Separations Division 2013

Core Programming Area at the 2013 AIChE Annual Meeting: Global Challenges for Engineering a Sustainable Future

San Francisco, California, USA
3 – 8 November 2013

ISBN: 978-1-63439-054-5

Printed from e-media with permission by:

Curran Associates, Inc.
57 Morehouse Lane
Red Hook, NY 12571

Printed by Curran Associates, Inc. (2014)

For permission requests, please contact AIChE
at the address below.

AIChE
3 Park Avenue
New York, NY 10016-5991

Phone: (203) 702-7660
Fax: (203) 775-5177

www.aiche.org

Additional copies of this publication are available from:

Curran Associates, Inc.
57 Morehouse Lane
Red Hook, NY 12571 USA
Phone: 845-758-0400
Fax: 845-758-2634
Email: curran@proceedings.com
Web: www.proceedings.com

TABLE OF CONTENTS

(9a) Selective Recovery From Interstitial Spaces of Leaf Tissue

Kingsbury, N., University of California, Davis
McDonald, K. A., University of California, Davis

Originally presented on: 11/3/2013 15:30:00 - 15:48:00

Title: Selective Recovery of Apoplast-Targeted Recombinant Proteins from Plant Leaf Tissue

Nathanial J. Kingsbury and Karen A. McDonald

Department of Chemical Engineering and Materials Science

University of California, Davis, CA 95616

Transient production of recombinant proteins in plant leaf tissue using vacuum infiltration with recombinant *Agrobacterium tumefaciens*has advantages over other approaches including speed, scalability, safety and cost-effectiveness. This method is being considered for large scale production of vaccines, biodefense agents and industrial enzymes and other products requiring a eukaryotic host. Recovery of functional protein from leaf tissue such as tobacco using traditional methods of homogenization and extractioncan be challenging due to high amounts of fiber, phenolics and proteases. Targeting the product to the apoplast has advantages in terms of product quality, particularly glycoform type and uniformity, as well as ease of product recovery, since the secreted product can be removed from the leaf tissue without tissue disruption. To recover transiently expressed apoplast-targeted recombinant proteins, a vacuum infiltration-centrifugation methodis applied to *Nicotiana benthamiana* and *Nicotiana tabacum* leaves. Leaves are submerged in a buffer solution and a moderate vacuum is applied for several minutes and then quickly released, allowing the buffer solution to infiltrate the interstitial spaces of the leaf. The leaf is then centrifuged to remove the liquid which is enriched with components from the apoplast and cell wall matrix. The collected fluid, called the apoplast wash fluid (AWF), is highly concentrated up to 16-fold and purified as much as 50-fold compared to a homogenate extract. Multiple rounds of vacuum infiltration-centrifugation, applied to further remove secreted protein from the residual tissue, can result in recovery ofover 90% of the secreted protein of interest. The ratio of air phase to liquid phase in the interstitial space, called the dilution factor, is calculated and related to the rate of recovery from each round of vacuum infiltration-centrifugation.

Leaves subjected to the vacuum infiltration-centrifugation method showed no signs of damage after recovery of AWF and continued producing the recombinant protein of interest. For example, when leaves expressing a recombinant cellulase enzyme, endoglucanase catalytic domain (E1cd) from *Acidothermus cellulolyticus*, were subjected to daily rounds of vacuum infiltration-centrifugation starting at two days post-agroinfiltration, the overall amount of E1cd collected fromthe leaves after six days post-agroinfiltration was as much as four times greater than in leaves where the E1cd was allowed to accumulate in the apoplast unperturbed. Possible mechanisms for this improvement in expression level are improved stability of the E1cd in the AWF compared to the leaf apoplast or improved secretion kinetics upon diluting the E1cd concentration in the apoplast and by extension the endomembrane system as a whole. The results therefore demonstrate a novel technique for boosting expression levels in transgenic leaf tissue.

Out of the model proteins used these studies, those with smaller molecular weights were recovered with superior yields, suggesting that the porosity of the cell wall matrix may be limiting removal by the vacuum infiltration–centrifugation method. Pretreatment of the leaves with low concentrations of pectinase enzyme improved the yield and concentration of an 82kDa recombinant protein, human butyrylcholinesterase, four-fold while improving its protein purity two-fold in the AWF. These results demonstrate that even relatively large proteins can be efficiently recovered using the vacuum infiltration-centrifugation method.

(9b) Control of Intein-Mediated Recombinant Protein Purification In a Mammalian Cell Expression System

Han, T. C., The Ohio State University

Originally presented on: 11/3/2013 15:48:00 - 16:06:00

Control of Intein-mediated Recombinant Protein Purification

In a Mammalian Cell Expression System

Tzu-Chiang Han

Dr. David W. Wood

Inteins are genetic elements that can self-excise from various host proteins in a post-translational protein splicing process. This activity can be modified by a simple mutation to yield isolated, controllable cleaving activity at either the N- or C-terminus of the intein. By combining self-cleaving inteins with conventional affinity tags, self-cleaving affinity tags have been developed. Once the tagged target protein is purified, the affinity tag and intein components are removed easily through the controlled self-cleavage of the intein without adding any protease. Although this efficient tag-removal method has been successfully applied in prokaryotic cell systems in our previous studies, the *in vivo* premature cleaving of the precursor protein has been a serious drawback for this method in eukaryotic cell systems. Here we present our most recent work in developing a next-generation intein for use in mammalian cells. In particular, we have engineered a new intein with enhanced control, where the control mechanism is compatible with our mammalian cell culture. This intein has been tested with several conventional affinity tags in mammalian cell culture, and the results will be presented. By controlling the *in vivo* premature cleaving in a mammalian cell culture context, we have generated a potential platform technology for the manufacture of non-antibody glycoprotein therapeutics at large scale. Additionally, this intein has been modified for high-throughput applications in drug and target discovery and development.

(9d) A Potential Platform Technology for Antibody Purification: Mixed-Mode Chromatography

Lin, D. Q., Zhejiang University
Yao, S. J., Zhejiang University
Zhang, Q. L., Zhejiang University
Tong, H. F., Zhejiang University
Wang, R. Z., Zhejiang University
Lu, H. L., Zhejiang University
Originally presented on: 11/3/2013 16:24:00 - 16:42:00

Mixed-mode chromatography (MMC) is a novel technology for bioproduct separation, especially antibodies. Mixed-mode ligands typically contain a combination of multiple binding modes like hydrophobic interaction, electrostatic forces, hydrogen bonding, etc.. High capacity, salt-tolerance, good stability and relatively low cost are the major advantages of MMC for direct capture processes. However, compared with Protein A capture, the separation selectivity of MMC for antibodies is still limited.

Based on series of tests with MMC using IgG, IgY and IgY(ΔFc), it was found that the control of loading pH or salt addition in the buffer was critical to improve antibody purification from albumin containing feedstocks. The control of caprylate addition might be another effective way for improving antibody purification. High purity and recovery can be obtained after the optimization of operation conditions. To better understand the ligand-protein interactions, a molecular simulation approach was established with combinative methods of molecular docking and dynamic simulation. The results indicated that at neutral conditions the ligand can bind onto some special sites of IgG surface which showed strong hydrophobicity. When the pH changed, the ligand departed quickly from the protein surface due to electrostatic repulsion. In addition, the ligand-net was constructed to simulate the protein adsorption on the porous surface of chromatographic resins. The MM/PBSA method was used to acquire the binding energy map which was used to scan the most suitable binding orientations of IgG on the ligand-net. The effects of ligand structure, ligand cluster, gel matrix and liquid conditions were also studied. The simulation results could provide useful information on the process development and the design of novel ligands for MMC.

For biopharmaceuticals such as monoclonal antibodies (mAbs), downstream processes require highly selective and robust technologies to achieve extremely high purity. The complexity of the interactions between the mixed-mode ligand and the desired protein might be considered as a challenge to improve the adsorption selectivity. Due to the flexibility of operation development and the innovative design of novel ligands, MMC is likely to become a potential platform for antibody purification with high efficiency and low cost.

* This work was supported by National Natural Science Foundation of China and the Zhejiang Provincial Natural Science Foundation of China.

(9e) Modeling the Synergies of Saline and Organic Modifiers On Rpc Separation

Johansson, K., Lund University
Søndergaard Frederiksen, S.,
Degerman, M., Lund University
Breil, M. P., Technical University of Denmark
Mollerup, J.,
Nilsson, B., Lund University
Originally presented on: 11/3/2013 16:42:00 - 17:00:00

During the last decades, great advances have been made regarding both modeling and understanding of preparative ion exchange chromatography (IEC). The Langmuir mobile phase modifier (MPM) [1] and the steric mass action (SMA) [2] models each give an accurate description of the effect of the salt concentration and recent developments in high throughput screening [3] as well as molecular dynamics simulations [4] have greatly increased the knowledge of the underlying mechanisms. In comparison, the chromatographic modes based on hydrophobic interactions, i.e. hydrophobic interaction chromatography (HIC) and reversed-phase chromatography (RPC), have been left in the shade. The complexity of the interaction between fatty ligands and hydrophobic patches on proteins has made it elusive. Although some serious attempts have been made, e.g. using a thermodynamic approach [5], many aspects of these hydrophobic phenomena are still not well understood.

As a step towards a better understanding of RPC and a model that applies for both HIC and RPC, an RPC separation of three insulins has been studied. This is a continuation of a project which was initiated with a study of the corresponding HIC separation. In both these cases, the effects of salt as well as organic modifiers have been investigated. This study included two different resins with C4 and C18 ligands, respectively, both on silica backbone.

The linear range of the isotherm was mapped by isocratic pulse experiments, performed at various concentrations of salt and organic modifier. The concentration of organic modifier was limited by nature; either by the lack of retention or by the salt solubility at the higher end, and by low UV signal-to-noise ratio, due to very high retention, at the lower end. For either resin, a series of pulse experiments at varying load within the non-linear range was performed to investigate the capacity properties.

Based on the experimental data, a mechanistic model of the system is developed. Since the resin particles are rather small and since the adsorption kinetics are generally rate-controlling for RPC, the column is described by a reaction dispersive model. The adsorption is described by a combination of thermodynamics [5] and the Langmuir MPM model, with adjustments which account for the observed effects of the modifier concentrations. The model parameters related to these effects as well as Henry's constant were calibrated against the retention data from the linear range experiments. The model parameters concerning capacity and kinetics were tuned by hand to make the peak shape fit the overloaded and linear range experiments, respectively.

The final model can be used as a tool to design new separation processes, as well as to optimize already existing ones. It is also useful in robustness studies and for filing purposes. Apart from the obvious practical benefits, it provides a new step of insight into the underlying phenomena and gives a deeper understanding of the process.

References

1. Melander, W.R., S. El Rassi, and C. Horváth, *Interplay of hydrophobic and electrostatic interactions in biopolymer chromatography: Effects of salts on the retention of proteins.* J. Chromatogr., 1989. **469**: p. 3-27.

2. Brooks, C.A. and S.M. Cramer, *Steric Mass-Action Ion Exchange: Displacement Profiles and Induced Salt Gradients.* AIChE J., 1992. **38**(12): p. 1969-1978.

3. Nfor, B.K., et al., *Rational and systematic protein purification process development: the next generation.* Trends Biotechnol., 2009. **27**(12): p. 673-679.

4. Holstein, M.A., et al., *Effects of Urea on Selectivity and Protein−Ligand Interactions in Multimodal Cation Exchange Chromatography.* Langmuir, 2012. **29**(1): p. 158-167.

5. Mollerup, J.M., *A Review of the Thermodynamics of Protein Association to Ligands, Protein Adsorption, and Adsorption Isotherms.* Chem. Eng. Technol., 2008. **31**(6): p. 864-874.

(9f) Novel Thermo-Responsive Fucose Binding Ligands for Glycoprotein Purification By Affinity Precipitation

Arnold, L., Georgia Tech

Originally presented on: 11/3/2013 17:00:00 - 17:18:00

Carbohydrates are the most abundant organic molecules found in nature, and they are metabolized and synthesized by nearly all organisms. Recently, focus in understanding the structure and function of glycoproteins has expanded into the rapidly growing field of glycobiology. However, the complexity and microheterogeneity of these oligosaccharide chains have made identification, isolation, and analysis very difficult. To aid in this demanding field, sugar binding proteins, or lectins, have emerged as critical tools in glycobiology. Lectins and lectin-containing biomaterials have been used in separation and identification, such as lectin affinity chromatography. However, while amenable at bench scale for glycoprotein separation, scale up of this chromatographic process is notably inefficient and expensive. Column fouling, diffusion limitations, and flow rate constraints are just some of costly disadvantages of this time consuming process.

To address these problems, we designed a recombinant fusion protein that potentially could be used for affinity precipitation purification of glycoproteins without chromatography. This protein has dual functionality: a fucose binding bacterial lectin from *Ralstonia solanacearum* (RSL) for carbohydrate affinity, and an elastin-like protein (ELP) thermo-responsive biopolymer for precipitation. Affinity separation will be achieved by simple inverse temperature cycling (ITC), which consists of temperature oscillation, followed by recovery of the precipitate via filtration or centrifugation. Separation may also be achieved by fluctuation of the salt concentration, maintaining the integrity of possible target enzymes. Furthermore, a newly developed extracellular secretion method for high expression level in *E. coli* is utilized to establish a cost effective process for production and isolation of this protein, achieving a secretion yield of 35 mg/L over a 48 hour cultivation period.

The presentation will detail the design, overexepression and characterization of several lectin-ELP fusions. The successful application of glycoprotein purification using these novel proteins will be demonstrated with a model protein, horseradish peroxidase (HRP), which contains α1,3-fucosylated glycans. Isolation of this target glycoprotein from a protein mixture in an affinity precipitation process resulted in a pure glycoprotein and a recovery yield of 88%.

(9g) Elastin-Like Polypeptide Tag Length Effect On Protein Expression and Purification

Shakalli Tang, M., The Ohio State University
Coolbaugh, M. J., The Ohio State University
Lease, R. A., Ohio State University
Wood, D. W., the Ohio State University
Originally presented on: 11/3/2013 17:18:00 - 17:36:00

Recombinant protein technology has enabled the expression of fusion proteins that can be purified in bacterial systems with the use of purification tags. In particular, the elastin-like polypeptide (ELP) precipitation tag in the presence of a self-cleaving intein protein domain allows the non-chromatographic purification of a target protein with only a change in salt concentration and pH. Shortening the ELP tag length has the potential to increase expression by freeing the energy that would be required to synthesize the large ELP repetitive protein sequence. An increase in production yield of a given protein would thus be expected when decreasing the ELP tag length. Furthermore, the optimal ELP tag length needed may be affected by the size and solubility of the protein product fused with the tag. In this study, the effect of five different ELP tag sizes in product yield was examined in the expression and purification of several proteins, including β-galactosidase and β-lactamase. Sodium dodecyl sulfate polyacrylamide gel electrophoresis (SDS-PAGE) qualitatively suggested the success of each purification. The corresponding activity and Bradford assays were then used to confirm these results quantitatively. Finally, studies including a wider range of salt concentrations and a wider selection of target proteins with varying hydrophobicity can help elucidate the relationship between the target protein fused to the ELP-intein system and the ELP tag length required for its successful purification.

(9h) High-Throughput Bioprocess Development for Polymer-Polymer Aqueous Two-Phase Systems

Sorensen, E., University College London
Bracewell, D. G., University College London
Patel, N.,
Originally presented on: 11/3/2013 17:36:00 - 17:54:00

The market for therapeutic proteins is currently increasing at a remarkable pace. In many cases, therapeutic proteins, such as monoclonal antibodies, must be administered in relatively large dosages. This large dosage requirement (> 100 mg $dose^{-1}$) (Shire et al., 2004), coupled with the high demand for these drugs, has resulted in an increased pressure on pharmaceutical companies to deliver high quality drugs at sufficient quantities. As a result, there has been an increasing amount of research & development (R&D) aimed at improving upstream efficiency, which has resulted in increased cell titres during the cell culture/fermentation stage of protein production. To date, improvements in upstream technology have, however, not been matched by improvements in downstream technology. Consequently expensive chromatography columns must be scaled accordingly to cope with the increasing amounts of material generated, which is proving to be a process bottleneck due to their capacity and cost, and has highlighted the need to re-evaluate current downstream separation methodologies.

Aqueous two-phase extraction (ATPE) has been shown to be a promising alternative downstream separation technology (Rito-Palomares, 2004). Systems can be formed from combining two incompatible polymers or a polymer and salt above a critical composition. Although having many advantages such as low surface-tension, easy scalability and a gentle environment for biopharmaceuticals, there is still a lack of widespread use of such systems. This can be attributed in part to the difficulty involved in characterising the large experimental space required for optimisation and effective process development. Factors such as charge, hydrophobicity, solute affinity, polymer molecular weight and concentration all play an important role on partitioning in ATPE (Asenjo and Andrews, 2011) and characterising all of the above factors manually is very resource intensive.

The high-throughput characterisation of polymer-salt aqueous two-phase systems has been explored (Oelmeier et al., 2011, Bensch et al., 2007) however, polymer-polymer aqueous two-phase systems behave somewhat differently and have so far not been considered although they have a clear potential for high separation efficiency. To help address this issue, we present an approach that utilises automated liquid handling systems for the high-throughput determination of polymer-polymer phase diagrams and protein partition behaviour. Such an approach allows for rapid exploration of the large experimental space provided by such systems, which in turn will allow for easier and more rapid development of optimised separation systems. Conventional techniques used to determine the binodial curve can be cumbersome and time consuming. An automated approach for polymer-polymer aqueous two-phase systems based on the determination of cloud point by light obscuration measurements is presented, and demonstrated for a system of PEG and Dextran. In addition, phase diagram tie-lines are generated by integration of automated phase volume determination with the lever arm rule. Measurement of protein partitioning is determined by measurements of UV absorbance at 280 nm with an appropriate control system as a blank.

The developed methodology will allow for screening of the wide variety of factors known to affect aqueous two-phase systems, such as polymer molecular weight, salt type, salt concentration, pH etc., thereby allowing the potential benefits of ATPE to be more readily explored. Future work would include the integration of more sophisticated assays into the high-throughput methodology to characterise components in complicated feed material such as impurities and products of interest. In addition, the integration of the experimental methodology with mechanistic models via parameter estimation is of particular interest as this would further speed up process development.

Acknowledgement:

We gratefully acknowledge the part-funding and support from BioMarin.

References:

Asenjo, J. A. & Andrews, B. A. 2011. Aqueous two-phase systems for protein separation: A perspective. *Journal of Chromatography A,* 1218, 8826-8835.

Bensch, M., Selbach, B. & Hubbuch, J. 2007. High throughput screening techniques in downstream processing: Preparation, characterization and optimization of aqueous two-phase systems. *Chemical Engineering Science,* 62, 2011-2021.

Oelmeier, S. A., Dismer, F. & Hubbuch, J. 2011. Application of an Aqueous Two-Phase Systems High-Throughput Screening Method to Evaluate mAb HCP Separation. *Biotechnology and Bioengineering,* 108, 69-81.

Rito-Palomares, M. 2004. Practical application of aqueous two-phase partition to process development for the recovery of biological products. *Journal of Chromatography B-Analytical Technologies in the Biomedical and Life Sciences,* 807, 3-11.

Shire, S. J., Shahrokh, Z. & Liu, J. 2004. Challenges in the development of high protein concentration formulations. *Journal of Pharmaceutical Sciences,* 93, 1390-1402.

(29a) Economic Evaluation of Freshwater Conservation Strategies for Thermoelectric Power Plants

Walker, M. E., Illinois Institute of Technology
Theregowda, R. B., Carnegie Mellon University
Safari, I., Illinois Institute of Technology
Abbasian, J., Illinois Institute of Technology
Arastoopour, H., Illinois Institute of Technology
Dzombak, D. A., Carnegie Mellon University
Hsieh, M. K., Carnegie Mellon University
Miller, D. C., U.S. Department Of Energy, National Energy Technology Laboratory
Originally presented on: 11/3/2013 15:30:00 - 15:50:00

Thermoelectric power plants withdraw and consume large volumes of freshwater. In parts of the country where freshwater resources are scarce, this demand can be problematic. One option for conserving freshwater is to utilize degraded water sources as a replacement cooling medium for thermoelectric power plants. However, utilizing degraded water in cooling systems can be problematic; increased concentrations of particulates and salt forming species can lead to increased fouling on heat exchanger surfaces. A degradation of performance in the heat exchanger will translate to a loss of thermal efficiency and may require the plant to perform more frequent maintenance routines; it is therefore necessary to treat the degraded water source before introducing it to the system.

This presentation discusses a comprehensive methodology for the determination of the total cost of degraded water use in thermoelectric plants, including the cost of treatment, additional fuel use, additional downtime and lost power production that results. This methodology is then applied via a novel, cross-disciplinary case study to (1) estimate the freshwater conservation cost associated with secondary municipal wastewater use in the recirculating cooling loop of a thermoelectric power plant; (2) evaluate the freshwater conservation cost associated with secondary municipal wastewater against that of dry-air cooling technology and the U.S. Department of Energy's 2020 target; and (3) to determine the breakeven differential purchase price of freshwater to secondary municipal wastewater, beyond which the use of secondary municipal wastewater is economically beneficial.

(29b) Modeling Of Pollutant Transport in the Nile Delta, Egypt

Mostafa, M., University of Alabama at Birmingham
Peters, R. W., University of Alabama at Birmingham
Originally presented on: 11/3/2013 15:50:00 - 16:10:00

MODELING OF POLLUTANT TRANSPORT IN THE NILE DELTA, EGYPT

Mohamed Mostafa, Department of Civil, Construction and Environmental Engineering, University of Alabama at Birmingham, Birmingham, AL and Robert W. Peters, Department of Civil, Construction and Environmental Engineering, University of Alabama at Birmingham, Birmingham, AL

Egypt is an arid country with high temperatures and low rainfall, and depends mainly on the Nile River as its source of fresh water, where it supplies Egypt with more than 95% of its diverse water needs. The Nile River divides into two branches (Damietta and Rosetta) forming the Nile delta. Both branches flow into the Mediterranean Sea. The Nile delta extends nearly 200 km (124.27 mi) from north to south and over 300 km (186.4mi) from west to east. This research is being performed on the Nile delta. The primary objective of this paper is to develop a dynamic model of Nile delta pollution by using MATLAB software. The model is calibrated based on field and historical data. Problem concerning the water quality on the Nile delta involve toxic dumping, and water quality degradation. The lengths of Damietta and Rosetta branches are approximately 245 km (152.2 miles), and 239 km (148.5 miles), respectively.

The Nile delta receives pollution loadings from industrial, agricultural, and domestic activities which are located along its path. This model can be useful in assessing the downstream distance affected by pollutant releases and will aid as a decision tool on restrictions to be imposed on industrial, agricultural, and domestic activities. The analytical model is also a reliable tool for water quality assessment and management in affected areas, and can be used to predict water quality and will aid in the estimation of potential environmental damage. The investigation started with sample collection on December 2012 and *will last until August* 2013. Water samples are collected every 15 days *along the Nile Delta*, and from drains discharging into the Nile delta. Water samples are analyzed for different parameters during the winter, spring, and *summer* seasons. The environmental water quality parameters include pH, chemical oxygen demand (COD), biochemical oxygen demand (BOD), *total organic carbon* (TOC), total suspended solids (TSS), total dissolved solids (TDS), dissolved oxygen (D.O), and temperature. The main sources of pollution are studied to identify the most appropriate ways to reduce pollution and improve water quality in the Nile delta.

(29c) Evaluation of RO Concentrate Treatment for High Recovery Desalting of High Salinity Inland Brackish Water

Rahardianto, A., University of California, Los Angeles
McCool, B. C., University of California, Los Angeles
Faria, J.,
Cohen, Y., University of California, Los Angeles
Originally presented on: 11/3/2013 16:10:00 - 16:30:00

Development of new water sources is crucial in order to address the crisis of dwindling water supplies around the globe. Reverse osmosis (RO) desalination is a technology that is now in common practice for developing new water sources via desalting of seawater, brackish surface water and groundwater, as well as municipal wastewater. While RO is considered a mature technology, its application for desalination of inland brackish water remains challenging due to the high costs associated with residual RO concentrate disposal and membrane mineral scaling. The above challenge is particularly daunting in California's San Joaquin Valley with respect to the critical need for high-salinity agricultural drainage (AD) water management. Accordingly, in the present study, the technical feasibility of primary RO (PRO) desalting and its integration with PRO concentrate treatment was evaluated, via both experimental field tests and process analysis, to assess its potential use in enabling recovery enhancement via secondary RO (SRO) desalting. Two concentrate treatment options were considered: precipitation softening (PS) and a new chemical enhanced seeded precipitation (CESP) process. In the CESP process, initial partial lime treatment provides adsorptive removal of residual antiscalant from the PRO concentrate. This enables subsequent concentrate desupersaturation (with respect to the major scaling salt calcium sulfate) to occur via seeded gypsum precipitation, unimpeded by residual antiscalant. The desupersaturated PRO concentrate (~ 34% reduction in SIg from ~ 1.7 to ~ 1.1) can then be further desalted in an SRO step up to the limit that is feasible with antiscalant dosing for scale control. In order to assess the cost of various AD water desalination management options, on the basis of treated feed volume, a comparison was undertaken of the total cost (including brine disposal for all options) for: (a) PRO-CESP-SRO and PRO-PS-SRO at the maximum recovery limit of 93%, and (b) RO desalting up to the maximum attainable recovery of 62% without concentrate demineralization. Process and economic analysis of water salinity management options for the AD water source considered suggested that, when considering the value of the desalted product water and accounting for the disposal cost of residual brine, PRO-CESP-SRO desalting is the most economically attractive option. The break-even AD water desalination management cost (i.e., zero net cost) for PRO-CESP-SRO, for the present cost estimates, was at a product water value of 0.67–0.74 \$/m3 which is lower by 22–29% compared to that of the PRO-PS-SRO process (0.95 \$/m3). Admittedly, RO desalting (along with disposal of residual brine) can be more economical when brine disposal cost is below about ~ 0.44 \$/m3 brine, but this is unlikely to be a feasible or acceptable water management option given the environmental challenge of disposal and loss of valuable water resource.

(29d) Simulation and Sustainability Analysis for Produced Water Thermal Desalination Technologies

Chen, L., Lamar University
Lou, H. H., Lamar University
Originally presented on: 11/3/2013 16:30:00 - 16:50:00

Simulation and Sustainability Analysis for Produced Water Thermal Desalination Technologies

Abstract

Produced water is the largest waste stream generated in oil and gas industries. Characterized by high salinity, environmental concerns and the prospect of beneficial uses have driven intensive research into produced water management. Dominant thermal desalination technologies include Multi-stage Flash (MSF) and Multi-effect Distillation (MED). In this work, several multi-stage MSF and MED processes are simulated using Aspen Plus based on the plant data, and comparisons are made on multiple dimensions. In technological level, water recovery and gain output ratio (GOR) are the major indicators. For economic comparison, capital and operational costs (CAPEX & OPEX) are collected. The Waste Reduction Algorithm (WAR) is utilized for the evaluation of their environmental impacts.

(29e) Dispatching Electricity Generation to Increase Water Availability in Times of Drought: A Feasibility Study in Texas

Pacsi, A.,

Originally presented on: 11/3/2013 16:50:00 - 17:10:00

Over the past several decades, the frequency and intensity of droughts has increased throughout the United States, and this trend is predicted to continue. The use of water as a coolant fluid in thermoelectric power generation accounts for 40% of freshwater withdrawals and 3% of consumption in the United States, which leaves these facilities vulnerable during times of decreased water availability. Thus, drought can stress both the water and electricity generation resources within an affected region.

This work seeks to characterize the feasibility, cost, and air quality implications of using the grid as a virtual water pipeline to "deliver" increased water availability in regions with high drought using a case study of eastern Texas during the summer of 2006. This idea involves shifting power generation from parts of the grid which are under extreme drought conditions to areas with relatively more water availability, changing the magnitude and spatial extent of consumptions and withdrawals of freshwater resources for power plant cooling. This approach could be rapidly implemented using existing power plants and transmission lines and could respond to spatial changes in the location of drought during other periods. Changing the spatial location of power generation based on water availability will impact the cost of electricity generation by shifting generation to facilities which are lower on the dispatch order (and thus more expensive to operate) and may affect regional air quality by changing the location and magnitude of air quality pollutant precursors such as nitrogen oxides (NO_x) and sulfur dioxide (SO_2), which are related to regional ozone and fine particulate matter (PM) concentrations. Texas offers an interesting test bed for the interconnections of water and electricity resources since the state has its own self-contained electric grid (ERCOT) that has a fuel-mix that is a reasonable approximation of electricity resources within the United States as whole. In addition, Texas has 23 self-contained river basins that drain to the sea, reducing complications in modeling associated with downstream water users from other states. Also, projected population growth within the state of Texas will likely add further the strain on the electric grid and water resources in the next few decades.

This work is the first to demonstrate that ERCOT has the existing capacity to switch ~10% of its base case generation from a highly-drought stricken area to other parts of the grid. For the June 2006 episode, water-withdrawing electricity generation was re-dispatched away from areas classified as being under extreme and exceptional drought by the U.S. Drought Monitor, which were in South Texas during this period. While drought is a complex hydrological phenomenon and water planning decisions within this framework would likely be more complex than assumed in this work, the case study offers insight into the resiliency of the grid to respond to localized drought conditions. In addition, scenarios in which water consumption in severe drought regions was also limited were examined to limit the exportation of water shortages to borderline drought regions. The June 2006 episode is important for air quality planning in Texas as it was the period chosen for the State Implementation Plan (SIP) filed with the Environmental Protection Agency in response to non-attainment of the 8-hour ozone standards in the Dallas-Fort Worth area.

Coupled models of ERCOT power generation (Power World Simulator), power plant water withdrawals and consumption, and regional air quality (Comprehensive Air Quality Model with Extensions, CAMX) were utilized in this work. Factors for water consumption and use as well as air quality precursor emissions were developed for individual power plants within ERCOT on a per kWh basis so that these parameters changed with generation at the facility. A base case model that minimized the total cost of electricity generation in ERCOT without water-based constraints was utilized in order to compare the impacts of four water-consumption based scenarios. The first scenario involved turning off all water-withdrawing electricity generation in extreme and exceptional drought regions in South Texas and re-dispatching electricity generation from that region to other power plants on a least cost basis. Other scenarios in which water consumption in severe drought regions (the next most intense US Drought Monitor classification) was constrained to be equal to the base case, to be reduced by 5%, and to be reduced by 10% were also modeled in the work.

The cost of the drought-based re-dispatching scenarios was compared to the base case operation of the grid to determine the financial impacts of this strategy to increase water availability. The grid-based re-dispatching modeling indicated that water consumption for thermoelectric power generation in South Texas could be forgone during the 2006 episode, creating enough water-availability in the region for the personal use of 360,000 Texans. The cost of this strategy was compared to the installation and use of air cooling technologies at the subset of power plants in extreme and exceptional drought regions as well as to a pipeline project in Texas. The use of drought-based grid dispatching was found to be generally more expensive than either of these strategies but had the advantage of being able to respond to seasonal changes in the location of the most severe drought within ERCOT. Future work and costing strategies could potentially improve the economics of grid re-dispatching. In addition, drought-based grid re-dispatching could be used as a parallel strategy while more long-term strategies, such as pipelines or air-cooling, were implemented.

Drought-based re-dispatching of power generation also changed the spatial location of power plant emissions of ozone and PM precursors within Texas in addition to changing patterns of water use within the state. Impacts on 8-hour ozone and average PM concentration, which are the metrics related to federal air quality standards, were generally modeled to be small. The areas with the greatest net air quality benefits were generally down-wind from the South Texas drought area from which power generation was shifted. The locations with the greatest air quality dis-benefit were near coal-fired facilities with increased power generation during the episode.

(29f) A Novel Method to Manage the Concentrate Disposal of Desalination Units

Aghahossein Shirazi, S., New Mexico State University
Saraeian, A.,
Rastegary, J., New Mexico State University
Ghassemi, A., New Mexico State University, Institute for Energy and the Environment/WERC
Originally presented on: 11/3/2013 17:10:00 - 17:30:00

Scarcity of clean water is a great dilemma today and needs better management. About 1.2 billion people are living in areas where water is physically scarce and also 1.6 billion encounter economic water shortage. Population growth, climate change, and inappropriate use of available resources are the chief factors that worsen the current condition. Given that the amount of water on the earth has remained quite constant over time, it is crucial to desalinate saline and brackish waters. However, cost and concentrate stream of desalination units are two major issues that have restricted expansion of inland water desalination technologies. The salinity of concentrate stream which is sent to disposals increases with each succeeding separation step and causes environmental issues resulting in confining practical deployment of desalination units. Cultivation of microalgae in concentrate disposal is a new method which could contribute to solving these problems inasmuch as microalgae have been considered as excellent candidates being capable of removing ions and salts existing in concentrate stream. In addition, the derived dry biomass could be used as a feedstock for biofuel production. To examine this idea, a full factorial experiment with completely random design (CRD) was conducted aiming to investigate the growth of two strains of marine algae, *Dunaliella tertiolecta* and *Nannochloropsis oculata,* in three different media including concentrate, artificial seawater medium, and a 50-50 combination of these two under 16 hours of illumination in 25°C. Furthermore, Nitrate, Phosphate, Sodium, and potassium removal were measured everyday to characterize the trend of removing ions and salts from media. Results revealed that both strains of algae can grow fairly well in concentrate and have significant effect on removing salts and ions from concentrate disposal.

(33a) pH Transitions in Chromatography Columns

Carta, G., University of Virginia

Originally presented on: 11/4/2013 8:35:00 - 8:55:00

pH transitions, defined as temporary excursions from a steady pH value even when initial and feed pH are constant, can occur in chromatography columns that contain weak acid or weak base groups even when the solutions used are buffered by non-adsorbed buffering species. Such transitions can lead to very long equilibration times, altered chromatographic separation performance, damage to the stationary phase and/or to the species being separated, and other operational problems. In this paper, we review how the local equilibrium theory of coherent waves can be used to understand the nature of these pH transitions, predict their magnitude and duration, and devise ways to either avoid them, when undesirable, or use them to an advantage for pH-based chromatographic resolution. As illustrative examples, we consider pH transitions in weak acid or weak base ion exchange columns and in hydroxyapatite columns caused by constant pH buffer steps as well as pH transitions in mixed beds of weak and strong ion exchangers generated by pH steps designed for the chromatographic resolution of protein charge variants.

(33b) Rapid Cycling of Pure Gases in a Single Bed PSA Apparatus

Rahman, A., University of South Carolina
Ebner, A. D., University of South Carolina
Erden, H.,
Holland, C. E., University of South Carolina
Ritter, J. A., University of South Carolina
Originally presented on: 11/4/2013 8:55:00 - 9:15:00

Pressure swing adsorption (PSA) is a cyclic gas separation technique that is just over 50 years old. It has many successful commercial applications including gas dehydration, air separation, hydrogen purification, and solvent vapor recovery. More applications are also on the horizon, including CO_2 capture from flue gas

Traditional, commercial PSA systems utilize cycle times on the order of minutes to tens of minutes, with individual step times no shorter than 0.5 min (30 s) or so. However, since the mid 80's there has been a great deal of interest in trying to reduce the cycle time to make the columns smaller and thus increase the feed throughput, which is the amount of gas processed per unit time per unit mass of adsorbent. These so called rapid PSA processes have met with limited success, however, until only recently.

Within the past 10 years, SeQual Technologies in San Diego and QuestAir in Vancouver (among less notable others) developed the rapid PSA concept into viable, commercial processes using rotary valve and rotary bed technology, respectively. They both have commercial processes that operate with cycle times on the order of a few seconds, with individual step times of less than 1 s. SeQual has been focusing on medical oxygen technology, while QuestAir has been focusing on hydrogen technology.

However, a paucity of work has been done on other rapid PSA applications; and overall, the understanding of the rapid PSA concept is quite limited with little fundamental information available. This leaves considerable room not only for the development of new theories and mechanisms underlying the rapid PSA concept, but also for the development of new applications for rapid PSA processes. The goal of this project has been to shed some light on both of these notions.

One of the shortcomings that have apparently limited the development of rapid PSA processes is a fundamental understanding of the controlling mass transfer mechanism that governs the performance. It has become clear to Ritter and co-workers over the past few years that the kind of information needed cannot be obtained by conventional measurement techniques, such as analyzing a column breakthrough curve. However, it can be obtained by building a prototypic rapid PSA process and then calibrating a rapid PSA process simulator to the performance data by varying the mass transfer coefficient. A 1-bed, rapid PSA system has been constructed for this purpose. So far, this rapid PSA system has been characterized by carrying out dynamic cycling experiments with single gases such as He, CO_2 and N_2, both with an empty bed and with the bed filled with beaded 13X zeolite over a wide range of cycle times and process conditions. Pressurization and depressurization step times as short as 0.5 s are possible in this system under both positive and negative gauge pressures and at different temperatures. This presentation will discuss the latest results obtained with this unique 1-bed rapid PSA apparatus.

(33c) The State of MOF Research Over the Past Decade: Connecting Chemistry With Adsorption Science

Walton, K. S., Georgia Institute of Technology
Originally presented on: 11/4/2013 9:15:00 - 9:35:00

Metal-organic frameworks (MOFs) are inorganic-organic hybrids that represent a new direction in porous materials research. The rich field of coordination chemistry provides a versatile platform from which these materials may be assembled from an almost infinite set of building blocks. Flexible synthesis techniques could lead to the creation of designer-specific multifunctional materials for a variety of applications. Research on MOFs began in earnest in the late 1990s/early 2000s, with thousands of new structures reported in a short time. Compared to synthetic zeolites, which had a discovery-to-commercialization time of ~ 6 years, MOFs have yet to find its niche in the porous materials market. Major barriers to using MOFs in adsorption applications include (i) stability in the presence of water or acid gases, (ii) broad scale-up synthesis, and (iii) cost of manufacturing. This talk will discuss these barriers and provide potential solutions that may lead to commercialization of MOFs.

(33d) Frequency Response Methods for Mass Transfer Study On Porous Materials

Wang, Y., Vanderbilt University
Kortunov, P., ExxonMobil Research and Engineering Company
Paur, C., ExxonMobil Research and Engineering
Ravikovitch, P. I., ExxonMobil Research and Engineering
Originally presented on: 11/4/2013 9:35:00 - 9:55:00

Frequency Response Methods for Mass Transfer Study on Porous Materials

Yu Wang, Pavel Kortunov, Charanjit Paur, Peter Ravikovitch

ExxonMobil Research and Engineering Company, Corporate Strategic Research

1545 Route 22 East, Annandale, NJ 08801

Knowledge of mass transfer mechanisms and rates is important for understanding transport phenomena inside porous adsorbents and catalysts as well as for the design of adsorption and catalytic processes. By introducing harmonic perturbations at various frequencies of oscillation, frequency response (FR) permits the decoupling of individual mass transfer steps within a complex system. Thus, it is capable of discriminating among different rate-liming mechanisms.

FR techniques developed for pure components include volume swing FR for a batch system and pressure swing FR for a flow system. A general model is presented to unify these two approaches for analyzing various mass transfer resistances including macropore diffusion, micropore diffusion, and surface barrier represented by LDF model, either independently or in combination. Heat effects can be included for non-isothermal cases. Furthermore, the heterogeneous properties of adsorbent/catalyst such as pore size distribution or pore mouth size can be described by introduction of continuous distribution of mass transfer rates, such as LDF, micropore diffusion and macropore diffusion time constants. With continuous development of the library of various mass transfer models, frequency response techniques can be applied to more complicated systems to determine accurate mass transfer mechanisms and rates.

Examples will be given for diffusion of light hydrocarbons on commercial microporous materials. By incorporating volume-swing FR (covering very low pressure and fast frequency range), and pressure-swing FR (covering higher pressure and slow frequency range), we provide a thorough mass transfer study in a wide range of temperatures and pressures.

(33e) Adsorption of Gases for Storage and Separation Purposes

Bastos-Neto, M., Universidade Federal do Ceara
Torres, A. E. B., Universidade Federal do Ceara
Lucena, S. M. P., Universidade Federal do Ceara
Azevedo, D. C. S., Universidade Federal do Ceara
Cavalcante, C. L. Jr., Universidade Federal do Ceara
Originally presented on: 11/4/2013 9:55:00 - 10:15:00

In the field of gas storage and separation by adsorption the evaluation of equilibrium and kinetics data is essential for the selection of more efficient materials. In this study, adsorption experimental data of light hydrocarbons in several porous materials is presented along with an experimental investigation of charge and discharge cycles of natural gas in a prototype storage vessel filled with activated carbon demonstrating the effect of the gas composition on the adsorption capacity. Also, results on adsorption of CO_2 on nitrogen-enriched activated carbon and zeolite are presented along with multicomponent adsorption measurements of CO_2-rich mixtures on activated carbon. In order to provide further insight for the experimental measurements, molecular simulation is used for support in adsorbents characterization, and for helping to perform multicomponent adsorption calculations. For carbon materials, heterogeneity studies were applied using the triangular pore model and the etched surface model. For natural gas storage in MOF materials a new site environment was proposed and applied to PCN-14 structure. The simulation results will be presented and discussed along with experimental adsorption results.

(33f) Can We Really Discover New Adsorbents On the Computer?

Snurr, R., Northwestern University

Originally presented on: 11/4/2013 10:15:00 - 10:35:00

Molecular modeling has long played an important role in improving our understanding of adsorption. In recent years, researchers have also started to discuss openly the possibility that we could discover new, useful adsorbents by computational screening of materials. The large number of metal-organic frameworks (MOFs) and related materials reported in the past decade has accelerated this discussion in two ways. First, MOFs are synthesized in a building-block approach, which suggests that targeted synthesis is possible. Second, the number of possible MOFs is practically unlimited, so the need for computational screening has assumed greater importance.

In this talk, I will review efforts to use molecular-level modeling to screen and suggest new MOFs with desired properties. Approaches include modeling single structures with grand canonical Monte Carlo simulations, quantum chemical modeling of sorbate interactions with particular surface functional groups, and high-throughput computational screening of hundreds of thousands of MOFs. The talk will include work from the literature and new results for applications such as CO_2 capture and natural gas storage.

(33g) A New Approach in the Modelling and Characterization of Porous Carbon With Argon and Water

Do, D. D., University of Queensland
Fan, C., University of Queensland
Nguyen, V., University of Queensland
Zeng, Y., University of Queensland
Nicholson, D., University of Queensland
Originally presented on: 11/4/2013 10:35:00 - 10:55:00

Characterization of porous carbon to determine the pore size distribution and its derivatives (surface area, pore volume and average pore size) is traditionally carried out with the implicit assumption of a pore model that has slit-like pores of uniform length having both ends open to the surrounding external gas phase. This ideal model does not adequately describe the inherently complex structure of porous carbons where the presence of functional groups should be accounted for. A better model should be simple enough in its application, not only in the characterization, but also for the design of separation and purification processes. Specifically, the following factors should be accounted for: (1) energetic variation along the pore axis due to constrictions etc., (2) connectivity between adjacent pores and (3) the presence of functional groups. In this presentation, we report detailed computer simulation results for a new model for porous carbon, which includes pore size variation, constrictions, and functional groups, with argon as a representative of non-polar adsorbate and water as an example of strongly associating fluids. The new model is able to produce all types of hysteresis loops reported in the IUPAC classification, especially the Type H3 and H4, commonly attributed to carbon materials. Analysis of the experimental data with the new model gives a pore size distributions, different from those obtained by conventional analysis, and the concentration of the functional groups.

(54a) Fundamentals Of Gas and Water Purification Using Polymer Membranes

Freeman, B. D., The University of Texas at Austin

Originally presented on: 11/4/2013 8:30:00 - 8:55:00

Polymer membranes are critically important in addressing urgent global needs in the 21st century for energy efficient gas separations as well as reliable, sustainable, efficient access to clean energy and clean water. In the gas separation field, polymer membranes are now well established for air separation, hydrogen purification, and, increasingly, natural gas processing. Polymer membranes have also emerged as a leading technology to desalination water (e.g., reverse osmosis) and are being explored for energy generation in applications such as reverse electrodialysis and pressure retarded osmosis. Furthermore, efforts are under way to develop additional applications of membranes for water purification, such as forward osmosis and membrane-assisted capacitive deionization. In each of these applications, control of small molecule transport (gases, water and ions) across polymer membranes is critically important for optimizing performance of such membranes. This presentation focuses on the fundamentals of small molecule transport in polymers obeying the solution/diffusion model. Structure/property correlations are shown for a variety of polymers, including uncharged and charged materials. The role of free volume in governing diffusion of solutes through polymers is explored. Consistent with the so-called upper bound relations in gas separation membrane materials, the existence of a water-salt permeability/selectivity tradeoff relation is observed for polymers being considered for water purification and energy generation applications. Areas where the physics of water and ion transport are both similar to and different from those of gas transport in polymers are highlighted.

(54b) Zeolite vs MOF Membranes For Carbon Dioxide Separation: Which Is Best?

Carreon, M. A., University of Louisville

Originally presented on: 11/4/2013 8:55:00 - 9:20:00

Carbon dioxide separation from light gases is a very important environmental and energy issue. The state-of-the-art process for the purification of CO_2 from gases uses amine adsorption, which is a complex, and costly approach. Membrane separation technology is far less expensive and requires less energy consumption. Feasibility studies demonstrate that membranes are a technically and economically viable option for CO_2 capture from flue gas and natural gas treatment. This presentation focuses on two type of families of membranes that have shown promise to separate carbon dioxide from light gases: traditional zeolites, and emerging metal organic frameworks. Zeolite (inorganic crystalline structures with uniform-sized pores of molecular dimensions) membranes have highly desirable separation properties such as molecular sieving ability, high thermal, mechanical, and chemical stability. The molecular sieving ability of zeolite membranes has been exploited for a vast number of relevant gas mixture separations, including CO_2 separation from N_2 and CH_4. Although zeolite membranes can separate CO_2 from light gases, the development of superior performance membranes for gas mixture separations requires novel materials with fundamentally different structural, compositional, adsorption and transport properties than those of zeolites. In this respect, metal organic frameworks, have emerged as a novel type of microporous crystalline materials which combine highly desirable properties, such as uniform micropores, high surface areas, and exceptional thermal and chemical stability, making them ideal candidates for gas mixture separation applications. Several MOF compositions, in membrane form have demonstrated to be effective for separating CO_2 from other gases. In this presentation, the advantages and limitations of relevant zeolite and MOF membrane compositions for CO_2 separation are discussed. Finally, representative examples of zeolite vs MOF membranes for CO_2 separation are compared to answer the question: which is best?

(54c) New Advances In Membranes For Antibiotic Recovery and Controlled Release

Ho, W. S. W., The Ohio State University

Originally presented on: 11/4/2013 9:20:00 - 9:45:00

This presentation covers new advances in two types of membranes: (1) supported liquid membranes (SLMs) with strip dispersion for recovery of antibiotics and (2) nanoporous polycaprolactone membranes for controlled release. SLMs with their stability ensured using strip dispersion have been studied for antibiotic recovery. Cephalexin has been extracted and concentrated with high recovery. A mass transfer analysis has shown that the resistance from the extraction reaction is dominant. Initial results from a 4-component feed mixture derived from its enzymatic synthesis have shown selective Cephalexin recovery possible. This allows Cephalexin recovery from the mixture in conjunction with complexation without enzyme deactivation, which commonly occurs when complexation is conducted in enzymatic reaction mixtures.

On nanoporous polycaprolactone (PCL) membranes for controlled release, a constant drug release rate is of paramount importance to implantable drug delivery systems in finding the remedy against chronic diseases. We have synthesized nanoporous PCL membranes to achieve the zero-order constant release rate via the combination of thermally- and nonsolvent-induced phase separations. Pore connectivity was improved significantly when coagulation bath temperature was lowered from 35 to 5°C. By using a 5°C water coagulation bath, the average pore size reduced from about 90 nm to 55 nm while increasing the casting solution concentration from 15 wt% to 25 wt% PCL. Thus, by varying the polymer concentration of the casting solution, the drug release rate can be well controlled with the constant zero-order rate using the higher polymer concentration.

(54d) Functionalized Responsive Membranes for Environmental Applications (invited paper)

Bhattacharyya, D., University of Kentucky
Gui, M., University of Kentucky
Xiao, L., University of Kentucky

Originally presented on: 11/4/2013 9:45:00 - 10:10:00

The functionalization of microfiltration membrane pores with macromolecules undergoing conformational changes provides nano-scale interactions in a confined domain. The development of these types of biomimetic, tunable, nanocomposite membranes with stimuli-responsive properties provide opportunities in flux tunability and separation selectivity control. Depending on the types of functionalized groups (such as, chain length, charge of groups, biomolecule, etc.), these membranes could be used in applications ranging from metal (or oxyanions) separation to enzymatic catalysis. The use of microfiltration membrane-based sorbents containing polypeptides which provides nano-scale interaction is a novel technique to achieve very high (1 – 1.5 g/g sorbent) metal sorption and selective separations under convective flow conditions. In addition to toxic metal separations, one can reduce the sorbed metals directly to nanoparticles for environmental catalysis. These membranes (with functionalized polypeptides or other polyelectrolytes) are also used for obtaining nanofiltration type separations at pressure < 3 bar. For example, we have demonstrated 90 % separations of arsenic oxyanions with high water flux at pressure of about 1 bar. By taking advantage of helix-coil transitions one can get tunable separations by altering pH. In addition, electrostatic self assembly in pores (layer-by-layer, LBL) can also be achieved through alternate adsorption of cationic and anionic poly-aminoacids or polyelectrolytes under convective flow conditions. Non-stoichiometric immobilization of charged multilayers within a confined pore geometry leads to an enhanced volume density of ionizable groups in the membrane phase leading highly tunable separations. Multilayer assemblies of polyelectrolytes provide excellent platform for protein/enzyme immobilization by providing reusability and high reactivity. Our results with Glucose oxidase enzyme with glucose as substrate showed very high (70 to 80% of homogeneous activity) activity, and continuous synthesis of hydrogen peroxide and gluconic acid. The presentation will include the role of nano-domain interactions for selective separations, full-scale synthesis aspects, stimuli responsive permeability, polypeptide assembly in membrane pores, and integrated enzyme-nanoparticle systems for environmental applications. This research has been supported by NIEHS and NSF-IGERT program.

(61a) Membranes for Advanced Gas Separations-Key Compoents in Large Scale Energy Efficiency Strategies

Koros, W. J., Georgia Institute of Technology

Originally presented on: 11/4/2013 8:30:00 - 8:55:00

Improved separation processes can provide competitive advantages to early adopters through economical reductions in energy intensity by debottlenecking large scale processes for transforming raw materials into higher value-added components. This fact alone makes it worthwhile to pursue the topic of energy-efficient separations, and membranes are the vanguard in this movement. This presentation identifies goals and strategies being pursued to implement this approach. The collaboration with industrial partners focuses on different gas technology areas, including natural gas purification, olefin-paraffin separation and nitrogen enrichment of air.

As molecular size differences between desired permeating components vs. rejected components become more similar, *molecular scale* "sorption-diffusion" discrimination is required to allow selective permeation that enables separation using membranes. Conventional solution-processable polymers and inorganic materials both have important areas of applications, but each have limitations that prevent their use across the spectrum of membrane-based separation applications needed in the broad spectrum of gas separation opportunities. For polymers, penetrant size and shape-discriminating ability is often lower than desired, while cost of production of large scale modules is a serious hurdle preventing the use of zeolite and silica membranes.

Hybrid materials comprising both polymers and inorganic components offer a good compromise to address limitations of either of the individual components. Moreover, carbon molecular sieve membranes combine desirable characteristics of both polymers and molecular sieving inorganic membranes within the existing economical processes for membrane and module formation. A framework will be provided to understand a strategy for major innovations across a broad range of gas separation applications will be discussed. Key challenges and opportunities will also be presented to encourage discussion related to this strategy.

(61b) New and Novel Materials for Membrane Separations
Noble, R. D., University of Colorado
Originally presented on: 11/4/2013 8:55:00 - 9:20:00

Synthetic membranes are important for many separation applications, such as the separation and purification of atmospheric gases for medical and industrial use, and salt removal from seawater in regions lacking fresh water sources. Unlike mesh filters used to separate large solid particles from a fluid (phase difference), membranes used to separate molecular species from one another (i.e., a chemical separation), is a much more difficult process [given the size scale and level of mixing of the components].

Because of the benefits of membrane-based chemical separations over other methods (e.g., smaller device footprints, lower energy use compared to distillation), there has been recent interest in applying membranes to new separation problems, as well as improving membrane materials used in existing separations. Consequently, there is a need for better membrane materials in general (i.e., with better selectivity, better productivity, longer life and operational stability, increased operating temperature range, use in chemically challenging environments, and the ability to adjust materials properties and performance for a given application).

The new materials are room temperature ionic liquids (RTILs), zeolites and lyotropic liquid crystals (LLCs). Each of these materials has unique physiochemical properties that provide an advantage in certain applications. Each of these materials will be discussed and their use in gas and water treatment.

For researchers working on the design of new membrane materials, it is important to not only consider the design factors that afford better separation selectivity but to also consider the other factors that afford good productivity as well. Often it is these materials engineering factors that decide whether a membrane material is viable in practical separation or not.

(61c) Membranes for Bioprocessing Applications: Overview of Research At Penn State

Zydney, A., The Pennsylvania State University

Originally presented on: 11/4/2013 9:20:00 - 9:45:00

Membrane systems are used extensively for the purification and formulation of a wide range of important biotherapeutic products, including monoclonal antibodies, vaccines, and gene therapy agents. Recent developments in bioprocessing have created important new opportunities for membrane separations. This presentation will highlight some of the key work being done within the Zydney group on the development of membrane processes for the purification of high value biological molecules, focusing on the relationship between the underlying membrane properties (e.g., pore size and surface charge) and the membrane performance characteristics. Several specific applications will be discussed, including the development of membrane systems for the production and purification of PEGylated proteins and the use of ultrafiltration for the purification of plasmid DNA for gene therapy applications. Highly selective membrane systems can be developed for the purification of pegylated proteins by exploiting the high degree of retention of the large pegylated species arising from both electrostatic and steric interactions. Plasmid DNA isoforms can be purified by exploiting differences in the elongational flexibility of the different topological isoforms of the plasmid. These results clearly demonstrate some of the exciting opportunities for using membrane systems in bioprocessing applications.

(61d) Polymeric Membranes for Forward Osmosis (FO) and Pressure Retarded Osmosis (PRO) Applications

Chung, T. S., National University of Singapore
Originally presented on: 11/4/2013 9:45:00 - 10:10:00

Clean water, clean energy, global warming and affordable healthcare are four major concerns globally resulting from clean water shortages, high fluctuations of oil prices, climate changes and high costs of healthcare. Clean water and public health are also highly related, while energy is essential for prosperity.

Among many potential solutions, advances in membrane technology are one of the most direct, effective and feasible approaches to solve these sophisticated issues. Membrane technology is a fully integrated science and engineering which consists of materials science and engineering, chemistry and chemical engineering, separation and purification phenomena, environmental science and sustainability, statistical mechanics-based molecular simulation, process and product design.

In this presentation, we will introduce our efforts on FO and PRO to solve some of these issues and enhance earth sustainability. Our recent technology breakthroughs on membrane developments for FO and PRO membranes will be highlighted.

Acknowledgement

This research was funded by the Singapore National Research Foundation under its Competitive Research Program for the project entitled, "Advanced FO Membranes and Membrane Systems for Wastewater Treatment, Water Reuse and Seawater Desalination" (grant number: R-279-000-336-281) and was also supported by the Singapore National Research Foundation under its Environmental & Water Technologies Strategic Research Programme and administered by the Environment & Water Industry Programme Office (EWI) of the PUB. The author would like to thank Prof. D. R. Paul, Drs. K. Y. Wang, N. Widjojo, P. Sukitpaneenit, S. Zhang, Q. C. Ge, M. M. Ling, S. P. Sun, Ms. X. Li, Ms. R. C. Ong, Miss Y. Cui, Ms. H. H. P. Duong, Miss P. Z. Zhong, Miss X. Z. Fu, Mr. G. Han, Mr. P. Wang and Mr. F. J. Fu for their supports and suggestions. Thanks are also due to BASF, Eastman Chemicals and Mitsui Chemicals for their unique materials and final supports.

References:

(1) T. S. Chung, X. Li, R. C. Ong. Q. C. Ge, H. L. Wang, G. Han, Emerging forward osmosis (FO) technologies and challenges ahead for clean water and clean energy applications, *Current Opinion in Chemical Engineering*1, 246–257 (2012).

(2) T. S. Chung, S. Zhang, K. Y. Wang, J. C. Su, M. M. Ling, Forward osmosis processes: yesterday, today and tomorrow, *Desalination* 287, 78–81 (2012).

(61e) Novel Oxygen Ion Conducting Ceramic Membranes for Oxygen Production in Enabling the Clean Energy Delivery

Liu, S., Curtin University
Zhang, K.,
Originally presented on: 11/4/2013 10:10:00 - 10:35:00

Oxygen ion conducting ceramic membranes for oxygen production in enabling the clean energy delivery

Shaomin Liu, Kun Zhang

Department of Chemical Engineering, Curtin University, Perth, WA6845, Australia

Email: shaomin.liu@curtin.edu.au; Telephone: (61) 8-92669056.

Abstract

Oxygen production by air separation is of great importance in both environmental and industrial processes as most large scale clean energy technologies require oxygen as feed gas. Currently the conventional cryogenic air separation unit is a major economic impediment to the deployment of these clean energy technologies with carbon capture (i.e. oxy-fuel combustion). Dense ceramic membranes are envisaged to replace the cryogenics and reduce O_2 production costs and cut energy penalty significantly when integrated in oxy-fuel power plant for CO_2 capture. This presentation reviews the current progress and their problems of oxygen ion conducting ceramic membrane technology for oxygen production. Novel ion conducting ceramic membranes with shelled structure or with a short circuit has been discussed. With the strong chemical stability against acidic gases like CO_2 and the higher oxygen fluxes at lower temperatures, these novel membranes could possibly provide a new way for tonnage oxygen production to improve the viability of these clean energy technologies.

(61f) Development of Fixed Carrier Membranes for CO2 Separation: From Membrane Materials to Membrane Processes

Wang, Z., Tianjin University
Li, S., Tianjin University
Qiao, Z., Chemical Engineering Research Center, School of Chemical Engineering and Technology, Tianjin University
Wang, J., Tianjin University
Wang, S., Tianjin University
Originally presented on: 11/4/2013 10:35:00 - 11:00:00

The development of cost-effective CO_2 separation process is the key for clean energy supply and environmental remediation. Polymeric membranes containing fixed carriers show promising applications in CO_2 removal from energy gas (e.g. syngas, natural gas, and biogas) and flue gas, because the reversible reaction between carriers and CO_2 leads to the both high CO_2 permeance and high CO_2/gas selectivity of membranes. However, the fixed carrier membrane has not been commercialized and applied in practical processes due to the lack of materials possessing high separation performance and stability which could meet the purity, recovery and cost requirements of clean energy purification and CO_2 capture. In the recent years, several high-performance fixed carrier membrane materials with good stability have been developed by our group, and the scale up and practical applications of fixed carrier membranes are in progress.

In order to improve the gas separation performance and stability of fixed carrier membranes to meet the technological and economical requirements of practical applications, multiple permselectivities including diffusivity selectivity, solubility selectivity, and reactivity selectivity have been combined into membranes by modifying membrane structures in multiple levels, such as levels of atoms and groups, molecular chains, molecular aggregation, and macroscopic structures. By following this strategy, several high-performance membranes have been developed recently. For example, a multi-permselective membrane has been developed by interfacial polymerization with hexane-soluble trimesoyl chloride and water-soluble diethylene glycol bis(3-aminopropyl) ether and 3,3'-Diamino-N-methyldipropylamine. Crosslinked polyamide with both ether oxide groups and tertiary amino groups forms after polymerization, indicating the successful combination of multiple permselectivities. The multi-permselective membrane exhibits high CO_2 permeance and high selectivity of CO_2 over H_2, CH_4 and N_2. Another example is the polyvinylamine (PVAm)-piperazine (PIP)/ polysulfone(PS) membrane. PIP that contains carriers (secondary amino groups) was used to crosslink PVAm, a key membrane material with fixed carriers, by hydrogen bonds, which modifies the membrane structures in the level of groups and the level of molecular aggregation. Compared with pure PVAm, the amine group concentration of the PIP cross-linked PVAm has been increased, which enhances the reactivity selectivity of the membrane. Moreover, the hydrogen bond formed between the secondary amine groups of PIP and the primary amine groups of PVAm, which makes the molecular aggregation more densification and enhances the diffusivity selectivity of the membrane. The PVAm-PIP/PS composite membrane shows a state-of-the-art CO_2 separation performance.

Based on the novel materials mentioned above, the scale up and practical applications of fixed carrier membranes are in progress. The effects of technological parameters of large-scale synthesis of materials were investigated, such as reaction temperature, reaction time, the concentration of the monomer and the initiator. After that, the thin-film composite membranes were prepared on both hollow fiber support membranes and flat-sheet support membranes. The flat-sheet composite membranes with the width of 0.3m have been continuously produced by an auto-casting applicator. Industrial grade hollow fiber modules and spiral wound modules have been developed with effective areas of $4.5m^2$ and $5m^2$, respectively. The spiral wound module exhibits excellent CO_2 separation performance, for example, CO_2 permeance of 250~300GPU with CO_2/N_2 selectivity of 80, CO_2/CH_4 selectivity of 50, and CO_2/H_2 selectivity of 40. Recently, a demonstration unit based on fixed carrier membranes for removing CO_2 from biogas with largest treatment capacity of 1500 Nm^3/day has been developed in Hebei, China. The results indicate that the scale up and practical applications of fixed carrier membranes are feasible.

(100a) Effects of Alcohol Modifier Aggregation and Complexation On the Retention Factors of Chiral Solutes With An Amylose-Based Sorbent

Tsui, H. W., Purdue University
Franses, E. I., Purdue University
Wang, N. H. L., Purdue University
Originally presented on: 11/4/2013 12:30:00 - 12:51:00

Various stoichiometric displacement models in the literature have been widely used for understanding the competitive adsorption mechanisms of solutes and the polar modifiers of the mobile phase. The models were used to explain the often-observed linear log-log plots of the solute retention factor k versus the concentration C_I^0 of the modifier. The slopes of the plots were sometimes inferred to be equal to the number of the displaced modifier molecules upon adsorption of one solute molecule, or upon adsorption of the solute-modifier complex, and were generally found to be greater than 1. In this study, the retention factors and enantioselectivities of ethyl lactate (EL) and pantolactone (PL) enantiomer pairs were measured for the amylose tris[(S)-α-methylbenzylcarbamate] sorbent, or AS, with isopropanol (IPA) in n-hexane at 25 °C. The slopes range from less than 1, from 0.43 to 0.76, at $C_I^0 = 0.13$ to 1.3 M, to slightly more than 1 (from 1.0 to 1.25) at the higher concentrations used (6 to 7 M). The literature models cannot account for such slopes. To understand such slopes, five achiral monovalent solutes were chosen and studied in detail; see Tsui, Franses, and Wang, J. Chromatography A, 1279, 36-48 (2013). Infrared Spectroscopy (IR) and Density Functional Theory (DFT) simulations provided indications that there is significant IPA aggregation with an average aggregation number of *n*=3. A new retention model for the monovalent solutes has been developed, to take into account the IPA aggregation and the solute-IPA complexation. The small slopes can only be explained if the alcohol forms aggregates in solution. The model and the HPLC data for acetone show that the limiting slopes at high IPA concentrations approach the value of =0.67. The data for acetone were fitted to the model to determine the IPA aggregation number and equilibrium constant. More complex multivalent models have been developed for chiral molecules, accounting for multivalent solute adsorption, with an average number of binding sites x, multivalent solute-alcohol complexation with number of complexation sites y, solute intra hydrogen-bonding, monovalent alcohol adsorption, and, most importantly, alcohol aggregation in the mobile phase. The chromatographic data of EL and PL enantiomers were fitted to the model, to determine the retention factors at zero IPA concentration (for PL only, because for EL k was measured directly), the equilibrium constants of complexation and intra hydrogen bond formation K_{intra}, and the values of x and y. The equilibrium constant K_{intra} was higher for EL than for PL, as indicated by IR spectra, apparently because the EL molecule is more flexible. The values of y were the same for each pair of EL enantiomers, y=2.9, and for each pair of PL enantiomers, y=2.7. Hence there are about three binding sites, on average, for solute/IPA complexation. The value of x was higher for PL, 1.8, than for EL, 1.3. The limiting slopes, at the highest IPA concentrations used approach the theoretically expected values of $(x+y)/n$=1.4 for EL and 1.5 for PL The equilibrium constant for PL adsorption was also higher than that of EL. These results can account for the much higher retention factors measured for PL than for EL, and help explain the solute-IPA and the solute-sorbent interaction mechanisms. The method used here has the potential for estimating the average number of binding sites for any solute in a mobile phase containing a polar modifier.

(100b) Multi Component Adsorption of Heavy Metals Using Sodium Iminodiacetate Resin

Koong, L. F., The Hong Kong University of Science and Technology
Barford, J., The Hong Kong University of Science and Technology
Mckay, G., The Hong Kong University of Science and Technology
Originally presented on: 11/4/2013 12:51:00 - 13:12:00

Acid leaching and bioleaching are simple technologies in extracting the metals from solid waste. However, a mixture of metal species are leached out which can complicate downstream recovery. Separating these metals using selective adsorption has a good advantage as the metals can be recoverd individually at low concentration and reconcentrated at higher purity for reuse. A commercially available sodium iminodiacetate resin was used to investigate the multicomponent adsorption studies of Copper, Nickel and Zinc Sulfate which are commonly found in bioleaching solution of printed circuit boards. Single, binary and ternary components system were carried out to study the effect of competing metals on selectivity and adsorption capacities. The resin has adsorption capacity around 2mmol/g and has a selecitivty in the order of Cu>Ni>Zn and Cu could be separated completely from the ternary component system but not the remaining Ni and Zn. Ni however is adsorbed at 50% higher capacity than Zn. pH was found to have no effect on the order of selectivity of metal. The effect of metal ratio on selectivity was also carried out and it was found that at lower metal ratio of the dominant metal, the other competing metal is also adsorbed as there are vacant sites available for other metal uptake, indicating displacement mechanism which contributes to selectivity. Single component adsorption data are modelled with Langmuir, Freundlich, Redlich Peterson and SIPS. Binary component adsorption data are modelled with Extended Langmuir,Extended Freundlich, Extended Langmuir-Freundlich,Extended Redlich-Peterson, IAST-Langmuir and IAST-Langmuir-Freundlich. The ternary component adsorption data is also modelled with a modified Freundlich.

(100c) Desulfurization and Denitrogenation Of Silica-Based Adsorbents For CLEAN Fossile Fuels
Lee, C. H., Yonsei University
Originally presented on: 11/4/2013 13:12:00 - 13:33:00

DesuLfurization and denitrogenation of silica-based adsorbents for clean fossile Fuels

Chang Ha Lee

Department of Chemical and Biomolecular Engineering, Yonsei University,

50 Yonsei-ro, Seodaemun-gu, Seoul, 120-749, Korea
E-mail: leech@yonsei.ac.kr

Due to increasingly stringent environmental regulations for transportation fuels, the desulfurization and denitrogenation of fuels have become very important research topics in the production of clean automotive fuels with ultra-low sulfur levels. According to the new environmental protection agency (EPA) Tier II regulations, the maximum allowable sulfur content in US diesel fuel is less than 15ppmw. In addition, some fuel cells will require deep-desulfurized fuels with sulfur content <1 ppmw in order to avoid poisoning and deactivation of the reformer catalyst.

Heavy low-quality crude oil and natural gas are rich in highly refractory nitrogen and sulfur compounds. During catalytic hydrotreating, denitrogenation (HDN) and desulfurization (HDS) are performed simultaneously. The ability to adsorb nitrogen compounds is generally higher when compared to aromatic thiophene derivatives, which have to compete with nitrogen compounds for active sites on catalysts during HDS processes. In addition, the HDN is the most difficult hydrotreatment reaction and slower than the HDS. Therefore, nitrogen compound residue remains on active sites longer than sulfur compounds. In addition, the removal of sulfur-containing compounds from natural, municipal or shale gas has also traditionally been achieved by the catalytic HDS process operated at elevated temperatures and pressures.

As an alternative method, selective adsorption appears to be a promising approach to ultra-deep desulfurization and denitrogenation because the process can be run at ambient temperature and atmospheric pressure without consuming pressurized hydrogen or oxygen gas.

In the presentation, a class of highly sulfur- and nitrogen-compounds selective silica-based porous adsorbents (Zr-Si cogel, mesoporous YSP, magnetite-mesoporous silica composite, MCF-Li, MgO silica composite) is introduced. Adsorption and desorption equilibrium by using real residue hydrodesulfurization diesel (RHDS DSL) were studied at ambient temperature and pressure. The dynamic behaviors of adsorption and desorption for RHDS DSL and municipal gas will be also presented.

Keywords: Silica, Adsorption, desulfurization, denitrogenation, fossil fuel

References

1. A. Koriakin, Y.H. Kim, C.H. Lee, "Adsorptive Desulfurization of Natural Gas Using Lithium-Modified Mesoporous Silica," Ind. Eng. Chem. Res., 51 (2012), 14489.

2. S.H. Lim, E.J. Woo, H. Lee, C.H. Lee, "Synthesis of Magnetite-Mesoporous Silica Composites as Adsorbents for Desulfurization from Natural Gas," Applied Catalysis B: Environmental, 85 (2008) 71.

3. A. Koriakin, K.M. Ponvel, C.H. Lee, "Denitrogenation of Raw Diesel Fuel by Lithium-modified Mesoporous Silica," Chem. Eng. J., 162 (2010) 649.

4. J.M. Kwon, J.H. Moon, Y.S. Bae, D.G. Lee, H.C. Sohn, C.H. Lee, "Adsorptive Desulfurization and Denitrogenation of Refinery Fuels by Mesoporous Silica Adsorbents" ChemSusChem, 1 (2008) 307.

5. Y.S. Bae, M.B. Kim, H.J. Lee, J.U. Ryu, C.H. Lee, "Adsorptive Denitrogenation of Light Gas Oil by Silica-Zirconia Cogel at Ambient Condition," AIChE J., 52 (2006), 510.

(100d) Diffusion Mechanism of CO2 in 13X Zeolite Pellets

Mangano, E., University of Edinburgh
Hu, X., University of Edinburgh
Friedrich, D., University of Edinburgh
Ahn, H., University of Edinburgh
Brandani, S., University of Edinburgh
Originally presented on: 11/4/2013 13:33:00 - 13:54:00

A systematic study of the diffusion mechanism of CO_2 in commercial 13X zeolite pellets is presented. In order to gain a complete understanding of the diffusion process of CO_2, kinetic measurements with a Zero Length Column (ZLC) system and a volumetric apparatus have been carried out.

The ZLC experiments were carried out on a single pellet of zeolite 13X at 38 °C at a partial pressure of CO_2 of 0.1 bar, conditions representative of post-combustion capture. Experiments with different carrier gases clearly show that the diffusion process is controlled by the transport inside the macropores. In order to provide a more accurate analysis of the ZLC curves we created an automated tool by linking our adsorption simulator to a Multi-Objective Genetic Algorithm (MOGA). From the fitting of experimental data the kinetic and equilibrium parameters were extracted.

Volumetric measurements using a Quantachrome Autosorb system were carried out at different concentrations. Differently from the ZLC experiments in this case no carrier gas is present. The low pressure measurements show clearly Knudsen diffusion control in the pellet macropores, while in the short time region a contribution from the viscous flow of CO_2 in the uptake cell is also detected. At increasing CO_2 concentrations the transport mechanism shifts from Knudsen diffusion in the macropores to a completely heat limited process. Measurements using two different pellet sizes confirmed that the diffusion process is controlled by the transport in the macropores.

Finally, the diffusivities extracted from the ZLC and the volumetric experiments were used to calculate the tortuosity factor of the 13X pellets. The values of tortuosity obtained from the two experiments are consistent and in excellent agreement with the one estimated from independent mercury porosimetry measurements.

(100e) Cu-BTC, Polar Or Apolar MOF?

Van Assche, T., Vrije Universiteit Brussel
Duerinck, T., Vrije Universiteit Brussel
Calero, S., University Pablo de Olavide
gutierrez-Sevillano, J. J., University Pablo de Olivade
Denayer, J., Vrije Universiteit Brussel
Originally presented on: 11/4/2013 13:54:00 - 14:15:00

Cu-BTC, polar or apolar MOF?

Tom R.C. Van Assche[a], Tim Duerinck[a], Juan Jose Guttierez-Sevillano[b], Sofia Calero[b], Joeri F.M. Denayer[a]

[a] Department of Chemical Engineering, Vrije Universiteit Brussel, Pleinlaan 2, B-1050 Brussel, Belgium
[b] Department of Physical, Chemical and Natural Systems, University Pablo de Olivade, Ctra. Utera Km. 1, 41013 Seville, Spain

Copper-benzene-1,3,5-tricarboxylate, also known under the name Cu-BTC or HKUST-1 [1], is one of the most studied metal-organic frameworks (MOF). The material was first reported by Chui et al. in 1999 [1] and has since then received considerable attention in the fields of adsorption, gas separation, hydrogen storage, catalysis, etc. Despite the fact that the material is well studied and understood in relation to its structure and adsorption of mainly apolar hydrocarbons and light gasses, not much is known about the adsorption of more polar adsorbates. Among the polar adsorbates, water has received the most attention [2]. However, since the Cu-BTC structure has questionable water stability, it is difficult to compare adsorption data of water with other molecules. To further complicate the matter, the adsorption properties of the material tend to vary significantly (BET surface area of 692-1635 m^2/g) between studies.
Using commercially available Cu-BTC and strict storage and handling procedures, minimizing water contact, reproducible vapor isotherms were measured on the Cu-BTC framework, with very high saturation capacities for various polar and apolar adsorbates. Additionally, the negative effect of water exposure was measured. With exception of water, pore volumes exceeding 0.66 cc/g were found for small alcohols, acetonitrile, acetone, THF, N,N-DMF and other molecules. Surprisingly, small and polar adsorbates give rise to a distinct two-step adsorption isotherm (Type IV). The isotherm step shifts to higher loading and lower relative vapor pressure for increasingly apolar adsorbates.
Using Monte Carlo simulation, the two-step adsorption isotherm can be explained by the difference in the various Cu-BTC cages. Originally, the Cu-BTC material was ascribed two types of cages; small octahedral pockets (T1) and large central cages (L). However, a further distinction can be made for the large cages; those with copper unsaturated sites exposed (L3), and thus who do not have these copper sites (L2). The

copper sites result in a more polar behavior of the corresponding cages. The combined behavior of the cages is proposed to result in a two-step adsorption isotherm for polar molecules, as the L2 cages will yield a S-shaped isotherm, while the L3 cages display more Langmuirian behavior.
The adsorptive behavior of the various cages is also reflected in the isosteric heat of adsorption and the Fickian and Maxwell-Stefan diffusion coefficients, as studied in this work for methanol. Here, the diffusion coefficients of polar molecules such as alcohols are significantly smaller than for apolar molecules (alkanes).
The dual character of Cu-BTC, polar and apolar, is expected to reoccur for other MOF's. Understanding the adsorption of various adsorbates in the well-known Cu-BTC structure is beneficial to understand the adsorptive properties of other, both hydrophilic or hydrophobic MOF. For example, hydrophobic MOF without free metal sites, like ZIF-8, show a very similar adsorption behavior as compared to the Cu-BTC apolar cages.

Acknowledgements:
T.R.C. Van Assche is grateful to the Agency for Innovation by Science and Technology (IWT) Flanders for financial support.

[1] Chui, S.S.Y.; Lo, S.M.F.; Charmant, J.P.H.; Orpen, A.G.; Williams, I.D. A Chemically Functionalizable Nanoporous Material $[Cu_3(TMA)_2(H2O)_3]_n$. *Science* **1993**, *283*, 1148-1150.

[2] Küsgens, P.; Rose, M.; Senkovska, I.; Föde, H.; Henschel, A.; Siegle, S.; Kaskel, S. Characterization of metal-organic frameworks by water adsorption. *Microporous Mesoporous Mater.* **2009**, *120*, 325-330

(100f) Kinetic Water Stability of An Isostructural Family of Zinc-Based Pillared Metal-Organic Frameworks

Jasuja, H., Georgia Institute of Technology
Burtch, N. C., Georgia Tech
Huang, Y., Georgia Institute of Techonlogy
Cai, Y., Georgia Institute of Techonlogy
Walton, K. S., Georgia Institute of Technology
Originally presented on: 11/4/2013 14:15:00 - 14:36:00

The rational design of metal–organic frameworks (MOFs) with structural stability in the presence of humid conditions is critical to the commercialization of this class of materials. However, the systematic water stability studies required to develop design criteria for the construction of water-stable MOFs are still scarce. In this work, we show that by varying the functional groups on the 1,4-benzenedicarboxylic acid (BDC) linker of DMOF [$Zn(BDC)(DABCO)_{0.5}$], we can systematically tune the kinetic water stability of this isostructural, pillared family of MOFs. To illustrate this concept, we have performed water adsorption studies on four novel, methyl-functionalized DMOF variations along with a number of already reported functionalized analogues containing polar (fluorine) and nonpolar (methyl) functional groups on the BDC ligand. These results are distinctly different from previous reports where the apparent water stability is improved through the inclusion of functional groups such as $-CH_3$, $-C_2H_5$, and $-CF_3$ which only serve to prevent significant amounts of water from adsorbing into the pores. In this study, we present the first demonstration of tuning the inherent kinetic stability of MOF structures in the presence of large amounts of adsorbed water. Notably, we demonstrate that while the parent DMOF structure is unstable, the DMOF variation containing the tetramethyl BDC ligand remains fully stable after adsorbing large amounts of water vapor during cyclic water adsorption cycles. These trends cannot be rationalized in terms of hydrophobicity alone; experimental water isotherms show that MOFs containing the same number of methyl groups per unit cell will have different kinetic stabilities and that the precise placements of the methyl groups on the BDC ligand are therefore critically important in determining their stability in the presence of water. We present the water adsorption isotherms, PXRD (powder X-ray diffraction) patterns, and BET surface areas before and after water exposure to illustrate these trends. Furthermore, we shed light on the important distinction between kinetic and thermodynamic stability in MOFs. Molecular simulations are also used to provide insight into the structural characteristics governing these trends in kinetic water stability.

(100g) Capitalization of Breathing Phenomena of Flexible MIL-53 (Al) Solid to Enhance Its Carbon Dioxide Capacity and Selectivity

Mishra, P., Indian Institute of Technology Guwahati
Edubilli, S., Indian Institute of Technology Guwahati
Mandal, B., Indian Institute of Technology Guwahati
Gumma, S., Indian Institute of Technology Guwahati
Originally presented on: 11/4/2013 14:36:00 - 14:57:00

Metal organic frameworks (MOFs) are being widely investigated for their potential to efficiently remove of CO_2 from process streams. Some MOFs (DOBDC series for example) have shown good CO_2 uptake and selectivity over other gases like N_2, CH_4 etc. at the desired process conditions. But poor stability towards moisture makes them unrealistic for any practical separation application. On the other hand, some MOFs (like in the MIL-53 series) are readily stable towards moisture, but have poor CO_2 uptake and selectivity at low pressures. MIL-53 MOFs exhibit structural transformation from (large pore) **lp** domain to (narrow pore) **np** domain and vice versa (so called breathing phenomena) upon adsorption of certain guest molecules (like CO_2, H_2O etc.) or by change in temperature.

Breathing phenomenon of MIL-53(Al) metal organic framework is capitalized to tune its structure for increasing CO_2 loading and selectivity in the sub-atmospheric region. We report a procedure to tune the pores of parent MIL-53(Al) into narrow pore (**np**) domain at ambient temperature in absence of guest molecules. On one hand, this tuned sample exhibits a significant (greater than four-fold) increase in CO_2 capacity at about 0.17 bar (in flue gases partial pressure of CO_2 is typically between 0.12 - 0.20 bar) and 293 K. On the other hand, it shows almost zero loading for CH_4, N_2, CO and O_2 at the same conditions. In addition we also propose a method to retain the **lp** form of the sample for adsorption in 1 - 8 bar region, thereby enhancing its CO_2 capacity.

(101a) Advanced Treatment of Shale Gas Frac Water to Produce NPDES Quality Water

Dahlin, R. S., Southern Research Institute
Chatterton, B.,
Pakzadeh, B.,
Vaillancourt, A., Southern Research Institute
Originally presented on: 11/4/2013 12:30:00 - 12:52:00

Advanced Treatment of Shale Gas Frac Water to Produce NPDES Quality Water

Bob Dahlin, Bill Chatterton, Behrang Pakzadeh, Austin Vaillancourt

Southern Research Institute, 5201 International Drive, Durham, NC, 27712

Natural gas extraction by fracking produces two types of wastewater that must be managed: 1) Flowback water is produced when the fracking is complete and pressure is released. Approximately 10% to 30% of the water used for fracking flows out of the well over a period of about ten days, and 2) Produced water, a mixture of water that was trapped underground with the gas deposit and frac water, flows out of the well with the natural gas over the several years life of the well at a rate of 100 to 1,300 gallons per day.

All wastewaters produced by drilling and fracking are regulated by the Clean Water Act (CWA) and are subject to the technology-based regulations (40 CFR Part 435, Subpart C) which apply to onshore facilities ? engaged in the production, field exploration, drilling, well completion and well treatment in the oil and gas extraction industry.? The effluent guidelines at 40 CFR 435, Subpart C establish best practicable control technology currently available (BPT) requirements for onshore facilities and prohibits the discharge of waste water pollutants into navigable waters from any source associated with production, field exploration, drilling, well completion or well treatment.

Southern Research Institute, with project partners M2 Water Treatment, Inc. and BKT United, has been awarded a contract by RPSEA (Research Partnership to Secure Energy for America) to develop an innovative, cost-effective, and robust approach for treatment of shale gas frac water to produce National Pollution Discharge Elimination System (NPDES) quality water for discharge and/or reuse. The approach comprises optimizing the combination of four technologies, two for frac water treatment and two for treatment and disposal of residues (high solid slurry and membrane concentrate) from the frac water treatment. The frac water treatment technologies are magnetic ballast clarification (MBC) for removal of total suspended solids (TSS), metals, and naturally occurring radioactive materials (NORMs) , and vortex-generating and nano filtration membranes, for removal of TSS and/or total dissolved solids (TDS). The residue treatment/disposal technologies are hydrogel adsorbent for metals, NORMs, and trace element removal, and precipitation, solidification and stabilization.

The project began in May of 2013 and is anticipated to continue for a period of 24 months. The project tasks include bench-scale simulation, onsite field research/testing, and a feasibility analysis. The objective of the bench-scale simulations is to develop a method to allow for a quick evaluation as to whether waters from a specific well are amenable to treatment with MBC, membranes, and hydrogel media or precipitation/solidification/stabilization. The objective of the onsite field research/testing is to provide a linkage from the bench-scale simulations to what can be expected in the field. The objective of the feasibility analysis is to bring together the technical, economic and environmental data to assess the advantages and disadvantages of the proposed integrations compared to conventional technologies. If selected, Southern Research plans to present the detailed analysis and results from the bench-scale testing and any available preliminary data from the onsite field testing.

(101b) Recovering/Reuse of Sodium Chloride from Zero Discharge Desalination Concentrate Streams

Yetayew, A.,
Camacho, L. M., University of Texas at El Paso
Cappelle, M. A., University of Texas at El Paso
Davis, T., University of Texas at El Paso
Originally presented on: 11/4/2013 12:52:00 - 13:14:00

Zero Discharge Desalination (ZDD) technology is able to achieve up to 98% recovery by desalinating brackish groundwater. ZDD is typically comprised of a reverse osmosis (RO) or nanofiltration (NF) system and a unique variation of electrodialysis called electrodialysis metathesis (EDM). The ZDD technology increases the yield of fresh water from water supplies that contain enough $CaSO_4$ to cause precipitation problems when the water is treated with RO or NF alone. EDM can remove $CaSO_4$ and other troublesome salts, using NaCl as the exchangeable ions. NaCl recovery and return to the EDM system is important to advance the development of the EDM technology. Electrodialysis with monovalent selective anion and cation exchange membranes was explored to selectively remove sodium chloride from a feed stream saturated with calcium sulfate. Recovered sodium chloride was reuse in the EDM process. Experiments with simulated solutions and extended tests with the actual EDM concentrate streams were performed. Very good selectivity was observed for chloride over sulfate. The performance of the process with actual EDM concentrate streams shows great promise for rapid transfer to the pilot scale.

(101c) Advances in Electrosorption By Mesoporous Carbon Materials for Water Desalination

Sharma, K., Georgia Institute of Technology
Mayes, R. T., Oak Ridge National Laboratory
Kiggans, J., Oak Ridge National Laboratory
Yiacoumi, S., Georgia Institute of Technology
DePaoli, D. W., Oak Ridge National Laboratory
Dai, S., Oak Ridge National Laboratory
Tsouris, C., Oak Ridge National Laboratory
Originally presented on: 11/4/2013 13:14:00 - 13:36:00

Electrosorption of ions from aqueous solutions by porous carbon electrodes has been investigated for the desalination of saline water through the capacitive deionization (CDI) process. Mesoporous carbon materials of 10-nm average pore size, synthesized at Oak Ridge National Laboratory (ORNL), have been proven suitable for CDI because of their high surface area associated with an optimum pore size for ion transport. To study ion transport in mesoporous carbon electrodes, we employed the neutron imaging technique, which is a non-destructive in-situ method for providing information on the local concentration of ions in the electrodes as a function of time. Neutron imaging of ions in mesoporous carbon was conducted to determine the effective diffusivity of ions under different conditions of applied potential. It was also found that an oscillating voltage component superimposed on the applied direct-current (dc) voltage that is usually applied in capacitive deionization enhances the rates of adsorption and desorption of ions. Using neutron imaging, it was determined that the effective diffusion coefficient increased by 35% when the oscillating potential was superimposed onto the dc potential of 1.2 V. Thus, modulation of the electrical signal applied in CDI processes can be used to enhance the kinetics of CDI and increase the competitiveness of this process in real water-treatment and reuse applications.

(101d) Evaluation of Ceramic Membrane Filtration for the Removal of Dbp Precursors From Surface Water Supplies

Mathews, A. P., Kansas State University
Fann, S.,
Maurmann, L.,
Originally presented on: 11/4/2013 13:36:00 - 13:58:00

The presence of high concentrations of natural organic matter (NOM) in surface water supplies makes many such resources unusable as drinking water resources due to the formation of disinfection byproducts (DBPs) during disinfection of the water for pathogen control. Chlorine, which is an inexpensive and effective oxidant, has been traditionally used for disinfection in a majority of water treatment facilities. The presence of numerous chlorinated compounds that have adverse human health impacts from chlorination, have resulted in new regulations that set maximum limits on total trihalomethane (TTHM) and haloacetic acids (HAA5) in potable waters. Alternative disinfectants such as chloramine, ozone, etc., are being used at some facilities. But, they may have other problems such as growth of nitrifying bacteria in the distribution system, and the lack of disinfectant residual in the distribution system and the potential for recontamination. As such, the removal of DBP precursors from water supplies using membranes has received considerable attention in the last decade. Polymeric membrane filters using microfiltration (MF), ultrafiltration (UF), and nanofiltration (NF) have been studied for the removal of DBP precursors from water suppliers. Nanofiltration, although effective, is expensive to operate. Fouling and flux decline are the major problems with NF membranes. Considerable advances have been made in the manufacture of ceramic membranes (CMFs) in recent years. CMFs are resistant to chemicals, pH variations, can withstand high transmembrane pressures, and provide a robust platform for the removal of contaminants from water supplies and wastewaters. Bench-scale and pilot studies were conducted using 15 kDa to 3 kDa ceramic membrane filters for the removal of DBP precursors from a surface water supply at a Navy installation water treatment facility. The NOM content was analyzed for using measures such as TOC, and trihalomethane potential (THMFP). Size exclusion chromatography was used to characterize the molecular size distribution of the organic matter in the water. The membrane flux, TMP values, and the data for NOM from bench-scale pilot CMF studies will be presented.

(101e) Efficacy of Membrane Bioreactor Process Combined With Advanced Oxidation Processes for Organic Removal in Water Reclamation and Reuse

Kim, W., University of Southern California
Ravindran, V., University of Southern California
Pirbazari, M., University of Southern California
Originally presented on: 11/4/2013 13:58:00 - 14:20:00

An increasing number of public owned wastewater treatment and water reclamation utilities and several municipalities have recognized the importance of recycled water as a new water source, particularly in arid areas of the United States. These utilities have developed a wide variety of options for water reuse, and have considered water recycling water for indirect potable reuse, or discharge into surface waters or infiltration into subsurface environments for augmenting potential water supply sources. They have considered membrane technologies such as microfiltration (MF) followed by nanofiltration (NF) or reverse osmosis (RO) for this purpose. In the present context, our study compares the performance of membrane bioreactor (MBR) process using ultrafiltration (UF) membranes integrated with the application of powder activated carbon (PAC) as adsorbent.

The major concerns of the U.S. Environmental Protection Agency and local regulatory agencies are health problems arising from the introduction of pathogens (protozoa, bacteria, viruses and other parasites), anthropogenic organic or inorganic chemicals into groundwater or surface water sources. Additionally, organic matter or natural organic matter (NOM) already present in drinking water or those formed during wastewater treatment due to decomposition of organic matter could significantly affect water quality in terms of dissolved organic carbon (DOC) concentrations, or can act as precursors to the formation of disinfection byproducts (DBPs) such as trihalomethanes, haloacetic acids, aldehydes, ketones and others, many of which are well known carcinogens. Therefore, the proposed environmental regulations in the State of California (more stringent than those of the US EPA) include a goal of achieving TOC levels not exceeding 1 mg/L originating from reclaimed waters.

The purpose of this study was to evaluate the performance of the MBR process regarding the fate of organics in the treated effluent, when reclaimed water is used as the feed. The intention was to compare the performance of the MBR system (using the UF membrane) with those of other advanced membrane treatment systems such as NF or RO processes, that have been reported in several earlier studies. The present work considered the application of advanced oxidation processes (AOPs) such as ozonation, ozone-hydrogen peroxide oxidation and UV-hydrogen peroxide oxidation to render the organic matter more easily amenable to biodegradation in the MBR system. These evaluations also included the evaluation of permeate fluxes and effluent quality pertaining to the MBR system.

Extensive characterization of organics (or DOC) in the reclaimed water and the treated effluent

(MBR) effluent was necessary for the following reasons. Quality issues for residual compounds arise from their resistance to sorption and biodegradation. Some of these organic constituents exhibit potentials for forming DBPs during disinfection. Hydrophilic constituents are poorly sorbed on adsorbents and cannot be easily removed. Large molecular weight compounds are resistant to degradation, while low molecular weight compounds and polar compounds are sometimes not well rejected during membrane filtration (NF or RO processes), and some have significant health concerns. Hence RO filtration and size fractionation using resins is important for organics classification.

The organics characterization strategies require state-of-the-art analytical techniques including chromatographic and spectroscopic methods. These techniques focused on several aspects including the bulk organic carbon characteristics including hydrophobicity/hydrophilicity, molecular weight or size distributions, elemental analysis, organic carbon functional groups, potential formation of trace organics and DBPs, and identification of fluoropores. The gas chromatography/mass spectrometry (GC/MS) analysis provided the qualitative and quantitative assessments of the trace organics and DBPs present. The RO filtration and resin adsorption methods determine the hydrophobicity and polarity of the constituents. The size exclusion chromatography (SEC) methods provided details on molecular weight and size distributions of the product, while the carbon-13 nuclear magnetic resonance spectroscopy (^{13}C-NMR) after size separation techniques showed the functional group characteristics of the DOC as well as their polar and non-polar fractions. The UV absorbance (UVA) measurements provide information on the natural of NOM present and the types of products formed after the application of AOPs. These advanced techniques for organics characterization provided an in-depth evaluation of the MBR system with reference to the effluent characteristics and removals of various organic constituents.

(101g) Sanitation Shallow Water and Groundwater Contaminated With Petroleum Hydrocarbons

Cruz, Y.,
Castro, J. J.,
Rico, E.,

Originally presented on: 11/4/2013 14:20:00 - 14:42:00

In this work a study was made of the biodegradation of water contaminated synthetically with diesel and oil , and a water sample taken from a actual-estate of the Valley of Mexico, having this problem; using a microbial consortium containing microorganisms of the genus Pseudomonas. The involvement of native microorganisms in the biodegradation, when implementing a bioremediation process must be carefully evaluated since their presence also affects the efficiency of biodegradation, reaching strengthen it in the best case. It was carried out to characterize the actual water and preparing synthetic solutions with salts essential for bacterial growth. Adapting the bacterial consortium was performed by supplying the necessary essential salts and adjusting the optimum pH and temperature of development. The biodegradation of hydrocarbon substrates was followed via COD, respirometry, CO_2 absorption, UV-vis spectroscopy,IR/ATR infrared and gas chromatography-FID. Biodegradation profiles and bacterial growth curves are obtained. Degradation efficiencies were obtained 90 to 98 percent, varying the nature of the hydrocarbon substrate used in a treatment period of 5 days.

(129a) Membrane Desalination: Limitations and Opportunities
Belfort, G., Rensselaer Polytechnic Institute
Originally presented on: 11/4/2013 12:30:00 - 12:56:00

Meeting: AIChE Annual Meeting

Session: Membrane Research Activities Around the World"

Host: Zhongyi Jiang <zhyjiang@tju.edu.cn>

Date: 3-8 November, 2013

Location: SF, CA

"Membrane Desalination: Limitations and Opportunities"

Georges Belfort[1]

[1]Howard P. Isermann Department of Chemical and Biological Engineering, and

Center for Biotechnology and Interdisciplinary Studies

Rensselaer Polytechnic Institute

Troy NY 12180-3590, USA

Reverse osmosis (RO) has matured from a laboratory curiosity at the University of Florida and the University of California at Los Angeles in the early 1960s to an accepted industrial desalination process widely considered as the least energy intensive and most economically attractive technology for separating salt from water. Plans are now in progress in Japan to build a megaton plant (1,000 m3/day). With this wonderful success story, where are the opportunities for improvement? They include development of smooth, low fouling, chorine stable and long lasting membranes. Can we replace interfacial polarization as a mean of producing salt rejecting rough and chlorine susceptible membranes? To address this question fully, we need to understand the limitations of RO, consider fundamental principles of the molecular behavior of water and salt at solid interfaces (i.e. synthetic membranes), and critically assess the limitations such as concentration polarization and fouling. In this presentation, we will first summarize the major technological breakthroughs associated with the success of large-scale RO including the reduction in energy use. While fouling and concentration polarization offer inherent limitations, they are dealt with through pretreatment (removal of foulants), chemistry and hydrodynamics. Despite its successes, the molecular basis of salt separation by RO is still controversial: Gluekauff, Matsuura and Sourirajan, and Belfort have suggested molecular mechanisms that depend on dielectric, bound water and ion hydration arguments, respectively. Given the importance of water-polymer interactions (hydrogen bonding), polar membrane materials such cellulose acetate and aromatic polyamides are widely used. However, the lack of a molecular-level understanding means that a rational design strategy for improving RO efficiency is missing. Recent experiments with oxidized graphene and single walled carbon nanotubes, together with simulations of water near polar and non-polar interfaces suggests exciting possibilities for high-flux desalination membranes. These results indicate that water moves through non-polar nano-pores with little or no friction, with very high water fluxes using small amounts of energy. Recent attempts to take advantage of these ideas are limited by concentration polarization. Increasing the mass transfer coefficient through improved fluid mechanics will be needed to remove the polarized salt and reduce salt leakage. Nature-inspired transport, through the well-known aquaporin channels in biological membranes, uses a mixture of intermolecular forces to optimize selectivity. In summary, we plan to discuss how integrating recent molecular insights with decades of industrial advances will provide exciting opportunities for scalable desalination.

(129b) Research Into Molecular Separations Using Osn At Imperial College

Livingston, A. G., Imperial College London
Originally presented on: 11/4/2013 12:56:00 - 13:22:00

Organic liquids are ubiquitous in chemical science based industries, which range in scale from refining to pharmaceutical production. It is generally accepted that 40-70% of capital and operating costs in these industries are dedicated to separations; and a substantial fraction of this cost is related to processing of organic liquids, both as product streams and solvents. Membrane technology has the potential to provide game changing alternatives to conventional concentration and purification technologies such as distillation, liquid extraction, adsorption and chromatography. In order to achieve this potential, membrane modules must meet several challenges. They must be stable in organic solvents, offer attractive fluxes and rejections for systems of interest, and give reliable and predictable service lifetime and performance. The obvious benefits of membrane processing have been apparent for many years, and have attracted research, development and commercialisation efforts from academic groups, end users, and membrane suppliers. In the last few years these efforts have resulted in Organic Solvent Nanofiltration (OSN) emerging as a new membrane technology.

This presentation will describe some key research activities in the Livingston group at Imperial College London into OSN. These include formation and function of OSN membranes, imaging and characterisation of these membranes, simulation and process design of OSN systems, and applications of OSN in organic processes. Several polymers (polyimides, polybenzimidazole, PEEK) have been used to create membranes stable in organic solvents. A key step is the post-formation chemical crosslinking of the polymers, which imparts stability even in harsh polar solvents such as DMF, THF and DMSO. Recently developed membranes include thin film composites based on interfacial polymerisation (TFC-IP). By using inert supports, TFC-IP membranes having good flux and rejection in polar aprotic solvents have been developed. A further innovation is the development of OSN membranes that can withstand highly basic and acidic environments, which further widens the scope of the technology.

OSN membranes have permeation pathways in the nanometre range and so are hard to characterise. On-going work using contrast agents and high resolution TEM is yielding some interesting insights into membrane structure. Simulation and process design allied to validation using spiral wound module data allows the prediction of process performance. There are a growing number of applications of OSN to industrial separation problems, and two main areas will be highlighted, the purification of pharmaceutical API, and the formation of peptides and oligonucleotides using OSN technology.

(129c) New Membranes Structures and Compositions Effective for CO2 Separation and Capture

Chen, Y., The Ohio State University
Ramasubramanian, K., The Ohio State University
Zhao, L., The Ohio State University
Severance, M. A., The Ohio State University
Dutta, P. K., The Ohio State University
Ho, W. S. W., The Ohio State University
Originally presented on: 11/4/2013 13:22:00 - 13:48:00

This presentation covers new structures and compositions for carbon dioxide-selective membranes, including (1) polymer and polymer blend membranes, (2) mixed matrix membranes, and (3) inorganic/polymer composite membranes.

We have synthesized membranes with polyamide-polyethylene oxide (PA-PEO) copolymer blended with polyethylene glycol dimethyl ether (PEG-DME) or polyethylene glycol (PEG). We have also synthesized a polydimethylsiloxane (PDMS) membrane reproducibly with good separation performance, which can be used as a cover layer for inorganic/polymer composite membranes.

We have dispersed zeolite-Y nanoparticles in PA-PEO/PEG-DME and PDMS solutions to synthesize the mixed matrix membranes, which can be used as the cover layers for inorganic/polymer composite membranes. The membranes showed a higher CO_2/N_2 selectivity than PDMS alone.

We have designed and synthesized inorganic/polymer composite membranes, which consisted of a thin selective inorganic layer embedded in a polymer structure. Zeolite-Y/polymer composite membranes were synthesized successfully by coating a polymeric cover layer containing amines on the zeolite-Y nanoparticles deposited on a nanoporous polymer support. The resulting membrane showed a CO_2 permeance greater than 1100 GPU and a CO_2/N_2 selectivity above 400 at 102°C. Si-1/polymer composite membranes were also synthesized successfully by coating a PA-PEO/PEG-DME blend or PA-PEO/PEG blend as the cover layer on both Si-1 particles/polymer support and Si-1/PDMS/polymer support multilayer structures. The syntheses and transport properties of the membranes will be presented and discussed.

(129d) Preparation of Porous PTFE Hollow Fiber Membranes Through Mechanical Operations

Liu, G., Ocean University of China
Lv, J., The Institute of Seawater Desalination and Multipurpose Utilization, State Oceanic Administration
Gao, C., Ocean University of China
Guan, Y., The Institute of Seawater Desalination and Multipurpose Utilization, State Oceanic Administration
Originally presented on: 11/4/2013 13:48:00 - 14:14:00

Porous polytetrafluoroethylene (PTFE) hollow fiber membranes were prepared from fine powder through a series of mechanical operations. The morphological structures of the prepared hollow fiber membrane were well analyzed by SEM, which were characterized by a spatial periodicity of node (domain of agglomerated PTFE particles) and fibril domains. The fibrils were highly oriented in the direction of the stretching operation, providing pores in the polymeric membrane as split-like voids between adjoining fibrils. The unit size of the periodic structure varied depending on the heat-treatment, stretching and sintering conditions. The property of primary PTFE fiber was changed by heating temperature, the PTFE membrane porous and morphological structures were significantly changed by stretching rate, stretching ratio, and sintering temperature. It was found that heat-treatment process could improve mechanical strength of primary PTFE fiber, and benefit for the forming and growth of fibrils. Remarkably, the higher stretching rate tended to result in smaller pore size and narrower pore size distribution, the higher stretching ratio increased porosity and enlarged pore size. Simultaneously, the shrinkage ratio decreased with the increase in the sintering temperature, and became constant at around 2.5% above 330°C. According to the results, it is indicated that mechanical operation is feasible to gain excellent porous PTFE hollow fiber membrane.

(129e) Engineering Chemically Hierarchical Surfaces With Superior Antifouling and Self-Cleaning Properties

Jiang, Z., Tianjin University
Su, Y., Tianjin University
Zhao, X., Tianjin University
Originally presented on: 11/4/2013 14:14:00 - 14:40:00

Surfaces simultaneously bearing antifouling and self-cleaning abilities are broadly employed by living systems to sustain diverse powerful functions. Integrating these features into synthetic surfaces emerges is highly desirable and critically important in the development of various advanced surface-relevant technologies during medical, industrial, and environmental applications. We report here an in situ strategy to engineer amphiphilic porous membrane surfaces by free and forced surface segregation of amphiphilic polymer with low surface energy components: During the traditional nonsolvent induced phase inversion, hydrophilic segments freely segregate to the membrane surface controlled by the self-organization to minimize the interfacial between hydrophobic surfaces and water, and the non-polar hydrophobic segments are spontaneously dragged onto membrane surfaces by hydrophilic segments via forced surface segregation. The resulting external surface covered with hydrophilic and low surface energy polymer brushes can endow membranes with superior antifouling (fouling-resistant) and self-cleaning (fouling-release) abilities. When utilized for wastewater treatment, both irreversible and reversible membrane fouling by typical foulants, such as oil, protein and microorganism, can be remarkably suppressed. Hydrophilic brushes generate hydration layers constructing a protective screen to reduce the direct interaction between foulants and membrane surfaces, endowing the surface superior antifouling ability; low surface energy brushes form amounts of non-adhesive microdomians on the membrane surfaces to minimize the intermolecular forces of interactions between the foulants and the membrane surface, rendering the surfaces with desirable self-cleaning ability.

(129f) Advanced Polymeric Membranes With Tailored Barrier Properties

Ulbricht, M., UniversitÃ¤t Duisburg-Essen
Adrus, N., UniversitÃ¤t Duisburg-Essen
Behnke, S.,
Frost, S., UniversitÃ¤t Duisburg-Essen
Kuhn, C.,
Gajda, A.,
Pei, Z.,
Originally presented on: 11/4/2013 14:40:00 - 15:06:00

Advanced polymeric membranes with tailored barrier properties

Mathias Ulbricht, Nadia Adrus, Sven Behnke, Sven Frost, Christian Kuhn, Aleksandra Gajda, Zhaoqin Pei

Lehrstuhl für Technische Chemie II, Universität Duisburg-Essen, 45117 Essen, Germany; mathias.ulbricht@uni-essen.de

Membrane technologies have been established in a wide range of industrial processes. However, in advanced separation and reaction engineering, membranes applied as separator or contactor can offer yet many more distinct advantages [1]. Two examples are membrane applications for more aggressive than aqueous feed streams or membranes where the barrier properties can be switched by external stimuli.

The majority of synthetic membranes are made from polymers because barrier and surface properties can be varied in wide ranges with help of established scalable manufacturing processes. Significant efforts have been made to further improve membrane performance by focussing on barrier properties (high selectivity, high flux) or surface properties (antifouling, affinity). In this presentation, we will discuss how the synthesis of novel membane polymers or the development of adapted polymer synthesis methods toward preparation of composite membranes can significantly improve the membrane performance or enable novel applications [2,3]. Here, important recent trends in the field will be illustrated with examples from own research.

In the first part, we will focus on the synthesis of polymers which can be used to obtain ultra- and nanofiltration membranes which can be used in organic solvents. In all cases, cross-linking of the polymer during membrane manufacturing is crucial, for adjusting the barrier properties or to obtain a solvent-stable membrane or both. Three different systems will be discussed, *i)* novel block copolymers from polyethyleneglycol and polyacrylonitrile for conventional phase separation and post-crosslinking, *ii)* modified commercial polyimides with photo-reactive groups for phase separation in combination with photo-crosslinking, and *iii)* tailoring of conditions to process commercial precursors for polyurethanes to obain thin-film composite membranes. The resulting novel membranes will also be compared with established materials.

In the second part, strategies toward stimuli-responsive membranes will be discussed with focus on materials where selectivity in the ultrafiltration range can be switched. Well-controlled grafting of a thermo-responsive polymer (cf. [4]) from the surface of ultrafiltration membrane pores can yield membranes where the size-selectivity can be tuned by the reaction conditions and the fractionation of nanoparticles can be reversibly switched by temperature. Reactive pore-filling with a hydrogel of tailored mesh size via *in situ* cross-linking polymerization is an alternative approach [5]. For another generation of such membranes, it will be possible to control the permeability and selectivity instead of heating the feed by "remote control"; this is possible by the integration of superparamagnetic nanoparticles in the membranes and excitation of the membrane by an external magnetic field.

All these strategies have relevance to widen the scope of the membrane applications and to contribute to sustainable technologies.

[1] E. Drioli, L. Giorno (Ed.), *Membrane operations: Innovative separations and transformations*, Wiley–VCH, Weinheim, **2009**.

[2] M. Ulbricht, *Polymer* **2006**, *47*, 2217.

[3] Q. Yang, N. Adrus, F. Tomicki, M. Ulbricht, *J. Mater. Chem.* **2011**, *21*, 2783.

[4] A. Friebe, M. Ulbricht, *Macromolecules* **2009**, *42*, 1838.

[5] N. Adrus, M. Ulbricht, *J. Mater. Chem.* **2012**, *22*, 3088.

(130b) Gas Separation Membranes: Issues Beyond the Upper Bound

Paul, D. R., The University of Texas at Austin

Originally presented on: 11/4/2013 12:45:00 - 13:08:00

In 1991 L. M. Robeson proposed that the well-known trade-off between the permeability of gases and the selectivity of gas permeation had an upper bound beyond which no polymers existed at that time. Finding polymers that exceed this upper-bound relationship for certain gas pairs has become a challenge being addressed in many research laboratories around the world. Another research trend has been to develop novel polymer structures that lead to unusually high levels of free volume in order to achieve very high gas permeability, but usually with a compromise in selectivity. Generally these searches for new membrane materials involve testing pure gas permeability in relatively thick films (25-100 microns) that are relatively easy to prepare in the laboratory.

Of course, the development of asymmetric and composite membranes with very thin dense "skins" needed to achieve high fluxes enabled the commercial use of membranes for molecular level gas separations. It has been generally assumed that these thin skins, with thicknesses of the order of 100 nm or less, have the same permeation characteristics as films with thicknesses of 25 microns or more. There is a growing body of evidence that this assumption can be in significant error, and use of thick film data to select membrane materials or predict performance should be done with caution.

This presentation will address issues for some of the new materials being explored for gas separation applications that go beyond where thick film data place them on the trade-off relationship and its upper bound. These include differences in behavior between thick laboratory films and thin films needed for high fluxes; these differences result from a more rapid physical aging and a more exaggerated response to plasticization by highly soluble gases as films become thinner. One focus will be on recent data for some high free volume polymers.

(130c) Permeability and Solubility of CO2 in Glassy Polymeric Membranes With and Without Plasticization: A Thermodynamic Approach

Minelli, M., Alma Mater Studiorum - Universite di Bologna
Sarti, G. C., Alma Mater Studiorum - Universite di Bologna

Originally presented on: 11/4/2013 13:08:00 - 13:31:00

The description of permeability and diffusivity of gases and vapors in glassy polymeric membranes has been obtained using a simple fundamental approach, which has been tested on a variety of different systems, and proved effective in all cases, with and without plasticization. The model relies on a notorious expression for the diffusive flux as the product of a kinetic factor, mobility, and a thermodynamic factor, associated to the concentration dependence of the chemical potential of the diffusing species. The latter factor can be obtained experimentally from solubility isotherm data, or can be evaluated theoretically by means of the well-known predictive procedures, as for instance the nonequilibrium lattice fluid (NELF) model or other equation-of-state models following the Non Equilibrium Thermodynamics for Glassy Polymers. On the other hand, the mobility factor is endowed with an exponential dependence on penetrant concentration, following the usual trend commonly found experimentally. Consequently, the resulting general permeability model contains only the two adjustable parameters entering the expression for mobility.

The solubility of the penetrant gas in the glassy membrane can be evaluated in a straightforward way through the NELF model, which proved appropriate to describe all the existing experimental sorption data for all the systems considered in this work, also in the presence of important swelling effects.

The final result is a rather simple and general expression for penetrant permeability, which contains at most two adjustable parameters: the gas mobility at infinite dilution and the coefficient characterizing the concentration dependence of mobility itself (plasticization factor). The model describes rather carefully the pressure dependence of permeability of CO_2 and other penetrants observed in a variety of different glassy polymers under steady state permeation, even in the presence of the so-called plasticization effects.

Remarkably, the same model parameters not only describe steady state permeation, but also describe the mass uptake versus time resulting from transient diffusivity, in all the cases inspected.

The present model has been tested with success for CO_2 permeability in several pure polymers as well as in a variety of complex heterogeneous systems, such as blends, random copolymers, glassy matrices with different degrees of crosslinks, and, further, in matrices additioned with low molecular weight species.

The case of highly plasticizing penetrants, as for instance hydrocarbons, in glassy polymers is also investigated, and the model proved his ability in describing the experimental trends.

Finally, it is also observed that the two parameters of the model follow general trends useful for predictive purposes. In particular, CO_2 mobility at infinite dilution correlates very well with the fractional free volume of the glassy matrix, while the plasticization factor correlates very well with the penetrant-induced polymer swelling.

(130d) Non-precious Metal Nanoparticle Catalyst for Fuel Oxidation

Greenlee, L. F., National Institute of Standards & Technology

Originally presented on: 11/4/2013 13:31:00 - 13:54:00

Electrochemical oxidation of small molecule fuels holds tremendous promise as the development of polymer electrolyte fuel cells continues. The advancement of catalyst design to include nanoparticle morphology and core-shell multi-metal structure has allowed the tuning of catalyst performance. The explosion of research on nanoparticle-based catalysis has demonstrated new synthesis strategies and has allowed the field to gain a better understanding of the connection between nanoparticle design and catalyst performance. However, key challenges remain, including nanoparticle durability and the reduction and elimination of precious metals. Our current work focuses on these challenges with the development of a completely non-precious metal nanoparticle catalyst that is able to successfully catalyze the electrooxidation of small molecule fuels, such as methanol and ethanol, in alkaline solution. This presentation will discuss our initial results on nanoparticle characterization and performance.

(130e) Dilute Methane Utilization Using A Membrane Gas Exchanger

Lin, H., Membrane Technology and Research, Inc.
Daniels, R.,
Merkel, T. C., Membrane Technology and Research, Inc.
Wijmans, J. G., Membrane Technology and Research, Inc
Originally presented on: 11/4/2013 13:54:00 - 14:17:00

Methane is a valuable energy resource and also the second largest contributor to global warming, after carbon dioxide. With almost twenty-one times the global warming effect of an equivalent mass of carbon dioxide, methane accounts for 10% of U.S. greenhouse gas emissions. Significant amounts of these emissions are dilute mixtures of methane and an inert gas (most commonly carbon dioxide), with a methane content below 40%. These dilute methane streams are not combustible when mixed with air, due to the low Btu values, and therefore, they are often vented. A need exists for an efficient technology to convert these waste streams into useful fuel, reducing global warming emissions.

This presentation will describe the development of a new membrane process to convert dilute, non-combustible methane/carbon dioxide gas mixtures into methane products that can be mixed with air and used as fuel. The approach is based on an innovative membrane device to selectively exchange carbon dioxide with oxygen from air. The resulting gas has a higher oxygen content and a lower carbon dioxide content than the feed gas, and is combustible when mixed with ordinary air. Results are reported for new high-performance membranes and modules that were fabricated and tested under simulated industrial conditions for concentrating methane. The energy efficiency and economics of the new membrane gas exchanger are analyzed, and the fuel value of the gas produced more than offsets the cost of the new process.

(130f) An Experimental and Modeling Study of CO2-Selective Membranes for Syngas Purification

Vakharia, V., The Ohio State University
Ramasubramanian, K., The Ohio State University
Ho, W. S. W., The Ohio State University
Originally presented on: 11/4/2013 14:17:00 - 14:40:00

As the awareness on global warming is growing, electricity generation from the Integrated Gasification Combined Cycle (IGCC) power plants seems to be a favorable alternative for power generation with simultaneous carbon capture (CO_2). In this case, CO_2 needs to be separated from syngas for the H_2 purification, where the purified syngas is the feed to the Gas Turbine (GT) for power generation. For the CO_2 capture in these power plants, conventional absorption process using Selexol as the solvent has been considered by DOE/NETL as the baseline technology. Recent data published in the DOE baseline document suggests that incorporating this technology for the CO_2 capture and sequestration in the IGCC power plants would result in an increase in the Cost of Electricity (COE) by 37%. Syngas from the low temperature Water Gas Shifter (WGS) reactor is at pressures close to 50 bar and at temperatures close to 235 ^oC.

This work describes the transport properties of our CO_2-selective membranes under conditions (pressure from 2 bar to 15 bar and temperature 107 ^oC) and the potential it has for operating at higher pressures (even higher than 20 bar) for CO_2 capture from syngas and apply it to the process that is integrated with the IGCC power plants. These CO_2-selective facilitated transport membranes are composed of amine carriers dispersed in the crosslinked polyvinylalcohol (PVA) matrix. CO_2 permeability and CO_2/H_2 selectivity at different pressures investigated are discussed. These membranes have been used for the H_2purification in a project funded by the Office of Naval Research.

This study deals with the modeling and cost study of a promising syngas purification process for IGCC plants with the CO_2capture using the membranes mentioned above. A two-stage process has been considered for this application. For both the stages, the effects of operating parameters like the feed pressures and temperatures on the overall economics were analyzed to obtain an optimal condition of operation. We have also studied the effects of membrane performance, selectivity and permeance on the overall increase in the COE and the capture cost.

A CO_2 permeability of 1287 Barrers and a CO_2/H_2 ideal selectivity of 87 at about 15 bar pressure were obtained from our group. From a study in our group on the transport properties at lower pressures (2 bar), a CO_2 permeability of 5000 Barrers and CO_2/H_2 ideal selectivity of 250 was obtained. Based on the transport results, our cost calculations show an increase in the COE of about 15.3% and a capture cost about \$17/tonne of CO_2 removed for the two-stage membrane based process for achieving about 90% CO_2 capture (>95% purity) and greater than 99% H_2 recovery. This process operates at pressures close to 23 and 2 bar on the feed side and the sweep side of the Stage 1, respectively. The membrane with a CO_2 permeance of 162 GPU, CO_2/H_2 ideal selectivity of 49 and with a thickness of 5 microns was considered. The transport properties of the membrane at 23 bar have been estimated from a study on the membrane performance-property relationship, and they will be verified experimentally. The feed pressure on the Stage 2 is close to 3 bar and the sweep side at atmospheric pressure. The Stage 2 membrane has a thickness of 5 microns with a CO_2 permeance of 1000 GPU and a CO_2/H_2 selectivity of 250. This study shows the potential our membrane based technology can have for the IGCC application.

(130g) The Use of Membranes to Lower the Cost of Hydrogen Production With Concurrent Carbon Dioxide Sequestration

Baker, R. W., MTR
Merkel, T. C., Membrane Technology and Research, Inc (MTR)
Zhou, M., Membrane Technology and Research, Inc.
Thomas-Droz, S. J., Membrane Technology and Research, Inc.
Amo, K. D., MTR
Lin, H., Membrane Technology and Research, Inc.
Originally presented on: 11/4/2013 14:40:00 - 15:03:00

Methods of producing fuel and electric power without concurrently emitting carbon dioxide to the atmosphere are being sought. One approach is to produce syngas (carbon monoxide and hydrogen) from coal or natural gas, convert the carbon monoxide to carbon dioxide through the water gas shift (WGS) reaction, producing carbon dioxide and more hydrogen, and then separate the resulting carbon dioxide/hydrogen mixture. The separated carbon dioxide can be sequestered underground.

This process is the basis of the proposed hydrogen economy. A key issue in this scheme is the hydrogen/carbon dioxide separation step. Currently, most process developers use the Selexol process (carbon dioxide absorption into cold oligomers of ethylene glycol), which is a safe but expensive choice. This paper describes the development of new carbon dioxide/hydrogen- and hydrogen/carbon dioxide-selective membranes that can be used to simultaneously sequester carbon dioxide and provide a high quality hydrogen fuel. Very significant cost savings can be achieved. The technology is at the bench/pilot plant stage.

(226e) A Sustainable Chemical Industry May Imply Dispersed Manufacturing

Cussler, E. L., University of Minnesota
Tiffany, D., University of Minnesota
Originally presented on: 11/4/2013 12:30:00 - 13:00:00

A SUSTAINABLE CHEMICAL INDUSTRY MAY IMPLY DISPERSED MANUFACTURING

E. L. Cussler and Douglas G. Tiffany

Departments of Chemical Engineering and Materials Science, and of

Agricultural Economics

University of Minnesota, Minneapolis MN 55455

A "sustainable chemical industry" implies using renewable feeds, especially those derived from agriculture. The ease of doing so depends on the amount of chemical produced. For example, antibiotics like penicillin already use agricultural waste products as a feed. The amount of antibiotic produced is tiny relative to commodity chemicals like olefins, and the transportation cost of transporting this high value-added product to market is minor. Thus the production of antibiotics will remain centralized, as it is now.

In contrast, the production of fuels from biomass will involve huge amount of feed and of product of much less value per kilo than antibiotics. Transportation costs of moving both feed and fuel will be major. As a result, the production of fuels will probably involve many smaller plants, whose lower efficiency is balanced by lower costs of transportation.

Thus a sustainable chemical industry may involve dispersed manufacture, rather than the highly centralized plants characteristic of today's industry. This paper will discuss such a future for ammonia, the key chemical fertilizer, which can be made from air, water, and wind power. We will describe what such a chemical plant could look like, and the advantages and debits of such dispersed manufacture.

(142d) Advanced Polymeric Membranes With Tailored Surface Properties

Ulbricht, M., Universitet Duisburg-Essen
Birkner, M., Universitet Duisburg-Essen
Frost, S., Universitet Duisburg-Essen
Kumar, M.,
Li, M.,
Quilitzsch, M.,
Pieper, T.,
Stahra, N.,
Sulc, R.,
Originally presented on: 11/4/2013 13:00:00 - 13:30:00

Advanced polymeric membranes with tailored surface properties

Mathias Ulbricht, Marc Birkner, Sven Frost, Mahendra Kumar, Miao Li, Mathias Quilitzsch, Thorsten Pieper, Nico Stahra, Robert Sulc

Lehrstuhl für Technische Chemie II, Universität Duisburg-Essen, 45117 Essen, Germany; mathias.ulbricht@uni-essen.de

Membrane technologies have been established in a wide range of industrial processes. However, in advanced separation and reaction engineering, membranes applied as separator or contactor can offer yet many more distinct advantages [1].

The majority of synthetic membranes are made from polymers because barrier and surface properties can be varied in wide ranges with help of established scalable manufacturing processes. Significant efforts have been made to further improve membrane performance by focusing on barrier properties (high selectivity, high flux) or surface properties (affinity, antifouling). Membrane fouling is still a major obstacle because it reduces productivity and product quality and increases energy consumption and need of additional chemicals for cleaning; hence, efficiency and sustainability of membrane processes are reduced.

In this presentation, we will discuss how tailored surface functionalizations can significantly improve the performance of established membranes and can also contribute to the development of membranes using novel separation principles. With respect to the preparation/manufacturing, two general routes will be outlined, *i)* membrane formation from functional copolymers or blends of established membrane polymers with functional copolymer additives, and *ii)* post-functionalization of membranes by (reactive) coating or grafting methods (cf. [2,3]). Important recent trends in the field will be illustrated with examples from own research.

One focus will be on ultra- and microfiltration membranes with hydrophilic, low-fouling surfaces. Such membranes can be obtained by phase separation of polymer solutions (route *i)*), either using tailored block copolymer additives (e.g. [4,5]) to established membrane polymers or *de novo* synthesized copolymers [6]. If the hydrophilic functional blocks comprise also groups which can reversibly bind certain molecules or particles, the resulting membranes can be used for combined separations using both size-selective barrier and adsorber surface properties. Another combined separation mechanism is the charge-enhanced selectivity in ultrafiltration, for instance of proteins [6].

With respect to route *ii)*, we will discuss recent progress toward ultra- and nanofiltration membranes with reduced fouling and improved cleaning. Thin-film hydrogel composite membranes obtained by graft copolymerization of hydrogels initiated at the membrane surface had been tailored with respect to the important parameters selectivity, flux and antifouling, either for selective separation of proteins and other nanoparticles [7] or for fractionation of smaller molecules and ions [8]. With zwitterionic grafted polymers and their stimuli-responsive properties it is also possible to develop more active antifouling and cleaning strategies, for instance based on a reversible conformation of the grafted "protective" layer change between operation and cleaning conditions. Finally, we will demonstrate that well-controlled grafting of a thermo-responsive polymer from the surface of ultrafiltration membrane pores can yield membranes where the size-selectivity can be tuned by the reaction conditions and the fractionation of nanoparticles can be reversibly switched by temperature.

All these strategies have relevance to widen the scope of the membrane applications and thus to contribute to the implementation or improvement of sustainable technologies.

[1] E. Drioli, L. Giorno (Ed.), *Membrane operations: Innovative separations and transformations*, Wiley–VCH, Weinheim, **2009**.

[2] M. Ulbricht, *Polymer* **2006**, *47*, 2217.

[3] Q. Yang, N. Adrus, F. Tomicki, M. Ulbricht, *J. Mater. Chem.* **2011**, *21*, 2783.

[4] H. Susanto, N. Stahra, M. Ulbricht, *J. Membr. Sci.* **2009**, *342*, 153.

[5] E. Berndt, M. Ulbricht, *Polymer* **2009**, *50*, 5181.

[6] M. Kumar, M. Ulbricht, *RCS Advances* **2013**, *accepted*.

[7] P. D. Peeva, T. Knoche, T. Pieper, M. Ulbricht, *Ind. Eng. Chem. Res.* **2012**, *51*, 7231.

[8] R. Bernstein, E. Antón, M. Ulbricht, *ACS Appl. Mater. Interf.* **20**

(142b) New Material Development for Membrane Separations

Noble, R. D., University of Colorado

Originally presented on: 11/4/2013 13:30:00 - 14:00:00

Synthetic membranes are important for many separation applications, such as the separation and purification of atmospheric gases for medical and industrial use, and salt removal from seawater in regions lacking fresh water sources. Unlike mesh filters used to separate large solid particles from a fluid (phase difference), membranes used to separate molecular species from one another (i.e., a chemical separation), is a much more difficult process [given the size scale and level of mixing of the components].

Because of the benefits of membrane-based chemical separations over other methods (e.g., smaller device footprints, lower energy use compared to distillation), there has been recent interest in applying membranes to new separation problems, as well as improving membrane materials used in existing separations. Consequently, there is a need for better membrane materials in general (i.e., with better selectivity, better productivity, longer life and operational stability, increased operating temperature range, use in chemically challenging environments, and the ability to adjust materials properties and performance for a given application).

The new materials are room temperature ionic liquids (RTILs), zeolites and lyotropic liquid crystals (LLCs). Each of these materials has unique physiochemical properties that provide an advantage in certain applications. Each of these materials will be discussed and their use in gas and water treatment.

For researchers working on the design of new membrane materials, it is important to not only consider the design factors that afford better separation selectivity but to also consider the other factors that afford good productivity as well. Often it is these materials engineering factors that decide whether a membrane material is viable in practical separation or not.

(142c) Hybrid Membranes for Improved Molecular Separation in Gas and Liquid-Phase

Tsuru, T., Hiroshima University

Originally presented on: 11/4/2013 14:00:00 - 14:30:00

Hybrid membranes have been investigated for improved performance which cannot be achieved by using only one type of materials. In this talk, our recent progresses on hybrid membranes will be introduced, including organosilica where Si was hybrid with organic functional groups on the molecular level, and mixed-matrix-membranes where inorganic nanoparticles were hybrid with polyamide during interfacial polymerization.

Amorphous silica membranes show highly permselectivity for hydrogen; however, there still remain two technical challenges: pore size control and hydrothermal stability.[1)] In order to improve hydrothermal stability of silica membranes, Castricum et al.[2)] developed hybrid silica membranes using bis (triethoxysilyl)ethane (BTESE), which showed excellent hydrothermal stability in pervaporation of aqueous alcohol solutions because of incorporated organic linking groups (Si-C-C-Si unit) in silica networks. We recently proposed a new strategy to control pore sizes of silica membranes using bridged alkoxides.[3-8)] Bis (tri-ethoxysilyl) ethane (BTESE), consisting of ethane group between 2 silicon atoms (-Si-CH_2-CH_2-Si-) as the minimum unit in hydrolysis and condensation reaction, was found to lead to loose silica networks. BTESE silica membranes showed approximately one order magnitude high H_2 permeance (~10^{-5} mol/(m^2 s Pa)) compared with previously reported silica membranes using TEOS (tetraethoxysilane), and a high H_2 to SF_6 permeance ratio of more than 1,000 with a low H_2 to N_2 selectivity (~10).[3)] Applications, including gas separation (hydrogen, organic gas[3-6)]) and liquid phase separation (reverse osmosis[7)], pervaporation[8)]), will be also discussed. BTESE membrane showed quite high rejection for NaCl and high stability to chlorine in reverse osmosis.

Regarding polyamide (PA) membranes for reverse osmosis, we proposed spray-assisted, 2-step interfacial polymerization[9)]. We found the formation of multilayered ridge-and-valley structures via the first spray-assisted interfacial polymerization, leading enhanced water permeation due to increased surface area. Zeolite Y nanoparticles, which were embedded into PA layer in the first-spray process, were confirmed to increase water permeability.

References

1) T. Tsuru, *J. Sol-Gel Sci. Technol.*, 46 (2008) 349; *J. Japan Petroleum Institute* 54 (2011) 277. 2) H.L. Castricum, A. Sah, R. Kreiter, D.H.A. Blank, J.F. Vente, J.E. ten Elshof,, *J. Mater. Chem. 18 (*2008) 2150. 3) M. Kanezashi, K. Yada, T. Yoshioka, T. Tsuru, *J. Am. Chem. Soc.* 131 (2009) 414. 4) H. R. Lee, M. Kanezashi, T. Yoshioka, T. Tsuru, *AIChE J.* 57 (2011) 2755. 5) G. Li, M. Kanezashi, T. Tsuru, *J. Membr. Sci.* 379 (2011) 287. 6) M. Kanezashi, M. Kawano, T. Yoshioka, T. Tsuru, *Ind. Eng. Chem. Res.* 51 (2012) 944. 7) R. Xu, J. Wang, M. Kanezashi, T. Yoshioka, T. Tsuru, *Langmuir* 27 (2011) 13996. 8) T. Tsuru, T. Shibata, M. Kanezashi, T. Yoshioka, *J. Membr. Sci.* 421-422 (2012) 25-31 9) T. Tsuru, S. Sasaki, T. Kamada, T. Shintani, T. Ohara, H. Nagasawa, K. Nishida, M. Kanezashi, T. Yoshioka, *J. Membr. Sci.* submitted.

(142a) Nature-Inspired Geomimetic Separation Membranes

Tung, K. L., National Taiwan University

Originally presented on: 11/4/2013 14:30:00 - 15:00:00

There are many lessons that one could learn from, and/or mimic in nature. In the past decades, biomimetic concept has been widely adopted for biomaterials and polymeric materials designs by scientists and biologists. The biomimetic concept has promoted a fantastic progress in the science and engineering of bio/organic materials. Nevertheless, the geomimetic concept was astonishingly few to be taken into consideration for inorganic materials development. Recently, several new concepts to develop novel inorganic membrane materials have been proposed and new preparation methods from a fundamental understanding of formation mechanism and morphology of natural materials haven been designed. In this talk, several potential routes intentionally mimic existing geological processes to develop new inorganic membranes will be presented.

(146a) Development of Composite Hollow Fiber Membranes for Aqueous and Non-Aqueous Nanofiltration

Sun, S. P., National University of Singapore
Chan, S. Y., National University of Singapore
Chung, T. S., National University of Singapore
Originally presented on: 11/4/2013 12:30:00 - 12:55:00

Nanofiltration (NF) has been growing rapidly in pharmaceutical separation, wastewater treatment, seawater desalination and many other applications. This presentation focuses on the new developments of composite NF hollow fiber membranes from both co-extrusion and interfacial polymerization (IP). Firstly, a highly water permeable dual-layer NF hollow fiber membrane fabricated by the simultaneous co-extrusion of Torlon® polyamide-imide (PAI) and cellulose acetate (CA) polymer solutions through a triple-orifice spinneret in a dry-jet wet phase inversion process. As compared to conventional membranes, the dual-layer hollow fiber membranes developed here have a unique structure which consists of (1) controllable thin selective outer layer; (2) innovatively designed interface, which is finely connected between outer and inner layers, does not contribute significant transport resistance while still maintains good mechanical strength; (3) fully porous substructure with reduced transport resistance. The cross-section struction and the nanopores of dual-layer hollow fiber membranes were molecularly designed by controlling the phase inversion process with the aid of various non-solvent additives into the polymer solutions. The newly developed NF dual-layer hollow fiber membrane with methanol as additive has a relatively high pure water permeability of 11.93 $l\ m^{-2}bar^{-1}h^{-1}$ with a mean effective pore radius of 0.63 nm and a more than 96.5% rejection of Na_2SO_4.

Secondly, a novel thin-film composite NF membrane was fabricated by IP of hyperbranched polyethyleneimine (HPEI) and isophthaloyl chloride (IPC) on a dual-layer hollow fiber membrane. The substrate possesses a unique cross-section structure comprising a layer full of macrovoids located in the middle and sandwiched by two thin spongy-like layers. Such a sandwich-like sublayer structure provides minimal transport resistance and sufficient mechanical strength for water permeation under high pressures. After IP, the NF membrane possesses a negatively charged substrate and a positively charged selective layer with a mean effective pore radius of 0.36 nm, molecular weight cut off of 500 Da, and pure water permeability of 4.9 $lm^{-2}bar^{-1}h^{-1}$. Due to this double-repulsion effect, together with the steric-hindrance and the solute electro-neutrality effects, by adjusting the pH of cephalexin aqueous solution to modify the ionization states of this zwitterionic molecule, the NF membrane shows high rejections over a wide pH range. Furthermore, the latest preliminary data shows the membrane is very stable in harsh organic solvents such as dimethyl sulfoxide. Therefore, the NF membrane may potentially be useful to reduce waste, recycle valuable pharmaceuticals and reuse solvents for the pharmaceutical industry.

(146b) Self-Heat Recuperative Seawater Desalination Process Using Fluidized Bed

Mizuno, H., The University of Tokyo
Kansha, Y., The University of Tokyo
Ishizuka, M., The University of Tokyo
Tsutsumi, A., The University of Tokyo

Originally presented on: 11/4/2013 12:55:00 - 13:20:00

Recently, it is presumed that there will be a serious shortage of water in several years later due to the high growth rate of population and industrial development. To alleviate the water shortage, more attention has been paid to desalination technology. One of the typical desalination processes is a multi-stage flash desalination. This process is categorized as thermal desalination process because it produces pure water by evaporation. The thermal process has the following two major advantages; seawater quality does not seriously affect the operation condition and product water salinity is low enough for industrial usage. However, this process consumes a lot of energy, thus the remarkable energy saving technology of thermal process is in great demand. So far, to reduce the energy consumption, we proposed a novel thermal desalination based on self-heat recuperation, which is a method for energy saving. By a process simulator, it was revealed that the proposed desalination process energy consumption is reduced to 1/4 of conventional thermal desalination process. Furthermore, it was elucidated that the specific energy consumption decreased as the recovery ratio (= the product water flow rate / the feed water flow rate) increased. However, in the thermal desalination process with high recovery ratio, a problem of scale deposition, which leads to a reduction of heat transfer efficiency in heat exchanger occurs.

In this research we introduce self-heat recuperative seawater desalination process with fluidized bed for solving the problem. In the desalination process, the fluidized bed is employed as an evaporator. It might be seen that this process has the following advantages; good heat transfer properties and less scale deposition because of water evaporation taking place on heated particles in fluidized bed evaporator.

In addition, we investigated the influence on the specific energy consumption and feasibility to the society of the proposed desalination process.

(146c) Characteristics of Water and Salt Transport, Free Volume and Their Relationship With the Functional Groups of Novel Cellulose Esters

Ong, R. C., National University of Singapore
Helmer, B. J., Eastman Chemical Company
de Wit, J. S., Eastman Chemical Company
Chung, T. S., National University of Singapore
Originally presented on: 11/4/2013 13:20:00 - 13:45:00

Fundamental transport properties including salt and water partition coefficients, permeability and diffusivity of various newly synthesized cellulose esters were evaluated in this study in order to investigate the relationship among them as a function of molecular structure. Dense flat cellulose ester films were prepared and studied so that the influence of membrane fabrication, morphology and processing history on transport properties can be minimized. Annihilation lifetime spectroscopy (PALS) was employed to characterize the free volume of these cellulose esters. It was found that the transport and free volume properties can be correlated with the functional groups and their content in cellulose esters. A high hydroxyl content leads to a high hydrophilicity but results in a high water and salt partition coefficient. Hydrophilic cellulose esters suffer from a low water/salt selectivity. On the other hand, an increase in bulky and hydrophobic propionyl or butyryl groups generally enhances cellulose esters' selectivity but brings down the permeability. However, a very high content of bulky functional group causes a drop in selectivity due to poor chain packing and the enlarged free volume as confirmed by PALS. Experimental results suggest that cellulose esters with moderate content of both hydroxyl and bulky side groups have the best diffusivity selectivity and may be suitable as membrane materials for salty water separation. An upper bound relationship between solubility selectivity and water partition coefficient was also observed in these cellulose esters as suggested in previous literatures for other polymers.

(146d) Bioremediation of Cu(II) Contaminated Aqueous Solution Using Acinetobacter Sp

Majumder, S., Birla Institute of Technology & Science, Pilani
Gangadhar, G.,
Raghuvanshi, S., Birla Institute of Technology & Science, Pilani
Gupta, S., Birla Institute of Technology & Science, Pilani
Originally presented on: 11/4/2013 13:45:00 - 14:10:00

Bioremediation of metal wastes using enriched mixed consortium of microbial species is a very promising and viable alternative in environmental biotechnology. Present study deals with the removal of divalent copper [Cu(II)] compound from aqueous solution using *Acinetobacter sp.* (GenBank Accession Number: HQ538658.1), an indigenous bacterial strain which is isolated from enriched mixed culture. Enrichment of mixed culture is carried for a period of 6 days. At the end of the enrichment period, the biomass concentration obtained is 0.0187 g/L of solution. The effect of different operating conditions such as contact time, initial pH and amount of nutrient dosage on bioremediation of Cu(II) are studied in a batch mode. The effect of contact time is studied for an initial Cu(II) concentration ranging from 60-150 mg/L. Increase in initial Cu(II) concentration from 60 mg/L to 150 mg/L shows better removal of Cu(II). The effect of initial pH is also studied by varying the solution pH in the range of 2 - 12. The maximum biomass growth obtained is 0.112 g/L at pH = 7.5 for initial Cu(II) concentration of 150 mg/L. The maximum percentage removal of Cu(II) is obtained for the initial Cu(II) concentration of 60 mg/L at the solution pH value of 7.5. Percentage removal of Cu(II) value is also compared with reported value of percentage removal of Cu(II). This study also shows that the microbial growth for Cu(II) removal is significantly affected by nutrient dosage. The optimum biomass growth obtained is 0.0248 g/L at double of the standard composition of MSM.

Keywords: *Acinetobacter sp., Divalent copper; Bioremediation; Acclimated mixed culture, MSM.*

(146e) Calcium Ions in Gypsum (CaSO4·2H2O) Scaling in Forward Osmosis Processes and a Novel Design Strategy to Fabricate Polybenzimidazole Anti-Scaling Hollow Fiber Membranes

Chen, S. C.,

Su, J., National University of Singapore

Fu, F. J.,

Mi, B., The George Washington University

Chung, T. S., National University of Singapore

Originally presented on: 11/4/2013 14:10:00 - 14:35:00

Calcium ions in gypsum ($CaSO_4$°¤$2H_2O$) scaling in forward osmosis processes and a novel design strategy to fabricate polybenzimidazole anti-scaling hollow fiber membranes

Si Cong Chen[a,b], Jincai Su[b], Feng-Jiang Fu[b] , Baoxia Mi[c], Tai-Shung Chung[a,b,*]

[a] NUS Graduate School for Integrative Sciences & Engineering (NGS), National University of Singapore, 28 Medical Drive, Singapore 117456, Singapore

[b] Department of Chemical and Biomolecular Engineering, National University of Singapore, 4 Engineering Drive 4, Singapore 117576, Singapore

[c] Department of Civil and Environmental Engineering, University of Maryland, College Park, MD 20742, USA

* Corresponding author: chencts@nus.edu.sg

Abstract

Gypsum ($CaSO_4$°¤$2H_2O$) scaling phenomena have been studied on hollow fiber membranes with different physicochemical properties in forward osmosis processes. Three types of hollow fiber membranes were demonstrated i.e., (1) cellulose acetate (CA), (2) polybenzimidazole (PBI)-polyethersulfone (PES) and (3) PBI-polyacrylonitrile (PAN) - polyhedral oligomeric silsesquioxane (POSS). For the first time, we have found that calcium ions not only dominate but also take a leading role on gypsum ($CaSO_4$°¤$2H_2O$) scaling through ionic interactions with the membrane surface. A high level of calcium ions in scaling solutions can initiate gypsum crystallization more effectively than a high level of sulfate ions. Based on this finding, for the first time, a novel strategy to molecularly engineer PBI hollow fiber membranes with superior anti-scaling properties has been proposed and demonstrated. As shown in Figure 1c, the newly designed PBI-PAN (POSS) membranes have a positively charged and smooth membrane surface at the nanoscale level but possess a ridge and valley structure at the micrometer level. It shows slightly positive charge at a low pH and exhibits minimal scaling of only 5% flux reduction in contrast to CA membranes with 63% flux reduction. The adhesion force measured by atomic force microscopy (AFM) confirms a strong repulsive force between gypsum particles and PBI (Figure 1b) at pH values lower than the isoelectric point of 5.7. The repulsion force leads to the anti-fouling properties, while the novel ridge and valley morphology in the micrometer scale facilitates surface turbulence flow and effectively remove foulants.

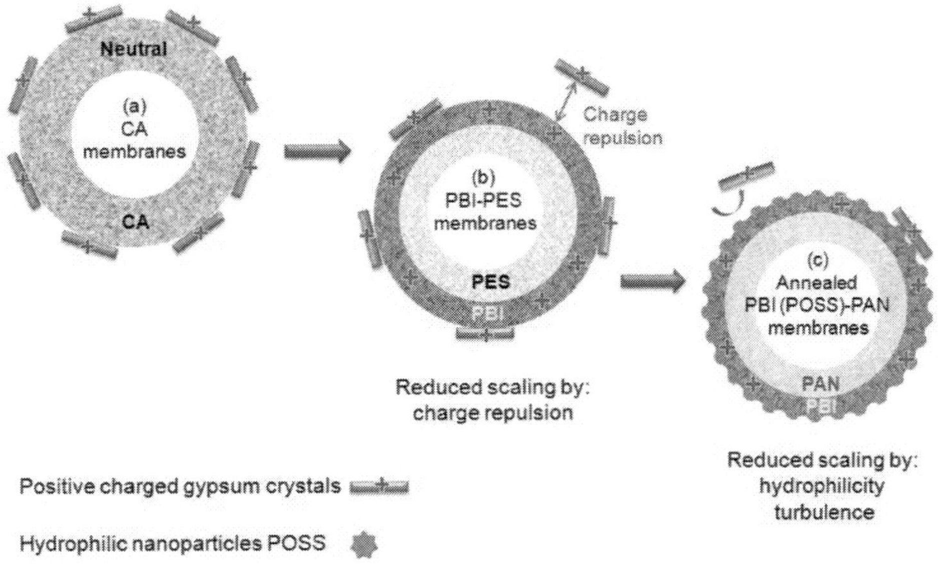

Figure 1: Fouling mechanisms of gypsum scaling on membranes

(146f) New Insights Into Hydrophobic/Hydrophilic Composite Hollow Fibers for Direct Contact Membrane Distillation

Zhu, J. M., Central South University
Jiang, L. Y., Central South University
Originally presented on: 11/4/2013 14:35:00 - 15:00:00

Hydrophobic/hydrophilic composite membrane is a new design of membrane in terms of structures and materials for DCMD. With dry-wet phase inversion process, a novel type of hydrophobic/hydrophilic hollow fiber membranes was fabricated in this research. The outer layer was made from polyvinylidene fluoride (PVDF), while the inner layer PVDF and polyvinyl alcohol (PVA) blend. Effects of dope composition, PVA/PVDF blending ratio and coagulants on the morphological, physiochemical and mechanical characteristics of the membranes thus generated were investigated. The pore size and porosity in respective layer of the dual-layer membranes were determined mainly by data obtained from SEM and corresponding software processing. The DCMD performance observed in experiments was compared with that obtained by mass and heat transport modeling of both single-layer and dual-layer DCMD membrane. Analysis relating the membrane structural features and the above comparison revealed that the inner layer characteristics in terms of both material hydrophilicity and pore size should be optimized to give a better membrane performance. The maximum flux in DCMD obtained was around $24.4 kgm^{-2}h^{-1}$ at 65℃.

(154a) Steam Stability of Alumina Supported Amine Materials for CO2 Capture From Ultra Dilute Gas Streams

Sakwa-Novak, M., Georgia Tech
Jones, C. W., Georgia Institute of Technology
Originally presented on: 11/4/2013 15:15:00 - 15:35:00

Solid supported amines are a promising class of adsorbents for CO_2 capture from large point sources such as post combustion flue gases, and also from ultra-dilute sources such as air. While much of the research community has focused on synthesizing materials with increased CO_2 capacity, addressing other practical issues, such as sorbent regeneration, remains a critical research need. Recently, we have shown that low grade steam can be used to regenerate supported amine sorbents, and that alumina is a promising support for amines in this application. Here, we further investigate the stability of alumina supported amine adsorbents under steaming conditions. Studies are conducted using a packed bed adsorption system capable of sorbent regeneration under flowing steam. We study both polymeric amines as well as aminosilanes grafted onto several alumina supports. Changes in the CO_2 capacities and materials properties of the adsorbents are presented after extended exposure to flowing steam, as well as exposure to flowing steam on the timescale of one desorption cycle. It is found that the alumina support shows structural changes after extended exposure to steam, and these changes appear to be influenced by the presence and type of amine incorporated into the material.

(154b) Separation of CO2 From Biogas, Natural Gas and Flue Gas Mixtures With Elastic Layered Metal-Organic Framework (ELM) Adsorbents

Lastoskie, C. M., University of Michigan
Trinh, T. D., University of Michigan
Sotomayor, F., University of Michigan
Gloumakov, R.,
Berendsen, S., Univ
Originally presented on: 11/4/2013 15:35:00 - 15:55:00

The suitability of elastic layered Metal-organic framework (ELM) as gas separators for CO_2-CH_4 and CO_2-N_2 mixtures from biogas, natural gas and flue gas was evaluated. Simulated CO_2 adsorption capacities on ELM-11 [Cu(BF4)2(bpy)2] and ELM-12 [Cu(OTf)2(bpy)2] above the gating pressure and CO_2 isosteric heat of adsorption agree well with experimental data. Calculated CO_2 selectivities from those mixtures based on pure-component adsorption isotherms generally agree with the Grand Canonical Monte Carlo (GCMC) simulations. GCMC simulated CO_2 selectivity from biogas mixture ranges from 10-45 and CO_2 selectivity from flue gas mixture ranges from 30-85 show that ELM materials are good candidates for gas separation.

Despite these promising results, only pure component data has been studied and no attempts for a real separation experiment in a fixed adsorbent bed have been made. In this presentation, the result of the first fixed bed packed with ELM-adsorbent for the separation of biogas mixture will be discussed. We also present density functional theory (DFT) results on the nature of the CO_2 and CH_4 separation in ELM-11 and ELM-12 framework. With the adsorption of CO_2 on framework the metal-anion bond in ELM framework was extended while that bond is virtually unchanged with CH_4 adsorption. Differences in the binding motif and energy of CO_2 and CH_4 with ELM frameworks using density functional theory calculation will also be discussed.

(154c) Immobilization of Carbonic Anhydrase On Ordered Mesoporous Alumina for Use in Biocatalytic Carbon Dioxide Capture

Salem, S., University of Cincinnati
Thiel, S. W., University of Cincinnati
Originally presented on: 11/4/2013 15:55:00 - 16:15:00

Carbon dioxide is considered to be a major greenhouse gas involved in global climate change, and the CO_2 concentration in the atmosphere has increased from its pre-industrial value of 280 ppmv to 455 ppmv [1]. This increased CO_2 concentration could, in turn, cause the average temperature at the Earth's surface to increase 1.8-4.0 K above the 1990 levels by the end of this century [2]. This warming is anticipated to cause sea level rise, increased intensity and frequency of extreme weather events, ice shelf disruption, and changes in rainfall patterns. Thus, much research is focused on developing methods to reduce CO_2 production and release into the atmosphere. CO_2 production can be reduced by using less carbon-intensive energy sources (wind, solar, and nuclear energy) and by increasing the efficiency of the energy production and use [3]. However, these strategies are not sufficient to meet CO_2 reduction targets.

Since combustion of carbon fuels will likely be a primary source of energy for the foreseeable future, it is critical to capture and sequester the resultant CO_2 [3, 4]. Capturing CO_2 from flue gas using absorption processes involves CO_2 dissolution in the aqueous phase, hydration, dissociation to form bicarbonate, and finally formation of carbonate; the hydration step is often rate-controlling. Fortunately, a catalyst for the rapid reversible hydration of CO_2 exists in biological systems: carbonic anhydrase (CA) [5-11]. This enzyme is so highly adapted to the hydration of CO_2 that reaction rates are almost diffusion limited [12].

CA functions by activating hydroxyl ions formed through ligation to a zinc atom bound within the enzyme [13]. The enzyme further holds CO_2 in the active site close to the activated hydroxyl group, while preventing water from gaining access to the hydroxyl group. The enzyme thus creates an extremely basic microenvironment that allows for the rapid formation of carbonate ions to occur without altering the pH of the bulk solution.

The broad objective of the research reported in this presentation to develop a biocatalytic membrane for the capture CO_2 from combustion flue gas; the membrane consists of an inert matrix in which carbonic anhydrase (CA) is immobilized. Specifically, previously presented work demonstrating the temperature tolerance, pH tolerance, and storage stability of CA adsorbed on ordered mesoporous alumina (OMPA) is extended to amine-functionalized alumina. The implications for the performance of a biocatalytic membrane incorporating CA are also explored.

References

[1] Rogner, H.-H., D. Zhou, R. Bradley. P. CrabbÃ©, O. Edenhofer, B.Hare (Australia), L. Kuijpers, M. Yamaguchi, 2007: Introduction. In Climate Change 2007: Mitigation. Contribution of Working Group III to the Fourth Assessment Report of the Intergovernmental Panel on Climate Change [B. Metz, O.R. Davidson, P.R. Bosch, R. Dave, L.A. Meyer (eds)], Cambridge University Press, Cambridge, United Kingdom and New York, NY, USA.

[2] Fisher, B.S., N. Nakicenovic, K. Alfsen, J. Corfee Morlot, F. de la Chesnaye, J.-Ch. Hourcade, K. Jiang, M. Kainuma, E. La Rovere, A. Matysek, A. Rana, K. Riahi, R. Richels, S. Rose, D. van Vuuren, R. Warren, 2007: Issues related to mitigation in the long term context, In Climate Change 2007: Mitigation. Contribution of Working Group III to the Fourth Assessment Report of the Inter-governmental Panel on Climate Change [B. Metz, O.R. Davidson, P.R. Bosch, R. Dave, L.A. Meyer (eds)], Cambridge University Press, Cambridge, United Kingdom and New York, NY, USA.

[3] Reichle, D., Houghton, J., Kane, B., and Ekmann, J. 1999. Carbon sequestration: Research and development. *U.S. DOE report.*

[4] Herzog, H. J. and Drake, E. M. 1996. Carbon dioxide recovery and disposal from large energy systems. *Annual Review of Energy and the Environment.* 21.145-166.

[5] Brown, R.S., in Aresta, M. and J.V. Schloss, editors, *Enzymatic and Model Carboxylation and Reduction Reactions for Carbon Dioxide Utilization, NATO ASI Series, Series C: Mathematical and Physical Sciences, vol. 314,* Kluwer Academic Publishers, Dordrecht (1990), 145-180.

[6] Dodgson, S.J., R.E. Tashian, G. Gros and N.D. Carter, editors, *The Carbonic Anhydrases: Cellular Physiology and Molecular Genetics,* Plenum Press, New York (1991).

[7]Pocker, Y., in Aresta, M. and J.V. Schloss, editors, *Enzymatic and Model Carboxylation and Reduction Reactions for Carbon Dioxide Utilization, NATO ASI Series, Series C: Mathematical and Physical Sciences, vol. 314,* Kluwer Academic Publishers, Dordrecht (1990), 129-143.

[8] Khalifah R.G., The Carbon dioxide hydration activity of carbonic anhydrase I. stop-flow kinetic studies on the native human isoenzymes B and C, The Journal of Biological Chemistry, 1971,246(8), 2561-2573.

[9] Bond G.M., Stringer J., Brandvold D.K., Simsek F.A., Medina .G., and Egeland G.2001. Development of Integrated System for Biomimetic CO2 SequestrationUsing the Enzyme Carbonic Anhydrase. *Energy & Fuels*, 15, 309-316. 7

[10] Cheng Li-H., Zhang L., Chen Huan-L., and Cong-Jie Gao Cong-J. 2008. Hollow fiber contained hydrogel-CA membrane contactor for carbon dioxide removal from the enclosed spaces. *Journal of Membrane Science* 324 33–43.

[11] Trachtenberg, M.C., Lihong, B., CO2 Capture: Enzyme vs. Amine, Proc. The 4th Annual Conference on Carbon Capture and Sequestration DOE/NETL (2005).

[12] Khalifah, R. and Silverman D.N., *Carbonic Anhydrase Kinetics and Molecular Function, The Carbonic Anhydrase*, Plenum Press, New York, 1991, pp. 49-64.

[13] Silverman D.N. and S.H. Vincent, “Proton Transfer in the Catalytic Mechanism of Carbonic Anhydrase.” *CRC Crit Rev Biochem*. 1983, 14, 207-55.

(154d) Adsorption Equilibrium Studies of Bio-Based Fermentation Broth (Acetone, Butanol, and Ethanol) Onto Immobilization of Potato Starch Sorbents

Indriani, D. W., Chung Yuan Christian University
Chung, T. W., Chung Yuan University
Originally presented on: 11/4/2013 16:15:00 - 16:35:00

Bio-based butanol has superior properties when compared to ethanol to be the gasohol and is gradually considered to be an important biofuel from the biomass fermentation. In this study, the high Si/Al ratio zeolites and self-made silica immobilized potato starch were used as sorbent. To evaluate the competitive adsorption behaviour of acetone, butanol and ethanol (ABE) solution, the gravimetric adsorption equilibrium apparatus and the dynamic adsorption system were used. The objectives of this research is propose an adsorption process to separate butanol from ABE solution and develop the regression model for adsorption isotherm of acetone, butanol and ethanol for further study The Langmuir model was used to fit the experimental data for both gas and liquid phase adsorption. The selected sorbents will be used to adsorb acetone, butanol and ethanol vapor for gas-phase single component adsorption equilibrium study and analysis by the gravimetric adsorption equilibrium apparatus. Then the breakthrough curves of each compound in ABE solution by the selected sorbents in liquid phase were obtained by using gas chromatograph.

Keyword: **Adsorption, Acetone, Butanol, Ethanol, Immobilization Starch Sorbent**

(154e) Bio-Ethanol Purification: A Comparison of the Performance of Fluoride-Mediated Silicalite-1 and Zeolitic Imidazolate Frameworks

Zhang, K., Georgia Institute of Technology
Chance, R. R., Georgia Tech
Koros, W. J., Georgia Institute of Technology
Nair, S., Georgia Institute of Technology
Originally presented on: 11/4/2013 16:35:00 - 16:55:00

Renewable energy has received increasing global interest due to rising oil prices. Among various alternatives, biofuels are generally environmentally benign and have the potential to reduce dependence on fossil energy sources. Since the biofuel products, either from corn-based or innovative algae-based technologies, are typically dilute alcohol-in-water solutions, an energy efficient alcohol–water separation technology is required to generate fuel-grade alcohols. Hydrophobic sorbents with high organic adsorption capacities are preferable for the separation of alcohol components while rejecting water molecules. The materials could be prepared as a membrane for a steady-state alcohol–water separation, or be integrated into a packed bed for low-throughput separations, or into a structured fiber sorbent configuration for high-throughput separations. It is therefore discernible that the key issue in this process is the innovation and realization of ideal hydrophobic materials.

We explored the performance of two promising hydrophobic materials, i.e. fluoride-mediated silicalite-1 (F-MFI) and zeolitic imidazolate frameworks (ZIF-8,-71,-90) for vapor phase bio-ethanol purification. For F-MFI, the introduction of mineralizing agent F^- mitigates the formation of internal silanol defects by forming ion-pairs with cationic structure directing agent that remain occluded within zeolite crystals, which results in the synthesis of pure-silica zeolites that are very hydrophobic. For ZIFs, their structures present an inherently hydrophobic nature due to the integral imidazolate linkers. Ethanol and water adsorption isotherms will be discussed in detail and the feasibility of applying these materials to the recovery of bio-ethanol is evaluated in terms of ethanol–water sorption selectivity. A comparison of distinct features for ethanol and water adsorption in F-MFI and ZIFs will be presented. The capability of biofuel purification under vapor mixtures is estimated by the ideal adsorbed solution theory (IAST) and binary breakthrough measurements.

(154f) Carbonyl Sulfide Removal From Fuel Reformates At Ambient Conditions During Hydrogen Production for Fuel Cells

Sujan, A., Auburn University
Tatarchuk, B., Auburn University
Originally presented on: 11/4/2013 16:55:00 - 17:15:00

PEMFCs operated at low to ambient temperatures in vehicular applications are gaining interest as the ambient temperature operation offers an advantage of lower pressure drop due to lower viscosities of the gas fed to the electrodes. Post reformer desulfurization is an important unit operation in fuel processing of hydrocarbon during the hydrogen production for PEMFCs. The major sulfur species in the reformate stream is H_2S and is effectively removed by ZnO based adsorbents. However, in addition to H_2S the reformate streams with low moisture content also contain considerable amount of CO, CO_2. At ambient conditions, the Carbonyl Sulfide (COS) formation occurs via the homogeneous reaction between CO and H_2S in the reformate stream. COS is less acidic and less polar than H_2S and ZnO is not efficient for removal. However, COS removal is essential to bring down the total sulfur concentration to sub-ppm levels for fuel cell applications. The total sulfur breakthrough on ZnO based adsorbents is determined by homogeneous COS formation (CO+H_2S) at low water concentrations. Even with hydro-treating, COS removal is particularly problematic as thermodynamic constraints dictate only partial conversion of COS to H_2S. COS is also formed via heterogeneous reactions between CO_2 and sulfided ZnO based sorbents. The focus of the current work is to understand the various circumstances under which COS formation takes places over wide temperature range (25°C – 400°C) in a simulated reformate stream with significant CO, CO_2 followed by a sorbent approach with several high surface area supports like SiO_2, TiO_2, Al_2O_3, PICA Carbon, ZrO_2 or in presence of moisture catalytically hydrolyze COS into H_2S which may then be removed by an efficient ZnO based sorbent (>0.5 gS/g sorbent).

(154g) Experimental Study On Hydrogen Sulfide Removal By Reactive Adsorption On Sulfatreat Sorbent

Pagliai, P., University of Genova
Di Felice, R., University of Genova
Originally presented on: 11/4/2013 17:15:00 - 17:35:00

EXPERIMENTAL STUDY ON HYDROGEN SULFIDE REMOVAL BY REACTIVE ADSORPTION ON SULFATREAT SORBENT

P. Pagliai and R. Di Felice

University of Genova, DICCA, via Opera Pia 15, 16145 Genoa, Italy;

tel. +39 0103532594, e-mail: paolo.pagliai@unige.it;

Passive adsorption on a porous bed of selected sorbents (e.g. activated carbons, zeolites) proved to be a cost effective technology in a wide range of industrial applications, which require the removal or separation of a few substances from a gas stream. However, in the energy production, the use of biogas derived from anaerobic digestion (AD) or collected from landfill (LFG) requires the sorbent materials to operate at conditions far from the ideal saturation capacity, thus raising the overall cost of the clean up process.

In fact, LFG and ADG contain a variety of trace compounds: more than 140 substances have been identified so far, which may account for about 0.15 % vol. (2000 mg/m^3) of the total amount. Such trace compounds may damage or hamper the normal operation of the power unit and therefore need to be removed down to tolerance levels.

Among these, Sulphur compounds, and particularly H_2S, can reach concentrations as high as 1000 ppm$_v$, which are several order of magnitude higher than the limit values (i.e. 0.1 ppm$_v$ H_2S for Solid Oxide Fuel Cells) required to avoid catalyst poisoning or equipment failure. Because of the substantial differences among biogas contaminants and their concentrations, it's very difficult to establish a standard method for biogas clean-up. Moreover, in sulphur removal by passive adsorption, the efficiency of the process is often lowered by the mechanism of competitive adsorption, which takes place between H_2S and other pollutants, water and carbon dioxide.

A commercially available porous sorbent, namely Sulfatreat HP 410, is claimed to be a particularly efficient (about 20% wt.) material for sulphur removal: sulphur compounds (H_2S, Mercaptans) are removed by reaction with a fine mixture of solids composed by iron oxides and noble metals and supported by a coarse size inert material.

However, as reported in a few experimental studies, the investigation on the removal efficiency by Sulfatreat at different operative conditions has been only partially fulfilled: particularly, the reproducibility of breakthrough experimental curves and the relevant modelling of the reactive process is not trivial because of the wide particle size distribution and packing of the tested material.

The objective of the experimental study here presented is to ascertain whether the reactive adsorption of H_2S can be described by considering two mechanisms, namely: the external mass transport of sulphur towards the particle surface and the irreversible reaction with the active superficial layer of the sorbent. Different flow conditions have been tested over a differential porous bed of sieved Sulfatreat particles in order to determine the relevant contribution of the external mass transfer. Experimental data are compared with theoretical data obtained by coupling mass balance and mass transport equations; the dependence of the reaction rate on H_2S concentration and the presence of humidity in the gas stream is also reported.

(156a) A Study On the Survival of Oral and Faecal Bacteria in Different Beverages

Mbachu, C., Middlesex University London

Originally presented on: 11/4/2013 15:15:00 - 15:37:00

Bacterial contamination of beverages has been a major concern and can cause many health issues. *Escherichia coli, Enterococci*, Coli forms and *Pseudomonas aeruginosa* are common known bacteria that can be found in beverages such as fruit juices and bottled water and these organisms can thrive well depending on the conditions (e.g. temperature and pH). These beverages sold at all public places and roadside shops and they can become a health concern depending on handling and storage or if the manufacturing process is unsafe. In this research, five different beverages (spring water, fountain water, fruit juice, lucozade and coke) were inoculated with *Pseudomonas aeruginosa* and *Escherichia coli* and survival rate monitored over time under two different temperature (4oC and 22oC) conditions. In this, the effect of storage time and temperature of different beverages on the survival growth of bacteria were studied to assess for the potential risk of food borne illnesses on different beverages. Analysis of Variance (ANOVA) of the data was conducted to determine and assess the differences in growth rate and for comparison purposes and to know which condition was significant. It was found that temperature was significant in the growth rate of *Escherichia coli* and both temperature and type of beverage used for the study was significant in the growth rate and survival of *Pseudomonas aeruginosa*.

(156b) Potable Water Production Via Microchannel Capacitive Desalination Technology (CDT) System Architecture

Loeb, C., Oregon State University
Jovanovic, G. N., Oregon State University
Originally presented on: 11/4/2013 15:37:00 - 15:59:00

As clean water availability diminishes throughout the world, an inexpensive, reliable, and scalable system is required to produce potable water from the world's vast supply of seawater. Taking advantage of the short diffusion lengths inherent in microchannel architectures, the research team at Oregon State University and MTEK Energy Solution Inc. is developing a parallel flow-through system capable of removing and segregating ionic species from concentrated salt water.

The application of a voltage potential to high surface area electrodes quickly removes ions from the salt-water feed to the electrode surface, allowing clean water to exit the CDT system while ions remain immobile. Segregated side channels quickly and efficiently elute concentrated brine solution during periods of unapplied electrical potential. Clean water production will be achieved through a series arrangement of CDT cells.

Validated computer simulations have provided the necessary insight to dramatically improve system design, operating conditions, and fluid flow relative to our demonstration prototype. The microchannel CDT system utilizes common materials of construction and fabrication processes available at the Microproducts Breakthrough Institute (MBI) in Corvallis, Oregon to quickly modify for enhanced performance. Increased throughput of seawater is readily achieved by 'numbering-up' parallel microchannel CDT architectures to match water production requirements. The CDT system allows for tremendous end-user customization that makes the CDT system the optimal alternative for any required process metrics, space limitations, and location.

(156c) Disinfection of Water By Adsorption Combined With Electrochemical Treatment

Hussain, S. N., University of Manchester
de Las Heras, N.,
Asghar, H. M. A., University of the Punjab
Brown, N. W., Arvia Technology Ltd
Roberts, E. P. L., University of Manchester
Originally presented on: 11/4/2013 15:59:00 - 16:21:00

The process of adsorption using graphite based adsorbents with electrochemical regeneration was developed at the University of Manchester for the removal of low concentrations of organic impurities from water. The technology employs a graphite intercalation compound (GIC) adsorbent that can be electrochemically regenerated. This process has significant potential application for water recycling since it does not involve the addition of chemicals and no solid or liquid wastes are generated. Safe storage of water for recycling often requires that the water is disinfected. To date there have been no studies of the effect of the adsorption / electrochemical regeneration process on any microorganisms present in the water. The aim of the present work was to investigate the removal and disinfection of microorganisms by adsorption on a GIC adsorbent followed by electrochemical treatment. The adsorption of an enteropathogenic bacterium, *Escherichia coli*, was performed using GIC particles followed by electrochemical treatment under a range of experimental conditions in an air agitated sequential batch reactor. The adsorption of *E. coli* cells on the GIC adsorbent was found to be a fast process and was capable of removing >99% *E. coli* cells from solution after 10 min with a ca. 6.5$\log_{10}$ reduction in the CFU concentration. The electrochemical treatment gave 100% regeneration efficiency showing that the capacity of the GIC adsorbent did not diminish over five adsorption and regeneration cycles . An ~8-$\log_{10}$ reduction of *E. coli* was achieved on each cycle. During electrochemical regeneration significant indirect electrochemical disinfection of *E. coli* present in the supernatant solution was observed. The process of adsorption using GIC adsorbents with electrochemical regeneration is capable of disinfecting a high concentration of E. *coli* at low current density and low energy consumption. The most likely mechanisms for electrochemical disinfection include electrochlorination, direct electrochemical attack and pH effects. In the absence of chloride, both the adsorption of *E. coli* on the GIC and the electrochemical regeneration were found to be less effective for water disinfection. In spite of this, effective disinfection was achieved in the absence of chloride by the appropriate selection of operating conditions. This observation indiates that the process of disinfection by adsorption with electrochemical regeneration could be employed for disinfection without the formation of chlorinated disinfection by–products.

(156d) Municipal Wastewater Treatment By An Osmotic Membrane Bioreactor: Use of Sodium and Magnesium Chloride As Draw Solution

Qiu, G.,
Ting, Y. P., National University of Singapore
Originally presented on: 11/4/2013 16:21:00 - 16:43:00

Osmotic membrane bioreactor (OMBR) technology, which integrates forward osmosis (FO) and biological treatment process, has received increasing attention. In this study, an OMBR was developed for municipal wastewater treatment, using two draw solutes (DS), i.e. $MgCl_2$ and NaCl. The potential and limitations of this innovative treatment process was investigated with respect to pollutant removal efficiencies, flux stabilities and membrane fouling. During a 100-day operation, high removal of organic matter (98%), nitrogen (98%) and PO_4^{3+}-P (99%) were achieved with both DS, with the effluent TOC, NH_4^+-N, total nitrogen (TN) and PO_4^{3-}-P being lower than 3.0 mg/L, 0.5 mg/L, 1.5 mg/L and 0.1 mg/L, respectively. Significant aerobic denitrification (up to 98%) was observed in the OMBR, owing to the high rejection of the FO membrane and the prolonged retention of NO_2^--N and NO_3^--N within the bioreactor which facilitated their aerobic removal. However, salt leakage from the DS and the retention of electrolytes from the feed wastewater resulted in significant salt accumulation in the bioreactor, which not only reduced the water flux considerably but also adversely affected pollutant removal efficiencies. Nitrification was affected acutely with both the DS. Marginal inhibition of denitrification was also observed with NaCl. However, for both DS, TOC removal was not evidently affected. Compared to $MgCl_2$, NaCl gave 20% higher initial water flux but resulted in more significant water flux decline (about twice), due to a more severe reverse transport of NaCl as the DS (about 6.5 times). Membrane fouling was controlled by periodically soaking and washing with tap water, and resulted in only 5% reduction in water flux decline over 60 days. Different rejection and transportation behaviors of NH_4^+-N, NO_2^--N and NO_3^--N were also observed with different DS, which was found to be governed by electrostatic interactions and the Gibbs-Donnan effect.

(156e) Scalability Analysis of An Open Channel UV Wastewater Disinfection Reactor Using CFD Simulation

Ray, M. B., The University of Western Ontario
Saha, R. K., University of Western Ontario
Zhang, C., University of Western Ontario
Originally presented on: 11/4/2013 16:43:00 - 17:05:00

The scalability of an open channel ultra-violet (UV) wastewater disinfection reactor is numerically investigated using a computational fluid dynamics (CFD) model. The CFD model consists of volume of fluid (VOF) method for capturing water-air interface, Lagrangian particle tracking for calculating microbial particle trajectory and the discrete ordinate (DO) model for resolving the UV intensity field in the reactor. The realizable k-ϵ turbulence model was adopted to simulate the turbulent flow, and the enhanced wall treatment was used to resolve the turbulence near the walls. Different scaling down procedures were tested to establish similarity of the hydrodynamics and the radiation field between the scale model and the full size reactor. To ensure similarity of the hydrodynamics, Froude number and the Reynolds number were taken into consideration. Scaling methodology for the UV radiation field was formulated to ensure the same reduction equivalent dose (RED) values can be achieved from the full size and scaled down reactors. A parametric study was performed in terms of the flow rate in the reactor and water UV transmittance to characterize the range of scalability. Initially, the model was validated using experimental data on free-surface profile. The CFD model predicts that the open channel reactor is scalable, thus allowing one to predict the large scale behavior from the small scale laboratory test.

(156f) Performance Evaluation of a Low Cost and High Rate Vertical Flow Wetland for Residential Graywater Treatment and Reuse

Yu, Z. L. T.,
Rahardianto, A., University of California, Los Angeles
Stenstrom, M. K., University of California, Los Angeles
Cohen, Y., University of California, Los Angeles
Originally presented on: 11/4/2013 17:05:00 - 17:27:00

Water shortages in urban environments across the US and around the world have sparked increased interests in the development of sustainable water practices and technologies. Despite increased sustainability efforts, water reuse in the United States accounts for only ~1% of the nation's total water supplies. Water reclamation and reuse through centralized suppliers (i.e. municipalities or water agencies) represent a major challenge given the lack of infrastructure and operational costs for treatment and distribution system for reclaimed wastewater and its reuse. One possible approach that addresses, in part, the above challenges is to utilize point-of-use (i.e. onsite) treatment to generate non-potable water source for onsite reuse. Such onsite reuse activity will expand the nation's water recycling efforts driven by end-users or small communities (e.g. multifamily homes) and augment the existing water resources portfolio. Accordingly, it is argued that on-site treatment and reuse of graywater is a viable option in residential and non-residential premises. The key challenge of encouraging such a practice is the lack of affordable and simple treatment technology suitable for residential users. In order to meet the above challenge, the present study presents a low cost, high rate, compact vertical flow wetland (CVFW) system for graywater treatment. The developed small footprint graywater treatment (~1m^2) was field tested in a single family home in Los Angeles with the treated water being used for irrigation. The kinetics of the graywater treatment system was characterized using a composite model for the vertical wetland configuration and system operation was accordingly optimized. Results from the field study demonstrated that the CFVW system is capable of removing up to 90% of total organic carbon and up to 99% of turbidity. The present approach has a potential for deployment in multifamily homes for onsite treatment and reuse of graywater with an estimated return on investment within 2 years.

(156g) Coagulation and Ultrafiltration of High-Alkalinity Greywater

Amundsen, T. J., Mainstream Engineering Corporation
Zastrow, D. J., Mainstream Engineering Corporation
Wagner, A. L., Mainstream Engineering Corporation
Originally presented on: 11/4/2013 17:27:00 - 17:49:00

Water is perhaps the most essential natural resource. Unfortunately, fresh water is becoming increasingly scarce, and with the world population continuing to grow, the need for improved methods of water utilization, treatment and recycling is becoming ever more pressing. Greywater is produced by a number of household activities including dishwashing, showering, bathing and laundry. Water from these sources is characterized by the presence of a variety of organic and inorganic impurities such as soap, oil, grease and dissolved solids. Greywater produced during dishwashing poses a particularly difficult problem for treatment processes due to a relatively large concentration of pollutants, particularly soap. The soap solubilizes nonpolar oils and greases in the water while simultaneously increasing the alkalinity of the water. Many treatment processes are very sensitive to pH and require either neutral or acidic pH to operate effectively. Traditional coagulants based on iron or aluminum do not perform well in an alkaline environment and can require permeate post-processing to remove dissolved metals. We investigated the use of novel electrode materials in an electrocoagulation process to treat dishwashing water without any pH adjustment. The coagulated water was then further purified by ultrafiltration. We report the impact of the pre-coagulation on membrane performance and longevity.

(182a) Economic and Engineering Assessment of Combined Heat and Power Fuel Cell Systems Installed in Commercial Buildings

Brooks, K., Pacific Northwest National Laboratory
Pilli, S.,
Originally presented on: 11/4/2013 15:15:00 - 15:37:00

PNNL support of the DOE's Fuel Cells Technology Office is to identify near term niche markets where fuel cells have potential, work with the DOE and stakeholders to develop activities in those areas, analyze the business case, and present the results to the community. Accelerating fuel cell use in these key early markets will create jobs in an industry that needs high volume purchases to ramp up production, decrease costs, support commercialization, and enable a domestic supplier base. PNNL has contracted ClearEdge Power to demonstrate fifteen 5 kW combined heat and power fuel cells in a wide variety of business in California and Oregon. These fuel cells are being remotely monitored for such things as electrical and thermal energy, overall efficiency, and availability. The gathered information will be analyzed, to provide "real-world" data from units "in the customer's hands" to validate performance, durability and reliability; installation, operations, and maintenance costs; and identify remaining barriers to widespread commercialization. As part of this assessment, a business case has been developed estimating of the size of the market and its growth potential, identifying possible niche markets, and comparing the CHP FCS with its alternatives in terms of economics, engineering and environmental impact. The results of this assessment will be presented.

(182b) Systematic Approach for Energy Efficient Eco-Industrial Parks

Noureldin, M. B., Saudi Aramco
Qahtani, S. A., Saudi Aramco
Owaidah, M., Saudi Aramco
Nutaifi, A., Saudi Aramco
Originally presented on: 11/4/2013 15:37:00 - 15:59:00

Sustainability since its original conceptualization is considered to have four dimensions; social; economic; environmental and institutional. It is simply the improved management of natural resources within a business establishment to provide economic and social benefits to the business and its surroundings. Eco-industrial parks (EIP) can serve a significant role in realizing economic, environmental and social benefits both to individual plants and to network of companies. Even though EIP are often publicized as a mean of reducing environmental damage through reduced waste, they can have number of benefits at different levels. Currently the major driving factor for the creation of most of it is to attain financial gain.

Industrial ecology concept may promote a new path of local development through the transition from industrial clusters to EIPs via exploiting common features of both models. Implementing Eco-Industrial Parks' (EIPs) principles in an existing industrial cluster or in the planning of new ones, represents a significant opportunity for its revitalization. IEP could exploit synergies from industrial clusters and non-industrial activities to easily create new production models in which the economic and environmental dimensions are symbiotic.

The well planning of new EIP or transformation of conventional industrial complexes to a one can brings significant value to energy efficiency and energy-based GHG emissions reduction. This task is a huge multi-variable multi-dimensional optimization problem in which the total eco-industrial park network depends on as small as stream condition and as big as the whole park functionality. The customized integration among many industrial plants and non-industrial ones in adjacent geographical locations can bring in more degrees of freedom to the "waste energy recovery" optimization problem and consequently presents new horizon to the radical energy-based GHG emissions reduction to levels never thought of before. Fortunately, many of the reasons which were hindering the application of inter-systems integration are nowadays addressed in cost-effective ways. However, to date EIPs key success has been a sequence of independent economically driven actions. Such evolutionary pattern may not be easily transferred to others. A holistic approach in addressing the problem using systematic techniques in the implementing of necessary modifications in either the brownfields or in the planning of Greenfields can bring huge value to energy resources conservation and GHG emissions reduction and is desperately needed today. It is very difficult with the current methods and tools to magically manufacture EIPs to work from scratch. First, there should be the basic ingredients in place, namely the desire of plants to actively participate and second the correct EIP membership and structure of firms. These basic ingredients can then be enhanced and improved upon, with correct support structure in place. The key to developing a successful EIP is to determine the best connections among different industrial plants and its surrounding communities in an EIP. Specifically it is the material and energy flows relationship among the different members in the alliance of plants which permit establishing optimal linkage to form a fruitful inter-dynamic structure. If such structure does not exist a successful EIP cannot be realized.

This paper addresses the energy efficiency of EIP via new systematic approach. It is different in concept than the current approach in which only the "transfer of waste heat" from one firm to a nearby one is conducted after exhausting all the possibilities to optimize the stand-alone plant in the firm. The total site targeting based on the pinch technology, which is the most widely used to date, starts with intra-systems integration which means again that only the waste heat of one member in the EIP is shared with other parks' members. Certainly, such level of sharing philosophy is not the ideal proactive form of true cooperation.

This paper introduces new method that achieves ambitious levels of energy efficiency and GHG emissions reduction via identifying best matchings' among EIP members; generating technically viable energy efficient alternatives; and defining the best generation and allocation of energy utilities systems. It also presents two new mathematical programming and constraints logic propagation tools to systematically find the best and the second best and so on matchings' among several plants for optimal eco-industrial park membership forms under all possible combinations of plants-specific design and operations modifications. The paper also introduces new graphical tool that enables the EIP designer to gain insights and create initial solutions for optimal energy loads for exchange among the members to best enhance the energy efficiency and GHG emissions of the whole park ("One for all and all for one philosophy"). The advantages of the new method and tools are illustrated and demonstrated by two case studies.

(182c) The Potential Of Osmotic Membrane Dehumidification

Kesten, A. S., Nanocap Technologies LLC
McCutcheon, J. R., University of Connecticut
Girelli, A.,
Blechner, J. N., Nanocap Technologies LLC
Originally presented on: 11/4/2013 15:59:00 - 16:21:00

THE POTENTIAL OF OSMOTIC MEMBRANE DEHUMIDIFICATION

Arthur Kesten, Jeffrey McCutcheon, Ariel Girelli and Jack Blechner

An osmotic membrane dehumidifier can use a flexible, semi-permeable membrane to facilitate capillary condensation of water vapor and the transport of condensed water through the membrane into a salt solution by osmosis. Here a humid gas stream is brought into contact with a semi-permeable membrane, which separates the gas stream from an osmotic (e.g., salt) solution. Some of the pores of the membrane are small enough to permit capillary condensation. Liquid formed within these pores can connect with liquid formed in adjacent pores, collectively forming continuous paths of liquid. These ?liquid bridges' extend across the thickness of the semi-permeable membrane and provide paths by which water can travel across the membrane. Because the membrane is so thin, water concentration gradients across the membrane can be large. This can provide a large driving force for water transport between the humid air and the osmotic fluid. The flexibility of the polymeric membrane allows for considerable design flexibility that enhances the potential for retrofit with any cooling system. An illustration of this two-step spontaneous process is given below.

Liquid desiccants are particularly effective as osmotic agents because water entering the desiccant solution is bound to the salt as water of hydration.Â This enhances the water concentration gradient across the membrane.

Laboratory testing of membranes and draw solutions under different environmental conditions is conducted using a cell comprised of two halves separated by a membrane. On the top half, the humid air is passed above the membrane. The draw solution is pumped through the bottom half of the cell; Â the draw solution is contained in a reservoir that is placed on a digital scale that measures mass to a hundredth of a gram. Humid air contacting the membrane condenses by capillary condensation in the pores of the membrane and the condensed water is drawn by osmosis into the draw solution. Mass changes versus time are recorded with the digital scale that is capable of measuring a maximum of 4,000 grams.

Osmotic dehumidification begins with condensation of water molecules, followed by the osmotic draw of the liquid water out of the pores into a draw solution. If the rate-limiting step is the osmotic de-swelling of the membrane, the highest osmotic pressure should be used to maximize flux. However, if capillary condensation is the limiting step, a threshold is reached beyond which increased osmotic pressure does not increase flux. To understand these limits, we tested three concentrations of magnesium chloride with a cellulose acetate membrane under identical conditions. The water flux varied directly with concentration suggesting that the limiting step for this membrane was the osmotic removal of liquid water as opposed to capillary condensation.

A separate experiment was constructed to compare the effectiveness of <u>no</u> membrane to our capillary condensation/osmotic dehydration system. Having no membrane in the system is the equivalent of exposing liquid desiccant directly to a humid air stream. Measured water removal rates for that system were less than one fifth of the rates for the membrane/draw solution tested here. And, of course, with no membrane between the humid air and the desiccant, there is always the potential for entraining desiccant in the air stream.

Cooling of the osmotic solution results in more effective dehumidification. Lower water vapor pressure leads to capillary condensation at lower relative humidity. For modest levels of cooling, the relative humidity exiting the Nanocap process will be around 50%.

Scale up of the process to larger surface areas of membrane can be accomplished with plate and frame systems, spiral wound devices and bundles of cylindrical fibers. Results with a plate and frame system demonstrate the effectiveness of a mathematical model as a basis for design of a larger system.

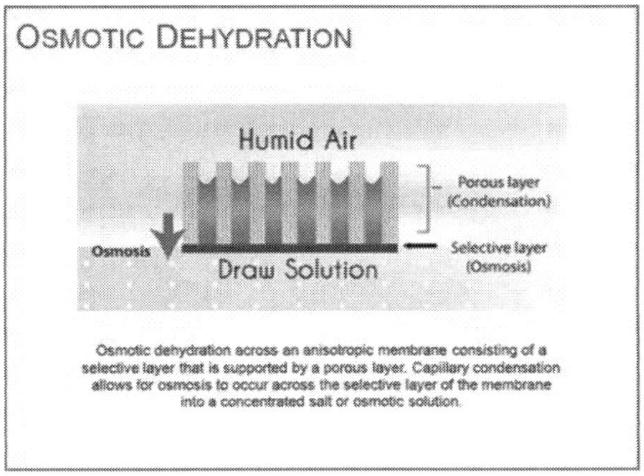

Osmotic dehydration across an anisotropic membrane consisting of a selective layer that is supported by a porous layer. Capillary condensation allows for osmosis to occur across the selective layer of the membrane into a concentrated salt or osmotic solution.

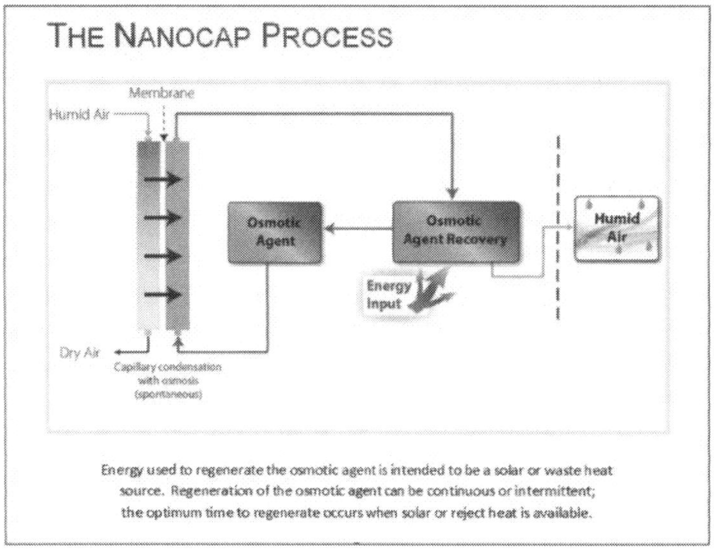

Energy used to regenerate the osmotic agent is intended to be a solar or waste heat source. Regeneration of the osmotic agent can be continuous or intermittent; the optimum time to regenerate occurs when solar or reject heat is available.

(182d) Improvement of Distillation Column Efficiency By Integration With Organic Rankine Power Generation Cycle

Sladkovskiy, D. A., St.Petersburg State Institute of Technology (technical university)
Originally presented on: 11/4/2013 16:21:00 - 16:43:00

Distillation as main separation technology consumes greater than 40% of the energy used by the chemical and petroleum process industries but the energy efficiency of a distillation column is low. It is well-known that significant energy savings can be achieved if complex column configurations and heat integration are used, although these improvements withdraw the attractive advantages of conventional distillation column such as flexibility, low capital investment and low operational risk.

This theoretical research focuses on improvement of distillation column vapor generation procedure by expansion turbine installation.

The liquid stream from the column bottom is pressurized by a pump and the high pressure liquid is vaporized in the reboiler. The resultant high pressure vapor stream drives the expansion turbine, which is coupled to the electricity generator or another rotating equipments. The turbine exhaust vapor enters the column bottom.

Such distillation column with power generation cycle (DCPGC) is kind of combination of conventional distillation column and organic Rankine cycle (ORC) power plant. The ORC is a cost efficient and proven method of converting low temperature waste heat to mechanical and/or electrical energy.

Such chemical components and its mixtures as propane, i-butane, n-butane, i-pentane et al. which occurs in distillation columns bottoms are known to be utilized or considered as ORC working fluids. So DCPGC is seem to be practically feasible.

DCPGC reboiler duty is about 10-20% greater than the conventional column reboiler duty but almost whole of extra heat transforms to electricity power. The theoretical DCPGC electricity generation efficiency is more than 90% (the heat required for the distillation remains the same as in conventional column and is not included in the efficiency calculation).

Only thermodynamic properties and thermal stability of column's bottom product impact on DCPGC feasibility, but not the temperature difference between top and bottom of the column. This is an advantage of suggested technology to compare with heat integrated column with heat pumps where temperature difference is quite important.

This theoretical research describes practical and economical feasibility of DCPGC for separating propane/propylene, i-butane/n-butane, i-pentane/n-pentane, butane-pentane, benzene/toluene, xylene mixtures. Crude oil and naphtha fractionation are also investigated. For instance DCPGC for i-butane/n-butane binary mixtures separation with 25 tonne/h feed flowrate can produce 1 MW of electricity power.

(182e) High Surface Area Zeolite Membrane for Efficient Bio-Fuel Production

LIU, W., Pacific Northwest National Laboratory

Originally presented on: 11/4/2013 16:43:00 - 17:05:00

Distillation and adsorption has been commonly used for respective enrichment and polishing of ethanol fuels in today's ethanol plants. In this presentation, development of thin zeolite/metal sheet membranes of surface area packing density comparable to polymeric sheets is discussed to enhance ethanol dehydration energy efficiency and lower capital costs. The mechanical structural integrity of this new membrane product concept is substantiated with stress modeling. A relatively simple fabrication process that renders its low-cost manufacturing is delineated with experimental results. Separation process studies are conducted along with the membrane product development, which include (i) performance tests with real-world feeds and (ii) process design and simulation of membrane separation integrated with whole plants. The present membrane is highly permeable to water that significant one-pass dehydration can be obtained within a short residence time (seconds). However, the membrane performance can be drastically affected by the feedstock. Proper pre-treatment is necessary to achieve stable separation performances with the real-world feed. Based on experimental results, a two-stage membrane separation system is proposed to conduct ethanol enrichment and purification by replacing respective distillation and adsorption processes. More than 90% energy saving is shown using the membrane separation. Due to high permeance and high membrane area packing density, the membrane module estimated for a 40-million gal/year ethanol plant can be fairly compact, about the size of plate heat exchangers. This membrane technology is promising for retrofitting of existing corn ethanol plants and for building of new cellulosic ethanol plants in the future as well.

(182f) Syngas Mass Transfer Analysis in a Hollow Fiber Reactor

Orgill, J. J., Brigham Young University
Lewis, R. S., Brigham Young University
Atiyeh, H. K., Oklahoma State University
Originally presented on: 11/4/2013 17:05:00 - 17:27:00

Biofuel production via fermentation is produced primarily by fermentation of simple sugars. Besides the sugar fermentation route, there exists a promising alternative process that uses syngas (CO, H_2, CO_2) produced from biomass as building blocks for biofuels. Although syngas fermentation has many benefits, there are several challenges that still need to be addressed in order for syngas fermentation to become a viable process for producing biofuels on a large scale. One challenge is mass transfer limitations due to low solubilities of syngas species. The hollow fiber reactor (HFR) is one type of reactor that has the potential for achieving appropriate mass transfer rates for biofuels production, but a better understanding of mass transfer limitations associated with all syngas species is critical for assessing syngas usage during fermentation.

Mass transfer coefficients for O_2 and CO_2 have been typically assessed for HFRs using dissolved gas probes. In limited cases, CO has been studied using myoglobin assays. Though there are many reports on mass transfer coefficients in HFRs, studies are lacking that compare mass transfer coefficients of syngas species, including comparisons obtained from different measurement methods. Such comparisons are important to verify the accuracy of the experimental methods and to provide an increased understanding of the competing mass transfer limitations associated with syngas species.

In this study, we measured the mass transfer coefficient for O_2 in a PDMS hollow fiber reactor as a function of Reynolds number (flow rate) using a dissolved O_2 probe. The data was fit to a model characterizing O_2 mass transfer and the model was used to predict the mass transfer coefficients for CO and H_2. In order to verify the model predictions for these latter species, a second measurement technique using GC analysis was developed to measure CO and H_2 mass transfer coefficients. The GC technique was validated by comparing the mass transfer coefficient of O_2 from both GC and dissolved O_2 probe analysis. The mass transfer coefficients of both H_2 and CO in a HFR will be compared and the impact of such coefficients on syngas utilization will be presented. This study will greatly enhance the knowledge necessary to assess syngas usage and mass transfer limitations in a HFR used for syngas fermentation.

(182g) An Innovative Methanol Synthesis Process Using Self-Heat Recuperation

Kansha, Y., The University of Tokyo
Ishizuka, M., The University of Tokyo
Tsutsumi, A., The University of Tokyo
Originally presented on: 11/4/2013 17:27:00 - 17:49:00

The demand for methanol will continue increasing, since methanol is attracting as a fuel for fuel cells, and is an intermediate raw material of hydrogen and dimethyl ether (DME) which are categorized green energy source. As well as other chemical processes, many methanol synthesis processes have reactant recycle system with product separation due to the low conversion ratio of the reactor. For this product separation from reactant recycle stream, a gas-liquid separator or distillation process has often been used. However, this gas-liquid separator and distillation processes are well known as an energy consuming process. Thus, many chemical engineers and investigators managed to produce a catalyst which achieves high conversion from reactant to production in the reactor for whole process optimization.

Authors have developed self-heat recuperation technology, in which not only the latent heat but also the sensible heat of the process stream can be circulated without any heat addition, leading to reduction of energy requirement in several chemical processes. This self-heat recuperation technology is suitable for thermal and separation processes and is applied to several chemical processes for energy saving as case studies.

In this research, we investigated the feasibility of applying self-heat recuperation technology to the methanol synthesis process and developed an innovative process for methanol synthesis process from the energy saving point of view. By installing the self-heat recuperation technology to the methanol synthesis process, the energy consumption of the process can be greatly reduced. Thus, it can be said that installation of self-heat recuperation technology is the other option of process design method to achieve high conversion in the reactor.

(186a) Influence of a Biofouling Layer On Membrane Distillation

Chew, J. W., Nanyang Technological University
Krantz, W. B., University of Colorado
Originally presented on: 11/4/2013 15:15:00 - 15:37:00

Membrane distillation (MD) is a relatively new technology that enables the production of potable water without the need to use the high pressures required for reverse osmosis. MD is a process whereby liquid water on the feed side is vaporized by heat supplied from the hot feed stream; subsequently, the water vapor diffuses through the pores in a hydrophobic membrane and is condensed on the cold permeate side. MD is attractive for several applications. In particular, it provides a means for producing potable water from saltwater while avoiding the high pressures associated with reverse osmosis. In addition, it provides a use for low quality waste heat available from many industrial operations. As such, it is a very attractive Green Technology.

Unfortunately, the optimal performance of MD is compromised owing to concentration polarization and fouling [1]. Models for the MD process developed to date have not incorporated all the effects of a biofouling layer [2]. Moreover, it is not clear how to use existing models for the MD process in order to infer the presence of a biofouling layer. The focus of this presentation is to present a new model for the MD process that incorporates the principal effects of a biofouling layer. The manner in which this model can be used to infer the presence of a biofouling layer will be outlined.

There are two principal effects of a biofouling layer on the MD process: (1) a resistance to heat conduction on the feed side of the membrane that can cause a reduction in the evaporation rate; and (2) a direct reduction in the vapor pressure of the evaporating liquid owing to the Kelvin effect introduced by the small pores in the biofouling layer. The latter effect was recently documented by Goh et al. [2] who used evapoporometry [3] to show that a biofouling layer can have pores ranging from 3.5 to 47.5 nm that can reduce the vapor pressure by as much as 36% and the MD water flux by as much as 50%.

The MD model advanced by Schofield et al. [4] is adapted to include both possible effects of a biofouling layer. This model permits constructing a plot based on the measured values of the MD water flux, feed temperature, and permeate temperature in order to determine if a biofouling layer is influencing the MD process. Moreover, it permits assessing if this biofouling layer offers a significant heat-transfer resistance. It also provides a direct means to determine the influence of the Kelvin effect on lowering the vapor pressure driving force for MD. A useful metric for assessing the effect of a biofouling layer is the temperature polarization coefficient, which is the ratio of the temperature difference across the membrane divided by the difference between the feed and permeate temperatures. A generalized design plot is developed in terms of the temperature polarization coefficient as a function of a dimensionless parameter that provides a measure of the Kelvin effect reduction in the vapor pressure.

[1] A.G Fane, S. Chang, Techniques to enhance performance of membrane processes, in A.K. Pabby, S.S.H. Rizvi, A.M. Sastre (Eds), Handbook of Membrane Separations: Chemical, Pharmaceutical, Food, and Biotechnological Applications, CRC Press, 2008, 193-232.

[2] S. Goh, Q. Zhang, J. Zhang, D. McDougald, W.B. Krantz , Y. Liu and A.G. Fane, Impact of a biofouling layer on the vapor pressure driving force and performance of a membrane distillation process, J. Membr. Sci. (2013) http://dx.doi.org/10.1016/j.memsci.2013.03.023.

[3] W.B. Krantz, A.R. Greenberg, E. Kujundzic, A. Yeo, S.S. Hosseini, Evapoporometry: A novel technique for determining the pore-size distribution of membranes, J. Membr. Sci. (2013) http://dx.doi.org/10.1016/j.memsci.2013.03.045i.

[4] R.W. Schofield, A.G. Fane, C.J.D. Fell, Heat and mass-transfer in membrane distillation. J. Membr. Sci. 33(3) (1987) 299-313.

(186b) Produced Water Treatment Using Direct Contact Membrane Distillation

Singh, D., New Jersey Institute of Technology
Sirkar, K. K., New Jersey Institute of Technology
Prakash, P., Pennsylvania State University
Originally presented on: 11/4/2013 15:37:00 - 15:59:00

De-oiled produced water in current produced water treatment facilities for water reclamation from steam-flooded oil exploration activities is available at temperatures between 160-200^{0}F (71-93^{0}C). Desalination of this de-oiled produced water is accomplished currently by reverse osmosis after a number of steps including cooling. Membrane distillation processes can utilize this sensible heat to achieve desalination. Direct contact membrane distillation studies using a hollow fiber membrane module with the brine in cross-flow have been successfully conducted using a variety of produced waters such as Chevron A (Pre-RO, Post WEMCO), Chevron B1 and Chevron B2 having total dissolved solids (TDS) respectively of 7622, 12040 and 35000 mg/l. The brine feed temperature was varied between 50 and 85^{0}C. The effect of brine feed flow rate variation was also studied. At 80^{0}C, the highest observed water vapor flux at a higher feed brine velocity was 15 kg/m^2-hr. The permeated waters were substantially pure with very low TDS value in each case indicating very high salt rejections. These conclusions held even when the feed de-oiled produced water was concentrated via batch recirculation to recover as much as 80% of the feed produced water in the distillate stream.

(186c) Membrane-Assisted Crystallization of L-Ascorbic Acid

Lakerveld, R., Delft University of Technology
Goddrie, S. F. M.,
Anisi, F., Delft University of Technology
Stankiewicz, A. I., Delft University of Technology
Kramer, H. J. M., Delft University of Technology
Originally presented on: 11/4/2013 15:59:00 - 16:21:00

The design and operation of solution crystallization processes still pose many challenges despite its wide application in chemical industry. The performance of the crystallization process and final product are closely related to specific properties of the crystals such as purity, crystal size distribution (CSD), and shape. Many interacting physical phenomena occur in solution crystallization processes, involving for example primary nucleation, growth and attrition of crystals. Supersaturation is a key driving force for these physical phenomena. Therefore, precise control over the generation of supersaturation is of crucial importance. In addition, improved yields and reduced energy consumption for the generation of supersaturation are of interest to improve sustainability of crystallization processes. The latter is especially relevant for evaporative crystallizers.

Membranes offer an interesting opportunity to generate supersaturation as alternative to conventional evaporative crystallization.[1,2] They can be used to increase the surface area available for evaporation per unit crystallizer volume and there is no restriction related to entrainment. Furthermore, in the design phase local gradients in supersaturation can be anticipated as a designer can choose the locations where solvent removal takes place. Also, membranes can be used for thermal labile compounds as they can be applied at low temperatures and outside boiling conditions. Finally, membranes offer an opportunity to reduce energy consumption, which is particularly true for application of reverse osmosis. However, before industry can fully embrace membrane-assisted crystallization processes, several challenges have to be met. First, the risk of membrane fouling due to crystallization on the membrane surface has to be mitigated such that stable operation for sustained periods of time can be achieved. Secondly, validated dynamic models for systems that integrate membranes and crystallization are required for selected case studies such that optimal designs and operational policies can be compared to conventional processes.

The objective of this study is to compare the performance of membrane-assisted crystallization of two model systems that both require a different membrane technology. A mini pilot-plant setup has been constructed featuring a separate crystallization vessel and a membrane module. Both sections of the pilot plant can be operated at different conditions, which allows for contacting the membrane surface with an undersaturated solution to prevent membrane fouling while operating the crystallization section at supersaturated conditions. The first model system is an adipic-acid water system for which membrane-assisted crystallization with reverse-osmosis membranes (MaC-RO) is feasible.[3,4] The second system studied is crystallization of L-ascorbic acid from water. The high osmotic pressure of this system prohibits application of MaC-RO; hence the experimental setup was modified to apply membrane distillation (MaC-MD). The measured flux over the membrane for MaC-MD was an order of magnitude lower compared to the measured flux for MaC-RO. Therefore, a much larger surface was installed for MaC-MD (1.4 m^2 for MaC-MD compared to 0.024 m^2 for MaC-RO). The experimental data has been used to model both the tested MaC-RO and MaC-MD processes including crystallization kinetics. Subsequently, the model has been used to reveal optimal operating conditions to obtain a desired supersaturation profile in the crystallizer while minimizing energy consumption. A comparison of the results for both case studies illustrates typical strengths and weaknesses of MaC-RO compared to MaC-MD and provides guidelines for the design and operation of future membrane-assisted crystallization processes.

References

1. Curcio, E.; Criscuoli,A.; Drioli, E. Membrane crystallizers. Ind.Eng.Chem.Res. 2001, 40, 2679–2684.
2. Gryta, M..Concentration of NaCl solution by membrane distillation integrated with crystallization. Sep.Sci.Technol. 2002, 37, 3535–3558.
3. Kuhn, J.; Lakerveld, R.; Kramer, H. J. M.; Grievink, J.; Jansens, P. J. Characterization and Dynamic Optimization of Membrane-Assisted Crystallization of Adipic Acid. Ind.Eng.Chem.Res. 2009, 48 (11), 5360-5369.
4. Lakerveld, R.; Kuhn, J.; Kramer, H. J. M.; Jansens, P. J.; Grievink, J. Membrane assisted crystallization using reverse osmosis: Influence of solubility characteristics on experimental application and energy saving potential. Chem. Eng. Sci. 2010, 65 (9), 2689-2699.

Acknowledgement

This work was supported by the European Commissions Framework 7 program through the OPTICO consortium

(186d) A Novel Membrane-Based Technology for Post-Combustion CO2 Capture and Recovery

Mulukutla, T., New Jersey Institute of Technology
Obuskovic, G., New Jersey Institute of Technology
Sirkar, K. K., New Jersey Institute of Technology

Originally presented on: 11/4/2013 16:21:00 - 16:43:00

Conventional method for CO_2 capture from post-combustion flue gas generally employs an absorption and stripping process. The CO_2 is absorbed for example in an aqueous solution of amine e.g., monoethanolamine and then stripped out in a separate tower with steam at 120^0C. Such a process is highly energy intensive. Further the amine used is volatile. The low partial pressure of CO_2 in the flue gas inhibits the application of CO_2-selective membranes unless methods are employed to increase the CO_2 partial pressure in the flue gas to be treated. Adsorption processes relying on adsorbent regeneration require among others the need to switch the flue gas between different columns. A number of other techniques are being studied to reduce the capital and operating costs for CO_2 capture from post-combustion flue gas. We will describe a novel membrane-based technique to potentially bypass most of the shortcomings of many existing approaches. We have employed a simulated humidified flue gas containing around 14% CO_2 and demonstrated successful removal of bulk of the CO_2 and its recovery in a CO_2-concentrated stream. We have characterized the performance of the membrane-based device used. We will deliberate on the strengths and weaknesses of this technology.

(186e) Vacuum Membrane Distillation of CO2-Rich Alkanolamines

Yang, S., Seoul National Univ.
Jeong, Y. S., Seoul National Univ.
Cho, S. P.,
Han, C., Seoul National University
Originally presented on: 11/4/2013 16:43:00 - 17:05:00

To mitigate climate change around the world, lots of technologies are on the competition. Chemisorption using alkanolamine is considered as the most competent technology especially when applied for the flue gas. However, the capture of carbon dioxide in a packed column has a very weak point in its energy intensiveness. Also, gas and liquid flow velocities are strictly restricted due to operating problems such as flooding, channeling, foaming and entrainment. [Kumar, 2002] Membrane gas/liquid contactor is known to be relevantly free from these problems, and now researched rapidly.

Membrane contactor can be used as absorber and stripper. Membrane distillation for regenerating amine solvent is energy-saving and also good for process intensification due to high surface to volume ratio. Especially, sweeping gas membrane distillation (SGMD) and vacuum membrane distillation (VMD) are the technologies that are frequently used in the experimental level. SGMD papers used nitrogen or argon as the sweeping gas which is hard to separate from carbon dioxide. [Khaisri, 2011] [Naim, 2012] So, vacuum membrane distillation using low quality steam as a sweeping gas can be applied instead of packed column carbon capture amine scrubbing.

Still, this VMD system is not well-simulated mathematically. In this presentation, some experimental results of vacuum membrane distillation data from [Fang, 2012] [Wang, 2013] [Yan, 2012] are validated with over-all mass transfer model using local composition model. And membrane contactor modularity for the scale-up is checked toward real application. Also some operating problems like membrane-wetting are considered.

This research was supported by the LNG Plant R&D Center funded by the Ministry of Land, Transportation and Maritime Affairs (MLTM) of the Korean government

Reference List

- Fang M, et al, CO2 desorption from rich alkanolamine solution by using membrane vacuum regeneration technology, *International Journal of Greenhouse Gas Control* **9**(2012) 507-521

- Khaisri S, deMontigny D et al, CO2 stripping from monoethanolamine using a membrane contactor. *Journal of Membrane Science* **376**(2011) 110–118

- Kumar PS. Development and Design of Membrane Gas absorption Processes. PhD Thesis, University of Twente, Enschede, The Netherlands.

- Naim R et al, Preparation of microporous PVDF hollow fiber membrane contactors for CO2 stripping from diethanolamine solution, *Journal of Membrane Science* **392– 393**(2012) 29– 37

- Wang Z et al, Amine-based absorbents selection for CO2 membrane vacuum regeneration technology by combined absorption-desorption analysis, *Chemical Engineering Science* **93**(2013) 238-249

- Yan S, et al, Regeneration performance of CO2-rich solvents by using membrane vacuum regeneration technology: Relationship between absorbent structure and regeneration efficiency, *Applied Energy* **98** (2012) 357-367

(186f) PTFE Hollow Fiber-Based Contineous-Effect Membrane Distillation for Deep Concentration of Aqueous Solutions At High Temperatures

Qin, Y., Chembrane Research & Engineering,Inc.
Jin, J.,
Wang, B., Tianjin University
Originally presented on: 11/4/2013 17:05:00 - 17:27:00

Membrane distillation (MD) is a relatively new membrane separation process during which only vapor molecules from the feed stream (hot side) are transported through porous hydrophobic membranes and then re-condenses by strip stream (cold side). Membrane distillation usually can be classified broadly into four categories according to the nature of the cold side of the membrane: direct-contact membrane distillation (DCMD), vacuum membrane distillation (VMD), sweeping gas membrane distillation and air gap membrane distillation (AGMD). The advantage of membrane distillation over traditional separation processes include lower operating pressures, 100% (theoretical) of ions and other non-volatile constituents are rejected, and so on. Membrane distillation was introduced in the 1960s and developed from 1980s onwards, but did not yet attain commercial status for limiting of membrane with the characteristics most suitable and thermal efficiency. Membrane distillation is usually operated at temperature of <90 oC. However, permeation flux (J) and thermal efficiency are much lower when dealing with concentrated salt solutions as a result of reduction of vapor pressure as well as concentration and temperature polarization effects. Even for low concentration solution, the performance ratio (PR) of conventional MD process typically ranged from 0.2 to 1, while the PR of MSF and MED usually ranged from 7 to 15. Another problem with MD is wetting or fouling of microporous membrane surface, which leads to the decrease of permeate flux and leakage; leakage further leads to the contamination of the permeate product by the impurities in the feed. Continuous-effect membrane distillation (CEMD) is based on the use of AGMD process, which has distinct characteristics of high PR value and internal recovery. Continuous-effect membrane distillation (CEMD) is based on the use of AGMD process, which has distinct characteristics with latent heat recovery and with a PR obviously more than 1.0. It has several advantages such as (1) completely eliminating evacuation operation, which implies that the AGMD-based CEMD module can minimize the leakage problem and thus has a long operation stability; (2) much higher PR than DCMD-based CEMD process since heat conductivity loss in an AGMD process is much less than that in a DCMD process; (3) dealing with feed streams containing acid, base, oxidant, toxic or volatile solutes, since the distillate will not directly contact with the cooling water stream. In the present study paper, continuous-effect membrane distillation (CEMD) is explored in the range of 85-115 ℃for deep concentrating feed of NaCl and MgCl2 solutions. The CEMD membrane was used polytetrafluoroethylene (PTFE), which could operate at high temperature. Under optimized operational conditions, The experimental result indicated that the J and PR were 3.1 L?m-2?h-1 and 15.2 for the 5% NaCl solution and 2.9 L?m-2?h-1 and 14.4 for the 5% MgCl2 when the value of T3 was 100 oC; the values of both J and PR decreased with the increase of feed concentration, however, their values could still achieved 1.88 L?m-2?h-1, 7.2 and 1.62 L?m-2?h-1, 6.4 respectively for the 25% NaCl and 30% MgCl2 solution when the value of T3 was 108 oC, while the salt retention rate could reach more than 99.95%. Also, a long-tern operational stability of the MEMD process was tested used 15% NaCl solution for 60 days. The experimental results indicated that CEMD process showed in good operation stability In conclusion, the study showed that CEMD was feasible to deal with brine for the advantage of high PR and thermal efficiency at high temperature. It has a great potential value for development and application.

(186g) Water Desalination By Vacuum Membrane Distillation

Chaurasia, S. P., Malaviya National Institute of Technology
Upadhyaya, S.,

Originally presented on: 11/4/2013 17:27:00 - 17:49:00

Water Desalination by Vacuum Membrane Distillation

S. P. Chaurasia, Sushant Upadhyaya, Kailash Singh

Department of Chemical Engineering, Malaviya National Institute of Technology Jaipur

J.L.N Marg, Jaipur-302017 (India)

Corresponding author, Email: chch1962@gmail.com

Co-authors Email: Sushant.cipet@gmail.com, ksingh.mnit@gmail.com.

Abstract

In this work, the desalination experiment was performed through VMD set up and the idea was focused on the removal of multi-ions basically chloride, fluoride and sodium ions, which were present in ground water of rural and urban areas by simulating the secondary solution containing the multi-ions above a permissible limit. The feed solution containing 3000 ppm of sodium chloride and 1000 ppm of sodium fluoride was pumped to feed side of flat sheet micro-porous hydrophobic Poly-tetra-fluoro-ethylene (PTFE) membrane. The performance of PTFE membrane for removal of multi-ions under VMD has been studied. Effect of various parameter like feed flow rate, feed bulk temperature, vacuum pressure and feed salt concentration on permeate flux and rejection have been investigated. The mathematical model [Upadhyaya et al., 2010] was simulated for various parameters. The experimental results are in good agreement with that predicted by model. The maximum purity of 99.8% approximately with permeate flux of 26.6 kg/m^{2}hr was obtained at feed flow rate of 120 liter/hr, vacuum pressure of 5.5 kPa and feed temperature of 333 K. This indicates the strength of VMD to produce safe drinking water from ground water in rural and urban areas.

Keywords: vacuum membrane distillation (VMD), drinking water, PTFE, fluoride removal.

(187a) The Development of Polymeric Membranes for Water, Gases and Fuel Cells

McGrath, J. E., Virginia Tech

Originally presented on: 11/4/2013 15:15:00 - 15:40:00

The speaker spent eight years at the Union Carbide Corporation, during which he contributed to the development of the poly(arylene ether) family, typified by the polysulfones (1). This basic work has been continued in a controlled fashion for the last 37 years since he joined the faculty of Virginia Tech. In the early days we focused largely on the generation of engineering thermoplastics, both amorphous systems and semicrystalline materials, such as the poly(arylene ether ketones). With the beginning of our National Science Foundation Center on Structural Adhesives and Composites, we focused attention on the synthesis of especially amino terminal functional oligomers based upon materials now known as Udel and related systems. The amino systems were prepared quantitatively and were demonstrated as early as the 1980s to be capable of significantly toughening the epoxy resin matrices used for carbon fiber and glass composites. They were differentiated from their analogous blends by the fact that the chemical reactions developed a controlled morphology and chemical resistance, which led to very significant fracture-toughness improvements in crosslinked epoxies and carbon fiber composites. Interactions with industrial organizations led to the commercialization of the materials, which are now present in toughened epoxies in both the Airbus and Boeing products (2).

The last decade in this area has turned toward the development of membranes, first for fuels cells (3) and more recently for water (4) and gas separations (5). In the first two cases, sulfonated copolymers have been investigated, including linear, random, and multiblock systems, which in the salt form can be water purification membranes; whereas in the acid form they are leading candidates for both portable power and vehicle hydrogen/air systems. Variations on these materials have led to crosslinkable systems, which have potential for further reduced swelling and enhanced selectivities for, as an example, seawater desalination. The comparison of the earlier studied post-sulfonated systems with that prepared by a one-step directly sulfonated comonomer will be presented and the advantages of the latter will be emphasized. Gas separation membranes also built upon earlier experience in high performance materials, such as polyimides, polyamideimides, and polybenzoxazoles. Illustrations of these and typical properties will be provided. Currently, further efforts are underway to generate water and gas separation membranes that contain both ionic and non-ionic covalently bound moieties to develop a controlled hydrophilicity needed.

REFERENCES

1. A.G. Farnham, L.M. Robeson and J.E. McGrath, J. Appl. Polym. Sci., Symp. 26, 373 (1975).
2. Guo, R. and McGrath J.E. (2012). Aromatic Polyethers, Polyetherketones, Polysulfides, and Polysulfones. In: MatyjaszewskiK and Moller M. (eds.) Polymer Science: A Comprehensive Reference, Vol. 5, pp. 377-430. Amsterdam: Elsevier BV.

3. M.A. Hickner, H. Ghassemi, Y.S. Kim, B. Einsla and J.E. McGrath, "Alternative Polymer Systems for Proton Exchange Membranes," *Chemical Reviews* (2004), 104(10), 4587-4611.
4. Water purification by membranes: The role of polymer science Geise, Geoffrey M.; Lee, Hae-Seung; Miller, Daniel J.; Freeman, Benny D.; McGrath, James E.; Paul, Donald R. *From Journal of Polymer Science, Part B: Polymer Physics* (2010), 48(15), 1685-1718.
5. Effect of Free Volume on Water and Salt Transport Properties in Directly Copolymerized Disulfonated Poly(arylene ether sulfone) Random Copolymers. Xie, Wei; Ju, Hao; Geise, Geoffrey M.; Freeman, Benny D.; Mardel, James I.; Hill, Anita J.; McGrath, James E. Macromolecules (2011), 44(11), 4428-4438.
6. Synthesis and Characterization of Multiblock Semi-crystalline Hydrophobic Poly(ether ether ketone) Hydrophilic Disulfonated Poly(arylene ether sulfone) Copolymers for Proton Exchange Membranes Yu Chen, Chang Hyun Lee, Jarrett R. Rowlett, and James E. McGrath, *Polymer 53* (2012). Pp. 3143 – 3153.
7. Gas permeability, diffusivity, and free volume of thermally rearranged polymers based on 3,3'-dihydroxy-4,4'-diamino-biphenyl (HAB) and 2,2'-bis-(3,4-dicarboxyphenyl) hexafluoropropane dianhydride (6FDA). Sanders, David F.; Smith, Zachary P., Ribeiro, Claudio, P., Jr.; Guo, Ruilan, McGrath, James E.; Paul, Donald R.; Freeman, Benny D. Journal of Membrane Science (2012), 409-410, 232-241.

(187b) Elucidating the Origins of Rechargeability Limitations in Li-O2 Batteries

McCloskey, B. D., IBM Almaden Research Center
Originally presented on: 11/4/2013 15:40:00 - 16:05:00

Li-air batteries have received significant attention as a potential high specific energy alternative to current state-of-the-art rechargeable Li-ion batteries. However, numerous scientific challenges remain unsolved in the pursuit of attaining a coulombically efficient Li-air battery with high capacity. This presentation will highlight current efforts to understand the nature of Li-O_2 electrochemistry occurring at the cathode. A quantitative *ex-situ* peroxide analysis is coupled with quantitative Differential Electrochemical Mass Spectrometry (DEMS) to better understand factors that control, and hinder, the desired rechargeable cathodic reaction (*i.e.*, $2Li^+ + O_2 + 2e^- \leftrightarrow Li_2O_2$). These measurements allow a comparison of Li_2O_2 formation/oxidation to O_2 consumption/evolution, giving us the ability to decouple parasitic processes occurring during discharge and charge in Li-O_2 cells. These measurements are employed to more completely understand factors that influence both electrolyte and cathode stability.

(187c) Salt Transport Structure/Property Relationships in Polymer Membranes for Water Purification and Power Generation

Geise, G. M., The University of Texas at Austin
Freeman, B. D., The University of Texas at Austin
Paul, D. R., The University of Texas at Austin
Originally presented on: 11/4/2013 16:05:00 - 16:30:00

Providing sustainable supplies of water and energy is a critical global challenge. Polymer membranes dominate desalination and could be crucial to power generation applications. These desalination and power generation applications include reverse osmosis (RO), nanofiltration (NF), forward osmosis (FO), pressure-retarded osmosis (PRO), electrodialysis (ED), membrane capacitive deionization (MCDI), capacitive reverse electrodialysis (CRED), and reverse electrodialysis (RED). Improved membranes with tailored water and salt transport properties are required to extend and optimize these technologies. Water and salt transport structure/property relationships must be used to provide the fundamental framework for optimizing polymer materials for water and salt transport property-critical applications.

A tradeoff relationship between water/salt selectivity (i.e., the polymer's ability to separate water and salt) and water permeability has been reported where polymers that have high water/salt selectivity tend to have low water permeability and vice versa. This tradeoff relationship is consistent with free volume theory suggesting that attempts to prepare materials with favorable combinations of both high selectivity and high water permeability should rely on approaches that do more than simply vary the polymer's free volume. One strategy is to functionalize an otherwise hydrophobic hydrocarbon polymer with fixed charge groups that can ionize upon exposure to water. Sulfonated hydrocarbon polymers are an example of such an approach. These polymers offer an additional advantage from a desalination perspective because they have been reported to resist degradation by oxidizers such as the chlorine-based chemicals that are widely used in water treatment applications.

The presence of fixed charge (e.g., sulfonate) groups in a polymer influences that polymer's salt transport properties. Salt permeability, sorption, and diffusion properties are sensitive to the uncharged or charged (e.g., sulfonated) nature of the polymer. Poly(ethylene glycol) (PEG) hydrogels, sulfonated poly(arylene ether sulfone) random copolymers, and sulfonated styrenic pentablock copolymers were studied to probe the influence of polymer charge, water content, polymer structure, and micro- as well as nano-scale structural heterogeneity on salt transport properties. Salt sorption can be modeled using a combination of Donnan exclusion theory and salt partitioning to further describe the influence of fixed charge concentration, water content, and micro- as well as nano-scale polymer structure heterogeneity on salt sorption and transport properties.

(187d) Highly Water Permeable Anti-Fouling Coatings for Water Purification Membranes

Na, Y. H., IBM Almaden Research Center
Vora, A., IBM Almaden Research Center
Dubois, G., IBM Almaden Research Center
Freeman, B., University of Texas at Austin
Nassar, M., King Abdulaziz City for Science and Technology
Al-rasheed, R., King Abdulaziz City for Science and Technology
Originally presented on: 11/4/2013 16:30:00 - 16:55:00

Water is a constrained resource today and will likely become more of a concern in the future as the world population grows. This global water crisis has lead to great interest in improved water purification technologies and innovative materials for water treatment. Here at IBM's Almaden Research Center, the materials, processes and computational models originally developed for nanotechnology are currently being leveraged for application in desalination and water purification. For example, in partnership with the King Abdulaziz City of Science and Technology (KACST, Saudi Arabia) and the University of Texas-Austin (Professor Benny Freeman, US), we have developed highly water-permeable coating materials that demonstrate significant anti-fouling efficiency during ultrafiltration (UF), nanofiltration (NF) and reverse osmosis (RO) processes which are widely used for waste water and drinking water treatment. In the presentation, bifunctional hydrogel materials designed for porous UF membranes, and ultra-thin hydrophilic methacrylate polymers designed for dense RO membranes will be introduced. The materials showed excellent antifouling efficiency in cross-flow filtration tested using an oil-water emulsion or a bovine serum albumin (BSA) solution as the feed. The antimicrobial activity of the anti-fouling materials was also demonstrated by a case study employing *E-coli*.

(187e) Determination of the Effect of Humic Acid Aggregation On Membrane Fouling By Zeta Potential and Dynamic Light Scattering

Rabbani Esfahani, M., The Center for the Management, Utilization and Protection of Water Resources,Department of Chemical Engineering , Tennessee Technological Universsity
Stretz, H. A., Tennessee Technological University
Wells, M. J. M., EnviroChem Services
Originally presented on: 11/4/2013 16:55:00 - 17:20:00

While natural organic matter and specifically humic acids (HA) have been studied extensively, the aggregation of these materials in solution is not well understood. In this study, the effects of time, concentration, and filtration; Dissolve Organic Carbon – Total Organic Carbon; (through a 0.45 micron Millipore filter) were examined by monitoring changes in zeta potential and dynamic light scattering. The effects of concentration were studied at 2, 5, 8, 45, 100, 250, and 500 ppm. At lower concentrations of Humic Acid higher variation of aggregation size was observed compared to higher concentration. Residence time does not affect the stability of aggregate specifically at higher concentration. Zeta potential decreased—became more negative—as concentration increased. Aggregates of total organic carbon (before filtration) are more stable over time than dissolved organic carbon (after filtration). Finally, the effects of concentration and humic acid size (DOC-TOC) at fouling on synthetic Polysulfone ultrafiltration were determined

(187f) Hybrid Composite Asymmetric Bi-Layer PVA-Cellulose Ultrafiltration Membranes for Water Desalination

Miles, J. II, Prairie View A&M University
Thompson, A., Prairie View A&M University
Nave, F., Prairie View A&M University
Originally presented on: 11/4/2013 17:20:00 - 17:45:00

In this study asymmetric hybrid Poly (vinyl alcohol) (PVA) hydrogels were functionalized with iminodiacetic acid (IDA) and Cu^{2+} to optimize the flux, selectivity, and anti-biofouling capabilities of membranes used for water desalination. The use of asymmetric membranes allows for a small mesh size at the surface to eliminate the transport of unwanted species as well as a mesh size gradient along the membranes thickness to minimize transport resistance. In addition to the advancement in transport capabilities by an asymmetric membrane, functionalizing it with a metal such as copper, which interferes with microorganisms' DNA replication, eliminates the accumulation of microorganisms that are likely to congregate at the membranes surface. Thus, the proposed membranes will reduce the amount of biofouling which tends to alter the membranes flux. Therefore, functionalizing membranes with an IDA-Cu^{2+} complex should maintain flux and decrease biofouling. The three types of composite asymmetric bi-layer membranes (CABMs) were formed using 1.25% (v/v) and 2.5% (v/v) of the cross-linking agent Glutaraldehyde (GA); (type 1) Nitrogen gas was applied to an ultrafiltration device to impregnate cellulose acetate support membranes with asymmetric PVA layer, (type 2) coated cellulose acetate support membranes with asymmetric PVA layer using a spin-coater and (type 3) manually coated cellulose acetate support membranes with asymmetric PVA layer. Characterization on CABMs using Fourier Transform Infrared spectroscopy (FTIR), Atomic Absorption spectroscopy (AAS), and Scanning electron microscopy (SEM) showed differences in the degree of cross-linking, functionalization by metal attachment and impregnation respectively. SEM micrographs indicate that membranes developed using the Ultrafiltration cell allowed for two homogenous gels to amalgamate leading to a more uniform asymmetric membrane. As an indirect measure of functionalization AAS results suggest that type 3 membranes attached the most Cu^{2+}, however the stability of the membranes were compromised. FTIR spectra confirm impregnation in support membranes with asymmetric PVA layer. The membrane performance was evaluated using a dead-end filtration, where preliminary flux data suggest that CABMs have a 66% decrease and 5% increase in flux compared to support membranes coated with a 1.25% and 2.5%(v/v) GA concentration homogenous layers respectively.

(225a) Fundamental Developments and Economic Feasibility of AOPs Involving Ultrasound for Environmental Remediation

Adewuyi, Y. G., North Carolina A&T State University
Peters, D. R. W., University of Alabama at Birmingham
Originally presented on: 11/5/2013 8:30:00 - 8:45:00

Recently, advanced oxidation processes (AOPs) have found world-wide applications for water treatment and environmental remediation and control of air toxics. AOPs involving sonochemistry (i.e., ultrasound), combinative sonophotochemical, sonophotocatalytic and sonoelectrochemical processes, hydrodynamic cavitation, persulfate and/or activated persulfate and other advanced oxidation processes are of particular emerging interests in the research community. These processes offer the potential for shorter reaction cycles, cheaper reagents, and less extreme physical conditions, leading to less expensive and perhaps smaller plants. AOPs have promise to destroy many emerging organic contaminants in water and are being considered in potable water treatment, wastewater treatment, site remediation, and industrial applications. Numerous aqueous emerging organic micropollutants such as perfluorinated surfactants and perfluroalkyl acids (PFAAs), pharmaceuticals, hormones, steroids, antibiotics, personal care products, disinfection by-products and other emerging pollutants; have been studied. This paper will provide a state-of-the-art and unified fundamental chemistry and reaction kinetics involved in the innovative advanced oxidation processes via cavitation, activated aqueous persulfate and other oxygenated oxidants; and will evaluate the performance and economics of AOPs involving ultrasound and hydrodynamic cavitation against conventional water treatment technologies. It is intended to advance our understanding of these emerging technologies and outline directions for future research.

References

Mahamuni, N.N.; **Adewuyi, Y.G.** Advanced Oxidation Processes (AOPs) Involving Ultrasound for Waste Water Treatment: A Review with Emphasis on Cost Estimation. Ultrason. Sonochem. **2010**, 17, 990-1003 http://dx.doi.org/10.1016/j.ultsonch.2009.09.005

Adewuyi, Y.G. Sonochemistry in Environmental Remediation. 2. Heterogeneous Sonophotocatalytic Oxidation Processes for the Treatment of Pollutants in Water. Environ. Sci. Technol. **2005**, 8557-8570. http://dx.doi.org/10.1021/es0509127

Adewuyi, Y.G. Sonochemistry in Environmental Remediation. 1. Combinative and Hybrid Sonophotochemical Oxidation Processes for the Treatment of Pollutants in Water. Environmental Science & Technology. **2005**, 39, 3409-3420 http://dx.doi.org/10.1021/es049138y

Adewuyi, Y.G. Sonochemistry: Environmental Science and Engineering Applications. Industrial & Engineering Chemistry Research. **2001**, 40, 4681-4715 http://dx.doi.org/10.1021/ie010096l

(225b) Advanced Oxidation of Estrone in Water and Wastewater

Sarkar, S.,
Nakhla, G., University of Western Ontario
Ray, M. B., The University of Western Ontario
Originally presented on: 11/5/2013 8:45:00 - 9:00:00

Natural estrogen, estrone (E1) is one of the most potent endocrine disrupting compounds (EDC) which is of primary concern due to its widespread presence in the aqueous environment and also due to its adverse impact both on aquatic life and potential risk to human health. Conventional waste water treatment plants are not designed to remove it, and most of it remains adsorbed on the activated sludge, while part of it is being discharged with secondary effluent, requiring tertiary treatment for water recycle, surface water protection, as well as surface water augmentation. Advanced oxidation processes (AOPs) are seen as promising technologies for removal of estrogens from environment. In this work, a comprehensive research was conducted to determine the relative performance of various advanced oxidation processes such as O_3, H_2O_2, UV, and combinations of UV/ O_3, UV/H_2O_2, ultrasonication, and Fenton agent for the removal of E1 from pure water and secondary effluent. In addition to the parent compound (E1) removal, performance of the AOPs was characterized using removal of total organic carbon (TOC) and estrogenicity in the effluent. Intermediate analysis indicates that UV photolysis produces intermediates, which absorb radiation at wavelength lower than 254 nm making them recalcitrant causing slow removal of TOC. Energy order calculations were performed and cost analysis indicates that ozonation to be the best oxidation process for the removal of E1, TOC and estrogenicity. About 0.5 mg/L of E1 degradation was complete within 30 minutes of continuous purging of water using 1 mg/L of ozone in water. Although, the initial rate of E1 degradation in secondary effluent was decreased by 50% compared to that of pure water, the final removal of E1 was not significantly affected in secondary effluent.

(225c) Analysis of Ozone Absorption and Reaction Kinetics in the Oxidation of Organics in Effluents in a Stirred Tank Reactor

Mathews, A. P., Kansas State University
Panda, K. K., Kansas State University
Originally presented on: 11/5/2013 9:00:00 - 9:15:00

The presence of high molecular weight organic compounds such as dyes, pharmaceuticals, and endocrine disruptors in receiving waters is of concern due to their toxicity and carcinogenicity to humans and aquatic organisms. Ozone is a powerful oxidant that can be used to oxidize such compounds in effluents. However, the use of ozone in these applications is somewhat limited at present due to the relatively high cost of ozone and ozonation systems. This is in part due to mass transfer limitations, and the lack of understanding of fundamental factors that contribute to poor utilization of applied ozone. Previous studies have concluded that mass transfer of ozone form the gas to the liquid phase is the main rate limiting step in ozonation. However, with high molecular weight compounds, the diffusion and transport of the compound to the reaction plane, the production of byproducts and secondary reactions, can also influence the overall oxidation kinetics. In this study, the mass transfer and reaction kinetics regimes are modeled based on fundamental principles and applied to analyze the oxidation of a high molecular weight dye in a stirred tank reactor with a bubble diffuser. Ozonation reactions are mediated primarily by ozone molecules at low pH values (pH ~3), and by hydroxyl radicals at higher pH values. The oxidation kinetics model for the dye is based on the consideration of two parallel reactions in the reactor. The first step involves a fast decomposition reaction that results in the production of primary products. In the second step, the primary products react further with ozone at a slower rate to produce secondary end products. Depending on the relative rates of reaction and mass transfer, the gas-liquid reactions can occur in the liquid film, the bulk liquid, or a combination of both. The decolorization reaction is moderately fast, and is assumed to occur in the film and the bulk liquid. The second reaction is slow, and is assumed to occur in the bulk liquid only. This paper will present experimental studies conducted at pH ~7 for the oxidation of high molecular weight dyes along with model description, verrification, and sensitivity analyses to optimize the performance of stirred tank ozonation systems.

(225d) Electrochemical Wastewater Treatment: Applications and Scale-Up

Woisetschlager, D., Graz University of Technology
Siebenhofer, M., Graz University of Technology
Originally presented on: 11/5/2013 9:15:00 - 9:30:00

In recent years great progress has been made in the research of electrochemical wastewater treatment (EAOP) with the aim to develop a universal method for the treatment of persistent water constituents. Especially anodic oxidation of organic substances with boron-doped diamond (BDD) electrodes has demonstrated its high potential for the degradation of persistent pollutants.

In this project the principles of electrochemical oxidation in wastewater treatment were developed and the applicability of this technology was verified. The degradation trends were characterised for synthetic and industrial wastewater effluents. The influence of the water matrix and the operating parameters were monitored and analysed. Boron-doped diamond electrodes prove to be a highly appropriate electrode in terms of oxidation power, application and stability. With this electrode configuration the purification of a variety of industrial wastewaters was successful. Side reactions and the formation of side products due to chloride, carbonate or ammonium load were also investigated. In the experiments the formation and prevention of chlorinated organic reaction products (AOX) was evaluated as well. The results of the research activities confirm that electrochemical wastewater treatment can establish as highly reliable technology.

(225e) Case Study - In-Situ Oxidation of Ground Water Contaminated By Petroleum Hydrocarbons, Acetone and Chlorinated Hydrocarbons At a Former Industrial Facility in Northern California
Qadir, R. M., Enpro Solutions, Inc.
Originally presented on: 11/5/2013 9:30:00 - 9:45:00

In-situ treatment of dissolved petroleum hydrocarbons and solvents presents many unique challenges due to variability in the subsurface hydrogeology, pH, and chemistry of the soil and ground water matrix. In addition, the petroleum hydrocarbon contaminants present a suite of chemicals in a variety of carbon chains, alkane lengths, and aromatic structures. Proposed in-situ treatments must be carefully tailored to the subsurface characteristics, land use, site-specific cleanup standards, and time constraints. This presentation describes the design and implementation of in-situ ground water remediation by injecting a mixture of Advanced Oxygen Release Compound (ORC-Advanced®) and Regenox® (a solid alkaline oxidant) to ground water contaminated by petroleum hydrocarbons, acetone, and chlorinated hydrocarbons at a former acetylene gas manufacturing site located in Northern California.

Nine ground water monitoring wells were installed and used to track water quality parameters and ground water contaminant concentrations within the treatment zone during remediation; and up-gradient and down-gradient to the treatment zone. Baseline ground water samples were collected and analyzed to estimate the chemical dosage and design the amendment mixture. The amendment consisted of a mixture of Regenox® for rapid, direct oxidation of contaminants and ORC Advanced®, for slow controlled release of oxidants for long-term bio-stimulation. A total of 10,023 gallons of amendment containing over five tons of oxidant was injected at 45 locations using multiple direct-push drill rigs. Eight quarters of post-treatment ground water monitoring showed that concentrations of all dissolved contaminants in ground water were successfully reduced to below cleanup standards. For example, acetone was reduced from a pre-remediation maximum concentration of 54,000 micrograms per liter (μg/L) to 1,600 μg /L.

Challenges encountered during implementation included surfacing of the injected material that reacted vigorously with naturally occurring carbon and anthropogenic contaminants. Subsurface conditions varied widely across the treatments area, including the presence of lime sludge and subsurface obstructions such as buried railroad ties, pipelines and concrete slabs. In addition, since site access constraints required that all work be performed on weekends, multiple drill rigs and support vehicles operated within a fairly small area, increasing logistical and safety concerns.

(225f) Engineering Challenges of Implementing the Peroxone Process for in-Situ Oxidation of Ground Water Contaminated By Acetone and Petroleum Hydrocarbons At a Former Industrial Facility

Qadir, R. M., Enpro Solutions, Inc.

Originally presented on: 11/5/2013 9:45:00 - 10:00:00

In-situ treatment of dissolved chemicals of concern (COC) presents unique challenges due to variability in the subsurface hydrogeology, pH, and chemistry of the soil and ground water matrix. At this former acetylene manufacturing facility, very high concentrations – currently up to 15 x 10^{6} micrograms per liter (μg/L) of acetone are present in the ground water! Other COCs include gasoline range petroleum hydrocarbons and benzene, toluene, ethylbenzene and xylenes (BTEX). The Peroxone process – simultaneous ozone (O_3) and hydrogen peroxide (H_2O_2) injection - was selected for in-situ chemical oxidation of dissolved species in the ground water. This advanced oxidation process produces highly reactive hydroxyl (OH^-) free radicals during the spontaneous decomposition in the ground water of O_3 to oxygen (O_2). Adding H_2O_2 increases the OH^-concentration and the overall oxidation rate. In addition, both O_3 and H_2O_2 break down into O_2 which leads to increased bio-stimulation.

In 2010 six dual-completion (two well casings in the same well bore) injection wells were installed to deliver the oxidants to the subsurface. Six ground water monitoring wells were used to track water quality parameters and ground water COC concentrations within the treatment zone during remediation. In addition, one well is placed up-gradient to the treatment zone and another off-site sentinel well is placed down-gradient to the site. Baseline ground water samples were collected and analyzed to estimate the mass-in-place of COCs dissolved in the ground water and adsorbed onto subsurface soils. The natural chemical oxygen demand (COD) based on naturally occurring organic materials and minerals present in the soil and ground water was estimated. The total COD is many times greater than that required to react with the COCs alone. The optimal oxidant dosage and flow rates required to overcome natural COD and consume the COCs over a period of several years were estimated. A packaged remediation system of appropriate capacity was installed to produce ozone from compressed air and deliver a mix of ozone, enriched O_2 and H_2O_2 to the subsurface through injection wells. The Peroxone process control system software allowed individually varying process variables such that injection flow rate, injection time and other parameters could be customized for each individual well. A complete injection cycle consisted of delivering oxidant to all six wells in turn for the injection times determined by the individual set point at each well. On site monitoring well data show that source area COC concentrations are being reduced. The off-site sentinel well data show that COC concentrations down-gradient to the source area have been reduced by as much as 90% since the Peroxone system was installed.

Challenges encountered during system installation included the presence of subsurface obstructions such as buried utilities and concrete slabs; and logistical and safety concerns due to limited site access and restricted work hours. A major operating challenge was process alarms and shutdowns caused by apparent leaks that turned out to be migration of the injected ozone from the sub-surface through secondary containment piping back to the process equipment trailer. Additional challenges include monitoring remediation progress and optimizing process variables and injection cycles over time.

(225g) Heterogeneous Fenton Oxidation of Synthetic Dye in Water With Surface-Modified Iron-Amended Activated Carbon

Kim, J. R., University of Hawaii at Manoa
Santiano, B., University of Hawaii at Manoa
Kim, H., University of Hawaii
Kan, E., University of Hawaii at Manoa
Originally presented on: 11/5/2013 10:00:00 - 10:15:00

The present study aims to develop effective adsorption and oxidation of synthetic dye in wastewater by using newly synthesized nano iron catalyst-deposited activated carbon. Recently synthetic dye-containing wastewater has gained more attention due to its mass discharge, high toxicity and low biodegradation. For enhancing adsorption of dye and oxidative regeneration of dye-exhausted activated carbon, the novel synthesis of nano iron-deposited granular activated carbon (GAC) was developed. It was to amend ferrous iron onto the acid-pretreated GAC followed by Fe amendment onto the acid-pretreated GAC when pH of iron solution was higher than the pH at point of zero charge (pH, $_{pzc}$) of the GAC. Methylene blue (MB) in water was adsorbed onto the iron-amended GAC (Fe-GAC) followed by single or multiple applications of H_2O_2. Batch experiments were carried out to study the adsorption isotherm and kinetics at at 20°C and pH 6 with the initial concentration of 20-200 mg/L and the solution pH of 6.0. It showed that the adsorption isotherm for MB onto Fe-GAC followed Langmuir isotherm, yielding maximum monolayer adsorption capacity (q_m) of 238.1 ± 0.78 mg/g which is higher than the virgin GAC with q_m of 175.4 ± 13.6 mg/g at 20°C and pH 6. The kinetics of adsorption results indicated that pseudo-second order kinetics represented adsorption of MB. The heterogeneous Fenton oxidation of MB in the Fe-GAC showed increasing the H_2O_2 concentration from 7 to 140 mmol H_2O_2/mmol MB led to enhancing the oxidation efficiency of MB in the GAC from 62.6% to 100%, respectively. The half-life of the H_2O_2 showed the range between 1.4 to 39.4 h depending on the H_2O_2 concentrations at a fixed iron loading of 6 mg Fe/g GAC. The oxidation of MB in the GAC increased at the higher initial concentration of H_2O_2 and Fe^{2+} because of the increased generation of hydroxyl radicals. Further enhancement of oxidation of MB in the Fe-GAC was made by the multiple injection of H_2O_2 while minimizing OH radical scavenging often occurring at high concentration of H_2O_2. The re-adsorptive capacity of the Fe-GAC showed 87-92% over the repeated multiple cycle of adsorption and oxidation of MB indicating effective treatment of MB in wastewater. Further works will include development of diffusion-reaction model and the optimization of the operating conditions over the repeated cycles of adsorption and oxidation.

(225h) Reaction Kinetics and Mechanistic Studies of Nitric Oxide Removal By Combined Aqueous Persulfate and Ferrous-EDTA Systems

Adewuyi, Y. G., North Carolina A&T State University
Khan, M. A., North Carolina A&T State Uuniversity
Originally presented on: 11/5/2013 10:15:00 - 10:30:00

Nitrogen oxides (NO_x – mainly NO_2 and NO) and sulfur oxides (SO_x – mainly SO_2) are the most prominent acid gases emitted from the burning of fossil fuel especially from the coal fired power plants, and are responsible for widespread problems of air pollution, health hazards, acid rains etc. Of these oxides NO_2 and SO_2 are readily soluble in water and can be scrubbed easily from the exhaust stream but nitric oxide (NO) is very sparingly soluble in water and cannot be separated easily. It is well known that the commonly practiced methods of removing NO such as selective catalytic and non-catalytic processes have high capital costs and undesirable problems of high temperatures and handling of harmful chemicals, and alternative cost-effective and environmentally friendlier processes are of ardent interest. It is also well known that suitable oxidizing agents can increase the solubility of NO significantly in water and our group's previous studies involved detailed experimental and mechanistic studies of the removal of NO by aqueous oxidizing agents, of which, peroxydisulfate or persulfate ($S_2O_8^{2-}$) activated by temperature showed significant removal (up to 90% at 90^0C). Later work involving the simultaneous activation by thermal and ferrous ion showed further 10% increment in the NO removal at all temperatures. But the energy requirement for high temperature process to obtain higher removal rates could be one of the main stumbling blocks for further developments. A number of reports have suggested NO removal by iron chelating agent especially ferrous ethylenediamine tetraacetic acid (Fe^{II}-EDTA) where NO is combined with Fe^{II}-EDTA and separated from the solution. The main drawbacks this process are removal of Fe^{II}-EDTA(NO) complex and fast oxidation of Fe^{II}-EDTA to inert Fe^{III}-EDTA and high cost of EDTA. The problem was somewhat overcome in BiodNO_x process where Fe^{II}-EDTA(NO) is broken down by denitrifying bacteria and Fe^{III}-EDTA is reduced by iron reducing bacteria and the resultant Fe^{II}-EDTA is recycled to the reactor again. But the process was impeded by the slow action of the biological process and uncertainty involving in bacteria culture and the additional cost of another unit. It is widely reported in the literature that Fe^{III}-EDTA can also be reduced by persulfate ion ($S_2O_8^{2-}$) and sulfate radical ($SO_4^{\bullet -}$) and the resulting Fe^{II}-EDTA used for the remediation of other chemical pollutants. However, to the best of our knowledge the simultaneous and synergistic application of peroxydisulfate ($S_2O_8^{2-}$) and Fe^{II}-EDTA for the removal of NO has never been studied and is the main focus of this work.

NO removal by combined aqueous persulfate and ferrous EDTA was studied in a bubble column reactor in semibatch mode. Firstly, the optimum Fe^{II}:EDTA ratio was determined for different temperatures and the

1:1 ratio was found to give the highest NO removal rates. The experiments were carried out at different temperatures and the results were compared to the thermal- and iron-activated persulfate alone removal experiments. It was shown that the combined persulfate-ferrous-EDTA system resulted in significantly higher NO removal at low temperatures, almost 30% higher than the thermal- and iron-activated systems. Also, the pH for optimal NO removal was found to be around neutral (pH 7.0). The concentrations of the ferrous, ferric ion and iron-EDTA complex species were measured by spectrophotometric methods and total iron material balance verified. Finally, kinetics and mechanistic reaction pathways for the complex process were investigated using a physicochemical mathematical model involving the use of pseudo-steady-state-approximation technique and film theory of mass transfer. The model was solved numerically using fourth order RK method in Matlab to find the theoretical concentration profile and results used to validate the experimental data. We are currently investigating the effects SO_2 and the various process parameters using design of experiments (DOE) methodologies.

References

Adewuyi, Y.G., Khan, N.E. Modeling the Ultrasonic Cavitation-Enhanced Removal of Nitrogen Oxide (NO) in a Bubble Column Reactor. AIChE J. **2012**, 58, 2397 - 2411. http://onlinelibrary.wiley.com/doi/10.1002/aic.12751/pdf

Khan, N.E.; **Adewuyi, Y.G.** Absorption and Oxidation of Nitric Oxide (NO) by Aqueous Solutions of Sodium Persulfate in a Bubble Column Reactor. Ind. Eng. Chem. Res. **2010**, 49 8749-8760. http://pubs.acs.org/doi/abs/10.1021/ie100607u

Adewuyi, Y.G., Owusu, S.O. Ultrasound-induced Aqueous Removal of Nitric Oxide from Flue Gases. Effects of Sulfur Dioxide, Chloride and Chemical Oxidant. J. Phys. Chem. A. **2006**. 110, 11098-11107. http://dx.doi.org/10.1021/jp0631634

Owusu, S.O; **Adewuyi, Y.G.** Sonochemical Removal of Nitric Oxide from Flue Gases. Ind. Eng. Chem. Res. **2006**, 45, 4475-4485. http://dx.doi.org/10.1021/ie0509692

Adewuyi, Y.G.; Owusu, S.O. Aqueous Absorption and Oxidation of Nitric Oxide with Oxone for the Treatment of Tail Gases: Process Feasibility, Stoichiometry, Reaction Pathways and Absorption Rate. Ind. Eng. Chem. Res. **2003,** 42, 4084-4100. http://dx.doi.org/10.1021/ie020709+

Adewuyi, Y.G.; He, X.; Shaw, H.;Lolertpihop, W. Simultaneous Absorption and Oxidation of NO and SO_2 by Aqueous Solutions of Sodium Chlorite. Chem. Eng. Commun., **1999**, 174, 21-51. http://dx.doi.org/10.1080/00986449908912788

(225i) Sonication Treatment of Sump Water Containing Diesel Fuel

Peters, D. R. W., University of Alabama at Birmingham
Bessette, R., Sound Environmental Systems

Originally presented on: 11/5/2013 10:30:00 - 10:45:00

Pervasive use of diesel fuel around rail yards impacts stormwater, by increasing petroleum releases to surface waters. On-site treatment of diesel sump water generally involves gravity oil-water separation. However, oil-water separators cannot meet National Pollution Discharge Elimination System (NPDES) stormwater discharge standards for oil and grease because they best remove "free" oil, not emulsified oil. Oil-water gravity separation treatment is least effective during stormwater runoff events. An on-site, cost-effective treatment technology treatment technology of diesel sump water is lacking.

This research had as its objective to determine sonication's potential to treat diesel laden stormwater. Sonication effectively oxidizes and pyrolyses a wide range of hydrocarbons in aqueous solution. A mass balance, using gravimetry and chromatography, determined the physical-chemical fate of diesel range organics (DRO). The rate of DRO decrease as a function of sonication time was investigated. Bench-scale sonication results from an initial diesel fuel concentration of 203 mg/L achieved the NPDES discharge level of 15 mg/L within 10 minutes treatment time. Sonication of diesel fuel produced carbon monoxide, short-chain hydrocarbon gases, short-chain hydrocarbons in aqueous solution, solid hydrocarbon precipitates, and residual diesel range organics. After 20 minutes treatment of No.2 diesel fuel (198 mg/L, 203 mg/L, and 2033 mg/L) all produced carbon monoxide and short-chain carbon compounds (< C6) in the gas phase. This indicated that lysis and oxidation of the diesel fuel hydrocarbons occurred. Evidence of diesel fuel hydrocarbon lysis was found in the sequence of changing chromatograms from each initial concentration's subjugation to sonication over a period of 20 minutes. The chromatograms showed a decrease of hydrocarbons in the diesel fuel range and a simultaneous shift of carbon compounds appearing the shorter (< C10) region of the chromatogram and "shouldering" into the solvent peak. There was a shift of the longer carbon chains to shorter carbon chains as sonication time increased from 0 to 10 minutes.

Sonication was more energy intensive per unit mass treated than oil water separation. However, sonication's [potential to eliminate peripheral costs required by oil-water gravity separation may prove cost effective.

(225j) UV-Based Treatment of Fluoroquinolone Antibiotics in Wastewater

Snowberger, S., University of Maryland, Baltimore County
He, K., University of Maryland, Baltimore County
Blaney, L., Lehigh University
Originally presented on: 11/5/2013 10:45:00 - 11:00:00

Over the past few decades, pharmaceutical use has become significantly more prevalent due to advances in technology and medicine; however, this rise in pharmaceutical consumption has been correlated with increases in the concentrations of pharmaceuticals detected in municipal wastewater. Several reports have expressed concerns over the presence of antibiotics in wastewater due to the potential for inhibition of select microorganisms and the development of antibiotic resistance. Furthermore, low concentrations of novel biologically active compounds may pose threats to aquatic ecosystems. Trace concentrations of pharmaceuticals persist through wastewater treatment plants and are discharged to surface water, compounding the issue.

This research examines transformation of pharmaceutical antibiotics using UV irradiation and the UV-H_2O_2 advanced oxidation process. Five fluoroquinolone antibiotics, namely ciprofloxacin, enrofloxacin, levofloxacin, norfloxacin, and sarafloxacin, were investigated. Transformation occurred through direct photolysis (UV), as well as through reaction with hydroxyl radicals (UV-H_2O_2). The reaction kinetics were examined by quantifying the quantum yield and the fluence-based rate constants. For each experiment, 10^{-7} M of the antibiotics were dosed into 2 L of 10 mM phosphate buffer (for pH control). In the case of UV-H_2O_2 testing, 10^{-5}-10^{-7} M hydrogen peroxide was added to the experimental solution. The solution was then run through a batch-recycle reactor containing a low-pressure UV lamp that emits 2.55×10^{-4} Einstein/min at 254nm. Experiments generally lasted ten minutes for direct photolysis and five minutes for UV-H_2O_2; samples were withdrawn at appropriate time intervals. Fluoroquinolone concentrations in experimental samples were analyzed using high performance liquid chromatography with fluorescence detection (HPLC-FLD), which allowed for measurement of all five fluoroquinolones in one sample run.

To examine the effect of complex water matrices on transformation, combinations of multiple fluoroquinolones (total concentration of 10^{-7} M) were simultaneously dosed into the reactor at varying molar ratios. Data resulting from these experiments verified the quantum yield and fluence-based rate constants for all five fluoroquinolones. In addition, the impact of real wastewater matrices was examined by spiking fluoroquinolones into influent and tertiary effluent from a regional wastewater treatment plant. The resulting transformation efficiencies were used to quantify the impact of wastewater-derived organic matter on treatment of fluoroquinolone antibiotics. The impact of dissolved organic matter on UV photolysis and UV-H_2O_2advanced oxidation was determined and provides important insight into transformation of pharmaceuticals by direct photolysis and hydroxyl radicals.

The results of this research include the following: fluence-based rate constants for the five fluoroquinolones at wastewater treatment-relevant pH; the impact of complex wastewater matrices on transformation kinetics (*i.e.*, multiple fluoroquinolones and wastewater-derived organic matter); and the improvement in transformation kinetics as a function of H_2O_2 concentration. As more wastewater treatment plants utilize UV disinfection processes, the results of this research will prove useful in developing effective solutions to the emerging threat from pharmaceuticals.

(228a) Crude Distillation Sequences

Sloley, A. W., CH2M HILL

Originally presented on: 11/5/2013 8:30:00 - 8:55:00

Crude distillation is the initial refinery processing step. Crude distillation produces the major cuts that determine overall refinery yield and sets the composition to downstream units. Properly operating crude units make refineries profitable. Poorly operating crude units make refineries lose money. For example, some refineries operate effectively with high diesel recoveries from the atmospheric unit while others have diesel recovery from the vacuum tower. In contrast, many refineries downgrade diesel to FCC feed, consuming FCC capacity, dropping FCC yields, and degrading middle distillate pool blending quality.

Some simple crude units have evolved into complex, difficult to understand, and nearly impossible to operate sequences using up to five towers. Material shifts between towers may have costs or benefits far from where the material was shifted. Understanding how material shifts change heat duties, yields, and tower capacities provides the basis for designing, operating, and revamping crude units.

The most common approach to understand yield shifts uses the fundamentals of superheat and partial pressures to illustrate how yields change in different crude sequences. The simpler approach of using overall heat balances, while less common, gives equivalent conclusions.

Rigorous analysis compares the partial pressure and heat balance approaches to analyzing crude sequences. Yield shifts for common sequences under different operating conditions are shown. Comparisons include light crude feed versus heavy crude feed for topping units; conventional atmospheric-vacuum distillation; atmospheric preflash towers; gas oil tower configurations; and vacuum preflash towers. Both conventional operating conditions and high-diesel recovery operations are shown for selected options.

(228b) EOM (EQUATION ORIENTED MODEL) For Crude Distillation Unit

Dave, D., British Petroleum

Originally presented on: 11/5/2013 8:55:00 - 9:20:00

Crude distillation unit is at heart of any refinery and processes all the crude that refinery purchases. The complexity in the modelling of crude distillation unit is enormous. The first challenge is the characterization of feed stock. Crude oil is composed of thousands and probably millions of components, with the current analytical techniques – it is impossible to measure them all. Typical approach is to model composition with the boiling point based pseudo components, which has its own short comings such as accuracy of results depend upon the choice of the lumping scheme and correlations used for determining their physical properties. In the present paper, author provide a novel lumping scheme using molecular type and carbon number based surrogate components which provides consistent properties (such as paraffinic component will have lesser gravity and critical temperature compared to aromatics component with same carbon number). For VLLE (Vapour – Liquid – Liquid – Equilibrium) calculations, author adopt an approach presented by Kamath et. al. (2010) to model Peng- Robinson equation of state (1976) calculations with API- Water calculations for handling the aqueous phase. A rigorous tray by tray distillation model is presented with rigorous (MESH) equations and correlation for pressure drop. Pump around in rigorous tower is modelled using the simplified UA based exchanger where a correlation is used to adopt UA values according to process conditions. Pump around heat exchangers are linked with preheat exchangers with consideration of pressure drop. A simple correlation was developed to model furnace which heats up the crude to the flash zone temperature. The numbers of stages in crude distillation column were adjusted to match the observed temperatures and the column model was solved with product purity (distillation end point, cloud point and flash point). An entire EO (equation oriented) model - EOM was solved with IpOpt Solver (2012). The results were compared with published literature and industrial studies. It was found that EOM approach is internally consistent (because of molecular approach), fast (because of equation oriented nature compared to simulation based approach which requires a lot of recycle) and reliable. An illustrative example is also presented where EOM model was used in refinery planning and scheduling compared to the traditional approach – the results suggests a better crude purchase decision is obtained resulting in savings in purchase cost. To the best of author's knowledge, this is the first molecular based, entirely heat integrated and fully equation oriented model presented for crude distillation unit.

Key words: Crude Distillation Unit, Equation Oriented, Molecular characterization, Heat Integration

References:

1. Ravindra S. Kamath, Lorenz T. Biegler, Ignacio E. GrossMann (2010); "An equation-oriented approach for handling thermodynamics based on cubic equation of state in process optimization", Computers and Chemical Engineering 34, 2085–2096
2. IpOpt version 3.0 downloaded from website(2012): https://projects.coin-or.org/Ipopt
3. Peng, D. Y., & Robinson, D. B. (1976). A new two-constant equation of state. Industrial and Engineering Chemistry Fundamentals, 15, 59–64.

(228c) Application of Exergy Analysis of Petroleum Crude Oils Fractions in Crude Distillation Unit

Pulgarin Leon, J., Universidad Nacional de Colombia

Carmona Otalvaro, J., Universidad Nacional de Colombia

Velasquez Arredondo, H. I., Universidad Nacional de Colombia

Molina, A., Universidad Nacional de Colombia – Sede Medellín, Facultad de Minas, Bioprocesos y Flujos reactivos

Originally presented on: 11/5/2013 9:20:00 - 9:45:00

Application of exergy analysis of petroleum crude oils fractions in crude distillation unit

J.Pulgarín-León, J. Carmona-Otálvaro, H. I. Velásquez, A. Molina[*]

[1]*Universidad Nacional de Colombia - Sede Medellín, Facultad de Minas, Bioprocesos y Flujos reactivos*

A crude distillation unit (CDU) consists of three parts, the prefractionator tower (PDU), the atmospheric distillation unit (ADU) and a vacuum distillation unit (VDU), as depicted in Figure 1. On a refinery process, the crude distillation unit (CDU) demands nearly 25%-30% of the total energy of the plant (Wang et al. 2011). Optimization of resources in the CDU is obviously mandatory.

As a complement to the traditional energy analysis, exergy links the first and second laws of thermodynamics and compares the results with a reference environment to identify the possibility to do work through gradient differences of intensive properties (T, P, μ). An exergetic analysis for oil processing can be carried out with three different points of view: (1) global comparison, (2) specific comparison, and (3) profile comparison. The first one considers real industrial operation and defines mass and energy balances that allow an exergy analysis without, however, clear specific conditions of each process. The main advantage of this approach is that it provides valuable information related to the potential global exergy improvement of the system (Le Goff et al. 1996; Rivero et al. 1999; Rivero 2002; Rivero et al. 2004). While the third approach is mostly of the interest of CDUs manufacturers, the second approach, or specific comparison can improve the understanding of irreversibilities present in the distillation process.

This paper presents an ASPEN simulation of PDU, ADU and VDU operating with Cusiana (36.35°API) and Vasconia (24.5°API) Colombian crude oils that has results for energy and separation efficiencies similar to those reported in the literature for similar systems. User-designed subroutines that account for the exergy of oil streams were incorporated into ASPEN to model the exergy in the system. The oil was represented as a combination of pseudocomponents.

Figure 2 shows the calculated exergy for the different streams and for the two oils. As the quality of the stream decreases, the exergy content is smaller. Furthermore, the exergy content is higher for the lighter oil (Cusiana). A comparison of the exergetic efficiency of the three towers involved in the overall separation process showed that the total irreversibility losses are 31.6 MW (23.1% PDU, 51.0% ADU and 25.9% VDU) for a crude flow rate of 154.8 kg/s (100 000 bbl/day). Although the conclusions obtained based on the exergy analysis agree with those obtained from a simple energy balance for the complete plant, the exergy concept identifies irreversibilities that an energy balance cannot detect.

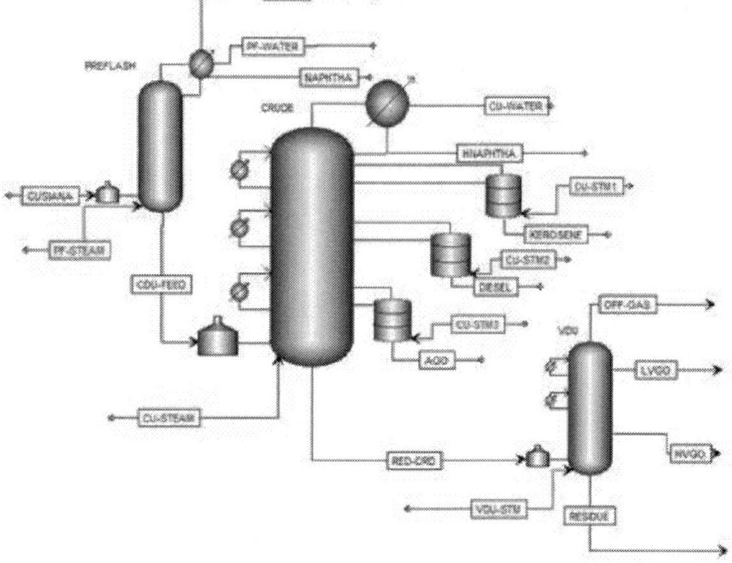

Figure
1. Process flow diagram for the crude distillation unit (CDU).

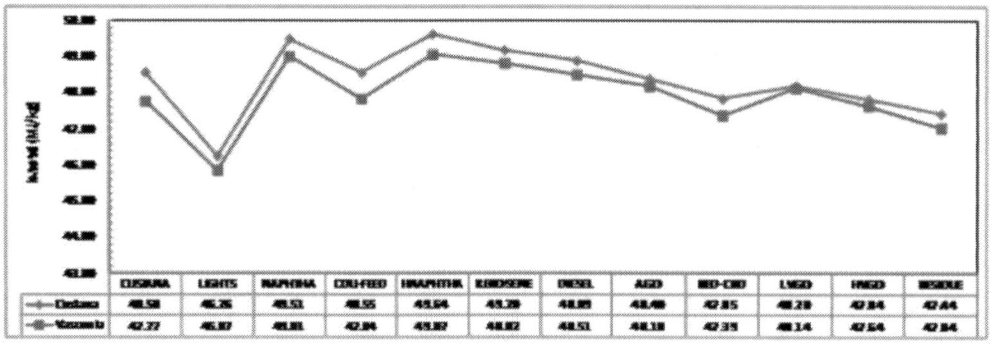

Figure
2. Exergy for flow diagram for the crude distillation unit (CDU).

REFERENCES

Le
Goff, P., et al. (1996). "Exergy analysis of distillation processes." Chemical Engineering & Technology 19(6): 478-485.

Rivero, R. (2002). "Application of
the exergy concept in the petroleum refining and petrochemical industry." Energy Conversion and Management 43(9-12): 1199-1220.

Rivero, R., et al. (2004). "Exergy

and exergoeconomic analysis of a crude oil combined distillation unit." Energy 29(12-15): 1909-1927.

Rivero, R., et al. (1999). The Exergy of Crude Oil Mixtures and Petroleum Fractions: Calculation and Application.

Wang, Y., et al. (2011). "Selecting the Optimum Predistillation Scheme for Heavy Crude Oils." INDUSTRIAL & ENGINEERING CHEMISTRY RESEARCH 50(18): 10549-10556.

---

[*] Corresponding author: amolinao@unal.edu.co

(228d) Separation of Complex Mixtures of Partially Hydrogenated Organic Compounds and Isomers By Distillation

Stark, K., University of Erlangen-Nuremberg
Muller, K., University of Erlangen-Nuremberg
Arlt, W., University of Erlangen-Nuremberg
Originally presented on: 11/5/2013 9:45:00 - 10:10:00

Liquid Organic Hydrogen Carriers (LOHCs) are novel media for the storage and carrying of energy. Amongst others, N-ethylcarbazole is a very promising candidate which is currently investigated with respect to its application as LOHC[1]. During the hydrogenation or dehydrogenation reaction the formation of several partially hydrogenated, stable intermediates has been proved experimentally, e.g. tetra-, hexa- and octa-hydro-N-ethylcarbazole[2]. Experimental studies by GC-MS, shown in this presentation, reveal that furthermore plenty of diastereomers are found turning the LOHC system into a complex multicomponent mixture. However, only several physicochemical property data for N-ethylcarbazole are available in a sufficient extend. To determine physicochemical properties, e.g. melting points or vapor pressures of the partially and fully hydrogenated isomers, those have to be separated from the mixture and purified.

In the current contribution the separation of a mixture of N-ethylcarbazole isomers by batch distillation is shown on lab scale (2-250 ml) as well as on pilot plant scale (5-15 l). Because of the different isomers having close boiling points a high number of separation stages is required. Published vapor pressure data for N-ethylcarbazole show that the vapor pressures of the components have to be considered to be rather low. Therefore the distillation was performed under vacuum conditions preventing distillation temperatures above 250°C at which thermal decomposition of the molecules is possible. Basic studies are performed using a lab scale micro spinning band column with a stage number of up to 50 per meter to fractionate a mixture of perhydrogenated N-ethylcarbazole isomers. The fractions are characterized by GC-MS and NMR. The purities of the fractions are up to $x = 99\%$. For further process design, modeling and scale up vapor pressures are determined using different methods e.g. a GC retention time method or a transpiration method. For the scale-up a pilot scale distillation column with a Sulzer structured gauze packing DX with the height of 2 meter is chosen with respect to the separation of close boiling mixtures. The presentation shows results for basic plant data obtained by experiments with binary test mixtures recommended for close boiling substances with low vapor pressures, e.g. cis/trans-decaline, focusing on the stage efficiency of the packing. Concluding, the separation of complex LOHC mixtures, based on the results of the test mixture experiments and model based process optimization using Aspen Plus®, is discussed.

[1] Teichmann, et al, Energy & Enviromental Science, 2012, 05 (10), 9044-9054.

[2] Eblagon, et al, International Journal of Hydrogen Energy, 2010, 35(20), 11609-11621.

(228e) Isoamyl Propionate Production By Reactive Distillation

Leyva, F., Universidad Nacional de Colombia
Orjuela, A., Universidad Nacional de Colombia
Kolah, A. K., Michigan State University
Lira, C. T., Michigan State University
Miller, D. J., Michigan State University
Rodriguez, G., Universidad Nacional de Colombia
Originally presented on: 11/5/2013 10:10:00 - 10:35:00

One of the by-products of the bio-ethanol industry is the fusel oil, which is principally composed by isoamyl alcohol. To improve the economics of the bio-ethanol industry, this work studied the production of isoamyl propionate by reactive distillation (RD) based on the direct esterification of isoamyl alcohol with propionic acid using Amberlyst 70® ion exchange resin as catalyst. From previous experimental works, models for the phase equilibria and reaction kinetics were obtained and used into the design of RD processes. A conceptual design was performed with the construction of residual curves maps (RCM), including and without including the chemical reaction effects. From the phase-equilibria and reaction kinetics models, with the principal conclusions of the RCM, simulations of possible continuous and batch DR operations were performed in the software Aspen Plus 7.3. From the more promising simulated conditions for high reagents conversions (over 95%) and high products purities (over 98%w), several experiments in a continuous pilot-scale RD column and a laboratory-scale batch RD equipment were performed, using Mellapack® as continous column packing. The experimental results agree accurately with the simulated predictions concluding the RD viability for this reactive system.

(228f) Simulation and Preliminary Design of a Reactive Distillation Column for Removal of Water From Ethanol-Water Mixtures

An, W.,
Lin, Z.,
Chen, J.,
Zhu, J.,
Originally presented on: 11/5/2013 10:35:00 - 11:00:00

Anhydrous ethanol, also known as absolute ethanol, is an excellent alternative clean-burning fuel to conventional hydrocarbon-based fuels such as gasoline, etc. Driven by several world energy crises that have taken place since the 1973 oil crisis, demand for anhydrous ethanol continues to be high. Because of the minimum-boiling azeotrope of ethanol (95.6 wt %) and water (4.4 wt %), special processes have to be employed to remove the remaining water from the mixture. Azeotropic distillation and extractive distillation are two widely used processes, in both of which a third component is introduced into the mixture to purify the ethanol. Besides, technologies of pervaporation and adsorption are also commercially used. However, no matter distillation or non-distillation technologies, all of the above processes involve either considerable energy consumptions or huge capital investments.

This work proposes a novel approach to remove water from a near-azeotropic ethanol-water mixture by reaction with ethylene oxide in a reactive distillation (RD) column. RD has been proved to be a very high efficient process with enhanced conversion, improved selectivity, and the overcoming of azeotropic limitations, since it combines the chemical reaction and separation into one unit. The proposed approach exploits the advantages of RD by forcing the water in the mixture to react with ethylene oxide, and simultaneously separating purified ethanol from the reaction products. The feasibility is verified by simulation studies. Sensitivity analyses are performed on key operating parameters, such as the column pressure, water-to-ethylene oxide ratio, distillate rate, reflux ratio, holdups, feed location, etc., – which are essential to the performance of the system. Appropriate variables are selected with the purpose to gain higher conversion of reactants, selectivity of ethylene glycol, yield of ethanol and desired product purity. By implementing selected parameters (i.e. operating pressure is 0.8 MPa; the molar feed ratio of ethylene oxide to water is 35:20; the reflux ratio is 9; the number of stages is 22), a high-grade anhydrous ethanol (99.6 wt %) can be produced continuously, plus a byproduct of the mixture of ethylene glycol, diethylene glycol, etc., which can be further used as a solvent for nitrocellulose, resins, dyes, or other applications. Especially, the heat released by the exothermic reactions is directly utilized by the distillation. Compared to the conventional processes, the energy consumption can be reduced by 30-70% in addition to the savings in the capital investment and added values of the worthwhile byproduct. In conclusion, the proposed approach shows its promising applications for removing water from a near-azeotropic ethanol-water mixture with great potentials for reduced energy and capital cost over traditional approaches.

(241a) Mathematical Modeling and Design of Layer Crystallization in a Concentric Annulus With and Without Recirculation

Zhou, L., Massachusetts Institute of Technology
Su, M., Massachusetts Institute of Technology
Benyahia, B., Massachusetts Institute of Technology
Singh, A., Massachusetts Institute of Technology
Barton, P. I., Massachusetts Institute of Technology
Trout, B. L., Massachusetts Institute of Technology
Myerson, A. S., Massachusetts Institute of Technology
Braatz, R. D., Massachusetts Institute of Technology
Originally presented on: 11/5/2013 8:30:00 - 8:52:00

In pharmaceutical manufacturing, the drug product is often separated from a liquid phase by crystallization, washing, filtration, and drying—each of which reduce the overall product yield. Poorly filtering crystals can result in bottlenecks in the downstream processing and can affect crystal product purity or reduce yield due to additional washes. These observations have motivated efforts to design continuous crystallization processes that are reliable for manufacturing a product that meets purity, yield, shape, and size requirements, and at the same time ensures efficient downstream processing.

One way to avoid slurry handling is to constrain crystal growth to occur on fixed surfaces so that the amount of solid particles in the bulk is minimized. Layer crystallization has been characterized as being a process in which coherent crystal layers grow on cooled surfaces of a specially designed tube or plate heat exchanger.[1,2] Layer crystallization from melts has been widely applied for the purification of organics and inorganics such as sucrose and milk fat, and in seawater desalination.[3-6] For layer crystallization, the liquid phase is always at a higher temperature than the solid phase, so that in melts, the temperature is required to be higher than the melting point, but layer crystallization from solutions can be operated at much lower temperatures with less energy consumption.[7]

This work proposes a layer crystallization process from a solution in a concentric annulus that avoids the need for washing, filtration, and drying.[8] A continuous flow of drug product could be generated by employing multiple cyclic annular crystallizers in parallel. Many past studies on mass[9-12] and heat transfer[12,13] in annular reactors have concentrated on obtaining correlations of dimensionless number, but these methods are limited by the configurations and operating conditions[13,14] and they are not able to predict the entire concentration or temperature field that is important for crystallization. To our knowledge, this work for the first time thoroughly examines heat and mass transfer of a crystallization process with solid-layer growth in an annulus with and without recirculation. The mass and/or heat transfer model with solid-layer growth on a cylindrical surface can be employed for many other applications, such as fouling in heat exchangers,[15] biofilms in petroleum pipelines,[16] and melt layer crystallization in a tank.[17,18]

A dynamic model is developed in the form of coupled partial differential equations to (1) predict the crystal layer thickness as a function of time and axial position, (2) update the nonuniform boundary temperature at the solid-liquid interface, which is affected by the local thickness of the crystal layer, and (3) calculate the temperature and concentration fields in the annulus. The model predictions are shown to closely track experimental data that were not used in the model's construction, and also compared to an analytical solution that can be used for quickly obtaining rough estimates when there is no recirculation loop. The effects of modifying the operating conditions are investigated, including variation in initial solute concentrations, cooling liquid jacket temperature, inlet solution velocities, core liquid temperature, and extent of mixing in the recirculation loop. The model can be used to optimize product yield and crystal layer thickness uniformity, with constraints on the supersaturation to avoid bulk nucleation by adjusting cooling temperatures in the core and jacket. Additionally, the model could be extended to predict molecular purity of the crystal layer, by considering more components.

References

1. Glade H, Ulrich J. *Melt Crystallization: Fundamentals, Equipment and Applications*. Aachen, Germany: Shaker; 2003.

2. Ulrich J, Bülau HC. 7 - Melt crystallization. In: Myerson AS, ed. *Handbook of Industrial Crystallization*. 2nd ed. Woburn: Butterworth-Heinemann; 2002:161-179.

3. Parisi M, Chianese A. The crystal layer growth from a well-mixed melt. *Chemical Engineering Science*. 2001;56(14):4245-4256.

4. Rich A, Mandri Y, Mangin D, Rivoire A, Abderafi S, Bebon C, Semlali N, Klein J-P, Bounahmidi T, Bouhaouss A, Veesler S. Sea water desalination by dynamic layer melt crystallization: Parametric study of the freezing and sweating steps. *Journal of Crystal Growth*. 2012;342(1):110-116.

5. Flesland O. Freeze concentration by layer crystallization. *Drying Technol*. 1995;13(8-9):1713-1739.

6. Peters-Erjawetz S, Ulrich J, Tiedtke M, Hartel R. Milk fat fractionation by solid-layer melt crystallization. *Journal of the American Oil Chemists' Society*. 1999;76(5):579-584.

7. Aumock NM. *A Process for Separation by Semi-Continuous Counter-Current Crystallization*. Cambridge, MA: Dept. of Chemical Engineering, Massachusetts Institute of Technology; 2011.

8. Zhou L, Su M, Benyahia B, Singh A, Barton PI, Trout BL, Myerson AS, Braatz RD. Mathematical modeling and design of layer crystallization in a concentric annulus with and without recirculation. *AIChE Journal*. 2013;59(4):1308-1321.

9. Ross TK, Wragg AA. Electrochemical mass transfer studies in annuli. *Electrochim. Acta*. 1965;10(11):1093-1106.

10. Mobarak AA, Farag HA, H.Sedahmed G. Mass transfer in smooth and rough annular ducts under developing flow conditions. *J. Appl. Electrochem*. 1997;27(2):201-207.

11. Ould-Rouis M, Salem A, Legrand J, Nouar C. Etude numÃ©rique et expÃ©rimentale des transferts de matière et de quantitÃ© de mouvement dans un Ã©coulement annulaire laminaire non Ã©tabli. *Int. J. Heat Mass Transfer*. 1995;38(6):953-967.

12. Ghosh UK, Upadhyay SN. Mass transfer to newtonian and non-newtonian fluids in short annuli. *AIChE Journal.* 1985;31(10):1721-1724.

13. Viskanta R. Heat transfer with laminar flow in a concentric annulus with prescribed wall temperatures. *Appl. Sci. Res.* 1964;12(6):463-476.

14. Mitrović J, Maletić B. Effect of thermal asymmetry on heat transfer in a laminar annular flow. *Chemical Engineering & Technology.* 2005;28(10):1144-1150.

15. Kirkpatrick JP, McIntire LV, Characklis WG. Mass and heat transfer in a circular tube with biofouling. *Water Res.* 1980;14(2):117-127.

16. Al-Darbi MM, Agha KQ, Islam MR. Estimation of the substrate concentration profiles in petroleum pipelines containing biofilms. *Pet. Sci. Technol.* 2006;24(11):1249-1265.

17. Chianese A, Santilli N. Modelling of the solid layer growth from melt crystallization—the integral formulation approach. *Chemical Engineering Science.* 1998;53(1):107-111.

18. Guardani R, Neiro SMS, Bülau H, Ulrich J. Experimental comparison and simulation of static and dynamic solid layer melt crystallization. *Chemical Engineering Science.* 2001;56(7):2371-2379.

(241b) Cocrystals: Latest Advances in 'in Silico' Methods for Predicting Cocrystal Coformers and Their Properties

Gallagher, D. A., CAChe Research
Loschen, C.,
Originally presented on: 11/5/2013 8:52:00 - 9:14:00

The formulation, design, and implementation of active pharmaceutical ingredients (API) is an area of great interest. Changing the structure and composition of an API by cocrystallizing it with a coformer can have a significant influence on the properties and bioavailability of the drug, and can even offer the potential for new patents and associated revenues. For example, changing the solubility can influence the formation of crystals, cocrystals or solvates, as well as the pharmacokinetics through dissolution during drug release.

The traditional experimental approach to screening for possible new coformers and measuring their properties is time-consuming and expensive. However, *in silico* methods are potentially orders of magnitude faster and are becoming more accurate, such that it is now possible to prescreen huge libraries of potential coformers in minutes or hours on a laptop computer, to determine the most likely candidates. This enables the chemical engineer to take into account a much wider range of possible candidates and then focus the experimental work on the most promising ones, so greatly improving the chances of success in the lab.

This presentation will review the capabilities and limitations of some of the available *in silico* tools for predicting cocrystal coformers and their properties, and compare them with experiment, including group contribution methods, quantum chemistry, and statistical thermodynamics. In addition, the latest algorithmic and method advances for improving the prediction of compounds that can form cocrystals and their properties in a range of solvents, will be discussed.

(241c) Determination of Critical Seed Loading From Nucleation Kinetics

Tseng, Y. T., National Taiwan University
Ward, J. D., National Taiwan University
Originally presented on: 11/5/2013 9:14:00 - 9:36:00

It has been observed experimentally that in at least some cases, secondary nucleation in batch crystallization can be suppressed by using an adequate seed loading. Kubota and coworkers[1–3]call this seed loading the critical seed loading and propose a correlation to predict the critical seed loading as a function of seed size. So far, critical seed loading has only been determined experimentally for a limited number of systems. In this contribution, a method is presented for preparing the critical seed loading diagram knowing only the nucleation kinetics for the system in question. This is advantageous because it is much less time-consuming than conducting experiments or developing and solving a complete batch process model including solubility expressions and growth and nucleation rate kinetics.

The results for 43 systems taken from the literature show that there is a great deal of variability in the critical seed loading, and a single correlation is inadequate to describe the relationship between critical seed loading and seed size. For a seed size of 10 microns, seed loading can effectively suppress nucleation in 92% of cases, while for seeds of size 50, 100 and 200 microns, seed loading can suppress nucleation in 74%, 61% and 39% of cases respectively.

Furthermore, the effects of changing process parameters (total batch time and the net crystal yield) can also be readily studied, although they are found to have a small effect compared to seed size. Finally the possibility of using a critical seed length or critical seed surface area for process development instead of critical seed mass is also investigated, however these alternatives are not found to be more attractive than the critical seed mass.

References

1. Doki N, Kubota N, Yokota M, Chianese A. Determination of critical seed loading ratio for the production of crystals of uni-modal size distribution in batch cooling crystallization of potassium alum. *J Chem Eng Jpn.* Jul 2002;35(7):670-676.
2. Doki N, Kubota N, Sato A, Yokota M, Hamada O, Masumi F. Scaleup experiments on seeded batch cooling crystallization of potassium alum. *AIChE J.* Dec 1999;45(12):2527-2533.
3. Jagadesh D, Kubota N, Yokota M, Sato A, Tavare NS. Large and mono-sized product crystals from natural cooling mode batch crystallizer. *J Chem Eng Jpn.* Oct 1996;29(5):865-873.

(241d) Design and Optimization of Continuous Crystallization Processes: Cascades of Mixed Suspension Mixed Product Removal Crystallizers

Vetter, T., ETH Zurich
Doherty, M. F., University of California, Santa Barbara
Originally presented on: 11/5/2013 9:36:00 - 9:58:00

Crystallization processes in the pharmaceutical industry have been carried out for decades as batch processes. It is recognized that this type of operation suffers from product variability from batch-to-batch and comparably high manufacturing costs [1]. While continuous processing is a proven technique in many large scale industries for overcoming both these deficiencies, the pharmaceutical industry was reluctant to adopt such an approach in the rapid growth era of the twentieth century. However, competition and regulatory demands are ever-increasing and there is now more focus on reducing manufacturing costs while maintaining high product quality.

While the design and optimization methodology for batch crystallization processes are relatively well understood for various combinations of cooling, anti-solvent or reactive crystallization [2] and for different optimization objectives [3], there are yet relatively few studies targeted on the design and optimization of continuous crystallization processes that keep the specific challenges posed to the pharmaceutical industry in mind [4,5].

In this work we present a comprehensive investigation of cascades of mixed suspension mixed product removal (MSMPR) crystallizers based on population balance equation models. The influence of the number of stages, temperature and residence time in each crystallizer on the particle size distribution of the product is ellucidated. Specifically, we show that for any number of MSMPRs and a constant production rate, there exists a clearly defined attainable region in a diagram of mean particle size vs. total residence time in the MSMPR cascade. This attainable region can be entirely traversed by altering the temperature and residence time in each MSMPR. We further report the influence of constraints (such as limitations in the maximally tolerated supersaturation in the MSMPRs and yield constraints) on the attainable region. By rephrasing well-established batch operating policies in the context of the attainable regoins, we highlight the underlying similarities between MSMPR cascades and batch processes and report design heuristics that can be used even in the absence of kinetic data. The concept of the attainable region and the associated heuristics could evolve into a convenient design tool for the conceptual design of continuous crystallization processes in a cost- and time-efficient manner.

Since the above mentioned investigations are dependent on the kinetics of crystal nucleation and growth, we report several case studies that employ different kinetic models and different ways to induce crystallization; among them the cooling crystallization of paracetamol from ethanol and the combined anti-solvent/cooling crystallization of aspirin from ethanol (solvent)/water (anti-solvent).

References:

[1] Randolph, A. D., Larson, M. A., Theory of Particulate Process: Analysis and Techniques of Continuous Crystallization, 2nd ed.; Academic Press: Toronto, Canada, 1988.

[2] Lindenberg, C., Krättli, M., Cornel, J., Mazzotti, M., Cryst. Growth. Des. 9, 1124-1136 (2009).

[3] Ward, J.D., Mellichamp, D.A., Doherty, M.F., AIChE J. 52, 2046-2054 (2006).

[4] Wong, S.E., Tatusko, A.P., Trout, B.L., Myerson, A.S., Cryst. Growth Des. 12, 5701-5707 (2012).

[5] Zhang, H., Quon, J., Alvarez, A.J., Evans, J., Myerson, A.S., Trout, B.L., Org. Process Res. Dev. 16, 915-924 (2012).

(241e) On the Fines Removal in a Continuous Plug Flow Crystallizer: Optimal Spatial Temperature Profile With Dissolution

Majumder, A., Loughborough University
Nagy, Z. K., Purdue University
Originally presented on: 11/5/2013 9:58:00 - 10:20:00

Fine crystals from a crystallizer can cause reduction of the efficiency of the downstream processes as well as final product quality. In this work, a systematic study is performed for obtaining the optimal temperature profile for the reduction of fines in a continuous plug flow crystallizer (PFC). The PFC consists of multiple modules or segments where the temperature of the each segment can be controlled separately. An optimization problem is formulated for a target crystal size distribution (CSD) (without fines) with the temperature of the PFC segments as decision variables. It is found that for the crystallization kinetics considered here, the optimal temperature profile also introduces dissolution steps so that the crystal fines due to secondary nucleation can be reduced significantly. A systematic study on the growth and dissolution kinetics is also performed. This study suggests that the key factor that determines whether the dissolution steps will be successful in reducing fines (without compromising the final size of the crystals from seed) is the size dependence of the growth and dissolution kinetics. If kinetics of the system is such that the larger crystals grow faster than the smaller ones and the smaller crystals dissolve faster than the larger ones, best results for fines removal is achieved. On the other hand, when both the growth and dissolution kinetics are independent of crystal size, fines removal found to be ineffective by temperature cycling.

(241f) Design and Operation of Crystallization-Based Enantioselective Processes: Enantioselective Crystallization of Chiral Compounds On An Enzymatic Reactive Environment

Encarnacion-Gomez, L. G., Georgia Institute of Technology
Bommarius, A. S., Georgia Institute of Technology
Rousseau, R. W., Georgia Institute of Technology
Originally presented on: 11/5/2013 10:20:00 - 10:42:00

Crystallization is a widely used separation and purification technique for a vast number of specialty chemicals and pharmaceutical compounds. However, its application on enantioselective separations can become troublesome. Enantiomeric compounds have the same physiochemical properties (other than optical rotation); as a result, an excess of the desired enantiomer is necessary to obtain enantiopure crystals.

To achieve the necessary enantiomeric excess, we are performing enantioselective enzymatic reactions on supersaturated solutions. To increase the productivity of the process, reaction and separation are performed in parallel in a single vessel. Preliminary work has shown the possibility of performing such a process utilizing chemoenzymatic reactions and kinetic resolutions. Present work is focused on using reaction and crystallization kinetics to design processes in which the operational time is minimized. As a case study, we are investigating the enantioselective deacylation of N-acetyl-DL-methionine by L-aminoacylase. The solubility of the DL-methionine has been measured as a function of substrate concentration, enantiomeric excess, and temperature. The information is being used to set the boundaries of the operation in which an enantiopure crystal can be recovered. The concentration in the reactive crystallizer is analyzed online by combining optical rotation, refractive index, and focused beam refractive measurements (FBRM). These signals can be used to monitor solution and crystals properties. Hence, we can control the chemical and enantiomeric purity of the crystals and their respective size distribution.

(241g) From Batch to Continuous Crystallisation: Nucleation and Growth Under Various Flow Conditions

Schacht, U., University of Strathclyde
Briggs, N., University of Strathclyde
Florence, A. J., University of Strathclyde
Sefcik, J., EPSRC Centre for Innovative Manufacturing in Continuous Manufacturing and Crystallisation University of Strathclyde
Originally presented on: 11/5/2013 10:42:00 - 11:04:00

Crystallisation processes are widely used for separation and purification in chemical and pharmaceutical industries. In order to develop continuous crystallisation of desired polymorphic forms, it is crucial to better understand the influence of mixing and flow conditions on nucleation of various polymorphs from supersaturated solution. For example, both supersaturation and post-mixing flow conditions can be used to control competitive nucleation of polymorphic forms in reactive crystallisation of L-Glutamic acid at constant temperature. We investigated various flow units for treatment of supersaturated L-Glutamic acid solutions prepared under identical initial mixing conditions in either reactive or antisolvent crystallisation.

We obtained respective crystal forms of L-Glutamic acid (either metastable α or stable β) by exposing supersaturated solutions to various flow units, such as a stirred tank, a magnetically stirred crystalliser, a peristaltic pump recycle loop or an oscillatory baffled crystalliser, operating at constant temperature. Using suitable flow units as nucleators to produce seeding suspensions of desired polymorphs, we designed a continuous crystallisation process by connecting a mixing unit (such as a static or in-line mixer) with a nucleator unit (a flow unit) and a continuous crystal growth unit (based on cooling crystallisation in a continuous oscillatory baffled crystalliser).

This set up was successfully used for continuous crystallisation of the stable β polymorph of L-Glutamic acid, achieving a good control over the final crystal size with a narrow particle size distribution.

(255a) Methane Steam Reforming in a Pd Membrane Reactor

Paglieri, S. N., TDA Research, Inc.

Abu El Hawa, H., Colorado School of Mines

Harale, A. X., Saudi Arabian Oil Company

Way, J. D., Colorado School of Mines

Originally presented on: 11/5/2013 8:30:00 - 8:55:00

Methane Steam Reforming in a PdPt Membrane Reactor

Stephen N. Paglieri[1], Hani Abu El Hawa[2], Aadesh X. Harale[3], J. Douglas Way[2]

[1]TDA Research, Inc., Wheat Ridge, CO 80033-1916, USA

[2]Department of Chemical & Biological Engineering, Colorado School of Mines, Golden, CO 80401, USA

[3]Saudi Arabian Oil Company, Dhahran, 31311, KSA

Competitively priced, widely available hydrogen is crucial for the transition to a hydrogen economy. The distributed production of hydrogen from abundant, domestic natural gas resources by methane steam reforming (MSR) currently requires very high temperature catalytic reactors coupled with bulky purification equipment. Demonstrable gains in efficiency for MSR in a membrane reactor can be obtained by the withdrawal of product hydrogen, which enables operation at much lower temperatures instead of the 750-900 °C that is required in conventional MSR reactors.

Thin film Pd and PdPt membranes (~5 μm thick)[1] supported on porous yttria-stabilized-zirconia/stainless steel substrates (ZrOD AccuSep, Pall Corp.) were used to carry out MSR over a commercial Ni-based reforming catalyst at temperatures up to 600 °C and pressures up to 2.9 MPa. Methane conversion was significantly higher than the thermodynamic equilibrium predicted for the feed composition; >90% conversion was obtained at 600 °C, a steam-to-carbon ratio (H_2O/CH_4) of 5 and a gas-hourly space velocity (GHSV) of 360 h^{-1} when the permeate hydrogen was evacuated, as shown in Fig. 1. For comparison, the equilibrium conversions at these conditions are <20% . The effect of parameters such as space velocity, steam-to-carbon ratio, membrane hydrogen permeance, temperature and pressure on methane conversion and hydrogen recovery will be discussed.

Reference

1. Lewis, A.E., et al., Pd?Pt/YSZ composite membranes for hydrogen separation from synthetic water?gas shift streams. *J. Membr. Sci.*, 2013. **437**: p. 257-264.

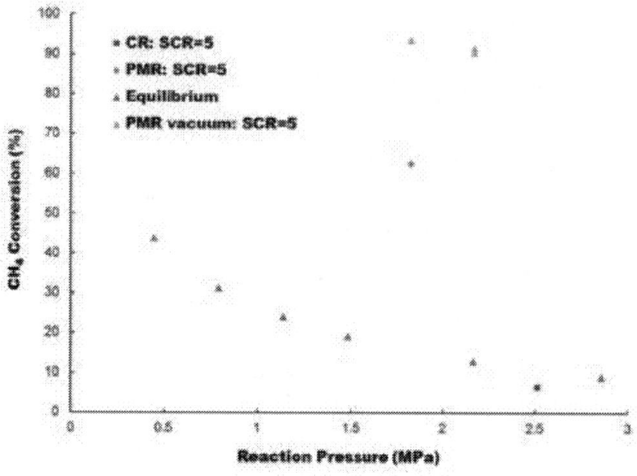

Figure 1. Methane conversion vs. pressure at 600 °C, $H_2O/CH_4 = 5$ and GHSV = 360 h^{-1} in the conventional reactor (CR) and the palladium membrane reactor (PMR). The permeate H_2 was evacuated in some runs (PMR vacuum). Thermodynamic equilibrium conversions are shown for comparison.

(255b) Methanol Synthesis in a Membrane Reactor

Soltani, S.,
Yu, X., University of Southern California
Sahimi, M., University of Southern California
Tsotsis, T. T.,
Originally presented on: 11/5/2013 8:55:00 - 9:20:00

Methanol Synthesis in a Membrane Reactor

SAHAR SOLTANI, XINHAI YU, MUHAMMAD SAHIMI and THEODORE T. TSOTSIS

University of Southern CALIFORNIA

LOS ANGELES, CA 90089

Methanol synthesis has undergone continuous improvements for nearly a century, as it represents the starting raw material for the production of a variety of other chemicals and solvents, including formaldehyde, methyl tertiary butyl ether, and acetic acid and fuel additives. Methanol has a number of advantages as a fuel and a source of chemical products, such as being more easily transportable than methane and other gaseous fuels, having a high energy density, needing no desulphurization, and its reactions (e.g., steam reforming) proceeding at moderate temperatures. Recent global energy shortages and more strict emission regulations have motivated research and development of new fuel cells, among which a direct methanol fuel cell is a prime candidate.

In the present work the CO/CO_2conversion into methanol in both a traditional reactor (TR) and a membrane reactor (MR) has been studied. The purpose of this study is to investigate the possibility of using a MR to increase the total carbon conversion into methanol relative to what a TR can convert.

MR with solvent purge, which is proposed in this work, incorporates the advantages of the existing reactive separation systems that result from the in-situ product removal. In our proposed process, we use a solvent as a sweep fluid. The membrane serves as an interface contactor for the selective permeation of methanol. The solvent is chosen in a way that the main product, methanol, has the highest solubility in it. Using TetraEthylene Glycol Dimethyl Ether as an inert, high boiling point agent, the alcohol will be selectively removed in-situ from the reactor using the membrane as an interface contactor between the methanol and the solvent. Due to the very low solubility of H_2, CO, and CO_2, they will remain in the reactor, since the solvent blocks the membrane's pores.

Prior to the start of the membrane reactor experiments, the surface of a membrane was successfully modified in order to increase membrane hydrophobicity and, therefore, to achieve higher mass transfer rate that, in return, results in higher methanol production. The modified membrane can be operated as a membrane contactor under the proposed operating conditions without the loss of the inert solvent. The kinetics of the methanol synthesis from CO, CO_2 and H_2 on a commercial $Cu/Al_2O_3/ZnO$ catalyst was investigated in an autoclave reactor. The effect of various $CO/(CO_2+CO)$ ratios, stoichiometric number, pressure, temperature and flow rate was investigated in the kinetic experiments.

Based on the experimental kinetics and the membrane surface modification results, we developed a model of the MR to explore the effect of several factors on the total conversion of carbon. The rate equation derived from the kinetic experiments was used in the model. Membranes with various thicknesses were simulated, and their performance was compared. Our results indicate great potential for application of membrane reactor in methanol synthesis using CO, CO_2 and H_2.

(255c) Synthesis of Nanostructured Flower-Like Zinc Oxide Through a Membrane Dispersion Micro-Structured Reactor

Zhilong, X., Nanjing University of Technology
Jian, H., Nanjing University of Technology
Feng, Z., Nanjing University of Technology
Weihong, X., Nanjing University of Technology
Originally presented on: 11/5/2013 9:20:00 - 9:45:00

- Acknowledgements

With the development of nanotechnology, Nanostructured ZnO has attracted extensive research attentions over the past decades in view of its numerous applications, such as gas sensors, desulfurizing sorbents and semiconductors. Many methods for synthesizing nanostructured ZnO have been proposed, including the chemical precipitation method, hydrothermal method, solgel techniques and so on4. The direct chemical precipition method is one of the most promising ways to fabricate nanostructured ZnO for its low cost, simple process and high purity. However, the mass transfer process is seriously limited in the typical precipitation process in which one feedstock is slowly added to another one dropwise, high supersaturation ratio is not available in whole, finally lead to the wide distribution of particle diameter. Consequently, it's significative to develop a micro-structured membrane dispersion reactor with high mass transfer efficiency. Figure 1 shows the membrane reactor experimental setup and the principle of membrane dispersion. NH4HCO3 and Zn(AC)22H2O were used in the reaction as precipitant and zinc salt respectively. Microporous tubular ceramic membrane with pore size of 50nm took the role of dispersion medium. Under the pressure of 0.4 MPa, NH4HCO3 solution was dispersed into tiny drops. The drops diffused rapidly under stirring and reacted with zinc salt, white precursor could be generated theno. Fllowing the process, the white precursor was washed by distilled water three times to remove foreign ions, and then it was dried at 70Ã¢?? for 6h and calcined at 300Ã¢?? for 1.5h.

Nanostructured flower-like zinc oxide with controlled size was successfully prepared though a novel membrane dispersion micro-structured reactor combined with the direct precipitation method. Figure 2 is the X-ray diffraction pattern. The white precursor obtained by the precipitation reaction is crystallized Zn5(CO3)2(OH)6 (No.19-1458). After the heat treatment, it matches well to the standard diffraction pattern of wurtzite ZnO (No.36-1451). Figure 3 is the SEM of Zn5(CO3)2(OH)6 and ZnO at different magnifications. Zn5(CO3)2(OH)6 precipitate demonstrates an interesting nanostructured flower-like morphology at micrometer size, which is assembled by nanoflakes. After the calcine process, similar morphology can be obtained but the nanoflake break into nanopaticles due to the heat treatment. The average crystallite size of ZnO is approximately 18.5 nm calculated by the Scherrer's formula.

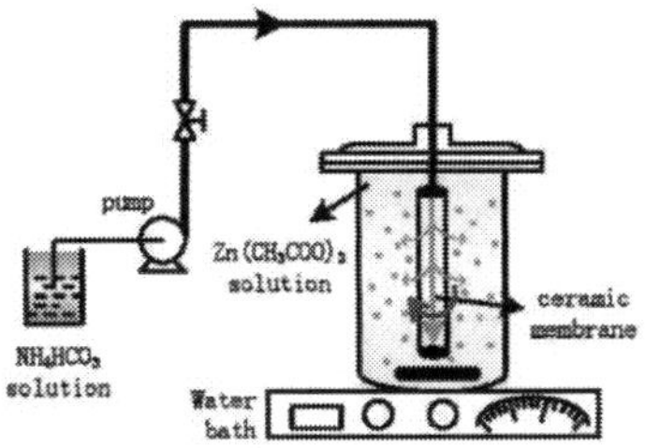

Figure 1. Schematic diagram of membrane dispersion microstructured reactor

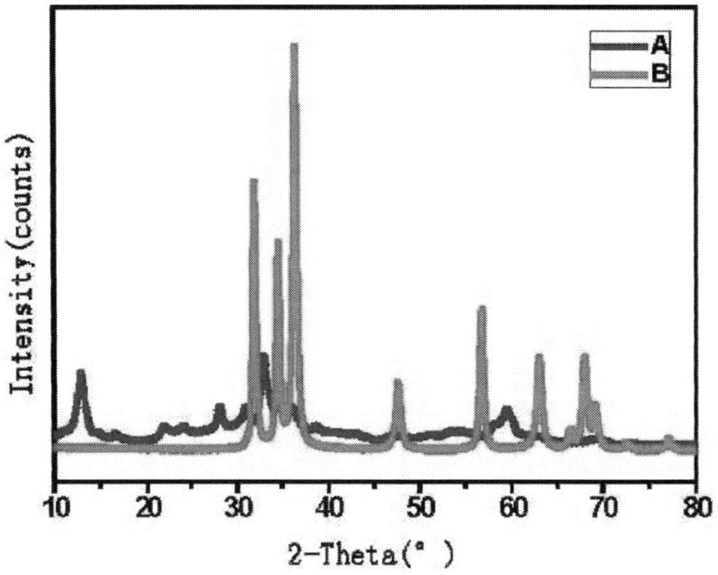

Figure 2.X-ray diffraction pattern of precousor and ZnO after calcination

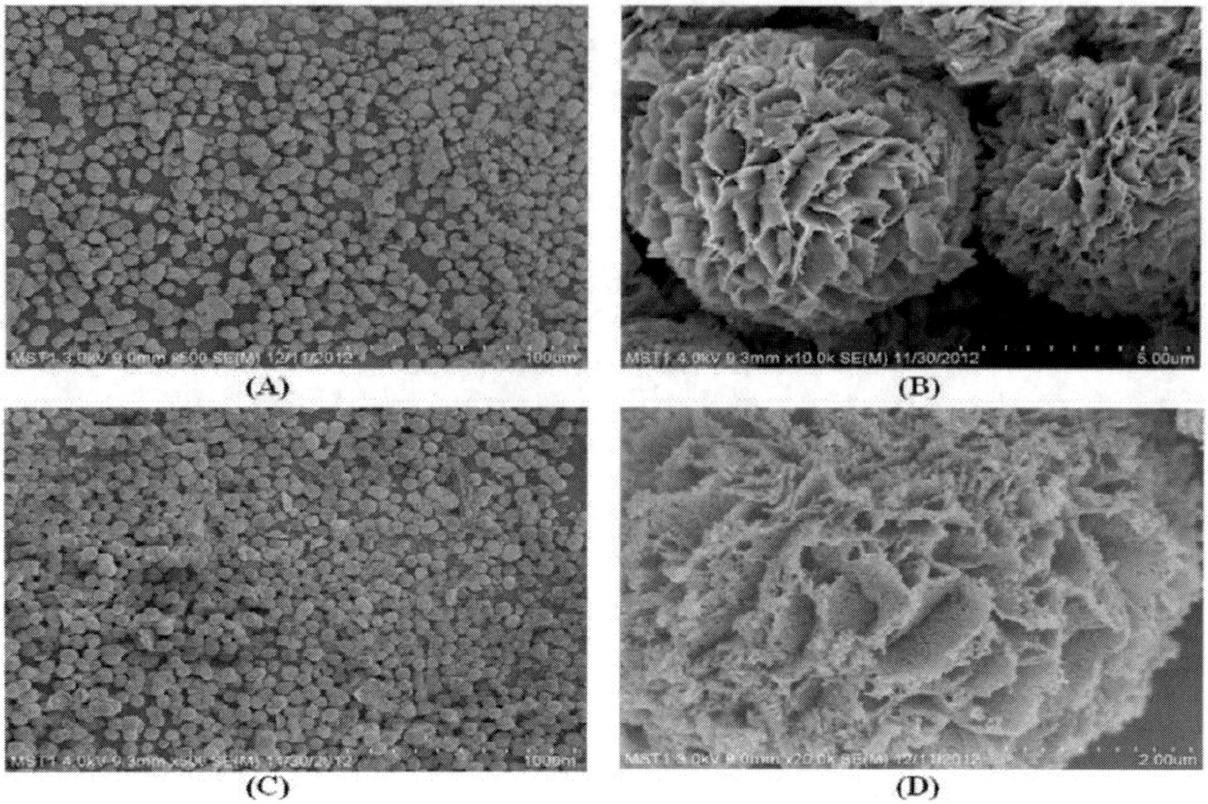

Figure 3.SEM micrographs of (A), (B) $Zn_5(CO_3)_2(OH)_6$ precursor (C), (D) ZnO at different magnifications

References

1.Chen M, Wang ZH, Han DM, Gu FB, Guo GS. High-sensitivity NO_2 gas sensors based on flower-like and tube-like ZnO nanomaterials. Sensor Actuat B-Chem. 2011,157(2):565-574.

2.Garces HF, Espinal AE, Suib SL. Tunable Shape Microwave Synthesis of Zinc Oxide Nanospheres and Their Desulfurization Performance Compared with Nanorods and Platelet-Like Morphologies for the Removal of Hydrogen Sulfid. The Journal of Physical Chemistry C. 2012,116(15):8465-8474.

3.Kripal R, Gupta AK, Srivastava RK, Mishra SK. Photoconductivity and photoluminescence of ZnO nanoparticles synthesized via co-precipitation method. Spectrochimica acta. Part A, Molecular and biomolecular spectroscopy. 2011,79(5):1605-1612.

4.Tiwari JN, Tiwari RN, Kim KS. Zero-dimensional, one-dimensional, two-dimensional and three-dimensional nanostructured materials for advanced electrochemical energy devices. Prog Mater Sci. 2012,57(4):724-803.

5. Medina A, Bejar L, Borjas SE, et al. Characterization of ZnO nanoparticles with short-bar shape pro-duced by chemical precipitation. Mater Lett. 2012,71:81-83.

Acknowledgements

Financial supports from the National Natural Science Foundation of China (No. 21125629), the National Basic Research Program (No.2009CB623406), the National High-Tech Research and Development Program of China (2012AA03A606)

(255e) Membrane Reactor With Water Selectve Zeolite Membrane for Esterification of Acetic Acid

Matsukata, M., Waseda University
Hirai, T., Waseda University
Originally presented on: 11/5/2013 10:10:00 - 10:35:00

Membrane reactoris expected to improve the level of esterification conversion, displacing equilibrium by selective removal of by-product, water. In this study, we performed a simulation study on the behavior ofthe membrane reactorforthe esterification of acetic acid (AcOH) with ethanol as well as the experiments using mordenite-type zeolite membrane, in which mordenite is a kind ofzeolite having both high stability in acidic media and hydrophilic nature. Recently, we successfully synthesized compact mordenite membrane [1], which selectively permeated and separated water from alcohol[2]. A compact mordenite membrane was prepared on the outer surface of an α-alumina tubular support by a secondary growth method, following the procedure described elsewhere [2]. The membrane tube was set inside a stainless steel module. A hundred milligramofsolid acid catalyst, Amberlyst-15, was placed on the outersurface ofmembrane. The exposure area of zeolite membrane was 6.28 cm^2. The feed for the esterification consisted of an equimolar mixture of AcOH and ethanol (both 0.1 mol). The retentate was circulated at a flow rate of 1 mL min^{-1}. The permeate side of membrane was vacuumed during the esterification.

Use of membrane increased the level of conversion from about 60 (equilibrium value)to >90 % at 403 K. Repeated use of the membrane gave reproducible results. In the simulation model, we integrated a perfectly-mixed batch reactor with a pervaporation unit. The expression of the rate of the esterification in the membrane reactor is controlled by both the rates of reaction and permeation. All the parameters for the simulation such as the rates of reaction and permeation were determined by the experiments: No fitting parameters were used. We confirmed that the simulation model developed in thisstudy sufficiently explainsthe experimentalresults. We carried out the sensitivity analysis of the model. Water removal from the reaction side contributed to improving the level of AcOH conversion by displacing the equilibrium. However, an increase of water permeance above 1.0×10^{-6} mol m^{-2} s^{-1} Pa^{-1}would not be effective to improve the level of conversion because the concentration of water in the reactorside is kept very low throughoutthe reaction. The simulation was done with different separation factors, namely water/AcOH, water/ehanol and water/ethylacetate. When the separation factor of water against AcOH or ethanol was too small, the reacants passed through the membrane, resulting in a decrease of attainable level of conversion. When the separation factor of water against ethyl acetate was too small, the yield of ethyl acetate has a maximum with time course, because produced ethyl acetate was lost to the permeate side. As a result, we concluded that separation factor should be larger than 100 for water/AcOH and water/ethanol, and larger than 1000 for water/ethyl acetate. Mordenite membrane developed satisfies such requirements for separation factors.

References

[1] Matsukata, M. et al.,J. Membr. Sci., 316, 18−27 (2008).
[2] Sawamura, K. et al.,J. Chem. Eng.Jpn., 41, 870−877 (2008).
[3] Li,G. et al.,J. Membr. Sci., 218, 185−194 (2003).

(255f) Enhancing Membrane Reactors By Water and Oxygen Removal

Shenvi, A., Compact Membrane Systems, Inc
Boruta, D., University of Delaware
Feiring, A., Compact Membrane Systes Inc
Fox, J., University of Delaware
Nemser, S., Compact Membrane Systems, Inc
Originally presented on: 11/5/2013 10:35:00 - 11:00:00

To improve product quality and reduce manufacturing costs of pharmaceutical end products, Compact Membrane Systems (CMS) has developed perfluoro membrane reactor technology that will provide a generally applicable tool for removing water and oxygen from pharmaceutical synthesis processes. This membrane system has the capability to withstand process temperatures of 110°C endowing it with universal applicability in almost any organic-water environment for water separation in chemical processes. Utility of this technology is demonstrated in a model reaction that will simplify the Buchwald reaction process used in the conversion of a aminoarylbromide into 4-methylpiperizyl derivative which is an advanced API intermediate used for the production of anti-anxiety drug AR-A2 .

In addition, an improvement in the reaction kinetics of nitration of 4-acetamidoanisole to the corresponding 3-nitro-4-acetamidoanisole which is a key intermediate used in the production of API Esomeprazole has been demonstrated. We will present reactor design and related results demonstrating advantages of water and oxygen removal in above mentioned reactions. In addition, we will discuss other applications.

(256a) Sulfonated Block Copolymer Technology for Water Treatment

Willis, C. L., Kraton Performance Polymers, Inc.
Wei, S.,
Originally presented on: 11/5/2013 8:30:00 - 8:55:00

Selectively middle-segment sulfonated pentablock copolymers[1], Nexar™ polymers, exhibit strong phase segregation in the solid phase as well as in the presence of many solvents. The phase separated structure of these polymers in solvent has been correlated with the morphology of the polymer in a cast membrane.

Two distinct micelle structures have been observed in solutions of these polymers[2]. In aliphatic hydrocarbon solvents like cyclohexane, a micellar structure having the sulfonated styrene segment in the core of the micelle is observed. On the other hand, polar solvents like N,N-dimethylacetamide, DMAc, afford micelles having the sulfonated styrene segment in the corona of the micelle.

In the cast membrane, the transport properties of the material are regulated by an ion microphase (sulfonated polystyrene segment) while the hydrophobic thermoplastic microphase (poly-t-butylstyrene and hydrogenated isoprene segments) accounts, in large part, for the mechanical properties of the product. The co-continuous nature of these two phases allows the generation of a material having simultaneously good transport characteristics and good mechanical properties. As the mechanical performance of the cast membrane is derived from the physical (not chemical) crosslinking of the poly-t-butylstyrene end segments of the block copolymer, thin membranes are strong under both wet and dry conditions. Water does not erode the strength mechanism that is holding these membranes intact.

The properties derived from this microphase separated structure of the cast membrane have been found to be useful in various applications where water and ion transport (reduced area resistance, elevated permselectivity) are valued in a membrane having strength whether wet or dry[3]. It is to be emphasized that the transport characteristics of these thin membranes are enhanced as the thickness of the membrane is reduced. Thin, strong membranes, in turn, have been achieved as a consequence of the structure of the phase separated block copolymer.

The relationship between Nexar polymer structure in solvent and performance of the cast Nexar membrane in various water treatment applications will be discussed in this presentation.

References

(1) Willis, C.L.; Handlin, D.L.; Trenor, S.R.; Mather, B.D.; US Patent 7,737,224 B2, 2010; ibid.; US Patent 7,919,565 B2, 2011; ibid.; US Patent 7,981,970 B2, 2011; ibid.; US Patent 8,003,733 B2, 2011; ibid.; US

Patent 8,058,353 B2, 2011.

(2) Choi, J.-H.; Kota, A.; Winey, K.I.; *Industrial and Engineering Chemistry*, 49, 12093-12097, **2010**; Choi, J.-H.; Willis, C.L.; Winey, K.I.; *Journal of Membrane Science*, 394-395, 169-174, **2012**.

(3) Willis, C.L.; Handlin, D.L.; Trenor, S.R.; Mather, B.D.; US Patent 8,084,546 B2.

(4) Geise, G.M.; Freeman, B. D.; Paul, D. R; *Polymer*, 51, 5815-5822, **2010**; Geise, G. M; Lee, H.-S.; Miller, D. J.; Freeman, B. D.; McGrath, J. E.; Paul, D. R.; *Journal of Polymer Science Part B: Polymer Physics*, 48, 1685-1718, **2010**.

(256b) Graphene-Derived Membranes for Gas and Liquid Separation

Park, H. B., Hanyang University

Originally presented on: 11/5/2013 8:55:00 - 9:20:00

Although graphene is a gas-impermeable, if pores could be properly engineered in graphene, its atomic thickness, and high mechanical strength would yield the thinnest-yet molecular filter. However, scalable fabrication of graphene-based membranes with subnanometre pores and high porosity is a great challenge. Here we report that gas molecules can diffuse through defective pores and slit-like interlayers in large-area, few-layered graphene or ultrathin graphene oxide (GO) membranes with nanometric thickness, and that gas diffusion through these membranes can be tuned by randomly stacking, engineering channels, and creating pores. These results are entirely counterintuitive because graphitic layers are usually believed to pose an effective gas barrier. Particularly, ultrathin GO membranes showed highly CO_2 permeable properties due to the strong affinity between CO_2 and GO, and are therefore excellent candidate materials for CO_2 separations. Our findings could potentially be extended to the nanomembranes fabrication for fast, selective mass transport of gases and liquids.

(256c) Pervaporative Purification of 1,3 Propanediol From Binary Solutions and Model Fermentation Broths: Design Optimization On Novel Functionalized Siloxane Membranes

Kanjilal, B., University of Connecticut
Noshadi, I., University of Connecticut
Parnas, R., University of Connecticut
Originally presented on: 11/5/2013 9:20:00 - 9:45:00

Among biorenewable chemicals, the market for intermediates such as 1,3 propanediol, though nascent, accounts for $575 million worldwide. The bio-fermentation of crude glycerol to 1, 3-propanediol is a profitable recourse to capitalizing on the current over capacity of glycerol. The purification of 1, 3 PD from the fermentation broth, in which its concentration is ~1.5%, amounts to almost 50 – 70% of the total production cost. Pervaporative membrane purification is a cheap and energy efficient alternative of upgrading the concentration from a very dilute starting solution. This work presents the design optimization of the pervaporative performance of novel pervaporation membranes based on functionalized Polysiloxanes. The optimal process parameters for the best possible separation factor, balanced by concurrent flux, were determined using a design of experiment by varying paramaters such as feed 1,3 PD concentration, temperature, cross flow velocity and the concentration of co metabolytes in model fermentation broths. Both the enrichment of 1,3 PD concentration and its selectivity over other co-metabolytes was monitored. Response surface methodology (RSM), an important branch of experimental design, was used to find the relationship between experimental parameters and the resultant separation factors and flux of each component in order to find the optimal conditions for the process. Pervaporation can reduce energy consumption in purification by ~80% compared to that required in evaporation / distillation. Additionally, pervaporation proffers a simple process which can be either used stand alone or coupled with other process as well as a continuous fermentation system for achieving high purity at low cost and is easily scalable for large scale operations.

(256d) Nanofiber Supported Thin Film Composite Membranes for Engineered Osmosis

McCutcheon, J., University of Connecticut
Bui, N., University of Connecticut
Originally presented on: 11/5/2013 9:45:00 - 10:10:00

Electrospun nanofibers are an emerging material in membrane separations today. Their high porosity combined with narrow pore size make them excellent candidates as supports for asymmetric composite membranes. In this work, electrospun nanofibers are used as supports for a novel type of thin film composite (TFC) membrane for Engineered Osmosis (EO) applications. These membranes exhibit excellent permselectivity while retaining the properties of membranes most suited for osmotic separations. These characteristics include having a thin, highly porous, and reasonably strong support layer. This presentation outlines a number of new nanofiber based TFC membranes developed at the University of Connecticut for use in osmotic separations. Their performance (flux, selectivity) greatly exceeds that of other membranes proposed in the academic literature as well as those membranes produced commercially. These new membranes may help enable EO to viably function in processes that involve water desalination, power production, and dewatering.

(256e) Small Molecule Transport in Polymers

Freeman, B., University of Texas at Austin

Originally presented on: 11/5/2013 10:10:00 - 10:50:00

This presentation will discuss the fundamentals of small molecule transport in polymers and then focus on applications that have been of particular interest to our group during the past 20+ years. Topics to be discussed include gas separation using polymers, water purification using polymers, and the development of polymer-based surface treatments to enhancing fouling reduction in oil/water separation. Areas where the materials concepts studied in my group have been reduced to practice will be highlighted. Much of the work has been collaborative in nature and cross-disciplinary, and specific examples of problems undertaken with the help of cross-disciplinary teams will be emphasized. Finally, some avenues for future work in the field will be discussed, as much remains to be done to fully understand the fundamentals and derive the most benefit possible from polymers or polymer-based materials for separations and barrier applications. Throughout the presentation, the contributions of collaborators as well as current and former students will be highlighted.

(285a) Tower Scanning With Packview™ - A Quantitative Tool for Diagnosing Liquid Maldistribution in Packed Beds

Pless, L., Tracerco

Originally presented on: 11/5/2013 12:30:00 - 12:55:00

The tower gamma scanning technique is well established in the process industries as a qualitative tool to help troubleshoot separation towers, both trayed and packed towers. In this presentation it will be shown that through careful and innovative application of the gamma scanning technique, quantitative information about the liquid distribution through packed beds in separation towers can be extracted.

For packed towers some form of a grid-scan, multiple vertical scans through packing, has been traditionally used to detect liquid mal-distribution through packing. A new technique will be introduced whereby gamma scan data is used to determine relative process densities through packing. Relative process densities can be converted into liquid holdup fractions, either of which can be used as a quantitative measure of liquid mal-distribution. The concept will be applied to case studies to illustrate how quantifying gamma scan data can provide more useful information than by the traditional qualitative analysis.

(285b) A Valve Tray Design Capable of Efficient Operation in Tilted Columns and Out-of-Level Tray Columns

Onyelucheya, O. E., Federal University of Technology, Owerri
Ndukwe, O. C.,
Okpala, K. O.,
Originally presented on: 11/5/2013 12:55:00 - 13:20:00

In this work, the operation and performance of a novel valve tray design capable of functioning without loss of efficiency when the column is operated in both vertical and tilted conditions is presented. The Tray Column used for the work is a Downcomerless liquid initiated and controlled Dual Flow Valve Tray with capacity to shut portions of the tray lacking liquid at any time. Portions of a Tray Column could be starved of liquid during operation when the tray is Out-of -Level and this may occur when Tray Columns are operated on moving platforms. The term “moving platform,” implies that the Column will not always be vertical, but can sway depending on the motion of the platform. These sways will result in tilting of the equipment mounted on the platforms, channeling of the fluids and a loss or total collapse of the efficiency of conventional Tray Columns subjected to such unique operating conditions. These conditions are encountered in many chemical processing operations that need to be carried out on floating platforms to recover “stranded gas” in remote locations, in stripping ships and even in columns mounted on land but exposed to strong wind forces and other adverse factors that cause the Tray column to be Out-of-Level.

A liquid controlled valve tray called “Plunger-Cap-Multifloat Valve Tray” having four floats per valve is examined. The “liquid control” aspect of the valve was achieved by exploiting the basics of the Archimedes principle of floatation and Upthrust. The fundamental aspect of this design was the determination of the optimum float size that will give the Upthrust required to operate the valve as desired. The design established that 0.5mm mild steel cylindrical floats with height/diameter of 50 mm will operate the valve when at least 75% of the float is submerged in water. The complete experimental rig used for this work consists of a 2-tray downcomerless column with 4 “Plunger-Cap-Multifloat Valves” per tray. Portions of the column shell were fitted with Perspex to permit observation of the workings of the tray internals during operation. Water supplied from a holding tank by a 0.5 Horsepower Electric pump was fed through the top of the column to the first tray. When the level of water on this tray was sufficient to lift the floats, the valves opened and allowed water to flow to the tray beneath. The response of the valves of the tray beneath was same as obtained with the tray above and the water flowed to the base of the column from where it was re-circulated by gravity to the holding tank. The air which was supplied by a 2.5 Horsepower Air Compressor was introduced beneath the second tray in the column. A head of liquid maintained at the base of the column served as a liquid seal and ensured that the air introduced flowed up through the open valve to the second tray first, then to the first tray contacting liquid in each of these trays. The air was released through a vent at the top of the column. The flow rates of both the water and air were measured by Rotameters. The column assembly was also fitted with valves for controlling the flows at the desired rates and the entire column is mounted on a rig which could be tilted and operated at various angles.

To evaluate the mass transfer performance of this column, an oxidation experiment was conducted and the percentage of Fe(II) oxidized to Fe(III) from the water by air at a fixed temperature (30^0C) was chosen as the system response. The impact of the various liquid and gas flow rates on this system response in both vertical and tilted conditions were examined.

Based on Factorial Design of Experiments, the effects of the Gas flowrate (x_1 in litres per minute), the Liquid flowrate(x_2 in litres per minute) and the Angle of tilt (x_3 in degrees) on the Response (y) were investigated. Both a First Order Linear experimental design and a Second Order Box-Wilson Rotatable design were used and four polynomial models tested.

The influence of these factors on the response was represented by the polynomial models and their effects were studied using the Student's t-test and Fischer's F-test for the Analysis of Variance (ANOVA) for each model. Out of the 4 models, the experimental data were found to be best represented by a linear polynomial model $y = 12.33 + 1.30x_1 + 0.68x_2 + 0.235x_3 + 0.178x_1x_2 + 0.023\,x_1x_3 + 0.043x_2x_3$ with Coefficient of Determination $R^2 = 0.9635$. The test of significance of the individual coefficients in the model based on the Student's t-test showed that only parameters x_1 and x_2 (i.e. the gas and liquid flowrates) are significant at 95% confidence level. The angle of tilt (with p=0.18) had no significant influence on the amount of Iron (II) oxidised. The ANOVA of this model showed that the model was significant as the calculated F-value (F-model = 17.60) exceeded the tabulated F-value (6.16), and also from its probability value (p=0.008) which is less than 0.05. From this test also, the model parameters x_1 and x_2 had significant influence as their calculated F-values (164.57 and 45.38 respectively) were much higher than their tabulated F-values (F=7.71). The parameter x_3 and their interaction effects were still not significant from this test as their calculated F-values (5.42 for x_3, 0.06 for x_1x_3 and 0.2 for x_2x_3) are less than their tabulated F-values of 7.71. These findings were further validated by the x-y scatter plots of the obtained data which showed that for all the liquid flowrates, the % Fe (II) oxidised increased with increasing gas flowrates but the impact of the angles of tilt was minimal. Also, for all the liquid flowrates within the range of our experiments, the change in the percentage Fe (II) oxidised between 0 degrees to 10 degrees column tilt and also between 0 degrees to 20 degrees column tilt is seen to be less than 1% at constant vapor rates.

The implication of these is that the operational efficiency of this novel Valve Tray Column will not be compromised by column tilt of up to 20 degrees from the vertical. This finding validates our earlier hypothesis that a liquid operated and controlled Valve Tray (such as the Plunger-Cap-Multifloat Valve Tray presented in this paper) will provide solution to the problems of liquid maldistribution and efficiency collapse occasioned by column tilt and motion.

(285c) Gas-Liquid Flow Modeling in Columns Equipped With Structured Packing Seen As Bi-Structured Porous Media

Soulaine, C.,
Quintard, M., Institut de Mecanique des Fluides
Originally presented on: 11/5/2013 13:20:00 - 13:45:00

Gas-liquid flow modeling in columns equipped with structured packing seen as bi-structured porous media

Structured packing play a large role in chemical engineering processes involving gas-liquid. The packings are made of an assembly of corrugated sheets where two adjacent sheets are respectively inclined by an angle and the opposite of this angle from the vertical axis. Each group of oriented sheets may force the flow in its preferential direction, hence the naming of such media as *bi-structured porous media*. In addition, such a device features the classical porous medium two-scales characteristics: the pore-scale and the packing scale or macro-scale. Due to this peculiar structured geometry, the flow modeling from a macroscopic point of view remains a challenging problem that has to be overcome to design enhanced devices. In particular, the macroscopic phenomena that leads to the spreading of a liquid point source at the top of a packing are still misunderstood, and the classical two-phase flow models in porous media failed to properly catch the liquid distribution within the column.

We have developed a mathematical model based on a multi-scale analysis to simulate gas-liquid flow through distillation columns. This model is partly inspired by the work of [1] and our recent research on single phase flow through *bi-structured porous media* [2]. The effective properties that appears in this model are evaluated from either simulations or analytical solutions of the flow at the pore-scale. Simulation results have been successfully compared to laboratory-scale experiments and industrial-scale measurements.

[1] CFD Modelling and Calculation of Dynamic Two-Phase Flow in Columns Equipped with Structured Packing *Chemical Engineering Research and Design, Volume 85, Issue 8, 2007, Pages 1112-1122* B. Mahr, D. Mewes

[2] A two-pressure model for slightly compressible single phase flow in bi-structured porous media *Chemical Engineering Science, Volume 96, 7 June 2013, Pages 55-70* Cyprien Soulaine, Yohan Davit, Michel Quintard

(285d) Experimental Investigation of a Multi-Stage Counter-Current Rotating Packed Bed for Distillation

Sudhoff, D., TU Dortmund University
Leimbrink, M., TU Dortmund University
Lutze, P., TU Dortmund University
Gorak, A., Department of Biochemical and Chemical Engineering, Laboratory of Fluid Separations
Originally presented on: 11/5/2013 13:45:00 - 14:10:00

Experimental Investigation of a Multi-Stage Counter-Current Rotating Packed Bed for Distillation

Authors: Daniel Sudhoff*, Mathias Leimbrink, Philip Lutze, Andrzej Górak

Keywords:

Rotating Packed Bed, Distillation, centrifugally-enhanced separation, process intensification

* Dipl.-Ing. Daniel Sudhoff

TU Dortmund University

Phone: +49 (0) 231/ 755 6002

Email: Daniel.sudhoff@bci.tu-dortmund.de

Web: www.fvt.bci.tu-dortmund.de

To meet future global challenges for the chemical and bio-chemical production processes such as lower lifetimes and the demand of innovative products but also to produce products more sustainable, the processes need improvements. Besides others, more flexible processes are necessary to be able to react to changing compositions of a bio-based feedstock, to meet changing market needs and to enable a quicker entry to market for new products. To fulfill these requirements processes are needed that either dynamically adapt the changing requirements quickly or can quickly be rearranged by modular devices and modular process parts. For both cases compact apparatuses are required that have a sufficient number of degrees of freedom and a wide range of operation.

The so called HIGEE-technology ("high gravity") is an example of a promising technology to meet these process demands for fluid separations. Therefore a high gravity field is created and the liquid and vapor are contacted inside this field. A promising possibility to realize a phase contactor applying a high centrifugal field are rotating packed beds (RPBs). In these machines a cylindrically shaped packing is rotating at a high rotational speed. The liquid is introduced to the center of this rotating packing and is therefore radially accelerated along the packing. The vapor is introduced to the outer casing and driven through the packing to the center of the rotor by a slightly enlarged pressure. The phase contact inside the packing occurs at applied forces that are a multiple of the gravitational force.

The high applied forces to the liquid enable a very high throughput of liquid and vapor as well as the use of high performance packings with a very high surface area. High turbulence with large interfacial areas are created and quick mass and heat transfer between the phases are enabled. This gives the opportunity for high capacities at short residence times and low space requirements. Additionally, the rotational speed that has a high influence on the separation efficiency of the RPB is an additional degree of freedom. Due to these advantages RPBs can be applied to meet the above mentioned requirements for chemical production plants.

Despite these advantages RPBs have not been established as phase contactors yet in the chemical industry. Although several studies, on the application of RPBs for absorption processes and for the production of nano-particles have shown its feasibility, no sufficient investigations of distillation processes in RPBs have been published yet. The few available results have been achieved with lab scale apparatuses with small radial lengths of the packing. Therefore, systematic investigations to understand energy, mass and impulse transport phenomena and the influence of equipment and operating parameters are lacking.

To fill this gap a novel rotating packed bed with three consecutive rotors at pilot scale for distillation has been developed (Multi-Stage Counter-Current Rotating Packed Bed, MSCC-RPB) and build. The details of the MSCC-RPB have as well been presented in a previous work as its hydrodynamic operating limits and an approach for the evaluation and comparison of the performances of RPBs for distillation [1].

In this work the hydrodynamic behavior with regard to pressure drop and hold-up have been investigated. Additionally, the separation performance of the MSCC-RPB for distillation has been characterized with as suitable water-alcohol test system. The influence of the parameters rotational speed, feed location and vapor and liquid load have been determined. The results are presented and evaluated on a basis of absolute reference values (integrated centrifugal acceleration, integrated liquid and vapor loads) as have already introduced in an earlier work [1]. Finally, the separation efficiency of the MSCC-RPB for distillation is compared to a pilot scale distillation column with structured packing existing at the same laboratory. For this purpose an established and in previous works successfully applied mass transfer model for the distillation column is used to determine the required packing height of a distillation column to achieve the same separation efficiency as the MSCC-RPB.

[1] Sudhoff, D.; Lutze, P.; Górak, A.: Multi-Stage Counter-Current Rotating Packed Bed for Distillation; presentation at AIChE Spring Meeting 2013, San Antonio, USA.

(285e) CFD Simulations of Rotating Zigzag Bed Hydrodynamics

Li, X., Tianjin University

Zhang, L., Tianjin University

Jiang, B., Tianjin University

Yang, N., Tianjin University

Originally presented on: 11/5/2013 14:10:00 - 14:35:00

The concept of process intensification was first proposed in the late 1970s. High gravity technology (abbreviated to HIGEE), which substitutes terrestrial acceleration for higher centrifugal acceleration (up to 1000 g) achieved by rotating a specially shaped bed, is a promising alternative to enhancing heat and mass transfer in the process intensification field. Compared to conventional distillation columns, the HIGEE can make dramatic reductions in equipment volume, and reduces the capital and operation costs accordingly.

Generally, an HIGEE device mainly consists of a rotor housed in a stationary casing and driven by a motor. The structure of rotor determines the characteristics of different HIGEE devices. There are several types of rotors, such as packing bed, wave form disk, helical bed, multi-staged spraying bed and split packing bed. Another type of HIGEE device, referred to as rotating zigzag bed (RZB), was invented by Ji et al. and patented in China and the United States. It has been successfully applied in continuous distillation service after unremitting improvements and tests. The simplified sketch of the RZB is shown in Figure 1. Tow series of concentric circular sheets are alternately fixed on the rotational (lower) and stationary (upper) disk, respectively; one series as rotational baffles and the other as stationary baffles. The clearance between rotational baffles and the stationary disk as well as between the stationary baffles and rotational disk offers zigzag flow channels for fluids. The gas is introduced into the casing and flows inward radially driven by pressure difference. The liquid is fed at the center of the rotational disk flowing outward countercurrent to gas by virtue of centrifugal force. In the interior of the rotor, the gas flows as continuous phase along a zigzag path and the liquid undergoes multiple dispersion-accumulation recycles. The liquid is dispersed in fine droplets by rotational baffles. These droplets are intercepted by stationary baffles and then accumulate to form liquid film which falls onto the rotational disk again. Then the liquid flows radially outward to the next rotational baffle and is dispersed, and so on. During this process, impingement and spray of the liquid in the rotor results in fast-renewed gas-liquid interface with very large specific surface area available for mass transfer.

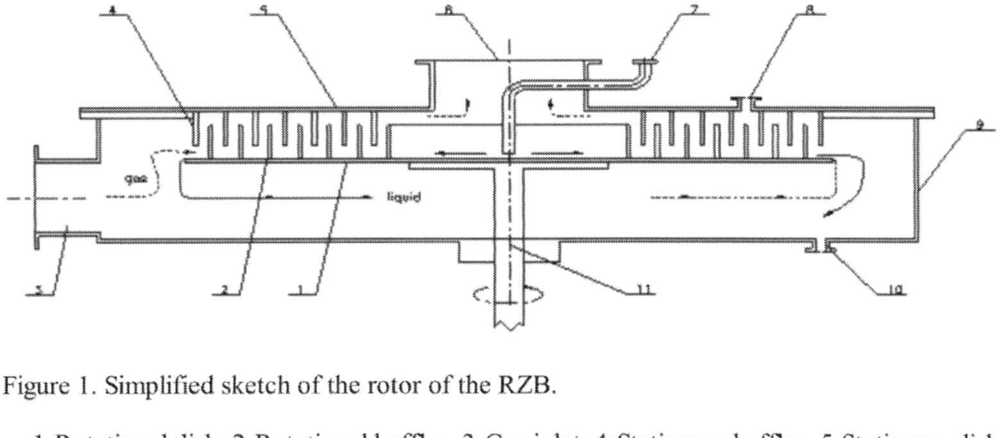

Figure 1. Simplified sketch of the rotor of the RZB.

1-Rotational disk, 2-Rotational baffles, 3-Gas inlet, 4-Stationary baffles, 5-Stationary disk, 6-Gas outlet,

7-Liquid inlet, 8-Intermediate feed, 9-casing, 10-Liquid outlet, 11-Rotating axle

Mass transfer performances of RZB and other HIGEE devices have been experimentally studied in the literature, but few studies on the hydrodynamics have been reported. The understanding of the two-phase hydrodynamic phenomenon in the rotor is helpful for the fine tuning and optimization of HIGEE device design.

Figure 2. 3D physical model of RZB.

Due to progresses in computer hardware and software and consequent increase of the calculation speed, the computational fluid dynamics (CFD) modeling technique becomes a powerful and effective tool for understanding the complex hydrodynamics in chemical engineering processes. In this study, a three-dimensional CFD model was developed to predict the hydrodynamics of RZB as shown in Figure 2. The model considered gas- and liquid-flow within the Eulerian-Eulerian framework in which both phase were treated as interpenetrating continuum having separate transport equations. With the model focusing on the interphase contacting region, the liquid phase has been taken as the disperse phase, while the gas phase has formed the continuous phase. Interphase momentum transfer term was employed for describing the interfacial forces between the two phases. Since the focus is on the hydrodynamics behavior of RZB, energy and mass transfers have not been considered in this study.

(285f) Research On Hydromechanics and Mass Transfer Performance of NS High Efficiency Packed Sieve Composite Plate

Yuanyu, T., Shandong university of Science and technology
Yingyun, Q., Taiyuan University of Technology
Taojiang, L.,
Kechang, X.,
Originally presented on: 11/5/2013 14:35:00 - 15:00:00

Characters of hydromechanics behavior such as pressure drop, entrainment and forth height and mass transfer performance of NS high efficiency packed sieve composite plate were studied totally,and were compared with ordinary dual-flow sieve tray. Effect of hydromechanics on mass transfer performance of the NS composite plate was discussed. Finally, a optimal operating range were determined, which provide the foundation for industrial applications of the NS composite plate. This special structure can fill the high efficiency, large flux, low pressure drop, large operation range and steady operation demand. “full-mixture” structure provide possibility for simplifying calculation model and improve modeling precision. Bubbles in NS efficient composite plate became smaller and favorable foam layer formed on the plate which made a large gas-liquid contact interface and large renew ratio. Compared with ordinary plates, mass transfer efficiency of NS plate was 20% more than ordinary ones. The efficiency was more than 80% and almost reach 100% at large gas velocity in the whole operation range.

(286a) Filtration of Sea Salt Particles (SSP) Using Composite Media At Different Relative Humidities (RH)

Zhao, P., Auburn University
Tatarchuk, B., Auburn University
Originally presented on: 11/5/2013 12:30:00 - 12:55:00

Solid Oxide Fuel Cells (SOFCs) are being widely used in recent times as a source of power on ships due to their high efficiency and lower waste emission. However, the exposure of several air contaminants like gaseous impurities and particulates is significantly detrimental to the performance of these fuel cells. In the marine environment, sea salt particles are the most common particulates and their removal from air stream fed into cathode side of fuel cell system is highly essential. To remove sea salt particles in a more efficient manner, our focus is on the design and optimization of a new composite filtration media known as Microfibrous Materials (MFM) and manufactured by the Center for Microfibrous Materials Manufacturing (CM3) at Auburn University. These composites consist of a three dimensional network of sinter-locked micron size fibers with high surface area. The diameter of the fibers is in the range of 3 to 15 microns and the fiber materials can be polymer, metal, activated carbon or their combination. The effects of important media parameters like fiber diameter, solid volume fraction (SVF) of certain fiber material, total voidage and media thickness on filtration performance characteristics, including filtration efficiency, pressure drop and quality factor were studied in this work. In addition, under the typical environmental conditions at sea, relative humidity (RH) is a variable factor that influences air quality. Therefore, the effect of different RHs (30%, 50% and 80%) on sea salt particles filtration performance was experimentally investigated using fiber material with different hydrophilicities and the results will be presented. SEM images and pore size distribution further help us to develop a better understanding of the structure of these materials. The summary of the work will include a discussion on the optimization of media filtration performance with highest efficiency and lowest pressure drop.

(286b) AC Field-Driven Dielectrophoresis in Concentrated Suspensions

Khusid, B., New Jersey Institute of Technology
Elele, E., New Jersey Institute of Technology
Shen, Y., New Jersey Institute of Technology
Originally presented on: 11/5/2013 12:55:00 - 13:20:00

AC dielectrophoresis is becoming one of the major techniques for micro- and emerging nano-scale systems because the use of an electric field requires no moving parts, employs polarization forces acting on a fine object that are insensitive to the object charge which is difficult to control and can be incorporated more favorably into micro-systems than more traditional methods. However, most of studies available in the literature consider dielectrophoresis in dilute suspensions. In contrast, our recent experimental and theoretical studies focus on fundamentals and applications of the field- and flow- driven phenomena in concentrated suspensions. We will consider the synergism of dielectrophoresis and electro-rheological effect in concentrated suspensions subjected to a strong field, which suggests a new method for field-driven fluid-particle separation.

(286c) Experimental and Computational Study of the Causes of the Fishhook Effect in a Mini-Hydrocyclone

Liow, J. L., University of New South Wales Canberra

Zhu, G., University of New South Wales Canberra @ ADFA

Originally presented on: 11/5/2013 13:20:00 - 13:45:00

Experimental and computational study of the causes of the fishhook effect in a mini-hydrocyclone

Guofeng Zhu, Jong-Leng Liow

School of Engineering and Information Technology, UNSW Canberra @ADFA, Canberra, ACT, Australia 2600

The ability of a hydrocyclone to separate finer material is governed by the cut size scaling proportionally to the square root of the hydrocyclone diameter. A mini-hydrocyclone of 5 mm was studied experimentally and through computational fluid dynamics (CFD) simulation for its ability to separate a feed of sub-100 mm particles. When the feed contains particles smaller than 5-10 mm, the separation efficiency of the finest particle to the underflow rises as the particle size decreases. The rise in the separation efficiency value leads to a shape similar to a fishhook, giving the term fishhook effect. The fishhook effect leads to poor separation of the fine particles as well as contamination of the underflow with fines. The exact cause and physics underlying the fishhook effect has not been fully elucidated and a series of experiments were conducted to investigate the effect of feed flow rate, particle size distribution in the feed, and particle sphericity on the fishhook effect in a mini-hydrocyclone where the 50% separation efficiency (d_{50}) is usually below 50 mm.

The mini-hydrocyclone was manufactured by micro-end milling to give a surface finish with a surface roughness less than 500 nm. A set of non-spherical beach sand samples was prepared by ball milling and screening and a set of spherical glass particles were obtained from Potter's Industries. Figure 1 shows the volume distribution of the samples. Figure 2 shows the separation of the 60 mm non-spherical sample at varying inlet velocities from 0.4 to 4.0 m/s, where the inlet Reynolds number (Re_{in}) varied from 600 to 6000, placing the flow in the transition to low turbulent regime. As the inlet velocity is increased, the separation efficiency of the larger particles (>5 mm) leads to a lower d_{50} value indicating that a higher inlet velocity leads to a finer cut size. For an inlet velocity of 0.4 m/s, there is a flat base in the separation efficiency between 2 to 20 mm, which is progressively narrowed as the inlet velocity increases. The fishhook dip, the point at which the separation efficiency is the lowest, occurs around 3 mm and is independent of the inlet velocity. However the fishhook peak, the maximum separation efficiency in the small particle region past the fishhook dip, increases with increasing inlet velocities.

It is proposed that the fishhook effect is due to the entrainment of fine particles within the wake of the larger particles and this is found to be more pronounced when there is a large fraction of large particles in the feed. The fishhook effect was found to be small or negligible when tests were carried with two feed streams, one with non-spherical particles with a d_{50}<18 mm and one with spherical particles with a d_{50}<8 mm. It was also found that as the concentration of large particles (>10 mm) decreased, the separation efficiency at the fishhook peak decreased as shown in Figure 3.

Earlier work has shown that the fluid flow within the mini-hydrocyclone is unsteady but not fully turbulent and up to 2 million grid cells were used with the ANSYS Fluent package to obtain grid independence (Zhu *et al.* 2012). A particle entrainment model was developed whereby the fine particles are entrained by larger particles within a finite cell volume around the larger particles. The fine particles trapped within this cell volume takes on the velocity of the larger particle, while fine particle outside the cell volume have a velocity based on its interaction with the fluid. A user added function was incorporated into the ANSYS Fluent package to model the particle separation. Twenty three logarithmic levels of particle sizes distributed over the 0.72 to 90 mm range with 8000 particles in each size fraction were tracked. The results of a simulation is compared to the experimental result with a 1.0 m/s inlet velocity for the 60 mm non-spherical sample and shows good agreement in predicting the separation efficiency and shown in Figure 4. A simulation without the particle entrainment model shows the monotonic decreasing separation efficiency with decreasing particle size. The differences between the experimental and simulated results in the larger particle size range (>10 mm) can be attributed to the non-sphericity of the particles. A spherical sample, labelled s2, with a d_{90}<50 mm and minimal particles below 8 mm was tested at the same inlet velocity and showed a separation curve in the larger particle size range closer to the simulation curve, confirming that the non-sphericity of the particles play an important role in the larger particle size range.

Reference

Zhu, G.F., Liow, J.L. and Neely, A. 2012 Computational study of the flow characteristics and separation efficiency in a mini-hydrocyclone. *Chemical Engineering Research and Design*, **90**(12), 2135-2147.

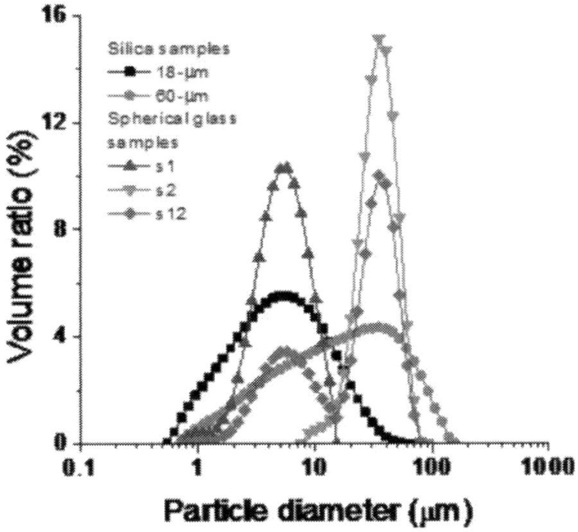

Figure 1: Particle size distribution of the particles used as feed into the mini-hydrocyclone

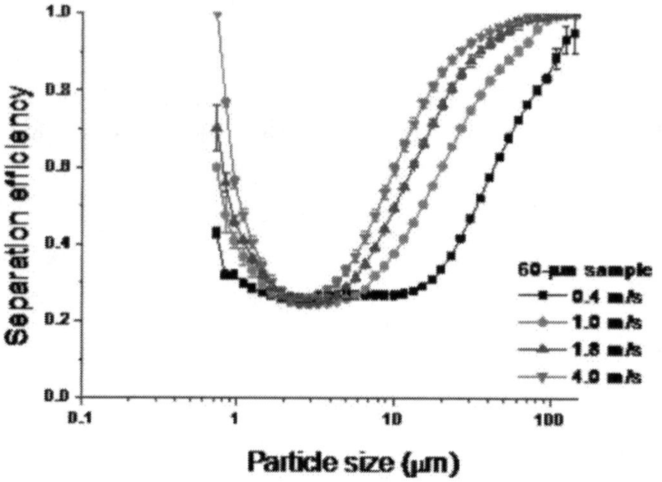

Figure 2: Separation efficiency curves of the 60 mm non-spherical sample for inlet velocities of 0.4, 1.0, 1.8 and 4.0 m/s. The standard deviation of the particle size analysis is of the order of the symbol size.

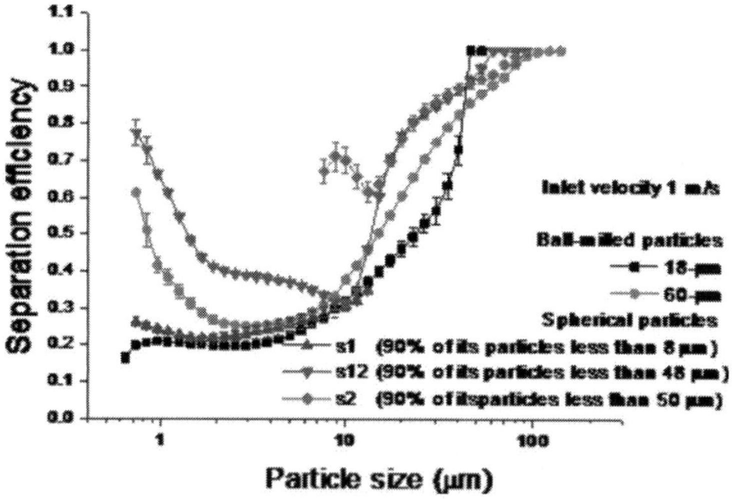

Figure 3: Separation efficiency curves for different particle feeds for an inlet velocity of 1 m/s.

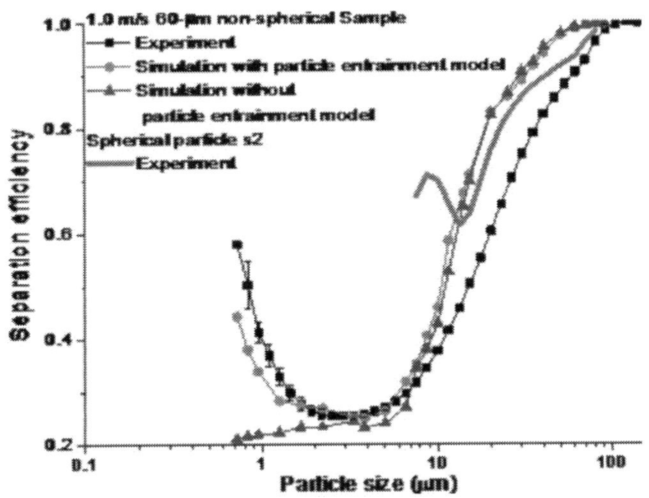

Figure 4: Comparison of the model results with the experimental results for an inlet velocity of 1.0 m/s and a 60 mm non-spherical feed sample. The spherical sample has a size range that has been screened to remove fine particle below 8 mm.

(298a) Applying Crystallization Modeling to Improve the Understanding of a Batch Cooling Process of An Agrochemical Active Ingredient

Brent, R.,
Parmar, M., Syngenta
George, N., Syngenta
Sillers, P.,
Bryce, M.,
Clelland, G.,
Mumtaz, H., Process Systems Enterprise
Mitchell, N., Process Systems Enterprise
Bermingham, S. K., Process Systems Enterprise
Originally presented on: 11/5/2013 12:30:00 - 12:52:00

The current production process of a Syngenta active ingredient isolates the batch by a cooling crystallization process. A major bottleneck of this process is the isolation time of the product after crystallization, which can vary significantly between batches. From observation, the key variable that dictates the isolation time of the product is the width of the particle size distribution of the crystals obtained; the wider the particle size distribution, the slower the filtration time observed. Typically for this product, two distinct environments of fine particles and large single crystals are observed within a single batch. In this work an experimental scale-down study was carried out to replicate the plant process and to assess whether the filtration properties could be improved by modifying the conditions in the crystallizer. Seeding the batch with different quantities of seed with a narrow particle size and modification of the cooling profile where carried out, with limited success in narrowing the particle size distribution significantly. Crystallization modeling using gCRYSTAL was employed to rationalize the experimental observations and to identify the key crystallization phenomena that are dominant in the process. It was observed that the controlling phenomena for this crystallization where the favorable kinetics for crystal growth and also attrition of the crystals, once attaining crystals that had grown to a critical size. Having identified this from the modeling, it was then possible to devise additional experiments to reduce the width of the particle size distribution of this material with the potential to improve plant operations.

(298b) Generic Modelling Framework for Design and Analysis of Crystallization Operations

Meisler, K. T., Technical University of Denmark (DTU)
Samad, N. A. F. A., Technical University of Denmark (DTU)
Gernaey, K. V., Technical University of Denmark (DTU)
Gani, R., Technical University of Denmark
Originally presented on: 11/5/2013 12:52:00 - 13:14:00

A systematic model-based approach to design and analysis of crystallization operations as a first step could result in reduction of time and experimental resources. To this end a generic and systematic model-based framework for the design of a process monitoring and control system to achieve the desired crystal size distribution (CSD) and crystal shape for a wide range of crystallization processes has been developed. This framework combines a generic multi-dimensional modelling framework, tools for design of supersaturation set-point, for design of PAT systems as well as options to perform the uncertainty and sensitivity analysis of the PAT system design. Through this framework, it is possible for a wide range of crystallization processes, to generate the necessary problem-system specific models, the necessary supersaturation set-point using an extended analytical CSD estimator and/or a response surface method (RSM) and a PAT system design including implementation of monitoring tools and control strategies in order to produce a desired product with its corresponding target properties. The problem-system specific models involve generic elements and system specific elements. The system specific elements, such as kinetic constitutive models (models for growth, nucleation, agglomeration and breakage) are stored in libraries, from which they can be extracted, analyzed, used and re-used. This allows models to be analyzed against other models for specific systems and the possibility of comparing models for different systems in a fast and flexible manner. An advantage of this is the possibility to study the effect of different constitutive model parameters on the kinetic phenomena associated with different chemical systems together with the crystallization operational parameters and employ this knowledge to plan and design crystallization operations. Uncertainty and sensitivity analysis is also possible within the framework, and can be carried out to determine, for example, Â theÂ kinetic model parameters in the growth rate model that are the most sensitive This presentation highlights the advantages of utilizing the generic modeling framework to support crystallization operation modeling and the use thereof in design and analysis of crystallization operations. This is illustrated through simulation studies of a potassium dihydrogen phosphate (KDP) and a paracetamol crystallization operation. The operation of KDP batch cooling crystallization is simulated with varying seeding policies and temperature profiles in order to establish the effects of different operational policies on the final product. This includes the effects of uncertainties in the input, for example in the kinetic model parameters. The advantage of using a framework for which models can be re-used is highlighted as the user is guided in specifying the needed data under the assumptions stated in the modeling problem through the formulation and solution of the model equations representing the operation scenario under investigation. The advantage of the generic framework is further highlighted when the chemical system is switched to paracetamol for a similar operation scenario. Here, switching the kinetic models within the operation scenario model allows the study of paracetamol crystallization operation, in one as well as two-dimensions.

(298c) Crystallization Process Optimization Using Derivative-Free Optimization Algorithms

Zhao, H.,
Vouzis, P., Carnegie Mellon University
Sahinidis, N., Carnegie Mellon University
Originally presented on: 11/5/2013 13:14:00 - 13:36:00

Polymorphism is of key importance in the pharmaceutical and fine-chemicals industry. Modeling of polymorphism phenomena is based on Population Balance Equations (PBEs), which are sets of partial differential equations whose analytical solution is not always available [1]. Moreover, when optimization is to be done over a simulator, it is often difficult to calculate derivatives of variables of PBEs either analytically or numerically. This renders most derivative-based optimization methods of little or no use. Derivative-free optimization (DFO) algorithms are specifically designed for this kind of problem [2]. In this work, we use g-CRYSTAL as a black box simulator to perform simulation-based optimization of crystallization processes. DFO is used to optimize one particular polymorph and minimize the time of producing a desired particle size distribution. This goal is achieved by optimizing process variables, including temperature and cooling profile of reactor, super saturation of seed molecule, flow rates, and concentrations of species. We present extensive computational results, where we compare the results of 28 DFO algorithms on several case studies.

References

[1] Rudiyanto Gunawan, Irene Fusman, and Richard D. Braatz. High resolution algorithms for multidimensional population balance equations. AIChE Journal, 50(11):2738{2749, 2004.

[2] Luis Miguels Rios and Nikolaos V.Sahinidis. Derivative-free optimization: A review of algorithms and comparison of software implementations. Journal of Global Optimization, DOI 10.1007/s10898-012-9951-y, 2012.

(298d) Stable Configurations in the Horizontal Ribbon Growth Process

Oliveros, G. A., Carnegie Mellon University
Ydstie, B. E., Carnegie Mellon University
Sridhar, S., Carnegie Mellon University
Originally presented on: 11/5/2013 13:36:00 - 13:58:00

In the Horizontal Ribbon Growth (HRG) process, a thin crystal ribbon is produced and extracted continuously from a molten substrate. The advantage of this technique is its capacity to achieve large rates of production compared to conventional crystal growth processes, such as the Czochralski and Bridgman. However, despite its potential and efforts to develop the technology for more than 50 years, many technical and experimental challenges have been reported regarding the difficulty in trying to maintain a continuous and stable ribbon production.

One of the principal causes of disruption is the freezing of the ribbon to the end of the crucible. The reason for this is that there is no melt present between the ribbon and the crucible. The solid surface of the crucible offers plenty of nucleation points for the ribbon to crystallize onto as it is being pulled. The opposite problem can also arise: an excess of melt flow that causes the melt to spill over the crucible. In an ideal operation, a space between the ribbon and the lip of the crucible must be maintained to avoid undesired nucleation points, and a meniscus must be pinned between the crucible lip and the ribbon to avoid melt spill-over.

In this work, we present a theoretical analysis to assess these technical difficulties. We analyze the conditions under which a meniscus exists between the ribbon and the crucible and stays pinned to its corner, as well as the allowable range of melt levels that give rise to stable menisci. Assuming that the existence of the meniscus is defined by hydrostatics, we use a theoretical model based on the governing Young-Laplace equation to solve analytically for the shape of the meniscus. Then, we show the range of operating conditions that are physically viable in a silicon system, and construct the possible stable configurations. We show that these stable configurations depend mainly on the pulling angle, the melt level, and the type and shape of the crucible.

(298e) Using ATR-FTIR to Track the Crystallization of Multiple Solutes

Griffin, D., Georgia Institute of Technology
Grover, M., Georgia Institute of Technology
Kawajiri, Y., Georgia Institute of Technology
Rousseau, R. W., Georgia Institute of Technology
Originally presented on: 11/5/2013 13:58:00 - 14:20:00

In-situ measurements of concentration play a large role in developing and implementing control schemes for crystallizers [1, 2]. In recent years Attenuated Total Reflectance Fourier Transform Infrared (ATR-FTIR) technology has been widely used to track the concentration profiles of a single solute during crystallization processes [3-5]. However, if the parent solution is complex and contains multiple potentially crystallizing solutes (e.g. nuclear waste) multiple solutes must be tracked simultaneously. This can be a challenging task if the characteristic infrared absorbance peaks for the individual solutes are overlapping. In this work a supervised learning approach was taken to construct a correlation between the infrared absorbance spectrum and the concentrations of multiple solutes in the solution. Furthermore, to avoid over-fitting and establish robust correlations, the training set was carefully selected and projection-based regression techniques [3, 6] were applied—of these, Partial Least Squares Regression (PLSR) performed the best.

This methodology was used to monitor crystallizations of complex electrolytic solutions representative of nuclear waste stored at Hanford, WA [7-9]. In particular, four solutes (carbonate, sulfate, nitrate, and nitrite) were tracked during batch cooling crystallizations. By monitoring a number of crystallizations, operating zones for crystallizing particular salts were identified. Furthermore, in future work, multi-solute concentration monitoring will be used to facilitate a direct design approach [10] to control the crystallization of clean (non-radioactive) sodium salts from nuclear waste mixtures.

1. Nagy, Z.K. and R.D. Braatz, *Advances and New Directions in Crystallization Control.*Annual Review of Chemical and Biomolecular Engineering, Vol 3, 2012.

2. Zhou, G.X., et al., *Direct design of pharmaceutical antisolvent crystallization through concentration control.* Crystal Growth & Design, 2006. **6**(4): p. 892-898.

3. Togkalidou, T., et al., *Solute concentration prediction using chemometrics and ATR-FTIR spectroscopy.* Journal of Crystal Growth, 2001. **231**(4): p. 534-543.

4. Lewiner, F., et al., *On-line ATR FTIR measurement of supersaturation during solution crystallization processes. Calibration and applications on three solute/solvent systems.* Chemical Engineering Science, 2001. **56**(6): p. 2069-2084.

5. Dunuwila, D.D. and K.A. Berglund, *ATR FTIR spectroscopy for in situ measurement of supersaturation.* Journal of Crystal Growth, 1997. **179**(1-2).

6. Xiaobo, Z., et al., *Variables selection methods in near-infrared spectroscopy.* Analytica Chimica Acta, 2010. **667**(1–2): p. 14-32.

7. Herting, D.L., *Clean Salt Process - Final Report*, E. Management, Editor. 1996.

8. Nassif, L., et al., *Pretreatment of Hanford medium-curie wastes by fractional crystallization.* Environmental Science & Technology, 2008. **42**(13): p. 4940-4945.

9. Rajbanshi, A., B.A. Moyer, and R. Custelcean, *Sulfate Separation from Aqueous Alkaline Solutions by Selective Crystallization of Alkali Metal Coordination Capsules.* Crystal Growth & Design, 2011. **11**(7): p. 2702-2706.

10. Fujiwara, M., et al., *First-principles and direct design approaches for the control of pharmaceutical crystallization.* Journal of Process Control, 2005. **15**(5): p. 493-504.

(298f) Operation and Optimization of Seeded Batch Crystallization When Crystal Growth Rate Is Size-Dependent

Ting, W. L., National Taiwan University
Wang, H. Y.,
Ward, J. D., National Taiwan University
Originally presented on: 11/5/2013 14:20:00 - 14:42:00

In this work, the operation and optimization of seeded batch crystallization processes is investigated when crystal growth rate is size-dependent using a dimensionless model[1] and model reported in the literature for the crystallization of potassium alum.[2] Many of the results are qualitatively similar to the case when growth rate is size independent: The seed mass and average seed size affect the results in a similar way. An approximate analytical trajectory that works well for size-independent growth also works well when the growth rate is size dependent. The optimal supersaturation trajectory is qualitatively similar to the case when growth rate is size independent. Finally, we show that an approximate, size-independent model can be used to determine the optimal supersaturation trajectory with little error. This is advantageous because optimizing the size-independent model requires much less computational effort.

References

1. Ward JD, Yu CC, Doherty MF. A New Framework and a Simpler Method for the Development of Batch Crystallization Recipes. *Aiche J.* 2011;57(3):606-617.
2. Tai CY, Lu EL, Yu KH. Determination of Secondary Nucleation Kinetics Using a Batch Crystallizer. *J. Chin. I. Ch. E.* 1991;22(4):231-239.

(298g) Effects of Several Heat Shield On the Crystallization Front of Sapphire Crystal Grown By Kyropoulos Method

Song, D. W., LG Siltron
Kim, H., University of Seoul
Ageyeva, S., CDT Equipment LLC
Originally presented on: 11/5/2013 14:42:00 - 15:04:00

In the growth of sapphire crystals, the critical density of dislocations determined by variations of the ratio of G/V affects on the quality of crystal structure. Where G is the axial temperature gradient and V is the velocity of crystallization front (CF). In general, it is not the initial temperature gradient is so important for structure control but rather the degree of its constancy maintaining during growth. However in Kyropoulos (KY) method the CF, has formed the inverse shape of cone, is gradually down to crucible bottom. Therefore it is more difficult to control its consistency during growth in KY method than other methods.

In our experiments, we have studied the influences of heat shield (HS) on the formation of low angle block or grain boundaries (LAGB) and micro-bubbles. We have estimated the angle and velocity of convex of CF in several HS configurations. The shapes of CF are determined by the patterns of micro-bubbles generated by the power down-and-up test (PDU) in quick time. And the defects such as LAGB and micro-bubbles are inspected in the vertical slug which had been cut in center line of grown crystals by polarized microscope. The velocities are measured in PDU-tests by the evaluation data of vertical slugs and calculated by SK Forecast (developed by Alyuda.com) and our simulators.

The defects free shape of CF is depended on the HS configurations with respect to time. It means that there exist an optimal tendency of CF in each HS. The LAGB is related to the acceleration of growth rate and the micro-bubble is related to the oscillation of growth rate that will be discussed in detail. The difference between our simulators and SK Forecast are come from the consideration of energy conservation and initial guessing value of angle and velocity of CF.

(299a) Solar- Powered Reverse Osmosis Technology for Brackish Water : The State of The Art

Abkar, L., Institute for Energy and the Environment New Mexico State University, Las Cruces, NM, USA.
Mallick, K.,
Alshukri, R.,
Ghassemi, A., New Mexico State University, Institute for Energy and the Environment/WERC
Loya, J., New Mexico State University,Institute for Energy and the Environment/WERC
Originally presented on: 11/5/2013 12:30:00 - 12:45:00

Water and energy are two of the most important factors for the development of any country. With increase in both population and industrialization, it's a challenge to produce ample freshwater with minimum energy cost. There are many places in the world that are turning into arid area due to water scarcity. Reverse osmosis desalination technology seem to be the best possible solution to this problem. With fossil fuel sources exhausting, renewable energy is most suitable to power RO for remote off-grid area. Solar powered– RO technology is applicable to area with ample solar resources. Many researches have been done for years to improve the PVRO system. Over the years this system has gone through many changes. For instance, batteries were taken down and direct connections were made between the PV panels and the pressure variable pump, turbines or hydraulic pumps were installed for energy recovery, etc. In this paper, a comprehensive review has been done on PVRO systems, its developments and water production cost in recent years.

(299b) Nanofiltration: High Potential, Poorly Defined and Understood

Paulson, D., Water Think Tank, LLC
Originally presented on: 11/5/2013 12:45:00 - 13:00:00

Nanofiltration: High Potential, Poorly Defined and Understood

The membrane mediated process known as nanofiltration is receiving much attention, but some of it due to the erroneous assumption that the technology came out of recent nano-materials advances, or that it performs based on the characteristics of nanoparticles. In reality, nanofiltration (NF) has been employed commercially for four decades, predating today's popular concept of nanotechnology. Even within the maturing membrane industry, nanofiltration is defined in varying manners. For an important technology to advance to full potential, such definitional incoherence is counterproductive.

The potential of nanofiltration is very high, and could be applied in more unique ways than its sibling membranes processes. To aid the research and industrial communities' understanding, this paper defines the current reality of the nanofiltration process, and addresses its history and its probable future. Few realize when the first NF membranes were developed, and for which applications and the origin of the term nanofiltration. Interestingly, the earliest NF membranes were involved in a patent suit that led to a seminal US Supreme Court decision, the only involving patent interpretation in a three decades span. These and other historical highlights are presented to help put the importance of this process in perspective.

The NF process is briefly described in relationship to the better understood membrane classes, reverse osmosis and ultrafiltration (RO and UF). Further definition is provided by description and comparison of NF's major applications in water treatment, industrial manufacturing and the more exotic emerging uses for medicines, fine chemicals and biofuels.

The lack of communication and common among the industries that use NF explains its incoherent definition. Lack of standardization and poor cooperation among standards setting organizations are described. Specific examples of misunderstandings that serve as roadblocks to the use of this process are given, including differences in definitions by standards groups, a membrane vendors and user groups (industries) are presented. Different views between the municipal water treatment community, industrial engineers, and the research community are described. A more universal definition, of both the process and its mediating membrane is proposed.

Whether defined as "nanotechnology" or not, separation of materials at the molecular level is a key to emerging technologies in several arenas of use. The paper concludes with a summary of recent advances, failed efforts, technical avenues to improve the membranes and the author's predictions for near and long-term improvements.

(299c) RO/NF Applications in Brackish Groundwater Desalination: Membrane Characterization

Vaseghi, G., New Mexico State University
Ghassemi, A., New Mexico State University, Institute for Energy and the Environment/WERC
Loya, J., New Mexico State University,Institute for Energy and the Environment/WERC
Originally presented on: 11/5/2013 13:00:00 - 13:15:00

Millions of people in the world suffer from the lack of potable and safe water. Water problem is practically the most important global issue nowadays, mainly because of decreasing fresh water resources and continuous deterioration of the quality of available resources. Water sources include oceans, brackish waters, and wastewaters. In the most arid and semi-arid regions of the USA such as New Mexico, California, Arizona, and Colorado, brackish water is an important source of feed water, and its most common quality problems are caused by suspended solids and hardness. Since desalination technologies are expensive, water shortage and scarcity pose significant threats for developing countries. Consequently, there is high interest in reducing the costs of water desalination.

Processes of desalination, or removing salt from water, include thermal processes, such as Distillation, and membrane processes, such as Reverse Osmosis (RO) or Electrodialysis (ED). Reverse Osmosis (RO) is one of the technologies used for desalinating brackish and saline waters to provide drinking water. RO treatment plants use semipermeable membranes and pressure to separate salts from water. They typically use less energy than thermal distillation, leading to a reduction in overall desalination costs, and allowing new brackish groundwater desalination facilities to use RO technology much more economically than distillation. Hence, this study improves the benefits of using RO by reducing the cost, time, and energy spending to find the best approach for different conditions. It focuses on the characterization of different types of membranes used in RO systems by testing effects of Temperature, pH, Conductivity, and Recovery on product. The objectives are accomplished by utilizing pilot plant experiments. A huge database collected from a full pilot-scale system, located at the Brackish Groundwater National Desalination Research Facility (BGNDRF) in Alamogordo, and operated by New Mexico State University are analyzed.

(299d) Investigating the Effecitive Parameters in Optimizing Reverse Osmosis Technology for Water Treatment

Abkar, L., Institute for Energy and the Environment New Mexico State University, Las Cruces, NM, USA.
Ghassemi, A., New Mexico State University, Institute for Energy and the Environment/WERC
Loya, J., New Mexico State University,Institute for Energy and the Environment/WERC
Originally presented on: 11/5/2013 13:15:00 - 13:30:00

As population growth, water scarcity become the major issue of concern. There are different types of desalination technologies such as thermal which includes multi- stage flash (MSF) and Multiple effect distillation (MED), and membrane based technologies such as reverse osmosis (RO) , electrodialysis (ED), electrodialysis reversal (EDR), membrane distillation (MD) and forward osmosis (FO) to provide fresh water. Among these processes Reverse Osmosis has shown the highest capabilities commercially. In this regard, optimizing RO system become a significant concern in order to decrease water production cost and make it more affordable. The process optimization depends on parameters such as temperature, pressure, membrane characteristics and feed water constituents. This paper presents a comprehensive review and assessment on the factors which affect the performance of the inland reverse osmosis system. A review on the major advancement in energy issues, water recovery, membrane characteristics, concentrate disposal, RO cost is done.

(299e) Carbon Electrode Driven Electrodialysis

Barber, J., GE Water & Process Technogies
Yang, H.,
Originally presented on: 11/5/2013 13:30:00 - 13:45:00

Electrodialysis processes require the application of an electric field in order to drive the migration of ions and cause deionization. The conventional approach is to utilize metal electrodes (Stainless Steel for the cathode and noble metals for the anode) and perform water electrolysis at the surface of the electrodes to create the electric field. The byproducts of water electrolysis, i.e. acid, base, and hydrogen, oxygen and chlorine gas, can cause system complexity, extra cost and environmental, health and safety concerns. To address these issues GE has develop a new capacitive carbon electrode to be used in electrodialysis reversal stacks which mitigates all of these concerns. In this presentation a detailed look at the electrodes design and operational parameters will be discussed, with reference to specific case studies from a number of field sites in rural and isolated communities.

(299f) Impact of Silica in Water Treatment Technology

Naderi Behdani, F., New Mexico State University,Institute for Energy and the Environment/WERC
Ghassemi, A., New Mexico State University, Institute for Energy and the Environment/WERC
Loya, J., New Mexico State University,Institute for Energy and the Environment/WERC
Originally presented on: 11/5/2013 13:45:00 - 14:00:00

In many parts of the world the use of water in reverse osmosis (RO), cooling, boiler, and geothermal applications is limited by the poor solubility of silica. Silica, the second most abundant element, is found in all natural water supplies. Silica solution chemistry is very complex and difficult to predict. In RO feedwaters, silica can exist in three different forms as follows: Monomer silica or silicic acid, polymeric silica that results from polymerization of silicic acid, and filterable silica. Ca^{2+} and Mg^{2+} ions have a strong influence on the formation of filterable silicate. It has been proved that the polymeric silica is mainly responsible for the membrane scaling. The challenge is how to exceed the range of silica solubility to facilitate the use of RO systems in areas where the available feedwaters have high silica concentrations. Silica solubility is well known to be influenced by both pH-value and temperature. This paper is a review on silica fouling problems and the use of various products for extending silica solubility.

(299g) Selectivity Comparison for Two Cation Exchange Membranes in the Electrodialysis Process

Karimi, L., New Mexico State University,Institute for Energy and the Environment/WERC
Ghorbani, A.,
Ghassemi, A., New Mexico State University
Loya, J., New Mexico State University,Institute for Energy and the Environment/WERC
Originally presented on: 11/5/2013 14:00:00 - 14:15:00

Electrodialysis is a membrane based separation in which charged species can be removed due to applied electrical potential. Although electrodialysis has been applied in very diverse industrial fields such as brackish and sea water desalination, salt production by concentering sea water or using brine discharged from a reverse osmosis seawater desalination plant, heavy metal recovery from waste water, recovering reverse osmosis concentrate stream, and citric acid separation, its high capability in brackish water desalination has been demonstrated. It is more applicable than reverse osmosis process for removing the impurities from saline water with high amount of silica. Saline water with high content of silica causes scaling problem in RO membranes, while ED is robust against scaling problems occurred due to silica content.

Depending on saline water resources, its composition can be very different and the levels of some ions can be more than the others. Therefore, they should be critically removed in order to reach the standard drinking water levels. Focusing on preferential ion removal in the ED operation makes it more purposeful to reduce the cost. Preferential separation of ions can lead to remove specific ions at appropriate rate and control the process cost. ED process is affected by several parameters; however, the ion exchange membranes have the most significant role in selective removal of ions in ED process. In fact, ion exchange membranes have significant role in selective removal of ions in this process. Various ion exchange membranes with diverse characteristics have different preference toward removing the charged species.

In this study the different operation of cation exchange membranes for CR67 and coated CR67 with Polyethyleneimine (PEI) were investigated. Their behavior toward preferential removal of different cations was verified. The experiments were done using a lab-scale ED equipped to a complete data acquisition which gives a significant level of confidence that both membranes were examined at same conditions. The water samples were analyzed using ion chromatography, IC 5000. The studied ion exchange membranes showed different behavior at toward different charged species. Sodium ion removal was considered as a reference ion in this study.

(299h) Developing a Mechanistic Transport Model for Electrodialysis/Electrodialysis Reversal Process

Ghorbani, A., New Mexico State University
Karimi, L.,
Ghassemi, A., New Mexico State University, Institute for Energy and the Environment/WERC
Loya, J.,

Originally presented on: 11/5/2013 14:15:00 - 14:30:00

Developing a Mechanistic Transport Model for Electrodialysis/Electrodialysis Reversal Process

Azadeh Ghorbani *, Leila Karimi, Abbas Ghassemi, Jim Loya

Institute for Energy and the Environment New Mexico State University, Las Cruces, NM, USA.
***azadeh12@nmsu.edu**

There is no doubt that the fresh water demand will continue to increases due to the growth of the world's population and the increase of the worldwide demand on consumer goods. Technologies that were originally developed to desalinate water such as Reverse Osmosis (RO) and Electrodialysis/ Electrodialysis Reversal, ED/EDR, are widely applied to remove contaminants other than salt from freshwater supplies.

ED/EDR is an Electrical-driven membrane technology that is effective with salt removal in which ions are transferred through ion exchange membrane. These process would cost less compared to the RO process but no complete model has been presented yet. Therefore, there is a definite need for presenting a model which would include mechanistic transport model.

Modeling is a critical step representing the process and the construction of system of interest. One of the significant features of a model is to enable the user to predict the effect of changes to the system without imposing cost.

Transport models relate fluxes through the active layers to driving forces and provide mechanistic descriptions of how material travels through the membrane media.

Mechanistic transport models may be used to predict how a particular membrane will perform in a new process, or may help to develop new membrane characteristics. Development of a reliable membrane transport model for describing mass transport phenomena in ED/EDR is a highly desirable goal since not only does this allow the analysis of performance characteristics of a membrane, but can also assist in the designing new membranes.

In this study, the existing models of RO were reviewed and investigated in order to develop a novel mechanistic transport model for ED/EDR processes.

(299i) Flexible Desalination Systems for Variable Salinity Sources

Chapman, M., Bureau of Reclamation
Leitz, F., Bureau of Reclamation
Tiffenbach, A.,

Originally presented on: 11/5/2013 14:30:00 - 14:45:00

Many water sources are less desirable as a potable water source because they vary in quality over time, such as in tidal reaches on coastal rivers. In other situations water needs to be treated intermittently so that either water from several sources is treated at a common treatment location or equipment is moved to the source of contaminated water such as in centralized produced water treatment and small scale industrial waste treatment. Irrigation return flows vary with the amount of precipitation and irrigation and fertilization schedule. Over a longer scale of time, communities that use ground water may find the salinity increase due to seawater intrusion or declining aquifer levels. In all of these scenarios the treatment process and equipment needs to be adaptable to a wider range of source water qualities.The Variable Salinity Desalination (VSD) project was designed to evaluate modifications to the Expeditionary Unit Water Purifier- Generation 1-A (EUWP), designed to treat any source up to 60,000 mg/L total dissolved solids, 100 NTU turbidity, and temperatures, at between 45-60 percent recovery, to allow higher water recovery operation when treating brackish water. The system was modified with a bypass around the pressure exchanger to attain 75% water recovery to convert the system from a one stage design to two stages. Operability, ease of transition between designs, power requirements, and water quality were evaluated using a brackish groundwater feed source. Performance was compared to operation in the original configuration at three other sites with a wide range of salinity. Power use and water quality were as would be expected for the higher recovery. Options for building in design flexibility and cost impacts are discussed.

(299j) An Innovative, Environmentally Friendly, Low-Cost Desalination System

Nossoni, G., Manhattan college
Markes, K., Manhattan college
Matinez, H., Manhattan College
Piazza, A., Manhattan College
Qasim, M., Manhattan College
Abulencia, J. P., Manhattan College
Originally presented on: 11/5/2013 14:45:00 - 15:00:00

The lack of clean drinking water is rapidly becoming one of the most important international health issues. Many remote regions in developing countries face water shortages and lack of potable water causing significant hardship to local people. One popular method used is the desalination of brackish water. The most widely used and commercially proven desalination technologies consist of thermal (evaporative), membrane based methods such as electrodialysis (ED) and reverse osmosis (RO). Unfortunately, these processes require a large amount of energy, which translates to being expensive. People in rural areas of developing countries do not have the means or resources for expensive water treatment systems

To this end, a multi-stage *hybrid ED-CON* (Electrondialysis/Concrete) process was designed to separate salt from seawater. The typical setup of an electrodialysis (ED) cell is adopted for the proposed *ED/CON hybrid* cell. However, rather than using a series of anion and cation exchange membranes that are arranged in an alternating pattern between anodic and cathodic electrodes, two concrete slabs of prescribed thickness are used as permeable membranes separating the anodic, cathodic, and desalination chambers. When a DC potential is applied between the two electrodes, positively charged cations (Na^+) move toward the cathode passing thorough the cathodic concrete membrane, and negatively charged anions (Cl^-) move toward the anode passing through the anodic concrete membrane therefore, the concentration of salt decreases as the salt water flows through each stage in the series of desalination cells. This design is unique and environmentally friendly because it can be produced using locally available recycled material and can be powered by a renewable energy source such as a solar panel.

Concrete is a mixture of Portland cement, water, and aggregate. It can be produced locally from abundant natural or recycled materials. As the water-cement ratio in a concrete mixture increases, the porosity of the cement paste increases. Concrete pores are interconnected and dissolved ions can move through concrete with water if there is a pressure gradient, or by diffusion due to a concentration gradient when a pressure head is absent.

Extensive testing was performed to optimize the process step by step. The parameters chosen for study were concrete porosity, the type of anodic solution, and potential used in the test. These variables were adjusted as the research progressed. The suitability of the parameters was judged based on the conductivity of the solution in the desalination chamber, which was used as a surrogate measure for the salinity. Preliminary results were promising and they show that the conductivity of salt water can be reduced by 10 ms/cm in 6 hours, which corresponds to about a 27% reduction in the salinity of water, in an optimized system. The optimized system used pervious concrete slabs, lime water in the anodic chamber, and a 10 V potential between the electrodes. The idea of using concrete as a membrane in water filtration is novel and addresses the issue of sustainability. This concept can be broadened and applied not only to the desalination of water, but also to the removal of other contaminants in ground or surface water by designing the concrete filters with suitable aggregate material and size gradations.

(313a) Parvoviral Filtration: Step Order Matters! Including Tools to Ensure Success of Worst-Case Process Characterization Studies

Mallaney, M., Genentech

Originally presented on: 11/5/2013 12:30:00 - 12:48:00

Title: **Parvoviral Filtration: Step Order Matters!**

Including Tools to Ensure Success of Worst-Case Process Characterization Studies

Parvovirus filtration is rapidly replacing retrovirus filtration as the industry standard for virus removal in monoclonal antibody cell culture manufacturing processes. The mass and volumetric throughput of monoclonal antibodies (mAbs) in parvoviral filters has been found to be significantly lower than on retroviral filters due primarily to the smaller pore size. A variety of solutions have been tested to improve throughput with some success. One seemingly easy solution is to place the parvovirus filtration step at the end of the purification process such that the feed is the most highly purified pool. However it may be the mode of separation of the chromatography step immediately preceding the filter that gives this improvement rather than the number of upstream purification steps. In addition, while the step order change can improve throughput, it does not rid the parvoviral filter of its inherent sensitivity to changes in product concentration, flow rate, and operating temperature. These parameters are tested at worst-case during process characterization studies, which can result in failure due to this sensitivity. Here we present strategies for implementation of a new virus filter type as well as a new location in a downstream process and for success when testing worst-case parameters during the process characterization of a parvovirus filtration step. Through case studies we also address the question of whether it is the mode of separation of the upstream chromatography step or the number of purification steps preceding the filter that has the greater effect on throughput improvement.

(313b) Ultrafiltration Optimization Via Experimental Design

Bhadouria, A., Rensselaer Polytechnic Institute, Troy
Sorci, M., Rensselaer Polytechnic Institute
Gu, M., Rensselaer Polytechnic Institute
Belfort, G., Rensselaer Polytechnic Institute
Hahn, J., Rensselaer Polytechnic Institute
Originally presented on: 11/5/2013 12:48:00 - 13:06:00

Separation and recovery of proteins are important processes in biotechnology. Various techniques are used for protein separation including chromatography, electrophoresis, and membrane ultrafiltration among others. Limitations of chromatography or electrophoresis include in some cases dilution of product, difficulty with high titer (i.e. highly viscous) solutions, scale up and the cost of the instrumentation. At the same time, membrane ultrafiltration has received a significant amount of attention because of its scalability and economics.

Membrane-based unit operations have been traditionally viewed as size-based separation processes where the size of the solutes differs by at least an order of magnitude [1]. However, factors other than molecular sieving can also affect the performance of ultrafiltration. For example, concentration polarization [2] and electrostatic interactions between proteins and proteins and membranes [3] have been investigated for the separation of proteins of similar size. The effect of pH on fractionation of proteins of similar size has been investigated by van Eijndhoven et al. [4, 5].

Optimization of factors affecting UF has been studied due to the strong dependence of UF performance on these factors [6,7,8]. The main challenge of fundamental approaches for this task is that they use models involving particle-particle interactions and resolving the resulting complex expressions requires significant computational effort. An alternative approach is to use experimentally generated data to predict membrane UF performance. However, the number of experiments to perform such optimizations can be large, resulting in significant time and financial commitments. This situation presents a clear need for optimal experimental design which can be used to either reduce the number of experiments or to maximize the information content from a pre-determined number of experiments.

In this work we present an empirical approach for optimizing operating parameters of UF. A polynomial model is fitted to available experimental data and optimal operating conditions are determined by optimizing the prediction of the model. The models include the factors that can be varied as inputs and they predict certain performance measures of interest. Special attention is paid to the order of the polynomial using cross-validation as the order is affected by the amount of data but also by trends in the data.

There are several performance measures that are commonly used, depending on how the separation process is used. The presented approach can compute operating conditions for different individual performance measures or some combination of them. The resulting expression for the performance measures can be numerically optimized to determine operating conditions that maximize the performance measures. This type of approach is in stark contrast to commonly used linear approaches where one factor is varied at a time. Here, using our nonlinear approach, we are able to account for nonlinearities and interactions among the factors.

The polynomial regression method is illustrated with two case studies. One study involves available experimental data from our lab [9] and another study uses a data-derived model from the literature [8]. The first case study illustrates that the optimal operating conditions can be significantly different from the ones found by individually varying the parameters, i.e. using a linear approach. The second case study illustrates that the model order should be chosen as low as possible, while still being able to fit the data well, if the main goal is to optimize a performance measure.

References

[1] Cherkasov A. N., Polotsky A. E., The resolving power of ultrafiltration, *Journal of Membrane Science* **110**, pp. 79-82 (1996).

[2] Porter M. C., Concentration polarization with membrane ultrafiltration, *Industrial and Engineering Chemistry Product Research and Development* **11** (3), pp. 234-248 (1972).

[3] Mahsa M. R., Zydney A. L., Role of electrostatic interactions during protein ultrafiltration, *Advances in Colloid and Interface Science* **160**, pp. 40-48 (2010).

[4] Eijndhoven R. H. C. M. van, Saksena S., Zydney A. L., Protein fractionation using electrostatic interactions in membrane filtration, *Biotechnology and Bioengineering* **48**, pp. 406-414 (1995).

[5] Saksena S., Zydney A. L., Effect of solution pH and ionic strength on the separation of albumin from immunoglobins (IgG) by selective filtration, *Biotechnology and Bioengineering* **43**, pp. 960-968 (1994).

[6] Box G. E. P., Wilson K. B., On the experimental attainment of optimum conditions, *Journal of the Royal Statistical Society* **13** (1), pp. 1-45 (1951).

[7] Bowen W. R., Williams P. M., Dynamic ultrafiltration model for proteins: a colloidal interaction approach, *Biotechnology and Bioengineering* **50**, pp. 125-135 (1996).

[8] Lin S. H., Hung C. L., Juang R. S., Effect of operating parameters on the separation of proteins in aqueous solutions by dead-end ultrafiltration, *Desalination* **234**, pp. 116-125 (2008).

[9] Sorci M., Gu M., Heldt C. L., Grafeld E., Belfort G., A multi-dimensional approach for fractionating proteins using charged membranes, *Biotechnology and Bioengineering* **110**, pp. 1704-1713 (2013).

(313c) Protein Fouling in Hollow Fiber Membranes

Teella, A., University of Massachusetts, Amherst
Zydney, A. L., The Pennsylvania State University
Zhou, H., GE Global Research
Olsen, C., GE Global Research
Misner, M., GE Global Research
Car, L.,
McCloskey, P.,
Originally presented on: 11/5/2013 13:06:00 - 13:24:00

Membrane fouling remains one of the critical factors limiting the performance of ultrafiltration and microfiltration systems for the separation and purification of high value biological products. Fouling is a complex phenomenon arising from both physical and chemical interactions between the proteins and membrane. Recent studies using flat sheet membranes with simple dead-end flow have provided important insights into the connection between membrane structure/morphology and the observed fouling characteristics. The objective of this work was to extend these results to hollow fiber membrane modules, specifically accounting for the more complex hydrodynamics in these tangential flow hollow fiber systems.

Fouling experiments were performed using solutions of human immunoglobulin (hIgG) with relatively large pore size polysulfone hollow fiber ultrafiltration membrane modules. The filtrate flux and protein transmission both declined significantly over the course of the experiment. For example, data obtained with 2 g/L solutions showed more than a 4-fold reduction in filtrate flux after only 60 min of constant pressure filtration, with the protein sieving coefficient decreasing from more than 95% at the start of the filtration to less than 50% at $t > 60$ min. The relative contributions of concentration polarization and fouling were evaluated from the difference in membrane hydraulic permeability before and after protein fouling. The extent of surface fouling was examined using streaming potential measurements. For example, the zeta potential of the membrane increased from -8 mV to -4 mV after fouling with hIgG, reflecting the coverage of the negatively-charged surface with bound protein. The pore size characteristics of the hollow fiber membranes was examined using dextran retention measurements before and after the fouling experiment. An initial theoretical model was developed that accounts for the effects of both fouling and concentration polarization on the transmembrane pressure. These studies provide important insights into the nature of the protein fouling phenomena in hollow fiber membrane systems.

(313d) Practical and Economic Implications of Filter Fouling Models and Sizing Methodologies
Patil, R.,
Originally presented on: 11/5/2013 13:24:00 - 13:42:00

Practical and economic implications of filter fouling models and sizing methodologies

Ranjeet Patil (ranjeet.patil@emdmillipore.com) ,Michael Felo (Michael.felo@emdmillipore.com)

Abstract:

In pharmaceutical industry, membrane filters are typically sized by running filtration tests with small scale devices/filter discs at constant pressure. The flux decay data from these tests is used to estimate the required filtration area for manufacturing scale processes. It is a common practice to use a filter fouling model to predict filter performance beyond small scale test throughput.

Most popular filter fouling model used in biopharmaceutical industry is gradual pore plugging or standard blocking model. Over past decades, other filter fouling models and filter sizing methods have been proposed. In some cases, predictions derived from these models and methods have been shown to be more accurate than gradual pore plugging model.

Although certain models have been shown to be more accurate than others, using a different model for filter sizing may not necessarily translate to better process control and economic gains. Actual pharmaceutical processes have a variety of feed steams with wide range of particle size distribution and employ different types of membranes. Marginal improvement in accuracy of prediction achieved due to a different fouling model may not translate to manufacturing scale due to aforementioned variables.

The objective of the proposed study is to understand the impact of using different filter fouling models and sizing techniques on filter sizing for manufacturing scale. In this study,

- · Accuracy of prediction of different fouling models and sizing techniques is determined at lab scale.
- · The small scale results are compared to a process scale device performance. .
- · Different feeds are used for filter sizing studies at lab scale to test the impact of feed variability on the accuracy of prediction.
- · Different filter meterials (e.g. PVDF, PES) and filter constructions (e.g. single layer, multi layer) are evaluated at lab scale to understand the effect on the accuracy of predictions of different filter fouling models and sizing methods.
- · Cost analysis is conducted for a generic Mab process (15000L scale) to evaluate the potential cost savings due to different sizing models/methods.
- · Based on the results, an optimum filter sizing strategy is proposed.

(313e) Membrane-Based Hybrid Method for Purifying Pegylated Proteins

Lam, S. F.,
Shang, X., McMaster University
Ghosh, R., McMaster University
Originally presented on: 11/5/2013 13:42:00 - 14:00:00

PEGylated proteins are value-added pharmaceutical products obtained by conjugating poly (ethylene glycol) or PEG with proteins. PEGylation improves the therapeutic efficacy of biopharmaceutical proteins by increasing their biological half-life, reducing their immunogenicity and increasing their stability. A range of synthetic methods is available for carrying out protein PEGylation After the reaction, the PEGylated protein is purified from unreacted protein and other species present in the reaction mixture, typically using chromatographic methods. While very pure products can be obtained using these chromatographic methods, they suffer from binding capacity and scalability limitations. The large-scale production capability for PEGylated proteins could be significantly enhanced through the use of a high-throughput, non-chromatographic purification step. However, there has been considerably less work done in the area of non-chromatographic purification of PEGylated proteins. PEG is an LCST polymer, i.e. it undergoes phase transition from a hydrophilic state to a mildly hydrophobic state when temperature is increased. Phase transition can be made to occur at ambient temperature by the addition of lyotropic salts such as sodium chloride and ammonium sulphate. Further increase in salt concentration leads to the formation of micelle like structures. Preliminary studies showed that a majority of these micelles were several microns in size. Based on this, a membrane-based hybrid technique involving formation, microfiltration, and dissolution of PEG micelles for purifying PEGylated proteins was developed. The results obtained are discussed.

(313f) Formation and Characterization of Functionalized Anion Exchange Membranes for Protein Separation

Ritchie, S. M., University of Alabama
Yang, L., University of Alabama
Originally presented on: 11/5/2013 14:00:00 - 14:18:00

The focus of this research is synthesis and characterization of functionalized polyethersulfone ultrafiltration membranes as anion exchange membranes for protein adsorption and separations. Functionalized membranes were synthesized by sulfonation, polymerization and amination. The initial step was to introduce sulfonic acid groups in a commercially available ultrafiltration membrane (300kDa) by sulfonation with 0.5 M H_2SO_4. The sulfonated membrane was subsequently polymerized with chloromethyl styrene by cationic polymerization. The final step involved amination with amine groups to impart positive charges to the grafted chains. Water flux and ion exchange capacity (IEC) studies were carried out to characterize the modified membrane. Decreased water permeability and increased IEC were observed after amination compared with the raw membrane indicating the presence of polymeric chains in the membrane flow path. ^1H NMR characterization confirmed the incorporation of poly(chloromethyl styrene) grafts. Sulfonation was also investigated in detail at room temperature by varying sulfuric acid concentration from 0.5 to 2 M. It was observed that initiator concentration increased from 0.014 to 0.027 meq/g when sulfuric acid concentration increased from 0.5 to 2 M. Amination was investigated with three aminating agents, diethylamine (DEA), para-aminobenzoic acid (PABA) and glycine. It was observed that glycine yielded the highest IEC. This was most likely due to the higher nucleophilicity of glycine compared with PABA. Results for protein adsorption will also be presented from whey protein rejection studies.

(313g) High Affinity Membrane Adsorbers for Protein Purification

Liu, Z., University of Arkansas
Qian, X., University of Arkansas
Wickramasinghe, S. R., University of Arkansas
Originally presented on: 11/5/2013 14:18:00 - 14:36:00

Specific protein-substrate recognition is critical for developing high affinity chromatography for protein purification. It provides an effective strategy to separate different proteins that have similar size and charge but different surface amino acid residues. Applying this molecular recognition strategy to modify membrane surface is promising to generate high performance membrane adsorbers for protein purification. Our work focuses on grafting a new polymer based ligand from the regenerated cellulose membrane (0.45μm pore size) surface by atom transfer radical polymerization (ATRP). Bisphosphonate was used as the functional monomer since it has a remarkable affinity to arginine residue. Other hydrophilic monomer such as N-(2-hydroxypropyl) methacrylamide (HPMA) and D-gluconamidoethyl methacrylate (GAMA) were used as the spacer monomer to enhance the flexibility of the polymer ligand and reduce the non-specific binding due to hydrophobic interaction. The copolymerization of bisphosphonate and the spacer monomers using ATRP was successfully achieved. Static and dynamic binding capacities were determined for the Arg rich Lysozyme and compared with other model proteins including BSA and IgG. The mechanism for the specific binding was explored using classical molecular dynamics simulations.

(313h) Scale Up of Membrane Affinity Chromatography Processes

Boi, C., Alma Mater Studiorum - Universite di Bologna
Dimartino, S., University of Bologna
Sarti, G. C., Alma Mater Studiorum - Universite di Bologna
Originally presented on: 11/5/2013 14:36:00 - 14:54:00

Membrane chromatography represents one of the emerging technologies for downstream processing in the biotechnology industry. This process is currently used in polishing steps for antibody manufacturing, while its application is still uncommon for the capturing step. To evaluate and to promote its application in large scale processes, it is crucial to use a reliable simulation tool able to describe the performance of membrane adsorbers at all scales in a predictive way.

In this work, the physical model for the description of protein purification with affinity membrane chromatography has been used to predict the performance of scaled up systems and compared with the simplified lumped model, in which axial dispersion is not accounted for explicitly but is rather lumped into the binding kinetic term. Two commonly used binding kinetics have been implemented in the models, namely the Langmuir and the bi-Langmuir equations.

It is known that by considering experimental data at bench scale only, both lumped model and physical model can describe equally well the observed behaviors. However the hypothesis to lumped all transport and kinetic processes into a single term leads to the use of "binding kinetic" parameters which are strongly flow rate dependent, in contrast with the usual physical properties of such terms. That throws motivated doubts to the validity of the lumped model for scale up purposes, which are further supported by the fact that the two models have different dimensionless equations.

In the present work we inspect the different scale-up predictions obtained by using the physical and lumped model, whose parameters have been assessed for the proper description of bench scale experimental data of h-IgG purification using B14-TRZ-Epoxy2 affinity membranes at different concentrations and different flow rates. The comparison of the lumped model with a validated physical model represents a valuable tool to assess the suitability of the lumped approximation. Although the two models describe equally well the experimental concentration profiles measured in a bench scale module, their up-scaled simulations at the size of commercially available adsorbers reveal a large discrepancy even at the early stages of breakthrough. The lumped model always overestimates the breakthrough point with respect to the physical model and, at 10% breakthrough, the difference in dynamic binding capacity evaluated by the two models presented is remarkably in the order of 23-28%, definitely impressive for the high added value of the antibody to be produced.

The physical model is based only on the actual physical phenomena present in the chromatographic module, the dispersion coefficient is measured separately and independently from the chromatographic cycle, and the conclusion is that it is the appropriate model for scale-up purposes.

Acknowledgements

This work was financially supported by MIUR, Italian Ministry of Education, University and Research (PRIN 2008) and by the University of Bologna, Italy.

(327a) Hybrid PSA-Distillation Process for Propane/Propylene Separation

Wu, F., University of South Carolina
Ebner, A. D., University of South Carolina
Ritter, J. A., University of South Carolina
Originally presented on: 11/5/2013 12:30:00 - 12:51:00

Polymer grade propylene is important and widely used in the manufacture of many chemicals and plastics; it is used especially as a monomer feedstock for polypropylene elastomer production. Its purity cannot be less than 99.5 mol%. The mixture of propane and propylene that results from the thermal or catalytic cracking of hydrocarbons is separated by distillation in a C_3 splitter.

In this traditional distillation, the relative volatility for this system is between 1.05 and 1.22 in the temperature range of 100-160 °F and in the pressure range of 189-454 psia. Thus, the separation is commonly performed in columns with more than 200 trays, with reflux ratios of about 13, and with high operating pressures of around 14 atm. The separation of propane and propylene mixtures represents one of the most energy consuming processes in the chemical process industries. This makes it highly desirable to develop new, more energy efficient processes for this very difficult separation.

Several pressure swing adsorption (PSA) processes have been developed by others for propane/propylene separation with several different adsorbents, such as 13X zeolite, 4A zeolite, carbon molecular sieve, and silica gel. However, the performance of a single PSA unit is not comparable with traditional distillation. This stems from it being difficult to produce two pure products from a single PSA process. The objective of this on-going project at USC has been to develop a hybrid PSA-distillation process for propane/propylene separation. Unfortunately, the energy savings could not be achieved, since none of the aforementioned commercial adsorbents work well enough, even in a hybrid process configuration.

With this in mind, a hypothetical adsorbent was conceived that has all the desirable and none of the undesirable properties of the commercial adsorbents already tested for this separation. Several PSA cycle configurations that utilize this hypothetical adsorbent under different operating conditions have been investigated via simulation. The results were very encouraging. They show that a hybrid PSA-distillation process is able to achieve significant energy saving compared to the traditional distillation process for propane/propylene separation. The goal now is to go to the lab and fabricate this novel adsorbent. The latest results of this work will be discussed during this presentation.

(327b) Surrogate Based Multi-Objective Optimisation for the Design of Pressure Swing Adsorption Systems

Beck, J., University College London
Friedrich, D., University of Edinburgh
Brandani, S., University of Edinburgh
Fraga, E. S., University College London
Originally presented on: 11/5/2013 12:51:00 - 13:12:00

Pressure swing adsorption (PSA) is a cyclic adsorption process for gas separation and purification with the potential for high productivity compared to alternative separation processes. The design of a cost-competitive and highly productive PSA process requires the optimisation of the process conditions with respect to the usually conflicting objectives purity, recovery and energy consumption. In the usual optimisation approach the governing hyperbolic/parabolic partial differential algebraic equations (PDAE) are simulated to cyclic steady state (CSS) for each parameter set generated by the optimisation routine. Since the system of PDAEs is usually large this approach is computationally very expensive. Thus most optimisation approaches either used simplified governing equations or optimised over a reduced set of parameters. Recently, reduced order and surrogate modelling approaches have been used for the optimisation of PSA systems.

Here we present a surrogate based multi-objective optimisation procedure for the design of PSA systems. This work is an extension of the previous single objective optimisation of the authors [1] to multiple objectives such as purity, recovery and energy consumption. The general idea is to build a surrogate model from a limited number of high-fidelity simulations and to perform the optimisation on the surrogate model. The surrogate is built and updated with design points calculated from our in-house PSA simulator which uses an accurate mathematical model for the dynamics of the PSA system. The NSGA-II multi-objective genetic algorithm is linked to the surrogate to perform the optimisation on the surrogate model; this reduces the computational complexity considerably because the surrogate is computationally less expensive to evaluate compared to the full PSA model. The solution obtained by the surrogate based optimisation procedure is validated against the full model and, if necessary, the NSGA-II optimisation procedure is repeated with an updated surrogate model; using the validation simulations with the full model to update the surrogate.

The resulting surrogate based optimisation outperforms a NSGA-II optimisation linked directly to the full PSA simulator in a case study optimising the separation of a binary gas mixture in a 2 bed, 6 step PSA system. The surrogate based optimisation reaches a more favourable Pareto front than the conventional optimisation with less than 25% of the high-fidelity model simulations; thus enabling the optimisation of full-scale PSA systems over a large set of parameters and using accurate mathematical models.

[1] J. Beck, D. Friedrich, S. Brandani, S. Guillas, and E. S. Fraga, "Surrogate based Optimisation for Design of Pressure Swing Adsorption Systems," in Proceedings of the 22nd European Symposium on Computer Aided Process Engineering, 2012, pp. 1217–1221.

(327c) Optimization of Ethanol Dehydration By PSA With 3A Zeolite

Rivera, A. F.,
Rodriguez, G., Universidad Nacional de Colombia
Bejarano, P. J.,
Originally presented on: 11/5/2013 13:12:00 - 13:33:00

Increasing demand for renewable fuels such as biodiesel and bioethanol had promoted research in separation processes. High water content in alcohol used for blending with gasoline could cause problems in engines. Anhydrous ethanol used as fuel must have purity 99.5%. Due to azeotrope formation, traditional separation process cannot be used. Adsorption processes has been widely used for industrial production of anhydrous ethanol.

Extractive distillation, heteroazeotropic distillation and pervaporation are alternative technology for ethanol-water separation. Comparison between alternative technologies and adsorption must be made with economical analysis.

In the present study, dehydration of ethanol by PSA using zeolite 3A was optimized with a sequential simplex algorithm. Adsorption processes are attractive alternative for its low energy consumption but it has high capital cost by equipment acquisition. Several studies had optimized adsorption process with recovery and product purity as objective function; however profits and cost are real comparative variables.

Objective function considers product sale, raw material cost, utilities and capital cost. Adsorption/desorption pressure, purge/feed ratio, bed dimensions and cycle time were chosen as independent variables for optimization. Auxiliary equipment such as compressor, condenser, vacuum system and utilities consumptions were estimated using algebraic equations but adsorption was modeled with a partial differential equation set. In all cases information for optimization was in the cyclic steady state (CSS). Start-up generated hot spot that could cause damage in adsorbent particles.

Differential equations were solved with method of lines, with upwind discretization in axial dimension and integrator ODE 15S of Matlab® for stiff problem. Model considers component balance for water, overall mass balance, "one-phase" energy balance, kinetic rate and equilibrium relation. Linear driving force model was used for kinetics and Langmuir isotherm for equilibrium. Ethanol was considered as an inert.

Numerical solution considers the following limits for optimization variable: Adsorption pressure; 1-4 bar, Desortion pressure; 0.1-1bar, purge/feed ratio; 0.05-0.2, diameter: 1-3m, length; 3-8m and time adsorption/desorption: 150-500s. Some fixed conditions were 393K for inlet temperature, 0.1 molar fraction of water in feed and constant feed of 200.000 l/day.

Increasing energy for compression or vacuum system allow to separate ethanol-water mixtures with a small bed on the other hand a largest column can achieve the separation with less energy for compression. Optimization process analyzed the effects and chose a better condition.

Separation with molecular sieves needs a previous process to concentrate ethanol near to azeotropic point. Azeotropic ethanol price was considered as a constant fraction of anhydrous ethanol price. Fraction of cost for raw material depends of several variables: sugar source cost, utilities in fermentative and concentration process and capital cost for reactor and distillation column. Optimization process considered several fractions of cost.

(327d) Evaluation of Oxygen Separation Performance Featuring High Temperature Pressure Swing Adsorption With Perovskite-Type Oxygen Sorbent

Izumi, J., ADSOTECH, Ltd.
Miura, N., Kyushu University
Fujimine, T., Tokyo Gas Co.
Ohi, M., Tokyo Gas Co.
Tazoe, M., ADSOTECH, Ltd.
Fukahori, K., ADSOTECH, Ltd.
Originally presented on: 11/5/2013 13:33:00 - 13:54:00

As indicated in the previous report, it was predicted that oxygen and nitrogen separation from air, featuring high temperature pressure swing adsorption using oxygen sorbent based on perovskite-type oxide ($La_{0.9}Sr_{0.1}Co_{0.9}Fe_{0.1}O_{3-\delta}$, LSCF1991), would demonstrate an extremely low electric power consumption rate.

In the present study, a small-scale testing apparatus was fabricated to measure the electric power consumption rate, as well as the oxygen production rate per loaded oxygen sorbent.

In this series of experiments, the adsorption pressure, desorption pressure, adsorption temperature, cycle time, etc. were varied, and the relationships between the afore-mentioned operational conditions and the electric power consumption rate and oxygen production rate per loaded oxygen sorbent were elucidated.

At the same time, assuming the future mass production of LSCF1991, an evaporation-to-dryness procedure of metal nitrates was adopted in this study to synthesize LSCF1991. The synthesized LSCF1991 powder was then formed into pellets as the oxygen sorbent for the small-scale testing apparatus.

Based on the data obtained, the electric power consumption rate and the oxygen production rate per loaded oxygen sorbent were posited as the medium-scale unit for 1,000 m^3N/h oxygen production.

To summarize the results of this experiment with a small-scale apparatus,

1) As the oxygen adsorbed amount of LSCF is proportional to the logarithm of oxygen partial pressure, the desorption pressure of 1-5 kPa (relatively higher vacuum conditions) gives the lowest electric power consumption rate and the greatest oxygen production rate per loaded oxygen sorbent.

2) With this isotherm, even when the adsorption pressure is increased, the oxygen production rate per loaded oxygen sorbent does show any significant rise. On the other hand, the electric power consumption

rate increases with rising compressor power consumption. Therefore, it appears that the adsorption pressure is sufficiently low to compensate for the pressure loss.

3) The adsorption temperature should be set higher than 500 deg-C to maintain a practical oxygen production rate per loaded oxygen sorbent.

4) At higher than 500 deg-C, as LSCF1991 shows a relatively high adsorption rate, and even during a shorter adsorption period such as 30 seconds, stable oxygen and nitrogen separation can be undertaken. For periods longer than 30 seconds, the oxygen production rate per loaded oxygen sorbent is inversely proportional to the adsorption time.

In conclusion, for this HTPSA-oxygen system, when a heat recovery ratio of 92% can be maintained, an electric power consumption rate of 0.2 kWh/m^3N is predicted.

With a relatively rapid operation of 100 seconds per 1 cycle (i.e., adsorption period of 40 seconds, desorption period of 40 seconds, and other operations requiring 20 seconds), an oxygen production rate per loaded oxygen sorbent of 30 m^3N-O_2/h/ton (500 kg/towerÂ □ A2 tower system) can be expected.

Compared with cryogenic separation, which delivers the lowest electric power consumption rate at the present time (0.32 kWh/m^3N), the HTPSA-oxygen system in this study has great potential to reduce the consumption rate in the near future.

The initial cost of the HTPSA-oxygen system in this study is the same as that for the PSA-oxygen system using nitrogen adsorbent, which is the representative medium-scale oxygen production unit available in the market.

(327e) H2 Purification By Pressure Swing Adsorption Using CuBTC

Silva, B., Laboratory of Separation and Reaction Engineering, Associate laboratory LSRE/LCM, Department of Chemical Engineering, Faculty of Engineering, University of Porto
Ribeiro, A. M., University of Porto
Chang, J. S., Korea Research Institute of Chemical Technology
Loureiro, J. M., LSRE - Laboratory of Separation and Reaction Engineering - Associate Laboratory LSRE/LCM
Rodrigues, A. E., University of Porto, Faculty of Engineering
Originally presented on: 11/5/2013 13:54:00 - 14:15:00

H_2 purification by Pressure Swing Adsorption using CuBTC

Bruna Silva (bnogueira@fe.up.pt), Ana M. Ribeiro (apeixoto@fe.up.pt), Jong-San Chang (jschang@krict.re.kr), JosÃ© M. Loureiro (loureiro@fe.up.pt), Alírio E. Rodrigues (arodrig@fe.up.pt)

Abstract

Hydrogen is one of the most important industrial and potential chemicals, used in many applications, such as hydrocracking, hydrogenation of oils and in the production of methanol and ammonia. Moreover, due to environmental concerns hydrogen has increasingly received attention as it is a promising energy source for electrical power generation and transportation fuel (Ribeiro *et al.*, 2009). Therefore, the demand of hydrogen is continuously increasing with the resulting worldwide attention and research motivation for advances in the field of hydrogen production and purification.

Nowadays, the catalytic reforming of natural gas combined with a water gas shift reaction step is the main and the most cost-effective process for hydrogen production at a commercial scale (Sircar *et al.*, 1999). The hydrogen production by steam methane reforming (SMR) originates a hydrogen stream containing several impurities, such as water vapor, CH_4, CO_2, N_2 and CO. The use of hydrogen in fuel cell applications requires a high purity (99.99+%), which is attained by removing these impurities (Ribeiro *et al.*, 2008). The commercial technology that is widely used for hydrogen purification and separation from steam methane reforming is pressure swing adsorption (PSA) (Ruthven *et al.*, 1994). Several studies have been done focusing on PSA processes for hydrogen separation and purification (Yang and Lee, 1998; Park *et al.*, 2000; Huang *et al.*, 2008; Lopes *et al.*, 2011; You *et al.*, 2012).

A new class of adsorbents named Metal Organic Frameworks (MOFs) represents an excellent alternative to the conventional materials used so far (Silva *et al.*, 2012). The diversity in the configurations of these materials results from co-ordination between inorganic metal atoms and organic ligands or linkers, to form highly porous network structures (Chowdhury *et al.*, 2012; Dasgupta *et al.*, 2012). The use of MOFs is advantageous compared to the utilization of zeolites, since their skeleton accepts almost all the cations of the periodic table (Ferey *et al.*, 2011). Among the different MOFs, CuBTC is one of the few materials that have already commercial availability. This material is composed of copper dimers coordinated to the oxygen atoms of benzene-1,3,5-tricarboxylate (BTC) linkers, forming a regular porous network with a large surface area (BET surface area up to 1600 m^2/g), high pore volume, high chemical stability and Lewis acidity (Plaza *et al.*, 2012). Because of these features, this material has a high potential for applications in the field of adsorption, such as gas separation and gas storage (Min Wang *et al.*, 2002; Dathe *et al.*, 2005; Millward and Yaghi, 2005).

In this work, hydrogen purification from mixtures that also contain CO_2, CO, CH_4, and N_2 was performed by PSA with CuBTC as adsorbent (supplied by KRICT). First, the equilibrium adsorption of each pure gas was measured in a magnetic suspension microbalance (Rubotherm, Germany). Second, single component, binary and ternary fixed-bed adsorption experiments were carried out. At the beginning of a breakthrough experiment, a stainless steel column filled with CuBTC and equipped with thermocouples, started to be fed with the gas mixture. During the adsorption and desorption steps, the temperature histories and the molar flow rate were recorded. Additionally, samples were collected in the loops of a multi-port valve for subsequent determination of the molar composition of the exit gas in the gas chromatograph. The same column, employed for the fixed-bed breakthrough curves described above, was used for PSA experiments. A PSA cycle of four elementary steps was performed. The PSA cycle started with the co-current pressurization with feed, followed by feed, blowdown and purge steps. In the purge step, a stream of pure hydrogen was used countercurrently at low pressure to regenerate the column.

The mathematical modelling of a multicomponent adsorption in a fixed bed involves the material, momentum and energy balances that govern the process, taking into account axial dispersion and mass transfer resistances. The mathematical model developed was applied for the simulation of the breakthrough experiments. This fixed-bed model establishes the basic procedure to simulate adsorption-based cyclic processes. The modeling of a PSA unit involves the same conservation equations used for the fixed-bed simulations coupled with the appropriate boundary conditions for each step. The previously developed model was validated against the results obtained in breakthrough experiments and PSA tests.

The results of the breakthrough experiments showed that the mathematical model describes well the dynamics of the adsorption processes that take place in the fixed bed. The adsorption and desorption showed a good fitting with the simulation results, since the molar fraction, molar flow rate and temperature histories were well predicted by the model. The typical roll-up that appears in multicomponent fixed bed adsorption is also observed and well predicted by the model. The increase of the molar flow rate of the adsorbed gases observed when the desorption step begins was well described by the model. The experimental pressure and temperature histories for all PSA experiments were well predicted by the model. The composition and the molar flow rate of each gas in the different experiments were generally well fitted by the model. In terms of PSA experiments with ternary mixtures, the best performance was achieved for a mixture with 78 % H_2, 18 % CO_2 and 3 % CH_4 and a total cycle time of 516 s: 262 s, 62 s and 192 s for pressurization + feed, blowdown and purge steps, respectively. For this non-optimized cycle, a hydrogen purity of 99.97 % was attained, with a recovery of 45.76 % and a productivity of 6.86 $mol.kg^{-1}.h^{-1}$ (simulation results).

In conclusion, this work demonstrates that the mathematical model developed is a powerful tool for the prediction of PSA experiments, which can save time and money, reducing the number of required experiments.

Acknowledgements

The authors acknowledge the financial support given by FCH JU, through FP7 EU Project Hy2Seps2 -

"Hybrid Membrane - Pressure Swing Adsorption (PSA) Hydrogen.

References

Chowdhury, P., S. Mekala, F. Dreisbach and S. Gumma, *Adsorption of CO, CO_2 and CH_4 on Cu-BTC and MIL-101 metal organic frameworks: Effect of open metal sites and adsorbate polarity*. Microporous and Mesoporous Materials, 2012. **152**: p. 246-252.

Dasgupta, S., N. Biswas, Aarti, N.G. Gode, S. Divekar, A. Nanoti and A.N. Goswami, *CO_2 recovery from mixtures with nitrogen in a vacuum swing adsorber using metal organic framework adsorbent: A comparative study*. International Journal of Greenhouse Gas Control, 2012. **7**: p. 225-229.

Dathe, H., E. Peringer, V. Roberts, A. Jentys and J.A. Lercher, *Metal organic frameworks based on Cu2+ and benzene-1,3,5-tricarboxylate as host for SO_2 trapping agents*. Comptes Rendus Chimie, 2005. **8**: p. 753-763.

Ferey, G., C. Serre, T. Devic, G. Maurin, H. Jobic, P.L. Llewellyn, G. De Weireld, A. Vimont, M. Daturi and J.-S. Chang, *Why hybrid porous solids capture greenhouse gases?* Chemical Society Reviews, 2011. **40**: p. 550-562.

Huang, Q., A. Malekian and M. EiÃ¦, *Optimization of PSA process for producing enriched hydrogen from plasma reactor gas*. Separation and Purification Technology, 2008. **62**: p. 22-31.

Lopes, F.V.S., C.A. Grande and A.E. Rodrigues, *Activated carbon for hydrogen purification by pressure swing adsorption: Multicomponent breakthrough curves and PSA performance*. Chemical Engineering Science, 2011. **66**: p. 303-317.

Millward, A.R. and O.M. Yaghi, *Metal−Organic Frameworks with Exceptionally High Capacity for Storage of Carbon Dioxide at Room Temperature*. Journal of the American Chemical Society, 2005. **127**: p. 17998-17999.

Min Wang, Q., D. Shen, M. Bülow, M. Ling Lau, S. Deng, F.R. Fitch, N.O. Lemcoff and J. Semanscin, *Metallo-organic molecular sieve for gas separation and purification*. Microporous and Mesoporous Materials, 2002. **55**: p. 217-230.

Park, J.-H., J.-N. Kim and S.-H. Cho, *Performance analysis of four-bed H_2 PSA process using layered beds*. AIChE Journal, 2000. **46**: p. 790-802.

Plaza, M.G., A.M. Ribeiro, A. Ferreira, J.C. Santos, U.H. Lee, J.-S. Chang, J.M. Loureiro and A.E. Rodrigues, *Propylene/propane separation by vacuum swing adsorption using Cu-BTC spheres*. Separation and Purification Technology, 2012. **90**: p. 109-119.

Ribeiro, A.M., C.A. Grande, F.V.S. Lopes, J.M. Loureiro and A.E. Rodrigues, *A parametric study of layered bed PSA for hydrogen purification*. Chemical Engineering Science, 2008. **63**: p. 5258-5273.

Ribeiro, A.M., C.A. Grande, F.V.S. Lopes, J.M. Loureiro and A.E. Rodrigues, *Four beds pressure swing adsorption for hydrogen purification: Case of humid feed and activated carbon beds*. AIChE Journal, 2009. **55**: p. 2292-2302.

Ruthven, D.M., S. Farooq and K.S. Knabel, *Pressure Swing Adsorption*. VCH Publishers, New York, 1994.

Silva, J.A.C., K. Schumann and A.E. Rodrigues, *Sorption and kinetics of* CO_2 *and* CH_4 *in binderless beads of 13X zeolite*. Microporous and Mesoporous Materials, 2012. **158**: p. 219-228.

Sircar, S., W.E. Waldron, M.B. Rao and M. Anand, *Hydrogen production by hybrid SMR–PSA–SSF membrane system*. Separation and Purification Technology, 1999. **17**: p. 11-20.

Yang, J. and C.-H. Lee, *Adsorption dynamics of a layered bed PSA for* H_2 *recovery from coke oven gas*. AIChE Journal, 1998. **44**: p. 1325-1334.

You, Y.-W., D.-G. Lee, K.-Y. Yoon, D.-K. Moon, S.M. Kim and C.-H. Lee, H_2 *PSA purifier for CO removal from hydrogen mixtures*. International Journal of Hydrogen Energy, 2012. **37**: p. 18175-18186.

(327f) Pressure Swing Adsorption Process for the Separation of Hydrogen, Methane and Ethane With the MOF

Okada, M., College of Industrial Technology, NIHON University
Ferreira, A., LSRE - Laboratory of Separation and Reaction Engineering - Associate Laboratory LSRE/LCM
Ribeiro, A. M., University of Porto
Lee, U. H., Catalysis Center for Molecular Engineering
Hwang, Y. K., Korea Research Institute of Chemical Technology
Chang, J. S., Korea Research Institute of Chemical Technology
Rodrigues, A. E., University of Porto - Faculty of Engineering
Originally presented on: 11/5/2013 14:15:00 - 14:36:00

The direct conversion of methane to hydrogen and/or higher hydrocarbons by plasma reaction is a recent field of interest. Indeed, a significant number of reports were published recently on this topic. Plasma reaction field can be generated by various discharge methods or equipment under atmospheric pressure. Yet, the non-thermal equilibrium plasma is one with highest potential, due its unique property to generate electrons with high temperature (high energy) at low gas temperatures. The available literature is mainly focused on product distribution, energy efficiency and reaction improvement strategies. As examples of improvement strategies we can find the integration of plasma field with catalysis (plasma assisted catalytic reaction), the study of different plasma generator configurations, and the influence of co-feed gases such as He, Ar, or even H_2. However, studies on downstream processes, such as the separation of unreacted methane are still insufficient from the engineering point of view, neglecting a possible improvement of the plasma reactor efficiency. The conversion of methane is strongly dependent on reaction conditions, such as discharge method, input power, composition of feed gas, structure of discharge equipment, among others. Usually, methane conversion of only 10% is obtained, when pure methane is fed to the plasma reaction field under ambient temperature and pressure. Therefore, downstream separation of the product gas mixture (H_2, CH_4, C_2H_6 ...) is very important in order to recycle the unreacted methane to the feed stream. This separation has similarities with the natural gas purification. Several different processes have been developed to produce pure methane from natural gas, by removing the carbon dioxide, nitrogen, hydrogen sulfide and a minor amount of hydrocarbons from natural source. Particularly, adsorption based processes, like Pressure Swing Adsorption (PSA), have been intensively studied, due to its high energy efficiency. Within those studies, different adsorbents have been proposed, such as molecular sieves, activated carbons, and more recently Metal Organic Frameworks (MOFs). In this work, a MIL-100(Fe) sample, synthesized and shaped into granulates by the Korea Research Institute of Chemical Technology (KRICT) is evaluated for methane purification, from the exhaust gas of the plasma reactor, by PSA. Firstly, the equilibrium of adsorption of hydrogen, methane and ethane was studied by means of a gravimetric method, at three different temperatures (303, 313, and 333 K). MIL-100(Fe) is ethane selective, presenting an ideal selectivity for C_2H_6/CH_4 of about 4, at 303 K and 5 bar. The calculated isosteric heats of adsorption of methane and ethane from this work, and the ones presented by Plaza *et al.*[1] for propane and isobutane, when plotted in function of the carbon atoms number present a linear behavior, with a value of 6.44 kJ per –(CH_2)-. Single component, binary, and ternary fixed-bed adsorption experiments were carried out, and these were used to validate the mathematical model by performing their simulation. From the breakthrough experiments it was confirmed the ethane selective behavior of the adsorbent, as predicted from the pure component isotherms. Finally, a PSA cycle was simulated, and experiments were carried out to validate the separation.

[1] Plaza, M.G., Ribeiro, A.M., Ferreira, A., Santos, J.C., Hwang, Y.K., Seo, Y.K., Lee, U.H., Chang, J.S., Loureiro, J.M., Rodrigues, A.E.: Separation of C3/C4 hydrocarbon mixtures by adsorption using a mesoporous iron MOF: MIL-100(Fe). Microporous and Mesoporous Materials 153(0), 178-190 (2012)

(327g) Verification of Two-Dimensional Simulation of Adsorption Process From Breakthrough Experiment With Varying Adsorption Pressure

Yingshu, L.,
Xingang, Z., University of Science and Technology Beijing
Originally presented on: 11/5/2013 14:36:00 - 14:57:00

In this study, the two-dimensional modeling of the pressure swing adsorption proposed before has been verified by breakthrough experiments, where the pressure and temperatures inside the bed as well as concentration of vent gas are measured. It is found that the varying adsorber pressure can be well predicted with a moderate mass transfer rate, which is essential for thoroughly simulating the PSA cycle process. The trends of the oxygen concentration in vent gas and temperatures inside the bed can also be described well, but due to the lumped mass transfer rate used in the model, there is some discrepancy in the temperature profiles. The temperature rise in dead volumes caused by compression can be found too. It is thus concluded that the two-dimensional model can give favorable prediction on the transport phenomena occurred inside the adsorber, but the accuracy of the model depends to a large extent on the expression used to calculate the mass transfer inside the sorbent particle.

In addition, the evolutions of process variables and the effects of particle size are briefly described and discussed in this paper. There are two mass transfer zones during the breakthrough process and the velocity distribution is in the form of staircase. The adsorber packed with very small particle will cause velocity increase induced by gas expansion and bring poor performance.

(330a) Adsorption of the Compounds Encountered in Monosaccharide Dehydration in Zeolites

Leon, M., University of Delaware
Xiong, R., University of Delaware
Swift, T. D., University of Delaware
Nikolakis, V., University of Delaware
Sandler, S. I., University of Delaware
Vlachos, D. G., University of Delaware
Originally presented on: 11/5/2013 12:30:00 - 12:52:00

Biomass carbohydrates are the most abundant renewable resources available and have the potential to replace petroleum as a source of both energy and chemicals, leading to more environmentally friendly processes and reducing the dependence on crude oil. A promising reaction in this field is the acid-catalyzed dehydration of monosaccharides, such as fructose and glucose, towards 5-hydroxymethylfurfural (HMF), an important intermediate for the production of valuable chemicals.[1] When zeolites are used as catalysts the preferential adsorption of reactants, products or byproducts could markedly affect the yield to the desired product. In this regard, studying the adsorption of the compounds present in the reaction media in the zeolite catalyst becomes crucial to understand the reactivity of the system and it is necessary to calculate the intrinsic reaction rate constants from kinetic data. Zeolites can also be used as adsorbents for the selective removal of the HMF from the reaction media (either in-situ or after dehydration). In this respect understanding the effect of the framework type, composition (e.g. Si/Al, extraframework cation), and presence of co-solvent (e.g. dimethylsulfoxide (DMSO)) on HMF adsorption are important for the design of an adsorptive separation process.

Adsorption loadings from solution are usually estimated from changes of the solute concentration assuming that the solution volume is not affected by adsorption. This quantity is called excess adsorption loading and sometimes it is quite different from the real loading. We formulated and used a method for the estimation of the real adsorption loading from the experimentally measured excess adsorption that can be applied to calculate the adsorption isotherms both in the case of single-solute and multi-solute mixtures.

The effect of temperature on the adsorption of fructose, glucose, HMF, levulinic and formic acid from aqueous solutions in the protonated form of zeolite beta with SiO_2/Al_2O_3 ratio of 18 (H-BEA) will initially be presented. The limiting heat of adsorption at zero coverage of all components was calculated from isotherms measured at 0, 25 and 40 °C. Adsorption experiments from multi-solute solutions were also carried out and the data were found to be in good agreement with predictions made using Ideal Adsorbed Solution Theory (IAST) and the single solute isotherm experimental data.

Adsorption of HMF was studied in more detail. We examined the effect of the framework type (BEA, MFI, FAU), composition (Si/Al) as well as the addition of DMSO as co-solvent on its adsorption uptake. It was found that in each zeolite type the adsorption uptake of HMF increases with the hydrophobicity (higher Si/Al ratio) of the zeolite. Experimentally measured isotherms were in good agreement with those calculated using Grand Canonical Monte Carlo - Expanded Ensemble (GCMC-EE) simulations. It is found that in the case of HMF/water binary mixtures, zeolites with larger pores (i.e. FAU), have higher HMF adsorption capacity but lower selectivity because large pores allow water adsorption due to favorable H_2O-HMF attraction. Therefore, dealuminated zeolite Y is less selective for the separation of HMF from water than silicalite-1 (siliceous MFI type zeolite) that has smaller pores sizes. In addition, experiments and simulations show that in HMF/DMSO/water ternary mixtures, the HMF loading decreases with increasing DMSO concentration. This decrease is due to the high solubility of HMF in DMSO, and to the fact that in the entropic contribution to the free energy of adsorption dominates over the enthalpic contribution. Analysis of our data shows that the ratio of the energetic interaction in the zeolite to the solvation free energy is a key factor in controlling separation from liquid mixtures.

To conclude, our findings could shed light on the acid catalyzed dehydration of monossacharides and the HMF separation, and serve as a hint to the rational design of microporous adsorbents for liquid-phase separations in biomass processing.

REFERENCES

1. Chheda, J. N.; Huber, G. W.; Dumesic, J. A. Liquid-Phase Catalytic Processing of Biomass-Derived Oxygenated Hydrocarbons to Fuels and Chemicals. *Angew. Chem. Int. Ed.* **2007,** *46*, 7164-7183.

2. Vandam, H. E.; Kieboom, A. P. G.; Vanbekkum, H. The Conversion of Fructose and Glucose in Acidic Media - Formation of Hydroxymethylfurfural. *Starch-Starke* **1986,** *38*, 95-101.

3. Kruger, J. S.; Nikolakis, V.; Vlachos, D. G. Carbohydrate Dehydration Using Porous Catalysts. *Curr. Opin. Chem. Eng.* **2012,** *1*, 312-320.

4. Kruger, J. S.; Choudhary, V.; Nikolakis, V.; Vlachos, D. G. Elucidating the Roles of Zeolite H-Bea in Aqueous-Phase Fructose Dehydration and HMF Rehydration. *ACS Catal.* **2013**.

5. Roman-Leshkov, Y.; Chheda, J. N.; Dumesic, J. A. Phase Modifiers Promote Efficient Production of Hydroxymethylfurfural from Fructose. *Science* **2006,** *312*, 1933-1937.

(330b) Elucidating Structure-Properties Relations for the Design of Highly Selective Carbon-Based HMF Sorbents

Yoo, W. C.,
Rajabbeigi, N., University of Minnesota
Mallon, E. E., University of Minnesota
Tsapatsis, M., University of Minnesota
Snyder, M. A., Lehigh University
Originally presented on: 11/5/2013 12:52:00 - 13:14:00

Continuously increasing energy demands in a fragile global market and in the face of growing environmental concerns motivate the drive to develop sustainable energy resources like plant biomass for replacing or augmenting current petroleum-derived fuels and chemicals [1-3]. One of numerous challenges facing the biorefinery is the selective and efficient production of important chemical products such as 5-hydroxymethyl furfural (HMF). HMF production from fructose dehydration in DMSO with various catalytic systems has shown superior selectivity and high yields [3, 4]. Yet, the subsequent need for energy-intensive separation of HMF from HMF/fructose/DMSO mixtures challenges the practical feasibility of this process. Microporous carbon materials have shown promise as substrates for alternative adsorption-based separations [5], but the origin of HMF selectivity and adsorption capacity, and thereby structure-properties relations enabling rational design of enhanced sorbents, has remained elusive. Through systematic quantification of the functionality and texture of various commercial (BP2000, Norit1240) and synthetic carbons (three-dimensionally ordered macroporous carbons; mesoporous carbons), we link, for the first time, adsorption capacity and HMF adsorption selectivity to two tunable materials properties of the carbon sorbents: micropore texture and oxygenate functionality (i.e., carbon polarity). In the process, we exploit these newly elucidated structure-properties relations for realizing a class of synthetic carbons with sub-micron particulate morphology that achieve more than 60% higher selectivity and more than 20% higher capacity for HMF over fructose as compared to the best performing commercial product, BP2000.

References

[1] P.Y. Dapsens, C. Mondelli, J. PÃ©rez-Ramírez, ACS Catal., 2 (2012) 1487-1499.

[2] G.W. Huber, S. Iborra, A. Corma, Chem. Rev., 106 (2006) 4044-4098.

[3] A. Corma, S. Iborra, A. Velty, Chem. Rev., 107 (2007) 2411-2502.

[4] X. Qi, M. Watanabe, T.M. Aida, R.L. Smith Jr, Ind. Eng. Chem. Res., 47 (2008) 9234-9239.

[5] N. Rajabbeigi, R. Ranjan, M. Tsapatsis, Microporous Mesoporous Mater., 158 (2012) 253-256.

(330c) Separation of Lignin From Hydrolyzates By Flocculation Using PEO Polymers

Ramarao, B. V., ESPRI, SUNY College of Environmental Science and Forestry
Yasarla, R., State University of New York
Gnanavel, V., SUNY ESF
Originally presented on: 11/5/2013 13:14:00 - 13:36:00

Hydrolyzates produced by pretreatment of lignocellulosic materials contain phenolic compounds, organic acids and aldehydes such as furfural in colloidal and dissolved form. Since these compounds can be potent inhibitors for fermentation, they need to be separated and isolated ahead of such biological treatments. We found that the application of a strongly binding non-ionic polymer, polyethylene oxide causes flocculation and rapid separation of lignin from the hydrolyzates, while preserving the carbohydrates in solution to a large extent. The effect of flocculation is measured by the reduction in turbidity and enhancement of the suspension settling velocity. The maximum flocculation was found to depend on the dosage of the polymer and its MW, but was independent of the pH of the hydrolyzates (from 2 through 9). PEO dosages were between 20 to 50 ppm at the maximum flocculation and reduction in suspension turbidity. The suspension temperature plays an important role, with temperatures below 21.5 C showing no flocculation.

The hemicellulose fraction in the supernatants was close to the raw extract showing that flocculation does not remove significant amount of fermentable sugars. Hence such pretreatment clarification is not likely to affect downstream bioproduct yields. Flocculation and clarification with PEO were also demonstrated on large pilot scale batches of 1000 kg of extracts. The flocculated extract was filtered and the filtrate showed a 99.5% reduction in turbidity, from 12000 NTUs in the raw extract to 50 NTUs in the filtrate. The impact of flocculation on solid liquid separations was quantified and flocculation was optimized with respect to dosage, suspension concentration, temperature and pH.

(330d) Fractionation and Recovery of Lignin and Xylan From Alkali Liquors - Characterization for Biorefinery Processes

Stoklosa, R. J., Michigan State University
Hodge, D., Michigan State University
Originally presented on: 11/5/2013 13:36:00 - 13:58:00

Understanding the behavior of plant cell wall biopolymers during alkaline biorefining processes and how their properties are affected by the processing conditions is important for maximizing value from woody lignocellulosic feedstocks. In this work, the fractionation of alkali pulping liquors was investigated to characterize the extractability, recoverability, and properties of the solubilized hemicelluloses and lignin during the alkali impregnation and heat-up stages of alkali organosolv pulping or soda pulping for three hardwood species: sugar maple (*Acer saccharum*), silver birch (*Betula pendula*), and a hybrid poplar (*Populus nigra x maximowiczii cv.* NM6). Recoverability of the solubilized biopolymers were determined using a variety of methods including aggregation and precipitation by acidification as well as ultrafiltration. The xylan showed a decreasing degree of polymerization (DP) with increasing temperature during reactor heat-up, which stabilized at a DP between 10-30. Lignin aliphatic and aromatic hydroxyl content was characterized by ^{1}H NMR and the lignin molecular weight distributions were determined by gel permeation chromatography (GPC). These properties showed a clear dependence on the extraction conditions and were clearly linked to recoverability of lignin by either acidification or ultrafiltration.

(330e) Novel Catalytic Membranes for Combined Biomass Hydrolysis and Sugar Recovery

Avram, A., University of Arkansas
Lei, J., University of Arkansas
Qian, X., University of Arkansas
Wickramasinghe, S. R., University of Arkansas
Beier, A., University of Arkansas
Originally presented on: 11/5/2013 13:58:00 - 14:20:00

Production of biofuels from lignocellulosic biomass is one of the leading candidates for replacement of fossil fuels. In the biochemical conversion platform, dilute sulfuric acid pretreatment is one of the leading technologies to hydrolyze hemicellulose, relocate lignin and expose cellulose. After pretreatment, a cocktail of cellulase enzymes is used to break down cellulose synergistically. The major problems associated with this technology are the slow and incomplete hydrolysis of cellulose and the degradation of monomer sugars during pretreatment. Furthermore, the cost of the enzymes has been an inhibitory factor for the commercialization of biomass conversion technology.

Here, a novel catalytic membrane reactor is designed to simultaneously hydrolyze cellulose and hemicellulose and separate the hydrolyzed sugars, thus minimizing product degradation. A polymeric solid acid catalyst is synthesized to replace the corrosive sulfuric acid and expensive enzymes, but operates at a higher conversion rate than the analogous natural enzymes and that costs far less to produce. We have grown adjacent nanostructure consisting of tethered polysulfonic acid and polyionic liquid chains from the surface of ceramic membranes. The feed consisting of cellulose and hemicellulose in a water / ionic liquid mixture is introduced on one side of the membrane. The polyionic liquid chains help solubilize the polysaccharides while the polysulfonic acid chains catalyze the hydrolysis of the polysaccharides. The membrane pore size is such that only water containing dissolved monomeric sugars and short chain oligomers can pass through. The ionic liquid and larger polysaccharides are rejected.

From preliminary hydrolysis experiments, our results indicate the potential of these novel membranes to hydrolyze polysaccharides, minimize degradation product and save on catalyst costs.

(330f) Optimization of An Extractive Distillation With Mixed Solvents As Entrainer

Garcia, L. C., Universidad Nacional de Colombia
Gil, I. D., Universidad Nacional de Colombia
Orjuela, A., Universidad Nacional de Colombia
Rodriguez, G., Universidad Nacional de Colombia
Originally presented on: 11/5/2013 14:20:00 - 14:42:00

Anhydrous ethanol is widely used in the biofuels industry, since it has demonstrated to be environmental friendly and to have excellent properties for clean-burning. In mixtures of ethanol and gasoline, alcohol must have purity higher than 99.5% in order to avoid phase separation. Among the most popular processes used in ethanol dehydration, extractive distillation has demonstrated to be an adequate alternative due to the low energy consumption compared with other techniques and its economic feasibility. In this work, an extractive distillation process was simulated with the Aspen Plus software platform, using the RadFrac module for distillation columns. A mixture of ethylene glycol and glycerol is proposed as entrainer. This enhanced entrainer combines the low viscosity and the ability of the ethylene glycol to break the binary azeotrope with the low prices of glycerol. Phase equilibrium of this quaternary system was calculated with the NRTL model, which correlates well these strongly polar mixtures. The nominal case consists in the dehydration column, the recovery column and the solvent recycle loop. In order to establish the most adequate operating conditions, it was studied for each column the effect of the molar reflux ratio, the feed stage and the feed condition in the ethanol and solvent compositions. For the dehydration column, it was also investigated the effects of the ethylene glycol-glycerol mixture composition, the separating agent feed stages, the separating agent split stream, and the azeotropic feed temperature on the separation. Subsequently, the thermal and hydraulic analysis of both columns was established, so that the system operates energetically efficient. Finally, the economic optimization of the system was achieved, using the maximization of the profit as objective function. In this case, the solvent to feed ratio, the reflux ratio and the number of stages in the extractive column were the optimization variables, while the design parameters of the recovery column were fixed.

(330g) Study of Physical Properties of Binary Mixtures of Sulfolane + 2-Butanol (or + 2-propanol) At Different Temperatures

Mesquita, F., Federal University of Ceara
Sant'ana, H. B., Universidade Federal do CearÃ¡
Feitosa, F., Universidade Federal do CearÃ¡
Aznar, M.,
Santiago-Aguiar, R. S., Universidade Federal do CearÃ¡
Originally presented on: 11/5/2013 14:42:00 - 15:04:00

Sulfolane and alcohols are classes of chemical compounds that are usually used as a solvent in different industrial recovery processes. The focus of the present work is to measure volumetric (densities) and transport (dynamic viscosity) properties of the following binary mixtures: sulfolane + 2-butanol (or + 2-propanol) at a range temperature from T = 303.15 to T = 343.15 K, at atmospheric pressure. Densities and viscosities were measured using a viscodensimeter (Anton Paar SVM 3000 Digital Oscillation U-tube). Viscosity deviations and excess volumes were obtained from the experimental data, by using a fitting procedure (Redlich-Kister type equation). The results show that excess volumes were positive over the entire mass fraction range for all binaries mixtures studied. Otherwise deviation viscosities are negative over the entire mass fraction range of the mixture. Excess volumes and deviation viscosities behaviors increasing with temperature.

(337a) Ambient-Air Plasmas for Frugal and Sustainable Water Disinfection in the Developing World

Pavlovich, M. J., University of California - Berkeley
Galleher, C., Univeristy of California - Berkeley
Ono, T., Tokyo Institute of Technology
Machala, Z., University of California - Berkeley
Clark, D. S., University of California
Graves, D. B., University of California - Berkeley
Originally presented on: 11/5/2013 15:15:00 - 15:37:00

Ambient-condition air plasmas have been shown to decontaminate water and aqueous solutions through the production of reactive oxygen and nitrogen species (RONS) that are toxic to bacteria and other microorganisms. One method of creating air plasma at ambient conditions is the "dielectric barrier discharge" (DBD) technique, which creates a small, confined plasma region near a pair of electrodes separated by a dielectric layer such as glass or ceramic; RONS diffuse away from the plasma region and into a contaminated aqueous volume. We have previously shown that DBD-based devices can disinfect water in two distinct "modes": a high-power, nitrogen oxides-rich discharge creates a persistent antimicrobial effect in water that can last at least one week [1], while a low-power, ozone-rich mode can rapidly decontaminate aqueous solution [2].

We describe a design and configuration of air plasma reactors for the novel application of water disinfection in low-resource settings, including the developing world. Ambient-air plasma is a promising method for low-resource sanitation because it requires only electricity and air to create a disinfectant. The plasma reactors used in this study are assembled from inexpensive materials including ceramic tubes and steel rods, and comparatively low-cost energy sources and power supplies like solar-energy rechargeable lead-acid batteries and neon-sign transformers are used to ignite the plasma. In this study, we compare the gas- and liquid-phase RONS chemistry created by these "frugal" devices to the chemistry observed by using more expensive and precisely tunable laboratory devices. In addition, we show how the disinfection efficacy, using *E. coli* as a model bacterium, depends on transformer voltage, dissipated power, air flow rate through the reactors, and reactor configuration. Our preliminary results indicate that the concentrations of some reactive species can differ dramatically based on the reactor geometry and material and the power source. However, the chemistry measured here agrees reasonably well with the established laboratory-scale plasma chemistry in some important aspects, including the ozone-nitrogen oxide "mode transition". Therefore, air plasma treatment remains a promising strategy for inexpensive and effective water disinfection.

References:

[1] M. J. Traylor, M. J. Pavlovich, S. Karim, P. Hait, Y. Sakiyama, D. S. Clark, D. B. Graves, *Journal of Physics D: Applied Physics* **2011**, *44*, 472001.

[2] M. J. Pavlovich, H.-W. Chang, Y. Sakiyama, D. S. Clark, D. B. Graves, *Journal of Physics D: Applied Physics* **2013**, *46*, 145202.

(337c) Chemical Degradation of Xanthate By Pulse Corona Discharges

Cerda, K. V., University Catolica del Norte
Arce, P. E., Tennessee Technological University
Originally presented on: 11/5/2013 15:59:00 - 16:21:00

This work is focused on the degradation of xanthates waste used in flotation process from copper mining companies, which are organosulfur compound useful in the extraction of certain ores, in this case sulfide ore. These wastes are often deposited on tailings - pile in areas surrounding the mine site, and sometimes they drain-down. For this reason, it will be useful to degrade them instead of being deposited. A very promising high oxidation technique based on electrical discharges will be used as the key method for the degradation of xanthate in this research project. This presentation will overview both oxidation technique as well as the plan to conduct preliminary studies of feasibility for the degradation of xanthate in aqueous solution. Preliminary results of the investigation will be discussed.

(337d) Pursuing a Viable Plasma-Based Water Treatment Process: Identifying Transport Limitations and Investigating the Effect of Reactor Design On the Degradation of Bisphenol A

Mededovic, S., Clarkson University
Bellona, C., Clarkson University
Dai, F., Clarkson University
Franclemont, J., Clarkson University
Holsen, T., Clarkson University
Stratton, G., Clarkson University
Zhang, C., Clarkson University
Originally presented on: 11/5/2013 16:21:00 - 16:43:00

To improve the feasibility of plasma-based water treatment technology, a study was conducted to identify and characterize the design parameters that influence treatment effectiveness and the physical phenomena that determine treatment efficiency. A compound's octanol-water partition coefficient (i.e., hydrophobicity) has been determined as the major factor that influences the compounds' relative ability to diffuse into the plasma channel. Experiments conducted with nine different solutes have shown that hydrophilic compounds tend to stay in the bulk liquid whereas hydrophobic compounds such as Bisphenol A (BPA) readily diffuse into the plasma channel where they react with OH and other radicals. Thus the electrical discharge as an advanced oxidation technology appears to be the most effective for the degradation of compounds with limited solubility in water.

The effects of several reactor design parameters were investigated, including electrode geometry, electrode spacing, reactor geometry, and the ground electrode plate area, on the degradation of BPA. The hybrid-series reactor in which the high voltage (plasma) electrode is in water and the grounded electrode is above the water's surface was about four times as effective as the reference configuration with both electrodes in water. The reactor geometry, particularly the distance between the high voltage electrode and the reactor wall, was observed to have a major impact on the characteristics of the electrical discharge, which in turn influenced the BPA removal rate. Tripling the reactor radius yielded a threefold increase in the peak current and thus the power deposited into plasma. Similarly, the area of the grounded electrode plate was found to affect the discharge characteristics with a twofold increase in the plate area corresponding to an increase of about 40% in the peak current and input power. The distance between the electrodes was identified as the key parameter. It was found that the optimum distance depends on the input voltage and solution conductivity, but is generally near the minimum distance at which no sparking between the electrodes occurs. At the same input power of 50 W, the optimization increased the extent of removal of a 5 mg/L BPA solution from 32% to 94% in less than one hour.

(337e) Enhanced Photocatalytic Degradation of Malachite Green By Visible Light Sensitive ZnO/ZrO2 Doped TiO2 Nanoparticles

Shende, A., South Dakota School of Mines and Technology
Shende, R. V., South Dakota School of Mines and Technology
Originally presented on: 11/5/2013 16:43:00 - 17:05:00

The large energy band gap of TiO_2 permits activation only in the UV portion of the solar radiation. To obtain visible light sensitive material ZnO/ZrO_2 doped TiO_2 nanoparticles were synthesized by the sol-gel method using Pluronic 123 surfactant. The structure and surface morphology of the photocatalyst was studied using X-ray diffraction, HRTEM, BET and diffuse reflectance spectroscopy. Photodegradation rates of malachite green (MG) in solar radiation using the $ZnO/ZrO_2/TiO_2$ and P-25 TiO_2 were compared. Effect of pH, catalyst load and dye concentration on the photoconversion efficiency were determined. The degradation products were identified by GC-MS. The toxicity of the degradation products were tested using E.coli.

(337f) Endogenous Hormones Degradation Using Nanotubular Oxide Layer Grown On Tiw Alloy By Electrochemical Anodization

Rodrigues, C. A., Universidade Federal de Sao Paulo
Oliveira, M. E.,
Barroso, B. L.,
Gomes, M. L. M.,
Bertazzoli, R., Universidade Estadual de Campinas
Moraes, M. D. L. L.,
Originally presented on: 11/5/2013 17:05:00 - 17:27:00

Advanced Oxidative Processes (AOP) are a significant green technology for application in water purification and organic pollutant degradation. The main mechanism of AOPs is the generation of highly reactive free radicals, such as the hydroxyl radical (•OH). Among different available AOPs, heterogeneous photocatalysis seems to be one of the most popular technologies for wastewater treatment. This process uses semiconductor materials as catalysts and UV light. The efficiency of the electrochemical and photoelectrochemical treatment is strongly dependent on the choice of the anode material and its morphology. The manufacture of Ti and Ti alloys oxide nanotube arrays by electrochemical methods has attracted a great deal of attention for applications in environmental photocatalysis due to its photoinduced reactivity[1]. In this work it is proposed the degradation of endogenous hormones, estrone (E1) and 17α-ethinylestradiol (EE2) by photoelectrocatalysis process using self-organized oxide semiconductor nanotubes on a titanium/tungsten alloy applying an anodization process. The growth of self-organized oxide semiconductor nanotubes on a Ti-x%W alloy (x= 0.5, 2.5 and 5.0% wt) was investigated as a function of the applied potential, electrolytic solution (aqueous or organic) and anodization time. The titanium alloy was prepared in an arc melting furnace using a non-consumable tungsten electrode and water-cooled copper crucible in a high purity argon atmosphere. The samples were solidified in an ingot form that was submitted to a rolling process to manufacture the alloy sheet X-ray fluorescence was used to determine the chemical composition of the alloy. The anodization experiments were carried out with a DC power supply and an acrylic electrochemical cell, with a platinum cathode and the Ti alloy anode separated by 2.5 cm. An aqueous solution of 0.2 M HF and two types of ethylene glycol solutions (0.2 M NH_4F and 0.2 M HF) were used as electrolyte solution. All the anodization experiments were carried out at 120 V, with applied time that varied between 5 to 120 minutes. SEM-FEG was used to characterize the nanotubes and X-ray fluorescence results confirmed that the composition of the alloys was as required. The best results of anodization were obtained when 0.2 M HF with ethylene glycol solution was used with the Ti-0.5w%W electrode, with an anodization time of 60 minutes at 120 V. Under these conditions the oxide nanotubular layer was homogeneous and continuous through the surface. After anodization, the nanotubular oxides on Ti0.5%W were annealed at temperature greater than 450°C in air atmosphere for 2 hours and then, cooled back to room temperature in a box furnace to obtain a crystalline structure. X-ray diffraction was used to characterize the crystallization behavior of oxide layer. The degradation of endogenous hormones was carried out in a photoelectrochemical cell. A Ni screen was used as a cathode, and the self-organized oxide nanotube on titanium alloy produced were used as anode. The efficiency of these treatment processes was monitored by following the TOC (Total Organic Carbon) and the E1 and EE2 concentration decay by HPLC technique (High Performance Liquid Chromatography). UV lamp was positioned to illuminate the anode electrode in order to induce the photodegradation of the organic compounds. The percentage of degradation and total organic carbon removal, and the energy consumed per volume of effluent treated were investigated during electrolysis in photoelectrochemical cell.

Acknowledgments:

CNPq (Proc. 483285/2011-0), Fapesp (Proc. 2011/51226-3), LNLS and UNICAMP

References:

[1] Nah, Y.C.; Ghicov, A.; Kim, D.; Berger, S.; Schmuki, P. *Electrocatalysis* 12 (2008) 16154 – 16155

(337g) Flame Aerosol Synthesized Cr Incorporated TiO2 for Visible Light Photodegradation of Acetonitrile

Inturi, S. N. R., University of Cincinnati
Boningari, T., University of Cincinnati
Suidan, M., University of Cincinnati
Smirniotis, P., University of Cincinnati
Originally presented on: 11/5/2013 17:27:00 - 17:49:00

The photocatalytic activity of flame aerosol made TiO_2 loaded with various amounts of Cr was studied as a catalyst for both liquid and gas phase photodegradation of acetonitrile (ACN) under visible light (400–800 nm). It was found that the optimal concentrations of Crand TiO_2 exist (Ti/Cr atomic ratio = 40) for the efficient oxidation of ACN in the gas phase.XRD patterns showed a decrease in the anatase phase with increase in the amount of Cr loading. Temperature programmed reduction (TPR) and UV–vis spectra profiles clearly indicated that the Cr^{6+} phase is dominant in the samples with the atomic ratio of Ti/Cr equal to 50, 40, and 30. The rest of the catalysts with Ti/Cr ratios of 100, 75, 20, and 10 were found to possess Cr^{3+} species along with Cr^{6+} species. Our XPS results illustrated that the relative atomic percentage value of Ti^{3+}/Ti^{4+} characterized by XPS was significantly high for the Cr/TiO_2 nanoparticles with Ti/Cr = 40 atomic ratio (Ti^{3+}/Ti^{4+} = 1.14, 42.1%), whereas, other samples demonstrated low atomic percentage value of Ti^{3+}/Ti^{4+} (Ti^{3+}/Ti^{4+} = 0.18-1.05). Moreover, Cr interacts with TiO_2 nanostructure in the interface of flame-made nanoparticles, bulk Cr oxide exists over the surface of TiO_2 nanostructure.The photodegradation ability of TiO_2/Crcatalyst with Ti/Cr atomic ratio of 40 was highly related to the existence of Cr^{6+}species which strongly interacted with TiO_2.Among all the catalysts tested, the system with Ti/Cr atomic ratio 40 demonstrated a superior catalytic performance with the rate constant of 0.812 $m^3g^{-1}mol^{-1}$ under visible light irradiation.The proposed route of the catalytic activity of the above material in visible light involves the reaction of dopant level electrons withsurface Cr^{6+}, which makes available valence band holes to performoxidation reactions.The photocatalytic performance of flame aerosol synthesized Cr incorporated TiO_2 material in both liquid and gas phase will be presented in the persentataion.

(342a) Immobilization of Recombinant Antibody Fragment Utilizing Material-Binding Peptides for Sensitive Immunoassay and Immuno-Sensors

Kumada, Y., Kyoto Institute of Technology
Kishimoto, M., Kyoto Institute of Technology
Originally presented on: 11/5/2013 15:20:00 - 15:40:00

Recombinant antibody fragments such as Fab, scFv and VHH are recently considered as next generation of diagnostic reagents because they can be immobilized with high density on the surfaces of materials, and consequently, antigen molecules would be efficiently captured by these ligand antibodies. However, such antibody fragments have not been utilized as a ligand antibody in enzyme immunoassay (EIA) because of difficulty in site-specific immobilization of antibody fragments with high remaining antigen-binding activities due to conformational change of antigen-binding domains after immobilization. In order to immobilize recombinant antibody fragments with high remaining activity, site-specific attachment of antibody fragments through the material-binding peptides is necessary. In this study, we originally identified material-binding peptides that can strongly interact with the surface of polystyrene (PS), polymethyl methacrylate (PMMA), and silicon nitride (SiN). By introducing these peptides to the C-terminus of recombinant antibody fragments, it was possible to increase both density as well as remaining activity of the fragments after passive adsorption to the solid materials. We also investigated efficient refolding method for peptide-tagged antibody fragments and thus, all of peptide-fused antibody fragments could be refolded with over 90 % of recovery, followed by site-specific immobilization. Our developed methods for refolding and immobilization of material-binding peptide-fused antibody fragments are considerably useful for preparation of high-performance of antibody-immobilized solid-support for immunoassay and immune-sensors.

(342b) Biomagnetic Nanomaterials for High Gradient Magnetic Affinity Separation

Lee, G. U., University College Dublin
Fields, C., 1School of Chemistry and Chemical Biology, University College Dublin
O'Mahony, J., UCD
Matias, V., UCD
Li, P., University College Dublin
Smith, L., UCD
Muzard, J.,
Originally presented on: 11/5/2013 15:40:00 - 16:00:00

The growth of biopharmaceutical and biomedical industries has created a demand for new technologies for purification of genetically engineered proteins. High gradient magnetic separation (HGMS) is a technology that has the potential to increase the speed of traditional affinity chromatography due to its use of an expanded bed of superparamagnetic microparticles (SPM). The efficiency of large-scale HGMS is greatly improved if SPMs offering high binding capacity, high magnetic susceptibility, and high magnetization are employed. In this communication, we will describe strategies for synthesizing SPMs that are composed of biomolecules and viruses assembled on a superparamagnetic core. Separation of specific proteins and antibodies from high-protein concentration solutions was performed in a rapid and single step. The dense SPM core of these particles makes them highly responsive to magnetic fields and the biomolecular coating produces a high binding capacity. These new biomaterials and the described technology appear to be well suited to large-scale HGMS separation and promise to be cost effective.

(342c) Theoretical Investigations of Multiple Weak Interactions in Multimodal Chromatography Using Molecular Dynamics Simulations and Protein Surface Characterization

Parimal, S., Rensselaer Polytechnic Institute
Woo, J. A., Rensselaer Polytechnic Institute
Srinivasan, K., Rensselaer Polytechnic Institute
Holstein, M. A., Rensselaer Polytechnic Institute
Garde, S., Rensselaer Polytechnic Institute
Cramer, S. M., Rensselaer Polytechnic Institute
Originally presented on: 11/5/2013 16:00:00 - 16:20:00

Multimodal (MM) chromatography has emerged as an extremely promising form of chromatography. Recent studies published by our group and others have shown that an appropriate choice of ligand chemistry and fluid phase modifiers can modulate the different modes of interactions present in MM systems to achieve enhanced selectivities as compared to traditional single-mode chromatography. However, exciting as the experimental results might be, there is still a lack of mechanistic details in these MM systems which is essential to optimize desired separations. The goal of the work presented here is to provide fundamental understanding of the nature of water-mediated protein-ligand interactions at a molecular level which can provide significant insight into the design of MM ligands, the roles of synergy and the modulation of selectivity using fluid phase modifiers (FPMs).

A combination of theoretical techniques has been used to understand MM interactions. Molecular Dynamics (MD) simulations have been employed to gain molecular insights into how multiple weak interactions in MM systems work together to create binding selectivity. All-atom explicit MD simulations have been performed with two different MM ligands interacting with proteins and the results are analyzed to understand binding preferences of these MM ligands in terms of the interaction chemistry and their conformational flexibility. Simulations have also been performed with different fluid phase modifiers to understand how co-solutes affect protein-ligand binding and to shed light on the different mechanisms which govern their behavior.

Protein surface characterization techniques have also been employed as a fast, coarse-grained approach to map proteins based on their potential to interact with different modes. Protein-ligand binding in MM systems involves a combination of electrostatic and hydrophobic interactions. To accurately measure the individual contributions of both these modes, and their possible synergistic effects to the overall binding behavior requires the characterization of the molecules involved in binding in terms of their electrostatic and hydrophobic potential. While electrostatic interactions between two molecules can be easily calculated, hydrophobicity has been difficult to understand. Hence, hydrophobic characterization of proteins has been carried out here using recently developed methods which incorporate the "context-dependence" of hydrophobicity. The results from these calculations, together with the molecular understanding gained from MD simulations, have been used to explain chromatographic retention behavior of proteins in MM systems on different resins and under different fluid phase conditions.

The work presented here demonstrates the use of theoretical techniques, especially MD simulations, to probe atomistic details of protein-ligand binding which can shed light on and indeed predict protein retention behavior in chromatographic systems. Finally, the knowledge base created using these simulations can be used to select appropriate combinations of MM ligands and modifiers to achieve unique selectivities for challenging protein separations.

(342d) Development of Adsorptive Membranes With Glycoligands for Targeted Lectin Binding

Chenette, H. C. S., Clemson University
Husson, S. M., Clemson University
Originally presented on: 11/5/2013 16:20:00 - 16:40:00

This presentation will describe our latest efforts to develop adsorptive membranes for use as protein purification media. The design approach used in these studies is to graft polymeric 'tentacles' from macroporous regenerated cellulose membranes by atom transfer radical polymerization (ATRP). Using one or more examples, we will illustrate how this design approach can be used to prepare high-productivity membrane adsorbers. In one case, we will discuss affinity membrane adsorbers that incorporate glucose-containing glycoligands. Studies have shown glycoligands selectively bind specific lectins [Yang et al., *Appl. Mater. Interfaces*, **2010**, *2*] mimicking the gluco-receptors found on cell surfaces that recognize and bind bacteria. The literature also reports advances in this field using model substrates to elucidate optimal polymerization conditions, and kinetic studies relating polymerization time to the degree of polymerization [Yang and Ulbricht, *Macromolecules*, **2011**, *44*]. The present study integrates these glycoligand adsorbers into a membrane system that is capable of scale-up to a column for chromatographic separation. A glucose-based ligand was grafted from regenerated cellulose membrane surfaces using ATRP. The model lectin, Concanavalin A, was used to evaluate membrane performance in bind-and- elute purification, using a low molecular weight sugar for elution.

(342e) Overload and Elute Chromatography for Enhanced MAb Purification

Nadarajah, D., Genentech, Inc.
Hohwald, S., Genentech, Inc.
Banerjee, D., Genentech, Inc.
Shomglin, K., Genentech, Inc.
St. John, R., Genentech, Inc.
Mehta, A., Genentech, Inc.
Originally presented on: 11/5/2013 16:40:00 - 17:00:00

Bio-pharmaceutical process development has increasingly been relying on platform processes that reduce development timelines and time to clinic. However, certain MAbs may not fit into the platform process thus requiring further development. This talk will focus on a set of MAbs that bind to the anion exchange (AEX) resin under flow-through conditions. To enable plant fit, improve yield and achieve adequate impurity clearance, a novel mode of chromatography was developed with a multimodal resin. The resin was operated in an overload mode with a step elution enabling significantly higher load density while obtaining the desired impurity clearance. Data on clearance of model viruses, host cell proteins and aggregates will be presented. High throughput screening was employed to rapidly identify process parameters for overload and elute chromatography. Data generated to develop a mechanistic understanding of product and impurity binding to the resin under the overload and elute mode of chromatography will also be presented.

(342f) Development of Production Method for Zeaxanthin By Fermentation Using Paracoccus Carotinifaciens

Nagai, H., Ajinomoto Co., Inc.

Originally presented on: 11/5/2013 17:00:00 - 17:20:00

Zeaxanthin is a yellow carotenoid contained in Corn, Paprika, Spinachi, Chinese matrimony vine, hen egg and so on. In human body, this carotenoid exists especially in the center of macular region on retina.

Age-related macular degeneration (AMD) causes the reduction of the content of Zeaxanthin on the macular of patients. Also, many researchers reported that most of these kinds of eye diseases are caused by oxidative stress. Since Zeaxanthin is a strong antioxidant, we can imagine that this carotenoid has a certain kind of prophylactic effect on human eyes.

In many situations, supplements are generally used for the uptake of Zeaxanthin into human blood, because the general foods contain the less content of Zeaxanthin. However, this process generally shows very low absorption efficiency while the concentration of zeaxanthin in the blood can be kept at high level (0.48□ 1.09μM). Since supplements are not inexpensive, this can be a barrier for the people who want to prevent falling to those eye diseases.

□However, it is possible to consider hen egg as a supply source of Zeaxanthin. In this case, Zeaxanthin concentration in the blood is very low(0.06□0.15μM), while absorption efficiency is very high. If we can increase the content of Zeaxanthin in hen egg, hen egg can be a candidate of the most efficient for Zeaxanthin supply for human.

Already some developers have used spirina or Chinese matrimony vine to produce Zeaxanthin enriched egg. However, the maximum its concentration of Zeaxanthin was less than 3.35mg/100g. In order to solve this provlem, we developed the dried Paracoccus cell containing high content of Zeaxanthin (Panaferd-ZX®). Addition of Panaferd-ZX into feed for hen shows higher content of Zeaxanthin in hen egg (3.5mg□ 10mg/100g).

In order to develop the production method, we selected some strains producing Zeaxanthin from mutant Astaxanthin production strains. Furthermore, we have developed fermentation process using these strains.

For the scale-up of fermentation process, we have faced to insufficient cell growth. From laboratory scale analysis, we have found that these strains showed insufficient growth under the pressure ranged from 0.02MPa to 0.08MPa. Since the large scale fermenter has the pressure up to 0.1MPa by the level of fermentation broth, it is possible that this is the reason of unsuccessful fermentation. To solve this problem, we examined the additives for fermentation media. Finally, we have found that citric acid can promote the cell growth under the pressure. This enabled us successful scale-up to 5kL scale fermenter and trial sales.

(342g) Molecular and Pore Diffusion Coefficients of DNAs and Modified Proteins for Process Chromatography

Ohshima, S., Yamaguchi University
Yoshimoto, N., Yamaguchi University
Yamamoto, S., Yamaguchi University
Originally presented on: 11/5/2013 17:20:00 - 17:40:00

As the performance of chromatography is governed by pore diffusion, it is important to know the pore diffusion coefficient especially for separation of large bioproducts such as proteins and DNAs.

However, compared with the diffusion coefficient of globular proteins (molecular weight 10,000- 100,000), the diffusion coefficients of DNAs and modified proteins have not been measured extensively. Consequently, it is difficult to predict the diffusion coefficient in the solution (molecular diffusion coefficient, D_m), which is needed for understanding the pore diffusion coefficient. D_s.

In this study, D_m was determined by the Taylor method, which uses a pulse response curve from a narrow tube at laminar conditions. Single strand and Double stranded DNAs were used as the sample. D_s was determined also by using a pulse response experiment with agarose-based ion-exchange chromatography particles, Sepharose HP under non-binding conditions.

D_m of DNA was correlated with the molecular weight M_W as follows.

Single strand DNA : $D_m = 1.94 \times 10^{-8} M_w^{-0.57}$

Double strand DNA : $D_m = 1.33 \times 10^{-8} M_w^{-0.49}$

Compared with the established correlation for proteins (D_m is proportional to $M_w^{-0.33}$), D_m changes much more strongly with M_W. This is because of the shape of DNA molecules.

D_s of single stranded DNA was proportional to $M_w^{-0.99}$, which indicates that diffusion of single strand is highly hindered in the pore. This also suggests that separation of DNA molecules with conventional porous particles is difficult and the use of convection-aided media such as monoliths is preferred.

(357a) Tower Scanning With Frothview™ - A Quantitative Tool for Diagnosing Trays

Pless, L., Tracerco
Nieuwoudt, I., Koch-Glitsch
Originally presented on: 11/5/2013 15:15:00 - 15:45:00

The tower gamma scanning technique is well established in the process industries as a qualitative tool to help troubleshoot separation towers, both trayed and packed towers. In this presentation it will be shown that through careful and innovative application of the gamma scanning technique, quantitativeinformation about trays in separation towers can be extracted.

For trayed towers this will be illustrated by comparing data taken under very controlled conditions on a hydrocarbon pilot plant tower equipped with windows, with the data obtained from scans of the tray active area. Three sets of trays, designed to flood under jet flood, downcomer backup or choke flood mechanisms, were tested and studied. The experimental observations, tray model predictions, and scan data for each of these sets of trays under various operating conditions will be compared. Based on this information it can be concluded that tower scanning can be used to generate quantitative information about tray performance.

(357b) Refinery Vacuum Towers - State of the Art

Sloley, A. W., CH2M HILL

Originally presented on: 11/5/2013 15:45:00 - 16:15:00

Refinery vacuum towers present difficult challenges for mass transfer internals. They combine severe mechanical constraints, difficult process conditions, and large economic consequences of being wrong in any direction. This is a service that often lacks a clear-cut conservative or safe answer. This paper reviews the state of the art for internals in refinery vacuum columns. Equipment must meet both service and mechanical requirements.

Services commonly include heat transfer (pumparounds), de-entrainment (feed entry and overhead) and stripping (trays). Less often, fractionation services may be included. In modern vacuum towers the dominant heat-transfer and fractionation choices include grid-packing and structured packing. The major driving force is the low pressure drop of the structured packing. Additional benefits also come from the good performance of packing at low liquid loads.

Mechanical difficulties come from large equipment diameters and severe mechanical loads during upsets.

Auxiliary equipment to make the packing works include collectors, liquid distributors, and vapor distributors. These must also meet the service and mechanical requirements imposed by the process.

(357c) Performance of Sieve Trays in a Heat Integrated Concentric Column

Olujic, Z., Delft University of Technology

Originally presented on: 11/5/2013 16:15:00 - 16:45:00

Results are presented on hydraulic and mass transfer performance of sieve trays in an internally heat integrated concentric distillation column (HIDiC), with hetat transfer devices installed above the active area of annular striping section trays. Experiemnts were performed with and without heat transfer between inner rectification section and outer stripping sections, indicating that annular stripping section trays perform in such a situation similar to conventional single path cross flow trays.

Results are presented on hydraulic and mass transfer performance of sieve trays in an internally heat integrated concentric distillation column (HIDiC), with hetat transfer devices installed above the active area of annular striping section trays. Experiemnts were performed with and without heat transfer between inner rectification section and outer stripping sections, indicating that annular stripping section trays perform in such a situation similar to conventional single path cross flow trays.

Results are presented on hydraulic and mass transfer performance of sieve trays in an internally heat integrated concentric distillation column (HIDiC), with hetat transfer devices installed above the active area of annular striping section trays. Experiemnts were performed with and without heat transfer between inner rectification section and outer stripping sections, indicating that annular stripping section trays perform in such a situation similar to conventional single path cross flow trays.

(357d) "Distillation Tray Efficiency Prediction: A Simpler Approach."

Garcia, A., Sulzer Chemtech USA Inc.

Originally presented on: 11/5/2013 16:45:00 - 17:15:00

In the last 30 years, the prediction of distillation tray efficiency has been extensively modeled, empirically and fundamentally. The efforts have been mostly focused on cross-flow sieve trays, and those models have been extended to trays with different contacting devices.

Experimental tray efficiency data from pilot and semi-industrial columns for several distillation systems will be used to present a simpler approach to estimate tray efficiency under distillation conditions.

The main goal is to present a simpler approach that process engineers can rely on to complete preliminary distillation column designs and/or to analyze industrial distillation columns mass transfer performance.

(357e) Dan the Tray Man

Resetarits, M. R., Fractionation Research, Inc.
Cai, T. J., Fractionation Research, Inc.
Chambers, S., Fractionation Research, Inc.
Originally presented on: 11/5/2013 17:15:00 - 17:45:00

Mr. Dan Summers' contributions to the distillation world are immeasurable. After graduating from SUNY at Buffalo, in 1977, Dan joined the Union Carbide distillation group in Charleston, West Virginia. In 1980, he relocated to Tonawanda, New York, remaining in Union Carbide but with their Linde Division. In 1988, Union Carbide's Process Equipment group was acquired by UOP and UOP soon named him as the Tonawanda Engineering Manager. In 1995, Dan moved to Houston to join Stone and Webster. In 1997, he moved to Tulsa to join Nutter Engineering, which was acquired by Sulzer Chemtech in late 1998. Dan is still in Tulsa with Sulzer Chemtech USA, Inc.

Mr. Dan Summers is highly proficient with process simulation, thermodynamics, distillation trays, column revamping and troubleshooting. While with UOP, their Process Equipment sales increased by a factor of 10. He is equally comfortable with crossflow, counterflow and cocurrent flow trays. Dan has written many articles and presented many papers, 52 to be exact, and the topics included the following: Beer columns, olefin towers, tray stability, 3-pass and 4-pass balancing, reboilers, fouling services, tray vibrations, push valves, V-Grid trays, dry tray pressure drops, performance diagrams and low-surface tension designs. Many of the articles/presentations were coauthored by engineers from operating companies who benefitted from Dan's distillation expertise.

Mr. Dan Summers is a PE in the states of New York and Oklahoma and he is listed on 2 US patents (one on bubble cap trays and one regarding fixed valve trays). He is the Chairman of FRI's Design Practices Committee and is a Fellow of AIChE. In addition, he co-chaired the recent rewrite of AIChE's Tower Testing Procedures.

(385b) Use of Single-Pass Tangential Flow Filtration for Decoupling Upstream and Downstream Processing

Brinkmann, A., Biogen Idec
Westoby, M. M., Biogen Idec
Elouafiq, S., Biogen Idec
Originally presented on: 11/5/2013 15:40:00 - 16:05:00

Higher cell culture productivity has led to a bottleneck during downstream purification. Various ways have been explored to decouple the upstream and downstream processes to allow for more flexible processing and enable the more efficient use of existing facilities. In this presentation, we will explore the feasibility of using tangential flow filtration to provide decoupling by concentrating and freezing centrate. However, traditional fed-batch tangential flow ultrafiltration (TFF) can be limited due to the size and minimum working volume of TFF systems. Single-pass tangential flow ultrafiltration (SPTFF) provides a viable solution by providing high concentration factors without the need of a recirculation vessel. This reduces overall system size and hold-up and enables concentration in-line with other operations. The SPTFF process was developed at bench scale to determine the appropriate operationg conditions for the desired concentration facter, and then demonstrated at pilot scale to show scalability and evaluate the effect on the downstream purification steps and product quality.

(385c) Clarification of Succinic Acid Fermentation Broth By Ultrafiltration and the Membrane Fouling Mechanism

Wang, C., Institute of Process Engineering, Chinese Academy of Sciences
Wan, Y., Institute of Process Engineering, Chinese Academy Of Sciences
Xing, J., Institute of Process Engineering, Chinese Academy of Sciences
Originally presented on: 11/5/2013 16:05:00 - 16:30:00

Ultrafiltration was investigated to clarify succinic acid fermentation broth in consideration of integrating fermentation and separation process and removal the product in situ. Different membranes(PES 100kDa, PES 30kDa, PES 10kDa and RC 10kDa) were used and two models were applied to analyze the fouling mechanism. Results indicate that ultrafiltration is feasible in clarifying succinic acid fermentation broth. Almost all the microorganism cells (99.6%) were removed from the fermentation broth. Proteins were also removed effectively by all membranes selected in this study. The removal rate was 79.86% for PES 100kDa, 86.43% for PES 30kDa, 86.83% for PES 10kDa, and 80.06% for the RC 10kDa. After ultrafiltration, clearer permeate was obtained comparing with the centrifugation. Among all tested membrane, PES 10kDa gave the best results which showed relative higher initial flux (18.86 L/m2 h), highest protein removal rate (87%), and less flux drop rate. Resistance-in-series model was applied to determine the main factor that caused the operation resistance. Results showed that most membranes selected in this study tended to be fouled by cake layer or concentration polarization. Hermia's model, which is composed of four individual sub-models, was used to analyze the predominant fouling mechanism for the used membranes. Results showed that the fouling of RC 10kDa and PES 30kDa was controlled by the complete blocking mechanism, while PES 100kDa was controlled by the intermediate blocking and PES 10kDa was controlled by cake layer. This conclusion was also proved by SEM photos. Membrane characteristics were analyzed before and after ultrafiltration by AFM and goniometer. Both contact angle and roughness of most membranes increased after ultrafiltration.

(385d) Effect of Filtration On the Concentration of Aqueous Ions in Process Water and Filter Cake Moisture During Iron Ore Beneficiation

Haselhuhn, H. III,
Kawatra, S. K., Michigan Technological University
Halt, J., Michigan Tech University
Originally presented on: 11/5/2013 16:30:00 - 16:55:00

The filtration processes within selective flocculation and dispersion iron ore beneficiation plants cause aqueous ions from the process water to concentrate within the filter cake moisture. These high concentrations of ions can lead to significant operational issues during downstream pelletization and reduction processes. This paper examines the ionic content of concentrated iron ore slurry that is filtered at an operating selective flocculation and dispersion hematite beneficiation plant and relates it to the ionic content of the moisture in the resultant filter cake. Water analyses were conducted to determine the sodium, potassium, calcium, magnesium, fluoride, chloride, nitrate, nitrite, carbonate, sulfate, and phosphate concentrations within the filter feed slurry and filter cake moisture of an operating hematite beneficiation plant. Results have shown that all tested ions within the process are more concentrated within the filter cake moisture than in the filter feed slurry. This has been attributed to the physical and chemical surface interactions of these ions with the iron ore particle surfaces. Of these ions, phosphorus concentrates the most with the filter cake moisture content having a phosphorus content that is 24.67 times that of the filter feed slurry.

(391m) Purification of 2-Amino-1-Phenylethanol Enantiomers By Stripping Crystallization

Shiau, L. D., Chang Gung University
Teng, H. G., Chang Gung University
Liu, K. F., Chang Gung University

Stripping crystallization (SC) is a new separation technique which combines distillation and crystallization. Basically, SC is operated at a triple-point condition, in which the liquid is simultaneously vaporized and crystallized due to the three-phase equilibrium. In essence, the process is continued until the liquid phase is completely eliminated and only the pure solid crystals remain in the feed. Thus, crystal washing is not required since only pure solid crystals remain in the feed and no impurities are adhered on the crystal surfaces at the end of SC. In the previous research, SC has been successfully applied to separate p-xylene from the mixed xylenes by Shiau et al. [Shiau *et al.*, 2005; 2006; 2008].

2-Amino-1-phenylethanol (2-APE) is an important basic skeleton in many adrenergic drugs and is often used as ingredients of cold medicines and anorectics. As the biological activity and pharmacokinetics of the enantiomers of a chiral drug can be different, the objective of this research is to apply SC to purify R-2-APE from the chiral 2-APE mixture. The results can provide important information in the pharmaceutical industry. A thermodynamic model is developed to simulate the three-phase equilibrium during the SC operation and to direct the batch SC experiments. The experiments show that no crystallization occurs during SC without seeding due to subcooling of the liquid enantiomers. On the other hand, seeding can eliminate subcooling during SC, leading to the R-2-APE crystals gradually formed preferentially on the seeds. Thus, seeding with magnetic stirring can be successfully employed to purify R-2-APE from the 2-APE enantiomers. The experimental purity of the resulting R-2-APE product is higher than predicted by the model although the experimental recovery ratio is slightly lower than the simulated recovery ratio.

References:

1. Shiau, L. D., Wen, C. C. and Lin, B. S. Separation and purification of p-xylene from the mixture of m-xylene and p-xylene by distillative freezing, *Ind. Eng. Chem. Res*, **44(7)**, 2258-2265, 2005.

2. Shiau, L. D., Wen, C. C. and Lin, B. S. Application of distillative freezing in the separation of o-xylene and p-xylene, *A.I.Ch.E. Journal*, **52(5)**, 1962-1967, 2006.

3. Shiau, L. D., Wen, C. C. and Lin, B. S. Separation of p-xylene from the multicomponent xylene system by stripping crystallization, *A.I.Ch.E. Journal*, **54(1)**, 337-342, 2008.

(391a) Selective Adsorption of Various Proteins By Cationic Spherical Polyelectrolyte Brushes

Wang, S., East China University of Science and Technology
Chen, K., Shanghai Jiaotong University
Li, L., East China University of Science and Technology
Guo, X., East China University of Science and Technology

Cationic spherical polyelectrolyte brushes (SPBs) were synthesized by photo-emulsion polymerization, consisting of a polystyrene core with a diameter around 80 nm and a poly(2-aminoethylmethacrylate hydrochloride) (PAEMH) shell with a thickness from 10 to 50 nm densely grafted on the core surface. Adsorptions of three proteins bovine serum albumin (BSA), β-lactoglobulin (BLG), and papain onto SPBs were observed and compared by turbidimetric titration, dynamic light scattering (DLS), zeta potential measurement, and isothermal titration calorimetry (ITC). Adsorption, aggregation, and desorption of proteins by SPB can be tuned by modulating pH value. The pH regions of adsorption for proteins BSA, BLG, and papain onto SPBs are markedly different and tunable by changing ionic strength and stoichiometry between protein and SPB. The adsorption energetic, affinity, and amount of various proteins onto cationic SPBs were determined by ITC. These findings can lay the foundation for SPB applications in protein separation and selective adsorption of proteins with very similar structures.

(391b) Ambient-Temperature Fusible Filter: A Preliminary Experiment and a Proposal for a Filtration Process

Nishioka, T.,
Teduka, M., Hokkaido University

To the best of our knowledge, there has been no report on the use of a fusible filter material (*FF*) for filtration apparatuses apart from our previous proposal (WIPO2010087055). We have been studying the use of ice crystals as an *FF*. In that study, squid oil contained in the visceral organ was separated from the oil emulsion. Here we briefly report the possibility of using an ambient-temperature fusible filter medium (*ATFF*) as a *FF*. An important characteristic with *ATFF* is that the melting or solidification process can be carried out within an ambient temperature range, resulting in a simpler and lower (in terms of running and capital) cost *FF* apparatus. A preliminary experiment was performed using trimethylolethane hydrate (TME-$3H_2O$) as an *ATFF*. The sieved TME-$3H_2O$ crystals formed the filter-bed. Rapeseed oil/water/surfactant-protein emulsion was artificially made as an alternative biomaterial, and centrifugal filtration was performed for 5 min at 400 G. 86% of oil (transparent) was recovered from the oil-contained *ATFF* by melting (heating) TME-$3H_2O$ at a temperature greater than the solidification point of 30 degree Celsius for the filter medium followed by gravity separation. For biomaterials, both *ATFF and FF* may solve a problem, integration of protein-coated oil emulsion droplets, which is accompanied with widely used method (sedimentation centrifuge). The reason seems that *ATFF or FF* has a coalesce function as deduced from the transparent state of the harvested oil. Recycle of both *ATFF* and *FF* may also solve an environmental problem that are associated with diatomaceous earth disposal.

(391n) Experimental and Simulation Study of a Universal Procedure for Crude Glycerol Purification From Different Feedstocks in Biodiesel Production

Xiao, Y., Purdue Univeristy
Xiao, G., Southeast University
Varma, A., Purdue University

With increasing supply and decreasing price, from the economic point of view, it is important to utilize crude glycerol, the main by-product of biodiesel production, to manufacture high value-added chemicals. Since crude glycerol typically contains less than 65 wt % glycerol and a variety of other components, its purification is the first step for its utilization. Owing to the wide variety of triglycerides, alcohols, catalysts and separation processes used in biodiesel production, the composition of crude glycerol from different biodiesel production methods varies widely. For this reason, glycerol purification processes are also generally different for the various crude glycerol feedstocks. In the present work, we have developed a universal procedure for crude glycerol purification, and utilized it for crude glycerol samples from two biodiesel production companies, experimentally upgrading both to >94 wt % purity. On Aspen Plus software platform, the purification procedure was simulated using a process model based on modeling of phase equilibrium behavior and modeling of NaCl solubility in ethanol-glycerol mixtures, to obtain good match with the experiments. Using the same model, two other glycerol purification processes described in the literature were also simulated and high purities were also obtained according to the simulation results. We thus demonstrate that the developed universal procedure is suitable for purification of crude glycerol obtained from different biodiesel production technologies.

(391c) CFD Simulation of Bubble Enhanced Hydrocyclone for Air-Water System Encountered in De-Oiling

Luo, X., Tianjin University
Sun, J., Tianjin University

De-oiling hydrocyclone has been used for years for its high capacity and separation efficiency. Recently, the applications are reported of air bubble enhanced hydrocyclone in removal of emulsified oil from water. This method, integrating hydrocyclone with flotation, is also worthy of consideration in biomass separations, such as algae harvest, especially oriented to bio-diesel collection. With the help of micro bubbles, the separation efficiency of oil increased about 15% at the reportedly optimal air-liquid ratio for this purpose. In this work, a 3D CFD model was built to investigate the flow patterns of water and air bubbles in a 35mm bubble enhanced hydrocylone. The simulated velocity distribution was compared with experimental data and it showed a good agreement. It was found that the air fraction at underflow increased and the bubble removal became lower when the air-liquid ratio increased from 0.025% to 0.3%. This result implied that more bubbles attached with oil could be in the underflow of the hydrocyclone, which provided one possible reason for the existence of the optimal air-liquid ratio.

Further more, this paper discussed the higher air-liquid ratio, i.e., the higher concentration of bubbles, which might increase bubble aggregation to form larger bubbles. Larger bubbles facilitate their removal because of higher rising velocities. To predict the process more accurately, CFD simulation including a population balance model is recommended.

(391d) Modeling and Simulations of the Liquefaction and Separation Process Via 3S Separator in Cryogenic Conditions

Sun, H., China University of Petroleum (Beijing)
Yan, F., China University of Petroleum (Beijing)
Liu, F., China University of Petroleum (Beijing)

Supersonic swirling separator, which is also called 3S separator, has been successfully applied in natural gas industry for water removing and heavy hydrocarbons removing. It could also been employed in the liquefaction of natural gas with proper pre-cooling of the feed natural gas. It has advantages such as high efficiency, simply process, low investment and no moving device, which means 3S separator will be appropriate for utilization in many fields.

The flow, liquefying and separation process of natural gas via 3S separator at cryogenic condition has different characteristics with that in the temperature range of the separation process of water and heavies. The modeling and simulation of the process is significant for understanding the liquefaction and separation process via 3S separator. It is also essential for the prediction the performance of liquefaction plant with a 3S separator.

Based on aerodynamics, thermal dynamics and nucleation theory, a theoretical model of the flow, liquefaction and separation process via 3S separator is established. The total process via 3S separator is divided into subsonic section, nozzle section, nucleation section, separation section and diffuse section. The assumption in each section is made based on the feature in this section and the theoretical model for the process in each section is set up. The natural gas at cryogenic condition should be retreated as real gas and its property is calculated using PR equations of stat. A modified CNT model is employed to describe the nucleation process of liquid droplet of natural gas when its temperature is below the saturate temperature. The Gyamathy model is used to describe the growing process of liquid droplet of natural gas when it expands further through the nozzle. The nucleation and growth of liquid droplet is rather important for the separation of LNG and gas. The numerical simulation is carried out by programming using MATLAB on the basis of the model.

A cryogenic experimental device with pre-cooling using liquid nitrogen is designed and set up. Experiments at different pre-cooling temperature are carried out and the experimental results confirm the practicability of the liquefaction of gas by proper pre-cooling. The pre-cooling temperature is an important factor which affects the liquefaction performance. The comparison between experimental data and prediction results of the model is done and the comparison results prove the validity of the model. The results also indicate the nucleation and separation is essential for the liquefaction of natural gas. The modeling is valuable to predict the performance of 3S separator at cryogenic condition and be useful for the development of LNG plant using 3S separator.

(391e) Microscopic Transport in Mixtures of Ionic Liquids and Carbon Dioxide By NMR

Hazelbaker, E., University of Florida
Wang, H., University of Florida
Vasenkov, S., University of Florida

Ionic liquids (ILs) show large sorption capacities for light gases such as carbon dioxide. This makes ILs highly attractive for potential applications as capture agents for carbon dioxide. Clearly, fundamental understanding of diffusion in mixtures of ILs and carbon dioxide is essential for an optimization of ILs for this application. Here we report results of diffusion studies of mixtures of selected ILs and carbon dioxide using pulsed field gradient (PFG) NMR technique at high magnetic field and high magnetic field gradients. This technique allowed obtaining first records of self-diffusivities of carbon dioxide inside ILs by any type of microscopic technique. Application of PFG NMR to a so-called "task specific" ionic liquid that chemically complexes with carbon dioxide resulted in the observation of an influence of exchange between the physisorbed and chemisorbed states of $CO2$ on diffusion in the mixtures of this IL and carbon dioxide. Exchange NMR was applied in combination with PFG NMR to quantify this chemical exchange process on the millisecond time scale. In addition to diffusion of carbon dioxide, also ion diffusion was investigated by PFG NMR in the same mixtures.

(391f) Multiobjective Optimization in the Design of a Process for Hydrodesulfurization of Diesel

Miranda-Galindo, E. Y.,
Segovia-Hernandez, J. G., Universidad de Guanajuato
Hernandez, S., Universidad de Guanajuato
Ramirez, N., Universidad de las Americas Puebla

The hydrodesulfurization (HDS) process is widely used to remove sulfur found in diesel. Conventional HDS processes basically consist of a reaction system where the organic sulfur compounds react with hydrogen to obtain organic compounds and hydrogen sulfide besides a separation system to remove compounds of diesel and a recirculation system. A technological alternative improvement for this process is the reactive distillation, which allows to perform the HDS operation and separation of products in a single unit process involving a considerable energy saving and cost investment. The distillation reactor-side (DSRC) also has been proposed to remove sulfur compounds of diesel. The design and optimization of a HDS process involves the selection of the configuration and the operating conditions to minimize the total annual cost, CO_2 emissions and the amount of sulfur compounds. In general, the optimal design of HDS process is a highly non-linear and multivariable problem, with the presence of both continuous and discontinuous design variables. In addition, the objective function used as optimization criterion is generally non-convex with several local optimums and subject to several constraints. The use of stochastic optimizers, which deals with multi-modal and non-convex problems, can be an effective way to face the challenging characteristics involved in the design of distillation columns. Stochastic global optimization algorithms are capable of solving, robustly and efficiently, the challenging multi-modal optimization problem, and they appear to be a suitable alternative for the design and optimization of complex separation schemes. In this work, we have implemented a stochastic global optimization method to obtain the design and optimization of three distillation reactor-side (DSRC) in the HDS process. In particular, we have used a multiobjective optimization based on differential evolution with tabu list coupled to Aspen ONE Aspen Plus. The results obtained in the Pareto Fronts indicate competition between total annual cost, CO_2 emissions and the amount of sulfur compounds of the HDS process. These results are useful to identify trends in the operational conditions in the DSRC process.

(391g) Conceptual Design of Distillation Colums for Ethanol Dehydration By Using Ionic Liquids

Ek, N., Universidad de las Americas Puebla
Ramirez, N., Universidad de las Americas Puebla
Jimenez, A., Instituto Tecnologico de Celaya

In the early 80's and 90's several authors studied the use of inorganic salts as entrainers for the ethanol dehydration. However, despite the benefits provided by salt-containing distillation, the design and implementation of this process faces some important drawbacks because of the limited solubility of such solid entrainers (Llano-Restrepo and Aguilar-Arias, 2003). In the last decade, Ionic Liquids (ILs) have been considered for several applications in chemical processes, performing roles as reaction media, solvents and separation agents (Shen et al 2011; Pereiro et al, 2012). Particularly, the utilization of ILs for the ethanol-water separation is gaining wider interest, since ILs integrate the advantages of a liquid entrainer and the benefits of a solid salt, as they promote an evident salting out effect over the vapor-liquid equilibrium properties. To determine the feasibility of this new technology, some modeling approaches have been recently proposed (Chavez-Islas et al, 2010; Meindersma et al, 2012).

The aim of this work is to present a methodology for the conceptual design of a distillation system for ethanol dehydration using an IL as entrainer. The proposed method is based on tray by tray calculations from the outside to the inside of the column, similar to the Boundary Value Method (BVM) proposed by Van Dongen and Doherty (1985) for ternary mixtures. In the BVM the feed stage is found where the liquid composition of both profiles intersect each other. However, although the system here under consideration is a ternary mixture, the IL is non volatile, such that its concentration remains essentially constant along each section, although it changes in the feed stage; as a result, the rectifying and striping profiles never intersect. In this work, the feed stage was located by using the minimum distance concept proposed by GutiÃ©rrez and JimÃ©nez (2007), which completed a successful design method for these types of systems. The effects of different feed composition and IL concentrations on the separation task were analyzed. In all cases, feasible designs were obtained. The behavior of the composition profiles is an interesting point worth of special discussion.

References

1. Llano-Restrepo, M., Aguilar-Arias, J. "Modeling and simulation of saline extractive distillation columns for the production of absolute ethanol", Computers and Chemical Engineering , Vol 27, 528-549, 2003
2. Shen, C., Li, X., Lu, Y., Li, C., "Effect of ionic liquid 1-methylimidazolium chloride on the vapour liquid equilibrium of water, methanol, and {water + ethanol} mixture", J. Chem. Thermodynamics, Vol. 43, 1748-1753, 2011.

3. Pereiro, A. B., Araújo, J.M., Esperança, J.M., Marrucho, I.M., Rebelo, L.P., "Ionic Liquid in separation of azeotropic systems – A Review", J. Chem. Thermodynamics , Vol. 46, 2-28, 2012.
4. Chávez-Islas, L.M., Vásquez-Medrano, R., Flores-Tlacuahuac, A., "Optimal Synthesis of a High Purity Bioethanol Distillation Column Using Ionic Liquids", Ind. Eng. Chem. Res., Vol. 50, 5175–5190, 2011
5. Meindersma, G.W., Quijada-Maldonado, E., Aelmans, T.A.M., Gutierrez-Hernandez, J.P., Haan, A.B., "Ionic Liquids in Extractive Distillation of Ethanol/Water: From Laboratory to Pilot Plant", Ionic Liquids: Science and Applications, *ACS Symposium Series*, Vol. 1117, 239–257, 2012
6. GutiÃ©rrez-Antonio, C., JimÃ©nez-GutiÃ©rrez, A., "Method for the Design of Azeotropic Distillation Columns", Ind. Eng. Chem. Res. 46, 6635-6644, 2007
7. Van Dongen, D. B.; Doherty, M. F. "Design and Synthesis of Homogeneous Azeotropic Distillations. 1. Problem Formulation for a Single Column". Ind. Eng. Chem. Fundam., 24, 454-463, 1985

(391i) Electrokinetic Separation of Heavy Metals in Porous Media: An Experimental Approach

Shpik, K.,
Tijaro, R., Universidad Arturo Prat
Stretz, H. A., Tennessee Technological University
Arce, P. E., Tennessee Technological University

Leaching of some industrial byproducts has been linked to health issues. This project will study the behavior of charged contaminants like copper, lead, and arsenic as they move through porous media. The project has specific relevance of heavy metal contamination of the ground water in areas where the metallic mining industry is prevalent. The objective is to explore the effects of different kinds of external fields (electric and a combination of both) on the diffusivity coefficients of the charged heavy metal particles in the saturated porous media. The electrolyte concentration will be varied in order to measure its effects on the diffusivity coefficients. By emulating the experimental design after a traditional Hele-Shaw cell, the first ever experimental results will be generated for radial diffusion of a charged solute. Measurements will be taken to find the effective mass diffusion coefficient in model soils to test the rate of contaminant spreading. The fundamental equation is Fick's Law, the first experiment will focused on the motion of contaminant along r, radial distance that the species travels over time. An image analysis software will be used to measure the charged particles in terms of concentration (intensity) versus position and time. Electrical fields can be imposed in these types of cells model low Reynolds number, constant flow, constant density, and one-dimensional flow are usually assumed.

(391k) Asphaltene-Stabilized Model Emulsion Studies: Organic Phase Composition and Aqueous Phase pH Effects

Machado, D., Universidade Federal do Ceara
Feitosa, F., Universidade Federal do Ceara
Sant'ana, H. B., Universidade Federal do Ceara
Santiago-Aguiar, R. S., Universidade Federal do Ceara

The asphaltenes and their aggregates have an important role in water-in-crude oil emulsion stabilization: under favorable solvent conditions and due to their surface-activity, they exhibit a colloidal behavior, accumulating at water/oil interface and forming a rigid film, they surround the water droplets, hinder the coalescence and hence the phase separation process. Therefore, the formation of emulsions with different stabilities represents the main technological challenge faced by oil producers during the primary processing stage of heavy oils, with their high content of asphaltenes. Our goal in this study is to evaluate the impact of changes in organic phase composition and aqueous phase pH on stability of model emulsion systems. Model emulsion stability was estimated by the percentage of water determined at the bottom of the bottle and it was performed by using a water separability tester, following the ASTM D1401 standard method. All tests were made at room pressure and T = 303.15 K, at 1700 rpm, and were executed by mixing 21 mL of synthetic organic phase, added to 500 ppm of asphaltenes, and 9 mL of synthetic formation water. The organic phase H/C molar ratio (a measure of aromaticity) varied between 1.4 and 2.0, being adjusted by different mixes of toluene and *n*-heptane. We measured a pH range of synthetic formation water was comprised between 3.0 and 11.0, when adding HCl or NaOH. Asphaltenes used in this work were extracted from a Brazilian crude oil, according to the IP 143 standard method. The results show, as expected, that the more stable emulsions under gravity setting are those where the H/C molar ratio of synthetic organic phase is greater (low aromatic content). We presented data of pH effects on emulsion stability from a range of synthetic organic phase formulated with the same asphaltenes, scarce in the literature.

(391j) Modeling and Simulation of Rotating Disc Contactor (RDC) for the Extraction of Aromatic Hydrocarbons From Lube-Oil Cut

Mehrkesh, A., University of Colorado Denver
Tavakoli, T.,
Hatamipour, M. S., University of Isfahan
Karunanithi, A. T., University of Colorado Denver

Previously, several mathematical models have been proposed for liquid-liquid extraction processes involving Rotating Disc Contactor (RDC) Column. Most of these models reveal that hydrodynamic and mass transfer phenomena are important for predicting the performance of the column. In this paper, a mathematical model, using a new scheme, was developed to acquire a simulation tool to predict the performance of an RDC column used in lubricating oil production. Field data obtained in an RDC column of from a commercial lubricating oil production company was used to evaluate the predictions. The model is used for parametric study of aromatics extraction from lube-oil cut in an RDC column to investigate the effects of operational data such as solvent and feed temperatures, solvent to feed ratio and disc rotation rate on the extraction yield.

The developed model compares more favorably with other existing back-flow models by consideration of temperature, physical property and flow rate variations across the column. Additionally, some new correlations for physical properties of the hydrocarbon compounds (aromatics and saturated) were developed. The accuracy of the mathematical model is checked with randomly selected field data of an industrial plant. The results show that the average error of the model in predicting the performance of the RDC performance is less than 10%. Parametric study of the extraction of aromatics compounds from lube-oil cut, carried out in an RDC column, reveals that by increasing the solvent to feed ratio, rotor diameter, discs rotating rate, and both feed and solvent temperatures, the extraction yield will be increased.

(393i) Isolation and Purification of Natural Antioxidants From Limnophila Aromatica By Using Low-Pressure Column Chromatography of Silica Gel

Do, Q. D., National Taiwan University of Science and Technology
Ju, Y. H., National Taiwan University of Science and Technology

In recent years, the ingestion of fresh fruits and vegetables rich in natural antioxidants has been associated with the prevention of cancer and cardiovascular diseases. *Limnophila aromatica* is used as a spice and medicinal herb in Vietnam and Southeast Asia. The aim of this study is to purify and identify antioxidant compounds from defatted *L. aromatica* by using low-pressure column chromatography of silica gel. Pure ethanol and aqueous ethanol (75% in water) were used for extracting antioxidant compounds in freeze-dried *L. aromatica*. Crude phenolic extracts were combined and used as raw materials to prepare high purity antioxidants in silica gel column chromatography (400 x 24 mm, 200–300 mesh) with different eluents and elution modes. TLC and DPPH (2, 2-diphenyl-1-pic-rylhydrazyl) were used to check fractions and their antioxidant activities. The purity of each fraction was determined by using HPLC. Six fractions had been obtained by applying sequentially two solvent systems: iso propanol:acetone:ethyl acetate:hexane (10:10:15:65) and methanol:acetone:ethyl acetate:hexane (10:10:25:55) and the purity of those fractions ranges from 27.682% to 96.447%. Nuclear magnetic resonance (NMR), FT-IR, ESI-mass spectrometry and elementary analyzer were used to determine the structure of purity compounds.

The economic and large scale method for isolation and purification of phenolic compounds from defatted *L. aromatica* had described in our work.

(393d) A Case Study of In-Process Modifications to the Unpaired Sulfhydryl of An N-Terminal Cysteine to Achieve Homogeneous Vaccine Antigen

Christanti, S., Merck & Co.
Woodling, M.,
Goerke, A. R., Stanford University

A protein antigen for a vaccine candidate produced in an insect cell system was found to contain one unpaired sulfuhydryl on the N-terminal cysteine residue. The sulfuhydryl was reactive with components in the cell culture media, such as glutathione and free cysteine, as well as to sulfuhydryls present on other antigens resulting in dimerization. Such propensity to react led to an undesirable heterogenous antigen population. In an effort to eliminate molecular heterogeneity without genetic mutation to eliminate the cysteine residue altogether, in-process modifications to the unpaired residue were incorporated during downstream processing. These modifications included removal of the cysteine residue by proteolytic digestion or capping with N-ethylmalemide (NEM) or free cysteine. The proteolytic digestion was performed by way of natural proteases present in the cell culture and was shown to yield consistent removal of the sulfuhydryl along with eight subsequent residues thereafter. Capping conditions were optimized using multi variable experimentation. Capping with cysteine was shown to be challenging due to slow reaction rate and the reversibility of the attachment. Conversely, capping with NEM was permanent but could lead to potential multiple NEM attachments if the reaction occurs outside of the optimal ranges. The use of TCEP as reducing agent for capping with NEM was found to be most suitable. The molecular properties of the resulting products were monitored using various analytical methods such as LC/MS, charge-based and size-based separations, and biophysical techniques to show the modification extent and stability. In addition, the biological activities of the modified products were also investigated in an animal model.

(393b) Separation of Pyruvic Acid:- Influence of Active and Inactive Diluents On Extraction Equilibria

Pal, D.,

Separation of Pyruvic Acid:- Influence of Active and Inactive Diluents on Extraction Equilibria

Dharm Pal & Dr. Amit Keshav

Assistant Professor,

Department of Chemical Engineering, NIT Raipur - 492010 C.G. (India)

E mail: dpsingh.che@nitrr.ac.in, Mob. +91- 95847 51651

Abstract

Recently, fermentation technology have started to replace conventional synthesis. Apart from central role in energy metabolism, pyruvic acid (PA) serves as an effective starting material for the synthesis of many drugs and agrochemicals and recently being utilized as a fat burner.

On an industrial scale, pyruvic acid is produced by dehydration and decarboxylation of tartaric acid. This process is simple to realize but not cost-effective. With the reception of biotechnological production of pyruvic acid, the recovery of pure acid from fermentation broth poses big hurdle in commercialization of biotechnological production.

In order to develop new synergistic recovery system of pyruvic acid, reactive liq.–liq. equilibrium with liquid ion exchanger in different diluents was carried out. The aim of present work is to investigate the effect of diluents on reactive extraction of pyruvic acid using phosphine based solvent Tri-n-butylphosphate (TBP, 10-70 Vol %) at ambient temperature ($T=25 \pm 0.1\ ^{o}C$). The physical and chemical extraction was performed at various initial aq. acid concentration (0.1 - 0.7 mol/m^3) to visualise the effects of different diluents (MIBK, n-heptane, toulene on complexation. Only (1:1) acid : extractant complex formation was observed with loading ratio ($Z < 0.5$). The higher extraction efficiency with active diluents is attributed to greater salvation and more stability of complex in organic phase.

(393j) Molecular Insight Into the Adsorption of IgG in Hydrophobic Charge-Induction Chromatography

Tong, H. F., Zhejiang University
Cavallotti, C., Politecnico di Milano
Lin, D. Q., Zhejiang University
Yao, S. J., Zhejiang University

The increasing demand of antibodies as therapeutics has led to the development of more efficient, robust and economical processes to purify antibodies from serum, ascitic fluid and cell culture broth. Protein A affinity chromatography is a typical platform technology for antibody purification for its high selectivity. However, it has some disadvantages, such as ligand leakage and CIP difficulty as well as high cost. Currently, hydrophobic charge-induction chromatography (HCIC) with 4-mercaptoethyl-pyridine (MEP) as the functional ligand has been developed as a novel technology for antibody purification. However, compared with protein A capture, the separation efficiency of HCIC still needs to be improved. Better understanding on the separation mechanism of HCIC in molecular scale is necessary for the process development and ligand design.

In the present work, molecular simulation methods were developed to investigate the interactions between MEP ligand and the Fc fragment of IgG. The forcefield parameters of MEP were first optimized via quantum chemistry, and then the hot sites on the surface of the Fc fragment for MEP binding were explored by molecular docking. The consensus binding site (CBS) was focused and some potential ligand-protein complexes were obtained. These complexes were adopted as the starting structure for the molecular dynamics (MD) simulation. The free binding energies were calculated using the WHAM-umbrella sampling as the reference interaction free energy to refit the linear interaction energy (LIE) parameters. Based on the MD results, four possible binding sites on a single chain and three binding sites on the crossover area of double chains were identified. Furthermore, the ligand-net was constructed to simulate the protein adsorption on the porous surface of chromatographic resins. The MM/PBSA method was used to acquire the binding energy map which was used to scan the most suitable binding orientations of Fc on the ligand-net. It was found that ligand density had significant influence on the binding of Fc, which is in good agreement with experimental results. Several potential binding conformations were taken and evaluated with the MD simulation. The multiple-point binding between Fc and MEP ligand-net was found to have an important role during the capture of Fc. The results indicated that hydrophobic forces dominate the binding of Fc on the ligand-net, and the synergistic actions of consensus binding sites (CBS) with other sites were observed. The simulations provided useful information about the molecular mechanism of IgG binding to the HCIC ligand.

With the computer-aided molecular simulation approaches, it is possible to evaluate the multiple interactions between target proteins and functional ligands at molecular level, which certainly helps to explore the molecular mechanism of bioseparation, and it can be used to optimize the separation conditions and develop new ligands for bio-product recovery and purification.

* This work was supported by National Natural Science Foundation of China, the Zhejiang Provincial Natural Science Foundation of China and special foundation for innovative talents training project of Zhejiang University "985 programs".

(393h) Physicochemical Properties of Highly Concentrated Monoclonal Antibody Solutions

Binabaji, E., The Pennsylvania State University
Rao, S., Amgen Inc.
Zydney, A. L., The Pennsylvania State University

The physicochemical properties of highly concentrated monoclonal antibody solutions can have a significant effect on downstream processing, drug formulation, product stability, and drug delivery. The objective of this work was to evaluate the osmotic pressure, second virial coefficient, and solution viscosity of a highly purified monoclonal antibody at concentrations up to 250 g/L over a range of pH and ionic strength. The osmotic pressure was evaluated using classical membrane osmometry. The second virial coefficient was evaluated independently using self-interaction chromatography, and the solution viscosity was evaluated using a RFS-II rotational viscometer. Independent measurements of the net protein charge were obtained using electrophoretic light scattering.

The osmotic pressure at pH 5 and low ionic strength was greater than 50 kPa for antibody concentrations above 200 g/L. The second virial coefficients determined from the oncotic pressure (after subtracting the Donnan contribution) were in good qualitative agreement with those determined by self-interaction chromatography. The second virial coefficient decreased with increasing ionic strength and increasing pH due to the reduction in intermolecular electrostatic repulsion. The third virial coefficient was negative under all conditions, suggesting that the multi-body interactions in this system are attractive. The solution viscosity increased significantly at antibody concentrations above 100 g/L, exceeding 20 cP under these conditions, although there was no direct correlation between the measured viscosity and the thermodynamic properties (osmotic pressure or second virial coefficient). These results have important implications for understanding the behavior of highly concentrated antibody solutions in bioprocessing applications.

(393k) Interaction of Carbohydrates With the Molecularly Imprinted Silica Particles

Das, S., University of Kentucky
Joshi, S., University of Kentucky
Rankin, S. E., University of Kentucky

Carbohydrates draw significant attention in cell and molecular biology due to their pivotal roles in molecular transport, interactions with proteins and responsibly for disease transmission [1][2]. As a result, significant advances are being made based on glycobiology in the fields of bio-interface engineering, glycomics, saccharide sensors, carbohydrate drug discovery and refinement, enzyme mimics[3][4] etc. These fields would benefit from technologies to recognize and separate specific carbohydrates from mixtures of similar compounds. Carbohydrate separation is challenging due to the similarity in their structures; a difference in stereochemistry at one carbon can be biologically very significant while chemically difficult to utilize for separation. Among all the approaches reported in the literature for high selectivity adsorption, molecular imprinting is possibly the most interesting and promising approach to separate valuable carbohydrates from mixtures due to the high affinity [5] and selectivity towards the target molecules possible by this technique - comparable to natural receptors. By molecular imprinting we create molecular binding sites in a stable framework with specific size, shape and functional group arrangement complementary to the target molecule. Here we report one simple inexpensive technique to synthesize precipitated silica particles (Stoeber particles) whose surface is imprinted using mixed surfactants where one surfactant has the target carbohydrate as a headgroup. We measure the adsorption kinetics and adsorption isotherms of different simple carbohydrates to demonstrate the selective binding of targeted molecules from carbohydrate mixtures due to molecular imprinting using nuclear magnetic resonance spectroscopy (NMR). Capitalizing on the availability of isotopically labeled carbohydrates (e.g. with one ^{13}C labeled) allows us to directly measure competitive binding from mixed solutions of carbohydrates to gain molecular-level insights into the interactions with the silica surface that lead to selective adsorption.

References

[1] Dinglasan, R. R.; Valenzuela, J. G.; Azad, A. F. *Insect Biochemistry and Molecular Biology* **2005**, *35*, 1.

[2] Mineo, H.; Hara, H.; Kikuchi, H.; Sakurai, H.; Tomita, F. *The Journal of Nutrition* **2001**, *131*, 3243.

[3] Fujie, T.; Matsutani, N.; Kinoshita, M.; Okamura, Y.; Saito, A.; Takeoka, S. *Advanced Functional Materials* **2009**, *19*, 2560.

[4] Di, L.; Wang, C.; Wu, J.; Wan, L.-S.; Xu, Z.-K. *Chinese Journal of Analytical Chemistry* **2011**, *39*, 592.

[5] G. Vlatakis, L. I. A., R. Moller, K. Mosbach. *Nature* **1993**, *361*, 645.

(393a) Electrostatics Potential in Annular Geometry: A Taxonomy of Models Useful for Applications to Particle Sorting, Tissue Scaffold, and Drug Delivery

Motamedilamouki, A., Tennessee Technological University
Golbayani, P., Tennessee Technological University
Sanders, J. R., Tennessee Technological University
Arce, P. E., Tennessee Technological University

Electrostatics Potential in Annular Geometry: A Taxonomy of Models useful for Applications to Particle Sorting, Tissue Scaffold, and Drug Delivery

ABBAS MOTAMEDILAMOUKI, PARVIN GOLBAYANI, J. R.SANDERS, PEDRO E. ARCE

Department of Chemical Engineering, Tennessee Technological University, Cookeville, Tennessee

Flow through an annular geometry has many applications in chemical, environmental, mechanical and bio-mechanical engineering [1]. A number of researchers have proposed combining electroosmotic flow (EOF) and pressure-driven flow as a means of controlling the motion and separation of bioparticles in an idealized geometry to sort particles using electrophoresis [2-3]. We present here a systematic investigation of the electrostatic potential distribution in an annular geometry. Our objective in this contribution is present a mathematical model for the electrostatic potential distribution in both straight and divergent annular geometry. The analytical solutions for the electric potential profile in the annulus are obtained by solving the 2D Poisson–Boltzmann equation with both long channel and Debye– Huckel approximations. This result is in preparation to the derivation of the Electrohydrodynamic Velocity profile. As a result of this investigation, one can assess the behavior of the electrostatic potential inside of annular channel. Three key parameters have been identified to describe the electrostatic potential behavior: the angle ($\propto$), ratio of up wall potential to the linear combination of both wall potentials, R, which handles the symmetrical/non-symmetrical aspects of the electrostatic potential, and the ratio of the width to the length (γ) that controls the "shape" of the channel section. Results of this study are illustrated by using a series of portraits that capture the key behaviors of the electrostatic potential with respect to the three parameters described above. Applications of this research maybe found in particle sorting, biomedical tissue scaffolds and drug delivery, among others. Illustrative examples related to these applications will be given.

1. Makinde, O. D., Sibanda, P., 2000 Steady Flow in a Diverging Symmetrical Channel: Numerical Study of Bifurcation by Analytic Continuation. *Quaestiones Mathematicae*. 23, 45–57.
2. Xuan X., Xu B., Li D., 2005 Accelerated particle electrophoretic motion and separation in converging–diverging microchannels. *Anal Chem*. 77, 4323–4328.
3. Ai Y., Joo S. W., Jiang Y., Xuan X., Qian S., 2009 Transient electrophoretic motion of a charged particle through a converging–diverging microchannel: effect of direct current-dielectrophoretic force. *Electrophoresis*. 30, 2499–2506.

(393c) N-Butanol Separation From Aqueous Solution By Pervaporation Using Composite PDMS Membrane

Liu, F., The Ohio State University
Yang, S. T., Ohio State University

Butanol is attracting more and more research interest as a second generation biofuel and due to environmental and energy concerns. Butanol can be produced through fermentation process, which is environmental friendly and sustainable, other than petrochemical process. Fermentative butanol usually exists in dilute aqueous solution. Therefore, we need to recover butanol from this dilute solution to get pure or high concentration butanol. There are several popular ways to separate butanol: pervaporation, gas stripping, adsorption and etc. Among them, pervaporation is an outstanding method as it offers higher selectivity and lower energy consumption compared to other techniques. Pervaporation is a membrane based technique and selective membrane is the key factor. Poly (dimethyl siloxane) (PDMS) membrane offers highest butanol separation factor and is widely used for recovering butanol from aqueous solution. However, butanol flux and selectivity of PDMS membranes can still be improved, especially butanol flux. In this work, efforts were made to improve butanol flux and separation factor of composite PDMS membranes. Different membrane thickness, filler material, and operating temperature were tested. Activation energy of PDMS membrane was also evaluated. Among those studied membranes, polytetrafluoroethylene (PTFE) supported PDMS membrane filled with zeolite offered a butanol separation factor of ~60 and butanol flux of 52 $g/m^2 \cdot h$. This membrane can concentrate a 15 g/L butanol solution to a ~420 g/L butanol solution with one single pass. This greatly reduces the butanol recovery cost and makes fermentative butanol production more economical.

(393e) Study of Protein Refolding in High Concentrations

Saremirad, P., Western University
RAY, A. K., University of Western Ontario

Inclusion Bodies (IBs) production is a widely used method in biopharmaceuticals due to very high expression levels. Large amounts of proteins are trapped in these insoluble aggregates which require solubilisation and purification. IBs are usually solubilized using high concentrations of suitable chaotropic salts and reducing agents to unfold the protein chain. However, the refolding of proteins into their compact structures is critical to confer biological activity on proteins allowing them to function as effective therapeutics. The simplest refolding method is batch dilution in which the denaturized protein is diluted in a suitable refolding buffer. However, the correct protein folding pathway often competes with misfolding and aggregation which substantially reduces refolding yields. Furthermore, presences of solubilized aggregates impurities in most biopharmaceutical processes are not acceptable as they may cause health problems in patients. Therefore, batch dilution has serious drawbacks during scale-up due to low product concentration and purity along with large process volumes necessitating additional concentration and purification steps. On-column chromatographic based refolding has shown to address challenges with product dilution by facilitating isolation of protein and chaotropic and reducing molecules while simultaneously purifying the protein after refolding. Since chromatographic protein refolding allows for significantly higher protein concentrations without drastic loss of refolding yield, compared to batch dilution, it is essential to investigate the system performance in industrially relevant concentrations. In this work, using Lysozyme from chicken egg white as a model protein, the effect of protein concentration (20-40 $mgml^{-1}$) and refolding buffer composition including its pH (8.1-9.5), ionic strength (0.2M NaCl) and Arginine presence as aggregation prevention additive (0.2M) on refolding yield, protein recovery and purity were evaluated and will be discussed in detail.

(393f) Development of Adsorptive Membranes for Targeted Protein Binding

Chenette, H. C. S., Clemson University
Husson, S. M., Clemson University

This presentation will describe our latest efforts to develop adsorptive membranes for use as protein purification media. The design approach used in these studies is to graft polymeric 'tentacles' from macroporous regenerated cellulose membranes by atom transfer radical polymerization (ATRP). Using one or more examples, we will illustrate how this design approach can be used to prepare high-productivity membrane adsorbers. In one case, we will discuss affinity membrane adsorbers that incorporate glucose-containing glycoligands. Studies have shown glycoligands selectively bind specific lectins [Yang et al., *Appl. Mater. Interfaces*, **2010**, *2*] mimicking the gluco-receptors found on cell surfaces that recognize and bind bacteria. The literature also reports advances in this field using model substrates to elucidate optimal polymerization conditions, and kinetic studies relating polymerization time to the degree of polymerization [Yang and Ulbricht, *Macromolecules*, **2011**, *44*]. The present study integrates these glycoligand adsorbers into a membrane system that is capable of scale-up to a column for chromatographic separation. A glucose-based ligand was grafted from regenerated cellulose membrane surfaces using ATRP. The model lectin, Concanavalin A, was used to evaluate membrane performance in bind-and- elute purification, using a low molecular weight sugar for elution.

(393g) Purification of a Novel Lipopeptide Biosurfactant By Green Recovery Methods

Marti, M. E., Iowa State University
Zainuddin, H.,
Schoborg, J., Iowa State University
Colonna, W., Iowa State University
Glatz, C. E., Iowa State University

FA-Glu is a novel biosurfactant that is produced by a genetically-modified strain of *Bacillus subtilis* on glucose-mineral salts media. It has unique properties compared to other biosurfactants and petroleum-based surfactants that make it suitable for several applications. In particular, its reduced toxicity to fish, low critical micelle concentration and high dispersant-to-oil ratio FA-Glu make it an attractive candidate for use in oil spill remediation.

Economic production of biosurfactants will improve their acceptability and augment their use, as well as reduce the need for synthetic surfactants. One approach for reducing production costs is through fermentation of low-cost feedstocks (preferably agricultural co-products) coupled with the use of efficient "green" recovery methods. Examples include membrane filtration and foam fractionation.

In this study, purification of FA-Glu from fermentation broth by froth floatation and ultrafiltration was explored. The effects of air flow rate, foam height and initial volume of the fermentation broth on enrichment and purification were tested. Additionally, membranes with different MWCO's were tested to optimize purification of FA-Glu from fermentation broth. Chemical analyses of product streams were used to identify processing parameters to maximize biosurfactant yields and purities. Purification strategies that may enhance the purity are presented.

(394a) Sublimation Mechanism for Small Molecule Organic Semiconductors

Zhang, Y., Kansas State University
Morgan, N. T., University of Minnesota
Holmes, R. J., University of Minnesota
Cussler, E. L., University of Minnesota

Sublimation Mechanism for Small Molecule Organic Semiconductors

Yi Zhang, Nathan T. Morgan, Russell J. Holmes, and Edward L. Cussler

Department of Chemical Engineering and Materials Science

University of Minnesota, Minneapolis MN 55455

Organic semiconductors continue to receive substantial application for next-generation display, photovoltaic and solid-state lighting technologies. Processes capable of producing high purity materials at large scales are desirable. Presently, the majority of these materials are purified using thermal gradient sublimation. We find that the sublimation rate of the archetypical, small molecule hole transport material N,N'-bis(naphthalen-1-yl)-N,N'-bis(phenyl)-benzidine (NPD), is controlled by a combination of viscous flow and physical vapor deposition. Other steps like diffusion within the solid feed, reaction at the feed particle surface, and mass transfer within the bed of feed particles do not significantly affect the sublimation rate. Such a mechanism is in contrast with other commercially practiced sublimations. The results obtained here for NPD suggest guidelines for the large-scale purification of active materials for organic electronic devices.

(394d) Discovery and Development of a Novel Hydrate of An API Via Controlled Solid-State Form Transformation

Reddy, Y. S., Dr. Reddy's Laboratories Ltd.
Velaga, J.,
Dhage, Y.,
Peddy, V.,
Pathivada, D., Dr. Reddy's Laboratories Ltd
Kumar, A.,
Ramakrishnan, S., Dr. Reddy's Laboratories Ltd.

Discovering novel forms can be a challenging task in the generic industry, with the search often resulting in solvates or unstable forms. This poster will present a case study wherein a stable novel form of an API was discovered by controlling the solid-state form transformation of an unstable methanolate. Compound A, which is a HCl salt, is reported to have several forms including a stable anhydrous form (Form-IV), stable mono-hydrate (Form V), mono-methanolate (Form XI), and several other solvates. While reactive crystallization of this compound in methanol lead to the reported mono-methanolate (Form XI), it was found that reducing the temperature of salt formation lead to a hitherto undiscovered form (designated as Form-P). However, Form-P, suspected to be an unstable methanolate, readily converted to a number of different forms depending on the process parameters during filtration: (i) The PXRD pattern of Form XI was obtained during filtration if the temperature increased or the wet cake was exposed to high humidity. Further investigation lead to the discovery that the Form-XI could be obtained by either formation of the mono-methanolate or by formation of a hydrate that was iso-structural with Form-XI. (ii) Interestingly, if Form-P was filtered quickly with exposure to mild humidity (40-60%RH), it converted to a second novel form, designated as Form-Q. (iii) Or, if filtered under pressure with dry nitrogen, it converted to a third novel form, Form-R.

Both Forms Q and R were found to be dihydrates, and Form-Q was selected for further development based on stability. The challenge then was to develop a robust process to control the conversion of Form-P to only Form-Q without any contamination of the other crystalline forms. This poster describes the development of a controlled solid-state transformation pathway, how challenges to scale-up the process were successfully addressed along with implementation of a QbD approach to ensure robustness.

Communication number: IPDO IPM - 00369

(394b) Effect of Fatty Acids On Cooling Crystallization of Glycine From Neutral Aqueous Solution

Kim, J. W., Sogang university
Noh, H. D., Sogang University
Koo, K. K., Sogang University

Polymorph control is significant process because different polymorphs induce different properties. In case of crystallization of glycine, it has been well known that α-glycine was crystallized from neutral aqueous solution. In the present work, it was found that γ-glycine could be obtained at neutral aqueous solution/fatty acid interface. γ-Glycine crystal was hanging at the neutral aqueous solution/fatty acid interface with a growth direction along [00-1]. It indicates that the –COOH head groups of fatty acid at the interface influenced the nucleation of γ-glycine because of interaction with NH3+-rich (001) face of γ-glycine. Furthermore, the variation of molecular arrangement of fatty acids with different hydrocarbon length was found to affect significantly the polymorphic selection of glycine. As a result, γ-glycine crystals were obtained from the hexanoic acid and octanoic acid, whereas α-glycine was induced from the fatty acid of which number of carbon is larger than 12 due to extension of molecular surface area at the interface (saturation temperature and cooling rate were 85 Â¢Âº C and 0.5 Â¢ÂºC/min, respectively).

(394c) Effect of Surface Energy Evolution On Particle Nucleation Under Gas Anti-Solvent Precipitation Conditions

Rosner, D. E., Yale University
Arias-Zugasti, M., Yale University

We address here the problem of particle nucleation and growth under so-called *Gas-Anti-Solvent-Precipitation* (GASP-) conditions. This complex environment is of considerable interest for several currenttechnologies. In the pharmaceutical industry[1], which motivated the present work, GASP environment is used to produce the controlled precipitation of an Active Pharmaceutical Ingredient (API) from an organic solvent, often sprayed into compressed CO_2. Of importance is the ability to predict and control the size distribution of the nucleated particles, with the usual goal of producing narrow distributions of ultrafine particles to facilitate rapid dissolution when administered. Our theoretical model[2] is illustrated using the system: phenanthrene [as API-surrogate] precipitated from liquid toluene using CO_2 as anti-solvent at ca. 60 bar and 298K). Our preliminary results demonstrate the need to account for the *time evolution* of not only the supersaturation (no surprise!) but also the *surface energy of the particle/fluid interface*(remarkably, rarely considered).

In our well-mixed single droplet theoretical model[2] we consider the time-evolution of particle nucleation- and condensation-growth-rates, fully coupled to the evolving environmental conditions associated with CO_2 uptake and solution volume change. Particle nucleation rates are conveniently approximated using *Classical Nucleation Theory* (CNT), and single particle growth rates are based on a molecular collision-theory-based kinetic model[3], including Gibbs-Kelvin(-Ostwald) curvature corrections for recently born particles. However, the numerical method used to solve the Population Balance Equation governing the particle number density distribution function (NDDF) is not restricted to any particular choice of rate laws for nucleation or growth, nor does it presume or restrict the mathematical form of the NDDF. Our model identifies the dependence of the characteristic particle size and nucleation/growth time scales of the problem as a function of the control parameters in the nearly isobaric/isothermal GASP-environment (mainly pressure level and initial API under-saturation). Besides revealing the relevant dimensionless groups governing the performance of such systems, we demonstrate the importance of *surface energy evolution* (SEE) in determining the critical nucleus size, and associated nucleation rates, in this variable composition "carrier fluid". Our model also reveals the effects of SEE on the ultimate particle NDDF. In this regard, we find that a non-monotonic time evolution of the critical nucleus size can give rise to multi-modal size distribution functions, without the need to involve physical processes other than particle nucleation. Our results also show that, for the system under consideration (phenanthrene precipitated from micron-size toluene solution droplets using CO_2 as anti-solvent) typical particle growth times are considerably longer than the corresponding characteristic nucleation times. In such cases post-nucleation particle growth introduces only a small correction, and the finally observed precipitated particle NDDF basically betrays the integrated time history of the critical nucleus size convoluted with log(supersaturation). As a corollary, controlling the time evolution of the critical nucleus size (*via*both effective surface energy and supersaturation) enables NDDF-control, which, in some cases, can even yield (normally undesirable) multi-modal API particle populations.

---

a Yale University, Department of Chemical and Environmental Engineering, New Haven, CT, 06520-8286, USA

b Departmento de Física Matemática y de Fluidos, UNED, Madrid, Apdo 60141, 28080, Spain

References:

1. Martin, A. and Cocero, MJ, "Micronization Processes With Supercritical Fluids: Fundamentals and Mechanisms", *Advanced Drug Delivery Reviews*(Elsevier),vol 60, pp 339-350(2008)

2. Rosner, DE and Arias-Zugasti, M, "Theory of Pharmaceutical Powder'Micronization' Using Compressed Gas Anti-solvent (Re-) Precipitation"; Paper # WG10S1O1, *European Aerosol Conference; Granada, Spain, September 6, 2012; *J Supercritical Fluids*(to be submitted) August , 2013

3. Rosner, DE, "Collision Theory Re-Interpretation of Kinetic Data for the Growth of Organic Crystal Surfaces; Part I. *Melt Growth*; Part II . Physical *Vapor–Growth*, Part III. *Solution–Growth*; *Implications for the Precipitation of Pharmaceuticals in Trans-critical Environments?*"; *Crystal Growth & Design* (ACS), (to be submitted); Summer 2013

(394f) Screening of Pharmaceutical Co-Crystals in Solution Crystallization

Lee, K. S.,
Kim, K. J., Hanbat National University

In recent years co-crystal formation has emerged as a viable strategy towards improving the solubility and bioavailability of poorly soluble drugs. The solution crystallization was used to produce co-crystals of active pharmaceutical ingredients (APIs): clopidogrel hydrogen sulfate, aripiprazole, bicalutamide, etc.. Co-crystals using the pharmaceutically acceptable co-former were produced from solution through evaporation and cooling crystallization methods.

Three unique solid phase utilizing six co-formers were generated in the screening experiments. The corresponding co-crystalline phases were screened by in-situ measurement and characterized by differential scanning calorimetry (DSC) and powder X-ray diffraction (PXRD). Phase solubility diagrams are used to illustrate how saturation of the reactants leads to supersaturation of the co-crystal. Kinetic and thermodynamic studies on co-crystal formation were carried out.

(394e) Recrystallization of Pharmaceutical Drug Particles Using Supercritical Fluid

Yim, J. H., Sogang University
Roh, H. D.,
Lim, J. S.,

In this research, recrystallization of adefovir dipivoxil particle was carried out by rapid expansion of supercritical solutions (RESS) process, one of the well-known supercritical processes. Supercritical carbon dioxide, a green solvent, was selected as a solvent. The effect of process parameters such as extraction temperature and pressure, pre-expansion temperature, spray distance, and nozzle configuration on the particles size of adefovir dipivoxil were evaluated and found that the process parameters influenced the particles size. Increasing the extraction temperature (308.15-328.15K) caused the particle size increasing of adefovir dipivoxil. The particle size decreased with increasing extraction pressure (20-30MPa). The particle size of adefovir dipivoxil was increased with increasing pre-expansion temperature. An increase in the spray distance (8-32cm) resulted in a decrease in the average particle sizes. The effect of the nozzle internal diameter on the size and shape of the precipitated particles was examined different nozzle internal diameters (152-762μm). When the nozzle internal diameter size was bigger, the particle size did not show significant change. Processed adefovir dipivoxil particles by RESS process were analyzed by SEM, XRD, DSC and FT-IR spectra.

(395a) Effect of ETS-10 Modification On Its Dielectric Properties and Microwave Regeneration

Shariaty, P., University of Alberta
Jahandar Lashaki, M., University of Alberta
Hashisho, Z., University of Alberta
Sawada, J., University of Alberta
Kuznicki, S., University of Alberta
Hutcheon, R., Microwave Properties North

Engelhard Titanosilicate molecular sieve (ETS-10) has unique selectivity and adsorption capacity for hydrocarbons separation and drying. Microwave heating was found to be a more efficient technique for regeneration of ETS-10 compared to conventional temperature swing regeneration (TSR). However, microwave regeneration of ETS-10 can be further improved by modifying its dielectric properties by exchanging the cations inside its structure. The objective of this research is to investigate the effect of ion exchange on the dielectric properties of ETS-10 as well as its adsorption and regeneration properties. Different monovalent and divalent cations (Li^+, K^+, Ca^{2+}, Cu^{2+}, and Ba^{2+}) were used for this purpose. All samples were characterized using Energy Dispersive X-Ray spectroscopy (EDXs) and micropore surface analysis to analyze the elemental composition and the porous structure of the adsorbents, respectively. The dielectric properties of ion exchanged samples were also measured and compared to the original ETS-10. The adsorbents were loaded by adsorbing a water/CO_2 mixture. The loaded adsorbents were then regenerated at 130°C using microwave heating. Higher adsorption capacities were found in case of Ca-ETS-10, Li-ETS-10, and Ba-ETS-10 compared to ETS-10. All ion exchanged samples yielded lower microwave energy absorbance and higher regeneration efficiency compared to ETS-10.

(395c) Synthesis of Hierarchical Hydrotalcite-Derived Mgal-Ldos for CO2 Adsorption: the Effects of Anions and Structure-Directing Agents
Cai, W., Wuhan University of Technology

Synthesis of Hierarchical Hydrotalcite-derived MgAl-LDOs for CO_2 Adsorption: the Effects of Anions and Structure-Directing Agents

Weiquan Cai[a,*], Pengpeng Shen[a], Jiaguo Yu[b] and Mietek Jaroniec[c]

[a]School of Chemical Engineering, Wuhan University of Technology, Luoshi Road 205#, Wuhan 430070, P. R. China

[b]State Key Laboratory of Advanced Technology for Material Synthesis and Processing, Wuhan University of Technology, Luoshi Road 205#, Wuhan 430070, P. R. China

[c]Department of Chemistry, Kent State University, Kent, Ohio, 44242, USA

[*]Corresponding author. E-mail address: caiwq@whut.edu.cn

Mg-Al layered double oxides (MgAl-LDOs), as one of LDOs, has a large number of Brønsted basic sites on their surface, and thus is considered as a promising candidate for CO_2 removal[1]. Herein, a series of hierarchical hydrotalcite-derived MgAl-LDOs with Mg/Al molar ratio of 2:1 were prepared by a simple hydrothermal method using chloride and nitrate as starting metal salts (SMS), respectively and urea as precipitating agent. Effects of structure-directing agents (SDAs) including Pluronic F127 and Na_2SO_4, and as mentioned SMS on the phase structures, morphologies, textural properties and adsorption performance towards CO_2 of the as-prepared MgAl-LDOs have been comparatively investigated by X-ray diffraction (XRD), scanning electron microscopy (SEM), nitrogen adsorption-desorption techniques, CO_2 temperature programmed desorption (CO_2-TPD) and CO_2 adsorption analyzer. With addition of Na_2SO_4, the morphologies of the samples transform from hierarchical assemblies assembled by hexagonal nanplatelets to hierarchical flower-like structure assembled by irregular nanoflakes. However, addition of F127 has no significant effect on the morphologies of the samples. The results also show that the adsorption capacity of these samples towards CO_2 relates with their surface basic site density resulted from different SMS and SDAs, and the sample MgAl-N-F prepared by using magnesium nitrate and aluminium nitrate as SMS, and F127 as SDA has a maximum adsorption capacity (1.18 mmol g^{-1}). This work provides a simple and an efficient way to prepare MgAl-LDOs with controlled adsorption capacity towards CO_2 from cheap magnesium salts and aluminum salts by using different SDAs.

[1] M. K. Ram Reddy, Z. P. Xu, G. Q. (Max) Lu, J. C. Diniz da Costa, Ind. Eng. Chem. Res., 45, 7504-7509 (2006).

(395u) Model Adsorption On Fixed Bed of Monolithic Carbon Aerogels, a Comparative Study Between Models of Barrier Resistance/Diffusion (CBRD) and Linear Driving Force (LDF)

Camargo-Trillos, D. A., Universidad Nacional de Colombia
Chejne, F. Sr.,

A base phenomenological model was developed to describe the behavior of adsorption systems at low and moderate concentrations. The model is based on partial differential equation of energy and mass balance to represent the kinetic adsorption and mass transfer at different space scales into the bed and porous material. The carbon aerogels were prepared previously by resorcinol-formaldehyde polymerization (RF), using acid and basic catalyst. Characterization of porous structure and morphology of the materials was obtained by N_2 (-196 ° C) and CO_2 (0 °C) gas adsorption. The specific surface areas BET up to 890 m^2/g and 850m^2/g were obtained for catalyst acid and basic respectively. The aerogels samples presented a high homogeneous mesopore and micropore size distribution.

Transient balance equations of gas phase into free mean of bed and particle were implemented to model the mass transport into the fixed bed of porous material. A comparative study between two adsorption kinetic models into the monolithic carbon aerogels was realized.

The first one, linear driving force model (LDF). This is based in that the rate of mass transfer is expressed through the global transfer coefficient k_p and the difference between the amount adsorbed q (mmol- adsorbed/g- adsorbent) in the particle and your equilibrium condition q^* (mmol- adsorbed/g- adsorbent)[1]. The mass transfer coefficient k_p was estimated by mutual addition of the mass transfer resistance like resistance to diffusion in the outer film in the particle, transport resistance by diffusion across the macropores and mesopores and kinetic adsorption on the micropores

The second one, model of barrier resistance/diffusion (CBRD). This model describes the mass transfer through of macropore and mesopore by introduction of the transient mass balance equation of gas at spherical particle porous. An unique mass flux at the interface of particle, that considered both the effect of kinetic adsorption on the outer surface and external film mass transfer was proposed. The kinetic adsorption in micropores of particle was represented through of Linear Driving Force model that introduced an additional transfer coefficient "k_d" for represent the specific rate adsorption by micropores.

The dynamic equations for the balance of material, absorption kinetics, and energy balance of the bed

are solved simultaneously with the finite volume method. The model was developed in the MATLAB software, using the function ODE23s and ODE23tb, which implement a numerical differentiation and implicit Runge-Kutta method respectively. The models were validated with experimental adsorption data of hexane on fixed bed of carbon aerogels, implemented the breakthrough curve methodology.

The two adsorption kinetic models allows obtaining a good fit with the experimental uptake data of adsorption Breakthrough curve of hexane on the samples of carbon aerogels. The LDF model presented a good fit with an experimental data of Breakthrough, when only the effects external mass transfer and interparticle transport diffusion on porous over the mass transfer coefficient k_p was considered. Similar results have been presented for activated carbon[2]. The mass transfer boundary condition implemented at CBDR model represent suitable adsorption process with experimental data that allow a very good mass conservation and good representation of mass flux between the mean free of bed and pores of particle. The CBDR model presents a larger sensibility with the textural change of the monolithic carbon aerogels and the effect over adsorption process than the LDF.

References.

1. Suzuki, M., *Adsorption Engineering*. 1990, Tokyo: Elsevier Science Publishers B. V.

2. Malek, A. and S. Farooq, *Kinetics of hydrocarbon adsorption on activated carbon and silica gel.* AIChE Journal, 1997. **43**(3): p. 761-776.

(395az) Development of a Three-Zone Simulated Moving Bed Process for Separation of Succinic Acid and Lactic Acid

Nam, H. G., Hanyang University
Park, C., Hanyang university
Hwang, H. J., Hanyang university
Mun, S., Hanyang University, Department of Chemical Engineering

In this study, a three-zone simulated moving bed process was developed for continuous separation of succinic acid and lactic acid, which came from the biotechnological process for production of succinic acid. First, an Amberchrom-CG300C resin was found to be suitable for the adsorbent of such SMB process. On the basis of the selected adsorbent, the intrinsic parameters of succinic acid and lactic acid were estimated at 40 °C through a series of multiple frontal experiments. The estimated parameters were then used in the next stage of optimizing the operating conditions of the three-zone SMB unit, which was assembled by our research group. Under the optimized conditions, the relevant SMB experiment was carried out at 40 °C and all the resultant samples from the product ports and column outlets were assayed, which verified that the continuous separation of succinic acid and lactic acid was performed successfully. The experimental data for the product concentration profiles and the internal concentration profiles were also in reasonable agreement with the model predictions.

(395d) Adsorption Desulfurization of Liquid Fuel Using Ag/TiO2 and Ag/Tiox-Al2O3 Assisted By Ultraviolet Irradiation

Sun, X., Auburn University

Organic sulfur compounds in liquid fuels are converted into SO_x during combustion, which cause severe environmental pollution such as acid rain and smog. Also sulfur is considered as a poison for catalysts in fuel cells. As a result, desulfurization has attracted much attention nowadays. The conventional hydrodesulfurization (HDS) process becomes more complicated for removing the refractory sulfur heterocycles from commercial fuels. Thus several alternative technologies have been investigated in order to supplement HDS. Among those technologies, adsorption desulfurization (ADS) is one of the simplest and efficient methods to achieve ultra-low sulfur level under ambient temperature and pressure. Previously, Ag/TiO_2 and Ag/TiO_x—Al_2O_3 adsorbents developed by our group showed high selectivity, capacity and regenerability. Our recent research interests focus on the photocatalytic property of TiO_2 material on desulfurization. In the present work, the desulfurization performance of these two adsorbents treated by ultraviolet (UV) irradiation was studied by both saturation and breakthrough tests using model fuels (e.g. 3500 ppmw S as benzothiophene in n-octane). Low power input UV sources at a longer wavelength (λ=365 nm) were applied to treat our sulfur adsorbents. Our research results showed that the desulfurization capacities of both Ag/TiO_2 and Ag/TiO_x—Al_2O_3 were improved apparently via photo-irradiation. The breakthrough capacity of blank TiO_2 increased from 1.83 mgS/g to 3.06 mgS/g with UV treatment. And the saturation capacity of 4wt% Ag/TiO_2 increased from 4.87 mgS/g to 6.42 mgS/g with UV treatment. UV source can produce more active sites on support materials' surface while loading Ag can further lower down the band gap of TiO_2 and also inhibit the the recombination of electron-hole pair. The characteristics of the adsorbents and possible mechanisms were investigated by X-ray Photoelectron Spectroscopy (XPS) and UV diffuse-reflectance spectroscopy (DRS). The UV photooxidation effects on organic sulfur compounds were also studied using gas chromatography (GC-PFPD).

(395t) Microwave Synthesis and Characterization of MOF-74 (M = Ni, Mg) for Gas Separation

Wu, X., New Mexico State University
Bao, Z., Zhejiang University
Yuan, B., New Mexico State University
Wang, J., New Mexico State University
Sun, Y., New Mexico State University
Luo, H., New Mexico State University
Deng, S., New Mexico State University

Isostructural MOF-74 (M = Ni, Mg) were successfully synthesized with both hydrothermal method (**1** and **3**) and microwave-assisted method (**2** and **4**). These MOF-74 samples were characterized with scanning electron microscopy for crystal structure, powder X-ray diffraction for phase structure, and nitrogen adsorption for pore textural properties. The experimental results showed that MOF-74 samples synthesized by the microwave-assisted method have a smaller particle size with relatively more uniform particle size distribution. The microwave effects also helped to produce a larger specific surface area and micropore volume, with a similar median pore diameter. Adsorption equilibrium and kinetics of various gases (CO_2, CH_4, N_2, C_2H_4, C_2H_6, C_3H_6 and C_3H_8) on these MOF-74 samples were determined at 298 K and gas pressures up to 1 bar. Adsorption equilibrium selectivity (α), combined equilibrium and kinetic selectivity (β), and adsorbent selection parameter for pressure swing adsorption processes (S) were estimated. The relatively high values of adsorption selectivity indicates the potential to separate CO_2/CH_4, CO_2/N_2, C_2H_4/C_2H_6, C_3H_6/C_3H_8 and C_3H_6/C_2H_4 pairs in a vacuum swing adsorption process using the MOF-74 adsorbent. The microwave-assisted method was found to improve MOF-74 with a larger adsorption capacity and higher selectivity for gas separation.

(395v) On The Use Of Ceramic Monoliths For IMAC Chromatography: Adsorption Kinetics and MASS Transfer Effects On The Separation Of BSA and Catalase

Vega, M., University of Salamanca
Martin del Valle, E. M., University of Salamanca
Cerro, R., University of Alabama in Hunstville
Galan, M. A., University of Salamanca

On the use of ceramic monoliths for IMAC chromatography: adsorption kinetics and mass transfer effects on the separation of BSA and Catalase.

Milena. A. Vega[1], **Eva.M. Martín Del Valle**[1], Ramon L. Cerro[2] and Miguel A. Galán[1]

[1] Department of Chemical Engineering, University of Salamanca, 37008, Salamanca, Spain.

[2] Department of Chemical Engineering, University of Alabama in Huntsville, Huntsville 37185, Alabama, USA.

The use of ceramic monoliths in affinity chromatography for laboratory applications is well established. The purpose of this research is to demonstrate the use of activated monoliths in several adsorption/desorption cycles, to device strategies for the scale up of bio-separation processes and demonstrate the economics of using monolith columns in industrial bio-separation processes. A separation process designed for industrial use requires prior knowledge of column lifetime, a realistic model for adsorbent and adsorption kinetics, and the ability to predict the column performance after reuse. These are important consideration for the development of industrial separation processes, because the economics of industrial processes is based on the continued use of the same column for several adsorption/desorption cycles without a significant loss in adsorption capacity.

On the basis of experimental data obtained on laboratory separation of BSA and Catalase a mathematical model was developed to predict the efficiency of a chromatographic column on multiple adsorption cycles. The mathematical/computational model describes the dynamics of adsorption of protein on a negligible porosity adsorbent such as agorose-coated ceramic monoliths. Since pore diffusion can be safely neglected, the adsorption process can be modeled as an equilibrium stage taking place on a flat two-dimensional surface. This model has been used to predict adsorption/desorption equilibrium, adsorption/desorption kinetics and mass transfer rates from the bulk of the circulation fluid to the monolith walls.

(395e) Phase Inversion Synthesis of ZIF-8/Polymer Spheres for Gas Adsorption

Li, L., Monash University
Yao, J., Nanjing University of Technology
Wang, H., Monash Unviersity

Carbon dioxide separation is a key step in carbon sequestration for preventing global warming with the increasing of the energy consumption. Adsorption is one of the most important methods to separate compounds from industrial gas mixtures. Thus, there is growing interest in developing technologies for the efficient adsorption of large quantities of carbon dioxide. The adsorbent materials are restricted to porous materials with a high surface/volume ratio. Metal organic frameworks (MOFs) are a new class of porous crystalline materials consisting of metal ions linked by organic bridging ligands. Zeolitic imidazolate frameworks (ZIFs), a subfamily of MOFs, have exceptional thermal and chemical stability; in particular, ZIF-8 is one of the most studied materials because of its functional applications in gas separation. However, one of the big issues to use ZIF-8 particles as adsorbent is how to shape them to a particular form. In the industrial gas separation processes, columns packed with adsorbents are mainly utilized, and the most common forms of adsorbents are the beads and the granules.

Here, we report a phase inversion spinning method for the preparation of the macroporous spheres through a single orifice spinneret. In principle, this method is a one-step method to synthesize ZIF-8 spheres with polymer as the binder, and the diameter of the spheres is controllable by changing the size of the spinneret. In a typical phase inversion process, the spheres are produced by extruding a concentrated polysulfone solution containing different amount of ZIF-8 through a single orifice spinneret. Surfactant F127 is pre-added in the polysulfone solution as a pore-forming agent. The morphology of the sphere shows a porous structure of the sphere with a pore diameter of around 1 μm, and the porosity of sphere is around 77% by the influence of the F127. The resulting ZIF-8/polymer composite spheres were analysed by nitrogen adsorption analysis, single carbon dioxide adsorption at different temperatures, and single hydrogen adsorption. The spheres containing 80% of ZIF-8 shows a high BET surface area which is appropriate 760 m^2/g, indicated the spheres are almost fully covered with ZIF-8 particles. The gas adsorption analysis of the ZIF-8/polymer spheres with nitrogen, carbon dioxide, and hydrogen show impressive value compared with the ZIF-8 powder; in addition, the spheres have the advantages of easy handling and recycling over ZIF-8 powder. This phase inversion method is a one-step method to synthesize spheres, which has potential to use for other adsorbent materials shaping.

(395ba) Supercritical Carbon Dioxide Regeneration of Activated Carbon for Industrial Re-Use

Ushiki, I., Tohoku University
Ara, T.,
Ota, M., Tohoku University
Sato, Y., Tohoku University
Inomata, H., Tohoku University
Ito, Y., DAI-DAN Co., Ltd.
Nakano, K., DAI-DAN Co., Ltd.
Nakamura, S., DAI-DAN Co., Ltd.

Activated carbon is widely used to remove hazardous substances such as volatile organic compounds (VOCs) by its high adsorption ability. Recently, regeneration processes of the used activated carbon are strongly demanded from the viewpoints of sustainability and reducing running cost. Supercritical carbon dioxide ($ScCO_2$) technology is regarded as a promising regeneration method because of its high diffusivity into microstructure of activated carbon and the operation capability at mild temperature (Critical point: 304 K). In this study, we have investigated the $scCO_2$ regeneration of activated carbon used at an industrial process.

We selected about 20 activated carbon samples used at an exhaust gas purification process of an electronic device factory in Japan. In experiment, $scCO_2$ was introduced to a regeneration column for 6h at 2.0L (SATP)/min, 20.0 MPa and 353 K. Regeneration ratio was defined as the recovering of specific surface area with a nitrogen adsorption method. The amounts of contaminants before and after regeneration were measured by a thermo gravimetric analysis (TGA). DTG (derivative TG) curve clearly showed bimodal shape corresponded to low (< 673 K) and high-boiling (> 673 K) components on the activated carbon.

The results showed higher regeneration ratio more than 90 % was achieved by $scCO_2$ cleaning in the cases of used activate carbon containing almost only low-boiling components. Furthermore, the regeneration efficiency linearly and dramatically decreased with increasing high-boiling components. This dependence indicated a predictive possibility of the regeneration ratio of the used activated carbon by means of the amounts of high-boiling contaminant.

(395bb) Re-Use Technology for Organic Gas-Removal Filters Using Supercritical Carbon Dioxide

Ito, Y., DAI-DAN Co., Ltd.
Nakamura, S., DAI-DAN Co., Ltd.
Nakano, K., DAI-DAN Co., Ltd.
Maeno, T., DAIDAN Co., Ltd.
Inomata, H., Tohoku University

We have studied supercritical carbon dioxide cleaning as a new regenerating technology for air filters, particularly organic gas removal air filters that had not be recycled and were forced to be disposed so far. Supercritical carbon dioxide ($ScCO_2$) is a solvent having both superior diffusivity and solubility for fine structures in filters and we utilized such ScCO2 features to develop a technology of removing VOC (Volatile Organic Compounds) components which adsorbed among the filter internal matrixes.

Organic gas removal filters are used in various fields and their main application are as follows;

- Electronic device factories: To prevent the organic matter contamination to the product

- Commercial tenant buildings: To deodorize the exhaust gas from the commercial kitchen.

- Research institute: To remove the noxious gas from the local exhaust ventilation (draft chamber).

On the other hand, the organic gas filters are generally disposed and replaced by new ones when their removal performance is decreases, leading to an increase of industrial waste.From the viewpoint of the conservation of global environment, our proposed methodology enables to re-use such used-filters and can contribute to the reduction of industrial waste by using our re-use technology.

In the poster, we introduce the performance in the case of applying this technology to organic gas-removal filter used in the draft chamber of the actual research institute.

(395w) Removal of Heavy Metal Ions From Lean Amine Solvent Using Chitosan Coated Ion-Exchange Resins in a Gas Sweetening Plant

Pal, P.,

Banat, F.,

AlShoaibi, A., The Petroleum Institute

1. Introduction

In natural gas sweetening plant aqueous methyldiethanolamine (MDEA) solution is used to scrub natural gas. Finally, the rich MDEA solution is fed to a regeneration column where heat is applied to strip the acid gas components out of the MDEA solution. Make-up water and MDEA are added to maintain the concentration of MDEA solution. The heavy metal ions present as contaminant in lean amine solvent may come from make-up water. Many researchers identified the metal ions present in the lean amine solvents and removed using electrodialysis and ion-exchange (IE) resins. IE process outperforms electrodialysis technique in terms of higher energy requirements and continuous supply of huge volumes of water [1]. IE processes are capable of achieving very low salt concentrations [2], operating and installation costs are less and the requirements for regeneration chemicals have reduced currently [3]. Cummings and Smith [4] used continuous adsorption studies passing the alkanolamine solutions over cation bed where positively charged contaminants such as ferrous ions and sodium ions were exchanged with protons. Audeh and Yan [5] showed that weakly acidic cationic resins with carboxylic acid functionality outperform strongly acidic resins with a sulfonic acid groups.

The basic objective of this research work was to remove heavy metal ions (iron, chromium, lead etc.) accumulated in lean MDEA solution in gas sweetening process. MDEA containing both amine and hydroxyl group chelate and adhere heavy metal ions strongly. Thus bare ion-exchange resins cannot always successfully remove heavy metal ions from this lean amine solvent overcoming the chelation. The novelty of this research work lies in the use of chitosan coated ion-exchange resin which contains both the above groups present in the amine solvent and could hence successfully remove heavy metal ions from MDEA.

2. Materials and Methods

Lean MDEA solvents (45 weights %) were obtained from the Gasco, Habshan, Abu Dhabi. Commercially available ion-exchange resins (ZGC151) were obtained from Hangzhou Zhengguang Resin Co., Ltd., China. Chitosan and all other chemicals used in this study were purchased from Sigma Chemical Co., USA. At first, the bare resins were coated with chitosan. 1.0 gram chitosan powder was added to 100ml 5% acetic acid solution and left overnight in stirring condition to dissolve completely. 10 grams of ion-exchange resin was added into the chitosan solution and stirred for six hours, decanted and kept at vacuum oven at 60°C for curing. The resin beads was then added drop wise to 100 ml 4% sodium hydroxide solution and stirred overnight for chitosan coating. Chitosan coated ion-exchange resins were treated to regenerative form before their use in ion exchange batch sorption experiments.

Elemental analysis was carried out using Inductively Coupled Plasma Optical Emission Spectrometry (ICP-OES, Optima 8000). The FTIR studies of chitosan coated ion-exchange resins were carried out using FTIR, Nicolet, iS10, Thermo Scientific instrument using OMNIC software (number of scans 64, resolution 4). Scanning Electron Microscopy (SEM) was carried out with FEI Quanta 200, The Netherland.

3. Batch and Continuous ion exchange experiments studies

Ion exchange experiments were conducted using mixed batch reactor technique for obtaining the equilibrium parameters data for the removal of heavy metal ions using chitosan coated cation exchange resins. The amount of heavy metal ions adsorbed per unit mass of the resin (q_e, microgram/gram) at equilibrium was calculated as:

$$q_e = (V/m)*(C_o?C_e) \qquad (1)$$

Where V is the volume of the lean amine solution, m is the mass of the ion exchanger resin; C_o and C_e are the respective ion concentrations in solution (microgram/liter) initially and at equilibrium. The equilibrium studies were conducted with lean amine solvent on ion exchange resins at pH 10.6, at 296 K.

The equilibrium data were fitted on to well-known established sorption isotherm models:

The Langmuir isotherm

The Langmuir isotherm [6] model and its linearized form are usually described by:

$$1/q_e = (1/q_{max}k_L)(1/C_e) + 1/q_{max} \qquad (2)$$

Where, q_{max} is the maximum sorption capacity of anions (microgram/gram of resin), C_e is the equilibrium heavy metal ion concentration (microgram/liter) and k_L is the langmuir isotherm constant (liter/microgram).

The Freundlich isotherm

The Freundlich [7] isotherm in its linearized form is expressed as:

$$\ln(q_e) = \ln k_F + (1/n) \ln C_e \qquad (3)$$

In this equation k_F(microgram$^{(1-1/n)}$L$^{(1/n)}$/gram) and n are the Freundlich constants.

For continuous adsorption experiment in column, a pyrex tube of 2.54 cm inner diameter and 10 cm height was used. Thirty grams of chitosan coated pretreated cation exchange resin (ZGC 151) was randomly packed in the column to a bed volume (BV) of 30 cm^3. The lean MDEA solution was fed to the bottom of the resin column by a peristaltic pump at a rate of 4.5 ml/min. Samples were taken periodically at the top of the resin column for individual and total heavy metal content and measured using ICP-OES.

4. Results and discussion

4.1 FTIR Analysis

FTIR analysis was carried out to observe the adherence of chitosan on resin surface. Figure 1 shows the FTIR spectra of the two structures. From the FTIR spectrum, it could be concluded that chitosan was successfully coated on resin surface to remove heavy metal ions from the lean amine solvent.

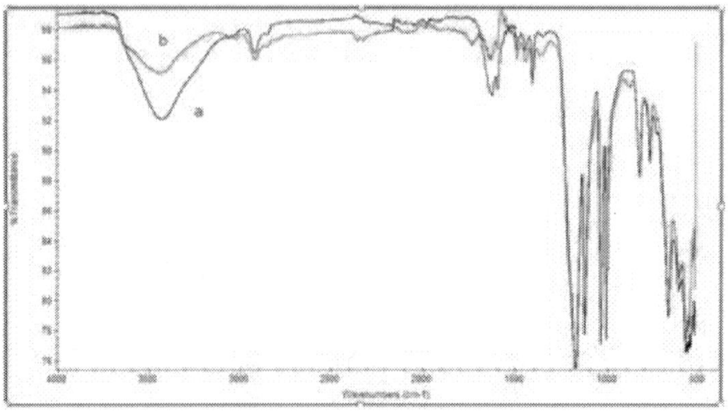

Figure1. FTIR Spectra of a) bare resin b) chitosan coated resin

4.2 SEM Analysis

Modification of the chitosan coated resin molecules followed by adsorption of MDEA was observed from the SEM images (Figure 2). Smooth chitosan coated resin surface was observed before adsorption (a) whereas surface becomes rough after adsorption reactions occurring between resin molecules and MDEA (b).

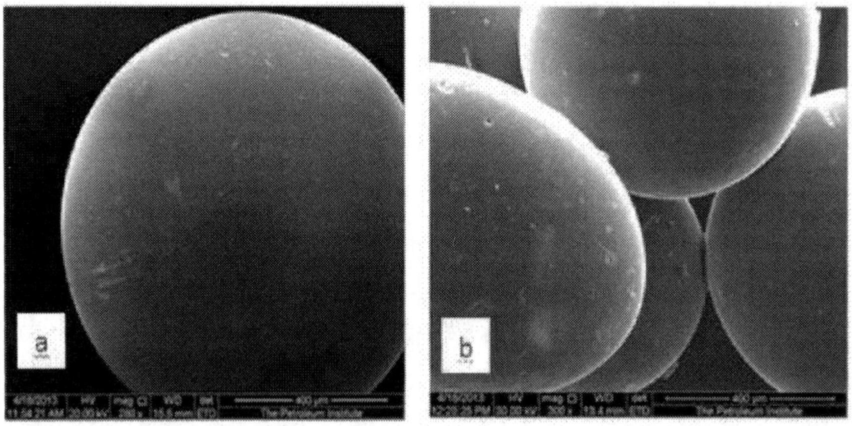

Figure2. SEM analysis of a) before and b) after adsorption studies

4.3 Heavy metal ions present in Lean amine samples

The major heavy metal ions present in amine samples [Chromium (479 ppb), Iron (967.1 ppb), Lead (1009 ppb), Magnesium (75.29 ppb), Manganese (111.7 ppb), Cadmium (48.32 ppb), Strontium (26.81 ppb) and Titanium (31.77 ppb)] were detected using ICP-OES.

4.4 Equilibrium Adsorption studies

The isotherm models gave an indication of the interaction between the heavy metal ions from solution with the ion exchange resin and their distribution over ion exchange surface. The corresponding isotherm models are shown in Figure3 and parameters are shown in Table1. Based on the data fit, major heavy metal ions follow a Langmuir type monolayer adsorption. The Freundlich isotherm predicted the constant, n calculated from linear fits falls in the range of favourable adsorption (n>1).

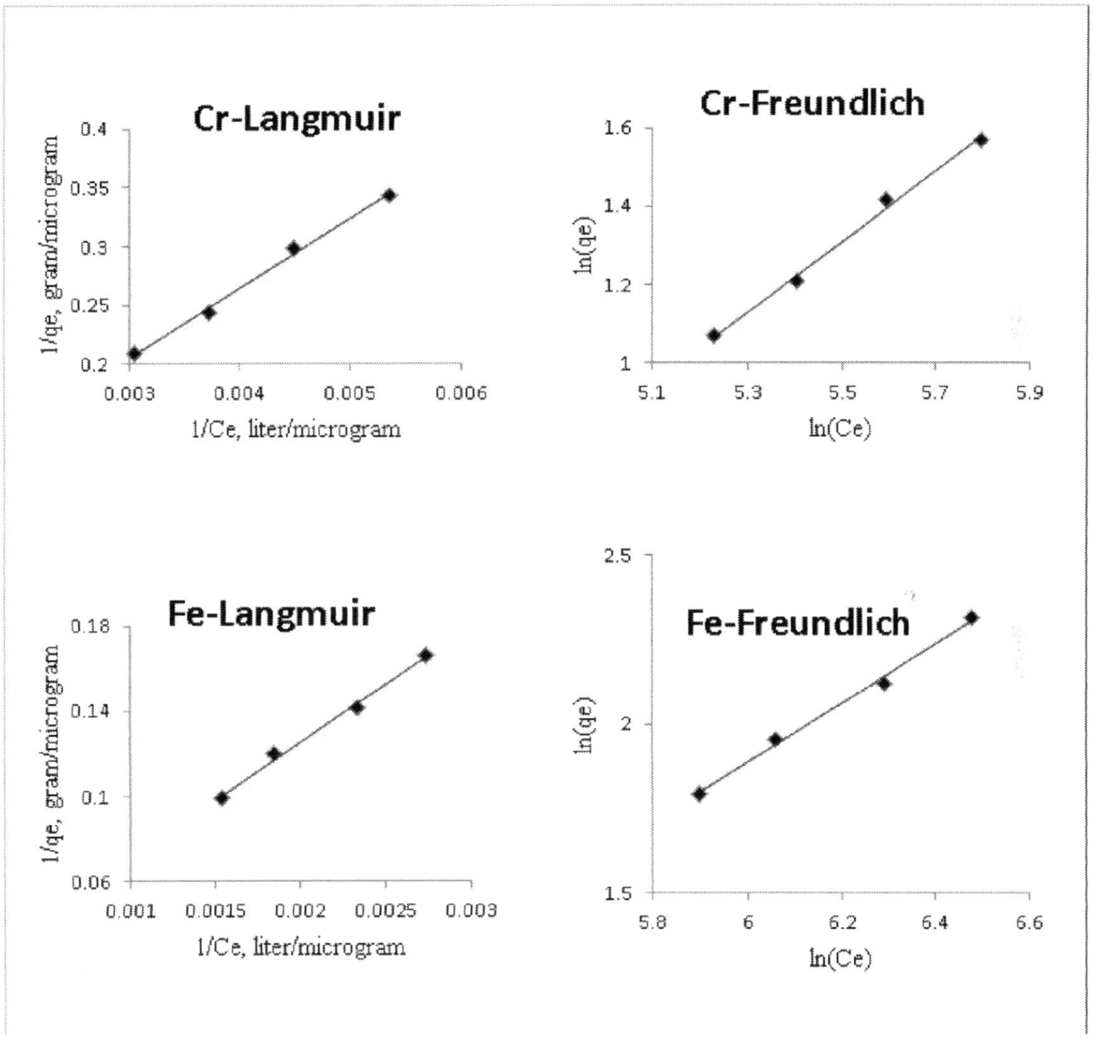

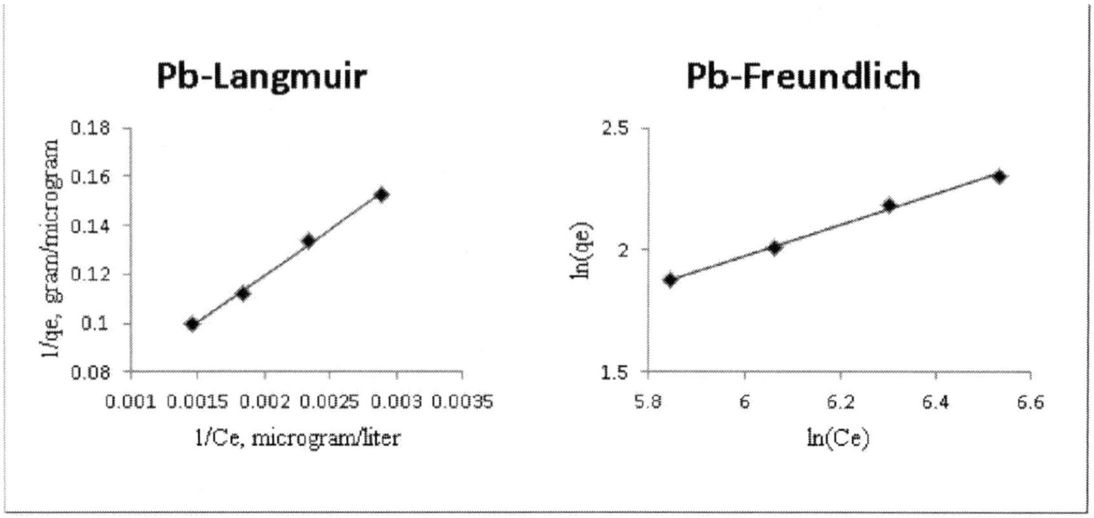

Figure3. Langmuir and Freundlich Isotherms for Chromium, Iron and Lead

Table1. Isotherm model parameters for the adsorption of heavy metal ions on ion exchange resin

Isotherm	Parameter	Chromium	Iron	Lead
Langmuir	q_{max} (microgram/gram)	38.022	59.88	22.47
	k_L (L/gram)	442.51	307.63	1181.9
	R^2	0.9951	0.9945	0.9972
Freundlich	k_F	0.0264	0.0352	0.1618
	n	1.112	1.146	1.579
	R^2	0.9945	0.9944	0.9952

4.5 Continuous Ion Exchange experiment studies of heavy metal ions from lean amine solution

A practical application of the experimental breakthrough curves is the determination of the breakthrough time, which helps establish the optimum operating conditions of the ion exchange process. It was observed that the breakthrough time was 8hours for same bed volume of the packed bed.

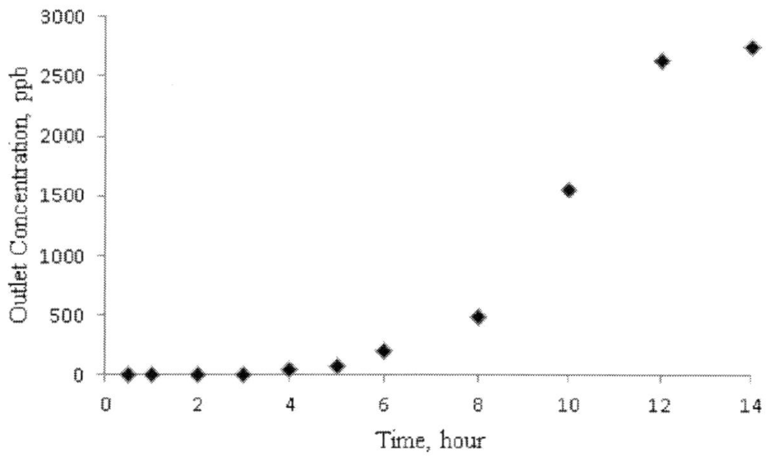

Figure4. The breakthrough curve for total heavy metal ions.

5. Conclusions

In this research, chitosan coated ion exchange adsorption was employed for efficient recovery of heavy metals from lean amine solution. Experimental tests were conducted to determine the equilibrium ion exchange capacities of the resins. Continuous adsorption experiment determined the breakthrough times of the ion exchange process and facilitates the process design by optimizing the operation conditions as useful to remove heavy metals from lean amine samples.

6. Acknowledgement

The authors are grateful to Gas Research Centre, The Petroleum Institute, Abu Dhabi for funding the projects. Sincere thanks to Mr. Alaa Abdul Aziz and Mr. Marwan Alawlqi, The Gasco, Habshan for their co-operation and support.

7. References

1. Hong, M., Shuang, Z., Chunxi, L., Liangshi, L., Removal of heat stable salts from aqueous solutions of N-methyldiethanolamine using a specially designed three-compartment configuration electrodialyzer, *Journal of Membrane Science*, 322, 436-440, 2008.

2. Burns, D., Gregory Jr., R.A., The UCARSEP: process for on-line removal of non-regenerable salts form amine units. In: Laurence Reid Gas Conditioning Conference, Norman, OK, 1995.

3. Cummings, A.L., Smith, G.D., Nelsen, D.K., Advances in amine reclaiming- why there's no excuse to operate a dirty amine system. In: Laurence Reid Gas Conditioning Conference, Dickinson TX, USA, 2007.

4. Cummings, A.L., Smith, G.D., Better alkanolamine system operations through chemical analysis. In: The Sulfur Recovery Symposium, Vail, CO, USA, 2010.

5. Audeh, C.A., Yan, T.Y., Clean-up of ethanolamine solution by treating with weak ion exchange resins.

United States Patent No. 802,586. Mobil Oil Corporation, Fairfax, VA., 1994.

6. Langmuir, I., The adsorption of gases on plane surface of glass, mica and platinum, Journal of *American Chemical Society*, 40, 1361, 1916.

7. Freundlich, H.M., Over the adsorption in solution, *Journal of Physical Chemistry*, 57, 385-470, 1906.

(395g) Electrochemical Synthesis of Metal-Organic Framework Based Microseparators

Van Assche, T., Vrije Universiteit Brussel
Campagnol, N., KU Leuven
Fransaer, J.,
Denayer, J., Vrije Universiteit Brussel

Electrochemical synthesis of metal-organic framework based microseparators

Tom R.C. Van Assche[a], Nicolo Compagnol'[b], Jan Fransaer[b], Joeri F.M. Denayer[a]

[a] Department of Chemical Engineering, Vrije Universiteit Brussel, Pleinlaan 2, B-1050 Brussel, Belgium
[b]Department of Metallurgy and Materials Engineering, KU Leuven, Kasteelpark Arenberg 44, B-3001 Leuven, Belgium

Microstructured devices offer the possibility of enhanced mass and heat transfer. Combining these devices with adsorbents allows for faster adsorption/desorption cycles using less material and enhanced chromatographic separations. Incorporating tailormade metal-organic frameworks (MOF) in such devices opens up new applications for specific separation problems, or by making use of their catalytic activity (i.e. unsaturated metal sites). In contrast to their inorganic counterparts (zeolites), MOF can be synthesized using a variety of synthesis strategies. From microdevice fabrication point of view, an interesting method is oxidative electrochemical synthesis of MOF.

During oxidative electrochemical synthesis, part of a metal electrode is electrochemically oxidized to form cations [1]. These cations subsequently coordinate with the linker present in the synthesis mixture. Advantages are the low temperature, short synthesis times, good atom efficiency and high process controllability. Under suitable conditions, the MOF can be grown as a well adhered and patterned film on the electrode, rather than a powder in the synthesis mixture. In this work, electrochemical MOF synthesis is combined with a facile microdevice fabrication technique to create an adsorption based microseparator [2]. This device is subsequently shown to outperform conventional packed bed technology in terms of adsorption kinetics.

A patterned MOF layer is achieved by either starting form a patterned conductive metal electrode, or more simply by using a patterned non-conductive mask on a metal electrode surface. By utilizing a self adhering patterned PEEK mask, microchannels can be produced, where the critical dimension is the PEEK layer thickness (200 or 500 μm). After 20-60 minutes of electrochemical synthesis, the masked electrode sheets can simply be clamped in a holder, utilizing the secondary function of the PEEK, as a gasket. The result is a rapid and facile method to produce a MOF-based microseparator, which was tested to 130°C and beyond.

For copper- benzene-1,3,5-tricarboxylate (HKUST-1), synthesis times are short and the synthesis can be performed at temperatures as lows as 30°C. The morphology can be controlled to a high degree, by varying the applied electrical potential, synthesis solvent and water content of the synthesis mixture. This allows the synthesis of HKUST-1 crystals of >0.5 µm - 50 µm diameter. By varying the temperature, octahedral or cubic crystals can be obtained.

For other MOF structures, well adhered layers are less evident to obtain by electrochemical synthesis. However, by combining the classical solvothermal synthesis methodology with an electrochemical cell, MIL-100 (Fe) coatings can be obtained on both electrochemically deposited (pure) iron, as well as steel [3]. Adsorption and separation properties of MIL-100 and HKUST-1 layers, were tested by fabrication of a microseparator device with a channel cross section of 2000 x 200 μm^2, and by performing breakthrough experiments.

For HKUST-1, the methanol adsorption kinetics on the MOF microdevice were improved compared to a conventional packed bed column (pellet size >0.6 mm). Here, the advantages of the microseparator are the absence of any large macropore diffusion resistance, low pressure drop and good heat transfer due to the intimate contact between the MOF layer and the conductive support. This fabrication methodology provides a facile, rapid manner to create MOF based microseparators and microreactors.

Acknowledgements:
T.R.C. Van Assche is grateful to the Agency for Innovation by Science and Technology (IWT) Flanders for financial support.

[1] Ameloot, R., Stappers, L., Fransaer, J., Alaerts, L., Sels, B.F., De Vos, D.E., 2009. Patterned growth of metal-organic framework coatings by electrochemical synthesis. Chem. Mater. 21, 2580–2582.
[2] Van Assche, T.R.C, Denayer, J.F.M, 2013. Fabrication and separation performance evaluation of a metal–organic framework based microseparator device. Chem. Eng. Sci. 95, 65-72.
[3] Compagnol', N., Van Assche, T., Boudewijns, T., Denayer, J., Binnemans, K., De Vos, D., Fransaer, J., 2013. High pressure, high temperature electrochemical synthesis of metal–organic frameworks: films of MIL-100 (Fe) and HKUST-1 in different morphologies. J. Mater. Chem. A 1, 5827-5830.

(395bc) Adsorption Characteristics of CO2, N2, CH4 and CO in CB6 Powder and Pellet

Moon, J. H., Korea Institute of Energy Research
Kim, H.,
Jin, G. T., Korea Institute of Energy Research
Park, Y. C., Korea Institute of Energy Research

The organic molecular porous material obtained by re-crystallization of cucurbit[6]uril (CB6) from HCl shows a high CO_2 sorption capacity at atmospheric condition. The adsorption characteristics, such as equilibrium and kinetics, of CO_2, N_2, CH_4 and CO molecules on a CB6 powder and CB6 pellet were investigated from 0 to 2 atm and from 283K to 343K using a gravimetric method. As adsorption mechanisms of CO_2, N_2, CH_4 and CO molecules on CB6, pore structures, molecular sizes/structures, adsorption affinities, steric hindrance, surface diffusion and equilibrium adsorption amount were considered. A magnetic suspension balance (MSB, Robotherm Co.) system was used for measuring transient adsorption amount and internal temperature. Various adsorption equilibrium models and kinetic models were introduced for comparing with experimental data. Diffusivity estimation tools and PDE (partial differential equation, 3D model) code for sorption kinetics were developed by using FORTRAN and MATLAB. Breakthrough tests of CO_2/N_2 mixtures were investigated in a CB6 pellet packed bed with varying CO_2 concentration, total flow rate, operating pressure and temperature.

(395bd) HCl Adsorption and Desorption Breakthrough Behaviors of Molecular Sieve 13X in a Fixed Bed Reactor

Moon, J. H., Korea Institute of Energy Research
Jo, S. H., Korea Institute of Energy Research
Jin, G. T., Korea Institute of Energy Research
Park, Y. C., Korea Institute of Energy Research

In this study, adsorption and desorption behaviors of the hydrogen chloride mixture (balance with nitrogen) using commercial sorbents (Molecular Sieve 13X) was studied with varying the flow rate, the concentration, the temperature and the pressure in a fixed bed reactor (15 cm tall bed with 0.5 cm I.d.).

As results, adsorption capacity of HCl was not affected by the inlet flow rate (0.5 l/min ~ 2.0 l/min) and the inlet concentration (200 ppm ~ 650 ppm). At 30Â°C ~ 90Â°C of the temperature ranges and 1 bar ~ 20 bar of the pressure ranges, the HCl sorption capacity decreases with increasing temperature, whereas it increases with increasing pressure. Since the MS-13X shows a physisorption behaviors for HCl removal, sorbent regeneration was possible. Cyclic operations with adsorption and desorption steps were also studied. On the basis of this result, such as the sorbent regeneration and the re-adsorption, we will develop thermal swing adsorption (TSA) process.

(395h) Propane/Propylene Separation in Ion-Exchanged ZMOFs

Demir, B., Istanbul Technical University
Ahunbay, M. G., Istanbul Technical University

Separation of propane/propylene mixtures formed during catalytic or thermal cracking of hydrocarbons is an important process in petrochemical industry. Conventional separation processes use widely cryogenic distillation, which requires high energy-consumption due to close relative volatilities of propane and propylene. Separation of propane/propylene mixtures via adsorption, such as PSA or TSA, is a less-energy demanding and environmentally friendly alternative. Conventional solid adsorbents such as zeolites are widely used in adsorption processes, but they suffer low capacity and selectivity in the case of propane/propylene separation. Metal organic frameworks (MOFs), on the other hand, can be good candidate as solid adsorbents. Zeolite-like MOFs (ZMOFs), a subclass of MOFs, have anionic structure and charge compensating extraframework cations. ZMOFs of rhombic and sodalite topologies, *rho*- and *sod*-ZMOFs, respectively, may exhibit customized gas separation performances due to their ion exchanging capabilities. [1]

In this study, propane/propylene separation performances of *rho*- and *sod*-ZMOFs were investigated at different temperatures and pressures via Grand Canonical Monte Carlo simulation method. Since the extraframework cation type is an important parameter in gas separation properties of ZMOFs, separation performances of Li^+, Na^+ K^+, Ca^{2+} and Mg^{2+}-exchanged structures of ZMOFs at different ratios were also evaluated by estimating adsorption selectivities and working capacities at different pressures.

Results showed that ZMOFs have strong potential for propane/propylene separation and their separation properties can be tuned through ion exchange. The extraframework cation type and ion-exchange ratio are other important parameters determining the adsorption capacity and selectivity of the materials for propane/propylene separation. Both the adsorption selectivity and the working capacity may be increased substantially by increasing the ion-exchange ratio. Furthermore, comparison with the separation performance of zeolites, such as LTA and 13X, suggests that the adsorption selectivities of ZMOFs may be as higher as an order of magnitude than those of the zeolites.

[1] Liu, Y.; Kravtsov, V. C.; Larsen, R.; Eddaoudi, *Chem. Commun.* **2006**, 1488-1490

(395i) Classical DFT for Large-Scale Screening of Nanoporous Materials

Fu, J., University of California, Riverside
Liu, Y., University of California, Riverside
Wu, J., UC Riverside

Metal-organic frameworks (MOFs) are porous crystalline materials formed by coordination bonding of organic ligands and metallic or organometallic centers. Different from typical intermolecular interactions that control the structure of a conventional crystalline solid, the coordination interactions are short-ranged and strongly anisotropic, much like covalent bonds underlying polymerization of monomeric units in a linear polymer. The strength and inherent stability of coordination bonds make the framework structure predictable from the coordination numbers and the geometry of the metal-ligand complexes. Such structure control at the molecular level renders unprecedented fast development of MOF-based materials with diverse and exceptional properties. However, the broad diversity of metallic nodes and organic linkers also imply large variations of MOF structures that overwhelm even most efficient synthesis strategies. Computational methods that allow fast and reliable predictions of physiochemical properties and *in silico* screening will drastically accelerate the process of MOF materials design and utilization. In this presentation, we introduce classical DFT as an accurate yet computationally efficient method for predicting gas adsorption in MOF materials. The numerical performance of this new method has been calibrated by extensive comparison with simulation results for gas adsorption in slit pores, spherical cages and various nanostructured materials including SWCNT and MCM-41. In addition, we compared the theoretical predictions with results from GCMC simulation at various temperatures and pressures for H_2 storage in a library of over a few thousands of MOF structures (Fig. 1). Because the classical DFT is much faster than GCMC simulation, it promises experimental scientists a convenient yet powerful tool for discovering novel MOF materials.

(395f) Annular Moving Bed Adsorber for Upgrading Natural Gas

Gupta, R. K., National University of Singapore
Thakur, R. S., GGV, Bilaspur
Kaistha, N., IIT Kanpur
Rao, D. P., Process Intensification Consultants

In the conventional fixed bed Pressure Swing Adsorption, most of the bed is inactive as the length of the mass transfer zone is many times smaller than the bed length. Further, the attainable driving force is small as the gas is in equilibrium with the adsorbent on either side of the mass transfer zone. Therefore the productivity and recoveries are low. However, these limitations can be overcome by using a 'simulated' moving bed (SMB) that ensures full utilization of the bed due to the countercurrent contact of the solids and the gas. Thermal swing offers an advantage over pressure swing as the thermal energy can be partially recovered by using heat regenerators that significantly reduce the cost of regeneration. Sivakumar and Rao (2012) proposed a thermal swing SMB with thermal regeneration using two fixed adsorbent beds and an inert bed for the recovery of thermal energy [1]. In this scheme too a large portion of the beds are inactive leading to low productivity.

We propose here an Annular Moving Bed adsorber (AMB) in which the inactive zones are eliminated altogether. The AMB is similar in construction to the rotary hearth furnace used in metallurgical industries. In addition, another concentric annular bed of inert solids, divided into two zones, is provided above the annular adsorption bed to facilitate the heating and cooling for the regeneration.

Simulation studies have been carried out to assess the efficacy of the AMB for upgrading lean natural gas over activated carbon. A non-isothermal model for the annular moving bed has been employed. The simulation studies show that for an operation at a pressure of 1 bar, the energy required for circulation of the gas for regeneration is more than that required for conventional PSA. However, if the operation is carried out above 10 bar, the cost of energy requirements is much smaller than conventional PSA and the productivity is several times higher than SMB and PSA. The AMB appears to be suitable for upgrading large volume of natural gas available at pressure above 10 bar.

(395x) Molecular Simulation Study of Argon and Krypton Confined Inside Slit Pores

Sengupta, A., Indian Institute of Technology, Bombay.

Adhikari, J., Indian Institute of Technology, Bombay.

Molecular simulation study of Argon and Krypton confined inside slit pores.

Angan Sengupta* and Jhumpa Adhikari+

Department of Chemical Engineering, Indian Institute of Technology, Bombay.

Mumbai ? 400076. India.

Grand canonical Monte Carlo (GCMC) simulations have been used extensively in recent years to study the adsorption and the phase behaviour of confined fluids inside various pore geometries. These molecular simulation studies enable an easy fundamental understanding of the distinct behaviour of a confined fluid which arises due to the presence of wall-fluid interactions in addition to the fluid-fluid interactions. We have investigated two mono-atomic fluids, viz. Argon and Krypton, confined inside the simplest pore model available, slit pore. The wall fluid interactions are given by equation 1[1].

$$\varphi_{wf}(z) = \begin{cases} -\varepsilon_{wf} & \left(-\frac{H}{2}+\frac{\sigma_{ff}}{2}\right) < z < \left(\frac{H}{2}-\frac{\sigma_{ff}}{2}\right) \\ \infty & \text{otherwise} \end{cases}$$

The fluid-fluid interactions in the system, have been modelled using a simple pair potential[1] energy function called the triangle-well (TW) model (equation 2)[2], which is more realistic than the square-well model; and being segmental in nature this potential is computationally faster than the Lennard-Jones model. Parameters of the TW model are: core diameter, σ_{ff}; well width, λ; and well depth, ε_{ff}.

$$\varphi_{ff}(r_{ij}) = \begin{cases} \infty & r_{ij} \le \sigma_{ff} \\ \left[\frac{\lambda\varepsilon_{ff}}{\lambda-1}\right]\left[\frac{r_{ij}}{\lambda\sigma_{ff}}-1\right] & \sigma_{ff} < r_{ij} \le \lambda\sigma_{ff} \\ 0 & r_{ij} > \lambda\sigma_{ff} \end{cases} \quad [2]$$

These parameter values for Argon and Krypton have been obtained by fitting the expression for the second virial coefficient determined from the Mayer's equation with the experimental data reported in literature[3,4]. Verification of the parameters have been done by performing the isothermal-isobaric ensemble simulations and determining the densities of both the fluids over a range of temperatures and pressures which were found to be in excellent agreement with the experimental data[5,6].

In the present work, GCMC simulations are carried out inside slit pores of heights 9 Å and 15 Å at 246 K and 300 K. The adsorption isotherms follow the same trends as reported by Hung and co-workers[7] from experiments as well as simulations, where Ar and Kr (modelled as LJ fluid) was confined inside meso-porous silica. For a given slit height, the change in the densities of both Ar and Kr; indicates a vapour ? liquid phase transition (a discontinuity in the adsorption isotherms) at higher values of ε_{wf}. The density changes continuously at relatively lower values of ε_{wf}, including for $\varepsilon_{wf} = 0$ (hard wall pores). The value of ε_{wf} above which this sudden change from vapour-like to liquid-like densities is observed depends on the type of fluid (Ar or Kr), the temperature and the pore volume. The density profiles show layering behaviour to occur perpendicular to the pore walls at progressively lower pressures as the slit height increases, at a given temperature. Also, noted is that the number of layers for Ar is less than that for Kr at the same temperature, pressure and pore height, though $\sigma_{ff}|_{Ar} < \sigma_{ff}|_{Kr}$. Inside the wider pores, enhanced adsorption increases the number of layers for both the confined fluids. Moreover, layering is also observed inside the hard wall pores at relatively higher pressures. Thus, the complex behaviour (layering and phase change during adsorption) of confined Ar and Kr, due to the effects of temperature, pressure, pore attractiveness and slit height are duly addressed using molecular simulation with two of the simplest potential models available in literature.

References:

1. Hamada, Y.; Koga, K. and Tanaka, H. Phase equilibria and interfacial tension of fluids confined in narrow pores. *Journal of Chemical Physics*, 127, 084908, 1 ? 9. (2007).

2. Feinberg, M. J. and Rocco, A. G. De. Intermolecular forces: the triangle well and some comparisons with the square well and Lennard-Jones. *Journal of Chemical Physics*, 41, 11, 3439 ? 3450. (1964).

3. Byrne, M. A., Jones, M. R. and Staveley, L. A. K. Second virial coefficients of Argon, Krypton and Methane and their binary mixtures at low temperatures. *Transactions of the Faraday Society*, 64, 1747 ? 1756. (1967).

4. Sevast'yanov, R. M. and Chernyavskaya, R. A. Virial coefficients of neon, argon, and krypton at temperatures up to 3000 K. *Translated from Inzhenerno-Fizicheskii Zhurnal*, 52, 6, 974 ? 977. (1986).

5. Van, I. A. and Verbeke, O. Density of liquid Nitrogen and Argon as a function of pressure and temperature. *Journal of Physica*, 26, 931 ? 938. (1960).

6. Trappeniers, N. J.; Wassenaar, T. and Wolkers, G. J. Isotherms and thermodynamic properties of Krypton at temperatures between 0° and 150^oC and at densities up to 620 amagat. *Journal of Physica*, 32, 1503 ? 1520. (1966).

7. Hung, R. F.; Bhattacharya S.; Coasne, B.; Thommes, M. and Gubbins, K. E. Argon and Krypton adsorption on templated meso-porous silicas: molecular simulation and experiment. *Journal of Adsorption*, 13, 425 ? 437. (2007).

+ *Corresponding author.* ***E-mail address:*** adhikari@iitb.ac.in

* *Presenting author.* ***E-mail address:*** *angan.sengupta@gmail.com*

(395y) Adsorption Equilibrium Properties of CO2 and Methane in MIL-100(Fe) Predicted By Monte Carlo Simulation

Goncalves, D. V., Universidade Federal do Ceara
Silvino, P. F. G., Universidade Federal do Ceara
Lucena, S. M. P., Universidade Federal do Ceara

MIL-100 is a mesoporous MOF material built by the connection of hybrid supertetrahedra which further assemble forming very large cylindrical pores. This particular framework gives high gas storage capacities. In this study, we investigated adsorption equilibrium properties of CO2 and methane in MIL-100 (Fe) using the Monte Carlo methodology in the grand canonical ensemble (GCMC). We develop a new consistent force field for both gases and compared with the general force fields. Adsorption isotherms were performed at room temperature using united atom (UA) and atom-atom (AA) gas models. Among the general force fields tested the Universal gave better results. The simulated isotherm using methane united atom model and CO2 atom-atom model gave the best agreement with the experimental data. We found that both gases did not adsorb into the supertetrahedra. NVT simulations at low temperature were carried out to estimate adsorption sites. Potential maps of selected sections of the framework were performed to give the energy of each site. Adsorption heat also was estimated.

(395j) Adsorption Of Xylene Isomers In MOF UiO-66 By Molecular Simulations

Granato, M. A., LSRE - Laboratory of Separation and Reaction Engineering - Associate Laboratory LSRE/LCM, Faculdade de Engenharia - Universidade do Porto

Martins, V. F. D., LSRE - Laboratory of Separation and Reaction Engineering – Associate Laboratory LSRE/LCM, Faculdade de Engenharia - Universidade do Porto

Ferreira, A., LSRE - Laboratory of Separation and Reaction Engineering - Associate Laboratory LSRE/LCM

Rodrigues, A. E., University of Porto - Faculty of Engineering

ADSORPTION OF XYLENE ISOMERS IN MOF UiO-66 BY MOLECULAR SIMULATIONS

Miguel Angelo Granato[1], *Vanessa F. Duarte Martins*[1], *Alexandre Filipe P. Ferreira*[1] *and Alírio E. Rodrigues*[1]

[1] *LSRE - Laboratory of Separation and Reaction Engineering ? Associate Laboratory LSRE/LCM, Faculdade de Engenharia, Universidade do Porto, Rua Dr. Roberto Frias, 4200-465 Porto, Portugal*

Abstract

This work presents results of molecular simulations on adsorption of the four xylene isomers in the porous zirconium terephthalate UiO-66(Zr). The grand-canonical Monte Carlo simulations (GCMC) are compared to multi-component adsorption equilibrium data obtained by breakthrough experiments (Bárcia *et al.*, 2011). Additionally, the adsorption behavior of each pure isomer was evaluated, in order to better understand the multicomponent adsorption. The adsorbate molecules were simulated using four different models:

- TraPPE-UA (Wick *et al.*, 2000): In this model, the aromatic pseudo-atoms [CH(aro) and R-C(aro) for the link to aliphatic side chains] are treated as a single interaction center as well as the CH_3-sp^3 and the CH_2-sp^3 aliphatic pseudo-atoms.

- OPLS (Jorgensen *et al.*, 1993): It is an all-atom (AA) model used for the substituted benzenes except the CH_3-sp^3 aliphatic groups which are treated as united atoms centered on the carbon.

- Extended TraPPE (Wick *et al.*, 2002): This new TraPPE?UA representation of the arenes contains a positive partial charge (q_{center} = +2.42 e), placed at the center of the ring, and two negative partial charges representing the p-electron clouds (q_{pi} = -1.21 e), placed at a distance of 0.785 Å from the ring center and on a line perpendicular to the arene plane.

- Charged-AUA model introduced by Nieto-Draghi *et al.* (2007a, b). This model describes the interactions inside the p-cloud of electrons by a positive partial charge (q_{center} = +8.13 e), and two negative partial charges (q_{pi} = -4.065 e), located at 0.4 Å above and below the ring plane on its hexagonal symmetry axis.

The simulations confirm that the experimentally observed *ortho*-selectivity is preferential in relation to the other isomers. Access to the small cages seems to be favored for *ortho*-xylene due to its more compact bulky structure that provides more advantages in terms of rotation degrees of freedom. Furthermore, it was found that there is a competition between the other three isomers (*para*-xylene, *meta*-xylene and ethylbenzene) for adsorption in the big cages. Molecular simulations allow prediction of single component adsorption isotherms, heats of adsorption, and selectivity properties of xylene isomers mixtures which are essential for the development of adsorption based separation processes.

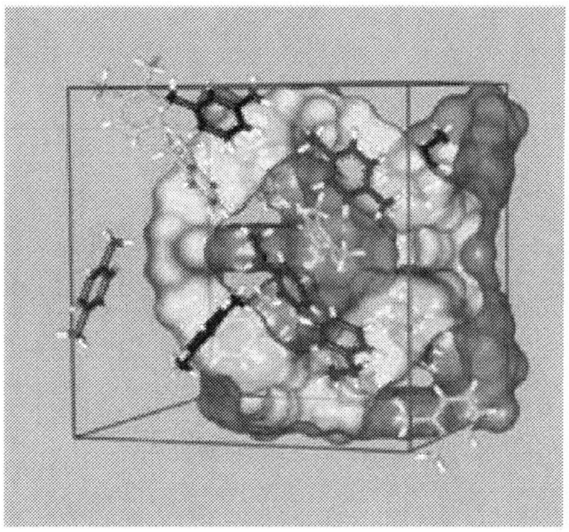

Figure1 *.Example snapshot of four components adsorption in UiO66. Green: accessible surface; light blue: o-xylene; blue: p-xylene; dark blue: m-xylene; dark red: ethylbenzene.*

References

Bárcia, P.S.; Guimarães, D.; Mendes, P.A.P.; Silva, J.A.C.; Guillerm, V.; Chevreau, H.; Serre, C.; Rodrigues, A.E. ?Reverse shape selectivity in the adsorption of hexane and xylene isomers in MOF UiO-66.? *Microporous and Mesoporous Materials,* **139**, 67-73 (2011).

Jorgensen, W.L.; Laird, E.R.; Nguyen, T.B.; Tirado-Rives, J. ?Monte Carlo Simulations of Pure Liquid Substituted Benzenes with OPLS Potential Functions.? *J. Comput. Chem.,* **14**, 206-215 (1993).

Nieto-Draghi, C.; Bonnaud P.; Ungerer, P. ?Anisotropic United Atom Model Including the Electrostatic Interactions of Methylbenzenes. I. Thermodynamic and Structural Properties?. *J. Phys. Chem. C,* **111**, 15686-15699 (2007a).

Nieto-Draghi, C.; Bonnaud P.; Ungerer, P. ?Anisotropic United Atom Model Including the Electrostatic Interactions of Methylbenzenes. II. Transport Properties?. *J. Phys. Chem. C,* **111**, 15942-15951 (2007b).

Wick, C.D.; Martin, M.G.; Siepmann, J.L. ?Transferable Potentials for Phase Equilibria. 4. United-Atom Description of Linear and Branched Alkenes and Alkylbenzenes.? *J. Phys. Chem. B.* **104**, 8008-8016 (2000).

Wick, C.D.; Siepmann, J. I.; Klotz, W. L.; Schure. M. R. ?Temperature effects on the retention of n-alkanes and arenes in helium?squalane gas?liquid chromatography Experiment and molecular simulation?. *J. Chromatogr. A,* **954**, 181?190 (2002).

(395aa) Ion-Exchanged Sod-ZMOFs for CO2 Capture Via Adsorption

Demir, B., Istanbul Technical University
Erdem, H. O., Istanbul Technical University
Ahunbay, M. G., Istanbul Technical University

Separation of CO_2 from streams containing CH_4 and N_2 is a still challenging problem in the era where low-energy consuming processes are studied extensively. CO_2 present in natural gas causes decrease in energy content and corrosion in aqueous medium. Therefore, removal of CO_2 from natural gas is still remaining a major problem in petroleum industry. Similarly, the discharge of CO_2-containing flue gas from coal-fired power plants is also a serious concern due to global warming, and to decrease CO_2 emission into the atmosphere remains a serious problem.

Amine absorption have been conventionally useds to capture CO_2 from CO_2/CH_4 and CO_2/N_2 mixtures. However, separation of CO_2 via adsorption or membrane-based separation processes, are less-energy demanding and environmentally friendly alternatives. Zeolites are widely used as solid adsorbents in adsorption processes, or as fillers in mixed-matrix membranes but they have limited separation capacity and selectivity. Recently, due their high free volumes and tunable porosities, metal organic frameworks (MOFs) have been proposed to replace zeolites in gas separation processes. A subclass of MOFs, zeolite-like metal organic frameworks (ZMOFs) have anionic structure and charge compensating extraframework cations, such as alkali earth metals.

In this study, separation of CO_2/CH_4 and CO_2/N_2 gas mixtures was studied in ZMOFs of sodalite topology (*sod*-ZMOFs) via Monte Carlo simulations. Gas separation performances of "as-synthesized" and partially and completely Li^+-, Na^+- and K^+-exchanged ZMOFs were evaluated on the basis of adsorption selectivities and working capacities at different operating pressures. The working capacity is an important parameter in evaluating adsorbents as a gas separation material and economics of gas separation processes such as pressure swing adsorption (PSA).

Results showed that both the adsorption selectivities and working capacities were enhanced by increasing the ion-exchange ratio. Thus, completely ion-exchanged *sod*-ZMOFs, particularly K^+-exchanged ZMOF showed the best performance in capturing CO_2 from CO_2/CH_4 and CO_2/N_2 mixtures.

(395k) Synthesis and Characterization of Ce-La Binary Oxide for Efficient Fluoride Removal

Li, G., National University of Singapore
Ma, Y., National University of Singapore
Chen, J. P., National University of Singapore

Abstract

In this study, a novel Ce-La binary oxide sorbent was synthesized by a co-precipitation method and the adsorption performance of the sorbent for fluoride removal was investigated. The best molar ratio of Ce/La was proved to be 1:2, and the co-precipitation pH was selected as pH 11.0. By potentiometric titration study, the pH_{zpc} of the sorbent was found to be around pH 7.8. Adsorption kinetics study showed that most fluoride uptake took place at the first 4 h followed by a relative slow process, and the adsorption equilibrium could be obtained within 10 h. The kinetic data correlated well with the pseudo-second-order kinetics model. The optimum pH range for the fluoride adsorption is 3-10. Isotherm study indicated that the Langmuir model can be well used for the description of the fluoride adsorption. These coexisting anions such as bicarbonate, phosphate and silicate significantly affected the uptake of fluoride by competitive adsorption.

(395l) Characteristics and Defluoridation Performance of Synthesized Zr-La Binary Oxide

Li, G., National University of Singapore
Ma, Y., National University of Singapore
Chen, J. P., National University of Singapore

Abstract

The Zr-La binary oxide synthesized under optimized conditions and its adsorption behavior for fluoride removal was investigated. The best molar ratio of Zr/La was proved to be 1:2, and the best co-precipitation pH was selected to be pH 11.0. By potentiometric titration study, the pH of zero point charge of the sorbent was found to be around 4.2. The kinetic study showed that the uptake of fluoride reached an equilibrium adsorption by 10 h. The kinetic data correlated well with the pseudo-second-order kinetics model and it indicates that the adsorption process might be chemical sorption. For fluoride removal, Langmuir model fitted the isotherm data better and the maximum adsorption capacity can be calculated as 88.06 mg/g. These coexisting anions such as bicarbonate, phosphate and silicate significantly affected the uptake of fluoride by competitive adsorption, while the coexistence of sulphate, arsenate and humic acid had a negligible influence on fluoride adsorption.

(395z) Modeling of Pb (II) Biosorption By Spent Coffee Ground in Fixed Bed Column

Davila-Guzman, N. E., Universidad Autonoma de Nuevo Leon
Cerino, F., Universidad Autonoma de Nuevo Leon, Facultad de Ciencias Quimicas
Soto, E., Universidad Autonoma de Nuevo Leon, Facultad de Ciencias Quimicas
Rangel-Mendez, J. R., IPICyT
Loredo, J. A., Universidad Autonoma de Nuevo Leon, Facultad de Ciencias Quimicas
Diaz-Flores, P. E., Universidad Autonoma de San Luis Potosi

Modeling of Pb (II) biosorption by spent coffee ground in fixed bed column

Biosorption of lead metal ions by spent coffee ground in a fixed bed column was predicted through a mathematical model based on the ion exchange between protons and calcium ions in the biomass and lead metal ions from aqueous solution. The breakthrough curves in a fixed bed column were obtained with a flow rate of 5.5, 7.3, 11 and 22 ml/min and a metal concentration of 0.1 M, the maximum removal efficiency was 43.4 % at 5.5 ml/min. The mathematical model that describes the dynamics of the breakthrough curve at 5.5 ml/min was obtained from the mass balance including a global mass transfer coefficient, axial dispersion coefficient and affinity constant. The parameters of the model were obtained from mathematical correlations and the affinity constant from the ion exchange multicomponent isotherm model. A sensitive analysis was used to determine the significant parameters of the model on the breakthrough time. The global mass transfer coefficient was not significant in the mathematical model. According to the lower error value, the mass transfer model for the biosorption of lead ions in a fixed bed of spent coffee ground was found to agree with the experimental data.

(395ab) Molecular Simulation of Heavy Alkanes Accumulation in Natural Gas Storage Cycles

Lucena, S. M. P., Universidade Federal do Ceara
Gomes, V. A., Universidade Federal do Ceara
Mileo, P. G. M., Universidade Federal do Ceara
Goncalves, D. V., Universidade Federal do Ceara
Silvino, P. F. G., Universidade Federal do Ceara

The storage efficiency of activated carbon systems can be hindered by trace higher molecular weight alkanes in the natural gas stream. Currently, deactivation of the carbon bed by alkanes is assessed through costly and time-consuming methods. In this study we propose a new methodology that uses only the carbon pore size distribution as input to predict the bed deactivation into storage cycles of natural gas. The method consists in approximating the pore size distribution with representative pores of each adsorption methane regimes, and performing mixture isotherms calculation using the Monte Carlo algorithm in the grand canonical ensemble. Bed deactivation for two species of activated carbons WV1050 and BPL were estimated at 17% and 39% while studies in the literature indicate values of 19% and 40% respectively. The methodology also allowed us to estimate the optimal pore size for maximum storage capacity and minimal retention of heavier fractions which is significantly useful to guide the synthesis of carbonaceous materials with improved performance for storage of natural gas.

(395ae) Adsorption of Benzene and Toluene in Supercritical CO2 On Molecular Sieving Carbon Using Stop & Go Method

Chihara, K., Meiji University
Shinsako, M., Meiji University
Yasuda, T., Meiji University
Ito, S., Meiji university

Chromatographic measurements were made for the adsorption of organics (benzene or toluene) on molecular sieving carbon (MSC) in supercritical fluid CO_2 mixed with organics. Supercritical CO_2 mixed with small amount organics (benzene or toluene) flowed as carrier in chromatograph. Adsorbates applied as impulse used same or different organics mixed with supercritical CO_2. Supercritical chromatograph packed with MSC was used to detect impulse responses of organics. Adsorption equilibria and adsorption dynamics parameters for organics were obtained by moment analysis of the impulse response peaks.

Stop & Go method was used as one of the simulation methods of perturbation chromatography which examine adsorption equilibrium and rate. Experimental adsorption equilibria and adsorption rate parameters for organics and CO_2 were obtained using Stop & Go simulation. Fig. 1 shows basic concept of this simulation which was N completely-mixed cell in axial direction of column. If the flow stopped only dt second, the adsorbate was made batch adsorption into each cells. After dt second, think material balance in adsorption and desorption of each cell within a stopped time frame.

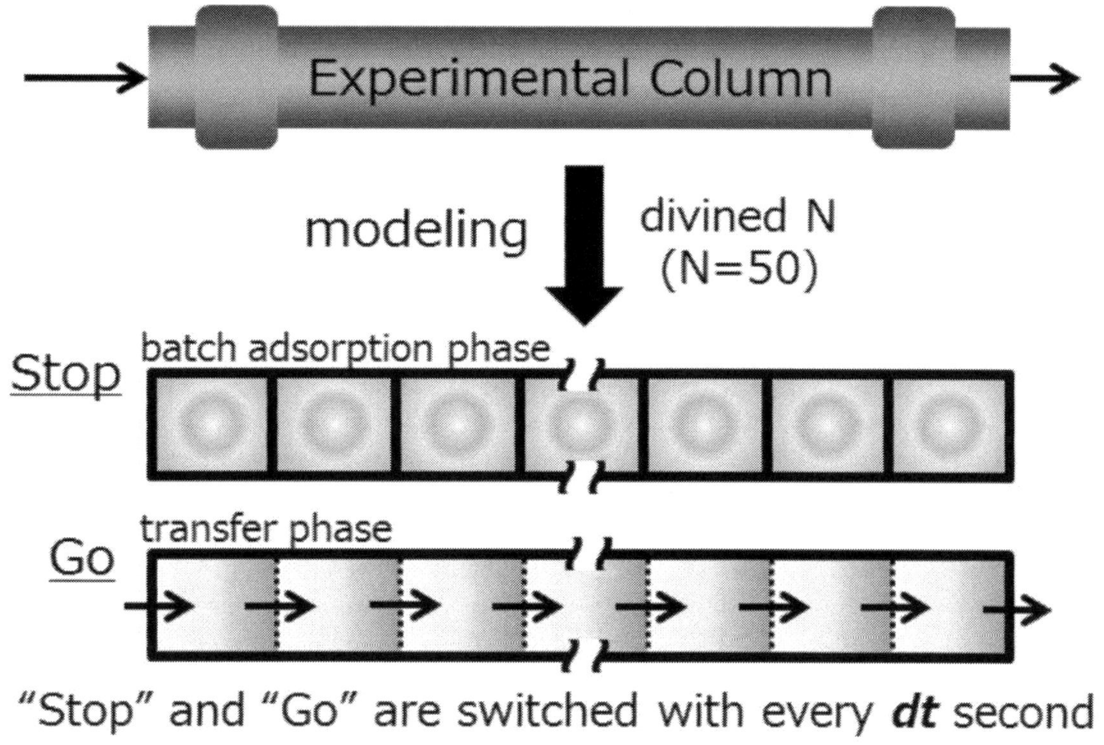

Fig. 1 Stop & Go method basic concept

(395ac) Free Energy Analysis of Adsorption-Induced Structural Transition in ZIF-8: A Molecular Simulation Study

Ohsaki, S., Kyoto University
Tanaka, H., Kyoto University
Yamamoto, D., Doshisha University
Watanabe, S., Kyoto University
Miyahara, M., Kyoto University

ZIF-8 is one member of a family of zeolitic imidazolate frameworks [1,2], and has attracted much attention in many applications because of its high thermal stability and intra-framework flexibility. Recent experimental and simulation studies have revealed that reorientation of the imidazolate linkers of ZIF-8 is induced by molecular adsorption, which lead to increase in the accessible pore volume of ZIF-8. However, the mechanism of the structural transition is still unclear because of the lack of knowledge about free energy change due to the reorientation of the imidazolate linkers. We have therefore focused attention on argon adsorption in ZIF-8, and performed the free energy analysis [3] with the aid of grand canonical Monte Carlo (GCMC) simulations to investigate the adsorption-induced structural transition of ZIF-8.

Adsorption isotherms of argon in ZIF-8 were measured at 79 K, 83 K, 87 K, and 91 K. A hysteresis loop was observed in all the adsorption isotherms, and its width increased with increasing temperature, which should provide important insight into the mechanism of the adsorption-induced structural transition of the argon/ZIF-8 system.

We constructed atomistic ZIF-8 models by rotating the imidazolate linkers about an axis through two nitrogen atoms from 0 ° to 30 °, and a series of adsorption isotherms of argon in the respective ZIF-8 models were obtained by the GCMC simulations. We used the following force fields with modifications: the universal force field for argon-ZIF-8 interactions and the Amber force field to calculate the potential energy of the ZIF-8 framework. The free energy of the system was calculated as a sum of the Helmholtz free energy of the ZIF-8 framework and the contribution of adsorbed argon obtained by integrating the simulated adsorption isotherm. In the adsorption process, with increasing pressure, the global minimum of the free energy shifts to a larger rotational angle of the imidazolate linker without any activation processes, and a second local minimum appears at around 25.5 °. Further increase in pressure provides a switch of the global minimum, and a spontaneous structural transition takes place when the energy barrier lying between the metastable state at the smaller rotational angle (10.5 °) and the global minimum at the larger rotational angle (25.5 °) becomes less than the energy fluctuation of the system of 0.51 kT per an imidazolate linker. The energy fluctuation of the system determined from the comparison with the experimental data is quite suggestive because its value is comparable in magnitude to the energy of one rotational degree of freedom of the imidazolate linker. Then, in the desorption process, the system stays at the local minimum at 25.5 °

regardless of the decrease in pressure, and finally an equilibrium structural transition occurs from 25.5 ° to 10.5 °. The fact that the equilibrium structural transition is accomplished in the desorption process differently from the adsorption process is an indication that the energy fluctuation of the system is larger than 0.51 kT/linker. This is reasonable because the energy fluctuation of the system should be proportional to the adsorption amount of argon. Actually, the absorbed amount before the equilibrium structural transition in the desorption process is about 1.2 times larger than that before the spontaneous structural transition in the adsorption process. Finally, we constructed a theoretical adsorption isotherm of argon in ZIF-8 by using the free energy analysis, and obtained a good agreement with the experimental isotherm over the entire range of pressures. This suggests that our scenario for the adsorption and desorption processes obtained from the free energy analysis is quite appropriate to describe the mechanism of the adsorption-induced structural transition of ZIF-8.

[1] X. C. Huang *et al., Angew. Chem. Int. Ed.* 45, 1557 (2006).

[2] K. S. Park *et al., Proc. Natl. Acad. Sci. U.S.A.* 103, 10186 (2006).

[3] S. Watanabe *et al., J. Chem. Phys.* 130 164707 (2009).

(395ad) Mechanism of Adsorption Hysteresis in Mesoporous Silica With Surface Roughness

Hiratsuka, T., Kyoto University
Tanaka, H., Kyoto University
Miyahara, M., Kyoto University

The siliceous mesoporous materials such as MCM-41 and SBA-15 have attracted much attention because of their potential use as catalyst supports, separation, and drug delivery, etc. These materials are also regarded as the most suitable model adsorbent for a fundamental study of capillary condensation. Vapor-liquid phase transitions of fluids in the mesoporous silica materials have been extensively studied by experiments, theory, and molecular simulations, and these studies reveal that adsorption hysteresis loop depends on pore diameter and temperature of the system. However, the mechanism of the adsorption hysteresis has still been a missing piece of the puzzle since MCM-41 appeared in 1992, because cylindrical pore models with smooth and structureless wall have been used to understand the capillary condensation behavior. Recent theoretical and simulation studies have shown that surface roughness of pore wall affects adsorption and capillary condensation of fluids in the siliceous mesoporous materials [1,2]. We therefore constructed an atomistic silica pore model mimicking MCM-41, which has molecular-level surface roughness, with the aid of the electron density profile (EDP) of MCM-41 obtained from X-ray diffraction data, and performed the molecular simulations to understand the mechanism of the adsorption hysteresis.

Molecular dynamics simulation was used for constructing a bulk fused silica with the BKS type potential. The system was equilibrated in the canonical ensemble at 4000 K, and subsequently, the fused silica was quenched from 4000 K to room temperature. To obtain an atomistic MCM-41 model, the amorphous silica block was carved out to reproduce the experimentally determined EDP of MCM-41 [3]. Adsorption isotherms of Lennard-Jones argon and nitrogen on the atomistic MCM-41 model at 75 K, 80 K, and 87 K were obtained by the grand canonical Monte Carlo (GCMC) method and the gauge cell Monte Carlo (MC) method [4]. Adsorption isotherms of argon and nitrogen on MCM-41 samples (pore diameter: 3.2 nm and 4.0 nm) at 75 K, 80 K, and 87 K were also measured by an adsorption apparatus consisting of a cryostat with a helium closed-cycle refrigerator and BELSORP-max (BEL, Japan) to compare with the simulation results.

The GCMC isotherms of argon and nitrogen on the atomistic MCM-41 model at 75 K, 80 K, and 87 K were in good agreement with the experimental data, which suggests that our model provide a good description of the surface roughness of MCM-41. Then, the simulated equilibrium capillary condensation pressures give close agreement with the experimental desorption branches, which indicates that the experimental desorption branch comes from thermodynamic equilibrium vapor-liquid transition. However, a large difference in the adsorption branch between the experiment and simulation was observed for all the cases. This suggests that the energy fluctuation of the system to overcome an energy barrier for the vapor-liquid transition is different between the experiment and the GCMC simulation. We therefore calculated canonical work function [5] by integrating a sigmoid adsorption isotherm obtained from the gauge cell MC method. The canonical work function provides a work required for a state change from a multilayer adsorption state to a capillary condensed state at a constant chemical potential. The obtained canonical work function shows that the energy barrier for the capillary condensation becomes small with increasing pressure, and if the system cannot overcome the energy barrier at the thermodynamic equilibrium condition, spontaneous capillary condensation should occur at a higher pressure than the equilibrium transition pressure. The energy barrier obtained from a comparison with the experimental adsorption branch of argon for MCM-41 (pore diameter 4.0 nm) at 75 K was 0.156 kT per adsorbed molecule. We assumed it as an energy fluctuation of the system at the spontaneous capillary condensation pressure, and predicted the capillary condensation pressure of argon on the atomistic MCM-41 model at 80 K and 87 K. The predicted spontaneous condensation pressures are in good agreement with the experimental data. We also determined the spontaneous condensation pressures of argon and nitrogen for the other atomistic MCM-41 model at all the temperatures, by setting the energy fluctuation of the system of 0.156 kT per adsorbed molecule, and succeeded in obtaining excellent agreement with the experimental adsorption branches for MCM-41 (pore diameter 3.2 nm).

[1] B. Coasne *et al.*, *Langmuir* **22**, 194 (2006).

[2] N. Muroyama *et al.*, *J. Phys. Chem. C* **112**, 10803 (2008).

[3] H. Tanaka *et al.*, *Adsorption* **19**, 631 (2013).

[4] A. V. Neimark and A. Vishnyakov, *Phys. Rev. E* **62**, 4611 (2000).

[5] A.Vishnyakov and A. V. Neimark, *J. Phys. Chem. B* **110**, 9403 (2006).

(395af) Multi-Component Gas Adsorption On MSC5A By Chromatographic Method and Simulation Study

Chihara, K., Meiji University

Nomoto, M., Meiji University

Koide, S., Meiji university

1. Introduction

The combination of chromatographic method and moment analysis of the response peaks is one of the useful techniques to study adsorption equilibrium and adsorption rate. Perturbation chromatography with the mixed multi component adsorbent gas carrier (two adsorbates) has been applied to several studies on adsorption. In this work, perturbation chromatography with multi component gas carrier (two adsorbates with inert gas) and non-equilibrium thermodynamics liner law was applied for discussion of the interference effect and the displacement effect (those are cross effects) on mass transfer in multi component gas adsorption as previous study for different gas mixture (He, CO_2, C_2H_4, CH_4 , N_2). The study of mixed gases (C_2H_4+CO_2,CH_4+N_2,CO_2+CH_4) were done in temperature of 313K, 323K,333K. Moment analysis method and stop & go simulation method were utilized to obtain each mass transfer parameters of adsorbate gases.

2. Experiment

The apparatus was similar to a conventional gas chromatograph. Adsorbent particles (molecular sieving carbon 5A, 20/30 mesh, Japan Enviro Chemical Ltd.,) were packed in a column. Carrier gas was a mixture of two or three components among He, CO_2, C_2H_4,N_2,CH_4. Perturbation pulse was introduced into the carrier gas stream. Introduction of pulses was performed by 6-way valve. The pulse size was 1cc, which meant injection period was 1.4 sec. Then pulse response was detected by TCD cell. Output signal of TCD was transmitted to a personal computer through RS232C. This signal was also transmitted to the personal computer. Simulated chromatogram by a personal computer can be overlapped on experimental chromatogram shown in the monitor screen. Further, moment of pulse response, which is shown in the monitor screen, can be automatically calculated by the personal computer. Also this simulated chromatogram can be compared with experimental chromatogram to determine the equilibrium and the adsorption kinetic parameters. Here Markham-Benton equation as for adsorption equilibrium and linear driving force (LDF) approximation as for adsorption kinetics were adapted for numerical calculation, which was based on stop & go method. Overall mass transfer coefficients (Ksav) for LDF model were determined.

3. Result and Discussion

Fig.1 shows an example of comparison of experimental chromatogram with simulated chromatogram for MSC5A to obtain *Ksav* for LDF model. Experimental conditions were 333 K, column pressure 5 atm, flow rate 25 cm/sec and He + CO_2 mixed gas carrier with CO_2 pulse. Here CO_2 concentration in the carrier gas was changed 10, 30, 50, 70 and 90 %. Retention time of simulation peak depend on langmuir parameter. Therefore experimental peaks differ from simulation peaks because of amount adsorbed from experiment and langmuir parameter differed.

Fig.2 shows experiment and simulation results in an example case of binary adsorbate carrier mixed with He and an adsorbate pulse. Experimental conditions were 333 K, column pressure 5 atm, flow rate 25 cm/sec and He(10%)+CO_2(30%)+CH_4(60%) mixed gas carrier with CO_2pulse. Simulation peak needs compensation becouse TCD signals different behavior from simulation's preassure change.

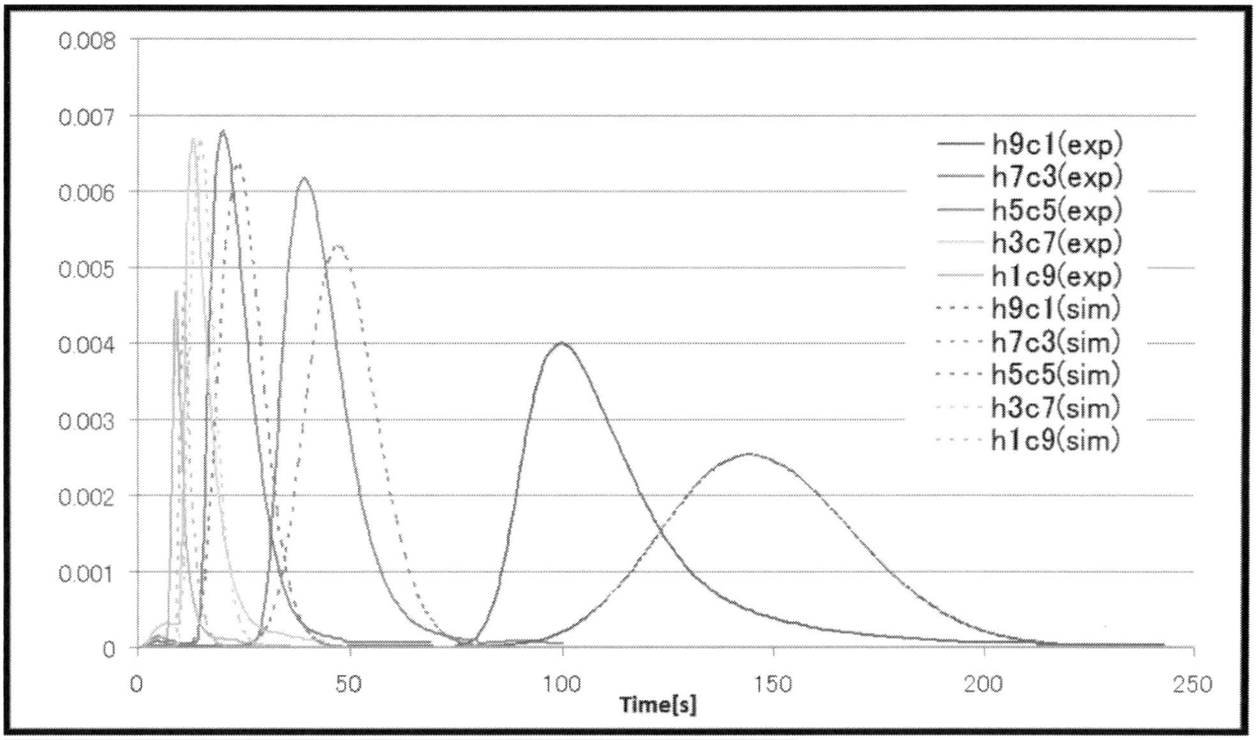

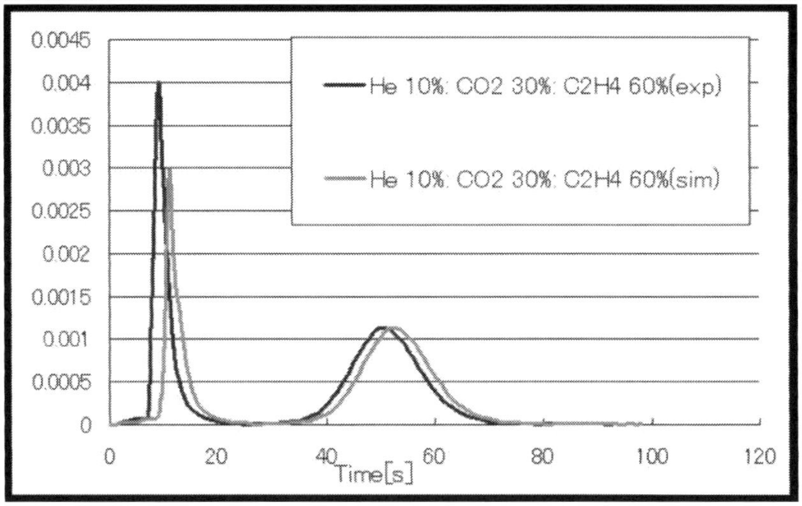

Fig. 1 Comparison of experimental peaks with simulation of He +CO_2 mixed gas carrier with CO_2 pulse at 333 K.

Fig. 2 Comparison of experimental peaks with simulation of He(10%)+CO_2(30%)+CH_4(60%) mixed gas carrier with CO_2 pulse at 333 K.

4. Conclusion

Good agreements between experimental chromatogram and simulated chromatogram, which were based on the modeling of Stop & Go method, were observed in case of peturbation chromatography with mixed adsorbate gas carrier. And micropore diffusivities obtained were interpreted by chemical potential driving force consideration based on non-equilibrium thermodynamics law.

(395aj) Solvent Recovery of 3 Component PSA Experiments and Influence of Water Vapor

Chihara, K., Meiji University
Kawahara, K., Meiji University
Kato, T., Meiji University
Kobayashi, J., Meiji University

When an organic solvent is drained from a factory , the organic solvent adversely affects the human body and the environment. Therefore , we study the adsorption method called the PSA method to collect the disposal gas. Water vapor obstructs the adsorption of an adsorbate and results in gradual deterioration of an adsorbent. We examined how water vapor affects the organic solvent recovery. As a method of removing the organic solvent , we are desired to use adsorption in the chemical field. It is PSA method. Adsorption of PSA depend on pressure , each operation pressure is alternately changed with the high pressure and the low pressure. Accoding to two kind of operation pressure , unit device of PSA is conducted adsorption and desorption at the same time.

In this study , water vapor content in the mixed gas was changed and the behavior of the density of the solvent in the product gas was examined with Stripping reflux PSA unit like Figure.1. As is Figure 1, flow gas repeats alternatively adsorption and desorption inturn in courses and , in each column which filled an adsorbent separate the gas well. CH_2Cl_2 and CCL_4 and water vapor were chosen as an adsorbate. Silica gel (CaRiACT Q-6) was chosen as an adsorbent.The recovery experiment of the organic solvent was conducted by PSA method. And , we tried to optimize the experiment method about Stripping-reflux PSA for solvent recovery by process simulation , named Stop & Go method. Process simulation using this method was able to compare the experimental data. In the same way as experiment, simulation operation sequence of PSA can carry out tentative experiment to process. And , to check how the column is adsorbed , we conducted a study to know the adsorption distribution. Compared with simulation (Stop & Go) , we considered whether there is placticability.

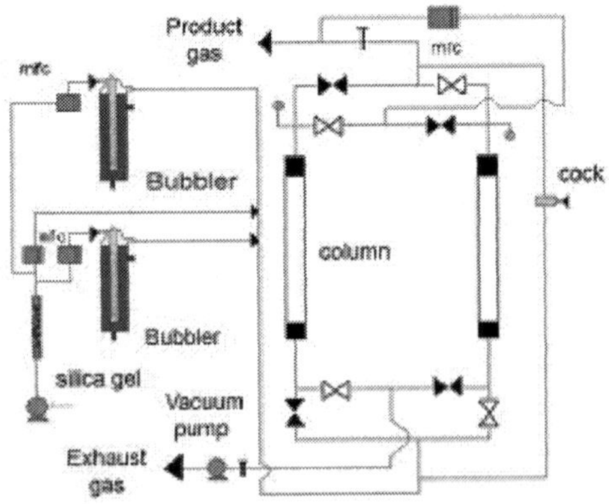

Fig.1 Stripping Re-flux PSA unit

(395ag) Azeotropic Adsorption Equilibrium of Mixed Organic Solvent On FAU Type High Silica Zeolite - Do's Method -

Chihara, K., Meiji University
Yamada, N., Meiji Univercity
Tatsuharu, T., Meiji University
Suzuki, Y., Meiji Univ.
Do, D. D., University of Queensland

In recent years, instead of the chlorofluorocarbon specified as the ozone-depleting substance, a chlorinated organic compound, hydrocarbon, alcohol, etc. are used as an alternative washing solvent, and they are used as a mixed solvent in many cases. Therefore, adsorption operation of binary systems is needed. Furthermore, since azeotropic adsorption may happen, in the case of the design of a solvent recovery system and operation, those adsorption equilibria are required as basic data. In this study, the equilibrium adsorption of the high silica FAU-type zeolite is obtained by using a molecular simulation device and a gravimetric adsorption experiments and analyzing the adsorption phenomenon. However, the IAS theory can't deal with the adsorption azeotrope. We applied the method of treating the azeotropic adsorption, that is, to treat the adsorbed phase as a combination of two independent adsorbed phases, and the IAS theory is applied for each adsorbed phase (Do, 2000).

The adsorbate used was ethanol and trichloroethylene, which are the system of azeotropic mixture. The experiments were carried out using a batch system with circulation flow of the adsorbate mixture around the adsorbent to prevent the concentration distribution in gas phase, changing the total concentration and composition of ethanol and trichloroethylene, by stepwise dosing of mixed solvent vapor. For this system, one azeotropic point appeared. Simulation module in Materials Studio (Accelrys Inc.) was used throughout MC simulations. It calculated by the Sorption module, and the adsorption isotherm was obtained.

Two kinds of adsorption site or adsorption phase were observed in the cell (25.08 x 25.08 "^{2}) by the Sorption simulation of binary adsorption equilibria for the system of EtOH-TCE. One phase containes EtOH and TCE (phase 1), and the other phase containes only EtOH (phase 2). The coordinates of two sites were determined. The ratio of EtOH amount adsorbed on both sites were counted from the simulated adsorbed cell. The ratio was found to be constant through the simulation. Using this constant ratio, the simulated or experimentally obtained amount adsorbed were divided for phase 1 and 2. (Fig.1). Then adsorption isotherms were determined for both phases. Phase 1 was expressed as extended Langmuir. Phase 2 was expressed as Langmuir. Total adsorption isotherm was observed as total of both phase shown as X-Y diagram as follows.Binary adsorption equilibria for the system of EtOH-TCE of total concentration of 70000ppm are shown in Fig.2 and Fig.3 in form of X-Y diagram (X is the mole fractions of gas phase concentration of

TCE. Y is the amount adsorbed of EtOH / total amount adsorbed). Fig.2 shows the experimental result. Fig.3 shows the simulation result.

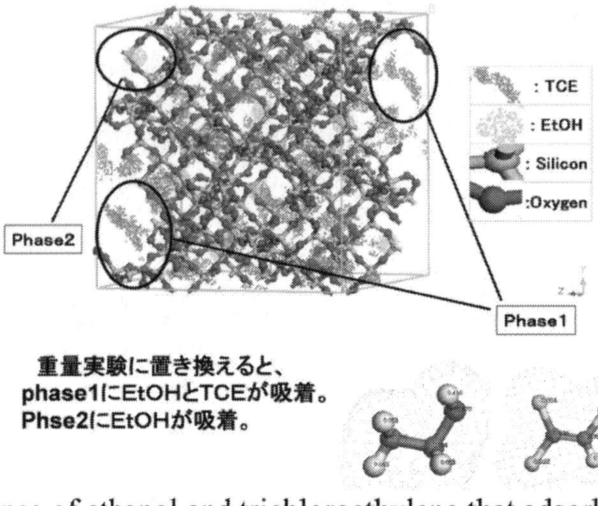

Fig.1 Appearance of ethanol and trichloroethylene that adsorbs zeolite

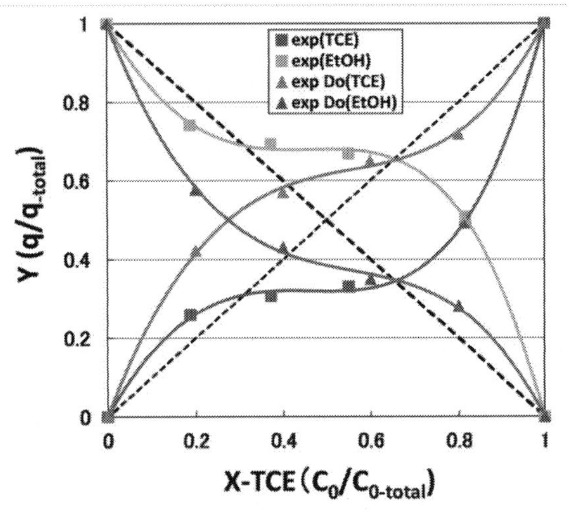

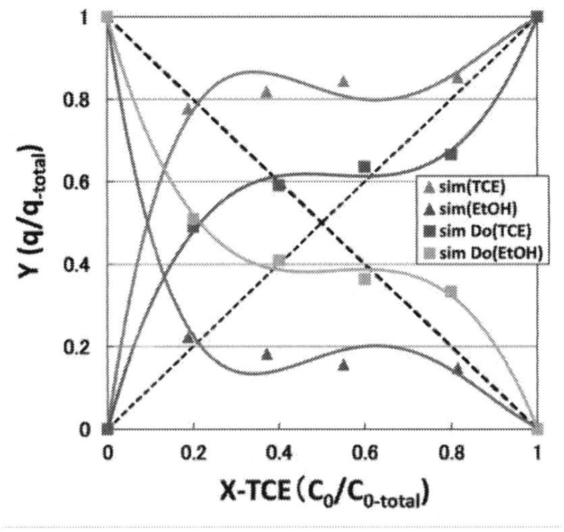

Fig.2 The experimental result(Do's method) Fig.3The simulation result(Do's method) Total concentration of70000ppm Total concentration of 70000ppm

References

1. Duong D Do (2000), "Adsorption Analysis:Equilibria and Kinetics," London , *Imperial College Press*, pp.219-221.

2. Chihara. K, Suzuki.Y, Tomita.S , Kabe.J (2011), "Adsorption equilibrium of the mixture of EtOH and TCE on FAU type high silica zeolite,"Adsorption 17, pp.201-203.

Acknowledgment: Special research fund of 2011 fiscal year by Meiji univ. is appreciated.

(395ah) Investigation of Adsorption On MSC5A in Supercritical CO2 â€“ Chromatography and Monte Calro Simulation â€“

Chihara, K., Meiji University
Yasuda, T., Meiji University
Shinsako, M., Meiji University

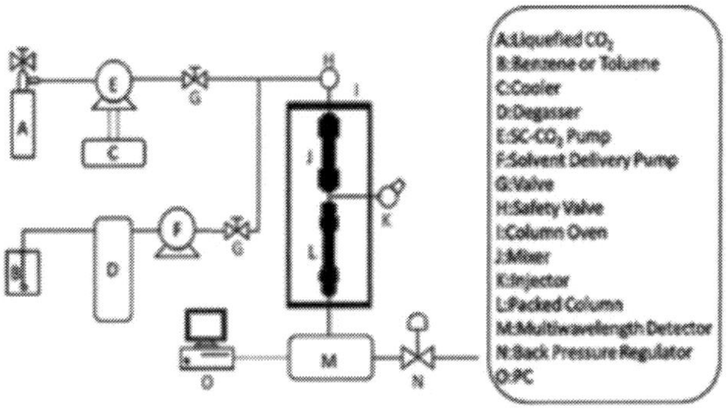

Chromatographic measurements were made for the adsorption of benzene or toluene on molecular sieving carbon (MSC) in supercritical fluid CO_2 mixed with organics. Supercritical chromatograph packed with MSC was used to detect pulse responses of organics. Adsorption equilibria and adsorption dynamics parameters for organics and CO_2 were obtained by moment analysis of the response peaks. Dependences of adsorption equilibrium constants, K*, and micropore diffusivity, D, on the supercritical CO_2 density were examined. To use Markham-Benton equation for obtaining the saturation adsorption amount, the adsorption isotherms was carried out the curve fitting method of least squares by the solver.

Fig.1. Experimental apparatus

Molecular simulation of multicomponent adsorption equilibria was performed, and compared the simulation results with experimental results. Simulation software is Materials Studio (Version5.5) made by MSI. The Grand Caronical Monte Carlo method (under constant chemical potential (Âµ), volume (V), and temperature (T)) was used to get the equilibrium amount adsorbed. The purpose of performing simulation is to capture visually the adsorption phenomenon and to elucidate an adsorption mechanism on the molecule level. So, we created Pseudo activated carbon sub-micropore model (MSC68-test021 model) as adsorbent. By performing molecular simulation, it is obtained average loading [molecular/cell] and is possible to determine the amount adsorbed therefrom.

(395al) CO2 Adsorption On Amine-Tethered MCM-41: Equilibrium Studies and Isotherm Modeling

Loganathan, S., Indian Institute of Technology Guwahati
Tikmani, M., Indian Institute of Technology Guwahati
Ghoshal, A. K., Indian Institute of Technology Guwahati

Recently, amine-tethered mesoporous silica materials have attracted considerable attention as a promising chemical sorbent for post combustion CO_2 capture applications. In the present work, we explored the influence of pore size distribution for amine-tethering application in order to enhance the CO_2 equilibrium uptake. MCM-41 samples with two different pore sizes (30 nm and 3.5 nm) were synthesized by sol-gel method. Then, a mono-amino silane compound named as "3-aminopropyl triethoxy silane" was allowed to react with silanol (Si-OH) groups of the MCM-41 samples through silane coupling chemistry. The adsorption capacity measurements of CO_2 on to mono-amine tethered MCM-41 samples were measured at four different temperatures (30, 45, 60, 75 ºC) and pressures ranging from 0-5 bar using Rubotherm magnetic suspension balance. Interestingly, it was observed that MCM-41 sample with pore diameter of 30 nm functionalized with monoamine silane exhibited highest CO_2 uptake of 1.2 mmol/g at 1 bar pressure and 30 ºC, when compared to that of other sample (0.7 mmol/g). From the outcome of CO_2 adsorption experiments, it can be seen that pore-expanded MCM-41 served as a promising support material for amine grafting applications. Ten adsorption-desorption cycles were conducted to check the stability of the adsorbent. No change in the performance of the adsorbent was noticed, indicative of reusable nature of the adsorbent and its applicability for multi-cycle operations. In addition to this, modeling for CO_2 uptake in amine-tethered MCM-41 was done using Freundlich and Toth isotherm models. The Freundlich model parameter *n* and Toth model parameters *t* and *n* significantly deviated from unity, indicative of heterogeneous nature of the surface. Surface heterogeneity is due to the presence of two active sites, which in turn indicates that CO_2 interacts with siloxane bridges through physisorption and amine groups through chemisorption. Isosteric heat of adsorption (~ΔH= 50 kJ/mol) calculated by Clasius-Clapeyrone equation revealed the dominance of chemisorption process in the current study.

(395n) Gas Separation By Adsorption On MIL-53(Al): Natural Gas and Biogas Application

Ferreira, A., LSRE - Laboratory of Separation and Reaction Engineering - Associate Laboratory LSRE/LCM
Ribeiro, A. M., University of Porto
Kulac, S.,
Rodrigues, A., LSRE - Laboratory of Separation and Reaction Engineering - Faculty of Engineering - University of Porto

Gas separation by adsorption on MIL-53(Al): natural gas and biogas application.

A.F.P. Ferreira[1], A.M. Ribeiro[1], S. Kulaç[1], A.E. Rodrigues[1]

[1] *Laboratory of Separation and Reaction Engineering, Associate Laboratory LSRE/LCM, Department of Chemical Engineering, Faculty of Engineering, University of Porto, Rua Dr. Roberto Frias, Porto, arodrig@fe.up.pt*

Keywords: MIL-53(Al), biogas, natural gas

Introduction

CO_2 and H_2S are two of the main polluting compounds in natural and bio gas. Therefore, the separation of CO_2 from CO_2 / CH_4 mixtures is a major process in the industry for upgrading the energy content of natural gas. [1] Adsorption, membrane and absorption processes are the three main technologies for CO_2/CH_4 separation [2]. Adsorption based separation is a key industrial technology due to its high energy efficiency, simplicity, low capital investment cost and ease of control. [1] Recently, several authors purposed the separation of CO_2 from CH_4 by adsorptive processes, using MIL-53(Al). In a recent work by Boutin and co-workers [3], the osmotic thermodynamic model was used in order to describe the flexible behavior of MIL-53(Al) upon adsorption of CO_2 and CH_4. Finsy and co-workers evaluated the effects of pressure and gas composition on the efficiency of the CO_2/CH_4 mixtures separation. [4] Stavitski *et al.* [5] studied the adsorption mechanism of CO_2 and CH_4 in MIL-53(Al), in order to explain its high efficiency for CO_2 capture. Shaped and powder forms of this material were compared by Heymans *et al.*, they obtained lower adsorption capacities for pellets than for powder, however similar separation selectivity was observed for both forms. [6] In this work, isotherms of CO_2 and CH_4 were obtained, based on single component breakthrough data. Adsorption data was fitted using Langmuir and Toth models. Single component breakthrough curves were simulated using gPROMS Software, allowing the global model verification. Furthermore, binary breakthroughs of CO_2/CH_4 at different feed pressures and concentrations were performed.

Experimental

The methane and carbon dioxide adsorption isotherms were determined using the breakthrough curves. A one column PSA set-up was used for the breakthrough experiments for a pressure range from 1.1 to 4.4 bar (total pressure). The same unit was used to perform the one column PSA cycle. This unit is described elsewhere. A back pressure regulator is used to control the pressure inside the set-up. Outlet gas composition was monitored using a Gas Data LMSxi Type G4.18 infrared gas analyzer (Gas Data Ltd, UK). A column with a length of 0.284 m and a diameter of 0.0212m was packed with 52.3 g of MIL-53(Al) pellets. Bed porosity is 0.36 and feed temperature is 303 K. The positions of the thermocouples are: 0.05 m, 0.118 m and 0.226 m. The MIL-53 (Al) adsorbent (BasoliteTM A100) was provided by BASF in the tablet form (2 x 2 mm), with a pellet density of 826 kg/m^3.

Results and discussion

Carbon dioxide and methane adsorption isotherms on MIL-53(Al) were obtained by breakthrough experiments, at the pressures of (0.11, 0.29, 0.6, 1.0, 1.6, 2.0, and 3.52) at 303 K. The adsorbed amounts were determined by integration of the experimental breakthrough curves. It was observed that the adsorption is reversible since the amounts of adsorption and desorption branches are in agreement. The adsorbed amounts are about 4 mmol g^{-1} and 2 mmol g^{-1} for CO_2 and CH_4, respectively, at 303 K and 3.4 bar. The isotherms assessed in this work are in good agreement the ones found in the literature, such as the results obtained by gravimetric technique in the work of Heymans *et al.* [6] The dynamic model describes adequately the adsorption/desorption breakthroughs, validating the model for its use in PSA design and simulation.

Conclusions

Methane and carbon dioxide isotherms were determined using the single component breakthrough curve experiments. Langmuir and Toth isotherm models were used to fit the adsorption data and it was observed that both models represent well the adsorption experimental data. Langmuir isotherm model was chosen to simulate CH_4 breakthrough curves. MIL-53(Al) is CO_2 selective, showing therefore a great potential for its application in the biogas and natural gas upgrade field.

Acknowledgements

The research leading to these results has received funding from the European Community's Seventh Framework Programme (FP7/ 2007-2013) under Grant Agreement No. 228862. MACADEMIA is a Large-scale Integrating Project under the Nanosciences, Nanotechnologies, Materials and New Production Technologies Theme in FP7.

This work is partially supported by project PEst-C/EQB/LA0020/ 2011, financed by FEDER through COMPETE – Programa Operacional Factores de Competitividade and by FCT – Fundação para a Ciência e a Tecnologia.

The authors thank BASF for providing Basolite A100 samples.

References

[1] Z. Xiang, X. Peng, X. Cheng, X. Li, D. Cao, CNT@Cu3 (BTC)2 and Metal-Organic Frameworks for Seperation of CO2/CH4 Mixture, J Phys Chem C, 115 (2011) 19864-19871.

[2] J. R. Li , Y. Ma , M. C. McCarthy, J. Sculley , J. Yu, H.-K. Jeong, Perla B. Balbuena b,c,∗ , Hong-Cai Zhou, Carbon dioxide capture-related gas adsorption and separation in metal-organic frameworks, oordination Chemistry Reviews 255 (2011) 1791–1823.

[3] A. Boutin, F.X. Coudert, M.A. Springuel-Huet, A.V. Neimark, G. Ferey, A.H. Fuchs, The Behavior of Flexible MIL-53(Al) upon CH4 and CO2 Adsorption, J Phys Chem C, 114 (2010) 22237-22244.

[4] V. Finsy, L. Ma, L. Alaerts, D.E. De Vos, G.V. Baron, J.F.M. Denayer, Separation of CO2/CH4 mixtures with the MIL-53(Al) metal–organic framework, Microporous and Mesoporous Materials 120 (2009) 221-227.

[5] E. Stavitski, E. A. Pidko, S. Couck, T. Remy, E. I J. M. Hensen, B. M. Weckhuysen, J. Denayer, J. Gascon, F. Kapteijn, Complexity behind CO2 Capture on NH2-MIL-53(Al), Langmuir 27 (2011) 3970–3976.

[6] N. Heymans, S. Vaesen, G. De Weireld, A complete procedure for acidic gas separation by adsorption on MIL-53 (Al), Microporous and Mesoporous Materials 154 (2012) 93-99

[7] F.V.S. Lopes, C.A. Grande, A.E. Rodrigues, Activated carbon for hydrogen purification by pressure swing adsorption: Multicomponent breakthrough curves and PSA performance. Chem Eng Sci 66 (2011) 303-317.

(395ai) Effect of Water Vapor On the Colum of a Pressure Swing Adsorption On Solvent Recovery - Process Simulation and MRI Analysis

Chihara, K., Meiji University
Kato, T., Meiji University
Kawahara, K., Meiji University
Kobayashi, J., Meiji University

Organic solvent discharged from the plant adversely affect the human body and the environment. Therefore, collection and removal is an essential operation in chemical engineering process. We investigated the effect when water vapor contained in the feed gas in the solvent recovery in the PSA process. We also analyzed the columns by MRI when the experiment was terminated at a certain time to investigate the adsorption distribution in the column. Respect to experimental results, create a STOP & GO simulations, we examined the reproducibility of the experimental results.

The experiments were carried out using the experimental apparatus of two-column PSA such as Fig.1, Which was operated by using a programmable controller, set the cycle time purge step (20 seconds) and pressurisation (160 seconds) and pressure reduction (20 seconds) and desorption step (160 seconds) , the organic solvent concentration at a constant cycle which was measured by gas chromatography. It also analyzed by MRI device the experiment was terminated at a constant cycle in, to obtain an adsorption distribution.

It is a calculation method to do the following with STOP & GO method (Fig. 2) . To think in the N cells the column, making the batch adsorption Stop the flow at a certain time. And I move to the next cell. It repeated this operation. It used three expressions of the material balance equation, equilibrium adsorption, adsorption rate expression to. Performs the simulation calculations, it is compared with the measured values.

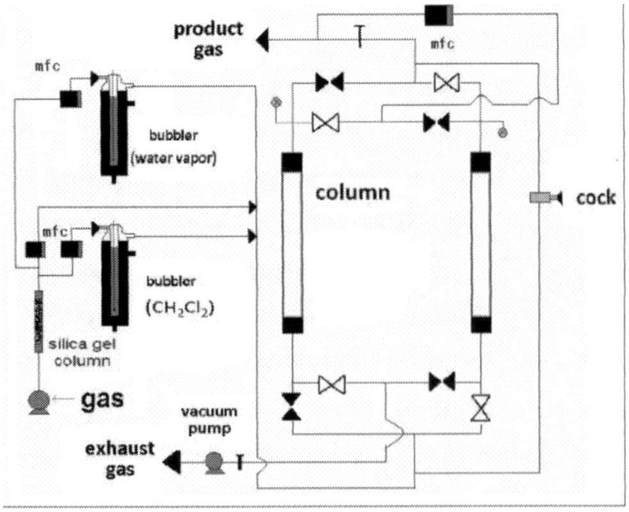

Fig. 1 Stripping reflux PSA unit

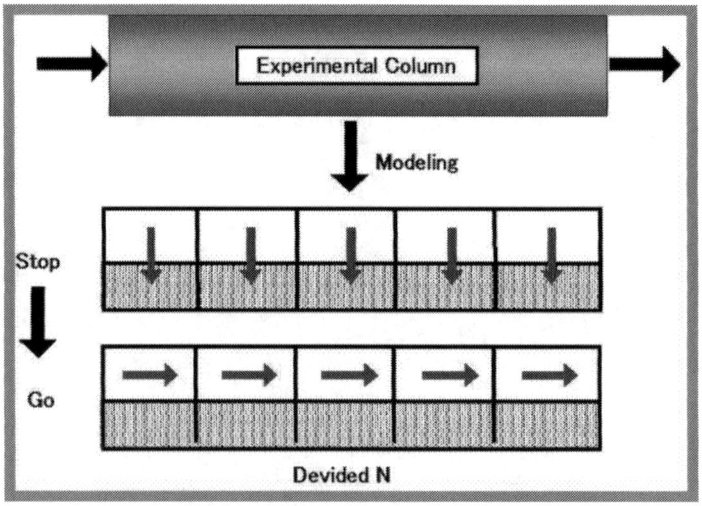

Fig. 2 Stop & Go method basic concept

(395ak) Experiment and Simulation Study of Multi-Component Gas Adsorption On MSC3A By Volumetric Method

Chihara, K., Meiji University
Koide, S., Meiji university
Nomoto, M., Meiji University

(1) Introduction.

MSC (Molecular Sieving Carbon) has slit like micropores, which separate different size molecules by molecule's different diffusivity.

In this study, the adsorption experiment was conducted by the volumetric technique. MSC 3A (KURARAY CHEMICAL CO., LTD) was used as adsorbent at 303K, 313K and 323K. Experimental adsorption of binary mixtures was compared with the simulation using the data obtained in equilibria and rates of adsorption of the single component.

(2) Study on gas phase adsorption by a volumetric method.

Equilibria and rates of adsorption of nitrogen, methane and binary N_2-CH_4 mixtures into molecular sieving carbon(MSC3A) were studied by the volumetric method.

Micropore diffusion, mass transfer coefficient are obtained by making an ideal curve and curve fitting of the pressure change about the analysis of the adsorption rate.

Equilibrium of single component was correlated by Langmuir equation and Markham Benton equation.

(3) Conclusion

A detailed value of the data obtained from mixed-gases experiments between experimental and simulated chromatogram, which were based on the modeling of Stop & Go method, are corresponding. And micropore diffusivities obtained were interpreted by chemical potential driving force consideration based on non-equilibrium thermodynamics law.

(395am) 'Currant-Bun' AmberliteTM-Silica Composites For CO2 Capture

Osei-Prempeh, G., West Virginia University Institute of Technology
Davari, A., West Virginia University Institute of Technology

Polymer-silica composites have the potential to provide adsorbent materials with large particle sizes and high surface area with accessible pores for enhanced CO_2 transport and capture. Amberlites^{TM} are commercially available polymeric adsorbents used for various separation processes. These materials possess fairly large pores which can be tailored with the right groups for CO_2 capture.

Prepared Amberlite^{TM}-silica composites resulted in materials with high silica content. Spherical materials with particle sizes greater than 200 μm were obtained. Amine functionalization of the composites results in good functional group incorporation and materials with fairly large pores for easy access to the functional groups. TGA analysis of CO_2 capture show improved capacity on amine functionalized materials. Sorption kinetics analyses were made and the stability of the materials investigated by taking them through several adsorption-desorption cycles. Comparison of the capture capacity and sorption kinetics of CO_2 on non-functionalized and functionalized Amberlite^{TM}-silica composites will be made. Gas transport properties through packed columns of the composites will be related to the capture capacity and sorption kinetics.

(395ao) Chiral Recognition Mechanism of Amide-Containing Chiral Solutes By Amylose Tris[(S)-Alpha-Methylbenzylcarbamate] Sorbent

Tsui, H. W., Purdue University
Wang, N. H. L., Purdue University
Franses, E. I., Purdue University

Although polysaccharide sorbents have been widely used for chiral separations, the recognition mechanisms have not been fully elucidated. In this study, we focus on one important commercial sorbent, amylose tris[(S)-alpha-methylbenzylcarbamate] sorbent, or AS. Four solutes were studied: ethyl lactate (EL), methyl mandelate (MM), benzoin (B), and pantolactone (PL), which have the same two key functional groups attached to the asymmetric carbon, O=C-C-OH. Each enantiomer pair of these solutes can be well separated by AS. The retention factors k_R and k_S and enantioselectivities ($S= k_R/ k_S$) were determined in n-hexane. Infrared spectroscopy (IR) data for polymer-solute, Density Functional Theory (DFT) simulations of the interactions of these solutes with the side chains of the polymer, and Monte Carlo (MC) and MD simulations lead to the following general hypothesis for the chiral recognition mechanism for these four molecules. First, a strong H-bond forms between the solute OH groups for each enantiomer and the sorbent C=O groups as the leading interaction, or "anchor" point. Then, a weaker H-bond forms between the solute C=O groups and the sorbent NH groups only for the R-enantiomer, but not for the S-enantiomer, which is prevented for steric restrictions. A third synergistic interaction may involve the C-O-C groups of the R-enantiomers of EL and PL with the sorbent NH groups, or the phenyl groups of the R-enantiomers of MM and B with the sorbent phenyl groups. Moreover, IR shows evidence of an intra H-bond for all four solutes, supported by DFT simulations. The retention factors increase with increasing strength of the inter H –bond, and decreasing strength of the intra H-bond. The intra H-bond strength correlates with the O=C-C-OH group molecular flexibility, as determined from a novel method involving the distribution of torsion angles of this group. The distributions were quantified with molecular dynamics simulations of the solute structure. Moreover, the retention factor and enantioselectivity increase with increasing molecular rigidity, which appears to facilitate the formation of the third synergistic interactions in the R-enantiomers. MD simulations of a left-handed 12-mer AS polymer rod model revealed that the polymer has orderly "grooves" and cavities. Simulations of AS with 200 n-hexane molecules indicate no effect of n-hexane on H-bonds in AS. Monte Carlo (MC) and MD "docking" simulations reveal certain chiral cavities which can lead to chiral discrimination. Molecular rigidity hinders the formation of strong intra H-bond and facilitates the formation of the synergistic third interactions in the R-enantiomers. Strong intra H-bond and molecular flexibility, on the other hand, hinder the formation of the three synergistic interactions in the R-enantiomers. The results support the proposed general mechanism.

(395an) Synthesis and Evaluation of Stable Oxygen-Selective BaO Sorbent for Air Separation

Park, J. H., Korea Institute of Energy Research
Jung, T., Korea Institute of Energy Research
Beum, H. T., Korea Institute of Energy Research
Jeon, S. G., Korea Institute of Energy Research
Kim, W. S., Kyung Hee University

Air separation unit is required for the production of pure oxygen for the oxy-fuel combustion. Cryogenic air separation is suitable for the large-scale air separation but requires high energy consumption. Nitrogen-selective adsorptive air separation is not appropriate for large-scale oxygen production and has poor selectivity. Oxygen-selective BaO/BaO_2 sorbent had been used for air separation as Brin process before cryogenic air distillation but it is known the BaO/BaO_2 sorbent has a poor stability at high temperature. In this study, to provide the multicylce operation capability, oxygen-selective BaO sorbent with uniformly mixed MgO which can stabilize BaO/BaO_2 particles at high temperature was prepared by precipitation of double carbonate and reduction to oxide. From thermogravimetric analysis, sorption isotherm interestingly showed an abrupt increase around the transition oxygen pressure. When the oxygen pressure was below 124 mmHg at 700°C, sorption of O_2 hardly occured. However, sorption isotherm curve sharply increased when the pressure increased above 124 mmHg. As the temperature increased, the transition O_2 pressure increased because decomposition of BaO_2 to BaO was favored at high temperature. Sorption capacity at about 600 mmHg O_2 was about 2 mmol/g at 700°C, which enables high productivity of O_2. The prepared sorbent exhibited a good stability during 10 times cyclic operations of sorption and desorption at 700°C because MgO acts as a barrier against sintering of BaO/BaO_2. The time for 80% of sorption and desorption at 700°C were around 1 to 2 min, meaning the uniformly mixed MgO/BaO sorbent has fast reaction kinetics.

(395ap) Denatured Bio-Ethanol Feedstock Desulfurization By Adsorption Onto a Nickel Containing Solid

Thomas, M., IFPEN
Montpeyroux, S.,
Surla, K., IFPEN

INTRODUCTION

IFP Energies nouvelles has developed a new hydrogen production system from denatured bio-ethanol liquid feedstock. Ethanol appears as a very promising hydrogen source due to its low toxicity, easy handling, transportation and storage and its wide availability. Moreover, using bio-ethanol produced from biomass sources has almost no impact on the greenhouse effect.

The objective is to produce pure hydrogen flow between 50 and 100 Nm^3/h from this renewable feed for medium scale industrial customers such as electronic, metals processing and oil hydrogenation industries. The conversion of bio-ethanol to hydrogen is performed using a catalytic auto-thermal reformer (ATR) working under pressure which combines the exothermic reaction of ethanol with oxygen (partial oxidation) and the endothermic reaction of ethanol with water (steam reforming). The reactions, leading to synthesis gas (syngas : mixture of H2, CO and CO2) production, can be described as follow :

$$C_2H_5OH + Â½O_2 => 2CO + 3H_2$$

$$C_2H_5OH + H_2O => 2CO + 4H_2$$

At the outlet of this reforming section, the hydrogen-rich effluent gas contains impurities, in particular carbon monoxide CO. In order to purify this syngas and to increase the hydrogen yield, the conversion of carbon monoxide is performed using the water gas shift reaction (WGS) :

$$CO + H_2O => CO_2 + H_2$$

A pressure swing adsorption unit is then used to purify the reformate gas. Using adsorption and desorption cycles over adsorbent beds (zeolite LTA 5A) allows to remove contaminants. Pure hydrogen stream (purity of 99.95 % vol.) is then produced.

For legal reasons, the ethanol that is not used in the food sector is denatured by the addition of a chemical substance, called the denaturing agent, and that makes it unfit for consumption. This denaturing agent can be a natural gas condensate, or, more often, a gasoline. In this case, this ethanol feedstock very often contains undesirable substances such as sulfides or mercaptans (condensate), or thiophenic compounds (gasoline). The presence of these sulfur-containing compounds is very disturbing because they can rapidly poison the catalysts used either in the ATR reactor or in the Water Gas Shift one. Therefore the feedstock has to be desulfurized before its reforming.

The aim of this study is to develop an adsorption process in liquid phase able to remove thiophenic compounds from the denatured bio-ethanol, by using adsorption over a nickel based solid.

EXPERIMENTAL

Typical denatured ethanol has to contain at least 92,1 vol. % of ethanol, a denaturant amount comprised between 2 and 5 vol. %, and no more than 1,7 vol. % of aromatics and 10 ppm mass of total sulfur. In this study, experiments have been carried out with synthetic model feeds, composed of pure ethanol (98 vol. %), toluene (2 vol. %) as the aromatic component of the denaturant and possible adsorption competitor, and thiophene (100 and 20 ppm "S") as the sulfur component.

All the experiments have been performed on a laboratory scale breakthrough curve unit in dynamic conditions, with a column of around 20 cm3, placed in an oven (350 Â°C max.). The synthetic feed is introduced under pressure, in order to be always in liquid phase, at a constant flow-rate at the column inlet containing the adsorbent, and the effluent is collected at the column outlet and then analysed by gas chromatography in order to determine the thiophene concentration versus time. A simple mass balance leads to the amount of thiophene adsorbed on the solid.

The solid used in this study is mainly composed of a mixture of nickel and nickel oxide deposited on a silica and alumina support, and is supplied in a stabilized form (cylindrical extrudates). The amount of nickel is around 55 mass %, and the specific surface area is 175 m2/g.

The main parameters which have been taken into account are:

- solid pre-treatment, consisting either in an activation step under nitrogen or a reduction step under hydrogen, at temperatures comprised between 120 and 300 Â°C

- adsorption temperature, comprised between 30 and 160 Â°C, with both activated and reduced solids

- influence of the presence or not of the aromatic inhibitor (toluene)

- thiophene concentration in the feed

RESULTS

a/ Adsorption temperature influence with an activated or reduced solid at 300 Â°C

A first set of experiments performed on a feed with toluene and 100 ppm "S" shows that a minimum adsorption temperature of 120 Â°C is necessary with a solid activated at 300 Â°C. The sulfur breakthrough point occurs after elution of 17 column volumes at 120 Â°C, and is not observed after elution of 34,0 column volumes at 160 Â°C. At lower adsorption temperatures, sulfur breakthrough is quite immediate.

A second set of experiments carried out with the same feed but with a reduced solid at 300 Â°C shows that the sulfur breakthrough occurs after elution of around 20 column volumes with adsorption temperatures from 30 to 120 Â°C, while it is not observed at 160 Â°C. Reduction appears therefore more efficient than activation.

b/ Activation or reduction temperature

With an activated solid at 120 or 150 Â°C, sulfur breakthrough is observed immediately or rapidly whatever the adsorption temperatures, 120 or 150 Â°C.

When the solid is reduced at 120 Â°C, sulfur breakthrough is also observed immediately at 120Â°C. Its efficiency increases dramatically when it is reduced at 150 Â°C, sulfur breakthrough being observed after elution of 23,6 column volumes with an adsorption temperature of only 80 Â°C, and after more than 34,0 column volumes at 150 Â°C.

c/ Toluene concentration

Toluene acts as a slight sulfur adsorption inhibitor. With a reduced solid at 300 Â°C and an adsorption temperature of 80 Â°C, the sulfur breakthrough occurs after the elution of more than 34,0 column volumes without toluene and 28,6 column volumes in presence of toluene (2 vol. %).

d/ Thiophene concentration

With a sulfur concentration of only 20 ppm "S" in the feed, an activation or reduction temperature of 300 Â°C and an adsorption temperature of 80 Â°C, sulfur breakthrough occurs after elution of larger feed volumes : 3,4 (100 ppm "S") and 32,5 (20 ppm "S") column volumes with the activated solid, 29,6 (100 ppm "S") and more than 34,0 (20 ppm "S") with the reduced solid.

CONCLUSIONS

The aim of this work was to study and develop a pre-treatment adsorption step in liquid phase in order to desulfurize a denatured bio-ethanol feedstock prior to ATR and WGS reactions in order to produce hydrogen.

Experiments carried out on a laboratory scale with a model synthetic feed have shown that the use of a solid containing both nickel and nickel oxide deposited onto an alumina-silica support was efficient for this purpose. Nevertheless the pre-treatment of the solid is very important in order to improve its efficiency in the sulfur removal. A reduction step at a temperature of 150 Â°C is necessary, and the adsorption temperature has to be at least of 80 Â°C. Higher temperatures of reduction and adsorption lead to a better efficiency. By comparison, a simple activation step does not lead to a great efficiency of the solid.

(395aq) Recovery of Phenolic Compounds From Pyrolysis Wastewater By Adsorption to Kenaf: Comparison of Different Kenaf Pre-Treatments

Holmes, W. E., University of Louisiana at Lafayette
Revellame, E. D., Mississippi State University
Hernandez, R., University of Louisiana at Lafayette
Buchireddy, P., University of Louisiana at Lafayette

There are many chemical and physical processes available for the removal of organic compounds from waste waters. To enhance the sustainability, energy efficiency, and yield of biofuel production processes it is desirable to recover and use these compounds using renewable raw materials. In this study we have used a plant material, Kenaf (Hibiscus cannabinus) for adsorption of small molecular weight phenolic compounds and poly aromatic hydrocarbons generated during the pyrolysis of biomass and contained in the waste water stream. Several pre-treatments of Kenaf were evaluated ranging from simple water washing to Ozone treatment or torrefaction. Torrefaction is gaining attention as an option to improve physical properties of physical and chemical properties of biomass prior to pyrolysis or gasification. This process could also increase the surface area, porosity and selectivity of biomass for adsorption applications. Kenaf was heated slowly in a reduced environment to a maximum temperature of approximately 200 - 300°C. This process can be considered a mild pyrolysis as it occurs at lower temperatures than normal pyrolysis process. The result of torrefaction of Kenaf are compared to other pre-treatments (e,g, ozonation, and washing). Carbon black was used as a positive control, since it is a commercially established adsorption technology and can remove up to 98% of phenol from waste water. Comparisons are based on mass reduction of organics in the waste water The resulting organic laden-Kenaf could be used within the pyrolysis process as an additional feedstock, resulting in a close system for wastewater treatment and an enhanced yield of fuels from biomass.

(395ar) Chemical Aspects of CO2 Adsorption On Mesoporous Silica Modified With n-Propylamines

Bacsik, Z., Stockholm Univerisity
Hedin, N., Stockholm Univerisity

Today, pilot plants for CO_2 capture use chemical absorption processes with various amines. These amines selectively absorb CO_2 from flue gases. The major drawbacks with such amines are the high energy demand for regeneration, corrosion of equipments, and the evaporation and leaching of the amine. Tethered amines on porous solid supports could possibly reduce the energy required for regeneration, limit the corrosion, and the evaporation of the amines.

We have studied the detailed chemistry that is involved when CO_2 reacts with tethered amines on model supports. Chromatographic silica and mesostructured silicas with cubic structures were modified with propylamine groups and studied by in situ FTIR spectroscopy and volumetric carbon dioxide uptake experiments. A range of equilibrium and transient spectra were recorded after and during uptake of CO_2 at different partial pressures. In addition, we will discuss how a trace component of carbonyl sulfide (COS) in the flue gas reacts and interacts with such propylamine groups. COS is known to interfere with the liquid phase chemistry of some amines used for CO_2 capture.

Numerous studies have suggested that certain ion pairs (ammonium carbamates) form with dry carbon dioxide, and that bicarbonates form under moist conditions. Complexities relating to analyzing and assigning the infrared spectra have rendered conflicting interpretations. We try to resolve such outstanding disagreement about the species that form during chemisorption of carbon dioxide on propylamines tethered on silica. We show that propylammonium-propylcarbamate ion pairs form for both dry and moist carbon dioxide. Significant amounts of propylcarbamic acid and its stabilized form of silylpropylcarbamate ester are formed by slow condensation on silanol groups under dry conditions. More ammonium carbamates are formed with moist carbon dioxide than with dry carbon dioxide, but no bicarbonates or carbonates were observed.

Infrared spectroscopy was also applied to directly quantify physisorbed and chemisorbed CO_2 on the propylamine-modified silicas. Irrespectively of the amine density, only a minor part of the CO_2 physisorbed at low (10 kPa) partial pressure of CO_2. At 101 kPa of CO_2 the relative contribution of physisorption of CO_2 to the total uptake of CO_2 increased to ~35 % at 101 kPa of CO_2 when the propylamine surface density was low or medium (0.87-1.67 NH_2/nm^2). When the amine density was >2.0 $NH_2/\ nm^2$, chemisorption dominated (up to 90 %) also at 101 kPa of CO_2. At low or medium amine surface densities, the concentration of ammonium carbamate scaled linearly with the density of amine, but the concentration of such ion pairs increased very significantly when the amine density was high. The significant increase in the formation of ammonium carbamate ion pairs is consistent with that sufficiently close propylamine groups are needed to form such ion pairs. This thesis was supported by the observation of a significant fraction of other chemisorbed moieties (propylcarbamic acid and silylpropylcarbamate) for sorbents with low amine surface densities, where significant amount of ion pairs could not form.

(395au) About The Interaction Between Cpls and CO2: A DFT and ONIOM-EE Study

Meza, P., University of Puerto Rico

Curet, M. C., University of Puerto Rico

ABOUT THE INTERACTION BETWEEN CPLs AND CO2: A DFT and ONIOM-EE STUDY

Paul de Jesus Meza-Morales, Radamés Irizarry-Castillo, and María C. Curet-Arana

Department of Chemical Engineering, University of Puerto Rico, Mayagüez, PR

Coordination polymer-ligands (CPLs) are porous-crystalline materials consisting of metal ions linked to organic bridging ligands as shown in figure 1. Because of the flexibility in their synthesis, a range of novel structures has been rationally designed with high surface area, large pores and distinct gas adsorption properties. The objective of this project is to study CO_2 - CPLs (CPL-2, CPL-4, & CPL-7) interactions using quantum mechanical (QM) and ONIOM-EE calculations, the last one is an integrated quantum mechanical molecular mechanical method linked via electronic embedding. We have used two representative models for the CPLs pores to reproduce binding sites in order to quantify energies of adsorption and characterize the nature of the interaction. For the QM calculations, we have compared two exchange/correlation functionals B3LYP and wB97XD, which has a specific treatment for non-covalent interactions. We have used a mix of 6-311+G and LANL2DZ for the basis sets. The latter basis set was used solely on the transition metal atom. For the ONIOM-EE calculations, we have used UFF to describe the low level part of the system. In addition, charges from electrostatic potential using a grid based method (CHelpG), natural bond orbitals population analysis and electric field gradient calculations were performed to describe the partial atomic charge distribution and electric field properties of the framework.

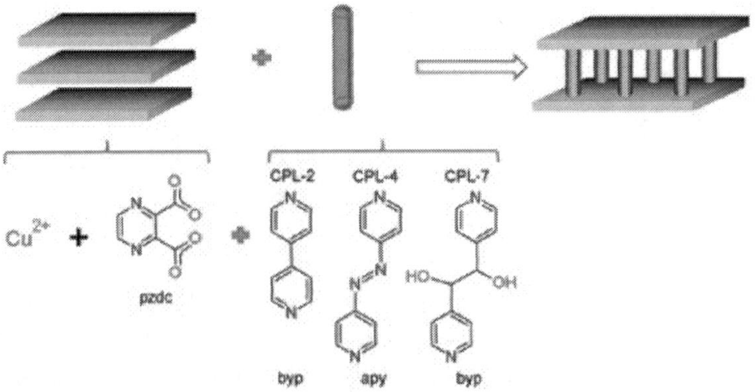

Figure 1. Schematic representation of CPLs structure

(395as) Kinetic Studies for the Removal of Copper From Industrial Wastewater Using Activated Neem (Azadirachta indica) Bark

Maheshwari, U., Birla Institute of Technology & Science (BITS)
Gupta, S., Birla Institute of Technology & Science, Pilani
M, B., Birla Institute of Technology & Science (BITS)

Water pollution from the industrial effluent containing toxic heavy metals is increasing globally. Environmental regulations for the drinking water are becoming more and more stringent. There is a need to develop an effective technique to process the industrial wastewater containing toxic metals before releasing into the water bodies. Cu(II) is one of the heavy metals used in a number of industries such as plating, electrolysis, metal cleaning, etc. A number of processes such as chemical precipitation, adsorption, solvent extraction, etc. are available for the treatment of industrial wastewater for the removal of Cu(II). Adsorption is proven to be an efficient technique at large flow rates and higher metal concentrations.

In the present study, a low cost adsorbent is developed using the neem bark. The neem bark is activated by proper chemical and physical treatment and is being used as an adsorbent for the Cu(II) removal. The kinetic studies for Cu(II) removal are carried out by varying the initial concentration of Cu(II) from 50 to 200 mg L^{-1}. The equilibrium time for the removal of Cu(II) using activated neem bark as an adsorbent is estimated to be about 40 hours. Pseudo first order and Second order kinetic model are tested and found that Second order model best describes the adsorption of Cu(II) from aqueous solution. Weber & Moris and Boyd model are used for the determination of mass transfer mechanism during the adsorption process. It is found that both are satisfactorily fitting with the experimental results. This confirms that the process is dependent on both, the film diffusion and intraparticle diffusion mechanisms.

(395o) New Insights On Sulfidation and Oxidizing Regeneration Reactions of Metal Oxides Used As Sulfur Sorbents

Chiche, D., IFP Energies Nouvelles
Neveux, L., IFP Energies Nouvelles
Girard, V., IFP Energies Nouvelles
Bazer-Bachi, D., IFP Energies Nouvelles
Baudot, A., IFP Energies nouvelles
Perez-Pellitero, J., IFP Energies nouvelles
Gay, A. S., IFPEN
Favergeon, L., Ecole Nationale Superieure des Mines
Geantet, C., IRCELyon
Pijolat, M., Ecole Nationale Superieure des Mines

Abstract

Fischer-Tropsch based XTL processes and IGCC technologies are attractive alternatives for future energy production. In order to prevent unit corrosion or FT catalysts poisoning, deep desulfurization of syngas is achieved with metal oxides able to react with H_2S to form sequestrated sulfide.

The present work focused on the solid-state transformations involved in metal oxides sulfidation and oxidizing regeneration reactions. Reactions kinetics were studied through thermogravimetric experiments. Regarding ZnO sulfidation, in opposition to common reaction modelings described in the literature, which assume an inward development of ZnS (shrinking core model), we propose a mechanism with ZnS outward growth, based on TEM and SEM observations. In agreement with the experimental data, an expression of the reaction rate has been determined.

The regeneration properties of various metal oxides were also studied. We will show that mixed metal oxides are for this purpose more adapted, and that lower regeneration temperature can be reached.

Introduction

Fischer-Tropsch (FT) based XTL processes and IGCC technologies are attractive alternatives for future energy production [1,2]. In order to prevent unit corrosion or FT catalysts poisoning, deep desulfurization of syngas is achieved with metal oxides, like zinc oxide, which reacts according to the reaction:

$$ZnO(s) + H_2S(g) \rightarrow ZnS(s) + H_2O(g)$$

The use of metal oxides as sulfur sorbents allows to reach very low H_2S levels in the treated gases, due to the highly favored sulfidation reactions. For IGCC applications, regenerable sulfur sorbents attract a great deal of interest, regarding the amount of H_2S to be removed from the gases and the process operating conditions. However the way back to the oxide is not favored at low temperature, as the formation of sulfate phases is usually observed during oxidative regeneration processes [3-5].

Although the ZnO sulfidation reaction has also been widely studied [6], little is known about the transformation into ZnS at the crystal scale, its detailed mechanism and kinetics. In this communication, we will present last advances in this field resulting from work performed at IFP Energies Nouvelles in collaboration with the Centre SPIN and IRCELyon laboratories. Especially, we will present the main results obtained in a study focused on the sulfidation of ZnO as model metal oxide. We will show evidences for an outward growth process of the resulting ZnS phase, which are in contradiction with the previously admitted assumption of an inward growth on ZnS with diffusion of sulfur species.

The regeneration properties of various metal oxides have also been studied, which will be presented. We will show that mixed metal oxides are for this purpose more adapted, and that lower regeneration temperature can be reached.

Experimental

The sulfidation reaction has been studied through thermogravimetric experiments (TGA) performed under variable H_2S partial pressures (0.001 to 1 bar) in helium, in a temperature range from 200°C to 400°C. TGA has also been used to study regeneration properties, under O_2 containing atmosphere (5-15% in N_2), between 100 and 700°C.

Materials characterizations have been performed by scanning and transmission electron microscopies (SEM, TEM), electron tomography (3D-TEM), X-Ray Diffraction (XRD), N_2 and Kr adsorption isotherm measurements, and real-time *in situ*characterizations, such as XRD, Extended X-Ray Absorption Fine Structure (EXAFS) and Fourier Transform Infrared spectroscopy (FTIR).

Results and Discussion

ZnO sulfidation study. Mono-crystalline ZnO materials have been synthesized and the sulfidation reaction has been studied through thermogravimetric experiments in order to establish a kinetic rate model. Kinetic tests based on thermogravimetric measurements have been performed and show that the kinetic rate can be expressed according to the following expression (rate-limiting step assumption) [7]:

$$d\alpha/dt = \Phi(T,P_i).S_m(\alpha)$$

where Φ (mol.m^{-2}.s^{-1}) depends on the nature of the rate-limiting step and is a function of temperature T and partial pressures P_i of the reacting gases, and $S_m(\alpha)$ (m^2.mol^{-1}) is a function of the fractional conversion related to the size and geometry of the reaction zone where the rate limiting step is spatially located.

SEM and TEM characterizations of sulfided ZnO particles reveal voids formation that evidence an external growth based mechanism for sulfidation reaction. Therefore the diffusion step is more likely to involve a diffusion of zinc and oxygen atoms from the ZnO/ZnS internal interface to the outer surface of the ZnS particle [8]. Related characterizations will be presented.

A reaction mechanism will be proposed that accounts for these results, and related possible rate laws determined upon approximation of the rate-determining step [9]. Thermogravimetry experiments performed in a wide range of H_2S and H_2O partial pressures have shown that ZnO sulfidation reaction rate exhibits a non linear variation with H_2S partial pressure, and that there is no significant influence of water vapor on reaction kinetics.

The formation of voids at the ZnO/ZnS internal interface, characterized by TEM and electron tomography, has been shown to influence the sulfidation reaction rate. The impact of the decreasing ZnO/ZnS internal interface on reaction kinetics, as well as the outward growth development of the resulting ZnS phase have been taken into account in a modeling, which will be presented in the communication.

Study of regeneration properties of mixed metal oxides. The sulfidation and regeneration properties of various single oxides have been studied. Based on the results obtained on single oxides, various zinc based mixed oxides were selected and synthesized, and their reactivity toward both sulfidation and oxidative regeneration was evaluated.

In particular after sulfidation step, Zn-Mo mixed oxides were shown to require lower regeneration temperature under O_2containing atmospheres than zinc oxide (from 150 to 200°C lower). Regeneration of sulfided mixed oxides was thus shown to exhibit synergetic effects resulting from exothermic oxidative reactions of molybdenum phases [10].

Moreover, textural modifications also occur during the sulfidation and regeneration processes. More characterizations on the modifications of the structure and texture of the mixed-oxide sorbents will be presented during the communication.

Conclusions

The researches performed at IFP Energies Nouvelles have allowed to bring new insights on the solid-state transformations involved during metal oxides sulfidation and regeneration processes. These results question generally admitted assumptions used for previous sulfidation reaction kinetics modeling. Kinetic model based on an outward growth development will be presented.

The sulfidation and regeneration properties of various single and mixed oxides have also been investigated. A decrease of 200°C of the regeneration temperature has been observed for some mixed metal oxides, and mechanisms allowing such temperature decrease will be presented.

In fine, this better understanding of the sorbents sulfidation and regeneration properties should allow to improve the efficiency of metal oxides sorbents-based desulfurization processes.

References

(1) D. Vallentin, *Energ. Policy*, 2008, **36**(8), 3198.

(2) O. Shinada, A. Yamada, Y. Koyama, *Energ. Convers. Manage.*, 2006, **43**(9-12), 1221.

(3) R.V. Siriwardane, S. Woodruff, *Ind. Eng. Chem. Res.*, 1995, **34**(2), 699.

(4) W.F. Elseviers, H. Verelst, *Fuel*, 1999, **78**(5), 601.

(5) R.V. Siriwardane, S. Woodruff, *Ind. Eng. Chem. Res.*, 1997, **36**(12), 5277.

(6) A. Samokhvalov, B.J. Tatarchuk, *PCCP*, 2011, **13**, 3197.

(7) M. Pijolat, M. Soustelle, *Thermochim. Acta*, 2008, **478**(1-2), 34.

(8) L. Neveux, D. Chiche, D. Bazer-Bachi, L. Favergeon, M. Pijolat, *Chem. Eng. J.*, 2012, **181-182**, 508.

(9) L. Neveux, D. Chiche, J. Perez-Pellitero, L. Favergeon, A.S. Gay, M. Pijolat, *PCCP*, 2013, **15**(5), 1532.

(10) V. Girard, D. Chiche, A. Baudot, D. Bazer-Bachi, C. Geantet, *To be published*

(395p) Monte Carlo Simulation of Methane and Carbon Dioxide Sorption in Clay-Like Slit Pores in Relation to Shale Gas

Jin, Z., University of California, Riverside
Firoozabadi, A., Yale University

As an important component of shale rocks, clay plays a significant role in the estimation species content and transport in shale gas reservoirs. Whereas previous investigations are mostly concerned with the gas sorption in carbonaceous materials, the dependence of the pore structure and chemical heterogeneity on the sorption behavior of gas in clay minerals needs further understanding. In this work, we present a grand canonical Monte Carlo method by using full atomistic molecular model for methane and carbon dioxide sorption in clay-like slit pores. Unlike previous computational and theoretical calculations on gas sorption in carbon slit-pores, the chemical heterogeneity of clay minerals and electrostatic interactions between gas-clay are explicitly considered. We demonstrate that both the pore structure and the chemical heterogeneity are important to the structural and sorption behavior of gas molecules resulting in different gas structural behavior in clay from that in the conventional carbon materials. Our work is believed to provide improved understanding toward the estimation species content in shale reservoirs.

(395av) Amine Functionalized Anion Tethered Ionic Liquid Based Substrates for Post Combustion CO2 Capture

Karunakaran, S., University of Cincinnati
Lee, D. J. Y., University of Cincinnati

Amino acid functionalized anion tethered Phosphonium or Imidazolium based Ionic liquids immobilized on various porous substrates are developed and studied for CO_2 adsorption at various operating conditions of temperature and pressure. The CO_2 adsorption takes place at 52ºC after wet flue gas desulphurization. Desorption from the CO_2 rich substrates shows faster kinetics when the temperature reaches 100ºC. The amine group facilitates the adsorption by chemical reaction whereas the ionic liquid physically adsorbs CO_2 by adjusting its free volume. Thus the CO_2 loading capacity is high for amine-functionalized ionic liquid substrates. In this study, the results for aminoacid- and blended amine-based ionic liquid adsorbents will be presented.

(395aw) Surface Functionalization of Micro and Mesoporous Sorbents For Post-Combustion CO2 Capture

He, J., Stanford University
To, J., Stanford University
Lyons, C. T., Stanford University
Haghpanah, R., Stanford University
Gary, B., Stanford University
Rupp, E. C., National Energy Technology Laboratory
Bao, Z., Stanford University
Stack, D., Stanford University
Wilcox, J., Stanford University

Solid sorbent technologies for CO_2 capture have several advantages over the traditional amine-based solvent absorption approaches. For instance, within an adsorption-based approach water is absent, which decreases the energy requirements associated with regeneration since heating a large amount of water is the greatest energetic expense associated with CO_2 capture using solvent-based approaches. Another benefit of using solid sorbents is the ability to tune the structural parameters of the material, as well as perform chemical modifications to enhance CO_2 uptake. In addition, materials such as carbon possess favorable heat conduction properties.

Mesoporous silica materials have been highly developed in materials chemistry, and allow for more extensive chemical modification for CO_2 capture. While developing chemistry amenable to both silica and carbon surfaces, mesoporous carbon-based framework development will allow for heat to be dissipated readily during the adsorption process, which will lead to maximum capacity, in addition to the ease of heat transfer into the system for regeneration.

We aim to use careful chemical modification of mesoporous silica- and carbon-based materials to achieve highly selective, recyclable, and efficient materials for CO_2 adsorption. Through careful tuning of the surface CO_2 adsorption chemistry, we can control the thermodynamics and kinetics of carbon dioxide adsorption to allow for efficient material development. Our inspiration for this idea comes from carbonic anhydrase (CA), an enzyme that can selectively adsorb and regenerate CO_2. CAs are found in the red blood cells of mammals, and are used to capture CO_2 as bicarbonate, which can easily dissolve in blood and transported to the lungs. The enzyme active site has been tuned such that the overall transformation is both kinetically fast and near thermodynamically neutral, allowing for both rapid and reversible CO_2 conversion. It is our aim to apply this design principle to sorbent-based technologies to enhance the mass transfer of CO_2 from the gas to adsorbed phase, in addition to enhancing the kinetics associated with adsorption and desorption (regeneration) processes.

(395r) Adsorption of Alkenes On 13X Zeolite: Heats of Adsorption and Henry Constants From the Gaussian Isotherm Model

Abouelnasr, D., American University of Shatjah
Loughlin, K. F., American University of Sharjah

Adsorption of Alkenes on 13X Zeolite:

Heats of Adsorption and Henry Constants from the Gaussian Isotherm Model

Adsorption data available in the literature for three alkenes (ethylene, propylene and 1-butene) on 13X zeolite is modeled using the Gaussian Adsorption isotherm. This isotherm model has three parameters: *qmax*, the saturation loading; the 50% adsorbate loading pressure ; and the standard deviation σ.

Methods: Values for qmax are calculated as specified in Al-Mousa [1]. For subcritical sorbates the sorbate density was calculated using Rackett's equation in combination with crystallographic data; for supercritical sorbates, a constant value was used. The remaining two parameters are deduced from a log plot of the isotherm pressure P versus the inverse Gaussian cumulative frequency distribution of Θ. The slope of this plot is σ and the y-intercept is $log\ P_{50}$. The heat of adsorption at 50 % loading is deduced from the slope of $log\ P_{50}$ versus $1/T$. Henry constants for supercritical isotherms are calculated using the Gaussian parameters in a method described in a forthcoming publication.

Z_H is a point within the range where the Gaussian model is consistent with Henrys Law. In this manner, the Henry constant can be deduced from isotherm data regardless of whether the experimental data extends into the Henrys Law region.

Results: The results were remarkably similar for all three alkenes. The standard deviation σ is observed to be strongly dependent upon temperature for subcritical adsorption, and only weakly dependent upon temperature for supercritical adsorption. The relationship between σ and reduced temperature is similar between the three alkenes.

The heats of adsorption $(-\Delta H_{50})$ are 9.02, 10.87, and 9.4 kcal/mole for ethylene, propylene, and 1-butene, respectively. These values compare well to literature values at zero loading $(-\Delta H_{0'})$.

The Henry constants for the supercritical isotherms range from 0.24 to 2.9 g/100 g-kPa for ethylene and from 0.3 to 3.6 g/100 g-kPa for propylene. There were no supercritical isotherms for 1-butene. The calculated Henry constants will be compared to values deduced from a virial plot of each isotherm.

[1] Alaa Al Mousa, K.F. Loughlin and D. Abouelnasr. Review and Analysis of Saturation Loadings for Subcritical Alkane Adsorption On 13X Zeolite, 2011 Annual Meeting of the AIChE, Minneapolis, 2011.

(395q) Synthesis of Titanosilicate Nanoparticles for Specific Adsorption of Biomolecules

Joshi, S., University of Kentucky
Knutson, B. L., University of Kentucky
Rankin, S. E., University of Kentucky

Many examples can be found in the literature of either porous surfactant-templated titanosilicates or nonporous non-imprinted core-shell titanosilicate nanoparticles. In this work, we propose to use sugar-based surfactants able to enter into complexes with titanium precursors to develop a method of synthesizing nonporous silica particles coated with imprinted titanium sites for adsorption of specific sugars.

The particles are prepared through a modification in the Stöber synthesis method for silica nanoparticles. Titania precursors complexed with specific saccharide surfactants are introduced to this method to provide surface functionalization and imprinting. It is hypothesized that addition of complexed sugar surfactant-bound titania along with cetytrimethylammonium bromide during early stages of the Stöber particle synthesis will lead to the creation of specific titanium-based adsorption sites on the surface of the particles. The titanium not only introduces the possibility of greater functionality and specificity of interactions with saccharides than silica alone, but also the possibility that non-specific binding sites on the silica surface can be blocked using silane chemistry with protection at the titanium site by the sugar template itself or other functional groups. Characterization of particles will be done using a combination of techniques including scanning electron microscopy, Fourier transform infrared spectroscopy, dynamic light scattering, transmission electron microscopy and nitrogen adsorption.

Adsorption experiments based on depletion of biological molecules from solution will be utilized to understand the specificity and selectivity of the titania-containing particles in comparison to silica along or particles prepared without surfactant imprinting. In particular, the ability of the particles to display enhanced binding to target saccharides (e.g. D-glucose vs. D-xylose) when imprinted using a surfactant with the same sugar headgroup (e.g. n-octyl-beta-D-glucopyranoside) will be discussed. Preparation of such imprinted titanosilicate particles is of potential importance for creating highly selective particles for separation of target saccharides and other biomolecules from mixtures of similar compounds.

(395s) Recovery of Biofuels From Dilute Aqueous Solution With Oxide-Supported Intrinsically Porous Oligomers

Thompson, A. B., Northwestern University
Scholes, R. C., Northwestern University
Notestein, J. M., Northwestern University

Fermentation-derived bioalcohols such as n-butanol and ethanol have potential as sustainable fuels, but their production is hindered by inefficiencies in separating from the dilute fermentation broth. The standard method of distillation is a major energy sink because of the low concentrations and unfavorable vapor-liquid equilibrium properties typical of ABE fermentation broths. More efficient methods are therefore needed in order to realize the potential of bioalcohols as renewable fuels. Adsorption is one such method, but conventional sorbents such as activated carbons have only limited ability to be fine-tuned at the molecular level, and as such, selectivities are typically low. Herein we investigate the use of intrinsically porous oligomers covalently immobilized to hydrophilic oxide supports as tunable adsorbents for separation of small organics from aqueous solutions. We previously synthesized these hybrid materials by grafting cup-shaped macrocyclic oligomers known as calixarenes, intended to act as the adsorption sites, directly to silica. We studied the effects of polarizing the calixarene cavities and found that the strongest adsorption occurs when both the adsorption site and the adsorbate are least polar, and that the adsorption energy decreases as the polarity of either the site or the adsorbate increases. These findings suggest that van der Waals forces drive the adsorption process and that permanent dipole-dipole interactions may hinder adsorption on calixarene-silica materials. Presently we investigate the effect of placing a Lewis acidic metal center such as Ti or Ta at the bottom of the calixarene cavity, which is intended to act as a weak coordination site for n-alkanols and allow for possible stronger, more directional interactions. In addition to their potential uses in separations, these materials enable the study of hydrophobic calixarenes in water, where they are otherwise insoluble. The synthesis schemes used here are generalizable and applicable for direct grafting of calixarenes, as well as other types of intrinsically porous oligomers, to different metal oxide supports. This leads to a large number of possible hybrid materials, and thus, the ability to fine-tune adsorbents for a given application.

(395bf) A New Flexible Cu-Based MOF for Separation of Linear and Branched C4-Hydrocarbons

Lange, M., aInstitut für Nichtklassische Chemie

Kobalz, M.,

Möller, A.,

Krautscheid, H.,

Staudt, R., University of Applied Science Offenburg

A new flexible Cu-based MOF for Separation of linear and branched C4-hydrocarbons

M. Langea, M. Kobalzb, A. Möllera, H. Krautscheidb, R. Staudtc

aInstitut für Nichtklassische Chemie e.V.,Permoserstr. 15, 04318 Leipzig, Germany bUniversität Leipzig, Institut für Anorganische Chemie, Johannisalle 29, 04103 Leipzig, Germany cUniversity of Applied Sciences Offenburg, Badstr. 24, 77652 Offenburg, Germany

Recently, a new Cu-based MOF material (overall formula 3 [(Cu2(H-Metrz-ia)2] **1**) with SBUs in *paddle wheel* structure was reported[1]. **1** possesses a three-dimensional pore system with two different windows of 0.27-0.35 nm and 0.44-0.50 nm as determined from X-ray single crystal structure analysis. The calculated solvent accessible pore volume reaches 50 % (PLATON[2]).
In this study, adsorption measurements were performed with both linear and branched alkanes and alkenes in order to assess the effect of structural flexibility on the separation of C4- alkanes/alkenes from the gas phase. The pure component adsorption of *n*-butane and isobutane on **1** was measured gravimetrically in a temperature range from 283 to 313 K and at pressures up to 300 kPa.
Sorption measurements show a stepwise pore filling, which is typical for flexible MOFs [3] (Fig. 1, left). By sorption of *n*- and isobutane, the framework opens and a maximum in loading is reached at 2.97 mmol g-1 and 3.12 mmol g-1, respectively, which results in similar pore volumes for both sorptives. In contrast, the equilibration times for *n*- and isobutane are significantly different (Fig. 1, right). Thus, the material has the potential to separate linear

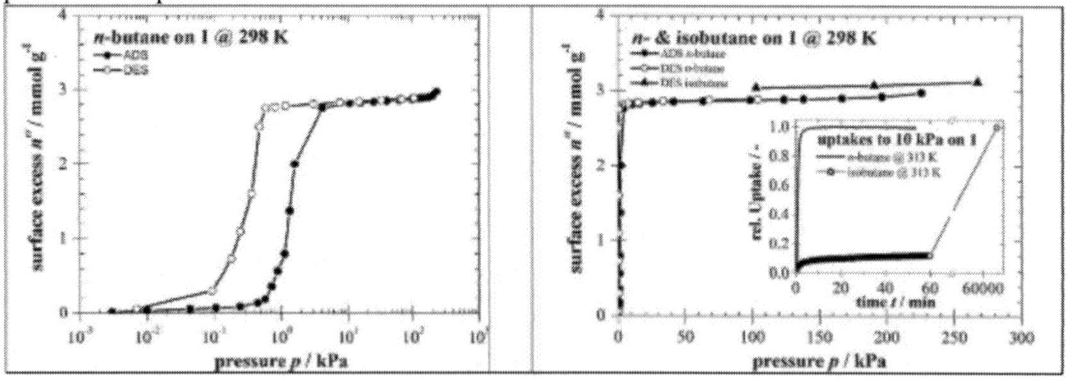

from branched C4-isomers on the basis of adsorption kinetics.

Fig. 1: Sorption isotherm of *n*-butane on **1** up to 250 kPa at 298 K (left) with a broad adsorption-desorption-hysteresis in the low pressure region, which is typical for flexible MOFs. The comparison of equilibria for *n*- and isobutane on **1** (right) show no thermodynamic separation effect, but the uptakes indicates a high potential for kinetic separation (insert).

Hence, kinetic measurements, like *uptake* experiments, breakthrough curves etc., will be presented to determine sorption rates. These are used as a basis for the calculation of effective selectivities. In addition, mixture measurements with time-resolved analysis of the gas phase composition will give detailed information about the adsorption process of *n*- and isobutane mixtures.

References:

[1] J. Lincke: Dissertation, Universität Leipzig, 2012. [2] PLATON, A. L. Spek: *J. Appl. Cryst.* (**2003**), *36*, 7.

[3] J. Lincke, D. Lässig, J. Möllmer, C. Reichenbach, A. Puls, A. Möller, R. Gläser, G. Kalies, R. Staudt, H. Krautscheid: *Micropor. Mesopor. Mater.* **2011**, *142*, 62-69.

(395ax) A Review On Adsorption Process for the Removal of Dyes From Textile Industry Effluent

Maheshwari, U., Birla Institute of Technology & Science (BITS)

There is a constant threat to the environment in terms of the pollution which is enhancing day by day. The impact of the pollution in the atmosphere, lithosphere and hydrosphere cannot be ignored. A lot of man made activities such as domestic, industrial, agriculture, shipping, radio-active, aquaculture waste activities are developing the negative impact on the water body by enhancing the contaminants present in the water. Textile industries are one of the type which are being counted as the top polluting industry. There are a number of harmful dyes which are present in the effluents from the textile industries and should not be discharged directly to the water body. These effluents are required to be treated for the removal of all the harmful dyes available in it before discharging it to the outer water body.

There are a number of processes available for the removal of textile dyes from the effluents such as such as trickling filtration, activated sludge process, oxidation ditch, oxidation pond, electrolytic precipitation & foam fractionation, membrane separation, ion exchange method, adsorption, etc. These processes can be further characterized into primary, secondary and tertiary category on the basis of the severity and effectiveness of the process. The processes other than adsorption are effective but are non co-operative in one or the other situation. Some are not effective at higher temperature, while others are not suitable at low temperature. Some processes are effective in cold conditions while others are more effective in hot environments. Adsorption is proven to be an efficient and cost effective method for the harmful dye removal. Adsorption is the only effective process which can work in a vast range of dye concentration. There are various adsorbents available for the dye removal such as fly ash, activated neem bark, china clay, silica, peat, etc. There be a requirement of developing a low cost adsorbent which will help in making the adsorption process more economical.

(395be) Multicomponent Adsorption Equilibria On Cu-Based MOF-Materials

Möllmer, J., Institut f. Nichtklassische Chemie e.V.

Möller, A., Institut f. Nichtklassische Chemie e.V.

Staudt, R., University of Applied Science Offenburg

Multicomponent Adsorption Equilibria on Cu-Based MOF-Materials

J. Möllmer, A. Möller, Institut für Nichtklassiche Chemie e.V., Leipzig/Germany

R. Staudt, University of Applied Sciences, Offenburg/Germany

Since the development of metal-organic frameworks (MOFs), the interest in studying specific adsorption interactions are gaining importance.[1] Starting with characterization of porous MOFs and followed by intensive study of gas storage applications, e.g., storage of hydrogen, carbon dioxide or methane, the current research interest is on multicomponent adsorption equilibria. In particular, for gas separation processes, the information about the adsorption equilibria of mixtures are of major importance. Thus, several groups have studied binary adsorption by dynamic methods, such as breakthrough curves or gravimetric method without a circulation pump.[3]
Here, three Cu-containing MOFs, i.e. the well known material Cu3(BTC)2 (copper - 1,3,5-tricarboxylic acid, HKUST-1)[2] and two MOFs with a triazole ligand, in particular the highly flexible MOF 3 [(Cu4(*Âμ*4-O)(*Âμ*2-OH)2(Me2trz*p*ba)4][4] and a new Cu-based MOF 3 [Cu(Me-*4*py-trz-ia)][5], were used for pure gas and mixed gas adsorption experiments using the volumetric-chromatographic method. More precisely, CO2 and CH4 pure gas adsorption experiments were carried out at temperatures between 273 K to 323 K and pressures up to 3 MPa. The IAST-model was used to calculate

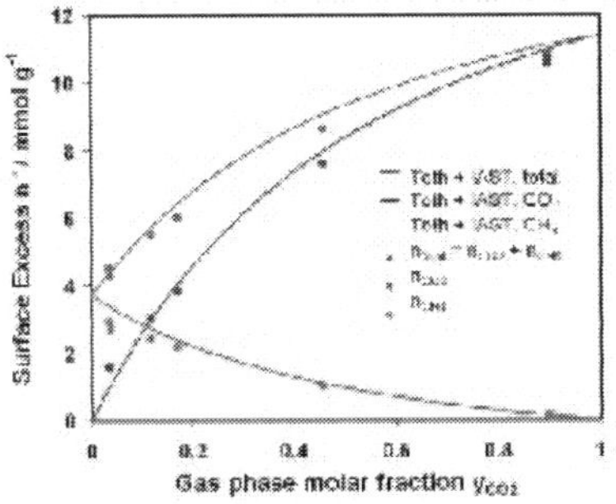

binary adsorption data from the pure gas isotherms. These data were further compared to experimental binary adsorption data, which were measured with a static volumetric system. Furthermore, the selectivity for adsorption of CO2 over CH4 was calculated and discussed in the context of the adsorption potential based on the accessible Cu-sites of these MOFs. The results provide evidence for the high potential of the investigated MOFs for gas separation applications, here, of CO2 separation from methane, especially at elevated pressures.

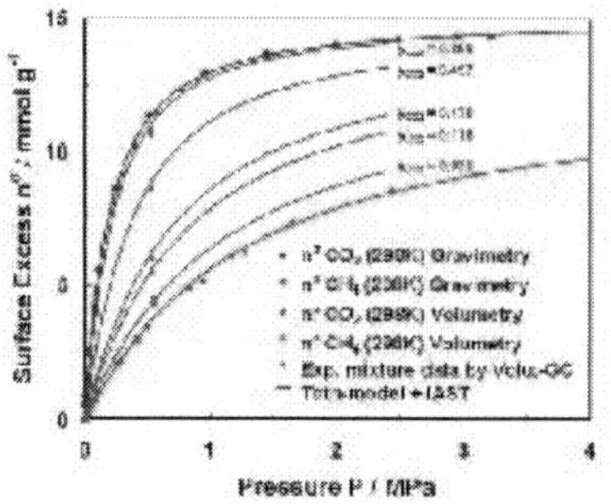

Tab. 1: Binary adsorption of CO2 and CH4 on Cu3(BTC)2 at 298 K and 0.54 MPa (left side -- adsorption isotherms at 298 K; right side - partial loading). Acknowledgements

The authors thank D. Lässig and J. Lincke (Institute of Inorganic Chemistry, University of Leipzig) for preparing the MOF

materials. The *Deutsche Forschungsgemeinschaft (DFG-Projekt SPP 1362 MOF STA428/17-1)* is also gratefully acknowledged. References

[1] O.M. Yaghi, H. Li, J. Am. Chem. Soc. 117 (1995) 10401.

[2] S.-Y. Chui, S. M.-F. Lo, J.P.H. Charmant, A.G. Orpen, I.D. Williams, Science 283 (1999) 1148. [3] L. Hamon, E. Jolimaitre, G.D. Pirngruber, Ind.Eng.Chem.Res. 49 (2010) 7497.

[4] J. Lincke, D. Lässig, J. Möllmer, C. Reichenbach, A. Puls, A. Möller, R. Gläser, G. Kalies, R. Staudt, H. Krautscheid, Micropor.Mesoporous Materials 2010, submitted.

[5] D. Lässig, J. Lincke, J. Möllmer, C. Reichenbach, G. Kalies, R Gläser, R. Staudt, H. Krautscheid, in preparation.

(396e) The Thermodynamics and Design of the Down-Stream Process of the Pervaporation With PDMS Composite Membranes

Fan, S., Sichuan University
Xiao, Z., Sichuan University

During a traditional membrane pervaporation process, in order to condense and capture the permeated vapors at the down-stream side of the membrane, a 'cold trap' unit at an extremely low temperature, e.g. -30°C, accomplished only with a refrigerating system, is needed, which takes a most part of the energy cost for the whole process. Based upon the thermodynamics of condensation and transportation of the vapors, a pervaporation process without a refrigerating system was designed, which can condense and capture the vapors at the normal temperature. In the procedure, a suction machine for the permeated vapor flowing, such as a vacuum pump or compressor was emphasized, and two condensers were set up at the inlet side and the outlet side of the machine, respectively. The condenser at the inlet side could be operated, by using the circulating water at the room temperature as the coolant, and the components with higher boiling point in the vapors could be condensed and collected here. The remaining vapors was discharged by the 'compressor' and condensed at the outlet side. The experiment was done for separating several organics-water mixtures in this new pervaporation system with a composite PDMS membrane. The system's stability and the membrane separation characteristics were tested, which demonstrated its good performance. The feasibility and the energy consumption of the process were evaluated, which demonstrated that the designed process had a lower-energy-cost, less than 26% of the energy consumption compared with that in the traditional process.

(396ac) Novel Cross-Linked Multi- Ammonium Poly (aryl ether oxadiazole) s Using Thermal Crosslink for Anion Exchange Membranes

Li, C.,

Novel cross-linked multi- ammonium poly (aryl ether oxadiazole) s using thermal crosslink for anion exchange membranes

Cunpu LI, Xiaofeng XIE *, Changsheng DENG

Institute of Nuclear and New Energy Technology, Tsinghua University, Beijing 100084, China

For a long time, the membrane which has been applied in the battery field, is basically Proton Exchange Membrane (PEM) that it also largely driven the development of the battery industry, but its limitations are increasingly exposed. Compared to PEM, anion exchange membrane technology is not mature. In this work, a novel cross-linked multi-ammonium poly (aryl ether oxadiazole) s ionomer (CMAPAEO) was synthesized. The 3-diethylaminophenol was used to get 4 tertiary amine groups to the backbone. Treated by bromomethane, the anion exchange groups were obtained. The anion exchange membranes were obtained by solution casting method. After thermal-crosslinking a more stable structure was obtained. The chemical structures and properties of the membranes were studied. The highest ionic conductivity of the membrane was achieved at 2.72×10^{-2} S cm^{-1} at 80 °C. The swelling ratio and TGA properties of the membranes revealed their good mechanical and thermal stability. Moreover, these membranes showed excellent solubility in low boiling point water-soluble solvents such as methanol, which would be helpful for membrane electrode assembly preparation. Also these membranes showed great potential for alkaline electrolyte membrane fuel cells (AEMFCs) applications.

(396y) High-Flux Reverse Osmosis Membranes With Fouling-Resistance for Seawater Desalination

Zhao, L., The Ohio State University
Ho, W. S. W., The Ohio State University

Reverse Osmosis (RO) is one of the most promising technologies for water desalination, which serves as an important solution to the global water shortage. However, the hydrophilicity of the aromatic polyamide (PA) thin-film-composite (TFC) RO membrane is insufficient, which results in a considerably limited water flux. In addition, traditional RO membrane is very vulnerable to fouling from large molecules.

A hydrophilic additive is defined as a chemical compound which contains at least one hydrophilic portion and one reactive portion that can react with either MPD or TMC during the interfacial polymerization. Therefore, the additive is chemically bound to the cross-linking network formed by MPD and TMC. It is believed that the hydrophilic additive incorporated in the TFC membrane can provide an additional pathway for the molecular transport of water. As a result, the incorporation of hydrophilic additive is able to improve the water flux of the RO membranes, thereby increasing the water productivity.

In this work, *o*-aminobenzoic acid-triethylamine salt was selected as the specific hydrophilic additive to investigate its effects on membrane separation performance under seawater desalination conditions for the first time. In addition, different membrane preparation conditions were investigated and optimized to further improve the water flux of the synthesized membranes. After optimization, the membrane samples were also tested by using the seawater from Port Hueneme, CA to determine their desalination performance for the real application. Furthermore, the fouling resistance of the high-flux membranes was compared with that of the standard TFC membranes.

The high-flux RO membranes were synthesized through advanced interfacial polymerization where *o*-aminobenzoic acid-triethylamine salt was added in the aqueous MPD solution. The membrane synthesized with the hydrophilic additive exhibited significantly improved desalination performance, in a side-by-side comparison with the membrane synthesized without the additive. After optimizing the concentration of the hydrophilic additive in the MPD solution, the water flux of the synthesized membrane was enhanced by about 62%, and the salt rejection remained the same. For instance, the membrane synthesized with 1.0 wt% *o*-aminobenzoic acid/triethylamine salt showed 39.2 gfd (gallons/ft^2/day) water flux and 99.46% salt rejection. The incorporation concentration of *o*-aminobenzoic acid-triethylamine salt was maintained at 1.0 wt% for the optimization of other membrane synthesis conditions to enhance water flux without compromising salt rejection. With 40 wt% isopropanol (IPA) in the aqueous MPD solution, the water flux increased to 44.4 gfd, and salt rejection was still above 99.4%. The effects of additional drying time after soaking polysulfone support in the amine solution and hydrocarbon removal time after interfacial polymerization were also investigated.

The high-flux membranes synthesized under the optimal conditions were tested by using the seawater from Port Hueneme, CA with an average turbidity of 0.032 Nephelometric Turbidity Units (NTU) and a silt density index (SDI) (15 min) of 2.2. The average water flux was about 41.8 gfd and the average salt rejection was above 99.4%. The lower water flux was found to result from the higher solid concentration in the real seawater compared to the synthetic seawater which contains 3.28% NaCl. In addition, the synthesized high-flux RO membrane exhibited an excellent stability in terms of water flux (average 40 gfd) and salt rejection (average 99.4%) during a 30-day test with the seawater from Port Hueneme, CA at 25°C and 800 psi applied pressure across the membrane. This good stability was verified by comparing FTIR spectra of the same membrane sample before and after the 30-day test. These results are important for the commercialization of this high-flux RO membrane for seawater desalination.

Sodium alginate was selected as the foulant to investigate membrane fouling, since it is a common compound in seawater from seaweeds. The membrane fouling resistance was significantly improved with the incorporation of the hydrophilic additive. For example, in the presence of 50 ppm sodium alginate in the feed solution, the water flux reduction of the membrane synthesized with the hydrophilic additive was 19.6% compared to 32.0% for the membrane synthesized without the additive.

(396z) CO2 Capture From Power Plant Flue Gas Using Zeolite/Polymer Composite Membranes

Ramasubramanian, K., The Ohio State University
Tong, Z., The Ohio State University
Zhao, L., The Ohio State University
Severance, M. A., The Ohio State University
Dutta, P. K., The Ohio State University
Ho, W. S. W., The Ohio State University

Membrane technology provides a promising approach for CO_2 capture from post-combustion flue gas due to its smaller footprint and potentially lower energy cost than the traditional amine scrubbing process. Thin inorganic membranes with high selectivity and flux have been conventionally supported by inorganic materials, which are expensive and difficult for scale-up in a continuous manner and fabrication into a high packing density in terms of high surface area/volume due to their brittle nature.

In this work, a thin zeolite seed layer (< 1 μm) has been successfully deposited on a flexible polymer support for the first time, which enables the continuous roll-to-roll fabrication of the membrane and the configuration of spiral-wound membrane modules with a high packing density. Zeolite-Y with pore size of about 7 Å was specially selected for CO_2/N_2 separation from flue gas, due to the preferential adsorption of CO_2 molecules with a larger quadrupole moment and higher polarizability. This is known as the "surface diffusion" mechanism along with the blocking of N_2 molecules due to the adsorbed CO_2 molecules. In addition, a polymer cover layer containing fixed polyamine and mobile amine carriers was applied on top of the zeolite seed layer by knife-casting to utilize the advantages of both inorganic and polymer membranes for CO_2 capture from flue gas. The thickness of polymer cover layer varied from 0.5 μm to 10 μm. The separation performance of the synthesized zeolite/polymer composite membrane was characterized under humid flue gas composition at close to atmospheric feed pressure and 102°C. The resulting membrane showed a CO_2 permeance greater than 1100 GPU and a CO_2/N_2 selectivity above 400. Based on the transport results, a process and cost modeling study will be discussed in view of achieving the DOE targets of < 35% increase on the cost of electricity and < $40/tonne CO_2 captured for 2025.

Considering the real application in a power plant, SO_2 tolerance of the aforementioned membrane was evaluated under different SO_2 concentrations in feed gas. When SO_2 concentration in feed gas was below 1 ppm, the separation performance of the composite membrane was not affected at all. When 5-ppm SO_2 was supplied, a reduction of CO_2 permeance was observed. However, the CO_2 permeance recovered back to about 80 Â¨C 85% of its original value within 24 hours after exposure to the feed gas without containing the SO_2, which could be crucial if the up-stream SO_2 polishing step fails.

(396aa) New Insights On Carbon Dioxide and Methane Transport in Carbon Molecular Sieve Membranes From Diffusion NMR Studies

Mueller, R., University of Florida
Kanungo, R., University of Florida
Kiyono-Shimobe, M., Georgia Institute of Technology
Koros, W. J., Georgia Institute of Technology
Vasenkov, S., University of Florida

Carbon molecular sieve (CMS) membranes are promising materials for energy-efficient separations of light gases of particular interest in energy applications. This work focuses on the separation of mixtures containing carbon dioxide and methane which are ubiquitous in natural gas and bio-gas streams. Here we present and discuss selected results of microscopic studies of self-diffusion of carbon dioxide and methane in samples of CMS membranes optimized for methane/carbon dioxide separations. These diffusion data were obtained by a pulsed field gradient (PFG) NMR method that combines advantages of using large (up to 30 T/m) magnetic field gradients and high (17.6 T) magnetic field. An application of large field gradients allows measuring relatively small diffusivities of gases in CMS membranes, while high field was essential for obtaining sufficiently large signal-to-noise ratios in diffusion studies. The microscopic diffusion data obtained by PFG NMR will be discussed in relation to previous macroscopic observations of membrane selectivities and permeabilities measured for the same gases in the same membranes.

(396a) Natural Gas Purification and Olefin/Paraffin Separation Using Thermal Cross-Linkable Co-Polyimide/ZIF-8 Mixed Matrix Membranes

Askari, M., National University of Singapore
Chung, T. S., National University of Singapore
Yang, T., National University of Singapore

Cross-linkable mixed matrix membranes represent an attractive technology that promises both outstanding separation properties and swelling resistance for the natural gas purification and olefin/paraffin separation. Using three 6FDA-based polyimides (6FDA-Durene, 6FDA-Durene/DABA (9/1), 6FDA-Durene/DABA (7/3)) and nano-size zeolitic imidazolate framework-8 (ZIF-8), we have fabricated mixed matrix membranes (MMMs) with uniform morphology comprising ZIF-8 as high as 40 wt% loading by directly mixing as-synthesized ZIF-8 suspension into the polymer solution. Permeability of all gases increases rapidly with an increase in ZIF-8 loading. However, the addition of ZIF-8 nano-particles into the polymer matrix increases the ideal CO_2/CH_4 selectivity of only 6.87%, while the ideal C_3H_6/C_3H_8 selectivity improves 134% from 11.68 to 27.38 for the MMM made of 6FDA-Durene/DABA (9/1) and 40 wt% ZIF-8. Experimental data show that the plasticization resistance and gas pair selectivity of MMMs are strongly dependent on the amount of cross-linkable moiety and annealing temperature. MMMs made of 6FDA-Durene do not show considerable improvements on resistance against CO_2-induced plasticization after annealing at 200 – 400 °C, while MMMs synthesized from cross-linkable co-polyimides (6FDA-Durene/DABA (9/1) and 6FDA-Durene/DABA (7/3)) show significant enhancements in CO_2/CH_4 and C_3H_6/C_3H_8 selectivity as well as plasticization suppression characteristics up to a CO_2 pressure of 30 atm after annealing at 400 °C due to the cross-linking reaction of the carboxyl acid (COOH) in the DABA moiety. The MMM made of 6FDA-Durene/DABA (9/1) and 40 wt% ZIF-8 possess a notable ideal C_3H_6/C_3H_8 selectivity of 27.38 and a remarkable C_3H_6 permeability of 47.3 Barrer. After thermally annealed at 400 °C, the MMM made of 6FDA-Durene/DABA (9/1) and 20 wt% ZIF-8 shows a CO_2/CH_4 selectivity of 19.61 and an impressive CO_2 permeability 728 Barrer in mixed gas tests. The newly developed MMMs may have great potential for industrial nature gas purification and C_3H_6/C_3H_8 separation.

(396g) Three-Component Mwcnts-Based Hybrid Membranes With Enhanced CO2 Capture Properties

Jiang, Z., Tianjin University
Wu, H., Key Laboratory for Green Chemical Technology of Ministry of Education, Sch Chem Engn & Technol, Tianjin Univ

Three-component hybrid membranes containing Pebax, PEO-based polymers and multi-walled carbon nanotubes (MWCNTs) were prepared and characterized. Pebax solutions were slowly added to PEO-based polymer solutions containing MWCNTs then evaporated to get the hybrid membranes. Scanning electron microscope (SEM) images presented the improved dispersion of MWCNTs in polymer matrix with the pre-dispersing of PEO-based polymers. The embedded MWCNTs disrupted the polymer chain packing and decreased crystallinity, which was confirmed by X-ray diffraction. Influences of PEO-based polymers and MWCNTs contents on CO_2, N_2 and CH_4 permeation were investigated. These results revealed that incorporating of high molecular weight PEO-based polymers favored for CO_2/CH_4 separation, while low molecular ones favored for CO_2/N_2 separation. For all these hybrid membranes, the enhanced CO_2 permeability was attributed to the more amorphous polymer structure. The incorporation of MWCNTs decreased CO_2/CH_4 selectivity while increased CO_2/N_2 selectivity for the size and solubility parameter differences of these two gas pairs. In particular, a selectivity of 108 and CO_2 permeability up to 743 Barrer at 10 bar and room temperature were documented, which showed promise for CO_2 capture from flue gas.

(396i) Modeling of Structure and Transport in Mixed-Matrix Membranes for Gas Separation

Petsi, A., FORTH
Burganos, V., FORTH/ ICEHT

Mixed matrix membranes combine the superior separation efficiency of inorganic fillers with the processability of polymeric materials and appear to be quite promising in the field of demanding separation problems. However, the resulting performance regarding selectivity is usually far less than the expected one, attributed among other factors to limited affinity between polymer and inorganic materials and the concomitant reduction of adherence of the two phases. This allows the formation of non-selective voids around the fillers, which can destroy the selectivity of the membrane, allowing the gases to follow pathways through these voids. The problem can be addressed by either filling the empty space with a suitable interphase material or eliminate it using appropriate functionalization at the interface. In both cases, the design and preparation of such composite materials relies heavily on the understanding and quantifying the relationship between the effective transport coefficient and the configuration of the material.

In this work, a numerical model is developed that allows the prediction of the effective diffusivity in composite membranes, taking into account the different diffusion and solubility constants in each phase. The algorithm involves computer-aided stochastic reconstruction of the membrane in two and three dimensions by randomly placing non-overlapping inclusions within the working domain, yet allowing free overlapping of the interphase layers. The size of the inclusions can be uniform or sampled from a given distribution, while their shape can be varied in two and three dimensions (spherical, elliptic, or circular). The width of the interphase layer can vary according to the physicochemical affinity between the inclusions and the polymer.

The permeability and selectivity of this composite membrane is numerically computed using a method that is adapted to the particular reconstructed material. Specifically, the resistor network approach is used following a suitable mapping of the composite material onto a regular network. It is found that the permeability of a composite membrane is a strong function of the filler-loading of the composite, the width of the interphase layer, and the diffusivities and solubilities in the three phases. The shape of the inclusions as well as their orientation in the case of elliptical particles are found to affect strongly the membrane permeability. Moreover, different spatial organizations of the particles in the polymer phase have been examined and it was found that the type of particle agglomeration can also have a significant effect on the membrane permeability. These results elucidate the utilization of the inclusions, which is evidently the key factor in the design and eventual performance of mixed matrix membranes.

(396j) An Experimental and Modeling Study of Membrane for Syngas Purification

Vakharia, V., The Ohio State University
Ramasubramanian, K., The Ohio State University
Ho, W. S. W., The Ohio State University

As the awareness on global warming is growing, electricity generation from the Integrated Gasification Combined Cycle (IGCC) power plants seems to be a favorable alternative for power generation with simultaneous carbon capture (CO_2). In this case, CO_2 needs to be separated from syngas for the H_2 purification, where the purified syngas is the feed to the Gas Turbine (GT) for power generation. For the CO_2 capture in these power plants, conventional absorption process using Selexol as the solvent has been considered by DOE/NETL as the baseline technology. Recent data published in the DOE baseline document suggests that incorporating this technology for the CO_2 capture and sequestration in the IGCC power plants would result in an increase in the Cost of Electricity (COE) by 37%. Syngas from the low temperature Water Gas Shifter (WGS) reactor is at pressures close to 50 bar and at temperatures close to 235 °C.

This work describes the transport properties of our CO_2-selective membranes under conditions (pressure from 2 bar to 15 bar and temperature 107 °C) and the potential it has for operating at higher pressures (even higher than 20 bar) for CO_2 capture from syngas and apply it to the process that is integrated with the IGCC power plants. These CO_2-selective facilitated transport membranes are composed of amine carriers dispersed in the crosslinked polyvinylalcohol (PVA) matrix. CO_2 permeability and CO_2/H_2 selectivity at different pressures investigated are discussed. These membranes have been used for the H_2purification in a project funded by the Office of Naval Research.

This study deals with the modeling and cost study of a promising syngas purification process for IGCC plants with the CO_2capture using the membranes mentioned above. A two-stage process has been considered for this application. For both the stages, the effects of operating parameters like the feed pressures and temperatures on the overall economics were analyzed to obtain an optimal condition of operation. We have also studied the effects of membrane performance, selectivity and permeance on the overall increase in the COE and the capture cost.

A CO_2 permeability of 1287 Barrers and a CO_2/H_2 ideal selectivity of 87 at about 15 bar pressure were obtained from our group. From a study in our group on the transport properties at lower pressures (2 bar), a CO_2 permeability of 5000 Barrers and CO_2/H_2 ideal selectivity of 250 was obtained. Based on the transport results, our cost calculations show an increase in the COE close to 15% and a capture cost about $17/tonne of CO_2 removed for the two-stage membrane based process for achieving about 90% CO_2 capture (>95% purity) and greater than 99% H_2 recovery. This process operates at pressures close to 23 and 2 bar on the feed side and the sweep side of the Stage 1, respectively. The membrane with a CO_2 permeance of 162 GPU, CO_2/H_2 ideal selectivity of 49 and with a thickness of 5 microns was considered. The transport properties of the membrane at 23 bar have been estimated from a study on the membrane performance-property relationship, and they will be verified experimentally. The feed pressure on the Stage 2 is close to 3 bar and the sweep side at atmospheric pressure. The Stage 2 membrane has a thickness of 5 microns with a CO_2 permeance of 1000 GPU and a CO_2/H_2 selectivity of 250. This study shows the potential our membrane based technology can have for the IGCC application.

(396k) On the Formation of Si-CHA Zeolite Layers and Their Secondary Growth for Continuous CHA Films

Kim, E., Korea University
Choi, J., Korea University

Despite its high potential for CO_2/N_2 separations in the post-combustion process, an all-silica CHA (Si-CHA) zeolite membrane has not been reported yet. In particular, the wide size distribution of conventionally synthesized near cubic CHA zeolite particles blocks the formation of a uniform Si-CHA layer.[1-3] In fact, the mono-dispersity of used zeolite particles is highly required to obtain the uniform layer.[4,5] Considering the fact that the uniform layer is critical in manufacturing desired zeolite films in the secondary growth method,[6] the formation of the uniform Si-CHA layer is an important task.

In this poster presentation, we will discuss how to achieve a uniform Si-CHA layer based on thin plate-like Si-CHA particles, though they exist in minor fraction in the mixture with the above-mentioned conventional near cubic CHA particles. In addition, we will demonstrate the formation of continuous CHA films via the secondary growth of the above-mentioned CHA layer and present the corresponding CO_2/N_2 separation performance. Specifically, we will show the CO_2/N_2 separation performance of the CHA film in the presence and absence of H_2O, which is third major component in the flue gas stream.

References

[1] Eilertsen, E. A.; Arstad, B.; Svelle, S.; Lillerud, K. P. *Micropor. Mesopor. Mater.* **2012**, *153*, 94.

[2] Olson, D. H.; Camblor, M. A.; Villaescusa, L. A.; Kuehl, G. H. *Microporous Mesoporous Mater.* **2004**, *67*, 27.

[3] Trzpit, M.; Rigolet, S.; Paillaud, J. L.; Marichal, C.; Soulard, M.; Patarin, J. *Journal of Physical Chemistry B* **2008**, *112*, 7257.

[4] Choi, J.; Ghosh, S.; Lai, Z. P.; Tsapatsis, M. *Angew. Chem., Int. Ed.* **2006**, *45*, 1154.

[5] Snyder, M. A.; Tsapatsis, M. *Angew. Chem., Int. Ed.* **2007**, *46*, 7560.

[6] Lai, Z.; Bonilla, G.; Diaz, I.; Nery, J. G.; Sujaoti, K.; Amat, M. A.; Kokkoli, E.; Terasaki, O.; Thompson, R. W.; Tsapatsis, M.; Vlachos, D. G. *Science* **2003**, *300*, 456.

(255d) Submerged Hollow Fiber Membrane Bioreactor for Symbiotic Microalgal Growth and Bacterial Wastewater Treatment

Vu, L., National University of Singapore
Loh, K. C., National University of Singapore
Praveen, P., National University of Singapore

Activated-sludge process is the most widely used biological method for the treatment of domestic wastewater. However, a main disadvantage of this process is the high energy cost for the mechanical aeration, which accounts for 45 – 75% of the plant energy consumption. The symbiotic microalgal-bacterial process, which incorporates microalgae and bacteria in one closed system, is a potential approach to solve this problem: in the presence of light microalgae can produce sufficient O_2 for bacteria to biodegrade organic matters in wastewater, while bacteria release CO_2 for microalgal photosynthesis. In order to efficiently apply this symbiotic microalgal-bacterial process to current activated sludge tank, a novel submerged hollow fiber membrane bioreactor (SHFMB) for symbiotic bacterial wastewater treatment and microalgal production has been developed. In this SHFMB, a microalgae tank is built above bacteria tank for effective light penetration, and the microalgal culture is circulated through hollow fiber membranes immerged in the bacteria tank. These hollow fiber membranes are used to isolate the bacteria from the microalgae, and for CO_2 and O_2 exchanges in both the microalgal and bacterial growth. Results showed that through the photosynthetic oxygenation, the BOD content of the synthetic wastewater was completely degraded, while the microalgae grew effectively in the absence of the external CO_2 supply. It is also feasible to operate the SHFMB in continuous mode. Hence, the developed SHFMB could be applied for existing activated sludge treatment process as it allows reduced energy cost for mechanical aeration and eliminates atmospheric CO_2 emission, while less or no additional land area for the construction of microalgal photobioreactor is required. Especially, the microalgal biomass obtained from the SHFMB could be used as sources of food, feed or other high-value compounds.

(396l) New Membranes Structures and Compositions for CO2 Separation and Capture

Chen, Y., The Ohio State University
Ramasubramanian, K., The Ohio State University
Zhao, L., The Ohio State University
Severance, M. A., The Ohio State University
Dutta, P. K., The Ohio State University
Ho, W. S. W., The Ohio State University

This presentation covers new structures and compositions for carbon dioxide-selective membranes, including (1) polymer and polymer blend membranes, (2) mixed matrix membranes, and (3) inorganic/polymer composite membranes.

We have synthesized membranes with polyamide-polyethylene oxide (PA-PEO) copolymer blended with polyethylene glycol dimethyl ether (PEG-DME) or polyethylene glycol (PEG). We have also synthesized a polydimethylsiloxane (PDMS) membrane reproducibly with good separation performance, which can be used as a cover layer for inorganic/polymer composite membranes.

We have dispersed zeolite-Y nanoparticles in PA-PEO/PEG-DME and PDMS solutions to synthesize the mixed matrix membranes, which can be used as the cover layers for inorganic/polymer composite membranes. The membranes showed a higher CO_2/N_2 selectivity than PDMS alone.

We have designed and synthesized inorganic/polymer composite membranes, which consisted of a thin selective inorganic layer embedded in a polymer structure. Zeolite-Y/polymer composite membranes were synthesized successfully by coating a polymeric cover layer containing amines on the zeolite-Y nanoparticles deposited on a nanoporous polymer support. The resulting membrane showed a CO_2 permeance greater than 1100 GPU and a CO_2/N_2 selectivity above 400 at 102°C. Si-1/polymer composite membranes were also synthesized successfully by coating a PA-PEO/PEG-DME blend or PA-PEO/PEG blend as the cover layer on both Si-1 particles/polymer support and Si-1/PDMS/polymer support multilayer structures. The syntheses and transport properties of the membranes will be presented and discussed.

(396m) Preparation and Characterization of Polyfluoroethylene Membranes and Pervaporation / Swelling Properties

Chen, X., Beijing Technology and Business University
Ye, H., Beijing Technology and Business University,
Zhang, T., Beijing Technology and Business University
Xia, Y., Beijing Technology and Business University, Beijing, China

The pervaporation is promising in the separation of phenol and water. The modification of membrane material is very important to obtain good selectivity and permeability. Polyfluoroethylene(PVF) and modified PVF membranes with the addition of 0.5wt% graphene were prepared from the solutions containing 2wt%-3wt% polymer. And the structures, crystalline properties and morphologies were characterized by FTIR, TGA, SEM and XRD. The swelling properties and pervaporation (PV) performances for separating 0.5wt% phenol/water mixture were investigated. The effects of feed composition on the flux and the selectivity were studied at 60-80°C. For PVF membranes, the results showed that fluxes increased and selectivity decreased with increasing temperature. The separation factors were from 12.04 to 48.36 at 60-80°C and fluxes were from 2.53 to 1.23 kg·μm·m-2·h-1 at 70-80°C. While for the modified PVF membranes, the fluxes decreased and selectivity increased with increasing temperature. The separation factors were from 4.99 to 10.08 and the fluxes were from 0.83 to 0.47kg·μm·m-2·h-1 at 60-80°C.

(780c) Forward Osmosis Hollow Fiber Membrane Incorporated With Ultrafiltration Bioreactor for Treatment of High-Strength Phenolic Wastewater

Nguyen, D. T., National University of Singapore
Praveen, P., National University of Singapore
Loh, K. C., National University of Singapore

Phenol and its derivatives are toxic compounds which can be found in the industrial wastewater from various manufacturing processes. These compounds have adverse effects on human health in terms of toxicity, persistence and bioaccumulation. Various methods are in place for removal of phenolic compounds, among which biodegradation is regarded as an environmentally friendly and cost effective treatment. However in such processes, substrate inhibition of microbial growth at high phenol concentrations poses a critical challenge.

In this project, a bioreactor incorporating forward osmosis and ultrafiltration is proposed to alleviate substrate inhibition in high-strength phenolic wastewater. Polyelectrolyte is added into the phenolic wastewater as draw solutes to effect forward osmosis, and for decreasing phenol to sub-inhibitory level, before biodegradation is effected in an ultrafiltration membrane bioreactor. The treated wastewater is circulated back to the system acting as dilution source, while polyelectrolyte is reused. Proof-of-concept will be a major focus of this project, followed by optimization of the operating conditions.

(519a) Co-Culture in Submerged Hollow Fiber Membrane Bioreactor for Bioethanol Production On Glucose and Xylose

Nguyen, D. T., National University of Singapore
Loh, K. C., National University of Singapore
Praveen, P., National University of Singapore

Lignocellulose is the most abundant renewable source of biomass available for conversion into bioethanol. In processing lignocellulosic biomass into bioethanol, major concerns are ethanol yield and fermentable sugars in the lignocellulosic hydrolysate. The major sugars in the hydrolysate are glucose and xylose. There is currently no wild-type microorganism that could efficiently accomplish the fermentation of this sugar mixture, so co-culture is a promising approach. However, the use of xylose-fermenting microorganism for co-culture in bioethanol production is limited due to their sensitivity to inhibitors, low ethanol tolerance and carefully regulated oxygen requirement. This project aims to develop a Submerged Hollow Fiber Membrane Bioreactor (SHFMB) with immobilized *Zymomonas mobilis* and *Pichia stipitis* for simultaneous co-fermentation of glucose and xylose in lignocellulosic hydrolysate. This SHFMB will facilitate efficient bioethanol production by shielding the cells from inhibitors present in lignocellulosic hydrolysate and from bioethanol inhibition in the fermentation broth. In addition, a proper regulating oxygen supply is applied to completely convert both sugars to ethanol with high yield. Proof-of-concept is a major focus of this project, followed by optimization of the operating conditions.

(396f) Fabrication of Silicon Carbide Sintered Supports and Silicon Carbide Membranes

Deng, W., University of Southern California
Yan, X., University of Southern California
Sahimi, M., University of Southern California
Tsotsis, T., University of Southern California

Fabrication of Silicon Carbide Sintered Supports and Silicon Carbide Membranes

Wangxue Deng, Xiaojie Yan, Muhammad Sahimi, and Theodore T. Tsotsis□

Mork Family Department of Chemical Engineering & Materials Science, University of Southern California,

Los Angeles, CA 90089-1211, USA

Abstract

Efficient separation of H_2 under high temperatures and pressures is fundamental to the development of clean energy industry, and has driven the research on inorganic membranes for two decades. Although several types of nanoporous membranes have been prepared, the fabrication of high-temperature and steam-stable inorganic membranes with high fluxes and large separation factors still remains a challenge. Of all the potential membrane materials, silicon carbide (SiC) is particularly promising due to its high temperature/corrosion resistance, high thermal conductivity, and excellent chemical/mechanical stability. In this work, we prepared highly permeable, mechanically strong porous SiC tubular supports for fabrication of nanoporous SiC membranes. We systematically investigated the effects of sintering temperature, the amount of sintering aids, the composition of the starting particles, and surface modifications on the transport characteristics of sintered SiC supports. The best SiC supports exhibit a He permeance of 5.82×10^{-5} $mol\bullet m^{-2}\bullet Pa^{-1}\bullet s^{-1}$, which is two orders of magnitude larger than the permeance of supports we previously prepared, and also 3~10 times larger than those of similar supports reported by other groups. These substrates are also mechanically strong, as manifested by their Vickers hardness tests. Moreover, we have fabricated nanoporous SiC membranes on such tubular supports by pyrolysis of thin allyl-hydridopolycarbosilane (AHPCS) films coated using slip-casting and dip-coating techniques, combined with a sacrificial polystyrene inter-layer method. These SiC membranes showed great improvement on both permeance and selectivity compared to our previous membranes.

(396b) Small Pore Zeolite Membranes for CO2 Capture

Carreon, M. A., University of Louisville

From the environmental and energy perspective, purification and recovery of carbon dioxide from flue gas and natural gas are of great interest. CO_2 must be separated from CH_4 because it reduces the energy content of the natural gas, and it is acidic and corrosive in the presence of water. Although polymeric membranes such as cellulose acetate, polyimide and polyaramide can separate CO_2 from CH_4 and H_2 mixtures, the high CO_2 pressures plasticize them and decrease their separation ability considerably. Zeolite membranes have significant advantages over polymeric membranes. They do not deform under high CO_2 pressures, have high thermal, mechanical, and chemical stability, and good erosion resistance. Of particular interest are small pore zeolites having three dimensional structures with pore sizes in the 3.5-3.6 Å range which represents an ideal pore window to molecular sieve CO_2 (3.3 Å) from CH_4 (3.8 Å) and N_2 (3.6 Å). Herein, we present our initial efforts to develop small pore zeolite membranes (including SAPO-56, DNL-6 and zeolite Rho) and evaluate its separation performance for CO_2/CH_4 gas mixtures. The membranes were supported on porous alumina tubes, and were grown via secondary seeded method. In particular, SAPO-56 displayed moderate CO_2 permeances and selectivities for equimolar CO_2/CH_4 mixtures.

(396p) Solvation of Nitrogen in Metallic Vanadium: A Theoretical and Experimental Study

Lee, K., Stanford University
Wilcox, J., Stanford University

N_2-selective membranes could be useful in air separation, ammonia synthesis in addition to the post-combustion CO_2 capture. High partial pressure of N_2 especially in the post-combustion gas streams might have benefits against the conventional processes directly capturing CO_2. We suggest a vanadium(V)-based metallic membrane for N_2 separation. Solvation of N in V is critical to evaluate the membrane properties. Numerous V-N systems have been identified for application to the precipitate materials for strengthening steel products. However, no attempts have been made to calculate the solubility of mobile N for membrane applications.

Atomic N absorbed to the pure V metal could form a solid solution or induce a phase transformation to vanadium nitrides in varying temperatures. In contrast to the hydrogen solvation in palladium in a traditional H_2-selective membrane system, N binding in bcc V is much stronger and V lattice distortion is more significant due to the large atomic size of N. The interaction between N and V was investigated by first principles calculations based on density functional theory. The optimized geometry, binding energy and electronic structure after N absorption were calculated to deduce essential implications to the N solubility in V. The solubility can be computed across different V-N systems using grand canonical Monte Carlo simulation. In comparison to the theoretical investigation, solubility of N in V was measured experimentally using a gravimetric sorption analyzer. The comparison can verify the previously suggested solvation models and the model can be applied to predict the properties of the N_2-selective membrane.

(396n) Fundamental Investigation On CO2 Permeation Mechanism of Amino Acid Ionic Liquid-Based Membranes

Otani, A., Kobe university, Center for Membrane and Film Technology
Kasahara, S., Kobe university, Center for Membrane and Film Technology
Kamio, E., Kobe university, Center for Membrane and Film Technology
Matsuyama, H., Kobe university, Center for Membrane and Film Technology

Prevention of global warming is a critical issue related to the conservation of the global environment. Carbon dioxide is one of the major contributors to the greenhouse effect. The CO_2 separation technologies have been researched in the world for prevention of global warming. In recent years, membrane separation has attracted attention because it has advantages including operational simplicity, high energy efficiency, simple process equipment and low capital cost. It is necessary to fabricate CO_2 selective membrane with high CO_2 permeability and selectivity for commercial viability. Many membranes for selective CO_2 separation have been developed during the last several decades. Especially, it is well known that a facilitated transport membrane (FTM) has high CO_2 permselectivity [1, 2]. FTMs show very high performance on CO_2 transportation because FTMs contain CO_2 carriers in membrane that react with CO_2. However, in the case of traditional FTMs, CO_2 carriers in FTMs cannot react with CO_2 without water. This restriction limits the field of practical applications of CO_2 selective FTMs. In our previous work, we have reported that amino acid ionic liquid (AAIL) based FTMs overcame this problem [3, 4]. Ionic liquids (ILs) are molten salts with unusually low melting point which have attractive properties such as negligibly small vapor pressure, tunable chemical and physical properties and so on. AAILs are one of the ionic liquids which have amino acid in their structure [5]. It has been reported that AAILs have high CO_2 absorption capacity because the amine group in the molecules react with CO_2[6-9].

We reported that AAILs in membranes worked as CO_2 carriers and diffusion mediums. AAIL-FTMs showed extremely high CO_2 permselectivity under various relative humidity at 373 K [3]. In addition, it was found that AAILs-based membranes facilitated CO_2 transport under dry conditions, whereas conventional FTMs cannot work in low humidity environments. On the other hand, contrary to great performance at elevated temperature, AAIL-FTMs showed critically low CO_2 permselectivity around room temperature under dry condition. To improve the CO_2 permselectivity around room temperature under dry condition, investigation on CO_2permeation mechanism of AAIL-FTMs were conducted.

The CO_2 permeation mechanism in AAIL can be divided into 3 steps. The first step is the formation of an AAIL-CO_2 complex via chemical reaction between amine group of an amino acid in AAILs and CO_2 at the feed side of the membrane. Subsequently, diffusion of the complex across the membrane occurs. Finally, the complex is dissociated at the permeate side of the membrane. It can be presumed that the diffusion of the complex is the rate-determining step, since generally the reaction rate is much faster than the diffusion rate in the liquid. Therefore, diffusivity is the key factor to control the CO_2 permeability of the AAIL-FTMs. That is to say, lowering the viscosity is promising to improve CO_2 permeability of the AAIL-FTMs. Regarding the viscosities of AAILs after CO_2 absorption, Goodrich et al. reported considerable phenomena [7]. For the primary AAILs, outstanding raise of viscosity was observed after they absorbed CO_2. On the other hand, the increase of the viscosity of the prolinate AAIL after CO_2 absorption was little. These results implied that the viscosity of AAIL-CO_2 complex would be a main factor to determine the CO_2 permeabilities of the AAIL-FTMs. In fact, in our previous work, we also reported the AAIL-FTMs containing the AAIL with prolinate anions showed higher CO_2 permselectivity than those with glycinate anions [4]. This would be due to the lower viscosity of proline based AAIL after CO_2 absorption. Glycine and proline have different amine group. Glycine has a primary amine, while proline has a secondary amine. From this point of view, understanding the differences of the CO_2 permselectivity as well as the viscosity among AAIL with different amine groups would be important to improve the CO_2 permselectivity of AAIL-FTMs around room temperature under dry condition. Among all the amino acids of the AAILs used in our previous research, only proline has secondary amine group [3, 4]. In this study, to investigate the affect of amine groups on the CO_2 permeability of AAIL-FTMs, AAILs with 3 types of amine group (primary, secondary and tertiary amine) were synthesized. The change of viscosities before and after CO_2 absorption of the newly synthesized AAILs was examined to understand the CO_2permeation mechanism of AAIL-FTMs.

In this work, viscosities and CO_2 permeabilities of a series of tetrabuthylphosphonium ionic liquids with amino acid as anions, [P_{4444}][AA] (AA= glycine (Gly), N-methylglycine (mGly) and N,N-demethylglycine (dmGly)), were measured. We investigated the effect of the viscosities of the AAIL-CO_2 complexes on the CO_2 permeation properties of the AAIL-FTMs. The viscosity measurement of the AAILs before and after CO_2 absorption indicated that the viscosities of the CO_2-saturated AAILs were higher than those of the pure AAILs. In addition, the viscosities of [P_{4444}][mGly]-CO_2 complex was lower than those of [P_{4444}][Gly]-CO_2 complex at elevated temperature. Regarding CO_2 permeability of the AAIL-FTMs containing [P_{4444}][Gly] and [P_{4444}][mGly], the CO_2 permeabilities of [P_{4444}][mGly] based FTM were higher than those of [P_{4444}][Gly] based FTM at elevated temperature. The trends of [P_{4444}][mGly] were same to proline based AAIL. On the other hand, viscosity enhancement of [P_{4444}][Gly] and [P_{4444}][mGly] were significant around room temperature. The phenomena would be attributed to formation of intermolecular hydrogen bonding among the AAIL-CO_2 complexes. In our examination, the viscosities of [P_{4444}][Gly]-CO_2 complex were higher than that of [P_{4444}][mGly]-CO_2 complex above 320 K, and vice versa below 320 K. If the viscosity of AAILs-CO_2 complex strongly affected to the CO_2 permeabirity of AAIL-FTM, the CO_2 permeabilities of [P_{4444}][Gly]-FTM and [P_{4444}][mGly]-FTM would be reversed at 320 K. As it was expected, the temperature dependence of CO_2 permeabilities revealed that the CO_2 permeability of [P_{4444}][Gly]-FTM was higher than [P_{4444}][mGly]-FTM below 320 K and vice versa above 320 K. From these result, it was strongly suggested that the diffusivity of the AAIL-CO_2 complex in AAIL-FTMs determines the CO_2 permeability of AAIL-FTMs. In addition, it was suggested that the increase of the viscosities of AAILs via CO_2 absorption would be coursed by the formation of hydrogen bonding between the formed AAIL-CO_2 complexes. In conclusion, it can be said that viscosity of AAIL-CO_2 is the most important factor that controls the CO_2permeability of AAIL-based facilitated transport membrane.

This research was supported by JST-ALCA.

[1] K. Okabe, N. Matsumiya, H. Mano, Stability of gel-supported facilitated transport membrane for carbon dioxide separation from model flue gas, *Sep. Pur. Tec.*, 57, (2007), 242-249

[2] C. A. Scholes, S. E. Kentish, G. W. Stevens, Carbon dioxide separation through polymeric membrane systems for flue gas applications, *Res. Pat. Chem. Eng.*, 1, (2008), 52-66

[3] S. Kasahara, E. Kamio, T. Ishigami, H. Matsuyama, Amino acid ionic liquid-based facilitated transport membranes for CO_2 separation, *Chem. Commun.*, 48, (2012), 6903-6905

[4] S. Kasahara, E. Kamio, T. Ishigami, H. Matsuyama, Effect of water in ionic liquids on CO_2 permeability in amino acid ionic liquids-based facilitated transport membranes, *J. Membr. Sci.*, 415-416, (2012), 168-175

[5] K. Fukumoto, M. Yoshizawa, H. Ohno, Room temperature ionic liquids from 20 natural amino acids, *J. Am. Chem. Soc.*, 127, (2005), 2398-2399

[6] J. Zhang, S. Zhang, K. Dong, Y. Zhang, Y. Shen, X. Lv, Supported absorption of CO_2 by tetrabutylphosphonium amino acid ionic liquids, *Chem. Eur. J.*, 12, (2006), 4021-4026

[7] B. F. Goodrich, J. C. de la Fuente, B. E. Gurkan, Z. K. Lopez, E. A. Price, Y. Huang, J. F. Brennecke, Effect of water and temperature on absorption of CO_2 by amine-functionalized anion-tethered ionic liquids, *J. Phys. Chem. B*, 115, (2011), 9140-9150

[8] Y. Zhang, P. Yu, Y. Luo, Absorption of CO_2 by amino acid-functionalized and traditional dicationic ionic liquids: Properties Henry's law constants and mechanisms, *Chem. Eng. J.*, 214, (2013), 355-363

[9] H. Yu, Y. Wu, Y. Jiang, Z. Zhou, Z. Zhang, Low viscosity amino acid ionic liquids with asymmetric tetraalkylammonium cations for fast absorption of CO_2, *New J. Chem.*, 33, (2009), 2385-2390

(396h) Kinetics Study of Formation of the Permselective Polyamide Layer in Thin Film Composite Membranes

Behera, S.,
Suresh, A. K., IIT Bombay

Kinetics study of formation of the permselective polyamide layer in Thin Film Composite membranes

Subhalaxmi Behera, A.K Suresh

Department of Chemical Engineering, Indian Institute of Technology Bombay, Mumbai - 400076

email: subha.b@iitb.ac.in, aksuresh@iitb.ac.in

Abstract

Thin film composite membranes, used extensively in reverse osmosis and other separation applications, depend for their function on a thin (100-200 nm) dense layer, which is formed on the surface of a macroporous support membrane by an interfacial polycondensation reaction. The properties of this dense layer depend on the conditions under which the interfacial reaction is carried out, through the kinetics. However, the kinetics of this process are extremely difficult to study, firstly because it occurs in a very thin region on the support membrane (and possibly inside the pores), and secondly because it is a very fast reaction with timescales of the order of seconds. A method to study the kinetics of the reaction, and hence understand the effect of preparation parameters on the rate and evolution of structure, would thus be of great value, not only for the membrane application, but also in other areas such as microencapsulation where such reactions are used.

Our approach to the study of interfacial polycondensation kinetics is developed in three parts. Firstly, we study the reaction in isolation in a liquid-liquid system, separated from the support membrane, to facilitate measurements. Secondly, we control the overall rate of monomer consumption by controlling the interfacial area available. These two steps are similar to our earlier studies of the polyurea microencapsulation reaction. Thirdly, we exploit the fact that the reaction involves (a) consumption of an alkaline monomer (a diamine) and (b) release of an acid byproduct (HCl), both of which should contribute to a decrease in pH as the reaction progresses, and hence use a fast-response on-line pH probe to follow the reaction. The method thus involves conducting the reaction in a dilute emulson, in which the aqueous phase forms the continuous phase, with the pH of the continuous phase being followed with time. A calibration curve, to allow measured pH to be translated to monomer conversion, is developed by measurements of pH of samples, which simulate different conversions. While the reaction configuration used is precisely that of microencapsulation, the determination of true kinetics requires a clean separation of reaction kinetic and transport influences on the rate. This is achieved here by conducting the reaction at high shear rates (5000 RPM in a homogenizer), so that the polymer formed is instantaneously stripped off the interface and transport limitations are eliminated.

The above methodology has been demonstrated in this work on the reaction between an aqueous solution of m-phynelenediamine (MPDA) and an organic solution of trimesoyl chloride (TMC). The variation of pH with conversion has been theoretically modeled and experimentally established. It is shown that the method does yield consistent and reproducible kinetic data. The method has been used to study the effect of surfactant concentration and monomer concentration ratio on the kinetics. Surfactant concentration is shown to influence the rate of the reaction, and hence the rates of the same reaction could be different when used for microencapsulation, from when used for TFC membranes. A limitation of the method is that it cannot be used when the aqueous monomer is in large excess, but this condition can be approached through kinetic modeling. The kinetic modeling is done assuming it's a reaction controlled process.

Keywords: Interfacial polycondensation kinetics, thin film composite membrane, Micro-encapsulation

(396r) Investigation of Adsorption-Deposition Behavior of Natural Organic Matter Fouling On Polysulfone Ultrafiltration Membrane

Rabbani Esfahani, M., The Center for the Management, Utilization and Protection of Water Resources,Department of Chemical Engineering , Tennessee Technological Universsity
Stretz, H. A., Tennessee Technological University

Natural organic matter (NOM), e.g. humic acid (HA), is a major foulant during ultrafiltration process of surface water. The research objective of this study is to investigate the differences between adsorption and deposition behavior of HA foulant on Polysulfone ultrafiltration process. Data were obtained by comparing the isothermal adsorption of powder polysulfone - humic acid with polysulfone membrane - humic acid. The experimental design was divided in to two sets of experiments. First set polysulfone powder with humic acid under different concentrations, ionic strength and time interval, was stirred and filtered. The filtrate was analyzed by UV-visible spectroscopy for HA. The second set was synthetic Polysulfone membrane in cross -flow and dead end filtration system with humic acid at the same conditions and same analyzing method. Results showed that there was not any adsorption of HA on polysulfone powder while HA fouling on polysulfone membrane was significant.

(396s) Facilitated Transport Membranes for Natural Gas Separation

Ansaloni, L., Alma Mater Studiorum - UniversitÃ di Bologna
Zhao, Y., The Ohio State University
Ramasubramanian, K., The Ohio State University
Giacinti Baschetti, M., Alma Mater Studiorum - UniversitÃ di Bologna
Ho, W. S. W., The Ohio State University

Facilitated Transport Membranes for Natural Gas Separation

Luca Ansaloni[1], Yanan Zhao[2], Kartik Ramasubramanian[2], Marco Giacinti Baschetti[1], W.S. Winston Ho[2,3]

[1]Dipartimento di Ingegneria Civile, Chimica, Ambientale e dei Materiali, Alma Mater Studiorum - UniversitÃ di Bologna, Bologna, Italy

[2]William G. Lowrie Department of Chemical and Biomolecular Engineering, The Ohio State University, Columbus, Ohio 43210, United States

[3]Department of Materials Science and Engineering, The Ohio State University, Columbus, Ohio 43210, United States

Nowadays, natural gas is an important source for energy production. Depending on the original source, raw natural gas contains various impurities in composition, which must be removed before delivery to the pipelines. Removal of acid gases, namely, CO_2 and H_2S, from natural gas is therefore an important industrial gas separation process.

For CO_2 removal from natural gas, polymeric membranes compete against conventional separation processes, such as amine absorption, due to superior advantages of lower costs, lower energy consumption, and lower waste production.[1] Several attempts have been made to develop CO_2/CH_4 separation membranes that match the pipeline specifications. Glassy polymers, in particular polyimides, have been reported to show the best performance for this specific separation.[2,3] The higher condensability and smaller kinetic size make CO_2 molecules more soluble and diffuse faster in the membrane matrix than methane, resulting in good selectivity and permeability. Recently, polyimide-based polymer with intrinsic microporosity (PIM)[4] and cyclodextrine grafted co-polyimides[5] demonstrated very high CO_2 permeability at high pressure (10 atm) but low CO_2/CH_4 selectivity, less than 25. In fact, at high pressures, these membranes suffer considerably from the plasticization problem caused by high concentration of the acid gas, resulting in decreasing membrane selectivity significantly. Crosslinking can help to minimize the CO_2-induced swelling; however, an increase in separation factor of a solution-diffusion polymeric membrane always accompanies a reduction in the penetrant flow due to the well-known trade-off relationship.[6]

Facilitated transport membranes hold great potential for natural gas CO_2 removal. This type of membrane is able to achieve high CO_2 permeability accompanied with high selectivities vs. CH_4, H_2, or N_2 due to the specific, prompt reaction between amine carriers and CO_2 molecules.[7,8] This work exploited the superlative mechanical property of multi-walled carbon nanotubes (MWNTs) to reinforce the mechanical strength of amine facilitated transport membranes, from which we successfully synthesized nanostructured polymer-MWNT membranes by incorporating MWNTs (about 18 nm diameter by 180 nm) into the crosslinked polyvinylalcohol-polysiloxane network containing amine carriers.

The synthesized membranes inherited the facilitated transport mechanism and thus achieved high CO_2 permeability and selectivity vs. CH_4 at high pressures of at least 15 atm and high temperatures of more than 100Â°C. At 15 atm and 107Â°C using a feed gas of 50% CH_4 and 50% CO_2, which simulated the natural gas composition, the membrane showed a high CO_2/CH_4 selectivity of about 100 accompanied with a CO_2 permeability of about 350 Barrers. This lower permeability was due to the higher degree of carrier saturation at the higher CO_2 concentration. It should be noted that this permeability is still higher than that of commercial membrane, e.g., cellulose acetate (about 5 Barrers).

In this work, we also synthesized acid-treated MWNTs and amino-functionalized MWNTs as advanced nanofillers for amine facilitated transport membranes. The acid-treated MWNTs were functionalized with hydrophilic groups including carboxylic acid and hydroxyl groups by strong acid oxidation. The amino-functionalized MWNTs were chemically modified with 3-aminopropyltriethoxysilane. The presence of these hydrophilic groups was confirmed by FTIR spectroscopy. The membranes incorporated with the modified MWNTs demonstrated significant improvement in CO_2/CH_4 separation performance, especially in term of separation factor. A CO_2/CH_4 selectivity of 180 along with CO_2 permeability of around 1000 Barrers have been obtained at 15 atm and 107Â°C for a feed gas composition of 59.8% CH_4, 20.2% H_2 and 20% CO_2. These improvements are attributed to the modification of MWNTs. With the hydrophilic surface groups, MWNTs were compatible with the network of crosslinked polyvinylalcohol and polyamines. Thereby, the MWNTs were able to insert between polymer chains and disrupted polymer chain packing more effectively, resulting in increased free volumes between polymer chains. In addition, the membrane performance at higher feed pressure up to 60 atm as well as the effects of membrane composition, temperature and membrane thickness, has been investigated.

Keywords: multiwalled carbon nanotube; mixed matrix membrane; facilitated transport; carbon dioxide separation; natural gas.

References:

1. Baker, R. W. & Lokhandwala, K. Natural Gas Processing with Membranes: An Overview. *Industrial & Engineering Chemistry Research* **47**, 2109?2121 (2008).

2. Wang, L. *et al.* Gas transport properties of 6FDA-TMPDA/MOCA copolyimides. *European Polymer Journal* **44**, 225?232 (2008).

3. Wind, J. D., Paul, D. R. & Koros, W. J. Natural gas permeation in polyimide membranes. *Journal of Membrane Science* **228**, 227?236 (2004).

4. Ghanem, B. S., McKeown, N. B., Budd, P. M., Selbie, J. D. & Fritsch, D. High-Performance Membranes from Polyimides with Intrinsic Microporosity. *Advanced Materials* **20**, 2766?2771 (2008).

5. Askari, M., Xiao, Y., Li, P. & Chung, T.-S. Natural gas purification and olefin/paraffin separation using cross-linkable 6FDA-Durene/DABA co-polyimides grafted with α, β, and γ-cyclodextrin. *Journal of Membrane Science* **390-391**, 141?151 (2012).

6. Robeson, L. M. The upper bound revisited. *Journal of Membrane Science* **320**, 390?400 (2008).

7. Bai, H. & Ho, W. S. W. Carbon dioxide-selective membranes for high-pressure synthesis gas purification. *Industrial & Engineering Chemistry Research* **50**, 12152?12161 (2011).

8. Xing, R. & Ho, W. S. W. Crosslinked polyvinylalcohol?polysiloxane/fumed silica mixed matrix membranes containing amines for CO2/H2 separation. *Journal of Membrane Science* **367**, 91?102 (2011).

(396t) The Promise of High Power Density Closed Loop Pressure Retarded Osmosis Via the Osmotic Heat Engine

Anastasio, D., University of Connecticut
Arena, J. T., University of Connecticut
McCutcheon, J. R., University of Connecticut

Pressure-retarded osmosis (PRO) is a means by which the energy contained within salinity gradients can be captured and converted into useable energy via a hydroturbine. Saline water solutions can be used as a means of energy storage, and this stored energy can be released using a closed-loop PRO process known as an osmotic heat engine (OHE). In this study, a cellulosic FO membrane was evaluated in closed-loop at a variety of temperatures, draw solution concentrations, and hydraulic pressures to determine how these parameters impact the flux of water and salt through the membrane. Furthermore, the impact of these test conditions on membrane power density, and thus the maximum energy captured per unit of membrane area, was also examined. While the membranes were only tested up to 21 bar, it is clear that increasing salt concentration increases peak power density. This observation is appropriate with respect to theory. The experimental results also suggest that for constant draw concentration, increasing temperature or decreasing hydraulic pressure will increase water flux and power density. Under these same conditions, salt fluxes were reduced by up to a factor of three, suggesting that significant external concentration polarization is impacting the performance of the membrane. Maximum power density seen throughout the trial was 18 W/m^2, which was seen at 40 C with a 1.5 M sodium chloride draw solution.

(396d) Novel Double Salt Membranes for Warm Syngas CO2/H2 Separation

Xing, R., Pacific Northwest National Laboratory
Li, X., Pacific Northwest National Laboratory
Zhang, K.,
King, D. L., Pacific Northwest National Laboratory

Membrane-based technology has shown advantages over other systems for warm syngas CO_2/H_2 separation due to its low capital investment, high process flexibility, energy efficiency, and ease of operation. When a CO_2-selective membrane is used, its benefits also include: (1) the generation of high purity H_2 on the high-pressure feed side, thus eliminating the recompression of H_2 for downstream use, and (2) lower membrane area can be used as compared to a H_2-selective membrane, thus saving capital cost. However, there has thus far been few membrane developed for practical use for CO_2/H_2 separation in the temperature range of 300-400C due to the limitations of existing membrane materials. MgO-based double salt crystals (e.g. Na_2MgCO_3), promoted by $NaNO_3$, show high CO_2 adsorption capacity and fast adsorption-desorption cycling properties at the above-mentioned temperature range by pressure swing and can potentially be developed as a CO_2-selective membrane. In this talk, we will present our recent progress on the development of double salt membranes for warm syngas CO_2/H_2 separation. This presentation will provide systematic discussion on impacts of both membrane composition and membrane separation conditions, such as temperature, feed pressure, and water steam content, on the membrane separation performances. The membrane separation mechanism will be discussed as well.

(396ab) Molecular Simulation of CO2 Sorption in Thermally Rearranged (TR) Polymers

Halitoglu-Velioglu, S., Istanbul Technical University
Ahunbay, M. G., Istanbul Technical University
Tantekin-Ersolmaz, S. B., Istanbul Technical University

Sorption induced volume swelling and plasticization is a serious concern limiting the use of polyimides as membrane materials for separations involving CO_2, H_2S, and condensable hydrocarbons. Thermally rearranged (TR) aromatic polyimides has recently been presented as a new class of membrane materials for gas separation. TR polymers are based on soluble aromatic polyimides with ortho-positioned functional groups; exposure of the polyimides to thermal rearrangement (TR) around 400°C leads to fully aromatic, insoluble polybenzoxazoles (PBOs) with exceptional thermal and chemical resistance characteristics. Modification of the polyimide chain during the rearrangement produces fundamental changes in molecular connectivity and conformation that alter chain packing, resulting in a narrow free volume distribution and hourglass shaped cavities [1]. These polymers show outstanding physical properties and high separation performance that exceeds the trade-off relationships for many gas pairs. Additionally, these polymers appear to be resistant to CO_2-induced plasticization, which is of great importance for separation of gas streams containing high levels of CO_2.

Research on TR polymers indicated no evidence of plasticization up to CO_2partial pressures of 20 atm [1-3]. Considering the difficulty of high pressure separation measurements, complementing experimental studies with molecular simulation predictions may be a powerful tool to understand the structure of and transport mechanisms in these promising materials. Although TR polymers are a hot topic in the membrane based gas separation area, there are only two simulation studies in the literature. Jiang et al. [2] determined the cavity size distribution and transport properties of six recently synthesized TR polymers and their precursors whereas Park et al. [4] analyzed the structure of TR polymers and their precursors. They carried out their simulations for a broad range of temperatures in order to clarify the effect of temperature.

In this study, molecular simulation techniques were used to estimate the degree of plasticization of 4,4-hexafluoroisopropylidene-diphthalic anhydride (6FDA)-2,2′-bis(3-amino-4-hydroxyphenyl)-hexafluoropropane (bisAPAF) membrane and its TR polymer (thermally rearranged polybenzoxazole) induced by CO_2 sorption. The structural properties, such as glass transition temperature (T_g), FFV and its distribution, d-spacing, radius of gyration, and cohesive energy density, along with sorption isotherms were investigated to understand the dynamics of TR polymer and its precursor chains. The simulations were compared to the experimental data available in the literature. The sorption simulations were carried out in the Grand Canonical Monte Carlo (GCMC) ensemble. To reproduce the CO_2-induced plasticization effect, sorption-relaxation cycles were applied until the CO_2 concentration converges. The increase in the fractional free volume (FFV) of the resulting polymer structure was considered as the extent of plasticization. Radial distribution functions (RDFs) were used for an in-depth analysis of interactions between $CO_2$and TR polymer or its precursor in order to obtain an insight to plasticization resistant structure of TR polymers. The oxygen in the hydroxyl groups of the diamine and the oxygen in the imide group were the most preferential sorption sites in 6FDA-bisAPAF, whereas they were not in its TR polymer.

[1] Park, H. B., Jung, J. H., Lee, Y. M., Hill, A. J., Pas, S. J., Mudie, S. T., Van Wagner, E., Freeman, B. D., Cookson, D. J., *Science*, 318, 254, 2007.

[2] Jiang, Y., Willmore, F. T., Sanders, D., Smith, Z. P., Ribeiro, C. P., Doherty, C. M., Thornton, A. W., Hill, Freeman, B. D., Sanchez, I. C., Polymer, 52, 2244-2254, 2011.

[3] Sanders, D. F., Smith, Z. P., Ribeiro, C. P., Guo, R., McGrath, J. E., Paul, D. R., Freeman, B. D., *J. Memb. Sci.,*409-410, 232-241, 2012.

[4] Park, C. H., Tocci, E., Lee, Y. M., Drioli, E., *J. Phys. Chem. Part B,* 116, 12864-12877, 2012.

(396v) Fabrication of Silica Nanoparticle-Stacked Ultrafiltration Membrane Using Spin-Assisted Layer-By-Layer Method

Saeki, D., Kobe University
Matsuyama, H., Kobe university, Center for Membrane and Film Technology

Ultrafiltration (UF) membranes are widely applied for industrial use to separate dispersed materials such as particles and colloids from water. Commercially available UF membranes are mainly prepared with organic polymeric materials by a phase-separation method. These conventional membranes have a thick separation layer over several hundred nanometers, therefore further improvement of water permeability has limitation. To overcome this limitation, a nanoparticle-stacked membrane was reported for ultrafiltration[1)]. In this study, we demonstrate a novel fabrication technique of a silica nanoparticle-stacked inorganic UF membrane with an ultra-thin separation layer by using a spin-assisted layer-by-layer method.

Silica nanoparticle-stacked UF membranes were prepared by a spin-assisted layer-by-layer method[2)]. A few drops of polydiallyldimethylammonium chloride (cationic polyelectrolyte) aqueous solution were deposited on an anodic aluminum oxide support membrane and then the excess solution was removed by the rotation at 3000 rpm using a spin coater. The membrane was rinsed with milli-Q water and dehydrated by ethanol to densify the membrane structure using the spin coater. Then, anionic silica nanoparticles prepared by hydrolysis and copolymerization reaction of tetraethyl orthosilicate were deposited on the membrane by the same operation as above. The nanoparticle-deposited membrane was calcined at 500°C at 5 h. The membrane performance was characterized by water permeability and dextran rejection.

The fabricated membrane had an extremely thin separation layer about several ten nm from FE-SEM observation. The thickness of the separation layer increased with the number of the stacked layer and decreased by the ethanol dehydration. The water permeability decreased with increasing the number of the stacked layer number while the separation performance was improved. The ethanol dehydration also remarkably improved the separation performance. These results indicate that the ethanol dehydration removed the hydration water of the polyelectrolytes and silica nanoparticles and densified the structure of the separation layer of the prepared membranes. The prepared membranes showed very high membrane performance than commercial organic or inorganic membranes.

Reference

1) X. S. Peng, *et al.*, *Nature nanotech.*, **4**, 353-357 (2009)

2) C. Jiang, *et al.*, *Langmuir*, **20**, 882-890 (2004)

(396w) Measurement of the Solubility of Water - Ethanol Mixtures in PDMS Membranes

Fuertes, L. A.,
Noriega, M. A., National University of Colombia
Fontalvo, J., Universidad Nacional de Colombia Sede Manizales
Gomez Garcia, M., National University of Colombia

The measurement of the solubility of the ethanol - water mixture into the PDMS polymer membranes is necessary to describe the flux of these components through the membrane; however little information is reported in the literature relating to such measurement. This paper develops an experimental technique for the determination of the solubility in polymer membranes of PDMS which is based on the quantification of the phase dispersion through photography, the experimental technique was used for the evaluation of the solubility of ethanol - water mixtures through PDMS membrane, the results have good reproduction and appears to fit well with the results reported in the literature.

(396q) CO2 Separation Performance of Poly(ionic liquid) Membranes Containing Branched and Cyclic Functionalities

Horne, W. J., University of Alabama
Shannon, M. S., University of Alabama
Moon, J., University of Alabama
Irvin, A. C., University of Alabama
Whitley, J. W., University of Alabama
Hindman, M. S., University of Alabama
Bara, J. E., University of Alabama

A series of poly(ionic liquid) membranes featured various branched and cyclic functionalities were synthesized and evaluated for CO_2/N_2, and CO_2/CH_4 separation performance. The membranes were formed via ultra violet-polymerization of various ionic liquid monomers. By functionalizing one of more carbon positions within the imidazole ring, we show that isomeric monomers can have very different selectivity and permeability behaviors when they are used to form poly(ionic liquid) membranes. The mechanisms behind this behavior may be a combination of free volume and polarity effects that can favor CO_2 separation. These new insights into poly(ionic liquid) membrane design reveal that functional groups have impacts that are greater than the sum of their parts. Polymers were analyzed via IR spectroscopy to determine the degrees of polymerization, and via scanning electron microscopy to observe surface features as well as polymer thickness. Once formed, membranes were tested in an experimental system to analyze the permeability and selectivity of the membrane in regard to N_2, CO_2, and CH_4.

(399a) Preparation of Macro-Porous Media By Reactive Gelation and Application to Chromatography

Brand, B., Swiss Federal Institute of Technology Zurich

Originally presented on: 11/6/2013 8:30:00 - 8:50:00

Reactive Gelation is a multi-step process to prepare mechanically stable, porous materials from the aggregation of polymer nanoparticles[1]. After preparing a dispersion of polymer particles in water by emulsion polymerization, these nanoparticles are swollen with monomer, aggregated and subsequently post-polymerized. The post-polymerization hardens the aggregates by forming covalent bonds between the primary particles. Depending upon the process conditions, either porous monoliths or micrometer-sized aggregates are formed. Both have been functionalized in several ways and applied in chromatography[2-4].

We are going to explain several distinct ways to carry out Reactive Gelation and analyze the obtained material using a variety of techniques including SEM, mercury intrusion, nitrogen porosimetry and microtomography. In particular, total porosity and pore size distribution are emphasized and similarities and differences between monoliths and micro-clusters are elucidated.

These media are then applied to high pressure liquid chromatography. The flow rate independent mass transport kinetics caused by their exclusively large pores will be shown using tracer peak moment analysis. After demonstrating possible functionalization methods both to anion and cation exchangers, several multi-species systems like isocratic standard monovalent anion separations, but also linear gradient elution separations of proteins are shown.

---

[1] Marti N., Quattrini F., ButtÃ© A., Morbidelli M., Macromol. Mater. Eng. 2005, 290, 221–229

[2] Bechtle M., ButtÃ© A., Storti G., Morbidelli M., J. Chromatogr. A 2010, 1217, 4675–4681

[3] Brand B., Krättli M., Storti G., Morbidelli M., J. Sep. Sci. 2011, 34, 2159–2163

[4] Lamprou A., Köse I., Soos M., Storti G., Morbidelli M., to be submitted

(399b) Soft-Templated Mesoporous Carbon From Natural and Renewable Precursor

Saha, D.,
Naskar, A. K.,
Originally presented on: 11/6/2013 8:50:00 - 9:10:00

Mesoporous carbon is a different variety of synthetic porous carbon which is different from its neighbor of activated carbon owing to its large mesopore volume that finds its typical applications in myriad technological sectors. Surfactant or soft-templating of mesoporous carbon is a better alternative over hard templating owing to lower cost, easy fabrication technique and involvement of less toxic chemicals. In this technique, the mesoporous carbon is synthesized by cross-linking a suitable polymeric resin in presence of micelles of sacrificial surfactants, which give typically zero to minimal char yield. The mesoporous carbon structure is formed by harnessing the hydrogen bond between surfactant and a phenolic precursor macromolecule followed by pyrolysis leading to the removal of the surfactant and subsequent carbonization in the same step. The key role of the surfactant is very similar to that of silica for the hard template based carbons, i.e., to dictate the mesoporous structure of the carbon. Most of the precursors employed so far are synthetic phenolic type of precursor, viz, phenol, resorcinol or phloroglucinol. In this work, we have utilized a natural precursor, lignin or lignocellulose based material as carbon source along with $[PEO]_a$-$[PPO]_b$-$[PEO]_a$ type of amphiphilic tri-block co-polymer (Pluronic) as structure dictating agent. Owing to the presence of plentiful of hydroxyl groups and ability to make intra- or intermolecular cross-linking make lignin based material an ideal natural and sustainable precursor for soft-templated mesoporous carbons. The resultant carbon possesses specific surface area of 200 to over 400 m^2/g along with dominating mesoporosity. The surface area and mesopore volume were further increased with physical and chemical activation techniques, viz, by CO_2 and KOH activations. The activation process increased the surface area up to 1100 m^2/g and total pore volume up to 1 cm^3/g. The resultant carbon was characterized for pore textural properties, electron microscopy and small-angle x-ray scattering (SAXS). Involvement of natural and renewable materials as sources of synthetic mesoporous carbons will benefit their sustainable manufacturing and serve as inexpensive alternative precursors.

(399d) Sulfur Tolerant MnOx/CeO2 Sorbents for Elemental and Oxidized Hg Removal From Flue Gas

Gunugunuri, K. R., University of Cincinnati
He, J., University of Cincinnati
Thiel, S. W., University of Cincinnati
Smirniotis, P., University of Cincinnati
Originally presented on: 11/6/2013 9:30:00 - 9:50:00

Mercury is a highly toxic element that is released both naturally and as a result of human activity. The U.S. EPA has identified mercury as a Hazardous Air Pollutant. Mercury vapor is emitted from power plants as a mixture of elemental mercury, oxidized mercury, and mercury adsorbed to particulates. In our previous studies we reported developments of Mn/TiO_2 and Mn/CeO_2-TiO_2 sorbents for the combined removal of the elemental and oxidized mercury. Addition of CeO_2 to the Mn/TiO_2 increased the Hg capacity and sulfur tolerance. X-ray photoelectron spectroscopy (XPS) results indicate that the adsorbed mercury is present as both Hg^0 and Hg^{2+} on these ceria-based materials. The XPS observations also suggest that the presence of both Ce^{3+} and Ce^{4+} oxidation states in the CeO_2-TiO_2 support and interaction between Mn and Ce is responsible for the HgO formation as well as sulfur tolerance. Hence, the present study mainly aimed at investigating Mn/CeO_2 materials for the elemental and oxidized mercury capture. We prepared $MnOx-CeO_2$ materials in different ways by using different synthesis methods like co-impregnation, sequential impregnation, co-precipitation, and deposition-precipitation methods. These materials were evaluated for the simultaneous capture of elemental and oxidized mercury from gas phase; the effects of manganese loading and gas-phase sulfur dioxide, carbon monoxide, and nitrogen monoxide were determined. To determine the structural and surface properties of the catalysts and to find out the elemental oxidation state of Mn, the fresh and spent catalysts were characterized by using XRD, TPR, TPD and X-ray photo electron spectroscopy techniques. These results will be presented in this study.

(399e) Ionic Liquid Thin Film Technology â€“ Advanced Materials for Gas Purification

Haumann, M., Friedrich-Alexander University Erlangen-Nuremberg
Wasserscheid, P., Friedrich-Alexander University Erlangen-Nuremberg
Roth, D.,
Popp, S.,
Kohler, F.,
Klefer, H.,
Eckle, I.,
Boehringer, B.,
Originally presented on: 11/6/2013 9:50:00 - 10:10:00

Novel Supported Ionic Liquid Phase (SILP) materials consist of an ionic liquid, dispersed as a thin film on the inner surface of a highly porous solid material. Since the ionic liquid is dispersed on the inner surface of the support, a dry solid material is obtained. These materials can be handled like classical adsorbents and can be applied in established fixed-bed reactors, thus making these novel materials highly attractive for large scale applications.

Due to the extremely low vapor pressure of ionic liquids, SILP materials are especially suited for continuous gas-phase reactions since no leaching of ionic liquid can occur via the gas-phase.SILP materials have been applied in dedicated gas purification systems, making use of the ionic liquids' tunable solubility for certain compounds in the gas stream. Latest results from gas-phase desulphurization studies in which the sulfur compound butyl-mercaptan is absorbed in the ionic liquid film of the SILP material indicate the high potential for industrial applications. Easy regeneration of the SILP material was achieved by pressure and temperature swing cycles. Additionally, N-containing compounds have been removed from feed gas, exemplifying the potential of SILP adsorbers for industrial, first responder and military gas filters.

It is anticipated that SILP technology will facilitate new and promising applications in gas-purification scenarios both for the removal of hazardous gas components from exhaust streams or for the purification of valuable gases in a highly energy efficient manner.

(399f) Activation of Metal Sites Within Carbon Silica Composites for Enhanced Gas Adsorption

Barpaga, D.,
LeVan, M. D., Vanderbilt University
Originally presented on: 11/6/2013 10:10:00 - 10:30:00

Carbon silica composites (CSCs) have been developed for their ability to target and adsorb a wide spectrum of light gases [1]. These materials take advantage of the specific interactions between the adsorbate and the adsorbent, which are dependent on the polarity of each [2,3]. CSCs are biphasic, with a nonpolar carbon phase that targets nonpolar gases and a polar silica phase that targets polar gases [2]. The silica phase of our CSCs is composed of mesoporous MCM-41, while the carbon phase is composed of carbonized furfuryl alcohol. CSCs prove to be promising adsorbents because optimization of the phases of the composite during synthesis enhances adsorption capacities compared to the independent precursors [2].

Active metal sites have been shown to promote the chemisorption of gas molecules by inducing polarity and increasing the attraction between the adsorbate and the adsorbent in silica materials [4]. For this reason, metal-containing CSC materials have the potential to show enhanced adsorption properties for many types of gas molecules without compromising the structural characteristics of the base composite. These materials are advantageous over single-phase adsorbents because they are biphasic and contain favorable attractions for both acidic and basic adsorbates. Such materials can be used for air purification applications to remove low concentrations of a variety of gases [1,2,4].

In this study, the CSC composite is modified by the consolidation of metal sites. The [Metal]-CSCs have been further enhanced by acid/base treatments to increase the adsorption of target acidic and basic adsorbates. Initial characterization and adsorption measurements for multiple series of [Metal]-CSCs (untreated/treated with acid/base) are presented. All synthesized materials are structurally characterized using nitrogen isotherms, powder X-ray diffraction, and thermogravimetric analysis.

[1] T. G. Glover, K. I. Dunne, R. J. Davis, M. D. LeVan, Microporous and Mesoporous Materials, 111, 1-11 (2008)

[2] A. M. B. Furtado, Y. Wang, M. D. LeVan, Microporous and Mesoporous Materials, 165, 48-54 (2013)

[3] F. de Clippel, A. Harkiolakis, X. Ke, T. Vosch, G. V. Tendeloo, G. V. Baron, P. A. Jacobs, J. F. M. Denayer, B. F. Sels, Chemical Communications, 46, 928-930 (2010)

[4] A. M. B. Furtado, Y. Wang, T. G. Glover, M. D. LeVan, Microporous and Mesoporous Materials, 142, 730-739 (2011)

(399g) Desulfurization of JP-8 Fuel Using Ag-Loaded Silica Supports Under Ambient Conditions

Oliver, S. R. J., University of California, Santa Cruz
St. John, J., Univeristy of California Santa Cruz
Tran, D. T., U.S. Army Research Laboratory
Chu, D., U.S. Army Research Laboratory
Originally presented on: 11/6/2013 10:30:00 - 10:45:00

The U.S. Army demand for mobile, portable and stationary fuel cells based on hydrogen has increased rapidly. To meet these power generation demands, the U.S. Army is interested in using JP-8 fuel as a hydrogen carrier. The reformation of JP-8 to hydrogen offers an energy dense feedstock via on-demand fuel processing. Unfortunately, organic sulfur compounds intrinsic to JP-8 pose a significant challenge because they can poison the reformer and fuel cell catalysts.

In this work, mesoporous silica and aluminosilicate adsorbents loaded with silver (and other metal) ions were synthesized, characterized and examined for the adsorptive desulfurization of JP-8 fuel. The adsorbents were evaluated with JP-8 fuel containing ~700 ppmw sulfur under ambient conditions. This talk will describe the preparation and characterization of these adsorbents and their sulfur adsorption characteristics for the goal of fuel desulfurization.

(399h) Surface Modification of Crosslinked Chitosan Beads Via Atom Transfer Radical Polymerization Technology for Effective Boron Removal: A Combined FTIR, XPS, and Quantum Modeling Study

Wei, Y. T., National University of Singapore
Zheng, Y. M., Natioanl University of Singapore
Chen, J. P., National University of Singapore
Originally presented on: 11/6/2013 10:45:00 - 11:00:00

As boron in water has potential adverse effect on plant species, it must be treated for irrigation utilization. In this study, a new approach was developed for preparation of an adsorbent with high boron sorption capacity, by surface modification of crosslinked chitosan beads via atom transfer radical polymerization. 2-bromoisobutyrate bromide as an initiator was first immobilized on the surface of CCTS through the interactions with hydroxyl and amine groups. Glycidyl methacrylate (GMA) polymer brushes were grafted from the surface-initiated chitosan by the ATRP. Epoxides of the grafted poly(glycidyl methacrylate) (PGMA) further reacted with N-methyl glucamine (NMDG) to obtain the sorbent (CTS-MG) for boron sorption. The morphological study of sorbents by optical and scanning electron microscopy showed that the chemical modification contributed to significant changes in both particle size and surface morphology. Fourier transform infrared and X-ray Photoelectron spectroscope analyses demonstrated that the oxygen in the form of secondary alcohol played an important role in boron sorption, and the tetrahedral boron complexes were present on the surface of the sorbent. Quantum modeling tool was used to illustrate the uptake of boron from aqueous solution.

(400a) Adsorption of Uranium From Seawater

Kim, J., Oak Ridge National Laboratory
Tsouris, C., Oak Ridge National Laboratory
Oyola, Y., Oak Ridge National Laboratory
Mayes, R. T., Oak Ridge National Laboratory
Janke, C., Oak Ridge National Laboratory
Dai, S., Oak Ridge National Laboratory

Originally presented on: 11/6/2013 8:35:00 - 8:55:00

Adsorption of Uranium from Seawater

J. Kim, C. Tsouris, Y. Oyola, R.T. Mayes, C.J. Janke, S. Dai

Oak Ridge National Laboratory, Oak Ridge Tennessee, 37831-6181, tsourisc@ornl.gov

G. Gill, L.-J. Kuo, J. Wood, K.-Y. Choe

Pacific Northwest National Laboratory

Polyethylene fibers functionalized with amidoxime ligands have been tested in laboratory and field experiments for uranium uptake. Uranium adsorption was scaled up from 5-gallon batch to 110-gallon flow-through columns with recycling prior to continuous flow-through column tests with seawater. Experimental data were analyzed by dimensional analysis to provide a better understanding of the rate-limiting step of the uranium uptake process. Continuous flow-through column experiments have been performed at the Marine Sciences Laboratory of the Pacific Northwest National Laboratory, Sequim, WA, the Rosenstiel School of Marine & Atmospheric Science of the University of Miami, Miami, FL, and the Woods Hole Oceanographic Institution, Woods Hole, MA. Packed adsorbent beds have been set up in different configurations to investigate the effects of flowrate and temperature on uranium uptake kinetics and equilibrium in flow-through systems. The maximum amount of uranium uptake from seawater tests was 3.3 mg U/g adsorbent after eight weeks of contact of the adsorbent with seawater. This amount was three times higher than the maximum adsorption capacity achieved in this study by a leading adsorbent developed by the Japan Atomic Energy Agency (JAEA), which was 1.1 mg U/g adsorbent at equilibrium. The initial uranium uptake rate of the ORNL adsorbent was 2.6 times higher than that of the JAEA adsorbent under similar conditions. Mathematical analyses have been performed to provide a better understanding of the uranium uptake behavior under transient and equilibrium conditions.

(400b) Modeling and Prediction of Sorption of Aqueous Contaminants

Fernandez, M. E., PINMATE
Bonelli, P., PINMATE
Cukierman, A. L., PINMATE
Lemcoff, N., Rensselaer Polytechnic Institute Hartford
Originally presented on: 11/6/2013 8:55:00 - 9:15:00

Industrial effluents are often contaminated and need to be processed before their discharge in bodies of water. One group of contaminants are the dyes used in textile, paper and plastic industries. In particular, most dyes used by the textile industry are synthetic and derived from either coal tar or petroleum-based intermediates. In order to develop and/or test adsorbents that can be used in the removal of these contaminants, probe molecules are normally used. Two of these dyes, which have been studied by several researchers, are methylene blue and rhodamine B.

In this work, a model was developed to predict the dynamic performance of biosorbents in the removal of binary mixtures of dyestuffs. Model parameters were obtained from single component adsorption data and from batch and dynamic studies on alkaline treated cypress cone chips. Numerical simulation studies of the sorption process were carried out using gPROMS, commercial computer simulation software. Langmuir multicomponent isotherm, modified to account for the interaction of the components, and a linear driving force model were used to predict the sorption of the binary mixture. The predicted breakthrough times and roll-up behavior are in good agreement with the experimental results.

References

M.E.Fernandez, G.V.Nunell, P.R.Bonelli, A.L.Cukierman, Bioresource Technology 106, 55-62 (2012)

M.Ghaedi, S.Hajati, B.Barazesh, F.Karimi, G.Ghezelbash, Journal of Industrial and Engineering Chemistry 19, 227-233 (2013)

(400c) In-Situ Isobutanol Recovery From An Autotrophic Fermentation in a Bioreactor for Incompatible Gases

Maiti, S. K., Michigan State University
Liu, Y. C., Michigan State University
Lira, C. T., Michigan State University
Worden, R. M., Michigan State University
Originally presented on: 11/6/2013 9:15:00 - 9:35:00

In-situ Isobutanol Recovery from an Autotrophic Fermentation in a Bioreactor for Incompatible Gases

For presentation at 2013 Annual American Institute of Chemical Engineers Conference, November 3-8, 2013

Soumen K. Maiti, Yangmu C. Liu, Carl T. Lira, and R. Mark Worden

Department of Chemical Engineering and Materials Science, Michigan State University

Isobutanol (IBT) is a drop-in biofuel that can be produced via heterotrophic fermentations on sugars or autotrophic "electrofuel" fermentations on H_2, CO_2, and O_2. Dilute aqueous solutions of IBT are highly non-ideal, with IBT exhibiting a substantially elevated relative volatility. We have recently developed a Bioreactor for Incompatible Gases (BIG) system that allows H_2 and O_2 to be simultaneously fed to IBT-producing cells without gas-phase mixing of the H_2 and O_2. The BIG generates both liquid and gas effluent streams, both of which may contain significant amounts of the IBT produced. The high relative volatility of the IBT vapor makes it difficult to condense completely without substantial refrigeration. An alternative method to recover the IBT vapor is adsorption onto an IBT-selective resin. Potential advantages of this approach are that it can be conducted at ambient temperature, and it can be used to recover IBT from both the gas and the liquid effluent streams. This paper describes studies to characterize recovery of IBT vapor from an air stream and dissolved IBT from a dilute aqueous stream. A styrene-based resin, Dowex Optipore SD-2 was found to effectively adsorb IBT from both gas and liquid streams of fermentation media at room temperature. Ethanol and methanol were effective desorbents, allowing nearly 100% IBT recovery from the resin. Methanol has the advantages of being less expensive and can be readily separated from the IBT via distillation. Reusability of resin has also been confirmed by using same resin for 20 times. This adsorption-desorption approach was used for in-situ recovery of IBT during autotrophic IBT production in the BIG system. A mathematical model describing IBT adsorption in a packed-bed column was developed to aid in process optimization and scale up.

(400d) Volatile Product Recovery From a Photobioreactor Using Modified Blue Green Algae Strains and Bio-Char Based Activated Carbon As Adsorbent

Gu, Z., South Dakota State University
Polin, J., South Dakota State University
Ramanathan, M., South Dakota State University
Halfmann, C., South Dakota State University
Johnson, T., South Dakota State University
Gibbons, W., South Dakota State University
Muthukumarappan, K., South Dakota State University
Originally presented on: 11/6/2013 9:35:00 - 9:55:00

Cyanobacteria (blue green algae) strains have been genetically modified to produce valuable biorenewable chemicals, long chain hydrocarbons and alcohols, including farnesene, myrcene, limonene, and linalool. This transformative concept is proposed to establish a continuous cyanobacterial platform to convert sunlight and CO_2 into energy dense fuels and organic chemicals which could be used for sustainable chemical production in the specialty fragrance industry. Furthermore, this cyanobacteria platform could fit into the current ethanol industry that produces surplus CO_2 throughout the Midwest and Great Plains.

Photobioreactors (PBR's) growing these algal strains require substantial CO_2/Air bubble flow for optimum growth, but at the same time, the air flow will generally volatilize the specialty chemicals and carry the bioproducts to the reactor's exhaust. In order to capture the volatized fraction of products, an adsorption column loaded with activated carbon was attached to the photobioreactor's exhaust for gas stripping. Prior to the adsorption column, the exhaust vapor proceeded through an ice water condenser to remove excess moisture. This effectively enhanced the chemical product's concentration before adsorption. The activated carbon column was thermally regenerated and coupled with a condenser to capture the desorbed products.

Different activated carbon materials were developed and compared to optimize the adsorption dynamic and equilibrium for each individual chemical product. The hydrophobicity and hydrophilicity of activated carbon were determined to be important factors for chemical product recovery. The regeneration characteristics, desorption point and heat of desorption, were measured using a Differential Scanning Calorimeter (DSC). Simultaneously, the lifetime of activated carbon adsorbents were quantified.

Developing an integrated, recirculating adsorption recovery system, which harvests volatile products by gas stripping the photobioreactor's exhaust, will be critical for expanding novel algal technologies. The CO_2 requirement for algae growth already limits its accessibility for widespread geographical applications, but a synergistic combination of PBR's with commercial ethanol plants could further improve the carbon efficiency and economics of ethanol production. This technology will allow us to exploit two under-utilized resources in the Midwest and Great Plains (sunlight and CO_2) by converting them into storable, transportable, energy-dense, biorenewable fuels and chemicals.

(400e) CO2 Removal From Natural Gas: Combined Materials Screening and Process Optimization

First, E. L., Princeton University
Hasan, M. M. F., Princeton University
Floudas, C. A., Princeton University
Originally presented on: 11/6/2013 9:55:00 - 10:15:00

The removal of CO_2 from natural gas is an important purification step to enhance energy content and meet pipeline specifications. There are vast natural gas reserves with high CO_2 content that are uneconomical to develop with current technology. Additionally, CO_2 removed at the wellhead is typically released into the atmosphere, making natural gas production the second-largest source of CO_2 emissions in the U.S (after fossil fuel consumption) [1]. Novel technology to separate and capture CO_2 from natural gas has the potential to open new opportunities for domestic energy production while reducing cost and CO_2 emissions.

Microporous adsorbents, such as zeolites and metal-organic frameworks, have been proposed for the selective separation of CO_2 from natural gas, due to their abilitiy to act as molecular sieves. We have developed the first computational approach to combine materials screening with process optimization to identify the most cost-effective sorbents for this separation based on their performance in a realistic pressure-swing adsorption (PSA) process.

Our hierarchical method [2-3] consists of material characterization through ZEOMICS [4] and MOFomics [5], filtering based on novel selectivity metrics [6-9], and PSA process optimization via a rigorous nonlinear algebraic and partial differential equation (NAPDE) model [10]. We have discovered novel materials for cost-effective processing of natural gas with various CO_2 contents up to 50-60%. We also consider the purification of landfill gas, created by the decomposition of waste, which is primarily a mixture of methane and carbon dioxide after drying. Natural gas is obtained with at most 2-3% CO_2, suitable for pipeline transport, while the captured CO_2 is compressed to 150 bar for transportation or storage.

References:

1. U.S. Energy Information Administration, Report Number DOE/EIA-0573(2009), 2011.
2. Hasan MMF, First EL, and Floudas CA. Cost-effective CO_2 capture based on *in silico* screening of zeolites and process optimization. Submitted for publication.
3. U.S. Provisional Patent Application #61/761,436 and #61/765,284.
4. First EL, Gounaris CE, Wei J, and Floudas CA. Computational characterization of zeolite porous networks: an automated approach. Phys. Chem. Chem. Phys. 13:17339-17358, 2011.

5. First EL and Floudas CA. MOFomics: Computational pore characterization of metal-organic frameworks. Micropor. Mesopor. Mater. 165:32-39, 2013.
6. Gounaris CE, Floudas CA, and Wei J. Rational design of shape selective separation and catalysis-I: Concepts and analysis. Chem. Eng. Sci. 61:7933-7948, 2006.
7. Gounaris CE, Wei, J, and Floudas CA. Rational design of shape selective separation and catalysis-II: Mathematical model and computational studies. Chem. Eng. Sci. 61:7949-7962, 2006.
8. Gounaris CE, Wei J, Floudas CA, Ranjan R, and Tsapatsis M. Rational design of shape selective separations and catalysis: Lattice relaxation and effective aperture size. AIChE J. 56:611-632, 2009.
9. First EL, Gounaris CE, and Floudas CA. Predictive framework for shape-selective separations in three-dimensional zeolites and metal-organic frameworks. Langmuir 29:5599-5608, 2013.
10. Hasan MMF, Baliban RC, Elia JA, and Floudas CA. Modeling, simulation, and optimization of postcombustion CO_2 capture for variable feed concentration and flow rate. 2. Pressure swing adsorption and vacuum swing adsorption processes. Ind. Eng. Chem. Res. 51:15665-15682, 2012.

(400f) Adsorption of Methane and Nitrogen On Activated Carbons: Numerical Derivation of Nitrogen Adsorption Affinity Coefficient

Wang, Y., Saudi Aramco
Ercan, C., Saudi Aramco
Hashim, M., Saudi Aramco
Othman, R. M.,
Originally presented on: 11/6/2013 10:15:00 - 10:35:00

In this paper, experimental investigations were carried out to determine the methane and nitrogen adsorption capacity of various activated carbons at the temperature range of 10 to 56 °C and pressure up to 50 bars. A temperature dependent isotherm equation, Toth equation, is used to model the experimental data of methane and nitrogen adsorption on an activated carbon labeled as AC5. The optimized parameters from Toth equation are used to calculate isosteric heat of adsorption.

Dubinin-Astakhov (D-A) equation is used to model the adsorption characteristic curve for methane adsorption. The derived micropore volume of 0.456 cm^3/g for AC5 from the optimized D-A equation matches the experimental value of 0.452 cm^3/g from nitrogen adsorption at 77 K. Heterogeneity parameter value of 1.250 indicates a broad micro-pore size distribution of AC5.

Using methane as the reference adsorbate, nitrogen affinity coefficient is calculated by numerically optimizing the D-A model for nitrogen adsorption data at 10 and 21 °C. The calculated nitrogen adsorption affinity coefficient is 0.857. Using this value in the D-A equation, the model predicts the nitrogen adsorption at 38 and 56 °C with an average relative error (ARE) of 6.67% and overall ARE for all the nitrogen adsorption data at 10, 21, 38 and 56 °C is 4.99%. This validates the nitrogen adsorption affinity coefficient calculated.

(400g) Sorption of Hydrogen Sulfide From Natural Gas Using Physicochemical Treated Fly Ash

Aslam, H. M. Z., King Fahd University of Petroleum & Minerals
Shawabkeh, R., King Fahd University of Petroleum & Minerals
Hussein, I., King Fahd University of Petroleum & Minerals
Al-Baghli, N., King Fahd University of Petroleum & Minerals
Originally presented on: 11/6/2013 10:35:00 - 10:55:00

Sorption of Hydrogen Sulfide from Natural Gas using Physicochemical Treated Fly Ash

[1]Zahir Aslam, *[1]Reyad A. Shawabkeh, [1]Ibnelwaleed Hussien, [1]Nadhir Al-Bagli

[1] Department of Chemical Engineering, King Fahd University of Petroleum & Minerals, Dhahran, Saudi Arabia

* AIChE members, corresponding Author, *rshawabk@kfupm.edu.sa*

Existing of acidic gases in natural gas causes major problems in gas processing industry. Hydrogen sulfide, for example, is certainly highly corrosive and hazardous material as it is flammable and poisonous to humans and animals, which can be lethal at 350 ppm level. Several attempts were made to remove H2S from upper- and down streams. The most widely used removal technique is chemical absorption and neutralization by amine solution. Although this technique is the most commonly used as gas separation process in the petrochemical industries, it suffers from numerous problems including: high-energy requirements, corrosion, and susceptibility to foaming. Alternatively, adsorption process is used with an efficient manner. Numerous adsorbents were used to clean up these toxicants from sour gases. Activated ash is intensively used in treatment of sulfur dioxide, carbon dioxide, and hydrogen sulfide in upstream gases [1]. Other adsorbents such as fly ash, titanium dioxide and synthetic and natural zeolite are also used [2,3].

In this work fly ash was chemically treated with different concentrations of mixed sulfuric, nitric and phosphoric acids, followed by physical activation in CO_2 atmosphere at 330 °C. Then the treated ash samples were refluxed over night with diethanol amine, filtered, dried and characterized according to their surface area, pore size distribution, surface morphology, surface functional groups, and acidity.

BET Surface area and pore volume using N_2 adsorption at 77 K showed an increase from 68 to 318 m^2/g and from 0.036 to 0.678 cm^3/g, respectively, with average pore diameter of 13.28. SEM morphology illustrated porous, spherical particles with an average particle size of 50 μm. FTIR analysis provided the existence of amine functional groups appeared at 3178, 2929 and 2770 cm^{-1}.

The treated fly ash was tested for adsorption of H_2S and found that both chemical and physical adsorption of H_2S takes place on the surface of the treated ash. Results of isotherm and kinetic studies will be presented and provided in the extended abstract (paper).

1. Wu J, Wang P, Pan W, Zhao L, Gu L, He P. Experimental study on flue gas emission removed by modified activated carbons. Paper presented at: 2010 International Conference on Advances in Energy Engineering, ICAEE 2010.

2. Irfan MF, Goo JH, Kim SD. Effects of NO, NO2, CO and SO2 on NO oxidation over Pt/TiO2 for hybrid fast SCR process. *Environmentalist.* 1-7.

3. Bai Xf, Cao Y, Wu W. Photocatalytic decomposition of H2S to produce H2 over CdS nanoparticles formed in HY-zeolite pore. *Renewable Energy.*

(402a) Stochastic- Based Method To Determine Biased Diffusion of Bio-Polyelectrolyte (Protein, Aerosole, Dust) Under Biased Forces

Golbayani, P., Tennessee Technological University
Machida, M., Tennessee Technological University
Sanders, J. R., Tennessee Technological University
Arce, P. E., Tennessee Technological University
Originally presented on: 11/6/2013 8:30:00 - 8:48:00

STOCHASTIC- BASED METHOD TO DETERMINE BIASED DIFFUSION OF BIO-POLYELECTROLYTE (PROTEIN, AEROSOLE, DUST) UNDER BIASED FORCES

PARVIN GOLBAYANI[1], MOTOYA MACHIDA[2], J. R.SANDERS[1], PEDRO E. ARCE[1]

[1]Department of Chemical Engineering, Tennessee Technological University, Cookeville, Tennessee

[2]Department of Mathematics, Tennessee Technological University, Cookeville, Tennessee

Key to this research is the determination of the effective diffusion (or biased diffusion) and effective mobility (under biased applied forces whose competition determine the quality of the separation for a given type of proteins. Under suitable conditions, this competition can be efficiently assessed by the time of separation [1-2]that incorporates the two effects in a convenient manner to study the impact of the two transport based parameters, i.e. when the effective mobility wins over the dispersion (or biased effects due to the electrical field) the device leads to an efficient separation. Otherwise, we will achieve an efficient mixing.

The novelty of this research is on the use of stochastic- based approach to obtain such coefficients. The analysis involves the solution of the equation of motion of a Brownian particle under both non-convective and convective media by using a Langevin equation [3]. In this research, separation time of Lysozyme (LYZ, MW: 14.4 KDa) and Cytochrome c (CYC, MW: 11.7 KDa) in presence of biased forces such as electrical and hydrodynamic based have been studied. The analysis involves the solution of the equation of motion of a Brownian particle by using a Langevin equation. Results of studying Couette flow from stochastic-based approach, as an example case, this investigation are consistent with results based on continuum mechanics. This, to the best of our knowledge, would be the first effort to determine biased diffusion coefficient directly from fundamental, i.e. stochastics-based principles by analytical methods. In summary, this project offers a very efficient path to obtain vital information to guide both experiments and new research relevant to both environmental proteomics and clinical diagnostics.

[1] Aris, R. "On the Dispersion of a Solute in a Fluid Flowing through a Tube." *Proceedings of* the *Royal Society of London. Series A, Mathematical and Physical Sciences*, 235, 67, **1956**.

[2] Taylor, GI. "Dispersion of Soluble Matter in Solvent Flowing Slowly through a Tube." *Proceedings of the Royal Society of London. Series A, Mathematical and Physical Sciences* 1137, 186, **1953**.

[3] Edward Nelson. Dynamical Theories of Brownian Motion, Second edition, Princeton University Press, **1967**.

(402b) Microfluidic Chip for Patient Sample Isolation Through Immiscible Phases

Wang, J., Brown University
Mani, Y., Brown University
Helfrich, M., Brown University
Tripathi, A., Brown University
Originally presented on: 11/6/2013 8:48:00 - 9:06:00

RNA isolation from a patient sample had been a laborious and time consuming process based on several extractions and centrifugation steps in which they were often limited by small yields and low purities of the separation products. This research project discusses the design of a PDMS microfluidic chip for nucleic acid isolation through two immiscible phases. The chip allows for a smaller sample size to be used, and for the isolation to occur rapidly by using a filtering barrier to eliminate washing steps, making the microfluidic chip a viable option for use in point-of-care diagnostics.

The isolation involves: (i) binding the RNA in the patient sample to magnetic beads in the presence of a lysis/binding buffer, (ii) transferring beads through immiscible hydrophobic liquid as a barrier to remove contaminants and purify the RNA, and (iii) the subsequent release of the RNA from the magnetic beads. As a proof-of-concept, the chip isolation efficiency was evaluated by transferring synthetic single stranded DNA through different width of channels. Microfluidic design was shown to be effective in transporting the nuclei acid bound to the magnetic beads through the chip. The influenza patient samples provided by Rhode Island Memorial Hospital will be further used for isolation and the results will be compared to commercial isolation kits.

(402c) Operational Characteristic Studies of Protein Refolding in Size Exclusion Chromatography

Saremirad, P., Western University
Ray, A. K., Western University
Originally presented on: 11/6/2013 9:06:00 - 9:24:00

Proteins are one of the most important biological compounds and many of them are used as drugs beneficial to human health. Proteins produced by recombinant DNA technology using bacterial hosts often form inactive aggregates called inclusion bodies. Large amounts of proteins are trapped in these insoluble aggregates which require solubilisation and purification. The solubilisation is possible through the introduction of Denaturising and Reducing agents (D&R) to unfold the protein chain. However, the refolding of proteins into their compact structures is critical to confer biological activity on proteins allowing them to function as effective therapeutics. Due to simplicity in design and operation, batch dilution refolding is commonly practiced in laboratories and industry. However, the correct protein folding pathway often competes with misfolding and aggregation which substantially reduces refolding yields. Furthermore, the presence of aggregates in the final product as impurities has many health concerns. Therefore, though ideal at laboratory scale, this technique has serious drawbacks during scale-up due to low product concentration and purity along with large process volumes necessitating additional post-refolding cost-intensive concentration and purification steps. Chromatographic based refolding has shown to address challenges with product dilution by facilitating spatial isolation of protein and D&R molecules while simultaneously purifying the protein after refolding. Since chromatographic protein refolding allows for significantly higher protein concentrations without drastic loss of refolding yield, compared to batch dilution, a multi-variable investigation of the system performance in industrially relevant concentrations is an essential step for optimization and intensification. In this work, using Lysozyme from chicken egg white as a model protein (129 amino acids, molecular weight of 14,307$gmol^{-1}$, four disulfide bonds and theoretical isoelectric point of 11.35), the effect of operating parameters on refolding yield, protein recovery and purity were evaluated. The parameters studied were protein concentration (20-40 $mgml^{-1}$) and refolding buffer composition including its pH (8.1-9.5), ionic strength (0.2M NaCl) and Arginine presence as aggregation prevention additive (0.2M). As expected, refolding buffer pH and protein concentration played an important role in refolding yield. In the tested range, pH further away from the protein isoelectric point (8.1) and low protein concentration (20$mgml^{-1}$) resulted in higher refolding yield. However, in the presence of Arginine, system performance was different which will be discussed in detail. Addition of sodium chloride resulted in considerable protein precipitation in column and flow blockage in high protein concentration. This observation is in contrast to Lysozyme refolding in lower concentrations, where the same type and concentration of salt is recommended to decrease the self and non-specific interactions.

(402d) A General Graphical Design Method for Affinity Chromatography

Ling, L., Purdue University
Kao, L. W., Purdue University
Wang, N. H. L., Purdue University
Originally presented on: 11/6/2013 9:24:00 - 9:42:00

In affinity chromatography, an adsorbent with a high selectivity for a target solute is used to isolate the target molecule from other impurities. Once captured, the target molecule can be eluted from the column in a purified state. Common applications of affinity chromatography include Protein A chromatography for antibody purification, Immobilized Metal Affinity Chromatography (IMAC) for protein purification, and many other capture processes based on high affinity sorbents. The well-known design method based on constant-pattern mass transfer zone analysis does not apply to small feed batches, which are insufficient to form constant-pattern frontal waves. Other literature design methods rely on simulation or experimental trials can be time-consuming and costly. In addition, it can be difficult to optimize the process to achieve desired purity, yield, and throughput. In this study, a convenient graphical design method based on intrinsic adsorption and mass transfer parameters and dimensionless groups is developed for affinity chromatography systems with Langmuir isotherms. Only a small number of experiments are needed to obtain the intrinsic parameters. The method is tested with literature data for Protein A chromatography for antibody purification. Close agreement is obtained. One can easily visualize from the graph the effects of material properties, capture yield, and throughput on column utilization. In addition, it can easily adjust to meet various design requirements and can take into account variations in the intrinsic parameters. Various sorbents can be evaluated for cost effectiveness based on the intrinsic parameters. This method is applicable to a broad range of affinity chromatography systems.

(402e) Pure Autologous Plasmin Obtained With Affinity Membranes

Castro, C., Dipartimento di Ingegneria Civile, Chimica, Ambientale e dei Materiali, Alma Mater Studiorum - UniversitÃ di Bologna, Bologna, Italy
Boi, C., Alma Mater Studiorum - UniversitÃ di Bologna
Sarti, G. C., Alma Mater Studiorum - UniversitÃ di Bologna
Originally presented on: 11/6/2013 9:42:00 - 10:00:00

Some eye conditions like diabetic rethinopaty, macular pukers, retinal detachment may benefit from vitreoctomy. Enzymatic vitreoctomy with plasmin is envisaged to augment or even replace conventional vitreoctomy by proposed means of less surgical risks, less surgeon time, lower costs. In addition, the use of autologous plasmin is beneficial since it avoids rejection problems and the search of compatible donors. Plasmin has the property to hydrolize a variety of glycoproteins, including laminin and fibronectin, by degrading the links between these components of the vitreoretinal interface and the inner limiting membrane, therapeutic posterior vitreous detachment has become possible.

Plasmin can be obtained by conversion of plasminogen using a variety of enzymes, including tissue plasminogen activator (tPA) and urokinase plasminogen activator (uPA). The purification of plasminogen from blood is normally performed with bead-based affinity chromatography.

However, for ophthalmology applications affinity membranes are ideally suited for the development of a disposable device to be used directly by the surgeon in the operating theatre, since they can be easily packed in small units providing a fast and economic process.

In this work we prepared affinity membranes for plasminogen purification using L-lysine as affinity ligand. To this aim regenerated cellulose membranes were used as a support for ligand immobilization. L-lysine was coupled by applying different binding protocols, with different buffers and reactants, testing also the effects of different temperatures.

The immobilization yield has been studied as a function of reaction time and L-lysine concentration. Ligand density was measured using a colorimetric assay with Orange 7 as indicator.

The membranes have been characterized in batch and in dynamic experiments using pure plasminogen, bovine and human sera. During chromatographic experiments with serum, fractions have been collected and analyzed with both HPLC and SDS-PAGE electrophoresis. In particular, from SDS-PAGE gel electrophoresis a well-defined plasminogen band can be shown in the eluate sample indicating that the affinity membranes obtained are suitable for the purification of plasminogen.

Acknowledgements

This work was financially supported by MIUR, Italian Ministry of Education, University and Research (PRIN 2008) and by the University of Bologna, Italy.

(402f) Hydroxamic Acids Production in Reverse Micelles and Recovery By Membrane Separation Processes

Bernardo, M., Instituto Superior de Engenharia de Lisboa
Pacheco, R., Instituto Superior de Engenharia de Lisboa (ISEL/IPL)
Minhalma, M., Instituto Superior de Engenharia de Lisboa
De Pinho, M. N., Instituto Superior Tecnico
Karmali, A., Instituto Superior de Engenharia de Lisboa
Originally presented on: 11/6/2013 10:00:00 - 10:18:00

Hydroxamic acids derivatives RC(O)NHOH present a high clinical potential, given that these compounds are efficient as antitumor, antituberculosis, antimalarial, anti-inflammatory agents and can be used in the treatment of diseases such as Alzheimer, AIDS and cardiovascular disorders [1]. Their synthesis and purification has been carried out through the use of chemical reactions from esters, in solution and on solid phase, but that involves complicated procedures [2]. Biocatalysis can be considered as an alternative approach, namely, when biocatalysis uses amidase (E.C. 3.5.1.4.) from *Pseudomonas aeruginosa*cells encapsulated in reverse micelles [3]. This approach offers advantages such as high specificity, mild conditions and shorter reaction times. The use of nonconventional systems of reverse micelles using a surfactant, water and an organic solvent are applicable in the biocatalysis of synthesis reactions, where these synthesis reactions are favored over hydrolysis reactions due to reduced water content, with an additionally enhanced activity and stability comparatively to buffer media.

The approach used involves a novel membrane bioreactor using encapsulated *Pseudomonas aeruginosa*cells for the synthesis of several hydroxamic acid derivatives using a range of different amidase substrates [4], according to the enzyme transferase activity illustrated below:

$$R\text{-}CO\text{-}R' + NH_2OH \rightarrow R\text{-}CO\text{-}NH\text{-}OH + R'H$$

where R': $-NH_2$or –OR''.

Intact cells from *Pseudomonas aeruginosa* strain L10 containing amidase were used as biocatalysts, both in free and immobilized form in a reverse micellar system. The apparent kinetic constants for the transamidation reaction in hydroxamic acids synthesis were determined using substrates such as aliphatic, amino acid and aromatic amides and esters, in both media. In reverse micelles, *Km* values decreased 2–7 fold relatively to the free biocatalyst using as substrates acetamide, acrylamide, propionamide and glycinamide ethyl ester. We concluded that overall the affinity of the biocatalyst to each substrate increases when reactions are performed in the reversed micellar system as opposed to the buffer system. The immobilized biocatalyst in general, exhibits higher stability and faster rates of reactions at lower substrates concentration relatively to the free form, which is advantageous. Additionally, the immobilization revealed to be suitable for obtaining the highest yields of hydroxamic acids derivatives. The optimization of work condition was then made using a factorial design methodology (CCD), through the study of the influence of some parameters: amide and hydroxylamine concentration, TTAB concentration (surfactant), pH and concentration of HEPES buffer. The results show that it is possible to obtain yields of 100% for all substrates.

The recovery of the hydroxamic acids derivatives produced can be achieved using membrane processes. Two pathways can be followed: i) use of solvent-resistant polymeric membranes, in this path the organic solvent/ micellar system can be fed into a flat sheet UF unit where the separation of the hydroxamic acids derivatives (in the permeate) from the micelles (retained on the concentrate) is envisaged or ii) a preliminary phase separation is carried out with an aqueous KCl solution, leading to the extraction of the hydroxamic acids into the aqueous solution, which can be then separated from residual surfactant and concentrated by membrane processes (UF and NF).

References:

[1] Codd R. (2008) Transversing the coordination chemistry and chemical biology of hydroxamic acids. Coordination Chemistry reviews 252:1387-1408.

[2] Devocelle M., McLoughlin B. M., Sharkey C. T., Fitzgerald D. J., Nolan K. B. (2003) A convenient parallel synthesis of low molecular weight hydroxamic acids using polymer-supported 1-hydroxybenzotriazole. Org. Biomol. Chem. 1:850-853.

[3] Pacheco R., Matos-Lopes M.L., Karmali A., Serralheiro M.L.M. (2005) Amidase encapsulated in TTAB reversed micelles for the study of transamidation reactions. Biocatalysis and Biotransformation 23: 407-414.

[4] Pacheco R., Karmali A., Serrallheiro M.L.M., Haris P.I. (2005) Application of FTIR spectroscopy for monitoring hydrolysis and synthesis reactions catalysed by a recombinant amidase. Analytical Biochemistry 346: 49-58.

(402g) Grafted Islands-in-the-Sea Nonwoven Fabrics for Enhanced Ion Exchange Binding Capacity

Heller, M.,
Pourdeyhimi, B., North Carolina State University
Gurgel, P. V., ProMetic Life Sciences and NC State University
Carbonell, R. G., Kenan Institute for Engineering, Technology & Science, NC State University
Originally presented on: 11/6/2013 10:18:00 - 10:36:00

Polybutylene terephthalate (PBT) nonwovens with poly(glycidyl methacrylate) (pGMA) grafts exhibit high protein binding capacities by ion exchanger. The three dimensional structure of the pGMA grafts allow capture of proteins at capacities many times that of monolayer coverage on the fiber surface. However, at large graft densities, protein diffusion into the grafted layer is slow resulting in long binding times to reach equilibrium and reduced binding capacities under flow conditions. To reduce these diffusional limitations, high surface area islands-in-the-sea PBT nonwovens were tested as base matrices for anion exchange capture of proteins. Islands-in-the-sea nonwovens are made of bicomponent fibers containing a removable polymer, "the sea", that encapsulates discreet fiber entities of a permanent polymer, "the islands", that are substantially smaller in diameter. These materials can be processed using traditional spunbonding techniques. This study used 50% polylactic acid/50% PBT islands-in-the-sea nonwovens consisting of 108 islands of PBT fibers. Once the polylactic acid sea was removed PBT fibers remained with diameters less than 1μm. Both commercially available PBT nonwoven fabrics with 3 μm fiber diameter, and islands-in-the-sea PBT were grafted with pGMA and derivatized with a tertiary amine to form weak anion exchangers. These materials were compared in terms of equilibrium protein binding capacity, and dynamic binding capacity for BSA. Islands-in-the-sea nonwovens require a minimum degree of grafting that is five times higher than the commercially available PBT due to the larger membrane intrinsic surface area. Additionally, the grafting rate is much faster for the islands-in-the-sea nonwoven due to the increased number of graft initiation sites on the material. The grafted islands in the sea matrix achieved equilibrium protein binding in approximately 5 min compared to the grafted commercially available PBT with a similar degree of grafting that required 24 hours to achieve equilibrium. In a dynamic format the grafted islands-in-the-sea nonwoven was capable of achieving higher overall binding capacities (420 mg/g at 5 cm/h) than the commercially available PBT (330 mg/g at 5 cm/h) for the same degree of grafting. At very high superficial velocities (300 cm/h) the islands-in-the-sea PBT maintained 80% of its binding capacity compared to the commercial PBT that maintained only 50% of its initial binding capacity.

(402h) Functional Magnetic Nanoparticles for Protein Capture

LIU, C.,
Guo, C., Institute of Process Engineering
Originally presented on: 11/6/2013 10:36:00 - 10:54:00

Magnetic nanoparticles have a wide range of applications in catalysis, separation, magnetic resonance imaging and drug delivery. This review focuses on the synthesis, functionalization and application of magnetic nanoparticles for protein separation. Great progress in synthesizing uniform superaramagnetic nanoparticles with high saturation magnetization has been achieved, and various functional magnetic nanoparticles with specific components have been fabricated for direct capture of various proteins with the help of an external magnetic field. Achievements in our research group together with recent progress around the world will be critically reviewed in this paper, and future perspectives in the field of magnetic bioseparation with robust magnetic nanoparticles will be outlined. An emphasis is highlighted on direct capture of laccase from *Trametes versicolor* fermentation broth using magnetic mesoporous silica nanoparticles with metal affinity ligands. The Cu^{2+}-chelated magnetic mesoporous silica nanoparticles with an average pore size of 14.5 nm provide 60.6-fold purification of laccase and 114.6% recovery yield of enzyme activity. The simple and efficient bioseparation method has great potential to be used for the robust and inexpensive preparation of purified laccase at the industrial scale.

(421a) Multicomponent Ppb Competitive Adsorption in the Presence of Varying Relative Humidity

Cheng, P., Auburn University
Tatarchuk, B., Auburn University
Originally presented on: 11/6/2013 8:30:00 - 8:50:00

Contaminant gas removal is a critical process for current and future power supply systems including fuel cells, gas turbines (GT) or a combination on ships. Jet fuels such as JP-5 are widely used in marine power supply systems, resulting in emission of volatile organic compounds (VOCs). The contaminants can affect the fuel cell performance of both anode and cathode, even exposed to an ultra-low concentration (<1 ppm). In this work, toluene and thiophene were introduced as gas phase contaminants at various initial concentrations (0.1 ppm, 0.2 ppm and 0.4 ppm). A packed bed consisting of activated carbon was used as an adsorbent operated at a face velocity of 0.9 m/s. Concentration of contaminants upstream and downstream was determined by GC/FID. The experiments for adsorption behavior (e.g. breakthrough time, saturation time and separation time) were carried out on different types of activated carbon with pretreatment. BET surface area, average pore volume and pore-size distribution were measured for each type of activated carbon in order to characterize their adsorption capacity. Furthermore, the effect of relative humidity (RH) was studied at such trace concentrations.

(421b) Removal of Low Vapor Pressure Perfluoroalkanes By Activated Carbon

Schindler, B. J., U.S. Army Research, Development and Engineering Command, Edgewood Chemical Biological Center
Buchanan, J. H., U.S. Army Research, Development and Engineering Command, Edgewood Chemical Biological Center
Mahle, J. J., U.S. Army Research, Development and Engineering Command, Edgewood Chemical Biological Center
Peterson, G. W., U.S. Army Research, Development and Engineering Command, Edgewood Chemical Biological Center
Glover, T. G., SAIC
Originally presented on: 11/6/2013 8:50:00 - 9:10:00

With increased use of highly fluorinated compounds has come an increased need for knowledge of the fundamental properties of these chemicals, such as vapor pressure, to determine how they will behave in the environment and how to remove them. A modified vapor saturation method was used to measure the ambient temperature vapor pressure of 3-(perfluorobutyl)propanol, (perfluorooctyl)ethylene , perfluorohexanoic acid, ethyl perfluorooctanoate, and perfluoro-3,6-dioxaheptanoic acid between 254.15 K and 309.15 K. The vapor pressure values are then used to generate a feed concentration at 139 PPM. The adsorption capacity of three of the chemicals on BPL activated carbon. At low relative humidity 3-(perfluorobutyl)propanol has a greater capacity than ethyl perfluorooctanoate, which has a greater capacity than (perfluorooctyl)ethylene. At high relative humidity the trend remains the same but the capacities differ.

(421c) The Design and Analysis of Activated Biochar Material for the Control of Elemental Mercury Emission From Industrial Facilities Into the Global Atmospheric Pool

Shewchuk, S. R.,
Dalai, A. K., University of Saskatchewan
Smith, D.,
De, M.,
Azargohar, R.,
Originally presented on: 11/6/2013 9:10:00 - 9:30:00

Mercury in the environment is a global environmental issue. One of the main sources of elemental mercury into the atmospheric environment is coal burning electricity generating power plants. Mercury is a trace contaminant in the burning of coal. Essentially, this anthropogenic elemental mercury is emitted into the earth's atmospheric environment. It then mixes globally. Once it deposits to the earth's surface, it is the aquatic systems of planet earth that are the most sensitive to the elemental mercury. The deposited mercury methylates the aquatic systems of the planet with the accumulation of this material occurring in selective fish tissue and then subsequently into the human being with the consumption of fish tissue.

Biological carbon or biochar has been prepared and used by the University of Saskatchewan Chemical Engineering Department, the Saskatchewan Research Council (SRC) and SaskPower in a research program over the years, to test its effectiveness as a capture agent for elemental mercury gas in Saskatchewan's coal-fired power plant flue gas streams.

The source of the biological carbon products for these programs has come from Advanced BioRefinery, which is a commercial firm that acquires its biological carbon from Canada's forest industry and other similar venues.

The biochar material is then activated at the University of Saskatchewan and prepared into an appropriate particle size range at SRC for injection into a coal-fired power plant's flue gas stream within the province of Saskatchewan, Canada. After activation, this biochar material is ground into a series of small aerosol particles in the size range of approximately 50 microns in diameter for direct in-stack injection into an operational coal-fired power plant's flue gas stream. Once in the flue gas stream, these biochar aerosols capture elemental mercury atoms which are subsequently sequestered from emission into the environment.

In this presentation we will discuss the need for the activation process of the raw biochar product. It has been discovered in this research that it is the surface aerosol microporosity that is the primary agent responsible for the elemental stack gas mercury capture efficiency. Various procedures have been employed to enhance the biochar elemental mercury capture effectiveness within an operational setting.

We have prepared two levels of aerosol microporosity described as an optimum level at 287 m^2/g and a reference level at 207 m^2/g. Once these various levels of biochar aerosol microporosity were injected into a coal burning power plan's flue gas stream, it was discovered that the optimum level of biochar aerosol microporosity reduced the elemental mercury emission rate into the environment by approximately 88 per cent. The reference level of aerosol microporosity reduced the elemental mercury emission rate in to the environment by approximately 60 per cent.

The primary conclusion of this research work with biochar is that it is the aerosol surface microporosity that is the main capture agent for in-stack elemental mercury atoms. Biochar aerosols can be prepared and sold into the international market as a sorbent-based emission control agent for in-stack elemental mercury emissions control for the coal burning power stations of the world.

(421d) Mercury Interaction With the Fine Fraction of Coal Fly Ash in a Simulated Coal Power Plant Flue Gas Stream

Jew, A. D., Stanford University
Rupp, E. C., National Energy Technology Laboratory
Jung, J. E., Stanford University
Farfan, G.,
Brown, G. E. Jr.,
Wilcox, J., Stanford University
Originally presented on: 11/6/2013 9:30:00 - 9:50:00

Mercury is a potent neurotoxin and a significant contaminant released through the burning of coal in coal-fired power plants. Besides Hg, coal-fired power plants produce large quantities of fly ash that interact with Hg contained in flue gas emissions. Because fly ash has the potential to be a sink for Hg and is used in products such as cements, it is necessary to understand how Hg is associated with the ash. This work focuses on both bulk and < 100 nm sized fly ash particles reacted with a simulated flue gas stream containing Hg vapor. Fly ash samples were analyzed by a variety of laboratory- and synchrotron-based techniques to determine the size fraction of fly ash responsible for the majority of Hg uptake, determine any correlations between Hg and fly ash components (both organic and inorganic constituents), and to determine the Hg phases present after fly ash reaction in the simulated flue gas stream.

The fine fraction of fly ash from a coal-fired power plant was separated to a particle size ≤ 100 nm. This fine fraction was reacted in a simulated flue gas containing CO_2, H_2O, O_2, NO_x, SO_2, HCl, and Hg. These gases were initially passed through a flame at ~1,100°C to simulate gaseous reactions found in a coal-fired power plant boiler, and then the gas stream was reacted with the fly ash in a packed bed reactor at ~140°C. Reacted and unreacted samples were analyzed by a large suite of techniques. Characterization of the ash was done using scanning electron microscopy (SEM), transmission Fourier transform infrared (FTIR) spectroscopy, x-ray diffraction (XRD), and electron microprobe. To understand how Hg correlates with different constituents of the fly ash, samples were analyzed by both electron microprobe and synchrotron-based x-ray fluorescence (s-XRF) mapping. X-ray absorption spectroscopy (XAS) was used to identify the Hg-bearing phases present in the samples. Micro x-ray absorption near edge structure (μ-XANES) spectroscopy was coupled with s-XRF to identify the form of Hg found in Hg hot spots identified during s-XRF mapping. Bulk extended x-ray absorption fine structure (EXAFS) spectra of the reacted fly ash were collected by slow cooling to 77K to identify the Hg-phases present in the sample, including elemental Hg if present.

Synchrotron XRF mapping of both the bulk and fine fraction reacted ash samples indicates that the fine fraction (≤ 100 nm) dominates the uptake of Hg in fly ash. XRF mapping also showed that Hg is present in two major regions: those with high Fe concentrations and Hg hot spots not associated with Fe. XRF mapping shows a correlation between Fe and Cl in Fe-rich regions, whereas Hg hot spots have associated S and Br as well as minor amounts of As and Se. Mercury was not found to be associated with Ti, alkali metals, alkaline earth metals, or the light transition metals in these samples. Electron microprobe analyses showed that some regions of the samples show a correlation between Hg and C, whereas in other regions there is no association of Hg with C. μ-XANES spectroscopy of the Hg hot spots indicates the presence of cinnabar (α-HgS) with no indication of the higher temperature polymorph, metacinnabar (β-HgS), being present. Bulk EXAFS analysis indicates that Hg phases present include Hg-Br, Hg-Cl, Hg bound to organics, Hg bound to Fe-oxides (presumably hematite), Hg-O (montroydite-like phase), and cinnabar. We did not detect elemental Hg in the fly ash, suggesting that Hg uptake by fly ash involves the oxidation of Hg(0) to Hg(II). FTIR analysis of the fly ash samples before and after reaction shows a significant intensity decrease of the carboxylic acid functional group absorption bands. This decrease is interpreted as suggesting that the carboxylic groups bind to Hg in the unburned carbon portions of the fly ash. Overall, these results show a complex interaction between Hg and coal fly ash. By understanding the speciation of Hg in fly ash, more refined model experiments using synthesized forms of phases present in the sample can be done to determine the stability of the Hg in the fly ash that is either disposed of in landfills or used in cement products.

(421e) Mercury Sorption Technology in Coal-Fired Power Plants

Meeks, N. D., Southern Company Services, Inc.
Originally presented on: 11/6/2013 9:50:00 - 10:10:00

Development of mercury sorption technology has recently emerged as an environmental challenge, most recently in response to the US EPA Mercury and Air Toxics Standards Rule, finalized in 2012. This talk wiill cover recent full-scale sorption trials of activated carbon, non-carbon sorbents, as well as pilot-scale studies of fixed structure mercury sorbents, including mixed matrix membrane technology. In particular, the talk will focus on overcoming challenges associated with conventional sorbents including Hg/SO3 selectivity. **(Most recent data will be included in the abstract closer to the meeting)

(421f) Gas-Phase Mercury Removal Using Sulfur Doped Porous Carbon

Reddy, K. S. K., The Petroleum Institute
Al Shoaibi, A.,
Srinivasakannan, C., The Petroleum Institute
Originally presented on: 11/6/2013 10:10:00 - 10:30:00

Gas phase mercury removal is a vital unit operation in gas processing industries. The present work attempts to prepare a sulfur impregnated carbon at optimized experimental conditions and compares its elemental adsorption capacity with the number of commercially available carbon based adsorbents. The effect of adsorption temperature on mercury adsorption capacity has been estimated for the prepared sulfur impregnated carbon. The adsorption capacity was found to increase with increase in adsorption temperature owing to the chemisorption nature of the adsorption. The adsorption isotherms were generated at three different temperatures and were found to close adhere to the Langmuir Isotherm model. The adsorption capacity was found to increase until 140°C, while decrease beyond, which was attributed to the softening and agglomeration of sulfur. The maximum adsorption capacity of 4325 µg/g was observed at a temperature of 140°C. A comparison of the relative adsorption capacity of various adsorbent at 30°C, reveled the adsorption capacity of the sulfur impregnated carbon prepared in the present work much higher than the commercially available carbons. The high adsorption capacities with simple preparation techniques favor the commercial mercury adoption process.

(421g) Study of H2S Removal At High Temperature With 20%Fe2O3/Al2O3 Sorbent Applying Factorial Experimental Design Method

Huang, C. Y., Institute of Nuclear Energy Research
Chyou, Y. P., Institute of Nuclear Energy Research

Originally presented on: 11/6/2013 10:30:00 - 10:50:00

In gasification-based energy systems, e.g., an IGCC (Integrated Gasification Combined-Cycle) system, the coal gasification process will generate large amount of high-temperature syngas, while release H_2S, COS, $CS_2$and other harmful gaseous pollutants. Many developed countries in the world were devoted to research and development of high-temperature dry desulfurization technology, which can improve the thermal efficiency as well as reduce the cost of power generation and environmental protection.

In this study, iron nitrate is impregnated on porous alumina and calcinated at 973K for 6 hours to prepare the desulfurization sorbent. The sorbent is used as a high-temperature desulfurization agent for the removal of hydrogen sulfide in a fixed-bed reactor. To improve the sulfur capacity of desulfurization sorbent, this paper applies the 2^5 factorial experiment designs that not only apply the ANOVA table and regression analysis to analyze the experimental data, but also verify the data completely by experiment step design before studying the important operational factors. Finally, the impact factors of desulfurization reaction of the main effects and the interaction effects can be identified. The 2^5 sets experimental data via conversion reduce the impact of interaction number. From analysis obtained that the main effect is the reaction temperature (positive) and CO_2 concentration (negative). The interaction effect is the CO_2 concentration & reaction temperature.

(430a) Impact of Acids On the Structure of Linde Type A Zeolites for Use in Reverse Osmosis Membranes for Wastewater Recovery

Jamieson, H., Arizona State University
Waller, A., Arizona State University
Yin, H., Arizona State University
Lind, M. L., Arizona State University
Originally presented on: 11/6/2013 8:30:00 - 8:55:00

Abstract:

Membrane separations and osmotic processes are attractive options for wastewater reclamation, and with growing concerns over a depleting global water supply such technologies are rapidly becoming more valuable. Previously, hydrophilic Linde Type-A (LTA) zeolites have been incorporated into polyamide reverse osmosis membranes, creating high flux membranes called Thin Film Nanocomposite membranes[1], [2]. In our ongoing work, we are exploring the use of zeolite nanocomposite membranes for the recovery of potable water from wastewaters. We are specifically interested in the wastewaters generated by NASA on the International Space Station (ISS). Currently, only 70% of the wastewater generated on the ISS is recovered; the remaining 30% of the water is stored for later disposal in the form of brine[3]. With the retirement of the NASA space shuttle fleet, the need to minimize use of ground-based resources is a high priority; as a result, NASA desires to maximize water recovery[4]. Our research focuses on using LTA zeolite nanocomposite membranes to recover water from the urine and urine brine wastewaters produced on the ISS. The small pore size of the LTA zeolite should make it ideally suited to allow passage of water, but rejection of small neutral organic molecules such as urea, a main component of urine[5].

In this talk, we will present our results evaluating the resistance of LTA zeolites to the components NASA uses to acidify the urine solutions. NASA treats the urine wastewaters with a multi-component acidic solution to inhibit bacterial growth. We have created synthetic urine brines in our lab and found that these have adverse effects on the crystal structure of the LTA zeolites. We produced solutions of potassium phosphate at pH 4.2 and 5, and sulfuric acid, and chromium trioxide solutions at pH values varying from 0.5 to 5 to identify the most damaging acid component. We exposed zeolites to the synthetic urine, and acid-specific solutions for 1 to 18 days to test the effects of pH and acid type over time. We characterized the samples with X-ray diffraction, scanning electron microscopy and Fourier transform infrared spectroscopy to evaluate any structural changes in the zeolites. Significant degradation was observed over a period of 9 to 18 days at pH values below 2 for all acids, and immediate dissolution occurred at or below pH 1. Potassium phosphate showed completely amorphous structure at pH 4.2, likely due to formation of aluminum phosphate. We will discuss the potential mechanisms for zeolite degradation, including acid hydrolysis at low pH. The results of this experimentation suggest that LTA zeolites are inappropriate for use in low pH environments.

References:

[1] Jeong, B. -H. *et al.* Interfacial polymerization of thin film nanocomposites: A new concept for reverse osmosis membranes. *Journal of Membrane Science*. **294**, 1-7 (2007).

[2] Lind, M. L. *et al.* Influence of zeolite crystal size on zeolite-polyamide thin film nanocomposite membranes. *Langmuir: the ACS Journal of Surfaces and Colloids*. **25**, 10139-10145 (2009).

[3] Carter, L., Marshall, N. & Flight, S. Status of the regenerative ECLS water recovery system. *AIAA International Conference on Environmental Systems; 11-16 Jul. 2010; Barcelona; Spain*1-11 (2010).

[4] Link, D. E., Carter, D. L. & Higbie, S. Development of an advanced recycle filter tank assembly for the ISS Urine Processor Assembly. *American Institute of Aeronautics and Astronautics*1-8.

[5] Database of Zeolite Structures. International Zeolite Association. http://www.iza-structure.org/databases/, IZA.

(430b) Arsenic Rejection By Nanofiltration Membranes: Experimental Results and Model Prediction

Deng, B., University of Missouri
Fang, J., Delon Hampton and Associates
Originally presented on: 11/6/2013 8:55:00 - 9:20:00

Nanofiltration (NF) membranes, NF270 and TFC-SR-2, were selected for arsenic rejection in dead-end stirred cell filtration equipment. The membranes were characterized using Atomic Force Microscopy, Attenuate total reflection-Fourier transform infrared spectroscopy, potentiometric titration, and contact angle measurements. The results showed that NF270 was rougher than TFC-SR-2, and both membranes contained typical carboxylic and amide groups. NF270 had a contact angle of 30.7°, and TFC-SR-2 was more hydrophilic with a contact angle at 20°. Potentiometric titration results showed that both membranes were negatively charged while pH>3.0, and carried more negative charges with increasing pH. The pore sizes of NF270 and TFC-SR2 were 0.49 and 0.46 nm, respectively, which were determined by fitting rejection data of uncharged solutes (glucose and maltose) into Spiegler-Kedem-Steric Hindrance Pore Model.

Arsenic rejection experiments showed that rejection of As(V) by NF270 and TFC-SR-2 membranes increased with increasing initial concentrations due to dilution effect and the rejection rate could be higher than 90%. The rejection of As(V) increased with increasing pH. As(III) rejection was not affected by pH while less than 9.2 because of its existence in neutral form. Arsenic rejection increased with increasing operating pressure and stirring speed. Ionic strength has influences on arsenic rejection due to shielding phenomenon. Co-occurring ions, such as sulfate, phosphate, and calcium, could reduce the arsenic rejection. As(V) rejection rate decreased with increasing concentration of co-occurring ions in the range of 0.0001M to 0.01M. Rejection rate of As(III) also reduced with the presence of co-occurring ions but was not affected by their concentration.

The As(III) rejection by NF270 and TFC-SR-2 membranes was fitted into Spiegler-Kedem-SHP model. The model calculated rejection was much higher than the experimental data. The poor fitting was possibly due to the affinity interaction between arsenite molecules and membranes. The formed hydrogen bond rendered the easy entry of arsenite into membrane structure, consequently, lowered the rejection of arsenite. The rejections of As(V) as a function of permeate flux by NF270 and TFC-270 were modeled using Donnan Steric Pore Model (DSPM) model combined with concentration polarization theory. The extended Nernst-Planck equation was solved numerically using a fourth order Runge-Kutta method. Membrane charge density and pore size are two important factors which have a major impact on membrane performance. The calculated rejections were in good agreement with the experimental results.

(430c) Fouling-Resistant Membranes for the Treatment of Flowback Water From Hydraulic Shale Fracturing: A Pilot Study

Miller, D. J., University of Texas at Austin
Huang, X.,
Li, H., Advanced Hydro, Inc.
Kasemset, S., University of Texas at Austin
Lee, A., University of Texas at Austin
Agnihotri, D.,
Hayes, T.,
Paul, D. R., The University of Texas at Austin
Freeman, B. D., The University of Texas at Austin
Originally presented on: 11/6/2013 9:20:00 - 9:45:00

Polyacrylonitrile hollow fiber ultrafiltration (UF) and polyamide spiral wound reverse osmosis (RO) membrane modules were surface-modified by contact with an aqueous solution containing dopamine to deposit polydopamine on the membrane surfaces and other wetted parts inside the modules. UF modules were further modified by grafting poly(ethylene glycol) (PEG) brushes to the polydopamine coating. Polydopamine and polydopamine-*g*-PEG coatings increase hydrophilicity of the membrane surfaces and have previously been shown to improve fouling resistance towards model oil/water emulsions in laboratory studies. In a pilot-scale test treating hydraulic fracturing flowback water from the Barnett Shale region of Texas, the fouling performance of modified UF and RO membrane modules was compared to that of unmodified analogs. UF modules were used to remove most of the highly fouling organic matter in the feed before desalination by a train of RO elements. Polydopamine-modified UF modules maintained higher flux, lower transmembrane pressure difference, and improved cleaning efficiency relative to unmodified modules. The polydopamine coating did not appear to improve RO fouling behavior, presumably because most of the organic foulants had been removed by UF pretreatment of the feedwater. However, higher and more stable salt rejection was observed in modified RO modules than in unmodified modules.

(430d) Microbial Adhesion to Advanced Water Treatment Membrane Materials: Implications for Biofouling

Kirschling, T. L., National Institute of Standards and Technology
Greenlee, L. F., National Institute of Standards & Technology
Originally presented on: 11/6/2013 9:45:00 - 10:10:00

Membrane based water purification will be essential for providing potable water from saltwater and wastewater as global water scarcity becomes more pressing. One of the biggest challenges to membrane based water treatment processes is biofouling which causes significant increases in transmembrane pressure drop and decreases in flux. This performance decline increases both energy requirements and the number of required cleaning cycles leading to significant cost increases. Novel process improvements and advanced materials have the potential to decrease biofouling, but an understanding of how these factors impact membrane associated microbial communities and their interactions with the underlying material is lacking. This study characterizes the microbial foulants recovered from a reverse osmosis (RO) pilot plant undergoing a six month study to assess the use of ozone pretreatment prior to the RO train. Pyrosequencing of the 16S rDNA gene was used to determine which microbial strains were prevalent in the biofoulant material. Sequencing data was compared with isolated bacterial strains from the membrane surfaces. Species of the genus *Mycobacterium* were found to be dominant in both ozonated and control trains. The adhesive properties of relevant isolates were studied along with their adhesion to various polymer films using atomic force microscopy (AFM) and quartz crystal microbalance (QCM) methods. The results of these studies with a *Mycobacterium sp.* isolate will be discussed.

(430e) Real-Time Monitoring Approach and System for Detection and Characterization of RO Membrane Fouling

Thompson, J. F., University of California, Los Angeles
Rahardianto, A., University of California, Los Angeles
Cohen, Y., University of California, Los Angeles
Originally presented on: 11/6/2013 10:10:00 - 10:35:00

Mineral scaling and membrane fouling remain significant impediments in reverse osmosis (RO) membrane desalination of various water sources, including inland brackish water, seawater and wastewater effluent for water reuse. In order to mitigate membrane fouling and mineral scaling it is imperative to be able to monitor the onset of the above in real time, particularly for water sources of high mineral scaling propensity. Accordingly, a novel membrane monitoring system was developed (for RO plant monitoring) consisting of a high-pressure transparent membrane cell and an automated detection system for high resolution optical scanning of the membrane surface. System pressure and feed flow rate are self-adjusted using a custom control system so as to set the desired level of concentration polarization in the membrane channel, at the desired level relative to that which is expected in the RO plant. Online image analysis software, which is integrated into the control system, analyzes the extent of mineral scaling to quantify the extent of surface coverage by particulate fouling or mineral scale and the number density of mineral crystals. Equipped with fluorescent marker injection and an inline spectrometer for monitoring, the integrated monitoring system also provides the capability for characterization of the feed-side mass transfer coefficient and marker rejection, thereby enabling estimation of the level of concentration polarization in the RO plant. The overall approach has been tested for operational scenarios in which calcium sulfate (gypsum), calcium carbonate (calcite), and silica are the mineral scalants of concern.

(430f) A Molecular Marker Approach for Characterization of RO Membranes and Solute Transport in Spiral-Wound RO Membrane Systems

Surawanvijit, S., University of California, Los Angeles
Thompson, J., University of California, Los Angeles
Rahardianto, A., University of California, Los Angeles
Cohen, Y., University of California, Los Angeles
Frenkel, V., ARCADIS U.S. Inc.
Originally presented on: 11/6/2013 10:35:00 - 11:00:00

Over the last few decades, there has been a marked growth in the application and integration of reverse osmosis (RO) with conventional processes for water production. RO processes are now used not only for seawater and brackish water desalination, but are also finding use in agricultural drainage water desalting, nitrate and color removal from impaired groundwater, as well as in water reuse applications. Even though RO is effective in rejecting solutes as small as monovalent ions, there are factors that may impact membrane performance (e.g. flux and rejection). These include membrane fouling and mineral scaling due to concentration polarization, inadequate hydrodynamic mixing in membrane modules, as well as loss of membrane integrity (e.g., due to prolonged exposure to disinfectants such as chlorine, as well as physical damage caused by manufacturing defects, faulty installation and maintenance, failure of components, and stress and strain on membrane surface from operating condition). In order to ensure robust process performance, it is paramount to obtain real-time information on solute transport parameters, hydrodynamics conditions, and membrane integrity. Accordingly, in the present study, an online fluorescent marker-based approach was developed for spiral-wound RO membrane systems for real-time monitoring of: (a) membrane mass transfer coefficient on the feed-side channel which enables assessment of concentration polarization conditions; (b) solute residence time distribution which provides information regarding hydrodynamic conditions in membrane modules and thus inference regarding potential structural changes within membrane elements, as well as membrane fouling and mineral scaling; and (c) transport parameters (e.g., solute solution-diffusion transport parameter and reflection coefficient) that enable characterization of membrane integrity. The approach involved a controlled introduction of a fluorescent marker dose to the feed stream and online monitoring of the marker concentration in the permeate and concentrate streams. Experimental studies were carried out, with an especially-designed online fluorescent marker detection system, using both intact and compromised (by exposure to chlorine and via mechanically induced damage). Results from the present study demonstrated the technical feasibility for real-time online assessment of concentration polarization, marker transport parameters, and highly sensitive detection of membrane integrity in spiral-wound RO membrane systems.

(433a) Simulation of Flow Through Spacers Used in Membrane Modules for Carbon Dioxide Capture

Lipscomb, G., University of Toledo
Lou, Y., University of Toledo
Originally presented on: 11/6/2013 8:30:00 - 8:52:00

Carbon dioxide (CO_2) emissions from coal-fired power plants may lead to significant global climate change if not controlled. CO_2 capture and sequestration (CCS) is one of many solutions being considered to mitigate emissions.

Membrane based processes potentially are a cost-effective option for CO_2 capture. Spacers are a critical component of membrane modules. Spacers create and maintain uniform flow channels during module fabrication and operation for the permeate and feed flows. Spacers also may be used to mix fluid within the channels to eliminate concentration polarization and increase mass transfer rates. Unfortunately, spacers increase pressure drop. This increase is strongly dependent on spacer geometry and is especially important in the membrane modules used for CO_2 capture. Process economics require the pressure difference that drives transport across the membrane to be as small as possible due to flue gas compression costs. Pressure drop that accompanies flow through a spacer will reduce this driving force and represents an additional parasitic energy load on the plant.

Computational fluid dynamics (CFD) is used to investigate the flow within a spacer-filled channel. The effect of mesh refinement and use of a periodic cell to represent the entire spacer are examined to determine the fidelity of the results. The results are compared to experimental measurements of pressure drop to validate the simulations.

The dependence of velocity fields and pressure drop on spacer geometry is examined to determine spacer configurations that offer the lowest pressure drops. Variables considered include diameter of each filament, filament spacing, angle between filaments, and fluid attack angle. The configurations considered include the low pressure drop configurations described in the patent literature. The results allow determination of the features that lead to the lowest pressure drops and the associated velocity fields.

(433b) The Design of a New Membrane Contactor for CO2 Capture

Hao, P., MTR
Kniep, J., MTR
Chan, K., MTR
Nguyen, V., MTR
Amo, K. D., MTR
Baker, R., MTR
Originally presented on: 11/6/2013 8:52:00 - 9:14:00

MTR is developing a new membrane process to capture and sequester CO2 from power plant flue gas. A key component in this process is a membrane contactor that selectively removes CO2 from flue gas exhaust stream and recycles it to the power plant boiler. Because the volume of the flue gas is very large and the driving force for CO2 permeate is very low, an economical and small footprint, very large membrane contactor is required. We are developing a plate-fame contactor for this application.

Counter-current sweep is a traditional and effective way to introduce sweep in plate-fame contactor but counter-current sweep contactor is mechanically very difficult to make and the pressure drop in the sweep side is another issue. We are developing a new design: cross-flow sweep. Computational Fluid Dynamics (CFD) is used to design the cross-flow sweep modules. Our base case was developed for use in a 1 MW_e power plant, which produces approximately 1 m^3 (STP)/s of flue gas. The feed gas contains 10% CO_2 at a pressure of 2 bar, and the module area is adjusted to produce a residual stream containing 2% CO_2 (80% CO_2removal). These are close to the conditions that this type of contactor would be exposed to when being used to treat coal power plant flue gas. Our base case module has a 1 m x 1 m module footprint.

Several design parameters are discussed:

1) Sweep ratio which is the volume ratio of the gas between two sides of membrane

2) Performance comparison between using counter-current sweep and cross-flow sweep

3) Multi streams of feed or sweep

4) Module with different ratios of length to width

5) Partial counter-current designs

In all the cases which have different design parameters the membrane contactor separation performance and pressure drop inside the module are compared.

(433c) Mathematical Modeling of Hollow Fiber Gas-Liquid Membrane Contactors for Acid Gas Removal

Makkuni, A., Gas Technology Institute
Li, S., Gas Technology Institute
Tamale, T., Gas Technology Institute
Zhou, S. J., Gas Technology Institute
Meyer, H. S., Gas Technology Institute
Bikson, B., PoroGen Corp.
Ding, Y., PoroGen Corp.
Originally presented on: 11/6/2013 9:14:00 - 9:36:00

The removal of acid gases such as CO_2, SO_2 and H_2S from industrial and natural gases is an important separation process. Membrane gas absorption is a hybrid approach wherein the advantages of both absorption and membrane processes are combined for cost-effective separation and removal of acid gas components from industrial and natural gases. Gas Technology Institute (GTI) and PoroGen Corporation are jointly developing novel PEEK (Poly-Ether-Ether-Ketone) based gas-liquid membrane contactors for CO_2 and H_2S removal. Extensive laboratory and bench-scale testing with different membrane module sizes are underway at GTI to bring this technology to commercialization. In these membrane contactors, the acid gas stream passes through the tube side of the hollow fiber while a chemically reactive solvent flows through the shell side of the module. The acid gas components permeate through the membrane and are absorbed in the selective solvent by chemisorption. The rich solvent is then thermally regenerated in a second hollow fiber contactor operated in reverse mode and recycled to the absorber module. The coupled mass and heat transport equations reflecting the absorption and desorption processes are formulated and solved both in one-dimensional (1-D) and two-dimensional (2-D) space coordinates. The models use plug and parabolic velocity profiles for the gas flow in the tube side. Velocity profiles for the shell side solvent flow include both plug flow and Happel's free surface model. The coupled differential equations are solved using an Excel VBA platform combined with process simulators (Aspen Plus® and Aspen HYSYS®) yielding the concentration and temperature profile of the fluid streams. In this scheme, the mathematical calculations are performed in the Excel modules with the process simulators supplying the required fluid properties.

(433d) Molecular Simulation of Carbon Dioxide Transport in Carbons-Highlighting the Importance of Potential and Structural Models

Kirchofer, A., Stanford University
Firouzi, M., Stanford University
Wilcox, J., Stanford University
Originally presented on: 11/6/2013 9:36:00 - 9:58:00

Understanding carbon dioxide (CO_2) adsorption and transport is crucial to the development of successful carbon capture and storage technologies, in addition to assisting in enhancing gas recovery from the natural systems such as coal and shale. Non-equilibrium molecular dynamics (NEMD) simulations were conducted to investigate pore entrance effects on transport through carbon pores. Pure and hydroxyl-functionalized slit and step pores are considered, to investigate pore entrance effects and tradeoffs between pore size, chemistry, capacity, and transport. To determine the significance of electrostatic interactions on CO_2 transport through pore entrances, simulations were conducted using potential models of varying levels of sophistication. The relative importance of 1) CO_2 geometry and flexibility, 2) electrostatic interactions, and 3) pore surface chemistry is investigated.

The transport investigations of CO_2 through carbon pores suggest that fluid density, diffusivity, and permeability are sensitive to the potential model used to describe molecular interactions, pore size, pore geometry, and surface chemistry. The differences in capacity and dynamic properties between the step and slit cases show the importance of investigating transport using models that account for different types of mass-transfer resistance. The current work highlights the importance of the potential and structure models in molecular simulations of adsorption and dynamics. These fundamental studies indicate the significance of molecular-scale phenomena in carbon dioxide transport in the confined spaces for carbon capture and sequestration applications.

(433e) Toward a Process-Based Molecular Model of SiC Membranes. III. Thermo-Mechanical Properties and Application to Transport and Separation of Gaseous Mixtures in SiC Membranes

Naserifar, S., University of British Columbia
Tsotsis, T., University of Southern California
Sahimi, M., University of Southern California
Originally presented on: 11/6/2013 9:58:00 - 10:20:00

We have developed a quantum-mechanically trained reactive force field, ReaxFF, to carry out reactive molecular dynamics simulation of hydridolycarbosilane (HPCS) to form amorphous SiC nanoporous membranes, mimicking the experimental pyrolysis process used for fabrication of the membranes. To test the validity of the generated membrane layer, extensive molecular dynamics (MD) simulations are carried out to compute the self-diffusivities of H2, CO2 and CH4 gases in the SiC membranes at various temperatures. Atomistic models are generated by inserting gaseous molecules in the structure of SiC model using energy minimizations, MD simulations, and the universal force field. The computed self-diffusivities of H2 gas are much larger than other gases and increase with increasing temperature. Next, we use the model and non-equilibrium molecular dynamics simulations in order to study transport and separation in the membrane of two binary gaseous mixtures, namely, H2/CO2 and H2/CH4, and test the accuracy of the results by comparing them with our own experimental data. The model is demonstrated to provide accurate predictions for various properties of interest, and in particular for the separation factors of the mixtures. The effect of the temperature and pressure drop applied externally to the membrane is also described. Finally, we also study the structural, mechanical, and thermal properties of the amorphous SiC film. The bulk properties, such as the thermal expansion coefficient and elastic moduli are calculated and compared with experimental data. In addition, the morphology of the SiC film is characterized by its accessible free volumes. The distributions of the cavity volumes are computed using the Voronoi tessellation method.

(433f) A Comprehensive Model for Gas Permeation and Separation in Asymmetric Hollow Fiber Membranes Considering Non-Ideal Conditions

Hosseini, S. S., Tarbiat Modares University
Krantz, W. B., University of Colorado
Roodashti, S. M., Tarbiat Modares University
Originally presented on: 11/6/2013 10:20:00 - 10:42:00

Abstract

Prediction and analysis of the gas permeation and separation performance in membranes require development of robust models that can account for the non-idealities and complexities involved in a real process. Pressure losses within the membrane module, real behavior of the gas molecules in the mixture, concentration polarization, temperature change due to permeation and its subsequent effect on the permeation performance, and temperature-, pressure- and concentration-dependence of the fluid viscosity are among the major deviations from ideal conditions. These parameters have been shown to play important roles in determining the accuracy of the models for performance evaluation. However, to our best understanding, despite numerous studies focused on this subject and various models presented in the literature, no specific study could be identified encompassing the role and influence of all these parameters in a single model.

This study aims to offer a comprehensive model that accounts for all of the above parameters in order to improve the accuracy of the performance prediction for hollow fiber gas separation membranes. Fundamental equations were based on the permeation, mass transfer and momentum equations for a feed containing a two-component ideal gas mixture. Then the effect of real gas behavior was included by considering fugacity coefficients and using the SRK equation of state. The effect of concentration polarization was considered by accounting for the relationship between the bulk and membrane surface concentrations. The temperature drop due to permeation was considered through calculation of the Joule-Thomson coefficient and its effect on the equations. The effects of pressure drop on both the shell and lumen side of the hollow fibers was accounted for by using the governing equations of fluid mechanics. The effects of temperature, pressure and concentration dependency of the fluid viscosity were accounted for at both the low and high pressure ranges. All these parameters and their effects were appropriately accounted in a model that was solved using MATLAB codes. This model was then extended to account for the calculation of the membrane performance for multi-component gas mixtures. The influence of non-ideal effects on the module efficiency were investigated and verified using available experimental and/or plant data from various processes.

(433g) Fractionation of Organic Fuel Precursors From Electrolytes With Membranes

Rickman, M., University of Colorado
Pellegrino, J., University of Colorado
Davis, R. H., University of Colorado
Originally presented on: 11/6/2013 10:42:00 - 11:04:00

Efficient membrane fractionation of small, neutral organics from electrolytes could improve the

sustainability of biofuel production from microorganisms, but the separations achieved are often

insufficient. To rationally develop improved membrane materials for these fractionations, we

need to better understand the mechanisms that govern solute transport through different types of

materials. To this end, we have surveyed the transport properties of a variety of commercially-

available and newly-synthesized membranes that belong to major classes of polymeric materials.

We determined that membranes belonging to a particular class have signature transport

properties—for example, fully-aromatic polyamides (FA-PA) are slightly more permeable to

glycerol than NaCl, while the converse is true for semi-aromatic polyamides (SA-PA)—although

all glycerol/NaCl separation factors are near unity. Selected membranes were further probed with

different reduced-carbon/electrolyte combinations. The cellulose-acetate (CA) membrane

achieves the greatest separation between ethanol and electrolyte (NaCl, LiCl, Na2SO4), and the

SA-PA membranes are better at fractionating larger reduced-carbons (glucose and sucrose) from the monovalent electrolytes. Meanwhile, an order-of-magnitude improvement in separation factor was found for the challenging glycerol/electrolyte fractionation with the SA-PA membranes when a divalent anion is used. The main transport mechanisms are interpreted based on Donnan exclusion and thermally-activated transport through the polymer. The CA and FA-PA membranes separate electrolytes through a predominantly steric-based mechanism, while the SA-PA membranes are more sensitive to anion valency. Furthermore, while solute size clearly plays a role in determining neutral solute transport, the precise role of solubility (i.e. polymer-permeant interactions) versus polymer free-volume in the transport of small organics through the different materials remains unclear.

(437a) Gas Purification and Separation With Ionic Liquids

Zhang, X., Institute of Process Engineering, Chinese Academy of Sciences
Dong, H., Institute of Process Engineering, Chinese Academy of Sciences
Huang, Y., Institute of Process Engineering, Chinese Academy of Sciences
Zhang, S., Institute of Process Engineering, Chinese Academy of Sciences
Zhang, X., Institute of Process Engineering, Chinese Academy of Sciences
Originally presented on: 11/6/2013 8:30:00 - 8:55:00

Gas purification and separation with Ionic liquids

Xiangping Zhang, Haifeng Dong, Ying Huang, Xin Zhang, Suojiang Zhang

Beijing Key Laboratory of Ionic Liquids Clean Process, State Key Laboratory of Multiphase Complex System, Institute of Process Engineering, Chinese Academy of Sciences, Beijing, 100190, China

Abstract

In the past decade, ionic liquids have emerged as a promising alternative to be used in the gas purification and separation due to their remarkable properties, such as negligible volatility, high chemical/physical stability, higher solubility and selectivity, and tenability which allows design of task-specific ionic liquids (TSILs). In industries, gas separation or processing was regarded as an important operation to remove harmful and toxic gas emissions from waste gas in order to reduce ecologic destroy and environmental pollution from an environmental perspective; on the other hand, purifying gases by separating impurities from mixtures can improve the quality of the gas to satisfy the requirement of gas products considering the industrial relevance. There are some typical gas separation processes with ionic liquids, such as CO_2 capture, desulfurization of flue gas, volatile organic compounds (VOCs) control, as well natural gas sweetening, have shown promising potentials for industrialization.

In this work, an Industrial Application-Oriented research framework was proposed and conducted to develop new processes to purify and separate gases taking ionic liquids as main solvents. The following studies have been focused on: 1) the structure-properties relationship of candidate ionic liquids was studied using the molecular dynamics simulation and the quantum mechanics, combining the results of spectrum characterization. Studying the interaction force between the gas and ionic liquid would be helpful to understand the absorption mechanism and then to determine the criterion of the designing and screening promising ionic liquids; 2) new thermodynamic models were developed and used to predict the physicochemical properties of ILs on the base of the fragment contribution-corresponding states (FCCS) method. The FCCS method has advantages to deal with both the molecular and ionic systems, and it is more precise than traditional method with an average absolute relative deviation less than 4% for more than 600 kinds of ILs investigated. The binary and ternary-compound gas-liquid phase equilibrium data were measured in the lab, and the SRK equation of state was used for predicting the solubility gas in ILs. The binary parameters for VLE of compounds were regressed from experimental data; 3) the transportation properties of gas and liquid were important for reactor design. The multi-bubble/single behavior in ionic liquids was studied by using high speed image pick-up system. The bubble behavior, such as bubble size distribution, gas holdup, Sauter diameter and interfacial area were investigated. New correlations based on the experimental data was proposed for the prediction of bubble Sauter diameter in gas-ionic liquid systems; 4) process simulation was used to determine the operational parameters and the flowsheet structure. The whole flowsheet of gas separation was established considering real industrial constraints. 5) techno-economic assessment procedure is also important to evaluate the new process. On the base of the models proposed and experimental data, the key parameters (e.g. solvent concentration and flow rate, stage numbers and lean loading) were optimized aiming at minimizing the energy requirement of the solvent regeneration. Capital costs and operating costs were estimated in this work to assess the economic performance.

Several ionic liquid-based gas separation processes were studies on the base of the above research framework. The possibilities and the feasibilities of the new processes would be expected to be obtained. The relevant study not only provides new gas separation technologies to replace the traditional energy-extensive and polluted ones, but also indicates the advantages and drawbacks before these new processes come into commercialization. Importantly, the obtained information will be helpful to avoid the failures of industrialization and reduce the risk of investment in the future.

Acknowledgements

We would like to acknowledge the support from Key Program of National Natural Science Foundation of China (No.21036007), the National Natural Science Foundation of China (No. 51274183) and the National High Technology Research and Development Program of China (No.2011AA050606).

(437b) Are Ionic Liquids "Drop-In" Replacements for Conventional CO2 Capture Solvents in Packed Absorption Towers? Capital Considerations

Minnick, D. L., University of Kansas
Weatherley, L. R., University of Kansas
Scurto, A. M., University of Kansas
Originally presented on: 11/6/2013 8:55:00 - 9:20:00

Ionic liquids are actively being pursued as absorption solvents for CO_2 capture. Some ionic liquids have been found to have some of the largest CO_2 absorption capabilities of any known solvent. However, the relatively slower mass transport properties of ionic liquids have been mentioned as potential challenges to their industrial implementation. This presentation will illustrate the performance of the known and hypothetical ionic liquids for their use in absorption towers. The size and hydraulics of absorption towers using ionic liquids will be compared with conventional alcohol-amine solvents. The interplay of the thermodynamic solubility with the mass transfer rates has large effects on the dimensions and hydraulics (capital costs) of packed absorption towers. Our findings suggest that even ionic liquids with the highest known CO_2 capacities can yield capital costs in excess to conventional solvents due to the packed tower specifications necessary to overcome their slower mass transfer rates. Generally, improvements in the mass transport properties often yield smaller absorption towers compared to improvement in the thermodynamic CO_2 solubility. This may indicate that other process intensification techniques may need to be considered to take advantage of some of the real thermodynamic advantages of ionic liquids. Molecular design parameters are suggested for the improvement in ionic liquids for CO_2-capture.

(437c) Post-Combustion CO2 Capture With Ionic Liquids Involving Phase Change

Seo, S., The University of Notre Dame
Simoni, L. D., University of Notre Dame
DeSilva, A., University of Notre Dame
Massel, M., The University of Notre Dame
Stadtherr, M. A., University of Notre Dame
Brennecke, J. F., University of Notre Dame
Originally presented on: 11/6/2013 9:20:00 - 9:40:00

Ionic liquids (ILs) are organic salts that are liquids below the boiling point of water. Non-volatility, good thermal stability and high carbon dioxide (CO_2) solubility make ILs an attractive replacement for commercially available volatile CO_2 absorbents. A new concept for CO_2 capture that uses ***phase change ionic liquids (PCILs)*** offers the potential to significantly reduce parasitic energy losses incurred from capturing CO_2 from flue gas. PCILs are solid ionic materials that react stoichiometrically and reversibly with CO_2 (one mole of CO_2 for every mole of salt at typical post-combustion flue gas conditions) and form a liquid when they react with CO_2. This allows for a novel process that uses the heat of fusion (ΔH_{fus}) to provide part of the heat needed to release CO_2 from the absorbent, reducing the total energy required. In addition, the phase transition yields almost a step-change absorption isotherm, which would require smaller pressure/temperature swing in the stripper. We present here detailed measurements of the CO_2 uptake, heat of absorption (ΔH_{abs}) and phase transition behavior for selected PCIL compounds as well as the development of an absorption isotherm model to capture the underlying thermodynamics of the CO_2 absorption and phase change process. Additionally, the performance of lab scale absorption unit and regeneration unit with a PCIL will be discussed.

(437e) Permeability of Carbon Dioxide Through Ionic Liquid Membrane With Polyvinylidene Fluoride

Shimoyama, Y., Tokyo Institute of Technology
Komuro, S., Tokyo Institute of Technology
Jindaratsamee, P., Tokyo Institute of Technology
Originally presented on: 11/6/2013 10:00:00 - 10:20:00

Membrane separation is expected to be candidate for carbon dioxide capture. Several materials, such as organic polymer, ceramics and hybrid of organic – inorganic, have been used for the membrane separation of carbon dioxide. A supported liquid membrane is a porous membrane filled with liquid. For carbon dioxide separation, amine solvents are often used as the liquid for the liquid membrane, but there is a problem of the low membrane stability due to the evaporation of the solvent. Ionic liquid can be suitable for the liquid membrane solvent more than the amines because their vapor pressure is ultra low. It is important for the process design of the membrane separation to understand the permeability of carbon dioxide through ionic liquid membrane. In this work, the permeabilities of carbon dioxide through ionic liquid membranes were investigated. Polyvinylidene fluoride (PVDF) was used as the porous membrane. Six species of ionic liquid were used as follows: 1-ethyl-3-methylimidazolium tetrafluoroborate ([emim][BF4]), 1-butyl-3-methylimidazolium bis(trifluoromethylsulfonyl) imide ([bmim][Tf2N]), 1-butyl-3-methylimidazolium tetrafluoroborate ([bmim][BF4]), 1-butyl-3-methylimidazolium hexafluorophosphate ([bmim][PF6]), 1-butyl-3-methylimidazolium trifluoromethanesulfonate ([bmim][OTf]) and 1-butyl-3-methylimidazolium dicyanamide ([bmim][dca]). The carbon dioxide permeabilities were measured by a sweep gas method at 303 to 343 K. Nitrogen gas was used as the sweep gas. The effects of temperature, anion species on carbon dioxide permeability though ionic liquid membrane were studied. The permeabilities of carbon dioxide increase with temperatures. The ionic liquid membrane with [bmim][Tf2N] results in the highest permeability of carbon dioxide of all the species in this work. The permeabilities of carbon dioxide through [bmim][PF6] and [bmim][Tf2N] membranes with water vapor were also measured at 303 to 333 K. It is found that the addition of water vapor at feed and permeate sides makes an enhancement of the permeabilities of carbon dioxide through ionic liquid membranes.

(437f) Phase Equilibria and Separations Involved in the Synthesis of the Ionic Liquid 1-Hexyl-3-Methylimidazolium Bromide ([HMIm][Br]) in Acetone

Minnick, D. L., University of Kansas
Scurto, A. M., University of Kansas
Originally presented on: 11/6/2013 10:20:00 - 10:40:00

A vast amount of research on ionic liquids appears in the open and patent literature. For the future industrial applications, large quantities of ionic liquids will be required. The scale-up of organic processes has been a longstanding mission of chemical engineering. However as ionic liquids are touted as potentially environmentally-benign solvents, ILs themselves must also be made in a sustainable manner. Our previous studies have measured the kinetics and initial human/environmental impact of the production of a model ionic liquid, 1-hexyl-3-methylimidazolium bromide ([HMIm][Br]) in a variety of solvents, including compressed CO_2. The conventional solvent acetone was determined to be a leading convetional solvent both from a reaction engineering standpoint but also an environmental. Here, the experimental vapor-liquid equilibrium and liquid-liquid equilibrium involved in the synthesis mixtures of [HMIm][Br] from 1-bromohexane and 1-methylimidazole in acetone have been performed. The experimental data is modeled using both the Peng-Robinson equation of state and excess Gibbs/activity coefficient models. These models are then used to simulate potential separation methods with particular attention on the energy demands on producing the ionic liquid at a given purity.

(437g) A Promising Electrolyte of Ionic Liquids for Aluminium Deposition
Wang, Q., Institute of Process Engineering, Chinese Academy of Sciences
Originally presented on: 11/6/2013 10:40:00 - 11:00:00

A Promising Electrolyte of Ionic Liquids for Aluminium Deposition

Qian Wang[a], Bin Chen[a], Qinqin Zhang[a,b], Suojiang Zhang[a,*]

[a]Beijing Key Laboratory of Ionic Liquids Clean Process, Key Laboratory of Green Process and Engineering,

State Key Laboratory of Multiphase Complex Systems, Institute of Process Engineering, Chinese Academy of Sciences,

Beijing 100190, P.R. China

[b]College of Chemistry and Chemical Engineering, University of Chinese Academy of Sciences, Beijing 100049, P.R. China

[*]Corresponding author. E-mail: sjzhang@home.ipe.ac.cn Tel.: +86 10 82627080

Ionic liquids (ILs) become the preferred electrolyte for aluminum electrodeposition due to their low melting point, negligible vapor pressure, high conductivity, and wide electrochemical window [1]. The research work on electrodeposition of aluminum from ILs began in 1951 [2], since then researchers conducted a series of studies [3-5], inwhich the types of ILs can be divided into four categories. They are mixtures of anhydrous aluminium halides with alkylpyridinium halides, alkylimidazolium halides, alkylpyrrolidinium halides and aralkylammonium halides, respectively. Usually, the anhydrous aluminium halide is anhydrous aluminium chloride ($AlCl_3$).

In this study, four typical ILs of the above four systems, 1- butyl -3- methylpyridinium chloride-$AlCl_3$, 1-butyl-3- methylimidazolium chloride-$AlCl_3$, 1-butyl-1-methylpyrrolidinium chloride-$AlCl_3$, and phenyltrimethylammonium chloride-$AlCl_3$, were chose using as the electrolyte for aluminium deposition. Firstly, the physical properties of the four ILs were measured. At room temperature, 1-butyl-3-methylimidazolium chloride-$AlCl_3$ has the highest conductivity and the lowest viscosity. Meanwhile, they are similar in density. Then aluminium coatings were prepared from them at the same current density, respectively. By comparing the quality of the obtained coatings, it was found that compact and smooth aluminium coatings can be prepared from the IL formed by mixing 1-butyl-3- methylimidazolium chloride with $AlCl_3$ at 30℃. But in the other three systems, only when the temperature is higher than 60℃, good quality aluminium coatings can be obtained. Finally, the cyclic voltammetry of the four electrolytes was determined in a three-electrode cell at different temperature, also showing that the 1-butyl-3-methylimidazolium chloride-$AlCl_3$ has the best electrochemical property at a lower temperature. All this proves that 1-butyl-3- methylimidazolium chloride-$AlCl_3$is a promising electrolyte for aluminium deposition.

References

[1] Brown L. Electroplating with non-aqueous solutions [J]. *Trans. Inst. Met. Finish.,* 2010, 88,122-123.

[2] Hurley F H, Wier T P. The electrodeposition of aluminum from nonaqueous solution at room temperature [J]. *J.Electrochem.Soc.*, 1951, 98, 207-212.

[3] Yue G, Zhang S, Zhu Y, et al. A promising method for electrodeposition of aluminium on stainless steel in ionic liquid [J]. *AIChE J.*, 2009, 55, 783-796.

[4] Liu Q, Liu K, Han Q, et al. Study on electrodeposition of Al from trimethylphenylammonium chloride (TMPAC)-aluminum chloride ($AlCl_3$) ionic liquids + benzene. *Chin. Surf. Eng.*, 2010, 23, 34-39.

[5] Giridhar P, El Abedin S Zein, Endres F. Electrodeposition of aluminium from 1-butyl-1-methylpyrrolidinium chloride/$AlCl_3$ and mixtures with 1-ethyl-3-methylimidazolium chloride/$AlCl_3$ [J]. *Electrochim. Acta*, 2012, 70, 210-214.

(450a) Bench Scale Shale Shaker Performance

Chase, G. G., The University of Akron
Benis, S., The University of Akron
Originally presented on: 11/6/2013 8:30:00 - 8:55:00

Shale shakers separate coarse solid particles from drilling muds in well bore-hole drilling operations. Drilling muds are engineered with rheological properties needed for efficient transport of coarse solid particles generated by the drill cutting into the soil and rock. Experimental measurements on full scale equipment are difficult, expensive, and inconvenient.

In this work experimental measurements are taken from a bench scale shale shaker. Experimental results show the effects of vibration frequency and amplitude, deck angle, mud flow rate, and mud concentration. Experimental results are compared with a shale shaker model.

(450b) Composition of Heavy Oil From Speed of Sound Data

Riazi, M. R., Kuwait University
Dashti, H., Kuwait University
Originally presented on: 11/6/2013 8:55:00 - 9:20:00

As reserves for light and conventional oils are in decline, demand on heavy and unconventional oils is on the rise. Production and processing of such oils require knowledge of their composition for proper characterization and design of related processing equipment. Heavy oils are usually rich in asphaltene which upon precipitation may cause significant flow problems in production and processing units. In this presentation we examine several measurable properties such as speed of sound and dipole moment which may be used for characterization of heavy oils.

Speed of sound in oils can be accurately measured under different conditions of temperature and pressure. In this presentation we show some recently measured data for the speed of sound in pure hydrocarbons, crude oil and deasphalted oils. Analysis of data show that sonic velocity may be used to determine amount of asphaltene and other heavy components in oils. In addition a predictive method to determine speed of sound from bulk properties is presented for cases that data on the speed of sound is not available.

(450c) A Case Study Of Debottlenecking Strides In A Petrochemical Plant

Okechukwu, C., INDORAMA ELEME PETROCHEMICALS LTD
Shrivastava, B.,
Dada, E. A., ChemProcess Technologies (CPT), LLC
Originally presented on: 11/6/2013 9:20:00 - 9:45:00

Process plant debottlenecking is aimed at increasing the plant's efficiency, optimizing raw material consumption, energy and other utilities, capacity utilization, on-stream availability, safety rating, competitiveness and profitability among others. To be competitive, the promoters of a process plant has to pro-actively adapt to ever changing realities and templates caused by equipment obsolescence, raw materials vulnerability, technological changes and even stringent local legislations and environmental laws.

Nigerian's government quest to attract foreign and local direct investment into key sectors of the economy in general and the oil and gas sector in particular to create jobs, contribute to GDP growth, facilitate the exploitation of vast natural resources are among the driving force for debottlenecking of existing facilities in the Petroleum and Gas sector for optimization and process improvement.

Indorama Eleme Petrochemical Ltd (IEPL), sited in the Niger Delta region of Nigeria has pursued and implemented a robust and integrated debottlenecking master plan to modernize its plants and facilities. In this presentation, we shall examine the giant debottlenecking strides made by IEPL Eleme which is fast transforming Nigeria into a large petrochemical hub in Africa. These debottlenecking strides include the modification for catalysts separation and regeneration. We shall also discuss some of the resources added by IEPL to consolidate the robust maintenance programs, upgrading of manufacturing facilities, streamlining of feedstock availability constraints and implementing a safety culture all of which have led to capacity improvements at a cost effective and profitable manner.

(450d) Experimental Measurements and Modelling of the Dissociation Conditions of Clathrate Hydrates for (Refrigerant + NaCl + Water) Systems

Ngema, P. T.,
Petticrew, C. N., University of KwaZulu-Natal
Naidoo, P., University of KwaZulu-Natal, Howard College Campus
Mohammadi, A. H., MINES ParisTech
Ramjugernath, D., University of KwaZulu-Natal
Originally presented on: 11/6/2013 9:45:00 - 10:10:00

Abstract

Globally, gas hydrates have been proposed as a potential means for many positive applications, *e.g.* wastewater treatment and desalination, CO_2 capture and separation, separation of close-boiling point compounds, hydrogen/methane storage, refrigeration and air conditioning industry, food industry[1,2,3,4,5]. Javanmardi and Moshefeghian[4]; Chun *et al.*[6]; Seo and Lee,[7] and Eslamimanesh *et al.*[8] have reported that the use of refrigerants for the formation of gas hydrates in water desalination processes is promising when compared to traditional desalination processes. The use of refrigerants to form gas hydrates is attractive because hydrates as host molecules may occur at ambient conditions. According to Eslamimanesh *et al.*[9], the dissociation of gas hydrate results in the production of pure water and the release of refrigerant which can be recycled. However, there is a lack of research regarding the use of hydrate technology for the treatment of industrial wastewater and desalination processes, particularly using fluorinated refrigerants as the hydrate former.

The focus of this study is to generate accurate hydrate dissociation data which will be used to design a wastewater treatment and desalination processes using gas hydrate technology. In this study, three refrigerants 1,1,1,2-tetrafluoroethane (R134a), (0.5 mass fraction difluoromethane + 1,1,1,2,2-pentafluoroethane) (R410a), and (0.5 mass fraction 1,1,1-trifluoroethane + 1,1,1,2,2-pentafluoroethane) (R507)were investigated in the absence and presence of a NaCl aqueous solution at various salt concentrations. Hydrate dissociation data was measured using an isochoric pressure-search method.

The binary systems, in this study, which consisted of {R134a, or R410a, or R507} + water were measured in the temperature range between (276.2 to 291.8) K and pressures ranging from (0.114 to 1.106) MPa. The ternary system R134a + water + NaCl, at three salt concentrations of (5, 10, and 15) %, was measured in the temperature range between (268.1 to 280.6) K and pressures ranging from (0.086 to 0.340) MPa. For the ternary systems comprising of {R410a or R507} + water + NaCl, at two salt concentrations of (5 and 10) %, measurements were undertaken in the temperature range between (275.1 to 290.3) K and pressures ranging from (0.269 to 1.170) MPa.

Modeling of the measured data was undertaken using a combination of the solid solution theory of van der Waals and Platteeuw[10] for the hydrate phase, the Aasberg – Petersen *et al.*[11] model for the electrolyte aqueous system, and the Peng-Robinson[12]equation of state with classical mixing rules for the liquid and vapour phases. The correlated results show good agreement with the experimental dissociation data.

The results of this study show that the presence of the NaCl in the aqueous solutions has a thermodynamic inhibition effect on refrigerant gas hydrates. The {R134a or R507} Hydrate–Liquid water–Vapour ($H–L_w–V$) equilibrium phase boundary is shifted to lower dissociation temperatures compared to the dissociation temperatures and pressures for R22 + water + NaCl systems measured by Chun et al.[6]. Also, it is higher than propane + water + NaCl systems measured by Chun et al.[13]. The hydrate dissociation temperatures for R410a + water system are close to ambient conditions. In addition, quadruple points at which the four phases ($H–L_w–L_{R134a}–V$) coexist were determined for the water + R134a + NaCl system. Further research is being investigated to identify suitable promoters with the systems discussed to determine feasible operating conditions the desalination processes.

Keywords:Gas hydrate, Refrigerant, Salt, Dissociation data, Model, Desalination.

***Corresponding authors**: Deresh Ramjugernath, e-mail: *ramjuger@ukzn.ac.za*,

Amir H. Mohammadi, e-mail: *a.h.m@irgcp.fr*

References

(1) Mohammadi, A. H.; Richon, D. Phase equilibria of methane hydrates in the presence of methanol and/or ethylene glycol aqueous solution, *Ind. Eng. Chem. Res,* **2010,**49, 925-928.

(2) Tumba, K.; Reddy, P.; Naidoo, P.; Ramjugernath, D.; Eslamimanesh, A.; Mohammadi, A. H.; Richon, D. Phase equilibria of methane and carbon dioxide clathrate hydrates in the presence of aqueous solutions of tributylmethlyphosphonuim methylsulfate ionic liquid, *Journal of Chemical and Engineering Data,* **2011**, 56, 3620-3629.

(3) Sloan, E. D.; Koh, C. A. *Clathrate hydrates of natural gases*, 3rd Edition, CRC Press, Taylor & Francis Group, London, New York, 2008.

(4) Javanmaardi, J.; Moshefeghian, M. Energy consumption and economic evaluation of water desalination by hydrate phenomenon, *Appl. Therm. Eng*, **2003**, 23, 845-857.

(5) Huang, C. P.; Fennema, O.; Poweri, W. D. Gas hydrates in aqueous-organic systems: II. Concentration by gas hydrate formation, *Cryobiology*, **1966**, 2, 240-245.

(6) Chun, M-K.; Lee, H.; Ryu, B. J. Phase equilibria of R22 ($CHClF_2$) hydrate systems in the presence of NaCl, KCl and $MgCl_2$, *J. Chem. Eng. Data*, **2000**, 45, 1150-1153.

(7) Seo, Y.; Lee, H. A new hydrate-based recovery process for removing chlorinated hydrocarbons from aqueous solutions, *Environ. Sci. Technol.* **2001**, 35, 3386-3390.

(8) Eslamimanesh, A.; Mohammadi, A. H.; Richon, D. Thermodynamic model for predicting phase equilibria of simple clathrate hydrates of refrigerants, *Chemical Engineering Science*, **2011**, 66, 5439-5445.

(9) Eslamimanesh, A.; Mohammadi, A. H.; Richon, D.; Naidoo, P.; Ramjugernath, D. Application of gas hydrate formation in separation processes: A review of experimental studies, *Journal of Chemical Thermodynamics*, **2012**, 46, 62-71.

(10) Van der Waals, J. H.; Platteeuw, J. C. Clathrate Solutions, *Adv. Chem. Phys,* **1959**, *2*, 1-57.

(11) Aasberg-Petersen, K.; Stenby, E.; Fredenslund, A. Prediction of high pressure gas solubilities in aqueous mixtures of electrolytes, *Ind. Eng. Chem. Res*, **1991,**30, 2180-2185.

(12) Peng, D. Y.; Robinson, D. B. A new two constant equation of state. *Ind. Eng. Chem. Fundam,* **1976**, 15, 59-64.

(13) Mohammadi, A. H.; Afzal, W.; Richon, D. Gas hydrate of methane, ethane, propane and carbon dioxide in the presence of single NaCl, KCl and $CaCl_2$ aqueous solutions: Experimental measurements and predictions of dissociation conditions, *J. Chem. Thermodynamics,* **2008,** 40, 1693-1697.

(451a) Engineering Co-Crystals: Solving a Three Dimensional Problem

Chadwick, K., Massachusetts Institute of Technology
Ling, J.,
Originally presented on: 11/6/2013 8:30:00 - 8:50:00

Enhancing the bioavailability of poorly water soluble drugs represents an outstanding challenge in pharmaceutical manufacturing. Engineering a crystal form of the Active Pharmaceutical Ingredient (API) that enhances dissolution and/or solubility is often a necessity for optimizing bioavailability. API co-crystals have the ability to greatly enhance the solubility and dissolution rate of the drug without altering its biological function. Current strategies for engineering co-crystals are limited to the design of desired intermolecular interactions between API and co-former. However, this approach is often unsuccessful as no consideration is given to whether the molecules can pack efficiently into a 3D structure. To our knowledge, the effect of the size and shape of the API and potential co-formers on co-crystal formation **has never been studied**.

In this presentation we shall describe the utilization of the Cambridge Structural Database to study how variation in the molecular weight and volume of the co-crystal components affects the probability of successfully obtaining a co-crystal. The effect of optimal co-former size on co-crystal packing density is discussed. Finally, an engineering strategy incorporating both intermolecular interactions and 3D packing affects has been devised. The success of which has been measured by successfully screening for co-crystals with APIs that have not previously been tested.

(451b) Ab Initio Prediction of Crystal Structure and the Effects of Temperature On the Relative Stability of Enantiotropic Polymorphs

Vasileiadis, M., Imperial College London,Center for Process Systems Engineering
Adjiman, C. S., Imperial College London, Centre for Process Systems Engineering
Pantelides, C. C., Imperial College London,Center for Process Systems Engineering
Originally presented on: 11/6/2013 8:50:00 - 9:10:00

A given molecule may lead to more than one crystal structure and the term *polymorph* describes each of the many possible -long ranged- spatial arrangements in which the molecule may crystallise [1]. Different polymorphs have different physical properties, rendering polymorphism a phenomenon of industrial interest since most solid products of practical interest are in crystalline form e.g. pharmaceuticals or agrochemicals. As a result the identification of possible polymorphs and their relative stability is an important challenge which must be addressed in order to engineer products and processes.

In recent years, there has been a growing effort to predict the number and structure of the polymorphs a compound forms. Successful predictions have been achieved for several small [2] molecules and more recently for large molecules exhibiting extended flexibility [3]. Most of the currently used crystal structure prediction techniques are based on lattice energy minimisation algorithms [4], thus restricting calculations to 0K and limiting our ability to assess reliably the relative stability of different polymorphs. Furthermore, based on those approaches, it is not possible to study enantiotropic polymorphs, i.e., polymorphs whose stability order is a function of temperature.

In this study, a methodology to include quantum and temperature effects within an existing crystal structure prediction methodology [3], based on the CrystalPredictor [5] and CrystalOptimizer [6] algorithms, is presented. It is based on fully atomistic lattice dynamics under the harmonic approximation [7], using an energy model that is consistent with the CrystalOptimizer algorithm. The approach is applied to tetracyanoethylene (TCNE), which is known to exhibit enantiotropic polymorphs [8,9]. TCNE possesses two resolved polymorphs, the cubic form, which is stable at low temperatures, and the monoclinic form, which is stable at more elevated temperatures. The free energies of the low lattice energy structures of TCNE identified by CrystalOptimizer are calculated at various temperatures using. The intra-molecular interactions are computed using isolated molecule quantum mechanical calculations and the electrostatic interactions are modeled using ab-initio derived distributed multipole moments [10]. An Ewald type summation technique is employed for divergent and poorly convergent sums. The effect of temperature on the predicted landscape is investigated and a reversal of stability is correctly predicted between the two enantiotropic polymorphs at higher temperatures. The impact of the approximations inherent in the harmonic approximation is assessed through comparison with experimental data.

[1] J. Bernstein, "Polymorphism in Molecular Crystals". *Clarendon Press*, Oxford, (2002)

[2] G.M. Day, T.G. Cooper, A.J. Cruz-Cabeza, K.E. Hejczyk et.al. , *Acta Cryst. B*, 65, 107-205

[3] A.V. Kazantsev, P.G. Karamertzanis, C.S. Adjiman, and C.C. Pantelides et.al., *International Journal of Pharmaceuticals*, 418, 168-178, (2011)

[4] G.M. Day, *Cryst. Rev.* 17, 3-52, (2011)

[5] P.G. Karamertzanis and C.C. Pantelides, *Molecular Physics*, 105(2-3), 273-291 (2007)

[6] A.V. Kazantsev, P.G. Karamertzanis, C.S. Adjiman, and C.C. Pantelides, *J. Chem. Theory Comput.*, 7, 1998-2016 (2011)

[7] M.T. Dove, "Introduction to lattice dynamics", Cambridge University Press, (1993)

[8] J. Murgich and S. Pissanetzky, *J. Chem. Phys.*, 62, 92-93 (1975)

[9] R. Mukhopadhyay, S.L. Chaplot and K.R. Rao, *Physca Status Solidi A: Applied Research,* 92, 467-474 (1985)

[10] A.J. Stone, *J. Chem. Theory Comput.*, 1, 1128-1132, (2005)

(451c) Polymorph Prediction: A Kinetic Approach

Singh, M. R., Purdue University
Parks, C.,
Tung, H. H., AbbVie Inc.
Bordawekar, S., AbbVie
Ramkrishna, D., Purdue University
Originally presented on: 11/6/2013 9:10:00 - 9:30:00

Polymorphs are characteristic identities of crystalline materials that stem from their crystal structures and can offer them significantly different properties such as solubility, morphology, hardness, cleavage, optical properties, heat conductivity, electrical conductivity and piezoelectric effect. The synthesis of crystals of desired polymorphs, based on the applications of interest, requires understanding of the self-assembly of molecules under different crystallization conditions. The self-assembly of molecules to form crystals depends on the atomic configuration of the molecule, flexibility of the molecule, solvent, co-solvent, additives, thermodynamic and hydrodynamic conditions. The molecules are known to self-assemble in 230 space groups or less, where the energy associated with each assembly can be used to infer stability or likeliness of the appearance of the polymorph under equilibrium. One of the methods to compute energies of different polymorphs is the Lattice Energy Minimization[1]which does not consider the effect of crystallization conditions on polymorph formation. The non-equilibrium approaches such as Kinetic Monte-Carlo and Molecular Dynamics shows limitations in treating dense systems and longer time scales. The objective of this article is to develop a mechanistic framework to predict polymorphs and associated nucleation kernels as a function of crystallization conditions. The application of this framework to predict polymorphs of Tetrolic acid will be shown.

According to the kinetic theory of nucleation, the critical crystal nuclei are formed by successive transitions of molecules to dimers, trimers, and so on. Clearly, the crystal structure is determined by the self-assembly of molecules to form dimers. The direct relation between the growth synthons[2] (structure of dimers) in solution with the structural synthons[2] (molecular building blocks) in the crystal is known as *Link Hypothesis*. The Link Hypothesis was experimentally verified for various crystalline materials including form I of Tetrolic acid.[3]According to the Link Hypothesis the structure of growth synthons will determine the crystal structure. We present a stochastic model to predict the formation of growth synthons under different crystallization conditions. The stochastic model is based on the constrained Brownian motion of a pair of molecules in an infinitesimal volume. The successful implementation of this framework will provide a tool for model-based screening of crystallization conditions for synthesis of desired polymorphs.

References:

(1) Day, G. M.; Chisholm, J.; Shan, N.; Motherwell, W. D. S.; Jones, W., An assessment of lattice energy minimization for the prediction of molecular organic crystal structures. *Crystal growth & design* 2004, 4, (6), 1327-1340.

(2) Desiraju, G. R., Supramolecular synthons in crystal engineering—a new organic synthesis. *Angewandte Chemie International Edition in English* 2003, 34, (21), 2311-2327.

(3) Parveen, S.; Davey, R.; Dent, G.; Pritchard, R., Linking solution chemistry to crystal nucleation: the case of tetrolic acid. *Chemical communications* 2005, (12), 1531-1533.

(451d) Modelling the Effect of Process Variables On the Polymorphic Transformation of L-Glutamic Acid

Mitchell, N., Process Systems Enterprise
Bermingham, S. K., Process Systems Enterprise
Mumtaz, H., Process Systems Enterprise
Originally presented on: 11/6/2013 9:30:00 - 9:50:00

This work concerns the modelling of the Solution-Mediated Polymorphic Transformation (SMPT) of L-glutamic acid in aqueous solutions. L-glutamic acid exhibits two polymorphs, the metastable alpha form and the stable beta form. For downstream processing the metastable α form is preferred due to its block-like habit, instead of the β form, which forms needles with a high aspect ratio. The modelling of the SMPT of L-glutamic was conducted using Process Systems Enterprise's (PSE) crystallization modelling software, namely gCRYSTAL. This software provides the capability for multiphase population balance modelling in a user-friendly GUI-based environment. Numerous process variables can affect the rate of the SMPT. In the case of L-glutamic acid, the effects of seed mass, seed PSD, seed purity and agitation rate on the SMPT will be examined in this work. The model utilised the experimental data of Cornel et al.[1] to evaluate the crystallization kinetics for both polymorphs. This work involved the seeded isothermal transformation experiments from the α form to the β form in aqueous solutions, saturated with respect to the α form. Therefore, the α form could only undergo dissolution, while the β form was found to exhibit secondary nucleation and crystal growth. Using experimental data for the evolution of the solution concentration and the polymorphic fraction, eight parameters including the level of β impurity and size in the α seed crystals, were employed to fit the experimental data.

The developed model of the SMPT process was able to accurately describe the effects of seed mass, PSD, purity and agitation rate on the transformation rate. The inclusion of parameters to describe the impurity of the alpha seeds employed in the experiments was validated by two blind tests, not utilised in the estimation of kinetic parameters, where a known amount of α and β seeds were added initially. The model was able to accurately describe the effect of the presence of the β seed impurity in these cases, which caused a large reduction in the time necessary for the SMPT process. The time necessary for the transformation process was found to reduce with increasing alpha seed mass, agitation rate and impurity concentration. These observations led to the conclusion that the rate limiting step of the SMPT process was the rate of secondary nucleation on the surface of the β crystals, which increased with agitation and available seed surface area. A contradictory finding for the effect of seed size was attributed to an increased impurity concentration of the β crystals in the larger α seed sizes.

[1]Cornel, J., Lindenberg, C., Mazzotti, M. (2009) 'Experimental characterization and population balance modelling of the polymorph transformation of L-glutamic acid', *Cryst. Growth & Des.*, 9(1), 243-252.

(451e) Application of Microfluidics in the Solid Form Screening of Pharmaceutical Parent Compounds

Goyal, S., University of Illinois at Urbana Champaign
Horstman, E., University of Illinois at Urbana Champaign
Economou, A., University of Illinois at Urbana Champaign
Lee, G., University of Illinois at Urbana Champaign
Gong, Y., AbbVie Inc.
Zhang, G. G. Z., AbbVie Inc.
Kenis, P. J. A., University of Illinois at Urbana-Champaign
Originally presented on: 11/6/2013 9:50:00 - 10:10:00

We present on the application of solvent resistant microfluidic platforms to screen solid form (salts, co-crystals, and polymorphs) of pharmaceutical parent compounds (PC). Success in finding a crystalline solid form of a PC with optimized CMC properties using conventional screening procedures is limited due to the limited experimental conditions that can be investigated while utilizing small quantities of PCs available (≈ 10 mg) in the early stages of drug development. Microfluidic technology capable of screening with reduced sample volumes of PC and precipitants (salt or cocrystal formers or antisolvents) solutions [1] can enable screening of multiple conditions using the limited amount of PC. This will enable early identification of all possible solid forms of PCs and help reduce the time and money invested in the solid form development.

To date, we have reported on hybrid polymer-based microfluidic platforms that permit combinatorial mixing of PC and precipitants solutions on-chip in arrays of 24/48, 50 to 200 nL wells [3-6], which is a drastic reduction in the volume of PC needed per condition compared to traditional approaches (~5 to 100 ul per condition [2]). These platforms were compatible with mild organic solvents and water, and enabled identification of the solid form crystallized on-chip via on-chip Raman spectroscopy. We screened for solid forms of PCs employing free interface diffusion (FID) mixing [3,5], antisolvent addition (AS) [4], and solvent evaporation [6] modes of crystallization employing the polymer-based platforms.

In this work, we present on the application of the aforementioned solvent compatible platforms to grow single, isolated crystals of PC solid forms (*esp.* cocrystals) that are hard to distinguish via Raman spectroscopy, employing *on-chip seeding methodology* followed by solid form identification via *in*-situ single crystal X-ray diffraction. In addition, we present the design, fabrication, and application of newly developed solvent resistant microfluidic platforms for solid form screening of PCs. These hybrid polymer-based microfluidic platforms are compatible with a much wider range of solvents including chloroform, toluene, and hexanes, and thereby enable investigation of a much wider range of conditions on-chip as well as allow a better control on the supersaturation levels attained on-chip. We employed these platforms to screen for solid forms of model compounds including theophylline, piroxicam, and carbamazepine. On-chip Raman and X-ray analysis was employed to identify and distinguish different solid forms crystallized on-chip. These microfluidic platforms have the potential to enable solid form development at the early stages of drug development.

References

[1] C. Hansen, E. Skordalakes, J. Berger, S. Quake, Proc. Nat'l Acad. Sci., 2002, 99, 16531-6.

[2] S. Talreja, P.J.A. Kenis, et al. J. Appl. Crystallography, 2005, 38, 988-995.

[3] M.R. Thorson, S. Goyal, P.J.A. Kenis, et al. Lab Chip, 2011, *11, 3829-3837*

[4] M.R. Thorson, S. Goyal, P.J.A. Kenis, et al. CrystEngComm, 2012, 14, 2404-2412

[5] S. Goyal, P.J.A. Kenis, et al. Crystal Growth & Design, 2012, 12 (12), 6023-6034

[6] S. Goyal, P.J.A. Kenis, et al. Lab Chip, 2013, 13, 1708-1723

(451f) Form Selection and Methods to Improve the Hygroscopic Behavior of the Selected Form

Shi, B.,
Lapina, O., Gilead Sciences, Inc.
Fung, P., Gilead Sciences, Inc
Wang, F., Gilead Sciences
Casteel, M.,
Zia, V.,
Stefanidis, D.,
Macleod, P.,
Lam, L.,
Originally presented on: 11/6/2013 10:10:00 - 10:30:00

Polymorph and salt/cocrystal screens were conducted in order to select the most suitable form of an active pharmaceutical ingredient (API) for development. Three forms including a hydrate, a channel solvate, and a cocrystal were discovered. The channel solvate was the most promising form, although it showed unusual hygroscopic behavior during dynamic vapor sorption (DVS) analysis: when humidity increased from 50% to 60% relative humidity (RH), the crystal form quickly absorbed water and then slowly lost water and the residual organic solvent. Various methods were explored to further understand the DVS behavior of the channel solvate and improve its physical properties. It was shown that exposing the channel solvate crystals to water removed the residual organic solvents, reduced the water uptake, and improved its hygroscopicity. The improved form was selected for further development. The process to prepare the form with improved properties was successfully scaled up to multi-kilo batches.

(451g) Integrated Approach Toward API Form Selection and Process Development for a Complex Reactive Crystallization Step

Carra, E., Gilead Sciences
Wang, X., Gilead Sciences
Brown, B., Gilead Sciences
Wang, F., Gilead Sciences
Originally presented on: 11/6/2013 10:30:00 - 10:50:00

Integrated Approach toward API Form Selection and Process Development for a Complex Reactive Crystallization Step

Ernest Carra, Xiang Wang, Brandon Brown, and Fang Wang

Gilead Sciences, Foster City, CA 94404

A suitable salt form of an active pharmaceutical ingredient (API) has been identified and selected for further development. This new salt form was discovered through extensive form screening work. The screening work produced several crystalline forms: a hemi-succinate salt and individual hemi-solvates of the parent compound with each of the several solvents, including acetone, acetonitrile, ethanol, isopropanol, and MTBE. Among all the newly discovered forms, hemisuccinate proved to be the most suitable form for further development based on its superior stability, manufacturability, and impurity control capability..

A successful manufacturing process was designed and developed using reverse addition of the API sodium salt solution to an aqueous succinic acid solution. This process was designed to overcome challenges presented by the competing crystallization of various solvated forms of the parent compound. Single crystal X-Ray structure data provided key understanding for the competitive crystallization during process development. This integrated approach towards form screen and crystallization process development has proven to be highly efficient as it is based on fundamental understanding of structural information of various API forms.

(461a) US Department Of Energy Critical Materials Strategy and R&D

McKittrick, M.,

Originally presented on: 11/6/2013 12:30:00 - 12:55:00

In 2010, the US Department of Energy (DOE) released its first Critical Material Strategy – which examined the role of rare earth elements and other materials in the clean energy economy. The Strategy, which was also updated in 2011, highlighted five elements as being "critical" to the deployment of clean energy technology – dysprosium, europium, neodymium, terbium and yttrium. Building off this work and leveraging R&D investments in the area, in 2013 DOE established the Energy Innovation Hub for Critical Materials. This Hub, led by Ames Laboratory, will conduct R&D across the entire lifecycle including materials discovery and design, feedstock supply, processing, manufacture, use, recycling, and re-use. A summary of the DOE Critical Materials Strategy and the Critical Materials Hub will be presented.

(461b) Research Gaps and Needs: Critical Materials Separations R&D Workshop

Schuster, D., Institute for Sustainability
Hunt, C. T., Emerging Technologies, Rohm and Haas Company
Chin, K., AIChE
Diallo, M., Caltech
Originally presented on: 11/6/2013 12:55:00 - 13:20:00

Report from recent workshop on research 'gaps and needs' in separation science and technology for sustainable extraction, recycling, and re-use of critical materials and alternatives will be presented. The enabling role of Separations Science and Engineering is critical in ensuring sustainable supply and utilization of critical materials and the alternatives.

(461c) The Availability Of Indium and Tellurium For Thin-Film Photovoltaic Materials

Eggert, R.,
Lokanc, M.,
Redlinger, M.,
Originally presented on: 11/6/2013 13:20:00 - 13:45:00

Indium and tellurium provide essential properties in emerging thin-film photovoltaic materials. These elements are geochemically scarce. Both are produced today almost exclusively as byproducts of zinc (in the case of indium) and copper (in the case of tellurium) production. If demand for these elements increases significantly, to what degree will new supplies be able to meet this demand? Will increased supplies come from expanded byproduct production at costs similar to recent prices? If not, where might incremental supplies come from and at what costs?

(461d) Rare Earth Concentration, Extraction, Separation and Reduction

Anderson, C.,

Originally presented on: 11/6/2013 13:45:00 - 14:10:00

Rare Earth Concentration, Extraction, Separation and Reduction.

Dr. Corby G. Anderson

Abstract

The key characteristics that deem the Lanthanides to be called Rare Earths focus on their inherent difficulty and challenges in beneficiation, separation and reduction.

This presentation will address the fundamentals behind these processing challenges along with the existing and potential technologies available to remediate them.

(461e) Critical Materials Recycling & Substitutes For Critical Materials

Jones, E., Lawrence Livermore National Laboratory

Originally presented on: 11/6/2013 14:10:00 - 14:35:00

The DOE Critical Materials Institute (CMI) is focused on increasing supply and reducing demand for rare earth elements in clean energy technologies. Approximately 60% of rare earths resourced from mining end up back in landfill or construction materials. This wastage comes from the manufacturing processes or disposals from the consumer reservoir. Thus, finding substitute manufacturing processes and increasing recyclability and reuse can substantially impact the sustainability of rare earth applications. This talk will discuss the R&D approaches that CMI is taking with respect to recycling and substitute technologies, as well as system analytics to improve the lifecycle sustainability of critical materials.

(461f) Critical Materials Recycling For Resource Sustainability

Mishra, B.,

Originally presented on: 11/6/2013 14:35:00 - 15:00:00

Rare earth metals, including yttrium and scandium, are being increasingly used in clean energy technologies, colored phosphors, lasers and high intensity magnets. There are important defense applications such as fighter jet engines, missile guidance systems and space based satellite and communication systems based on these metals. The commitment to clean energy technologies by governments and the projected growth in power and transportation sectors across the globe ensure that the demand for rare earth metals and compounds would continue to escalate. This demand implies that, to ensure unhindered technological innovation, it is essential to possess secure supply chains for rare earth elements. The United States continues to be one of the largest consumers and importer of rare earths and the trend is expected to continue as the demand increases. In order to ensure secure rare earth supply and attenuate supply-demand imbalance post 2014, it is not only necessary to encourage and support exploration of newer reserves, build a rare earth stockpile, but it is also of utmost importance to look at opportunities to recycle and reuse Rare Earth Elements from secondary sources, such as post-consumer and manufacturing process wastes. This research describes the technological developments made to convert these valuable resources into functional manufactured materials for lighting industry, automotive and petroleum refining catalysts, and high density permanent magnets. In addition, production of rhenium from advanced aerospace alloys will be discussed that is recovered for introduction in turbine alloys.

(462a) Steady-State Behavior and Control of Kaibel Distillation Column for Aromatics Separation

Ling, H., East China University of Science and Technology
Lin, X., East China University of Science and Technology
Wu, H., East China University of Science and Technology

Originally presented on: 11/6/2013 12:30:00 - 12:55:00

Kaibel column permits the separation of a four-component mixture into four pure fractions in a divided-wall column. This work studies the separation of a four-component mixture of benzene, toluene, xylene and tri-methyl-benzene in a Kaibel column. The optimum economic design of the Kaibel column was obtained first, and then seventeen cases of various feed compositions were simulated to investigate the steady state behaviors of the Kaibel column. Results showed that xylene should be controlled at the top of the prefractionator, while toluene should be controlled at the bottom of the prefractionator. The heavy impurities are much larger than the light impurities in the two side streams. These behaviors are very similar to the performances of three -product divided-wall column, but the extra freedom, the xylene side stream, makes it difficult to control. At last, a control structure with five composition control loops was implemented. Simulation showed that it can handle 10% disturbances of flow rate and feed composition, but the xylene side stream sometimes had small purity deviations.

(462b) The Influence Of The HEAT Transfer Between The Wall On The Separating Effect and Energy Consumption In The DWC

Fang, J., Hebei University of Technology
Li, C., Hebei University of Technology
Wang, H., Hebei University of Technology
Hu, Y., Hebei University of Technology
Originally presented on: 11/6/2013 12:55:00 - 13:20:00

For the past few years, the dividing wall column(DWC) showed huge potential application and broad prospects in the conventional separation technology.

Dividing wall column is a special structure of fully thermally coupling distillation. There is a dividing wall separating the prefractionation from the main column section. At present, most of the research on the dividing wall column were based on the assumption that the dividing wall is adiabatic. But in fact, there is a obvious temperature difference between the two side of the dividing wall. This temperature difference will cause horizontal heat transfer. The influence of the heat transfer to the separating effect and energy consumption was studied.

Two kinds of different structure dividing wall columns were designed. A lab-scale experiment was conducted on the two columns. The influences of reflux ratio and distribution ratio on the concentration of the overhead n-Pentane and side withdrawing n-Hexane were also obtained.

After builded mathematical models, the rectification processes was simulated by the chemical process simulation software. A comparison between the simulative results and the experimental results was carried out and it is found that they are in agreement well. That is to say, the mathematical models were suitable.

The rectification processes of the two kinds of dividing wall columns were simulated with the specified reboiler duty. The results showed that the heat transfer can the separation efficiency. With the same reflux ratio□distribution ratio and reboiler duty, the concentration of the overhead n-Pentane and side withdrawing n-Hexane were improved because of the heat transfer. The Increase amplitude could be amounted to 30%.

(462c) Simulation Based Approach to Optimal Design of Dividing Wall Column Using Random Search Method

Ge, X.,
Yuan, X., Tianjin University
Originally presented on: 11/6/2013 13:20:00 - 13:45:00

Simulation based approach to optimal Design of dividing wall column using random search method

X. L. Ge, X. G. Yuan (Corresponding author: yuanxg@tju.edu.cn)

State Key Laboratory of Chemical Engineering, School of Chemical Engineering and Technology, Tianjin University, Tianjin 300072, China

ABSTRACT

Optimal design of dividing wall column (DWC) involves multivariable optimization, which includes both continuous and discrete variables. And these variables interact with each other and need to be optimized simultaneously. Because of the more freedoms in terms of the numbers of stages as well as decision parameters of DWC comparing to conventional simple columns, the optimization problem turns out to be in large scale and combinatorial, and can be expressed as a mixed-integer nonlinear programming (MINLP) problem. However, the cost estimation and the material and energy balances of DWC are complex, and as a result, the objective function and constraints of the optimization problem are difficult to be formulated explicitly and directly as functions of the mixed variables. Thus, the optimal design of DWC is usually based on the detailed analysis, rather than a systematic approach.

In this work, a systematic optimization method was developed for optimal design of DWCs. The proposed method is based on the combination of a radial basis function neural network (RBF-NN) and genetic algorithms (GAs). The RBF-NN was built (trained) by a series of rigorous simulations of DWC for different sets of stage numbers with available tool provided in commercial software, *ASPEN PLUS* for example. In each of the simulations, the utility cost of the DWC for a specific set of stage numbers was optimized by adjusting the split ratios of respectively the liquid and vapor side draw streams from the main column to the overhead and the bottom of the prefractionator. Then, a GA was applied to optimize the stage numbers of the DWC with an objective (cost) function evaluated with the results obtained by the RBF-NN. The proposed method was validated with optimal design of DWCs for IDEAL, *BTE* and *EPB* systems respectively. The results were shown encouraging compared with those found in the literature. As a random search based approach, the proposed method showed its strength in finding the optimal solution by just evaluating small portion of the possible combinations of the stage numbers, and thus can be known as a promising method for optimal design of DWC.

Keywords

Distillation, Dividing wall column, optimal design, simulation, random search

(462d) Design and Optimization of a Fusel Oil Separation System Using a Dividing Wall Column

Mendoza Pedroza, J. D. J., Universidad de guanajuato
Orjuela, A., Universidad Nacional de Colombia
Segovia, J. G.,
Originally presented on: 11/6/2013 13:45:00 - 14:10:00

Design and optimization of a fusel oil separation system using a dividing wall column

JosÃ© de Jesús Mendoza Pedroza, Alvaro Orjuela, Juan Gabriel Segovia H,

Fusel oil is a mixture of alcohols obtained as a side stream during distillation of fermentation-derived ethanol with yields around 1 to 10 liters per thousand liters of fuel ethanol. According with GC-MS characterization, sugarcane derived fusel oil is mainly composed of isoamyl alcohol, ethanol, water, with minor concentrations of n-butanol, isobutanol and other alcohols, acids, and esters. Currently fusel oil is used as fuel in steam boilers in the ethanol biorefinery, or mixed back with anhydrous ethanol and sold as fuel. However fusel alcohols can be transformed in higher added-value products such as esters for the fragrances and flavors industries.

Conventional separation of fusel oils is accomplished in batch distillation systems using different cuts where water and ethanol are removed first. After lights removal, isoamyl alcohol is distilled from top obtaining a 95% w/w product while a dark brown liquid is removed from the reboiler. This process is energy intensive and operating costs represent a major fraction of the final costs.

In this direction, this work describes the design and optimization of a novel continuous distillation system including a dividing wall column. Thermodynamics of phase equilibria were validated by regression of binary and ternary equilibrium data reports with a NRTL-HOC activity-based model. The method used for the optimization corresponds to a stochastic differential evolution and a tabu list, setting the objective function to minimize energy consumption and the purity of product streams. This optimization was implemented by connecting Aspen Plus^{TM} and Visual Basic^{TM} (included in EXCEL^{TM}). A preliminary design and optimization was developed including only the three major components: isoamyl alcohol (66% mol), ethanol (16 % mol) and water (18% mol). In a subsequent stage n-butanol, isobutanol and heavy compounds were included in the fusel oil feed stream. In both cases, a reduction in energy consumption of around 20 to 30% with respect to conventional continuous distillation was observed.

(462e) Application On NS Composite Co-Current Flow Trays of Tilt Strip in Ethylene Oxide Distillation

Yuanyu, T., Shandong university of Science and technology
Yingyun, Q., Taiyuan University of Technology
Originally presented on: 11/6/2013 14:10:00 - 14:35:00

NS composite co-current flow trays of tilt strip (NS CTTS) is a new kind of stereo trays with larger throughput and high efficiency. This technology is an improvement of the New VST technology according to its flow structure, combining the co-current injection structure and structured packing effectively. Not only the structures, operation principle and characters were introduced in the passage, but also its application on renovation of ethylene oxide refining tower in Maoming petro-chemical company and Dongfang petro-chemical company was described. After 7-year industrial examination, it's found that just renovating Ethylene Oxide Distillation with NS CCTS trays, Maoming petro-chemical company has decreased circumfluence ratio by 3 times, improved throughput by 5.5 times, indeed 7 times; made the EO product as the high quality product; saved the energy by 69% and handed in tax more than before 1.5 million a year .Its industrial application fully confirmed that with NS SCTTS, the production and separation efficiency can be improved significantly. Obviously, it is the best choice for industrial renovation and energy thrift, additionally, the potential applications of this technology has a splendid future.

(479a) Functionalization of Metal Organic Frameworks for Enhanced Stability Under Humid Environment for CO2 Capture Applications

Andirova, D.,
Lei, Y., Northeastern University
Choi, S., Northeastern University
Originally presented on: 11/6/2013 12:30:00 - 12:48:00

A need for efficient carbon dioxide capture and storage (CCS) technology heightened in the last few decades due to alarming increase in CO_2 atmospheric concentration which was ascribed to be a possible cause for global climate change. Decarbonization of flue gas from the post-combustion process via adsorption onto porous solid media such as metal organic frameworks (MOFs) have shown their effectiveness among a variety of carbon mitigation technologies.[1,2] Unique structural characteristics of MOFs including extremely large surface area coupled with nanopores, a number of MOFs have demonstrated competitive CO_2 adsorption capacity under the low-pressure conditions simulating the CO_2 concentration in the flue gases (10-15%). However, most of the previous CO_2 capture performance evaluations for MOFs have been carried out using dry CO_2, which is not realistic considering the practical composition of the flue gas. For instance, Mg/DOBDC, the MOF that presents the highest CO_2 adsorption capacity among other MOFs, exhibited significant decrease in its CO_2 capacity under the presence of humidity, possibly due to its strong hydrophilicity.[3]

Here we present the functionalization of MOF frameworks as a facile route to enhance the adsorbent stability and the CO_2 affinity under practical conditions involving humid CO_2 capture. Specifically, the surface of the nanopores in Mg/DOBDC is decorated by the functional groups such as ethylene diamine to prevent the direct contact of water molecules to the open metal sites of the MOF framework, using the post-synthetic functionalization techniques reported in our previous work.[4] The material stability of the functionalized Mg/DOBDC is evaluated via accelerated steam treatment experiments, while being tested by various characterization techniques including surface area and pore size analyzer, XRD, TGA, IR, etc. The CO_2 capture performance of the functionalized Mg/DOBDC upon exposure to humid environment will be also presented.

(1) Choi, S.; Drese, J. H.; Jones, C. W. *ChemSusChem* **2009**, *2*, 796.

(2) Sumida, K.; Rogow, D. L.; Mason, J. A.; McDonald, T. M.; Bloch, E. D.; Herm, Z. R.; Bae, T. H.; Long, J. R. *Chemical Reviews* **2012**, *112*, 724.

(3) Kizzie, A. C.; Wong-Foy, A. G.; Matzger, A. J. *Langmuir : the ACS journal of surfaces and colloids* **2011**, *27*, 6368.

(4) Choi, S.; Watanabe, T.; Bae, T.-H.; Sholl, D. S.; Jones, C. W. *The Journal of Physical Chemistry Letters* **2012**, *3*, 1136.

(479b) Evaluating Different Classes of Porous Materials for Carbon Capture

Obst, J. M., University of California-Berkeley
Lin, L. C., University of California-Berkeley
Berger, A. H., Electric Power Research Institute
Martin, R. L., Lawrence Berkeley National Laboratory
Haranczyk, M., Lawrence Berkeley National Laboratory
Bhown, A. S., Electric Power Research Institute
Reuter, K., Fritz-Haber-Institut der Max-Planck-Gesellschaft
Smit, B., University of Amsterdam
Originally presented on: 11/6/2013 12:48:00 - 13:06:00

Carbon Capture and Storage (CCS) is one of the promising ways to significantly reduce the emission of CO_2 from power plants. The separation and compression of CO_2, however, is currently considered the most energy consumptive part. It is therefore important to find adequate materials to reduce the energy load additionally imposed on a power plant, which necessitates the development of a suitable metric to rank those materials. The concept of parasitic energy is a potential means for the evaluation of CCS materials. By combining thermodynamic properties and optimizing the process conditions the parasitic energy requirement can be determined on a material-by-material basis. Ranking materials according to their energy efficiency in CCS not only provides a direct comparison to current state-of-art technologies like MEA but also enables the application to different capturing strategies.

Over 50 different materials, including metal-organic frameworks (MOFs), zeolitic imidazolate frameworks (ZIFs), porous polymer networks (PPNs), zeolites, and hypothetical materials were analyzed using the approach of parasitic energy and compared to the results of other suggested ranking strategies such as selectivity and working capacity. Furthermore, the influence of different regeneration techniques as well as the impact of the adopted prediction model for mixture adsorptions on the parasitic load were investigated.

(479c) Carbon Dioxide Separation Using Highly Selective Narrow Pore Zeolite ZK-4

Cheung, O., Stockholm University
Bacsik, Z., Stockholm Univerisity
Mace, A., Stockholm Univerisity
Krokidas, P., Stockholm University
Laaksonen, A., Stockholm Univerisity
Hedin, N., Stockholm Univerisity
Originally presented on: 11/6/2013 13:06:00 - 13:24:00

In recent years, zeolites were comprehensively studied for their potential as CO_2 sorbents in carbon capture and storage (CCS) processes. Zeolites offer high CO_2 capacity, selective CO_2 sorption, high thermal stability and low costs. Zeolite ZK-4, is a silicon rich version of a zeolite, which has the same framework topology as the Zeolite A. Previously, we studied and showed that the Zeolite NaKA could be tuned into a highly CO_2 selective sorbent. ZK-4 with its potential for high Si/Al ratios could potentially alleviate the disadvantages linked to the very hydrophilic nature of zeolite A. Here, we show that a zeolite ZK-4 (Si/Al =1.36) could be tuned into a highly CO_2 selective form by carefully tuning the cation ratio (K^+/(K^++Na^+))

When the cation ratio of K^+ content was over ~ 25 atomic %, the sorbent exhibited a molecular sieving of CO_2 over N_2 and was highly selective towards CO_2 (ideal CO_2/N_2 selectivity >200). We ascribe this manifold enhanced selectivity to the specific location of the K^+ cation in the pore windows, where their comparably larger size hinders the diffusion of N_2 more effectively than for CO_2. The CO_2 sorption capacity remained high on the CO_2 selective materials (~ 4.1 mmol/g 273K, 101 kPa). The heat of CO_2 physisorption varied between 32 – 45 kJ/mol and was partly related to the K^+ ion content. The rate of CO_2 physisorption was relatively fast. Infrared spectroscopy and molecular modeling gave an insight into the configuration of adsorbed CO_2 and how the cations affected the sorption of CO_2. In summary, zeolite ZK-4 offers the similar CO_2 adsorption characteristics as zeolite NaKA, but its more hydrophobic nature and the possibility for varying the Si/Al ratio allow higher flexibility for tuning this material for specific application. We are continuing these studies by increasing the Si:Al ratio of ZK-4 until the molecular sieving capacity of CO_2-over-N_2 is lost.

(479d) Understanding Carbon Dioxide Adsorption On Alkali Metal Exchanged Zeolite SSZ-13

Pham, T., University of Delaware
Lobo, R. F., University of Delaware
Originally presented on: 11/6/2013 13:24:00 - 13:42:00

Understanding Carbon Dioxide Adsorption on Alkali Metal Exchanged Zeolite SSZ-13

Trong D.Pham[1], Matthew R. Hudson[2], Craig M. Brown[2], Raul F.Lobo[1*]

[1]*Center for Catalytic Science and Technology, University of Delaware, Department of Chemical and Biomolecular Engineering, USA, lobo@udel.edu.*

[2]*Center for Neutron Research, National Institute of Standards and Technology, Gaithersburg, Maryland 20899-6102, USA.*

Keywords: CO_2, SSZ-13, adsorption, neutron diffraction

Introduction

The emission of anthropogenic carbon dioxide into the atmosphere due to the combustion of fossil fuels is one of the main causes of climate change. Low silica zeolites (types A, 13X, zeolite Y) are among the most commonly used CO_2 adsorbents in industrial gas separations[1,2], but are generally strong adsorbents, making regeneration difficult [3]. Because of lower Al content, high silica zeolites with a less hydrophilic pore environment are an interesting option with respect to the regeneration and the tolerance to polar impurities in the gas such as water vapor and sulfur compounds[4]. Here we report an investigation of CO_2 adsorption on high-silica zeolites SSZ-13 (CHA framework-type, Si/Al=6 and 12), examining the effects of framework Si/Al ratio and extra-framework cations (Li^+, Na^+, K^+) on CO_2 adsorption isotherms and heats of adsorption. The interaction of CO_2 with the SSZ-13 is investigated at the molecular level using Rietveld refinement of neutron diffraction patterns of adsorbed CO_2 in the host zeolite.

Experimental

SSZ-13 samples (Si/Al=6 and 12) were prepared by hydrothermal methods using N,N,N-Trimethyl-1-adamantanammonium ion as a structure-directing agent. Element compositions of synthesized and ion-exchanged samples were determined by Energy-dispersive X-ray spectroscopy (EDX). CO_2 adsorption isotherms of the SSZ-13 samples at various temperatures (273, 303 and 343 K) were measured on a Micromeritics ASAP 2020 device up to ambient pressure. Neutron powder diffraction (NPD) measurements were conducted on the high-resolution diffractometer BT1 at NIST, using a Ge(311) monochromator, λ = 2.0787(2) … for bare and CO2 gas-dosed alkali-exchanged SSZ-13.

Results and Discussion

Na-SSZ-13/6 is remarkable because it shows comparable equilibrium CO_2 capacity to 13X and Y, and higher capacity than NaA, three zeolites with much higher aluminum content (Table 1). This is because *all* the cations in the smaller cages of SSZ-13 are accessible to CO_2, and despite the fact that SSZ-13 has a lower cation concentration compared to other zeolite frameworks. Na-SSZ-13/12 adsorbed less CO_2 than the lower Si/Al samples under the same conditions, as expected, but shows much higher adsorption capacity than FER and STI with similar Si/Al ratio. This difference is attributed to the higher microporous volume of the SSZ-13 sample.

Table 1. Adsorption capacity of CO_2 on various sodium-exchanged zeolites at 303 K and 1atm.

Samples	Na-13X	NaY	NaA	Na-FER[6]	Na-STI [6]	Na-SSZ-13/6	Na-SSZ-13/12
Si/Al ratio	1.25	2.47	1	9	15	6	12
Uptake (mmol/g)	5.23	5.03	4.13	2.6	2.5	5	3.96

*ref [6]: The experiment was carried out at 298K and 750torr

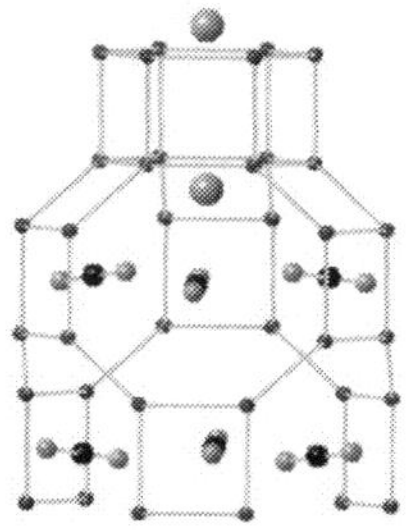

Figure 1: Schematic view of cation positions in SSZ-13 framework (left) and CO_2 positions in Na-SSZ-13 (Si/Al~12) (right)

Rietveld refinement of neutron diffraction on bare adsorbent SSZ-13/12 shows that Li^+, Na^+ is located right above the center of the hexagonal prism (site SII), and the larger cation K^+ is in the middle of 8-membered rings. With 4 K^+ cations/cage in K-SSZ-13/6, K^+ could reduce the effective adsorption capacity of the zeolite by blocking access to some of the SSZ-13 cavities. Difference Fourier maps were used to locate the positions of CO_2 molecules in the zeolite. At a loading of 0.66 molecules CO_2 per 8-ring window on Na-SSZ-13/12, we found that CO_2 is located in the center of the 8-membered ring window instead of directly interacting with extra-framework cations, which is similar to the results reported by Hudson et al[5] for Cu-SSZ-13. Na^+ is bound to 3 oxygen atoms of the 6-ring with a bond length of 2.35, and the closest distance of $O(CO_2)$-Na is 3.71. This distance is too long for efficient interaction coupling of the CO_2 and the cations suggesting that adsorption is driven by the electrostatic interactions of the framework oxygen atoms in the 8-ring and the carbon atom in CO_2.

Conclusions

SSZ-13 has a high potential for CO_2 capture, reaching an adsorption capacity of 5.1 mmol CO_2 per gram of zeolite at ambient temperature and pressure, a value comparable with the best-known zeolite adsorbents (X, Y). The adsorption characteristics of SSZ-13 are strongly dependent upon the cation concentration, the cation-type, and the CO_2 loading amount.

Acknowledgements

This work was supported by the National Science Foundation through grant CBET-940768 and National Institute of Standards and Technology Center for Neutron Research (NCNR). Trong Pham acknowledges support from Vietnam Education Foundation (VEF). MRH acknowledges support from the NRC Postdoctoral Research Fellowship.

References

[1] Yang, H.; Xu, Z.; Fan, M.; Gupta, R.; Slimane, R. B.; Bland, A. E.; Wright, I., *Journal of Environmental Sciences-China* **2008,** *20* (1), 14-27.

[2] Walton, K. S.; Abney, M. B.; LeVan, M. D., *Microporous Mesoporous Mater.* **2006,** *91* (1-3), 78-84.

[3] Harlick, P. J. E.; Tezel, F. H., *Microporous Mesoporous Mater.* **2004,** *76* (1-3), 71-79.

[4] Bao, Z.; Yu, L.; Dou, T.; Gong, Y.; Zhang, Q.; Ren, Q.; Lu, X.; Deng, S., *J. Chem. Eng. Data,* **2011,** *56* (11), 4017-4023.

[5] Hudson, M. R.; Queen, W. L.; Mason, J. A.; Fickel, D. W.; Lobo, R. F.; Brown, C. M., *J. Am. Chem. Soc.* **2012,** *134* (4), 1970-1973.

[6] Zukal, A.; Zones, S. I.; Kubu, M.; Davis, T. M.; Cejka, J., *Chempluschem* **2012,** *77* (8), 675-681. ADDIN EN.REFLIST

(479e) Molecular-Level Insight Into Unusual Low Pressure CO2 Affinity in Pillared Metal-Organic Frameworks

Burtch, N. C., Georgia Tech
Jasuja, H., Georgia Institute of Technology
Dubbeldam, D., University of Amsterdam
Walton, K. S., Georgia Institute of Technology
Originally presented on: 11/6/2013 13:42:00 - 14:00:00

Fundamental insight into how low pressure adsorption properties are affected by chemical functionalization is critical to the development of next generation porous materials for post-combustion CO_2 capture. In this work, we present a systematic approach to understanding low pressure CO_2 affinity in isostructural metal-organic frameworks (MOFs) using molecular simulations and apply it to obtain quantitative, molecular-level insight into interesting experimental low pressure adsorption trends in a series of pillared MOFs. Our experimental results show that increasing the number of non-polar functional groups on the benzene dicarboxylate (BDC) linker in the pillared DMOF-1 $[Zn_2(BDC)_2(DABCO)]$ structure is an effective way to tune the CO_2 Henry's coefficient in this isostructural series. These findings are contrary to the common scenario where polar functional groups induce the greatest increase in low pressure affinity through polarization of the CO_2 molecule. Instead, MOFs in this isostructural series containing nitro, hydroxyl, fluorine, chlorine, and bromine functional groups result in little increase to the low pressure CO_2 affinity. Strong agreement between simulated and experimental Henry's coefficient values is obtained from simulations on representative structures, and a powerful yet simple approach involving the analysis of the simulated heats of adsorption, adsorbate density distributions, and minimum energy 0 K binding sites is presented to elucidate the intermolecular interactions governing these interesting trends. Through a combined experimental and simulation approach we demonstrate how subtle, structure-specific differences in CO_2 affinity induced by functionalization can be understood at the molecular-level through classical simulations. This work also illustrates how structure-property relationships resulting from chemical functionalization can be very specific to the topology and electrostatic environment in the structure of interest. Given the excellent agreement between experiments and simulation, predicted CO_2 selectivities over N_2, CH_4, and CO are also investigated to demonstrate that methyl groups also provide the greatest increase in CO_2 selectivity relative to the other functional groups. These results indicate that methyl ligand functionalization may be a promising approach for creating both water stable and CO_2 selective variations of other MOFs for various industrial applications.

(479f) Adsorption of CO2 Over Supported Amine Adsorbents in the Presence of SOx and NOx Impurities

Rezaei, F., Georgia Institude of Technology
Jones, C. W., Georgia Institute of Technology
Originally presented on: 11/6/2013 14:00:00 - 14:18:00

For supported amine adsorbents to be useful in practical CO_2 capture application, it is crucial to exhibit a stable performance and a long lifetime in the presence of flue gas impurities such as SOx and NOx. Here, the stability of supported aminosilica adsorbents to SO_2 and NOx impurities in CO_2 capture process is evaluated by conducting equilibrium and dynamic adsorption experiments. The effect of temperature, feed concentration and adsorbent amine loading are investigated by performing the adsorption experiments in TGA and fixed bed using different SO_2, NO and NO_2 concentrations. In equilibrium measurements, the single-component adsorption of CO_2, SO_2, NO and NO_2 onto supported amine adsorbents is carried out by directly exposing the material to the various gases in carrier gas. The CO_2 capacity is measured before and after exposure to SO_2, NO or NO_2 to evaluate any CO_2 capacity loss due to irreversible binding of SO_2, NO or NO_2 to the supported amine adsorbents. Furthermore, to assess any synergies in co-adsorption of SOx or NOx and CO_2 gases, three gas mixtures comprising SO_2/CO_2, NO/CO_2 and NO_2/CO_2 are used in addition to single-component adsorption measurements. The experiments are performed by simultaneously exposing the adsorbents to CO_2 and SO_2 or CO_2 and NO/NO_2 and thereby their CO_2 uptake is measured at different temperatures.

(479g) Selective Dynamic CO2 Separations On Mg-MOF-74 Pellets At Low Pressures: A Detailed Comparison With 13X

Remy, T., Vrije Universiteit Brussel
Peter, S. A., Indian Institute of Technology Bombay
Van Der Perre, S., Vrije Universiteit Brussel
Baron, G., Vrije Universiteit Brussel
Denayer, J., Vrije Universiteit Brussel
Originally presented on: 11/6/2013 14:18:00 - 14:36:00

Selective separation of CO_2 from various gas streams is typically carried out via chemical absorption with amine solutions in the (petro-) chemical industry. Adsorption with porous solids such as metal-organic frameworks (MOFs) is an interesting alternative because this technique could be more energy efficient and thus cost reducing.

Different MOFs from the MOF-74 or MDOBDC family have large CO_2 capacities when exposed to pure CO_2 streams (M = Ni or Mg / DOBDC = 2,5 dioxido-1,4-benzenedicarboxylate). These MOFs contain open metal sites and experiments with small amounts of material indicate a large selectivity and capacity for CO_2 [1-2].

In this work, the potential of the promising Mg-MOF-74 material was studied in closer detail and compared to the benchmark zeolite 13X. Adsorption isotherms of CO_2, CH_4 and N_2 were recorded on MOF powder using the gravimetric technique at different temperatures (273 – 373 K). The uptake measurements indicate much lower affinity for CH_4 and N_2 and suggest a larger capacity and selectivity for the MOF in relevant low-pressure CO_2 separations.

To assess the dynamic separation potential we developed a new pellet preparation technique to ensure a minor capacity loss upon shaping of the powder. Breakthrough experiments with mixtures of CO_2 and CH_4 on a column packed with Mg-MOF-74 pellets at 308 K and 1 bar indicated a higher CO_2 capacity and separation efficiency for the MOF with respect to the benchmark 13X zeolite. The MOF regains 81% of its original capacity when purging with helium at 308 K during 10 minutes and can be fully regenerated at 353 K, which indicates the facile desorption of CO_2 on this material. However, the performance of Mg-MOF-74 severely deteriorates upon long-term exposure to relevant impurities in CO_2 separations at low pressures such as water and oxygen.

[1] P.D.C. Dietzel, V. Besikiotis, R. Blom, J. Mater. Chem., **19**, (2009).

[2] R. Krishna, J.M. van Baten, J. Membr. Sci., **377**, (2011).

(479h) Metal-Organic Framework With Optimal Adsorption Thermodynamics and Kinetics for CO2 Separations

Belmabkhout, Y., King Abdullah University of Science and Technology
Originally presented on: 11/6/2013 14:36:00 - 14:54:00

Metal-Organic Framework with Optimal Adsorption Thermodynamics and Kinetics for CO_2 Separations

Youssef Belmabkhout[1], Amy Cairns[1], Ryan Luebke[1], Patrick Nugent[2], Stephen D. Burd[2], Michael. J Zaworotko[2] and Mohamed Eddaoudi[1,2].

[1]KAUST Advanced Membranes & Porous Materials Center, Functional Material Design, Discovery and Development (FMD[3])King Abdullah University of Science and Technology
4700 King Abdullah University of Science and Technology Thuwal 23955-6900 (Kingdom of Saudi Arabia)

[2]Department of Chemistry, University of South Florida, 4202 East Fowler Avenue, Tampa, Florida 33620, USA.

**Corresponding author email: youssef.belmabkhout@kaust.edu.sa*

Metal Organic Frameworks MOFs (MOFs) is burgeoning class of materials are comprised of metals or metal clusters ("nodes") coordinated to multi-functional organic ligands ("linkers") and they offer myriad of surface areas up to 7000 m^2/g [1]. Furthermore, the modular nature of MOMs and their use of known coordination chemistry lead to enormous diversity of structures and properties [2]. Unfortunately, though extra-large surface area and pores facilitate high gravimetric uptake of gases at low temperature and/or high pressure, they are not conducive to highly efficient separations at the relatively low pressures necessary for the capture of CO_2 particularly for post-combustion. Unsaturated metal centers (UMCs) [3] or organic amines [2] that chemically interact with CO_2 can promote enhanced binding for CO_2 but there are drawbacks: high energy costs associated with activation, regeneration and recycling of the sorbent material, especially for amines [2]; competition with water vapor, especially for UMCs [4]; selectivity tends to monotonically decrease with increased loading of sorbate. Consequently, a sorbent with favorable CO_2 sorption kinetics and/or constant and moderate adsorption enthalpies over a wide range of CO_2 loading will permit efficient CO_2 capture from flue gas with low regeneration costs. We describe herein an approach to synergistically balance the kinetics and thermodynamics of CO_2 sorption selectivity through a crystal engineering and reticular chemistry strategy that affords control over pore functionality and size. The resulting MOFs are microporous sorbents with narrow one dimensional square channels lined by periodically arrayed SiF_6^{2-} (SIFSIX) anions. These MOFs exhibit exceptional CO_2 selectivity under conditions practically relevant to CO_2 separation and capture in the context of post-combustion (flue gas CO_2/N_2), pre-combustion (shifted synthesis gas stream CO_2/H_2) and natural gas upgrading (natural gas cleanup CO_2/CH_4) [5].

[1] Farha, O. K. *et al. J. Am. Chem. Soc.* 134, 15016-1502 (2012).

[2] Sumida, K. *et al. Chem. Rev.* 112, 724-781 (2012).

[3] Caskey, S. R., Wong Foy, A. G. & Matzger, A. J.. *J. Am. Chem. Soc.* 130, 10870-10871 (2008).

[4] Kizzie, A. C., Wong Foy, A. G. & Matzger, A. J. *Langmuir* 27, 6368-6373 (2011).

[5] Nugent, P., Belmabkhout, Y., Burd, S. et al. Nature (2013) 10.1038/nature11893.

(480a) From Fundamental Single Drop Analysis to the Description of Liquid/Liquid Dispersions - Part I: Coalescence

Kamp, J., TU Berlin
Villwock, J., TU Berlin
Kraume, M., Technische UniversitÃ¤t Berlin
Originally presented on: 11/6/2013 12:30:00 - 12:50:00

There are numerous applications in industry, where reactions and mass or energy transfer between two immiscible liquids are the major concerns. When trying to influence or predict the contact area one has to focus on the underlying principles that lead to the formation of the drops. The resulting drop size distributions (DSD) are thereby determined by drop breakage and coalescence. In order to predict the size distributions, both phenomena have to be understood and described properly. Although comprehensive scientific research identified a multitude of theoretical models describing both processes, the prediction of the DSD is only possible with restrictions when varying for example the power input, material and process parameters. Consequently, excessive and expensive experimental investigations are still necessary at different scales of process development. For example the design of extraction columns still requires pilot plants using high amounts of process liquids.

To reduce the amount of influencing factors of the whole process, it is necessary to reduce the problem to the fundamental behavior of single droplets. Therefore, the impact of influencing parameters for drop coalescence and breakup needs to be identified and quantified for each phenomenon separately. The gained knowledge of this microscopic behavior can be used to prove existing models and develop new ones. These models can either be semi-empirical or mechanistic. The obtained detailed models can be implemented into the framework of population balance equations (PBE), which includes separate kernels for drop breakage and coalescence. The PBE can be applied to describe the time dependent DSD in technical applications and therefore connects the microscopic to the macroscopic behavior of droplet swarms. The microscopically based modeling allows a determination of the model parameters using single drop experiments with only a small amount of liquid components in order to adapt the submodels to the present process.

This part is focused on the coalescence of two droplets and will discuss two characteristics of this phenomenon in detail: the influence of superimposed mass transfer on the coalescence probability and the coalescence inhibition due to electrostatic charge at high pH values.

The coalescence probability, which quantifies if two droplets coalesce or repulse each other after collision, is quantified systematically using single drop experiments. Therefore, a newly developed test cell is used where a rising drop collides with a pendant one recorded by a high speed camera. The drop generation, detachment, collision and the triggered recording of the process is fully automated. This allows carrying out serial examinations of droplet collisions under different system conditions (varying e.g. drop sizes, mass transfer direction and intensity). As the coalescence and contact time (and consequently the coalescence probability) of drops have a broad variance, a significant number of experiments has to be conducted to provide a statistically solid data base. Analyzing the recorded images of the coalescence process, important quantities like coalescence probability, contact and coalescence time, relative velocity, momentum and deformation are evaluated. Conducting these experiments, the EFCE recommended model system toluene / acetone / water is used.

Analyzing the influence of drop size ratio on coalescence probability shows the same tendency which commonly used models predict. If two drops with the same size collide, the coalescence probability is the lowest and rises with a shift in drop size ratio. Increasing the rising distance and therefore the relative velocity between the drops decreases the coalescence probability. This finding is contradictive to the so called energy models in literature, which assume a higher coalescence probability with higher kinetic energy (corresponding to a higher relative velocity). As mass transfer is inherent in extraction and is known to have a significant impact on drop coalescence, its influence on the coalescence probability is investigated systematically. Introducing the transferring component acetone into the system shows that the coalescence behavior is changed significantly over a broad range of driving concentration differences. A transfer direction from dispersed to continuous phase (d → c) leads to a decrease in the contact time of two drops until they coalesce. This results in a significant increase of coalescence probability to a complete coalescence of all drops independent from the drop size ratio. Reversing the mass transfer direction (c → d) generally decreases the coalescence probability, although this effect seems to be outweighed at distinct drop size ratios. Apparently, the gathered knowledge of this microscopic behavior offers a basis for model validation and future model development describing the influence of superimposed mass transfer.

A coalescence inhibition caused by electrostatic interactions was found in experimental investigations in stirred toluene/water systems at constant ionic strength of 0.1 mol/L with pH values higher than 11. Two designated models were used to simulate the transient DSD in a stirred tank, showing that the influence of droplet charge due to a change in pH value or ion concentration cannot be predicted satisfactorily by existing models. This finding motivated a new mechanistic modeling approach implementing the DLVO theory into the PBE framework. For that reason the coalescence efficiency is separated in a hydrodynamic and an electrostatic part. Being independent from each other, the hydrodynamic part of the efficiency can be described by common models from literature, whereas the developed electrostatic part is based on the ratio of the repulsing force due to the overlap of the electrical double layers of two colliding drops to the attractive van der Waals force between them. This model is able to describe the hindered coalescence at high pH values and the transient and steady-state Sauter mean diameters of the DSD fit well with the experimental values. Further investigations will focus on single drop experiments with a systematic variation of ionic strength, pH and ionic species and zeta potential measurements. These examinations will build a basis for further model development.

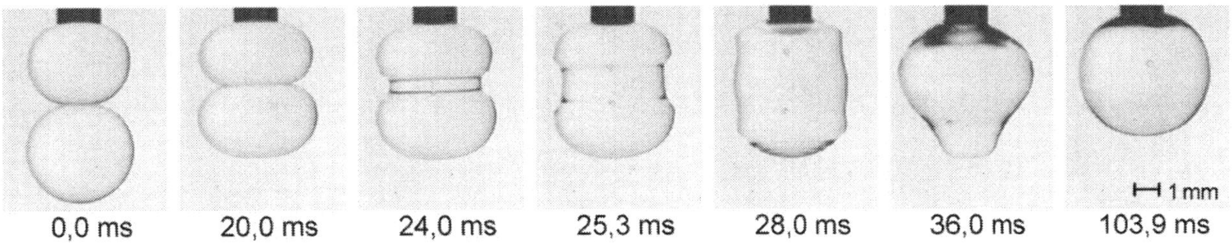

Fig. 1: Coalescence of two droplets

Partially founded by DFG project ?Coalescence efficiency in binary systems? and DECHEMA ?Max-Buchner-Forschungsstiftung?

(480b) From Fundamental Single Drop Analysis to the Description of Liquid/Liquid Dispersions-Part II: Breakage

Nachtigall, S., Technische UniversitÃ¤t Berlin
Kraume, M., Technische UniversitÃ¤t Berlin
Originally presented on: 11/6/2013 12:50:00 - 13:10:00

There are numerous applications in industry, where reactions and mass or energy transfer between two immiscible liquids are the major concerns. When trying to influence or predict the contact area one has to focus on the underlying principles that lead to the formation of the drops. The resulting drop size distributions (DSD) are thereby determined by drop breakage and coalescence. In order to predict the size distributions, both phenomena have to be understood and described properly. Although comprehensive scientific research identified a multitude of theoretical models describing both processes, the prediction of the DSD is only possible with restrictions when varying for example the power input, material and process parameters. Consequently, excessive and expensive experimental investigations are still necessary at different scales of process development. For example the design of extraction columns still requires pilot plants using high amounts of process liquids.

To reduce the amount of influencing factors of the whole process, it is necessary to reduce the problem to the fundamental behavior of single droplets. Therefore, the impact of influencing parameters for drop coalescence and breakup needs to be identified and quantified for each phenomenon separately. The gained knowledge of this microscopic behavior can be used to prove existing models and develop new ones. These models can either be semi-empirical or mechanistic. The obtained detailed models can be implemented into the framework of population balance equations (PBE), which includes separate kernels for drop breakage and coalescence. The PBE can be applied to describe the time dependent DSD in technical applications and therefore connects the microscopic to the macroscopic behavior of droplet swarms. The microscopically based modeling allows a determination of the model parameters using single drop experiments with only a small amount of liquid components in order to adapt the submodels to the present process.

This part focuses on the drop breakage phenomenon in turbulent flows. It is described in literature, that drop breakage occurs when the external forces exerted on particles by the continuous phase exceed the internal forces which hold the particle together. The Weber number describes the relation between both stresses. Thereby only the interfacial tension is considered for the stabilization of the drops. The stabilizing influence of the viscosity is described by the Ohnesorge number which sets the effect of the viscosity in relation to the effect of the interfacial tension. Applying Kolmogorovs theory of isotropic turbulence, correlations between the mean diameter of a turbulent liquid/liquid droplet swarm and the above mentioned dimensionless numbers can be derived.

In the present studies single drop breakage events, where coalescence can be neglected, are analyzed in a turbulent flow field comparable to the field of a stirred tank. A single stirrer blade, mimicking a part of a Rushton turbine is therefore mounted in a rectangular channel and induces drop breakage. Drop deformation and breakage is analyzed by high-speed imaging. Image data is evaluated by fully automated image analysis. For statistically relevant results, at least 1000 drops entering the vicinity of the stirrer blade are analyzed. The breakage probability and the breakage and oscillation times are determined. Additionally the resulting Sauter mean diameter and the number and the size distribution of the daughter drops are evaluated.

A variety of different droplet substances were investigated at constant droplet diameter of 1 mm. Thus, the physical properties (density, viscosity) of the dispersed phase and the interfacial tension between the two phases were varied over a wide range. Additionally, surfactants were added to increase the range of the investigated interfacial tensions. Consequently, the role of the stabilizing forces, which determine the Ohnesorge numbers of the investigated systems, can be quantified.

Furthermore the fluid flow velocity of the continuous water phase was varied (v=1.0 - 2.0 m/s) to also determine the influence of the deforming external forces and therefore the Weber numbers. These experiments allow proving if the drop breakage phenomenon can be described with the above depicted concept using two dimensionless numbers.

The influence of the dispersed phase properties on drop deformation and breakage can be described well with the Weber and the Ohnesorge number. For systems with low Ohnesorge numbers, the breakage probability and the Sauter mean diameter are only a function of the Weber number. The stabilizing influence of the viscosity has to be considered to describe these factors.

The breakage time is set to be the time between the beginning deformation of a droplet and the final splitting in two or more parts. To characterize the deformation, the two major axis of the particle were set into relation. In accordance with literature, droplets with deformation factors smaller than 0.9 were set to be deformed, while the droplets with greater values were defined to be spherical. For very low Ohnesorge numbers, the breakage time only depends on the interfacial tension. For higher Ohnesorge numbers, the breakage time is determined by the ratio of viscosity to interfacial tension which results in a proportional dependency thereof.

When combining both dimensionless numbers, the breakage times, probabilities and Sauter mean diameters of all the investigated systems with different dispersed phases can be described. The PBE breakage kernel includes the breakage time, probability and size distributions of the resulting daughter drops. Therefore, the gained detailed results from the single drop breakage investigations can be used to validate and improve existing mechanistic models from literature.

Partially founded by DFG project "Coalescence efficiency in binary systems" and DECHEMA "Max-Buchner-Forschungsstiftung"

(480c) Phase-Inversion and Coalescence in Turbulent Liquid/Liquid-Dispersions

Maab, S., SOPAT GmbH
Russo, M.,
Kraume, M., Technische Universitat Berlin
Originally presented on: 11/6/2013 13:10:00 - 13:30:00

Understanding the dispersions of immiscible liquids is still very challenging. A detailed understanding of the opposed phenomena of drop breakage and coalescence is necessary for a successful mathematical modelling. A further difficulty handling such dispersed systems is the possible danger of a phase inversion (PI). Does this phenomenon occur spontaneously during an industrial process, it can have fatal consequences for product and equipment [1]. Therefore also a deeper understanding and modelling of this phenomenon is desirable.

A turbulent liquid/liquid system was investigated. The operation parameters dispersed phase fraction and surfactant concentration have been varied systematically to reliably achieve PI. The transient drop sizes as the instable behaviour of the water and organic phase have been observed using an in-situ microscope [2]. The drop size was measured using a fully automated image analysis tool [3]. Furthermore, the simulation of the drop size was carried out using population balance equation (PBE).

By using the named measurement technique, it was possible to detect the starting point of the PI. Furthermore the occurrence of double emulsions was observed during the procedure. The drop size was fluctuating. The change of the droplet sizes with variation of the operating parameters mentioned was successfully simulated with the help of the PBE. A simulation of the occurrence of PI using PBE was not predictable with the used model.

[1] Nienow, A. W., *Adv. Colloid Interface Sci.* 2004, *108*, 95-103.

[2] Maaß, S., Wollny, S., Voigt, A., Kraume, M., *Exp. Fluids* 2011, *50*, 259-269.

[3] Maaß, S., Rojahn, J., Hänsch, R., Kraume, M., *Comput. Chem. Eng.* 2012, 45: 27-37.

(480d) Progress in RDC Design

Aksamija, E., Graz University of Technology
Siebenhofer, M., Graz University of Technology
Originally presented on: 11/6/2013 13:30:00 - 13:50:00

Taylor-Couette vortices are formed in the gap between concentrically rotating cylinders. They have been subject of investigations for decades and some attempts have been done for implementation in liquid-liquid extraction.

Problems with vortex instability at high axial flow rates, high angular velocity differences, high diameter differences of the cylinders and the presence of a second phase made this hydrodynamic phenomenon appear unsuitable for liquid-liquid extraction.

Experimentally validated CFD-Simulations of phase flow and flow patterns in the *Rotating Disc Contactor* (RDC) led to the suggestion that the characteristics of *Taylor-Couette* flow pattern can seemingly be obtained in a simplified RDC design without stator discs and appropriate rotor disc diameter. As long as the compartment height is large enough to accommodate two toroidal eddies, vortices are dynamically stabilized by the rotor discs, allowing smaller shaft diameters and larger active space in the apparatus. CFD-Simulations show less axial mixing for both phases compared to a state of the art RDC design with stator discs. One the one hand compartments are separated more efficiently by the enlarged rotor discs, on the other hand higher energy dissipation rates caused by the enlarged rotor discs lead to narrower drop size distributions and more uniform axial velocity of the dispersed phase. The geometrical stabilization of the toroidal flow pattern with the rotor discs also induces a straight path of the dispersed phase, minimizing the tendency for vortex inversion, which can be observed in state of the art RDC design.

Mass transfer experiments were performed in a pilot scale RDC100 to quantify optimization in terms of separation efficiency. The height of one transfer unit (HTU) was found to be 40% lower for the simplified internals geometry.

Beside lower investment costs and increased mass transfer efficiency several operation advantages in cleaning, maintenance and assembly may arise from statorless design of RDC extraction columns.

(480e) CFD Analysis of Flow of Two Immiscible Fluids in a Disk and Doughnut Pulsed Column

Saini, R.,
Bose, M., Indian Institute of Technology Bombay
Originally presented on: 11/6/2013 13:50:00 - 14:10:00

Solvent extraction is one of the key unit operations in the process industry. Among various
types of solvent extraction units, the pulsed columns are emerging as one of the best choices
because of their much smaller footprints compared to the most commonly used mixer-settler
type extraction units [1]. Optimal design of a pulsed column requires a thorough understanding of the multi-phase flow dynamics as the mass transfer efficiency directly depends on the inter-facial area which in turn, is influenced by the extent of the turbulent mixing in the unit.

The objective of the present work is to investigate the pulsatile flow of single and two immiscible liquids in a disk and doughnut pulsed column (DDPC) using the techniques of
computational fluid dynamics (CFD), in order to understand the effect of the geometry and
the operation conditions on the distribution of the drop size of the dispersed phase. To that
end, model geometry is selected based on the dimensional analysis of the large scale columns. A DDPC can be described as a circular column with an internal fitting that has several units stacked vertically in a repeated manner. Each unit comprises of a disk placed
between two ring shaped baffles (the doughnuts). The height between the ring and the disk
remains uniform throughout the column. All the geometric parameters such as the aperture of
the ring (d_r), the diameter of the disk (d_D), the height between the ring and the disk (h), are
non-dimensionalised by the diameter of the column.

These ratios (d_r/d_c, d_D/d_c, h/d_c) are modified to investigate the influence of geometry on the mixing. The effect of turbulence in the continuous phase is modelled using low Reynolds number k − ε model. Simulation results are validated against the experimental work [2]. The results obtained from simulation of the scaled down DDPC is in very good agreement with the experimental observation.

We have also determined the spatial distribution of the turbulent kinetic energy to estimate the maximum size of the droplet formed [3]. The estimated droplet size is used in the simulations of two immiscible fluids in a pulsed column using Eulerian-Eulerian approach, where, both the phases are considered to be interpenetrating continua. Effect of the column geometry, frequency and amplitude of pulsation and the location of the feeding for the secondary phase on the dispersion will be discussed in the presentation.

References:

1. Angelov, G., Gourdon, C., 2012. Pressure drop in pulsed extraction columns with internals of discs and doughnuts. Chemical Engineering Research and Design, 90 (7), 877-883.

2. Bujalski, J., Yang, W., Nikolov, J., Solnordal, C., Schwarz, M., 2006. Measurement and cfd simulation of single-phase flow in solvent extraction pulsed column. Chemical Engineering Science 61 (9), 2930 -2938, fluid Mixing VIII International Conference.

3. Angelov, G., Gourdon, C., 2009. Turbulent flow in pulsed extraction columns with internals of discs and rings: Turbulent kinetic energy and its dissipation rate during the pulsation. Chemical Engineering and Processing: Process Intensification 48 (2), 592 -599.

(480f) High Efficiency Liquid Liquid Extraction By Emulsion Formation and Separation in Robust Milli-Fluidic Devices

Jablonski, E. L., Bucknell University
Originally presented on: 11/6/2013 14:10:00 - 14:30:00

An improved method for accomplishing liquid-liquid extraction in milli-fluidic devices is presented. New devices have been fabricated that can process large emulsion volumes in parallel while providing increased contact time and driving force for extraction. Current devices can process several mL/min and have greater chemical resistance than prior designs. The fabrication technique has been modified to allow for improved manufacturability and greater yield of working devices with long operating lifetimes. Device geometry can easily be changed by laser cutting custom parts while retaining the same substrate. With the new configuration many devices can operate in parallel to contact the aqueous/organic phases through emulsion formation, allow optimal residence time controlled by channel/tubing length and flow rate, then extraction in a quasi-countercurrent system that first breaks the emulsion and subsequently utilizes stacks of alternating devices.

Previously, the continuous extraction of rhodamine 6G into water from solution in pentanone was achieved in both milli- and microfluidic devices. Continuous extraction was possible through cocurrent laminar flow of the organic and aqueous streams stabilized by a hydrogel defining one wall of the channel. Water flow rates of 10 ml/h to 40 ml/h were used with a pentanone flow rate of 40 ml/h. The mass transfer coefficient for this system was found to be $(1.7 \pm 0.8) \times 10\text{-}3$ cm/s.

Milli-fluidic LLE with cocurrent laminar flow of immiscible liquids provides a suitable geometry to accurately determine mass transfer coefficients due to the well defined interface between the phases. After accurate determination of the mass transfer coefficient for the system was achieved, droplet-based extraction was performed and a milli-fluidic method capable of continuous passive separation of the emulsions with an efficiency of ~90 % was utilized to recover the aqueous phase. The overall scheme is an efficient method for liquid liquid extraction. To achieve continuous passive separation, a device with opposing channel walls of disparate hydrophobicity is used to stabilize co-laminar flow of the solvent and water. The disparity in hydrophobicity of the channel walls is accomplished by defining one length of the channel with a hydrogel, in this case polyethylene glycol. Emulsion separation is facilitated by introducing the emulsion at the water/hydrogel interface. Advantages of performing separations at the millli-fluidic scale are presented. The scale up of this continuous, passive emulsion separation method to several mL/min was achieved.

To demonstrate the utility and robustness of the emulsion separation scheme, experiments used high oleic sunflower oil or mineral oil as the continuous phase and aqueous solutions of methylene blue, crystal violet, or dextran with methylene blue as the dispersed phase. The water-in-oil emulsion was created within a T-junction device and the emulsion was transported to a milli-fluidic device where passive emulsion separation was achieved by the co-current laminar flow of oil and water. Co-current laminar flow was stabilized by defining one channel length with a hydrogel layer. Introduction of the emulsion at the hydrogel–water interface induced emulsion break up. The maximum emulsion separation throughput was defined by a decrease in emulsion separation efficiency to below 80%, a loss of stable parallel flow, or the creation of an emulsion. A maximum emulsion separation throughput of 1.4 metric tons per year was achieved.

(480g) Liquid-Liquid Extraction: Comparison in Micro and Macro Systems

Talapuru, S., UNIVERSITY COLLEGE OF TECHNOLOGY, OSMANIA UNIVERSITY
S, L.,
Rao, U., IICT HYDERABAD
Originally presented on: 11/6/2013 14:30:00 - 14:50:00

Liquid – liquid extraction operation in micro channels and capillaries received attention with a focus on intensification on mass transfer. Yuechao Zhao et.[1] reported mass transfer characteristics of water, succinic acid and n-butanol in case of counter and cross-flow T-junction micro-channels and proposed correlations to predict the volumetric mass transfer coefficients, which are two or three orders of magnitude higher than those of conventional liquid-liquid contactors. Kashid et al [2] have investigated the mass transfer coefficient in liquid-liquid slug flow through a capillary micro-reactor for three different non-reacting systems. They have observed enhanced rates of mass transfer in micro-reactors as compared to the conventional reactors reported in literature. Yoshihito Okubo et al [3] have investigated three extraction operations, contact flow in a Y-shaped micro channel, segmented flow and emulsification. They noted that extraction using segmented flow is dependent on mass transfer by molecular diffusion and its rate is enhanced by the internal circulation flow generated inside slugs and extraction rate can be controlled precisely by adjusting operational parameters such as the flow rate and the flow rate ratio. Nobuaki Aoki et al.[4] have studied that to increase the maximum flow rate that enables slug flow formation in miniaturized channels, gas phase slugs were added in a liquid- liquid slug flow to form a gas-liquid –liquid slug flow. Effects of channel size, void fraction (ratio of volumetric flow rate or gas phase to that of total flow rate), and volume ratio of aqueous phase to organic phase on flow regime were examined. Usha Rao [5] have studied the flow pattern mapping and also dimensionless analysis of two phase liquid-liquid systems in glass micro channels of sizes varying from 500 pm to 1000um (ID) and varying shapes such as straight and sinusoidal. Thee liquid-liquid systems were studied viz., (i) water methyl ethyl ketone (ii) palm oil-methanol (iii) ethyl acetate-water, each having different properties. These fluids were fed into the micro channel through different junctions, Y, T and L at controlled flow rates ranging from 0.02-0.08 ml/min and slug flow pattern was seen for all the three systems irrespective of diameters and properties of the systems and concluded that slug flow pattern is good for operations in micro channels, as it gives good mixing and good contact between phases.

In the present study liquid-liquid operation was investigated in micro channels of different diameters and compared the results with that obtained in packed column using Raschig rings and structured sulzer packing. Water-acetic acid-ethyl acetate was the system chosen for the purpose. Ethyl acetate was used as solvent and dispersed into the aqueous phase containing acetic acid. Micro channels of different diameters namely, 0.7, 0.9 and 1.8 mm were used for extraction. The liquids were drawn using peristaltic pumps from reservoirs consisting of feed and solvent. The channel volume was kept constant (0.1708 ml) by decreasing the diameter of the channels and increasing the length. The acetic acid volume was 20 percent in water. Runs were made with flow rates such as 0.04 to 0.08 ml/min and operated with solvent to feed ratio, S/F = 1.

The continuous counter current extraction glass column with an internal diameter of 45mm and a height of 500mm had a packed length of 200mm and disengaging spaces at the top and at the bottom. Two dosing pumps were used to pump the solvent and the aqueous phase. Concentrations of acetic acid in water was 20 volume percent. Solvent flow rate was varied from 4 to 16 ml/min which gives S/F of 1 to 4.

Volumetric mass transfer coefficient, $k_L a$ and the number of transfer units, N_{tL} were found to increase with S/F ratio in the packed columns. $k_L a$, N_{tL} and percentage extraction were observed to be higher for structured sulzer packing. Slug flow or segmented flow is most promising flow patterns for good mass transfer in micro channels. Slug flow, which occurs at a relatively low flow rates, is characterized by elongated bubble flowing in the axial direction. For a fixed residence time $k_L a$ and percentage extraction were observed increase with decrease in diameter for S/F = 1. This is due to increase in surface to volume ratio which reduces the mean distance of the specific liquid volume to the channel walls or to the domain of the second liquid. As a consequence mass transfer performance with the second fluid is enhanced. The values of $k_L a$ were of the order of 3.56×10^{-4} in the packed columns and 8.98×10^{-3}in the micro channels. In micro channels $k_L a$ is almost 25 times than that of packed columns with S/F =1. The percentage extraction obtained at an S/F = 1 in micro channels, was almost same with S/F = 4 in the packed columns.

When different phases are injected as adjacent streams in one channel, one phase often preferentially wets the boundaries and encapsulates the second fluid as discrete drops. However, the flow can also become stable by the generation of a clear interface. With downscaling, the gravitational force becomes less important [6]. Multiphase-fluid dynamical responses are commonly successfully characterized by dimensionless numbers, Reynolds number, Re, Capillary number, Ca and Weber number, We. In micro channels Re obtained were less than 10 and hence laminar profile can be observed. The results indicate the influence of interfacial tension, characterstic of We and is found to be more than the inertial and viscous forces.

Flow patterns by micro Particle Image Velocitymetry indicated internal circulation of liquid in the organic slug caused due to two liquid flows. The diffusion of solute from aqueous to organic slug provides better mixing.

References

1. Yuchao Zhao, Guangwen Chen and Quan Yuan, "Liquid-Liquid Two-Phase Mass Transfer in the T-Junction Microchannels", AICHE.J.53.12.3042-3053,2007.
2. Kashid M.N., A.Renken, L.Kiwi-Minsker, "Gas-liquid and liquid-liquid mass transfer in microstructured reactors", Chem.Eng. Sci. ,66,3876-389,2011.
3. Yashihito Okubo, Taisuke Maki,Nobuaki Aoki, Teng Hong Khoo, Yoshikage Ohmukai,Kazuhiro Mae," Liquid-liquid extraction synthesis and separation by utilizing micro spaces". Chem. Eng. Sci., 63,4070-4077,2008.
4. Nobuak Aoki, Ryuichi Audo and Kazuhiro Mac, "Gas liquid –liquid slag Flow for improving Liquid-Liquid Extraction in Miniatured Channels", Ind.Eng.Chem.Res., 50, 4072-4677,2011.
5. Usha Rao H.,"Study of microscale fluid dynamics an input for micro reactor modeling and design "Ph.D., Thesis submitted to Osmania Universwity,2012.
6. Yuanhai Su, Guangwen Chen, Yuchao Zhao and Quan Yuan, "Intesfication of Liquid-Liquid Two Phase Mass Transfer by Gas Agitation in a Microchannel" AICHE J.,55, 8, 1948-1958,2009.

(484a) Selective Adsorption of CO2 On Flexible Ni/Dbm-Bpy MOF

Matranga, C., National Energy Technology Laboratory
Culp, J., National Energy Technology Laboratory
Shi, F., NETL
Originally presented on: 11/6/2013 12:30:00 - 12:50:00

Selective Adsorption of CO_2 on Flexible NiDBM-BPY MOF

Fan Shi[1,2,*], Jeffrey Culp[1,2] and Christopher Matranga[1],

(1)U.S. Department of Energy, National Energy Technology Laboratory, Pittsburgh, PA, USA (2)URS Corporation, South Park, PA, USA

* fan.shi@netl.doe.gov

To maintain fossil fuels as viable energy source, carbon capture and storage technology is essential. Recent novel sorbents for CO_2 capture [1-3], including Metal Organic Frameworks (MOFs), have attracted a lot of interests. Flexible MOFs have one attractive feature that their structure become dynamic only in the presence of suitable guests such as CO_2, CH_4, N_2, and H_2. These flexible materials can show interesting host-guest phenomena including crystal-crystal transformations and gated adsorption behaviors due to the structural transitions which occur between the guest free and guest-loaded states.[4-6] Such dynamic host-guest behavior is particularly attractive for applications involving sorbent-based gas separations and selective adsorption such as CO_2 and CH_4 separations.[7-9] In order to develop materials for these applications, a detailed understanding of the adsorption mechanisms of these flexible hosts is needed.

The material in this study is the linear chain coordination polymer catena-bis (dibenzoylmethanato)-(4,4′-bipyridyl)nickel(II), or referred as NiDBM-BPY. This neutral one-dimensional polymer was propagated by a bridging 4,4′-bipyridine ligand bound to the z-axes of the nickel complexes which were bischelated in a planar fashion by two anionic DBM ligands. CO_2 breakthrough from binary mixtures, including CO_2/N_2, CO_2/H_2, and CO_2/CH_4, was performed in a fix-bed reactor with ID of 0.375? and a length of 12?. The average particle size of NiDBM-Bpy was around 100 mm to minimize external mass transfer limitations. The gas composition of effluent was analyzed by a *Mass Spectrometry* (*Pfeiffer Vacuum*). The breakthrough curve of CO_2 from a fixed-bed reactor packed with NiDBM-Bpy showed a step-shape curve, which also indicated the structure expansion above the structure transition threshold pressure. The selectivity for CO_2

over N_2, H_2, and CH_4 was calculated and results confirmed the sorbent's affiliation to CO_2 than other gases.

References

1. Kauffman et. al. *Angew. Chem. Int. Ed.* **2011**, *50, 10888 ?10892*
2. Millward, A. R.; Yaghi, O. M. *J. Am. Chem. Soc.* **2005**, *127*, 17998.
3. Wang, B.; Cote, A. P.; Furukawa, H.; O'Keeffe, M.; Yaghi, O. M. *Nature* **2008**, *453*, 207.
4. Culp, J. T.; Smith, M. R.; Bittner, E.; Bockrath, B. J. Am. Chem. Soc. **2008**, 130, 12427.
5. Takamizawa, S.; Nakata, E.; Yokoyama, H. Inorg. Chem. Commun. **2003**, 6, 763.
6. Uemura, K.; Kitagawa, S.; Fukui, K.; Saito, K. J. Am. Chem. Soc. **2004**, 126, 3817.
7. Maji, T. K.; Matsuda, R.; Kitagawa, S. Nat. Mater. **2007**, 6, 142.
8. Shimomura, S.; Horike, S.; Matsuda, R.; Kitagawa, S. J. Am. Chem.Soc. **2007**, 129, 10990.
9. Snurr, R. Q.; Hupp, J. T.; Nguyen, S. T. AIChE J. **2004**, 50, 1090.

(484b) Amine Impregnated N-Doped Mesoporous Carbon for Selective CO2 Capture

To, J., Stanford University
He, J., Stanford University
Gary, B., Stanford University
Stack, D., Stanford University
Wilcox, J., Stanford University
Bao, Z., Stanford University
Originally presented on: 11/6/2013 12:50:00 - 13:10:00

Carbon dioxide is a major component of the greenhouse gases. It is expected that the atmospheric CO_2 level will continue to increase over the current concentration of approximately 390 ppm in the near future, as fossil fuels remain the major source utilized to meet global energy demand. Carbon capture and storage is a technology developed to prevent release of large quantities of CO_2 into the atmosphere from power generation by capturing CO_2 and securely storing it away from the atmosphere. However, current state-of-the-art technology for CO_2capture is to use liquid phase amine scrubbing, which is corrosive and energy intensive due to high heating capacity of water during solvent regeneration. On the other hand, solid sorbent material adsorbs CO2 via physisorption, which requires less energy for regeneration. Therefore, porous carbon with high thermal conductivity is desirable for such application. With improved kinetics in adsorption and desorption processes, separation column required is smaller and the overall cost of the power plant is lower.

Mesoporous carbons are promising for CO_2 capture due to chemical inertness, low cost, high surface area and tunable pore structures. Its porous structure allows addition of chemical functionality by grafting or impregnation. Amine chemistry tells us that nitrogen functionalization plays an important role in surface chemistry to achieve high CO_2 adsorption capacity. We report here an ordered mesoporous carbon made using co-assembly of modified-pyrrole and triblock copolymer. To increase the nitrogen content of the adsorbent, nitrogen is incorporated into porous carbon framework and further functionalized with amines. Another potential benefit is that the thermal conductivity of mesoporous carbon is much higher than its silica counterpart, resulting in a faster regeneration step with enhanced stability using temperature swing process. These overall properties of mesoporous carbon made from conducting polymer made it a desirable material for CO2 capture.

(484c) Effect of Alumina Doping On Structure and Stability of CaO Based Sorbents for High Temperature CO2 Capture

Gunugunuri, K. R., University of Cincinnati
Qullin, S., University of Cincinnati
Smirniotis, P., University of Cincinnati
Originally presented on: 11/6/2013 13:10:00 - 13:30:00

In an age where environmental concerns have become increasingly prevalent, efficient and carbon neutral methods for satisfying the global energy demand have become necessities. Increasing levels of greenhouse gases are widely regarded as a key factor in the shifts observed in global climate patterns. In this study, we reported that Flame synthesized doped Calcium-based novel refractory sorbents have shown excellent performance in CO_2 adsorption. In our previous studies we incorporated several metals (Hf, W, Y, La, and Al) into calcium oxide lattice to improve stability of the sorbent during the CO_2 adsorption experiments. Among these dopants, Alumina doped CaO with molar ratio of 10:3 showed very high adsorption stability during 100 cycles of operation and better adsorption capacity than Ca/Zr sorbents (our previous best system). We reported that the formation of perovskite type of structure in CaO-Al sorbent is responsible for the observed stability. The present study mainly focuses on the Ca-Al system and obtains the optimum combination during extended carbonation-decarbonation operation cycles in the presence of highly concentrated CO_2. Various molar ratios of Ca: Al were synthesized using flame spray pyrolysis technique. The sorbents were characterized by BET, XRD, SEM and TEM techniques and analyzed using TGA. The optimal sorbent were tested under severe condition of high temperature. These interesting findings will be discussed in the presentation.

(484d) Enhanced Performance of Polymeric Amines Supported On Heteroatom-Incorporated Silica in CO2 Adsorption/Desorption

Kuwahara, Y., Georgia Institute of Technology
Copeland, J. R., Georgia Institute of Technology
Sievers, C., Georgia Institute of Technology
Kamegawa, T., Osaka University
Yamashita, H.,
Jones, C. W., Georgia Institute of Technology
Originally presented on: 11/6/2013 13:30:00 - 13:50:00

Silica-aminopolymer composites are promising materials that can be used to adsorb CO_2 from dilute gases, with the most commonly studied compositions being simulated flue gas from power plants (ca. 5-15% CO_2) and ambient air (ca. 400 ppm CO_2). The CO_2 adsorption characteristics of prototypical, class 1 [1] poly(ethyleneimine)-silica composite adsorbents are shown here to be significantly enhanced by altering the acid/base properties of the silica support via heteroatom incorporation into the silica matrix. [2] To demonstrate this, a series of poly(ethyleneimine)-impregnated mesoporous silica SBA-15 materials containing heteroatoms (Al, Ti, Zr, and Ce) in the silica matrix are prepared and characterized in adsorption experiments under conditions simulating flue gas (10% CO_2 in Ar) and ambient air (400 ppm CO_2 in Ar). The structure of the composite adsorbents, including local information concerning the state of the incorporated heteroatoms and the overall surface properties of the silicate supports, are investigated to develop a relationship between the adsorbent structure and CO_2 adsorption/desorption performance. The CO_2 adsorption/desorption kinetics are assessed by thermogravimetric analysis and *in-situ* FT-IR measurements. These combined results, coupled with data on adsorbent regenerability, demonstrate a stabilizing effect of the heteroatoms on the poly(ethyleneimine) contained within the support mesopores, enhancing adsorbent capacity, adsorption kinetics, regenerability, and stability of the supported aminopolymers over continued cycling. [3] It is suggested that the CO_2 adsorption performance of silica-aminopolymer composites may be further enhanced in future work by more precisely tuning the acid/base properties of the support, an approach that has not yet been widely exploited in the literature.

References

[1] P. Bollini, S. A. Didas, C. W. Jones, J. Mater. Chem. 2011, 21, 15100.

[2] Y. Kuwahara, D.-Y. Kang, J. R. Copeland, N. A. Brunelli, S. A. Didas, P. Bollini, C. Sievers, T. Kamegawa, H. Yamashita, C. W. Jones, J. Am. Chem. Soc. 2012, 134, 10757.

[3] Y. Kuwahara, D.-Y. Kang, J. R. Copeland, N. A. Brunelli, S. A. Didas, P. Bollini, C. Sievers, T. Kamegawa, H. Yamashita, C. W. Jones, Chem. Eur. J. 2012, 18, 16649.

(484e) Hydrophobic Polymeric Solvents for the Selective Absorption of CO2 From Warm Gas Streams That Also Contain H2 and H2O

Enick, R. M., University of Pittsburgh
Luebke, D., National Energy Technology Laboratory
Nulwala, H., National Energy Technology Laboratory
Warman, S., University of Pittsburgh
Stevenson, C., University of Pittsburgh
Koronaios, P.,
Originally presented on: 11/6/2013 13:50:00 - 14:10:00

Hydrophobic Polymeric Solvents for the Selective Absorption of CO_2 from Warm Gas Streams that also Contain H_2 and H_2O

Robert Enick, David Luebke, Hunaid Nulwala, Samantha Warman, Craig Stevenson, Peter Koronaios

Hydrophobic polymers may provide an alternative to physical solvents based on the hydrophilic solvent polyethyleneglycol dimethylether (PEGDME) for the pre-combustion capture of CO_2 from a warm, high pressure, post-WGS reactor gas stream that also contains H_2O and H_2. The candidate hydrophobic polymers are designed to absorb as much CO_2 as possible, and as little H_2 and water as possible. There are three promising candidates: polydimethyl siloxane (PDMS or silicone oil), PDMS-PEGDME hydrid solvent developed by our team, and a commercial PDMS/Fe compound with the addition of iron as an anti-oxidant to enhance its high-temperature performance. PDMS is commercially available over an extremely wide MW range (up to 500,000+). The PDMS-PEGDME hybrid solvent is also easy to synthesize from commercially available starting materials. Two silicone oils containing both methyl and phenyl groups were also tested and found to be much less promising candidates.

In this work, the solubility of hydrogen, CO_2 and water in PEGDME, PDMS, PDMS-PEGDME and PDMS/Fe is presented for pressures up to ~2000 psi and temperatures to 120°C. The results indicate that PDMS is completely hydrophobic over broad ranges of temperature and pressure, and PEGDME is completely miscible with water in all proportions. PDMS-PEGDME and PDMS/Fe are hydrophobic at ambient temperature but the hydrophobicity of both have not yet been measured over a broad temperature range. These three hydrophobic solvents absorb comparable amounts of CO_2 (on a wt% basis). The solubility of H_2 in the solvents at ~25°C is roughly three orders of magnitude less than the solubility of CO_2, with PEGDME absorbing the least H_2 and PDMS absorbing the most H_2 (about 2.5 times more than PPGDME). Unfortunately, because the solubility of CO_2 decreases with increasing temperature as the solubility of H_2 increases, the CO_2-selectivity of the solvents (as estimated by the ratio of CO_2 solubility:H_2 solubility at the same conditions) decreases with increasing temperature.

(484f) Synthesis and Characteristics of Amine Functionalized Biopolymers for CO2 Capture

Fujiki, J., Research Institute of Innovative Technology for the Earth
Yogo, K., Research Institute of Innovative Technology for the Earth
Originally presented on: 11/6/2013 14:10:00 - 14:30:00

The increase in the atmospheric concentration of CO_2 is considered as one of the major causes of global warming. Separation and recovery of CO_2 from large CO_2 emission point source have been investigated. Recently, there are many reports on amine loaded mesoporous silicas having high CO_2 adsorption capacity. However, toxic and expensive chemicals are used to synthesize mesoporous silicas. Additionally, most reports on amine loaded mesoporous silicas discuss the CO_2 adsorption characteristics for "powder" form mesoporous silicas. Mesoporous silicas seem to be less reliable now because the adsorbent must be in the form of bead, granule or pellet to use in actual CO_2recovery processes. Therefore, alternative adsorbent "beads" made from low-cost and eco-friendly chemicals should be developed.

In this study, highly porous amine functionalized biopolymer beads were synthesized and CO_2 adsorption performance was evaluated. As a result, it was found that the polyethyleneimine functionalized beads had high CO_2 adsorption capacity with high adsorption rate. The CO_2 adsorption capacity reached 2.3 mmol/g in the absence of water vapor and 3.6 mmol/g in the presence of water vapor at 313 K and 15 kPa of CO_2.

In general, amine impregnated materials show high CO_2 adsorption capacities but lower adosorption rate. On the other hand, amine grafted materials have the advantage of physical and chemical stability of loaded amines, high amine efficiency and adsorption-desorption rate. The adsorption characteristics of our material show advantages of both amine impregnated and grafted materials, i.e., relatively high adsorption capacity and fast adsorption rate. The results suggest that this material has potential for effective CO_2 adsorbent.

(484g) Removal of Sulfur Compounds in Methane Or Nitrogen By Nano-Sized Magnesium Oxide and Its Composite

Lee, C. H., Yonsei University
Kim, Y. H., Yonsei University
Kim, J. J.,
Tuan, V. A.,
Originally presented on: 11/6/2013 14:30:00 - 14:50:00

Removal of Sulfur Compounds in Methane or Nitrogen by Nano-sized Magnesium Oxide and Its Composite

Yo Han Kim, Vu Anh Tuan, Jae Jeong Kim, Chang Ha Lee*

Department of Chemical and Biomolecular Engineering, Yonsei University, Seoul, Korea

*E-mail: leech@yonsei.ac.kr, Tel: +82-2-2123-2762

Metal oxide particles are considered as reactive adsorbents for the decontamination of toxic materials. The important feature is to enhance the surface chemical reactivity toward incoming adsorbates. It is expected that the adsorption efficiency of nano-sized MgO is greater than commercially common used MgO.

Nano-sized magnesium oxides (MgOs) and MgO-SiO_2 composite were synthesized by polyol-meditation thermolysis, hydrothermal, and aerogel methods to remove sulfur compounds from methyl mercaptan/methane and CEES/nitrogen mixtures. And the difference in sorption capacity among the synthesized MgOs and between dry and wet gas conditions was compared by using dynamic adsorption method at different temperatures. In addition, the effects of calcination condition on adsorption capacity by calcination condition were investigated. The synthesized MgOs were characterized by XRD, BET, TGA, and TEM.

Keywords: Magnesium oxide, Composite, Desulfurization, Sol-gel

Reference

[1] S. Utamapanya, K.J. Klabunde, and J.R. Schlup, “Nanoscale metal oxide particles/clusters as chemical reagents. Synthesis and properties of ultrahigh surface area magnesium hydroxide and magnesium oxide,” Chem. Mater., 3 (1991), 175-181,

[2] S.H. Lim, E.J. Woo, H.J. Lee, C.H. Lee, “Synthesis of magnetite–mesoporous silica composites as adsorbents for desulfurization from natural gas,” Applied Catalysis B: Environmental, 85 (2008) 71–76.

(486a) How Do Water and Protons Move Through Ionomers?

Benziger, J., Princeton University

Originally presented on: 11/6/2013 12:33:00 - 12:57:00

Ionomer membranes are the heart of Polymer Electrolyte Membrane Fuel Cells. The ionomer membrane transports protons from the anode to the cathode. Proton conductivity is a strong function of water activity. In Nafion, the archetypal PEM for fuel cells the proton conductivity increases by more than six orders of magnitude as water activity increases from 0 (0% humidity) to 1 (saturated vapor). Water transport in Nafion also increases by more than two orders of magnitude as water activity increases.

We have examined in detail the transport and properties of Nafion, sulfonated polyether etherketone (SPEEK) and sulfonated polystyrene (SPS) as functions of temperature and water activity. Ionomers separate into hydrophilic domains of sulfonic acid and water dispersed in a hydrophobic matrix. Water and protons are transported through the hydrophilic domains. The volume fraction of the hydrophilic phase increases with water activity, which also changes the structure of the hydrophilic domains. A percolation threshold of 0.1 vol fraction for proton and water transport is found for 1100 EW Nafion; above the percolation threshold water diffusion and proton mobility increase quadratically with water volume fraction. The percolation thresholds for SPEEK and SPS are larger, indicating different shapes of hydrophilic domains for different ionomers. We will show how ionomer structure is related to the shape and percolation of hydrophilic domains.

Water transport in PEMs involves the coupling of interfacial mass transport, diffusion and polymer swelling. We will show that interfacial transport resistance at the vapor/membrane interface is the limiting transport resistance at high water activity. The interfacial transport resistance is a key reason why Nafion is such a good PEM for fuel cells. We will show how the coupling of internal diffusional resistance and external interfacial transport resistances are couple in ionomers and how it affects the coupled water and proton transport.

(486b) Membranes With Sulfonated Perfluorocyclobutane (PFCB) Ionomer for Fuel Cells

Jiang, R.,
Fuller, T., General Motors Company
Gittleman, C., General Motors
Originally presented on: 11/6/2013 12:57:00 - 13:21:00

Membranes with Sulfonated Perfluorocyclobutane (PFCB) Ionomer for Fuel Cells

Ruichun Jiang, Timothy Fuller, Craig Gittleman

Global Fuel Cell Activities, General Motors Company

895 Joslyn Ave, Pontiac, MI 48340, USA

Ruichun.Jiang@GM.com

Polymer electrolyte fuel cells (PEFC) have been regarded as one of the most promising power sources for various applications. As the heart of PEFC, the proton exchange membrane, or PEM, plays a critical role in proton transport, electrical isolation and reactant gas separation between anode and cathode components of a fuel cell. Perfluorosulfonic Acid (PFSA) ionomers, such as Nafion$^{Â®}$ from Dupont, are the most commonly used materials for PEMs because of their favourable proton conductivity, and mechanical and chemical durability characteristics. There are, however, drawbacks associated with the use of PFSA membranes. These include high cost and high gas permeability which contributes to reductions in both fuel cell efficiency and durability. To overcome these deficiencies, hydrocarbon membranes with aromatic backbones have received considerable attention as PEM materials due to their lower gas permeability and decreased production costs.

A new class of hydrocarbon block copolymer membrane based on perfluorocyclobutane (PFCB) chemistry was developed at General Motors for fuel cell PEM applications. The PEMs with sulfonated PFCB ionomer demonstrate benefits of low production costs, good proton conductivity, and lower gas (H_2, O_2, N_2) permeability than PEM with PFSA ionomer. One of the challenges for sulfonated PFCB based PEMs is enhancing their mechanical stability. Several strategies, including blending PFCB ionomer with a poly(vinylidene fluoride) (PVDF) elastomer, and introducing reinforcement layer containing expanded polytetrafluoroethylene (ePTFE) have been investigated to develop composite PFCB PEMs with improved mechanical properties.

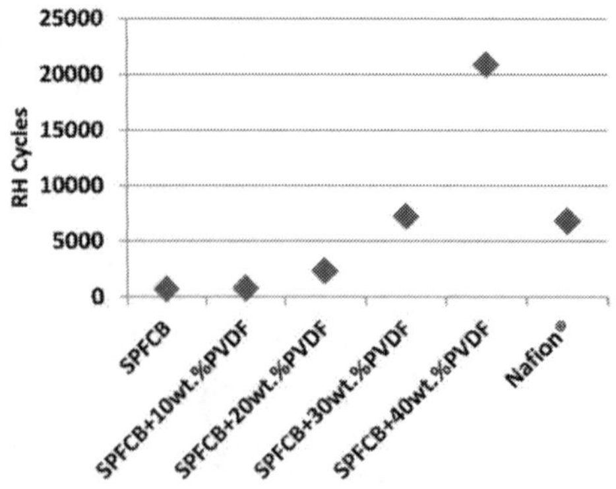

Performance and durability of fuel cell membranes based on PFCB ionomer are greatly influenced by the content of PVDF, as well as by the properties of ePTFE reinforcement layer. The figure on the right shows that the mechanical durability in RH cycling tests of the sulfonated PFCB/PVDF blend membranes were improved with increasing level of PVDF content. More performance and durability information of PFCB/PVDF and PFCB/PVDF-ePTFE types of fuel cell membranes will be reported and discussed.

(486c) Increased Hydrophobic Content Multiblock Copolymers for Use As Proton Exchange Membranes in Direct Methanol Fuel Cells (DMFC)s

Rowlett, J., Virginia Tech
Shaver, A., Virginia Tech
Kim, Y. S., Los Alamos National Laboratory
Zelenay, P., Los Alamos National Laboratory
McGrath, J. E., Virginia Tech
Originally presented on: 11/6/2013 13:21:00 - 13:45:00

Two series of increasing hydrophobic poly(arylene ether nitrile) and poly(arylene ether ketone) oligomers were successfully synthesized. These poly(arylene ether) hydrophobic oligomers consistent of different amounts of methyl groups on the bisphenol A co-monomer (bisphenol A, dimethyl bisphenol A, and tetramethyl bisphenol A), and were combined with either difluorobenzonitrile or difluorobenzophenone to from the hydrophobic segments. In addition to further increase the hydrophobicity of the moieties, small mole percent's of the bisphenol A monomer series (0-25%) were replaced by the fluorinated bisphenol A analog 4,4'-hexafluoroisopropylidenediphenol. Only up to twenty-five percent of the fluorinated monomer was used in the hydrophobic copolymers due to the concern over methanol cross-over, which has been shown to greatly increase with higher fluorine content.

Both series of hydrophobic oligomers were then reacted with a poly(arylene ether sulfone) hydrophilic oligomer (BPS100) via a low temperature coupling reaction to produce the hydrophilic-hydrophobic alternating multiblock copolymers. The resultant block copolymers formed a tough, durable, and transparent membrane when cast from solution, with structure confirmation and composition done using ^{1}H and ^{13}C NMR spectroscopy. Mechanical properties were examined using tensile testing, and used in combination with size exclusion chromatography (SEC) to confirm high molecular weight was achieved. In addition the conductivity, swelling ratio, and fuel cell performance was measured and compared against that of the nafion membrane. Membrane electrode assembly (MEA) and fuel cell performance was carried out by Los Almas National Laboratory (LANL) in collaboration with this work. The synthesis and characterization of these multiblock copolymer membranes will be discussed.

(486d) Controlling Mechanical and Transport Properties of Layer-By-Layer Electrospun Mat Composite Membranes for Fuel Cell Applications

Liu, D. S., Massachusetts Institute of Technology
Rutledge, G. C., Massachusetts Institute of Technology
Hammond, P. T., Massachusetts Institute of Technology
Originally presented on: 11/6/2013 13:45:00 - 14:09:00

The continued development of thin-film solid polymer electrolytes with improved mechanical and ion transport properties is critical for the further advancement of electrochemical energy devices. For hydrogen and methanol fuel cells, the proton exchange membrane (PEM) is the electrolyte and has to have high protonic conductivity, low fuel crossover, and be mechanically and chemically stable. More recently, there has been considerable interest in improving the mechanical properties and durability of the proton exchange membranes (PEMs) without sacrificing its transport properties. For methanol fuel cells in particular, there is a need for a cheap, mechanically robust PEM that also has high protonic conductivity and low methanol permeability.

LbL assembly allows for the controlled deposition of alternating polyelectrolytes at the nanometer scale. An LbL system composted of poly(diallyl dimethyl ammonium chloride) (PDAC) and sulfonated poly(2,6-dimethyl 1,4-phenylene oxide) (sPPO) has shown to have proton conductivity comparable to that of Nafion®, the industry standard, but with methanol permeability two orders of magnitude less than Nafion's. While this system by itself is mechanically deficient when hydrated, the mechanical properties can be greatly improved if the film is spray-coated on an electrospun fiber mat to form a composite membrane. With vacuum assisted spray-LbL assembly, the fibers can be individually coated, and the mat filled; mats as thin as 15 μm can be coated. Without vacuum, the spray-LbL assembly can form a continuous fuel-blocking layer with properties similar to the free-standing LbL film by itself. Combined, composite membranes with methanol permeability twenty times lower than Nafion and through-plane proton selectivity five times greater than Nafion could be fabricated.

As anticipated, the mechanical properties of the underlying electrospun mat were passed on to the composite membrane. The composite membranes were further strengthened by annealing the electrospun mat prior to coating. In addition, the swelling of the composite membranes in water was significantly lower than the LbL-only system and over a magnitude lower than Nafion. The mechanical hysteresis was also lower than Nafion under hydrated conditions. The composite membranes were tested in an operational direct methanol fuel cell with results showing a higher open circuit voltage and comparable cell resistances when compared to Nafion.

(486e) First-Principles Modeling of Proton Conductivity in Perfluorosulfonate Fuel Cell Membranes

Zhang, B., University of Tennessee
Edwards, B., The University of Tennessee
Originally presented on: 11/6/2013 14:09:00 - 14:33:00

Proton conductivity through a perfluorosulfonate polymer electrolyte membrane was investigated using an extension of a recent model developed for the purpose of quantitatively describing transport of charged species through the ionomer interface of a typical membrane [Kumar et al., JCP, 138, 064903, 2013]. In this study, the membrane was modeled as a collection of cylindrical pores with the anionic groups (i.e., $-SO_3^-$) assumed to be fixed at a prescribed periodic spacing along the pore wall, as determined by the material's equivalent weight and surface charge density. Microscopic potentials for the interactions between the various chemical species present within the pore were scaled up to a mesoscopic level, where transport equations were derived via the generalized bracket formulation of nonequilibrium thermodynamics in terms of the density and velocity of each species (hydronium ions and water) within the pore. These coupled transport equations were solved based on reasonable boundary conditions, and an expression of conductivity was deduced in terms of the following quantities: water content, equivalent weight, temperature, density, and the monomeric structure of the perfluorosulfonate chain. Theoretical predictions of this model were compared against experimental data of conductivity in three different membranes, Nafion 117 (EW = 1100), a 3M membrane (EW = 900), and a Dow membrane (EW = 800), for different values of water content and temperature. Theoretical predictions of the model matched quantitatively the experimental data without the necessity of using empirical fitting parameters.

(486f) Functionalized Inorganic Particle-Doped Composite Proton Exchange Membranes for Enhancement in Fuel-Rejecting and Proton-Conducting Property

Wu, H., Tianjin University
Zhao, Y., Tianjin University
Jiang, Z., Tianjin University
Originally presented on: 11/6/2013 14:33:00 - 14:57:00

Organic-inorganic hybrid proton exchange membranes have gained tremendous attention for use in direct methanol fuel cells due to their combined advantages originated from both the inorganic and organic components. A large variety of polymers have been blended with various inorganic particles by either physical blending or sol-gel process to generate inorganic phase in polymer matrix. The addition of carefully selected inorganic components leads to inhibited methanol crossover and improved stability. The inorganic dopants are usually functionalized to improve the organic-inorganic interfacial compatibility and to introduce additional pathways for proton transfer at the interface, thus increasing proton conductivity of the resultant hybrid membranes. Inorganic dopants such as titania and silica were applied widely for the ease of availability and functionalization. In our previous work various functional groups including phosphonic acid groups, sulfonic acid groups, carboxylic acid groups, amine groups and amino acid groups by means of chelation absorption of these functional groups onto titania. In this study, two kinds of phosphorylated silica submicrospheres are synthesized and incorporated into sulfonated poly(ether ether ketone) (SPEEK) to fabricate hybrid membranes. Phosphorylation is carried out by the reaction of phosphorus oxychloride and the epoxy groups on the silica surface introduced by glycidyl-silane (SiP-I) or poly(glycidyl methacrylate) (SiP-II). The length of the chains and the amount of phosphoric acid groups grafted on the silica surface are tuned by the above two different phosphorylation methods. The dispersion of SiP-II, anti-swelling and methanol resistant property of the SPEEK/SiP-II hybrid membranes are enhanced. And the proton conductivity is increased by doping with both kinds of phosphorylated silica particles, while, interestingly, the SPEEK/SiP-I hybrid membranes with less amounts of phosphoric acid groups show unexpectedly much higher proton conductivity than the SPEEK/SiP-II hybrid one at the same doping content. A proton conductivity of 0.335 S cm^{-1} at 60 ^{o}C and 100%RH is obtained. The result is tentatively discussed in terms of effective proton-conducting groups/sites instead of the absolute amount of acidic groups.

(496a) Modeling the Ultrafiltration Behavior of Highly Concentrated Monoclonal Antibodies

Binabaji, E., The Pennsylvania State University
Rao, S., Amgen Inc.
Zydney, A. L., The Pennsylvania State University
Originally presented on: 11/6/2013 12:30:00 - 12:55:00

Monoclonal antibodies are used extensively in the treatment of diseases and sometimes need to be delivered at very high concentrations to achieve the desired therapeutic effect. This creates unique challenges in the development and implementation of ultrafiltration processes for the concentration and final formulation of these antibody products. The objective of this study was to develop a new theoretical framework for describing the filtrate flux during the ultrafiltration of highly concentrated protein solutions.

Experiments were performed with a highly purified monoclonal antibody provided by Amgen Inc. Data were obtained for the protein osmotic pressure (using membrane osmometry), the second virial coefficient (using self-interaction chromatography), the solution viscosity (using a RFS-II viscometer), and the filtrate flux over a wide range of antibody concentrations. The filtrate flux attained a pressure-independent value at high transmembrane pressures, consistent with the concentration polarization model. However, plots of the filtrate flux versus the logarithm of the bulk protein concentration showed a distinct curvature, in contrast to predictions of the classical stagnant film model. The complex behavior at very high protein concentrations is due to the corresponding increase in the protein osmotic pressure and solution viscosity.

The filtrate flux data were analyzed using a theoretical framework that accounts for the effects of intermolecular interactions and the enhanced solution viscosity on protein transport in the concentration polarization boundary layer. Calculations were also performed to evaluate the effect of the increased axial pressure drop in the highly concentrated protein solutions on the local variation in the filtrate flux. These results provide important insights into the factors controlling the ultrafiltration behavior of highly concentrated protein solutions encountered in bioprocessing applications.

(496b) Permeability of Electrospun Mats Under Pressure Driven Flow

Choong, L. T.,
Rutledge, G. C., Massachusetts Institute of Technology
Originally presented on: 11/6/2013 12:55:00 - 13:20:00

Abstract

Electrospun mats are promising in filtration applications because of their high surface-to-volume ratio (10-500 m^2/g for 10-500 nm fiber diameter) and high porosity (>0.9), which results in high permeability constants [1,2]. However, electrospun mats are also highly compressible, hence their porosity decreases with increasing pressure. This compressibility of the mat can reduce the benefits of high porosity in filtration applications. An understanding of the extent of the reduction in permeance upon compression for electrospun mats is vital for performance comparison with other commercial filtration membranes in filtration processes.

Hydraulic permeability of electrospun mats under flow-induced compression has been modeled and verified experimentally. The permeation model accurately estimates the changes in solidity, and hence the permeability of the electrospun mats, over a range of pressure differentials. The model is based on Darcy's law using Happel's permeability equation coupled with a compression model of the form, $P_m = kE(\varphi^n - \varphi_0^n)$ where P_m is the compressive stress and φ is solidity, in accord with the analysis of Toll [3]. Hydraulic permeability of electrospun mats of bis-phenol A polysulfone (PSU) comprising fibers of different mean diameter, annealed at temperatures at and above the glass transition of the polymer, was measured for feed pressures ranging from 5 kPa to 140 kPa. By using kE value as a fitting parameter in the permeation model, permeation test may be used as an alternative tool to estimate the kE value, which is difficult to measure accurately from compression testing. The electrospun mats experience a decrease of more than 60% in permeability constant, compared to a 32% decrease for a commercial membrane, because electrospun mats are more compressible than the commercial membrane.

References

[1] Yoon K., Hsiao B., Chu B. (2008) *J. Mater. Chem.* **18**, 5326-5334.

[2] Burger C., Hsiao B., Chu B. (2006) *Annu Rev Mater Res* **36** (1), 333-368.

[3] Toll S. (2004) *Polymer Engineering & Science* **38** (8), 1337-1350.

(496c) Transport of Ions and Water and Lipid Flip-Flop During Ligand-Coated Nanoparticle Permeation in a Lipid Membrane

Oroskar, P.,
Song, B., University of Illinois at Chicago
Jameson, C., University of Illinois at Chicago
Murad, S., University of Illinois at Chicago
Originally presented on: 11/6/2013 13:20:00 - 13:45:00

Nanoparticles are generally considered excellent candidates for targeted drug delivery. However, ion and water transport and cytotoxicity induced by nanoparticle permeation is a potential problem in such drug delivery schemes because of the toxic effect of many ions. In this study, we have carried out a series of coarse-grained molecular dynamics simulations to investigate the water penetration, ion transport, and lipid molecule flip-flop in a protein free phospholipid bilayer membrane during ligand-coated nanoparticle permeation. The effect of ion concentration gradient, pressure differential across the membrane, nanoparticle size, length of ligand, and permeation velocity has been examined in this work. Some conclusions from our studies include (1) The number of water molecules in the interior of the membrane during the nanoparticle permeation increases with nanoparticle size and pressure differential across the membrane but is unaffected by the nanoparticle permeation velocity or the ion concentration gradient. (2) Ion transport is sensitive to the size of the nanoparticle as well as the ion concentration gradient between two sides of the membrane; no anion/cation selectivity is observed for small nanoparticle permeation, while anions are preferentially translocated through the membrane when the size of the nanoparticle is large enough. (3) Incidences of lipid molecule flip-flop increases with the size of the nanoparticle and ion concentration gradient and decreases with the pressure differential and nanoparticle permeation velocity. Finally, we will also report on the effect of ligand size on water and ion leakage and the frequency and mechanism of lipid flip-flop.

(496d) Solute-Solvent-Membrane Interactions in Organic Solvent Nanofiltration

Marchetti, P., Imperial College London

Butté, A., Lonza AG

Livingston, A. G., Imperial College London

Originally presented on: 11/6/2013 13:45:00 - 14:10:00

Solute-solvent-membrane interactions in Organic Solvent Nanofiltration

In recent years, nanofiltration (NF) has been applied in organic solvents more and more widely [1,2,3]. In the pharmaceutical industry, for example, NF has been proposed as an emerging technique to perform concentration, separation, and salt/solvent exchange of peptide solutions, as a consequence of the increasing interest in peptides as pharmaceuticals, which has been challenging the peptide industry to develop economically competitive methods for large scale manufacturing. Application of NF to peptide processes has permitted the improvement of the overall peptide downstream processes.

Since application of membrane technology in non-aqueous media is a recently emerging field, understanding the transport mechanism of organic solvent mixtures through NF and UF membranes is still an open research topic. Straight extension of steric and electrostatic separation mechanisms, typical of aqueous environments, to non-aqueous systems is complex, due to the significant differences in the structures and properties of the solvents [1,4,5,6]. It has been observed that the NF performance is much less predictable in the presence of organic solvents, than in aqueous solutions [4,5] and the molecular weight cut-off (MWCO) is an insufficient descriptor for the separation capability of the membrane in organic solvents [4]. In comparison to well-established aqueous processes, non-aqueous processes are characterized by the increase in the number of solute-solvent-membrane interactions [3], which play a determining role in the understanding of both solvent flux and solute rejection.

The factors affecting the molecular interactions at the nanoscale can be classified in terms of solvent-membrane and solute-membrane affinity interactions. Solvent-membrane affinity is a function of solvent nature and composition of the solvent mixture. Solute affinity is generally a function of four factors:

(i) chemical nature of the molecule, in terms of steric, electrostatic, hydrophilic/phobic and polar properties;

(ii) electrostatic interactions between solute and membrane charge, already extensively studied in water, as Donnan effects [6], and questioned for Organic Solvent Nanofiltration (OSN) [1,2,5];

(iii) composition of the solvent mixture. When a solute is dissolved in a mixture of two or more solvents, solute-solvent interactions may differ in strength for each solvent. This may cause the composition of the solvation shell of the solute to be different from that of the bulk solution. In this case, the solute is said to be preferentially (or selectively) solvated by one of the two solvents [7].

(iv) concentration of salts/ions, often present in a typical peptide process solution. On one hand, ions affect the charge interaction, by affecting the solution pH or ionic strength. On the other hand, ions can affect the solute solubility, depending on their kosmotropic ("order-maker") or chaotropic ("disorder-maker") nature [9], via salt formation or water-mediated ion pairs.

The aim of this work is to study the effect of solute-membrane vs. solvent-membrane affinities on the OSN performance. The work is divided in three parts.

Firstly, permeation of water, organic solvents and their mixtures is investigated through hydrophilic nanofiltration and ultrafiltration (UF) ceramic membranes (Inopor® Nano 450 Da, Inopor® Nano 750 Da and Sulzer 1000 Da, all composed of TiO_2/Al_2O_3, and Inopor® Ultra 2000 Da, composed of ZrO_2/Al_2O_3). Experimental results clearly indicate that the Hagen-Poiseuille viscous flow, which assumes the viscosity as the key parameter influencing permeation, fails in describing solvent permeation through NF membranes, whereas it is able to describe the experimental permeability through UF membranes. This is ascribed to the significance of surface effects against bulk effects (i.e. viscosity) for high specific surface systems (such as NF membranes), compared to low specific surface systems (such as UF membranes). Several correction factors are considered in this work, to account for the surface phenomena that arise in a nanotube due to solvent-membrane surface interactions. These correction terms (namely interfacial tension, polarity and molecular dimension) are included into an improved phenomenological model for solvent permeation [10]. As expected, they are functions of the pore size, and, namely, significant for small pores (cf. NF), and negligible for large pores (cf. UF).

Afterwards, permeation of single salts and acids (NaCl, KCl, LiCl, NaI, NaF, HCl and trifluoroacetic acid, TFA-H) in water and organic/water mixtures through the same NF and UF ceramic membranes is investigated. Salts/acids are chosen for three reasons. Firstly, salts are not affected by steric retention, and therefore electrostatic and molecular affinity effects can be isolated and analyzed separately. Afterwards, charge effects have been largely studied in water [6] and have been well explained by the Donnan theory. The extension of these considerations to organic solvents is therefore possible. Finally, preferential solvation data for common inorganic ions are either available in the literature or easily derivable from theoretical models. Availability of these data permits one to look for a correlation between permeation performances and solvation characteristics. The effects of pore dimension, nature of the organic solvent and nature of anion/cation for different salts and acids are presented and discussed in this paper. The effect of the type of solvent on the ion rejection is found to be a function of the Hansen solubility parameter of the solvent, the Hansen solubility parameter of the ion, and the preferential solubility parameters of the ion in the aqueous mixture. The solute-solvent competition has a more significant effect at low pore dimension and becomes more negligible for large pore dimension (UF range). This is shown in Figure 1, by the more significant negative rejection (i.e. enhanced solute transport vs. solvent transport) for NaCl in acetonitrile/water mixtures through the tighter membrane (Inopor Nano 450) than through the looser one (Inopor Nano 750).

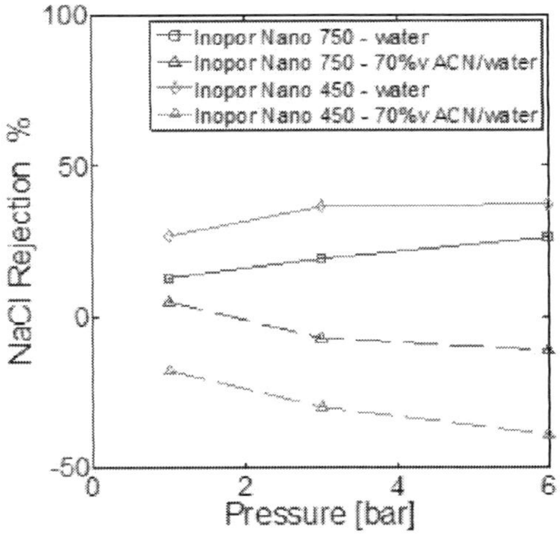

Figure 1. NaCl rejection during NaCl NF in water and 70%v ACN/water (C_{NaCl} = 1 mM).

This demonstrates that the affinity effects are more significant in the NF range, due to the high specific surface that characterizes these membranes, and confirms our previous findings from the investigation of pure solvent permeation.

Finally, rejections of one small organic molecule, Npys, and one model peptide, synthesised by Lonza (Visp, Switzerland) and named PEP1 in this study, through the same hydrophilic ceramic membranes, are studied as a function of solvent composition (%v ACN/water) and ion content (Na^+, H^+, Cl^-, and TFA^-,). The effects of solvent composition, ion nature (kosmotropic or chaotropic) and ion concentration affect the solute rejection, by affecting the solute preferential solvation, i.e. its affinity with the membrane. The effect of ions on the composition of the solute solvation shell is explained by the hydration/dehydration mechanism proposed by Collins [9].

In conclusion, this work demonstrates that surface effects and molecular affinity interactions influence permeation of solutes and solvents through NF membranes. For solute+solvent systems, additionally to steric and electrostatic retention mechanism, preferential solvation of organic molecules in solution affects the membrane transport, by changing the molecule hydration degree, and in turn the hydrophilic/hydrophobic interactions with the membrane. Solvent composition, ion nature and concentration are therefore significant additional effects for the pressure-driven permeation through ceramic NF membranes.

Bibliography

[1] Y. Zhao, Q. Yuan, *A comparison of nanofiltration with aqueous and organic solvents,* J. Membr. Sci. 279 (2006) 453-458.

[2] D. Bhanushali, S. Kloos, D. Bhattacharayya, *Solute transport in solvent-resistant nanofiltration membranes for non-aqueous systems: experimental results and the role of solute ? solvent coupling,* J. Membr. Sci. 208 (2002) 343-359.

[3] S. Darvishmanesh, J. Degréve, B. Van der Bruggen, *Mechanisms of solute rejection in solvent resistant nanofiltration: the effect of solvent on solute rejection*, Phys. Chem. Chem. Phys. 12 (2010) 13333-13342.

[4] J. A. Whu, B. C. Baltzis, K. K. Sirkar, *Nanofiltration studies of larger organic microsolutes in methanol solutions,* J. Membr. Sci. 170 (2000) 159-172.

[5] X. J. Yang, A. G. Livingston, L. Freitas dos Santos, *Experimental observations of nanofiltration with organic solvents*, J. Membr. Sci. 190 (2001) 45-55.

[6] W. Bowen, J. Welfoot, *Modelling the performance of membrane nanofiltration critical assessment and model development*, Chem. Eng. Sci. 57 (2002) 1121-1137.

[7] A. Bagno, M. Campulla, M. Pirana, G. Scorrano, S. Stiz, *Preferential solvation of organic species in binary solvent mixtures probed by intermolecular 1H NOESY NMR spectroscopy*, Chem. Eur. J. 5 (1999) 1291-1300.

[8] Y. Marcus, *Preferential solvation of ions in mixed solvents. 2. The solvent composition near the ion,* J. Chem. Soc. Faraday Trans. 1 84 (1988) 1465-1473.

[9] K. D. Collins, *Ions from the Hofmeister series and osmolytes: effects on proteins in solution and in the crystallization process*, Methods 34 (2004) 300-311.

[10] P. Marchetti, A. Butté, A. Livingston, *An improved phenomenological model for prediction of solvent permeation through ceramic NF and UF membranes*, J. Membr. Sci. 415-416 (2012) 444-458.

(496e) Rigorous Calculations of Permeation in Mixed-Matrix Membranes

Nair, S., Georgia Institute of Technology
Singh, T.,
Kang, D. Y., Georgia Institute of Technology
Originally presented on: 11/6/2013 14:10:00 - 14:35:00

Polymeric membranes face an intrinsic trade-off between the permeability and selectivity. A widely-taken approach to overcome this trade-off is to incorporate higher-performance nanoporous particles (e.g., zeolites, metal-organic frameworks) as fillers into polymeric membranes. Such membranes are also referred to as 'mixed-matrix' membranes (MMMs), and have been shown to yield enhanced separation performance (higher permeability, higher selectivity, or both).

Several analytical models have been developed to understand and predict the effective permeability and selectivity of MMMs, such as the Maxwell, Bruggeman, Pal, Lewis-Nielsen, and other models. The most significant – yet least discussed – limitation of the above models is that none of them consider the effects of the adsorption equilibrium at the polymer/filler interface on the effective permeability. This can lead to qualitatively and quantitatively erroneous interpretations of permeability data from MMMs when analyzed with the above 'permeability-based' models, since most polymer/filler interfaces will not have an interfacial equilibrium constant of unity. Furthermore, the models often cannot successfully interpret experimental MMM permeation data, or reconcile data from different sources, without the postulation of 'non-idealities' such as the presence of interfacial voids or rigidified polymeric regions at the interface with the filler. These hypotheses introduce additional fitting parameters and are difficult to verify by independent characterization.

With the advent of a large new class of nanoporous MOF materials as MMM fillers, a large quantity of permeation data is emerging that is increasingly difficult to interpret. A fundamental difficulty is that the permeability-based models do not allow one to separate the effects of interfacial equilibrium from the effects of hypothesized non-idealities. Thus, there is a requirement for rigorous predictions of permeation in ideal MMMs that capture the true dependence of the effective permeability on the diffusivities of the matrix and filler phases, the adsorption equilibrium at the polymer/filler interface, and the volume fraction of the filler. Such predictions can then be used for reliable interpretation of MMM data, assessment of the effects of the intrinsic material properties versus those originating from possible non-idealities, and selection of matrix and filler materials for desired separation properties.

We will discuss rigorous calculations of single-component permeation in mixed-matrix membranes (MMMs), and show their value in developing a reliable understanding of permeation behavior in these membrane architectures. We first develop methods for the construction of detailed and large-scale 3D mixed-matrix membrane (MMM) models, which are then solved by finite-element methods. Our models explicitly account for the effects of interfacial equilibrium between the matrix and the filler, in addition to the differences in molecular diffusivity between the two phases. Analytical models (e.g., Maxwell model) can only predict the MMM permeability when the interface equilibrium constant K=1. Most real MMMs do not satisfy this condition. It is shown that the individual values of the equilibrium constant K and the diffusivity ratio D_f/D_m, and not the combined permeability ratio $P_f/P_m = K.D_f/D_m$, determine the MMM permeability. We then use our 'exact' predictions to examine some of the current speculations regarding MMM permeation behavior. We use CO_2 solubility and diffusivity data on example filler and polymer materials to show that the use of permeability-based models to analyze MMM permeation data leads to spurious results. Our simulations reveal that the incorrect permeabilities obtained from the Maxwell (and other) models are likely to mislead researchers to hypothesize non-idealities such as matrix-dependent filler behavior or rigidification of the matrix polymer near the interfaces. Our simulations also indicate clearly that an ideal MMM shows no significant direct effect of filler particle size. Finally, we fit our computational data to an empirical correlation that can be used to calculate MMM permeabilities given adsorption and diffusion data for the matrix and filler phases.

(496f) Effective Medium Theory-Based Modelling of Transport in Disordered Nanoporous Membranes

Bhatia, S. K., The University of Queensland
Gao, X.,
Originally presented on: 11/6/2013 14:35:00 - 15:00:00

A fundamental understanding of the processes affecting fluid transport in nanoscale confinements is crucial to numerous emerging applications in nanotechnology, including those involving membrane separations. The Knudsen model has been our primary tool for modelling the transport of fluids in confinement; however, it has not been unequivocally validated for nanoscale pores due to the inability to make measurements at this scale. Its weakness is generally masked by the use of correlations which invoke tortuosity as a fitting parameter, while also overlooking complexities associated with structural heterogeneity. Indeed, theoretical and simulation-based results, as well as recent literature experimental data, show the Knudsen model to substantially overpredict the diffusivity in nanoscale pores.

In modelling transport in membranes it is common to consider only a single representative pore size in the support and in the membrane layers, thereby overlooking the role of structural heterogeneity in these layers. Experiments have been performed in our laboratory in which the transport in the support and the composite membrane are independently investigated for several gases, and it is shown that the use of a single representative pore size in the support is misleading. This leads to tortuosities that are dependent on the gas, and vary with temperature and operating conditions. A new effective medium theory (EMT) approach for modelling the transport in the support and membrane layers is developed, which utilizes the entire pore size distribution (PSD), and shown to adequately interpret our experimental data. The technique is free from the fitting parameter of tortuosity, but which can now be theoretically predicted, and demonstrates the importance of considering the entire PSD and network effects in the pore structure of the support. Our EMT results show that for macroporous networks different tortuosities are obtained in the Knudsen and viscous regimes, due to the different dependence on pore size. It is also found that the Knudsen model fails when applied to the mesoporous membrane layer when the modal pore diameter is below about 10 nm. On the other hand our novel transport theory considering the fluid-solid interaction and the adsorption field in the pore proves successful, when combined with the effective medium theory for such layers.

Our results clearly demonstrate that it is critical to independently investigate the resistances of the various layers, carefully decoupling these, for interpreting membrane transport data. Correct interpretation, free from artefacts, is only obtained when the full PSD and network connectivity effects are separately considered in each layer, such as by EMT, and when the effect of fluid-solid interaction on transport is considered in nanoscale pores smaller than about 10 nm in diameter.

(501a) Low-Temperature and Low-Pressure Fuel Hydrodesulfurization By Solid Catalyst Coupling With Ionic Liquids

Li, C., Institute of Process Engineering
Zhang, S., Institute of Process Engineering, Chinese Academy of Sciences
Originally presented on: 11/6/2013 12:30:00 - 12:52:00

A new hydrodesulfurization (HDS) process for the fuels utilizing a solid catalyst coupled with ionic liquids (ILs) at mild operating conditions is first proposed in this study. A novel Ce-Ni-Mo/γ-Al_2O_3 HDS catalyst with high catalytic activity was developed and coupled with different ILs to investigate the removal of dibenzothiophene (DBT) from model oil at low temperature and low pressure. Results showed that the sulfur removal efficiency follows the order: [BDMIM]BF_4> [BMIM]BF_4>[EMIM]BF_4> [BMIM]Cl. In addition, the highly efficient [BMIM]BF_4 coupled with the solid catalyst was selected to undergo a series of parallel experiments to systematically investigate the effects of temperature, hydrogen pressure and IL/oil mass ratio on DBT removal from model oil. To achieve optimal sulfur removal efficiency, model oil was conducted at the following conditions: reacton temperature, 150℃; hydrogen pressure, 3MPa; reaction time, 3h; rotate speed, 150rpm/min; and IL/oil mass ratio, 1:6. The process and mechanism of HDS were also studied using a molecular simulation software. This study shows that the HDS process using the solid catalyst coupled with ILs provides a new direction for producing cleaner industrial fuels.

(501b) Reactivity of Biopolymers in Ionic Liquids

Hallett, J. P., Imperial College
de Gregorio, G., Imperial College
Ismail, N. L.,
Originally presented on: 11/6/2013 12:52:00 - 13:14:00

Ionic liquids are gaining increasing acceptance as novel solvents for chemical transformations. This has included a recent focus on the use of ionic liquids for the separation of biopolymers (lignin and cellulose). During this process, the biopolymers are susceptible to multiple types of interactions with the ionic solvent, which can lead to depolymerisation (fragmentation or cleavage reactions), solvolysis or solvent addition reactions.

Our recently proposed Ionosolv process focusses on the use of highly acidic ionic liquids for biomass pretreatment. During pretreatment, all biopolymers in solution (lignin, cellulose, hemicelluloses) are subjected to harsh chemical consitions (non-aqueous, highly acidic, moderate temperature) which can induce novel reactivities. This paper will explore several potential reactions of the ionic liquid solvents with these biopolymers, including ether and ester cleavage and anion-induced substitution reactions. A variety of chemical characterisation techniques have been explored to ensure both a qualitative and quantitative measure of this reactivity.

We have also utilised a series of model compounds (esters, ethers and alcohols) to explore the reactivity of individual polymer subunits with the solvent under a variety of consitions. This portion focusses on electronic effects of the substrate and explores reaction kinetics and changes in mechanism with different functional groups.

(501c) Microwave-Assisted Ionic Liquid Pretreatment and Catalysis of Lignin to Phenol in a Biphasic Reaction System

Xu, J.,

Originally presented on: 11/6/2013 13:14:00 - 13:36:00

Microwave-assisted Ionic Liquid Pretreatment and Catalysis of Lignin to Phenol in a Biphasic Reaction System

Yongshui Qu, Hao Luo, Jian Xu*

National Key Laboratory of Biochemical Engineering, Institute of Process Engineering, Chinese Academy of Sciences, Beijing 100190, People's Republic of China

*Corresponding author: jxu@home.ipe.ac.cn

Abstract: Recent concerns about the gradual depletion of conventional fossil resources and the pressure from global climate change have accentuated the need for new alternative resources. As one of the main components in biomass, lignin is the second most abundant natural polymer after cellulose which has the potential to serve as a sustainable source of energy and organic carbon to replace petroleum-based chemicals. Efficient conversion of lignin into high value-added chemicals is a key to improve the economic feasibility of biomass refinery. In the present study, a biphasic reaction system assisted by microwave heating is employed in the conversion of lignin into phenol, a vital petroleum-based building blocks used in production of fine chemicals and plastics. In order to understand the relationship among the structure of the ionic liquid, its effect on the change of lignin structure and yield of phenol, a series of functional ionic liquids are prepared to pretreat the industrial lignin and the changes on the structures and characteristics of the lignin (functional groups, microcosmic morphology, etc.) in the pretreatment and its effect on the phenol production are investigated. The study on the directional conversion of the pretreated lignin to phenols in a biphasic system is subsequently carried out by using the microwave irradiation instead of conventional heating, replacing the common catalysts (acids or alkalis) with the ionic liquids. The result shows that the phenol yield of 8.14% is achieved with the microwave irradiation at 400 W for 60 min and the produced phenol can be separated easily by a separation-friendly biphasic system. In this biphasic system, the industrial lignin pretreated by 1-ethyl-3-methylimimidazolium acetate ([EMIM]OAc) is catalyzed into phenol by an alkaline ionic liquid. The phenol product is continuously extracted into an organic phase consisting of methylisobutylketone (MIBK) and 2-butanol.

Keywords: Lignin; Ionic liquid; Pretreatment; Phenol; Catalysis

(501d) Short- and Long-Term Thermal Stabilities Of Imidazolium-Type Ionic Liquids On Catalytic Metal-Oxide Supports

Akcay, A., Koc University

Balci, V., Koc University

Uzun, A., Koc University

Originally presented on: 11/6/2013 13:36:00 - 13:58:00

Ionic liquids (ILs) have been attracting tremendous attention in the last few decades because of their unique physicochemical properties, such as negligible vapor pressure, high thermal stability, high solvating capacity, and wide ranges of density and viscosity. With these tunable characteristics, they have become the first choice in many applications such as organic synthesis, electrochemistry, separation, and especially in catalysis.

In catalysis field, two directions of IL application recently discussed in the literature are particularly attractive: *supported ionic liquid phase* (SILP) catalysts and *solid catalysts with ionic liquid layer*(SCILL). Here, the utilization of ILs in these catalysts not only provides opportunities to control and stabilize the nuclearity and morphology of the active sites, but also offers the following opportunities to improve the catalytic performance: i) ILs can impose a positive influence on the "chemical" properties of the catalysts by acting as a co-catalyst, either as an acidic function in a bifunctional catalyst or as a modifier; ii) ILs can control the effective concentration of reaction intermediates and stabilize charged transition states or ligands to control the reaction pathways. SILP type catalysts, where active species dissolved in IL, act as a homogeneous system immobilized on a high surface area porous solid support offering the benefits of heterogeneous systems. Since support enables a higher surface area and IL enables the better utilization of catalyst particles in the pores, the system results in the higher selectivity and activity compared to traditional catalysts. Similar to the SILP catalyst concept in SCILL applications, a supported metal catalyst is coated with IL layer.

Although SILP and SCILL concepts offer improved activity and selectivity, applications of these IL-assisted supported catalysts are limited with the thermal stabilities of ILs. Thermal stability limits determined for pure ILs are not applicable for these systems. Because, ILs directly interact with the metal-oxide supports and these interactions influence the thermal stability of ILs significantly. Therefore, elucidation of structural factors determining the thermal stability limits of ILs on metal-oxide supports is crucial for the selection of optimum IL for the application conditions of IL-assisted supported catalysts. Here we report the *short-term* and *long-term* thermal stability limits of 34 different imidazolium-based ILs coated on commonly used metal-oxide supports, SiO_2, γ-Al_2O_3, MgO. Structures of imidazolium ILs were varied systematically and the acidity of the support material increased from MgO to γ-Al_2O_3 and to SiO_2.

Short-term thermal decomposition temperatures of ILs supported on metal-oxides were determined by thermogravimetric analysis (TGA) at a linear heating rate of 10 K/min from 298 to 1073 K from the onsets of the derivative per cent weight changes. Results show that thermal stability of pure ILs significantly changes upon coating on metal-oxides with different point of zero charges (PZC) such as < 2.82 for SiO_2, 6-9.3 for γ-Al_2O_3 and 11.6-12 for MgO. Pure *1-ethyl-3-methylimidazolium tetrafluoroborate* [EMIM][BF_4], for example, is reported to be stable up to 723 K; however, our data illustrate that it starts to decompose at 609 ± 3, 534 ± 3 and 446 ± 3 K when coated on SiO_2, γ-Al_2O_3, and MgO, respectively. In general, imidazolium type ILs lose their thermal stability significantly as the acidity of the support decreases from SiO_2 to MgO. However, ILs with high acidity levels such as those with *dicyanamide* $[DCA]^-$ anion, *1,3-dimethoxyimidazolium bis(trifluoromethylsulfonyl)imide* [DMOIM][NTf_2] *and 1,3-dihydroxyimidazolium bis(trifluoromethylsulfonyl)imide* [DHIM][NTf_2] and *1-ethyl-3-methylimidazolium hexafluorophosphate* [EMIM][PF_6] do not follow this general trend and show higher thermal stabilities on MgO than on SiO_2. Similarly, effects of increase in the length of alkyl groups attached to the imidazolium ring depend strongly on the anion type too. For instance, thermal decomposition temperatures of [EMIM][BF_4], *1-butyl-3-methylimidazolium tetrafluoroborate* [BMIM][BF_4], *1-hexyl-3-methylimidazolium tetrafluoroborate* [HMIM][BF_4] and *1-decyl-3-methylimidazolium tetrafluoroborate* [C_{10}MIM][BF_4] were determined on the SiO_2, γ-Al_2O_3, MgO metal-oxide supports. Results show that thermal stabilities of these ILs decrease from 609 ± 3 to 502 ± 3 K on SiO_2, from 534 ± 3 to 496 ± 3 K on γ-Al_2O_3 and from 446 ± 3 to 440 ± 3 K upon step-wise increase in the length of alkyl chain attached to the imidazolium ring from *ethyl* to *decyl*. However, when $[BF_4]^-$ anion is replaced with $[NTf_2]^-$, increase in the alkyl chain length of cation does not affect the thermal stability significantly. Similarly when the *butyl* chain was replaced with a nucleophilic *cyanopropyl* group on imidazolium ring with $[NTf_2]^-$anion, data does not indicate any effect on the thermal decomposition temperature on the same support.

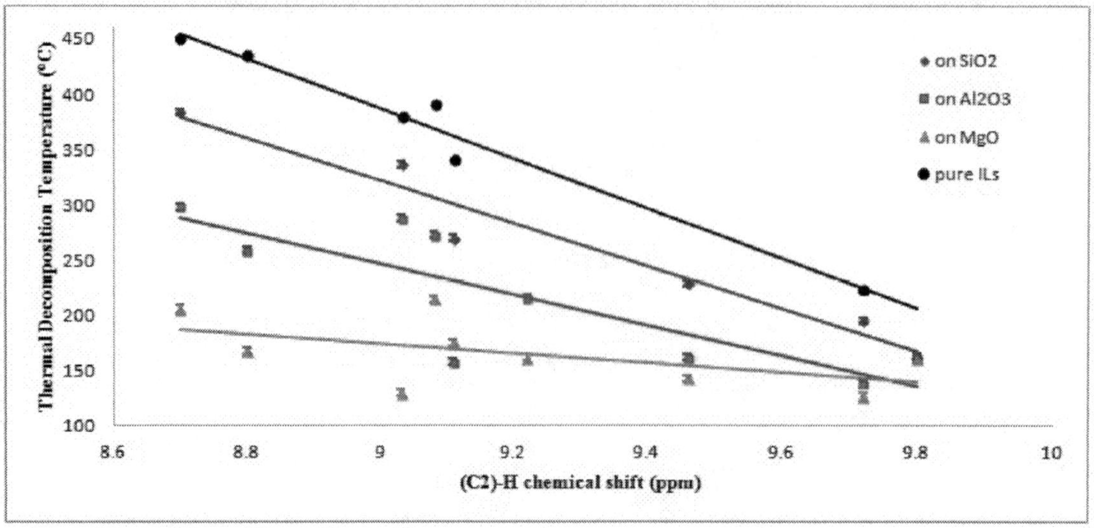

Figure 1. Thermal decomposition temperatures of several ILs with $[BMIM]^+$ cation as a function of (C2)-H chemical shift

As shown in Figure 1, thermal stabilities of ILs with $[BMIM]^+$ cation on various metal-oxides vary linearly with the chemical shift value of acidic (C2)-H on imidazolium ring. Moreover, slopes of these lines depend strongly on the PZCs of the metal-oxide supports. Thus, we infer that there is a strong correlation between the thermal stabilities of ILs with their acidity levels and the PZCs of supports.

In order to complement the TGA-based *short-term* thermal stability results, *long-term* thermal stability limits of ILs on metal-oxides were also investigated. For this purpose, ILs on various metal oxides were characterized by FTIR spectroscopy and SEM analysis with EDX mapping before and after 6 hour-long isothermal treatments at various temperatures. Results illustrated that some ILs start to decompose earlier than the temperatures determined by quick TGA scans while others remain intact up to the TGA-determined decomposition temperatures. As an example, on γ-Al_2O_3 metal-oxide support, $[EMIM][BF_4]$ is stable up to the 534 ± 3 K (temperature determined with TGA scans) while $[C_{10}MIM][BF_4]$ starts to decompose after 473 ± 3 K, earlier than the TGA-determined temperature.

In summary, results presented here provide the basis for selecting stable ILs for IL-assisted supported-catalysts based on their application conditions.

(501e) Cost-Efficient Ionic Liquid and Water Mixtures for Biomass Dissolution

George, A., Joint BioEnergy Institute
Tran, K.,
Sathitsuksanoh, N., Joint BioEnergy Institute
Sun, N., Joint BioEnergy Institute
Mac Dowell, N., Imperial College London
Singh, S., Joint BioEnergy Institute, Emeryville, CA 94608 and Sandia National laboratories, Livermore, CA
Simmons, B., Joint BioEnergy Institute
Welton, T., Imperial College
Hallett, J. P., Imperial College
Originally presented on: 11/6/2013 13:58:00 - 14:20:00

The adoption of lignocellulosic feedstocks in an industrial process used to commercially produce biofuels is contingent upon cost-effectively and efficiently disrupting the lignin present, as it inhibits downstream enzymatic-hydrolysis efficiency. There are numerous reports in the scientific literature on the capability of ionic liquids (ILs) to disrupt cellulose structure and dissolve lignin effectively. Ionic liquids offer exciting prospects due to properties such as their low volatility and "tunability". In this study a new class of ionic liquids were designed and tested for their capacity to improve the enzymatic hydrolysis efficiency of switchgrass. The ionic liquids tested were based upon a sulphate anion, whilst varying the cation. The number of alkyl chains on ammonium based cations were investigated. The ionic liquids were designed so that they can effectivly remove lignin and leave the cellulose intact, thereby enabling easier fractionation post pretreatment. Another benefit of this class of ionic liquids is their lack of sensitivity to H_2O, enhancing their handling properties. A focus was also placed on designing these ionic liquids from ubiquitous low cost chemicals, with the aim of mitigating reagent cost, thereby enhancing their industrial relevance. The metrics used to evaluate these ionic liquids were degree of lignin dissolution, effect on cellulose structure, ease of pretreated material fractionation and enzymatic hydrolysis efficiency.

(501f) Conversion of Residual Ethylene Oxide in the Ionic Liquid Catalyzed CO2 Cycloaddition Process

Sun, J., Institute of Process Engineering, Chinese Academy of Sciences
Cheng, W., Institute of Process Engineering, Chinese Academy of Sciences
Wang, J., Institute of Process Engineering, Chinese Academy of Sciences
Zhang, J., Institute of Process Engineering, Chinese Academy of Sciences
Zhang, S., State Key Laboratory of Multiphase Complex Systems, Institute of Process Engineering, Chinese Academy of Sciences
Originally presented on: 11/6/2013 14:20:00 - 14:42:00

Conversion of residual ethylene oxide in the ionic liquid catalyzed CO_2 cycloaddition process

Jian Sun, Weiguo Cheng, Jinquan Wang, Junping Zhang and Suojiang Zhang

Institute of Process Engineering, Chinese Academy of Sciences,100190, Beijing, PR China

**Contact information: jsun@home.ipe.ac.cn*

Recently, the synthesis of cyclic carbonates via cycloaddition of CO_2 to epoxides is much attractive not only because of its efficiency in CO_2 conversion, but also becuase of its widely derivative high-value chemicals. [1] Although a number of efficient catalysts including ionic liquids (ILs) have been developed, [2] attention on the trace residual epoxide in the reaction system, especially in gas phase, is rarely covered, which could cause a potential explosion of epoxide when unreacted CO_2 is compressed for recyle. As a continous work for CO_2 conversion,[3] herein, studies on the conversion of residual trace epoxide (0.2 wt%) in an IL catalyzed CO_2 cycloaddition process to produce ethylene carbonate (EC) were carried out. The effect parameters including pressure, temperature, time and solvents (e.g. H_2O, EC+ethylene glycol+H_2O, BmimCl, $BmimPF_6$, $BmimBF_4$, etc) were systematically screened, and a >99.9% conversion of EO could be realized by reduding the partial pressure of EO in the system. The results indicated that a vapor-liquid equilibriuma of EO in the reaction system has a more important effect on EO elimination rather than reaction time and temperature.

Refferences

[1] B. Schaffner, F. Schaffner, S.P. Verevkin and A. Borner, ***Chem. Rev***., 2010, 110, 4554.

[2] a) J. M. Zhang, J. Sun, X. C. Zhang, Y. S. Zhao, S. J. Zhang. ***Greenhouse Gas Sci Technol***., 2011, 2:142-159; b) M. North, R. Pasquale, C. Young, ***Green Chem***., 2010, 12, 1514.

[3] a) J.Q. Wang, J. Sun, W.G Cheng, S.J. Zhang. ***Phys. Chem. Chem. Phys.***, 2012, 14, 11021; b) J. Sun, L. J. Han, W. G. Cheng , J. Q. Wang, X. P. Zhang, S. J. Zhang. ***ChemSusChem***, 2011, 4: 502; c) J. Sun, J.Q. Wang, W. G. Cheng, J. X. Zhang, X. H. Li, S. J. Zhang, Y. B. She. ***Green Chem***., 2012, 14: 654; d) J. Sun, W. G. Cheng, W. Fan, Y. H. Wang, Z. Y. Meng, S. J. Zhang. ***Catal. Today***, 2009, 148: 361.

Acknowledgments

This work was supported by National Basic Research Program of China (2009CB219901), and National Science Fund of China (21006117).

(501g) Aggregation and Solubilization Phenomena During Ionic Liquid Co-Solvent Based Lipid and Phorbol Ester Extraction From Biomass

Bandlamudi, S., South Dakota School of Mines & Technology
Cooney, M., University of Hawaii
Benjamin, K. M., South Dakota School of Mines and Technology
Originally presented on: 11/6/2013 14:42:00 - 15:04:00

The use of ionic liquids to separate oils, proteins, and phorbol esters from biomass is a significant achievement in efficiently separating high quality jet and diesel fuels and value-added products from abundant ocean algae and other agricultural feedstock. Micro-algae and oil-seed biomass were suspended in co-solvents mixtures comprised of ionic liquids of varying polarity and polar-covalent molecules and subjected to mild heating under reflux at atmospheric pressure. Yields of bio-oils and phorbol esters extracted were determined by gravimetric analysis and HPLC, respectively. In addition, light scattering and fluorescence techniques were utilized to investigate polarization/anisotropy in the co-solvent mixture, and their effects on lipid solubility and rates of diffusion.

Molecular dynamics simulations have been employed to probe the fundamental interactions between solutes (lipids and phorbol esters) and ionic liquid and polar covalent molecules in solution. Simulations are capable of predicting the mixture density as a function of composition, and compare well with experimental data. Radial distribution functions (rdfs), potentials of mean force (pmfs), and simulation snapshots confirm the increased aggregation between IL cations and anions with increasing methanol concentration.

The aggregation behavior of triolein in IL/co-solvent mixtures was also explored. From the MD generated rdfs and potentials of mean force, we observe a clear correlation between the extent of triolein aggregation and the experimental extraction yields as a function of methanol composition. This indicates that the lipids are aggregating in these IL/co-solvent mixtures, a simulation finding that is also supported by light scattering experiments. Therefore, there is a clear correlation between the free energy of lipid aggregation and the overall extraction yield. Therefore, the dominant mechanism for the lipid auto-partitioning behavior is lipid aggregation, likely leading to the formation of micelles, which subsequently rise through the IL/co-solvent mixture via buoyancy (density difference).

In addition, density functional theory (DFT) calculations using the COSMO-RS implicit solvent model and the COSMO-THERM framework are used to predict phorbol ester solubilities in IL/methanol mixtures, which are subsequently compared to experimental ester extraction results.

Both experimental and simulation efforts provide an improved understanding of the relationship between molecular-level co-solvent structure and solute aggregation/solubilization, which will help in the design of tailored ionic liquid/PCM solvent mixtures to optimize lipid and ester extraction and separation.

(503a) Selection and Crystallization Development of a Practical Salt Form for a Pharmaceutical Compound

Yang, B. S., Boehringer Ingelheim Pharmaceuticals, Inc.
Li, Z., Boehringer Ingelheim Pharmaceuticals, Inc.
Jiang, M., Massachusetts Institute of Technology
Kim, S., Boehringer Ingelheim Pharmaceuticals, Inc.
Wang, X.,
Yee, N., Boehringer Ingelheim Pharmaceuticals, Inc.
Senanayake, C., Boehringer Ingelheim Pharmaceuticals, Inc.
Originally presented on: 11/6/2013 12:30:00 - 12:50:00

This paper presents a fully integrated solid form screen and crystallization development workflow for the development of an active pharmaceutical ingredient (API) at BI. Salt I of an in-licensed compound was initially used in the Phase I and Phase IIa studies. However, Salt I was later characterized as an ethanolate/hydrate and had several development issues, including the high and variable moisture content and unstable physical/chemical properties under various conditions (heat, humidity, mechanical and light). As a result, a new salt form that can provide better developability was required that triggered extensive salt/polymorph screen and selection activities, in order to select a new salt form to be used for Phase IIb and beyond.

After a thorough solid form screen and selection process, Salt II, with its superior physicochemical properties to the other crystalline salts identified, was chosen as the new development form. In the initial development, a workable process (Process I) was developed in EtOH/Heptane that was able to produce multiple kilo scale of Salt II to supply for the formulation development. The particle size was controlled by dry milling initially. Later, a more comprehensive crystallization development was conducted to address several development challenges, including the purification of the major impurity from the synthesis, hydrolysis of the compound in the crystallization system, color removal and powder property control, etc. A new modified process (Process II) was successfully developed, enabled by the selection of appropriate solvent system and design of crystallization condition that can inhibit the hydrolysis of the compound during crystallization. Process II used milled seeds and controlled crystallization to directly produce the desired particle size distribution without milling. Process II was successfully scaled up to multiple 30 kg batches in the Pilot Plant.

(503b) In-Situ Monitoring of Resolution of Chiral Compounds Using Process Analytical Technology

Chen, Z., National Engineering Research Center for Industrial Crystallization Technology
Gao, Y., National Engineering Research Center for Industrial Crystallization Technology, School of Chemical Engineering and Technology, Tianjin University
Wang, Y., National Engineering Research Center for Industrial Crystallization Technology, School of Chemical Engineering and Technology, Tianjin University
Yin, Q., National Engineering Research Center for Industrial Crystallization Technology, School of Chemical Engineering and Technology, Tianjin University
Zhang, M., Tianjin University
Hou, B., Tianjin University
Hao, H., National Engineering Research Center for Industrial Crystallization Technology, School of Chemical Engineering and Technology, Tianjin University
Originally presented on: 11/6/2013 12:50:00 - 13:10:00

In-situ Monitoring of Resolution of Chiral Compounds using Process Analytical Technology

Zhen Chen, Yuan Gao, Yongli Wang, Qiuxiang Yin, Meijing Zhang, Baohong Hou, Hongxun Hao*

National Engineering Research Center for Industrial Crystallization Technology, School of Chemical Engineering and Technology, Tianjin University, Tianjin 300072, P.R. China

* Corresponding author: Tel:+86-22-27405754 Fax: + 86-22-27405754

Email: hongxunhao@tju.edu.cn

Abstract

Since chirality was discovered and defined hundreds of years ago, the resolution and purification of chiral compounds has become one main challenge that scientist have to face. Most physicochemical properties of the enantiomers are nearly the same or pretty close. So it is quite difficult to find an appropriate method to resolve the chiral compounds, especially in industry scale. As one main resolution method of chiral compounds, preferential crystallization has been widely used not only in the labs but also in industries. To develop and design an optimum crystallization processes for chiral compounds, it is crucial to understand the thermodynamic and kinetic mechanism of the processes. Process analytical technologies (PAT) can be used to monitor the preferential crystallization processes in line and are very helpful to further understand the mechanism behind the phenomena. In this paper, several PAT tools were successfully used to in-situ monitor the resolution of L-lactide. Raman Spectroscopy and attenuated total reflectance Fourier transformation infrared (ATR-FTIR) were used to monitor the solid phase and liquid phase concentration respectively. Focused beam reflectance measurement (FBRM) and particle vision measurement (PVM) were used to monitor the particle size and morphology respectively. Some other offline analytical tools, such as X-ray diffraction (XRD), differential scanning calorimetry (DSC), Automatic polarimeter were also used to confirm the online analysis results. Solubility data of L-lactide were also measured experimentally. The effect of operating parameters, such as seeding, cooling profile and the initial concentration, on the preferential crystallization processes was also investigated. It was found that all of these operating parameters can affect the resolution results. The reasons and the mechanism of the influence were discussed in detail. Using these knowledge obtained in this paper, an reliable optimum resolation method with high output capacity and good repeatability for L-lactide were finally developed.

Keywords: Resolution, Chiral compounds, Process analytical technology, Preferential crystallization, L-Lactide

(503c) A Model for Crystal Growth for Solids Crystallizing From Solutions Containing Multiple Conformers and Evidence of Growth Inhibition By the "Wrong Conformer"

Derdour, L., Bristol-Myers Squibb co.
Sivakumar, C., Carnegie Mellon University
Skliar, D., Bristol-Myers Squibb Co.
Pack, S., Bristol-Myers Squibb Co.
Lai, C., Bristol-Myers Squibb Company
Vernille, J., Bristol Myers Squibb
Kiang, S., Bristol-Myers Squibb Company
Originally presented on: 11/6/2013 13:10:00 - 13:30:00

The validity of a model derived for crystal growth rate for solids crystallized from solutions containing multiple conformers is evaluated. A simple and straightforward method for determining crystal growth rates using FBRM and FTIR/ATR is utilized to maximize growth rate data collected from a single experiment. Crystal growth rate data obtained experimentally were in good agreement with model predictions. In particular, the data indicated that when the approach of the right conformer (Derdour et al. Chem Eng Sci., 66(1), pp. 88-102, 2011) applies, crystal growth rate is inversely proportional to crystal size. Further, the results indicate that growth rates determined from supersaturation measurements can be over-estimated because of surface nucleation and/or agglomeration that can result from supersaturation spikes during anti-solvent addition. Lastly, crystal growth data obtained at different temperatures indicated that the slow growth rate observed at low temperatures may be caused in part by surface poisoning by the wrong conformer.

(503d) API Particle Engineering for Drug Substance: Crystallization & Wet Milling to Control API Powder Properties

Schild, R., Bristol-Myers Squibb
Engstrom, J. D., Bristol-Myers Squibb Company
Hubert, M., Bristol-Myers Squibb Company
Jones, J.,
Qian, X.,
Wu, Y., Bristol-Myers Squibb Co.
Leane, M.,
Gamble, J., Bristol-Myers Squibb
Wang, S. S. Y., Bristol-Myers Squibb Company
Originally presented on: 11/6/2013 13:30:00 - 13:50:00

In the pharmaceutical industry, target drug substance powder properties, such as particle size and flow are defined for a given formulation to ensure robust drug product processing. Batch crystallizations are primarily designed to maximize the active pharmaceutical ingredient (API) yield and product purity and frequently produce API's with non-ideal size and morphology. In this investigation, the effect of seeding, supersaturation, and post crystallization wet milling were used to produce API batches with a range of particle sizes and flow properties to define the initial design space of acceptable content uniformity and powder flow. For laboratory and pilot plant scale-up experiments, the particle size was measured real-time during crystallization and wet milling using focused beam reflectance measurement (FBRM) and correlated to spot slurry samples analyzed by laser light scattering as well as image based particle characterization. These results were compared with empirical correlations to account for initial particle size, mill generator configuration, rotor or tip speed, milling residence time, and rotor/stator gap or tooth dimensions to aid successful crystallization and wet milling scale-up.

(503e) Hysteresis in Solid-Liquid Suspensions in Optimax®: Using CFD Simulations and Experiments With FBRM

Vedantam, S., National Chemical Laboratory

Ranade, V. V., National Chemical Laboratory

Originally presented on: 11/6/2013 13:50:00 - 14:10:00

Title: **Hysteresis in solid-liquid suspensions in OPTIMAX®: Using CFD simulations and experiments with FBRM**

*Authors: Vedantam, S., Ranade, V.V.**

Affiliation: Industrial Flow Modeling Group (iFMg), CEPD Division, National Chemical Laboratory (NCL), Pune, INDIA

**corresponding author: vv.ranade@ncl.res.in*

ABSTRACT

Introduction:

Crystallization is the most widely used separation and purification process employed throughout the chemical and pharmaceutical industry for chemical products that are solids at room temperature and pressure. Today, crystallization modeling tools are limited by the availability of methods to reliably predict solubility and super saturation for chemical systems of industrial importance. There is also limited capability for describing nucleation and crystal growth, and mixing and design scale-up for industrial crystallizers. Tools are needed to model and optimize crystallizer process performance with respect to product quality, throughput and efficiency. It is essential to understand and relate key thermodynamic, kinetic and hydrodynamic aspects to crystallizer performance not just in terms of yield but also in terms of product quality (characterized by particle size distribution, morphology, polymorphism and the amount of strain as well as the uptake of solvent or impurities in the crystal lattice). Process Analytical Technology (PAT) tools are also on the rise in these applications and published in the literature (Daneau and Steele., 2005; Samad *et al.*, 2013).

Hydrodynamic modeling, one of the key aspects, is the focus of the work presented in this abstract. In literature, Joshi et al (2011), presented a comprehensive review on CFD modeling studies in stirred tanks, stating that it is worth noting that dense suspensions require a correct modeling of both solid?liquid and particle? particle interactions. However, as an initial step, an automated reactor called OPTIMAX® (launched by Mettler-Toledo Ltd.,) is being used for the study of various aspects of crystallization and the quantitative effect of all key aspects on the process. OPTIMAX® (OM), a synthesis workstation provides a platform for product and process development and is coming into use in bio-pharmaceutical synthesis applications. OM is typically an automated reactor, offering heating electrically and cooling on the basis of Peltier technology. Crystallization is one of the wide hopes of applications to be carried out in this particular workstation. In this particular set-up, OM is designed to accommodate internals such as Focused Beam Reflectance Measurement (FBRM) probe (used for estimating the Chord Length Distributions (CLD) and thus, Crystal Size Distributions (CSD)); Temperature probe (Tr probe), Fourier-Transform Infra Red (FTIR) spectroscopy probe; turbidity probe; pH probe; impeller and baffles as well. However, as for this study, the internals used are FBRM probe; Tr probe and 4-bladed down-pumping PBT as impeller are used. It is understandable that the reactor geometry, operating conditions and the flow characteristics govern the hydrodynamic aspects of the crystallization process thus subsequently also affecting the kinetic aspects. Solid-liquid suspension studies on solids distribution and the effect of operating conditions has been simulated and experimentally verified.

Methodology, flow characterization and mixing time:

Commercial software Ansys Fluent (14.0) has been used for all CFD simulations. Multiple Reference Frame (MRF) approach is used for modeling the impeller. Multiple phases have been simulated using Eulerian approach. 1000 mL reactor volume has been used as the domain for simulations and experiments, with an inner diameter of 0.101 m, and a height of 0.137 m. The impeller used is a 45^0, four-bladed, down pumping PBT. A mesh size of 4, 60,000 cells has been used for the entire domain. Simulations are carried out for impeller speed ranging from 100 to 1000 rpm, and varying solids hold-up of the particles used. The geometry indicating the internals, tank and inner and outer regions of the domain, clearly demarked for MRF approach are shown in the Figure 1. The commonly used validation check points such as Power number (N_P, from turbulent energy dissipation rate and moments), Pumping number (N_Q) are estimated from the CFD predictions, and satisfactory validation has been obtained, which gives a go ahead for further studies on crystallization in this reactor.

Experiments are carried out using FBRM to estimate the distribution of particles, and verify the hysteresis likely to occur with the variation in impeller speed. Neutrally buoyant particles of 350 micron are used as solid particles. Hold-up has been varied from 1 % to 10 % in the reactor. Impeller speeds are varied from 100 rpm to 1000 rpm and the particle size distribution has been tracked using FBRM. Similar conditions have been simulated using CFD.

Conclusions:

Solid-liquid suspension studies in OPTIMAX® are carried out using simulations as well as experiments. The CFD study is validated with the established parameters such as N_P (from estimated turbulent energy dissipation rate and the torque), N_Q from the CFD predictions. Possibility of hysteresis has been verified using FBRM and CFD simulations.

References:

Samad, N.A.F.B.A., Sin, G., Gernaey, K.V., Gani., R. 2013, ?A systematic framework for design of process monitoring and control (PAT) systems for crystallization processes?, Computers & Chemical Engineering, 54, 8-23. http://dx.doi.org /10.1016/j.compchemeng.2013.03.003

Buwa, V. V. and Ranade, V. V., 2003a Characterization of dynamics of gas-liquid flows in rectangular bubble columns, AIChEJ, 50, 2394.

Deneau, E., Steele, G, 2005, An in-line study of oiling out and crystallization, Org. Proc. Res. Dev., 9, 943-950.

Joshi, J.B., Nere, N.K., Rane, C.V., Murthy, B.N., Mathpati, C.S., Patwardhan, A.W., Ranade, V.V., 2011, CFD Simulation of Stirred Tanks: Comparison of turbulence models (Part II: Axial Flow Impellers, Multiple Impellers and Multiphase Dispersions), The Canad. J. of Chem. Eng., 89, 754-816.

Ranade, V. V., Bourne, J. R. and Joshi, J. B., 1991. Fluid mechanics and blending in agitated tanks. Chem. Eng. Sci., 46, 1883.

Figures:

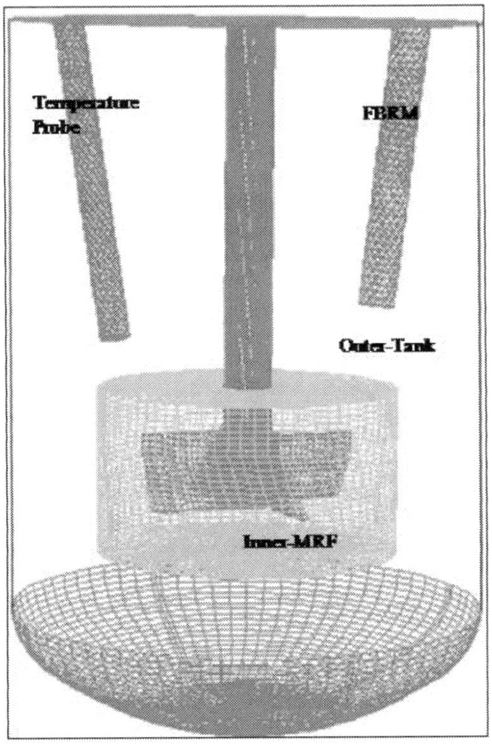

Figure 1. Computational domain showing inner and outer regions of MRF approach

(503f) Development of a Model for Anisotropic Crystal Growth From Solutions

Derdour, L., Bristol-Myers Squibb co.
Chan, E. J.,
Originally presented on: 11/6/2013 14:10:00 - 14:30:00

A model for anisotropic growth was developed. It is based on an isotropic growth model recently developed for crystal growth from solutions containing multiple conformers. The model presented herein implies that growth rates are constrained in one spatial dimension, thus only two linear growth dimensions, namely length and width, are considered. Crystal growth is assumed to occur via step propagation. The model takes into consideration step advance velocity dependence upon supersaturation and solvent/crystal interaction. An approach which uses two solute integration coefficients is introduced that describes crystal growth kinetics in each growth dimension. These coefficients are functions of the linear growth of the faces and are inclusive of solvent/crystal facet interactions. A parameter estimation algorithm to determine these coefficients is described. The algorithm requires experimental data of concentration variation during crystallization and final aspect ratio. The model's predictions of temporal variation of supersaturation and final crystal elongation factor were in agreement with experimental data for a Bristol-Myers Squibb proprietary compound. Finally, model simulations allow for the estimation of integration coefficients for any given aspect ratio.

(503g) A Straightforward Technique to Control the Oxidation of Amorphous Cobalt Nanoparticles During Synthesis Using DMSO

Duggan, J. N., Auburn University
Roberts, C. B., Auburn University
Chaturvedi, A.,
Suzuki, T.,
Originally presented on: 11/6/2013 14:30:00 - 14:50:00

Magnetic nanoparticles have attracted a great deal of interest lately because of their unique magnetic properties and have a wide range of potential applications. For example, nano-scale magnetic particles have enhanced magnetic susceptibility and saturation compared to their bulk-scale counterparts. Cobalt nanoparticles can exhibit a superparamagnetic behavior, which makes these particles particularly useful as contrast agents in Magnetic Resonance Imaging (MRI). However, many of the techniques commonly used to produce well-defined cobalt nanoparticles involve the use of harsh reagents and synthesis conditions, and may require days to produce a single sample of nanoparticles. Furthermore, it is often difficult to produce cobalt nanoparticles because they have a tendency to become overly oxidized, and can subsequently agglomerate and precipitate from the solution.

We have developed a single-step method to synthesize and stabilize cobalt nanoparticles using an FDA-approved, functional solvent, dimethyl sulfoxide (DMSO). Moreover, this research demonstrates that DMSO can be a multi-functional component in nanoparticle synthesis by serving concomitantly as a solvent and as a stabilizing agent that effectively suppresses nanoparticle growth and prevents excessive oxidation of the nanoparticle surface. The size and morphology of the particles were determined using TEM imaging analysis. In addition, we have performed a variety of different characterization techniques to investigate the interaction of the cobalt nanoparticles with the DMSO stabilizing molecules, including FT-IR, XPS, and XRD. Using these techniques, we found that both the sulfur and oxygen functional groups on the DMSO molecules interact with the nanoparticle surfaces via a resonance hybrid structure that stabilizes the nanoparticles. We have also found that these cobalt nanoparticles are amorphous when synthesized at room temperature. Ongoing studies in our lab are focused on subjecting the cobalt nanoparticles to a variety of different temperatures post-synthesis in order to determine the minimum temperature needed to induce oxidation and crystallinity within the particles. In addition, the magnetic behavior of these amorphous cobalt nanoparticles compared to the magnetic behavior of crystalline cobalt and cobalt oxide nanoparticles will be discussed.

(506a) Development of a Novel and Efficient Cell Culture Flocculation Process Using a Stimulus Responsive Polymer to Streamline Antibody Purification Processes

Felo, M., EMD Millipore Corp.
Kang, Y.,
Hamzik, J.,
Originally presented on: 11/6/2013 12:30:00 - 12:55:00

Recent advances in cell culture processes have significantly increased product titers with corresponding elevated levels of cell densities and cellular debris as well as process and product related impurities, necessitating improvements in downstream processing. In this study we have developed an alternative antibody harvest process that includes flocculation using a novel stimulus responsive polymer, benzylated poly(allylamine), followed by depth filtration. The process demonstrates high process yield, improved clearance of cells and cell debris, and efficient reduction of aggregates, host cell proteins (HCP) and DNA for multiple antibodies. A wide operating window has been established through design of experiments condition screening and optimization. This process has achieved residual levels of impurities in the Protein A eluate that potentially meet the requirements of drug substance and thus alleviate the burdens for further impurities removal in the subsequent chromatography steps. In addition, efficient clearance of residual polymer has been demonstrated using a florescence tagged polymer in the presence of a stimulus reagent. This novel and efficient process can be easily integrated into current mAb purification platforms, and may improve current mAb purification processes by overcoming downstream processing challenges.

(506b) Virus Flocculation in the Presence of Protecting Osmolytes

Tafur, M. F., Michigan Tech
Pearson, E., Michigan Technological University
Heldt, C. L., Michigan Technological University
Originally presented on: 11/6/2013 12:55:00 - 13:20:00

Virus Flocculation in the Presence of Protecting Osmolytes

Maria F. Tafur, Eric Pearson & Caryn L. Heldt

Department of Chemical Engineering, Michigan Technological University, 1400 Townsend Drive, Houghton, MI, 49931 USA

Viruses are best known for their ability to cause disease. However, viruses can be manipulated to become vaccines or engineered to deliver genetic material that help combat disease. There is a need to improve the purification of viruses for the production of therapeutic virus particles. The main unit operation for virus purification is chromatography. There are many disadvantages to chromatography, including the need for high product concentration, large solvent quantities, and the low yield of large particles. We have studied the use of protecting osmolytes in the flocculation of viruses for improved purification. Osmolytes are natural, organic compounds that stabilize intracellular proteins against environmental stresses, such as severe temperature or high osmotic pressure, by changing the water content of cells. Protecting osmolytes have the ability to fold proteins by structuring water around the protein surface, which can either stabilize or destabilize the protein. We are currently working with porcine parvovirus (PPV), a non-enveloped, icosahedral, single-strand DNA virus, to demonstrate the ability of osmolytes to flocculate viruses. The structuring of water around the hydrophobic virus is proposed as the method for PPV flocculation. PPV was flocculated with different concentrations of osmolytes, followed by filtration through a 0.20 μm filter. This large pore-sized filter is usually used to retain bacteria, not small viruses. Using high throughput screening, we have discovered that protecting osmolytes flocculate PPV and demonstrate a high removal (80-97%) with the large pore-sized filter. The virus aggregate particles are very fragile and can be broken with high agitation. When the pH of the osmolytes solutions was adjusted close to the isoelectric point of the virus, improved virus removal was achieved. We propose to use protecting osmolyte flocculation as a potential platform purification approach for vaccines and gene therapy vectors.

(506c) Use of Flocculation to Improve Clarification Efficiency in a Microbial Feedstream

Molek, J. R., GlaxoSmithKline
Goklen, K. E., GlaxoSmithKline
Originally presented on: 11/6/2013 13:20:00 - 13:45:00

E. coli is one of the fastest growing, least expensive, and highest product-to-volume systems available for the production of therapeutic proteins. Despite these advantages, efficient clarification of the product stream can be challenging due to its high solids content, propensity for cell lysis, as well as high concentrations of impurities such as DNA and host cell protein. While centrifugation can be used to remove larger particles, such as whole cells, many small particles, such as cell debris and protein DNA complexes, are difficult to sediment. Depth filtration can be used to remove fine particulates but capacities of the depth filters can be limited, especially by high DNA concentration. In this presentation, polycation precipitation is explored as a method to enhance the removal of cells, cell debris and DNA during centrifugation processes. Screening is performed using a combination of high throughput methods and bench scale experimentation. We examined the impact of polycations on processability, particle size distribution, DNA and host cell protein content. Subsequent scale-up work has demonstrated a minimum four-fold increase in depth filter capacity using optimized flocculation conditions.

(506d) Foam Separation for In Situ Biosurfactant Recovery

Winterburn, J., The University of Manchester
Martin, P., The University of Manchester
Originally presented on: 11/6/2013 13:45:00 - 14:10:00

Biosurfactants are naturally occurring amphiphiles with a range of unique properties; including biodegradability and tolerance of a wide range of temperature and pH, making their use in a range of novel applications, such as environmental bioremediation and enhanced oil recovery, attractive. At present the utilisation of biosurfactants outside of high added value niche applications is hindered by a lack of economic production routes, with issues common to many bioprocesses being encountered; low upstream productivity and the subsequent need for costly downstream separation. In aerobic fermentations for production of extracellular biosurfactant nuisance foaming creates further difficulties.

Our current research is focused on applying our previously developed[1] controlled foam separation process to the production of the biosurfactant surfactin from *Bacillus subtilis* and investigating the controllability and scale up of the process. The effects of foaming during batch and fed batch surfactin producing fermentations have been studied, with foam generation being monitored over time. Foaming is associated with the surfactin producing stage of fermentations, which begins at the onset of glucose limitation and lasts for several hours. Treating the biosurfactant production and separation stages of the process as separate unit operations means that optimal conditions can then be established for each process, such as the aeration and agitation rates needed to achieve the oxygen mass transfer required to meet the biological oxygen demand of the cell population. For the uncontrolled foaming case in batch fermentations surfactin recovery was high, > 99%, with a typical total surfactin production of 100 mg. Similarly in fed batch fermentations uncontrolled foaming recovered 92% of surfactin produced, enriched by a factor of 48.

[1]Winterburn J.B et al (2011) Integrated recirculating foam fractionation for the continuous recovery of biosurfactant from fermenters *Biochem. Eng. J* **54**

(506e) Impact of Clarification Treatment On Protein A Purification

Dupont, A.,

Originally presented on: 11/6/2013 14:10:00 - 14:35:00

Technology advances in the biopharmaceutical industry have resulted in significant increases in MAb titer and cell densities. Current clarification technologies such as centrifugation and depth filtration are struggling to keep up with more challenging feeds. There is a clear need for alternative clarification techniques and a focus has been on precipitation technologies. Precipitation technologies have the potential to remove both cells and impurities in a single step, thus compressing the downstream process. EMD Millipore has developed a smart polymer along with the Clarisolve™ filter for clarification of MAbs. The solution provides a disposable clarification option up to 2000 L and also provides significant DNA clearance. The impact of using smart polymer on subsequent purification steps was investigated. Results indicate that smart polymer offers advantages beyond clarification alone. Data will be presented which illustrates that the use of smart polymer provides enhanced Protein A purification and an improvement in the Protein A elution pool stability.

(506f) Characterizing the Effect of Cationic Polyelectrolyte Flocculants for Microfiltration of Cyanobacterial Culture

Ramanathan, M., South Dakota State University
Muthukumarappan, K., South Dakota State University
Gu, Z., South Dakota State University
Originally presented on: 11/6/2013 14:35:00 - 15:00:00

Membrane filtration of blue green algae is an efficient method of harvesting than the high energy consuming centrifugation only when high enough flux is achieved. Three different flocculants were investigated for their performance in response to total filtration time by improving the permeate flux for cyanobacterial culture. The concentration required to achieve flocculation was varied between the flocculants to identify the required concentration. Cellulose and nitrocellulose membranes were compared for their performance in microfiltration. Total filtration time is reduced significantly with the addition of flocculant compared to the control non flocculating culture suspension. Simple cationic polyelectrolyte resulted in larger floc formation, but not resulted in the shortest filtration time like poly amine resin. Cellulose membrane performed better than the nitrocellulose membrane for same flocculant type and concentration. In addition, the optical density of the permeate is measured to observe the turbidity formed by flocculating agent and to evaluate the filtering efficiency. Thus cellulose membrane with poly amine resin at concentration of 0.03% provides the improved permeate flux with least filtration time.

(522a) Absorption Global Standard - Standardisation of Mass Transfer Parameters in Absorption and Desorption

Muller, S., Ruhr University Bochum
Kopatschek, M., University of Bochum
Grunewald, M., Ruhr-Universitat Bochum
Kunze, A. K., Department of Biochemical and Chemical Engineering, Laboratory of Fluid Separations, TU Dortmund University
Lutze, P., TU Dortmund University
Gorak, A., Department of Biochemical and Chemical Engineering, Laboratory of Fluid Separations
Originally presented on: 11/6/2013 15:15:00 - 15:40:00

?Absorption Global Standard'

Standardization of mass transfer parameters in absorption and desorption

1. Motivation

Dimensioning of absorption and desorption columns is highly dependent on applied mass transfer parameters, which need to be determined experimentally. However, a partially significant deviation in the value of mass transfer coefficients determined in the past has been observed, thus their comparability is restricted. Based on deficient transparency, reasons for the variation between mass transfer coefficient are often unknown. Consequently, high safety factors for dimensioning of columns are necessary and following high construction costs. Therefore, a standardization of mass transfer measurements in absorption is essential for a consistent database of mass transfer parameters. Hence, increasing application of rigorous mass transfer models for absorption process design will be more effective. Due to the need of standardization the project Absorption Global Standard' was launched in fall 2011. In this project a consortium of universities, producers and users of packings and absorption columns are involved.

2. Goals

The aim of this project is to develop a method for consistent experimental determination of volumetric mass transfer parameters. In addition to the experimental standardization of measurements, a mathematical method is established that gives the opportunity to designate phase specific mass transfer parameters and effective interfacial area. Based on this, transparency in development of both methods is essential to guarantee global acceptance and comparability of determined mass transfer parameters. For this purpose a detailed description of experimental method will be published at the end of this project. Additionally, a standardized mathematical method is summarized in a tool and will be published along with the experimental method. Thus, database for mass transfer coefficients become more reliable and safety factors decrease. Hence, costs and resources for column design can be optimized. In this work status quo of the project will be shown, with focus on the current state of experimental method development.

3. Approach

The approach within this project can be classified into three periods. The first period is a qualitative and quantitative analysis of the influence of column configuration parameters and preparation parameters on hydrodynamic and mass transfer results. Thereby, the following configuration parameters are analyzed regarding their influence:

- Liquid distributor
- Gas distributor
- Packing Height
- Sampling
- Demister

Additionally, the following preparation parameters are analyzed:

- Pretreatment of packings
- Inserting random packings
- Water quality

In the second period of the project a standardized experimental method is established. Therefore, characterization experiments with respect to the results of the qualitative and quantitative analysis of the first period are arranged. Those are carried out with selected structure packings of Mellapak 250.Y and random packings of Pall Ring 25 M and 50 M at different facility plants. For the estimation of phase specific mass transfer parameters the standardized mathematical method is used. The results will show the comparability of the standardized methods and subsequently the determined mass transfer parameters.

In the third and last period an investigation of interactions between experimental data and calculation pathway of the mathematical method is performed. For this purpose, a detailed local and global sensitivity analysis shows the influence of chemical system data for the standardized calculation method regarding HTU-NTU evaluation.

This project is founded by the ?Forschungs-Gesellschaft Verfahrens-Technik e.V.?, IGF-Vorhaben 17116N.

(522b) Studies On a Newly Developed Modular Packed Absorption Column

Müller, S., Ruhr University Bochum

Grünewald, M., Ruhr-Universität Bochum

Originally presented on: 11/6/2013 15:40:00 - 16:05:00

Studies on a newly developed modular packed absorption column

1. Motivation

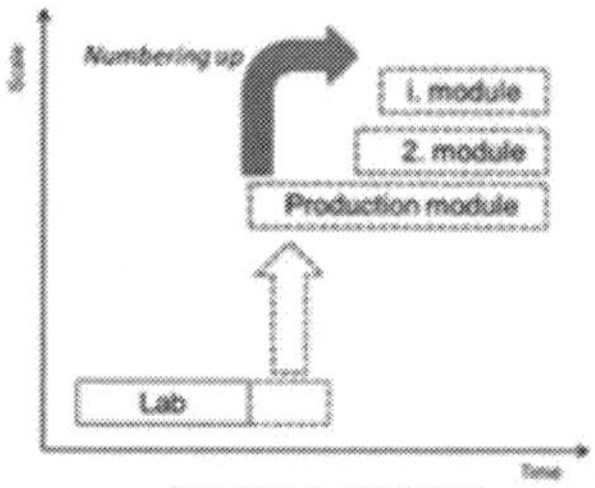

Increasing product differentiation and changing from commodities to specialized products are reasons for shortened product life circles in chemical engineering. Traditionally, large scale-plants possess inadequate flexibility in capacity and product portfolio based on their ?one purpose' development. Therefore apparatus with increased flexibility, like modular devices, becomes necessary to handle fluctuating customer demands. Due to modular assembly capacity can be increased either by equaling up modules from general structures or by numbering up equipment. Hence, a ?smart scale-up' of the complete modular apparatus becomes possible. In favor of smart scale-up only the operating parameters of the single module are necessary and thus reduced time-to-market periods for the products (see figure 1). In fact the challenge on scaling-up modular devices is an equal feed flow for every single module.

Modular devices are state of the art in chemical engineering. Apparatus types like plate heat exchangers or filter presses are established technologies with a long history of development. Nevertheless, most of other apparatus principles are not modular in terms of adaptable capacity and operational flexibility. Even continuously working multiphase reactors and separation units are scale-up by the traditional similarity analysis. Therefore our chair is running several research projects presently dealing with the modularization of multiphase units like packed columns and bubble column reactors for example.

In this contribution first experimental studies on a newly developed packed absorption module are compared with the operation behavior of a conventional packed column.

2. Experimental Studies

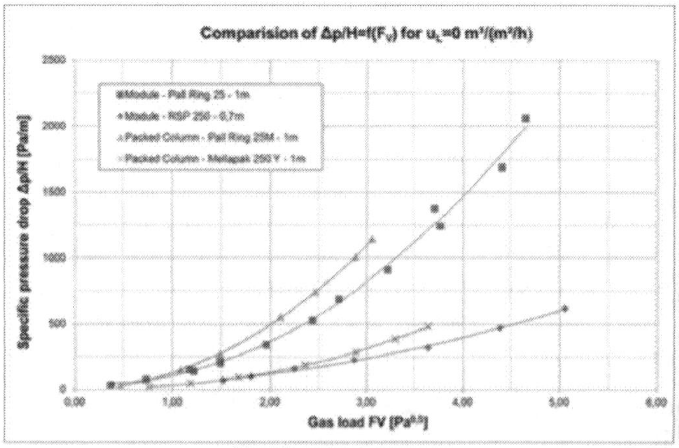

Figure 3: Comparison of pressure drops

Figure 2: Packed absorption column

For the experimental studies a packed absorption module with rectangular cross-section was developed (see figure 2). Firstly, experimental studies regarding hydrodynamics and mass transfer efficiency of the module have been performed. Those studies have been performed with dumped and structured packings for liquid and gas loads up to the flooding point. In detail, this work presents results of measurements with Pall Ring 25 M and Raschig Super-Pak 250. A comparison of experimental results between a packed absorption module and conventional packed columns shows the practicality of this new apparatus (see figure 3). With regard to identical packings, the module's results are comparable to conventional packed columns. Although the single module shows the expected losses in capacity for the gas and liquid load, it shows comparable results for the mass transfer efficiency in consideration of the reduced cross-section. Particularly with regard to the distribution of the fluids in the module, experimental studies have been performed to show the influence of the rectangular cross section and thus the shortened distance for liquid and gas to the wall. For qualitative and quantitative investigations of the wall effect a new in-situ measurement method for the thickness of the wall near liquid film in packed columns will be developed. By a detailed examination of the wall effect inside the module, internals will be designed to minimize the wall film thickness and increase the mass transfer efficiency.

For an investigation of the gas distribution, studies have been performed with fog to visualize the stream of the gas inside the rectangular absorption module. The tests with fog will show the need of a gas distributor and the behavior of gas stream inside the newly module.

3. Interconnection of the modules

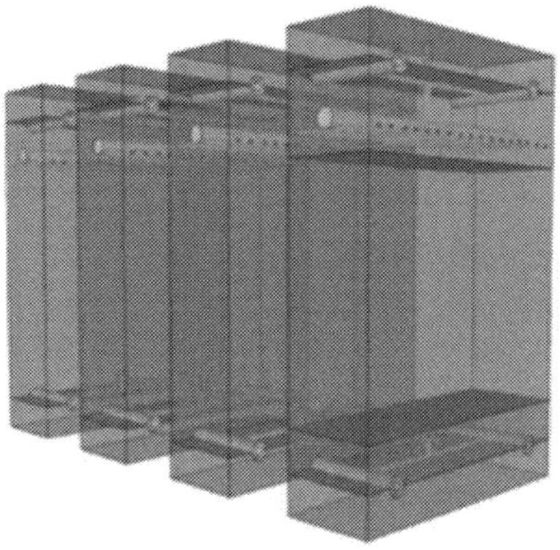

Figure 4.: Design and interconnection principle of a modular absorption unit

As aforementioned the utilization of the packed absorption module can be increased by a multiple connection of single modules, the so-called *numbering-up*. Purpose of this work is to show possible design studies to connect single modules (see figure 4). These studies indicate the conduction of gas and liquid flows and possible integration of the modules into existing modular concepts like frames of plate heat exchangers.

(522c) Separation of Small Paraffin-Olefin Mixtures With An Ionic Liquid. Thermodynamic Study Using COSMO-RS

Ruiz, E., University of California Berkeley
Afzal, W., UC Berkeley
Liu, X., Xi'an Jiaotong University
Yu, G., Research Laboratory of Green Chemistry and Technology, Institute of Process Engineering, Chinese Academy of Sciences
Ferro, V.,
Palomar, J., Universidad Autonoma de Madrid,
Prausnitz, J. M., UC Berkeley
Originally presented on: 11/6/2013 16:05:00 - 16:30:00

Separation of small paraffin-olefin mixtures with an Ionic Liquid. Thermodynamic study using COSMO-RS

Author: Elia Ruiz Pachón

ABSTRACT

Ionic liquids are nonvolatile solvents with desirable properties to be potential candidates for separation processes. The physicochemical properties of ILs can be tuned by different cation - anion combinations. In this paper, we use COSMO-RS to select the most promising ILs for the paraffin/olefin separation. We simulate the IL in two ways: [CA] and [C+A]. We estimate Henry's constant of ethane, ethylene, propane, propylene, butane, butene, pentane, pentene, hexane and hexene using [C+A] COSMO model. We compared our calculations with some experiments reported in literature. The paraffin or olefin solubility in the IL is higher as increasing in the long chain of the hydrocarbon. We calculate the relative volatility and the selectivity of ethane/ethylene, propane/propylene and butane/butene. Because the [C+A] COSMO predictions overestimate the experiments, we consider the [CA] COSMO model. [CA] model deviations are smaller than those of [C+A]. Finally, from a thermodynamically point of view, we propose new promising IL for capturing ethane, propane, and butane from ethylene, propylene and butene respectively.

(522d) Carbon Dioxide Absorption In Amine-Ionic Liquid Hybrid Solvents

Osman, K.,
Ramjugernath, D., University of KwaZulu-Natal
Coquelet, C., Ecole Nationale SupÃ©rieure des Mines de Paris
Originally presented on: 11/6/2013 16:30:00 - 16:55:00

The amount of carbon dioxide (CO_2) emitted into the atmosphere by various industries is of great concern due to the global warming effects of CO_2, and industries worldwide are faced with increasing pressure by environmentalists, and the public at large, to reduce their emissions of the gas. In 2008, nearly 30000 mega tonnes (Mt) of CO_2 was emitted into the atmosphere[1]. This is nearly a 100% increase since 1971. The increasing worldwide demand for relatively cheap electricity and the advances in coal-to-liquids technology has resulted in 42.9% of CO_2 emissions stemming from coal industries, with oil and gas industries accounting for 36.8% and 19.9% of CO_2 emissions respectively[1].

South Africa produces 224 million tonnes of coal annually and 25% is exported, making South Africa the third largest coal exporting country. It is estimated that South African coal reserves amount to 53 billion tonnes, enough to meet the country's electricity needs for nearly 200 years to come[2]. It is estimated that 53% of coal produced in South Africa is used to generate electricity, while 36% of electricity produced worldwide stems from the burning of coal[2].

The primary electricity utility in South Africa is Eskom Ltd.. While the company harnesses energy from a diverse range of sources, at least 77% of its electricity production stems from the burning of coal at its 14 coal-fired power plants[2]. South Africa enjoys cost effective electricity due to its abundant coal reserves, but coal energy also results in the country emitting substantial amounts of CO_2. Eskom's coal operations have made it the second highest CO_2-emitting company in the world.

The flue gas emitted by a pulverised coal power plant typically contains 13 vol% CO_2, 68 vol% N_2, 16 vol% water, 3 vol% O_2, and lower concentrations of other components, 200ppm SO_2, 60ppm NO_X, 60 ppm hydrocarbons[3]. The flue gas is typically emitted at 1-1.7 bar[4].

A conventional technique to remove CO_2 from the flue gas and hence prevent the emission of CO_2, is to pass the flue gas through an absorption column containing a solvent. The solvent selectively absorbs CO_2 while allowing other flue gas components to be emitted. The CO_2-loaded solvent is then heated in a regeneration column to release the CO_2 for compression and storage, while the solvent is recycled.

There are many solvents under investigation for the energy efficient absorption of CO_2. Conventional solvents are amine based including, Monoethanolamine, Diethanolamine, Methyl diethanolamine, Diglycol Amine, Triethanol Amine, Methyl Monoethanolamine, Amino Ethyl Ethanolamine, Ethyl Amino-ethanol, and Butyl Aminoethanol[5]. Amines achieve reactive absorption of CO_2 with a high absorption rate at low pressure, in comparison to physical solvents. A common shortcoming of amines is their high corrosivity, requiring amine solvents to be diluted with water at a composition 50 to 70% H_2O by mass. This not only limits the absorption capacity of the solvent, but also results in very high energy costs when heating the solvent in a regeneration column to isolate CO_2.

A relatively new approach is the study of CO_2 absorption in ionic liquids. Ionic liquids are physical solvents composed entirely of ions. They are chemically and thermally stable, and non-corrosive . In this research we investigated the effect of combining amines with ionic liquids instead of water. The effect was studied in terms of CO_2 solubility in the solvent. The amines investigated were Monoethanolamine (MEA), Diethanolamine (DEA), and Methyl-Diethanol (MDEA) amine. The ionic liquids studied were [1-Butyl-3-Methyl Imidazolium] [Tetrafluoroborate] ([bmim][BF_4),] and [1-Butyl-3-Methyl Imidazolium] [bis(trifluoromethylsulphonyl) imide] ([bmim][Tf_2N]).

The solvents studied contained 45 to 100% of the above ionic liquids (by mass), with the above amines occurring singularly or in combinations at compositions of 10 to 50% by mass.

All solvents were contacted with CO_2 gas in order to measure the equilibrium CO_2 loading in each solvent. CO_2 loading measurements were conducted at CO_2 partial pressures ranging from 0.5 to 15 bar, and at isotherms of 303.15, 313.15, and 323.15 K (approximately).

Measurements were conducted using an Intelligent Gravimetric Analyser (IGA-01) constructed by Hiden Analytical Ltd., which utilises a sensitive microbalance to track the absorption of CO_2 gas into the solvent by noting an increase in the mass of the solvent. The microbalance has a resolution of ±0.0001 mg or ±0.1 µg. The apparatus achieves a vacuum of 1e-7mbar and operates at pressure up to 20 bar. Temperature of 273.15K to 773.15 K is achievable using a refrigerated waterbath or cryofurnace. 80 to 100 mg solvent was required for loading into the stainless steel reactor.

The low quantity of sample required and the programmable nature of the apparatus, made absorption measurements efficient and less time consuming. The weight change of the sample as gas absorption occurs was calculated by taking into account all contributing forces including the mass of the counterweight and its associated attachments (hook and chain), the sample container and its associated attachments (hook and chains), the sample itself, and the buoyancy force exerted by the gas in the reactor.

Due to the sample being of liquid form, there is a significant change in the sample density upon absorption of the gas, thus making the sample significantly more buoyant and altering the actual weight reading of the sample. This buoyancy force was taken into account by first noting the weight reading of the sample at different gas densities using a non-absorbent gas[6]

CO_2 solubility measurements were presented in terms of CO_2 mole fraction versus CO_2 partial pressure for each sample and at each isotherm.

The data for the above systems are to be presented and discussed, along with the modelling of the data using the Soave-Redlich-Kwong (SRK) equation of state for CO_2 absorption into the ionic liquid component, and Deshmukh-Mather model for reactive absorption of CO_2into primary, secondary and tertiary amine components.

The measurements found that the combining of amines with ionic liquids was significantly beneficial to the absorption of CO_2 in the solvent, in comparison to the use of pure ionic liquids. All hybrid samples containing amines and ionic liquids achieved higher CO_2 absorption at low pressures of 0.5 to 4 bar, than conventional amine solvents diluted with water[7], as well as pure ionic liquid solvents measured.

However, at pressure greater than 4 bar, certain hybrid solvents, particularly those containing [bmim][BF_4] ionic liquid and higher compositions of secondary and tertiary amine achieved lower CO_2 absorption than pure ionic liquid solvents of [bmim][BF_4] and [bmim][Tf_2N], indicating superior absorption capacity of pure ionic liquids over hybrid solvents. Hybrid solvents containing the [bmim][Tf_2N] ionic liquid however, achieved superior CO_2absorption compared to pure ionic liquids even at high pressure.

Overall, the combining of amines with ionic liquids resulted in significantly higher CO_2absorption, indicating that this direction of research is certainly worth pursuing further.

References:

1. International Energy Agency (IEA), 2010, "Key World Energy Statistics – 2010", International Energy Agency, Paris Cedex, France.
2. Eskom Power Generation, 2011, "Eskom – Coal Power", Eskom Holdings Ltd., South Africa. Accessed 28/02/2011. http://www.eskom.co.za/live/content.php?Item_ID=279
3. Brennecke J.F. and Gurkan B.E., 2010, J. Phys. Chem. Lett., 1, 3459–3464.
4. National Energy Technology Laboratory (NETL), 2010, "Doe/Netl Advanced Carbon Dioxide Capture R&D Program: Technology Update", NETL, U.S.A. Accessed 25/5/2011. http://www.netl.doe.gov/technologies/coalpower/ewr/pubs/CO2%20Capture%20Tech%20Update%20Final.pdf
5. Ma'mun S., Nilsen R. and Svendsen H.F., 2005, J. Chem. Eng. Data, 50, 630-634
6. Macedonia M.D., Moore D.D., and Maginn E.J., Langmuir 2000, 16, 3823-3834.
7. Park S.H., Lee K.B. , Hyun J.C. , Kim S.H., 2002, Ind. Eng. Chem. Res., 41, 1658-1665

(522e) Theoretical and Experimental Investigation of An Absorption Refrigeration System Using R134/[Bmim][PF6] Working Fluid

Kim, S.,
Kohl, P. A., Georgia Institute of Technology
Originally presented on: 11/6/2013 16:55:00 - 17:20:00

Energy recycling is an overarching goal in many areas including high performance computing and data storage systems where all the electrical energy is converted into low-quality waste heat. To this end, the first experimental demonstration of an absorption refrigeration system operating on recycled energy using a safe refrigerant pair has been achieved. The refrigerant was 1,1,2,2-tetrafluoroethane (R134) and the absorption media was an ionic liquid (IL), 1-butyl-3-methylimidazolium hexafluorophosphate ([bmim][PF_6]). ILs are liquid salts at ambient temperature and are a promising class of absorbents in an absorption refrigeration system because they have tunable properties, high molar uptake for specific gases, zero vapor pressure, high thermal stability, and environmental safety.

There is limited information on the thermodynamic and transport properties of IL/refrigerant mixtures at the temperatures and pressures that are relevant to operation of the refrigeration systems. A Redlich-Kwong type equation of state model was employed to analyze potential working fluids and overall system performance. The two phase pressure drop equation and binary interaction parameters were introduced to improve the computational model. Using the model, the R134/[bmim][PF_6] fluid pair was compared to previous studies using 1,1,1,2-tetrafluoroethane (R134a) and the same IL. The R134/[bmim][PF_6] fluid pair had up to 92% greater cooling-to-total-energy efficiency than the R134a/[bmim][PF_6] fluid pair even though R134 and R134a are isomers and have similar physical properties. The coefficient of performance (cooling capacity/pumping work) of the R134/[bmim][PF_6] fluid pair was up to three times larger than that of R134a/[bmim][PF_6] fluid pair when only waste heat was used at the desorber. The maximum projected coefficient of performance was 161 for the R134/IL system.

A working refrigeration system with R134/[bmim][PF_6] was constructed. For enhanced performance and miniaturization, microchannel heat exchangers were fabricated on copper plates for evaporator, condenser, absorber, and desorber. The system performance was experimentally evaluated using the measured pressure, temperature and heater power values. The measurement of the system performance showed that R134/[bmim][PF_6] had 1.9 times larger cooling capability compared to R134a/[bmim][PF_6] at a desorber temperature as low as 63°C.

(522f) Salidroside Separation By Magnetic Polysulfone Microcapsules Containing Tributyl Phosphate

Shen, S., Hebei University of Science and Technology
Ren, S.,
Originally presented on: 11/6/2013 17:20:00 - 17:45:00

The active ingredients of Tibetan Rhodiola are salidroside and tyrosol determined by the HPLC method. Different extractants, such as tributyl phosphate (TBP) and N-octanol ,and the change of pH were investigated to recover the main component in the Rhodiola extract. A new method was developed for the salidroside separation from Rhodiola rosea by using magnetic polysulfone (PSF) microcapsules containing Fe_3O_4 particles and tributyl phosphate (TBP) (TBP/Fe_3O_4@PSF microcapsules) in this manuscript. The magnetic polysulfone microcapsules were prepared based on the phase-inversion method, and characterized by using Scanning Electron Microscopy (SEM), Infrared Spectroscopy (IR), Thermal Gravimetric Analysis (TGA) and Vibrating Sample Magnetometer (VSM). Magnetic microcapsules could provide easier separation in the external magnetic field. Results showed that salidroside could be separated from aqueous solution at pH 8.0. After three cycles, the magnetic microcapsules had almost the same extraction performance. These novel magnetic microcapsules have potential applications in the separation and purification of natural products.

(531a) Molecular Simulation Methods to Understand the Partition Behavior of Charged Solutes in Biomembranes

Jakobtorweihen, S., Hamburg University of Technology
Ingram, T., Hamburg University of Technology
Smirnova, I., Hamburg University of Technology
Originally presented on: 11/6/2013 15:15:00 - 15:33:00

Interactions of charged solutes with biomembranes (liposomes, lipid bilayers) are far less investigated than the interactions of neutral solutes. However, they are of outstanding interest for toxicological studies, environmental science, drug design and the food industry. In the drug industry for instance, more than 60% of pharmaceutical substances are ionizable. Furthermore, the technical separation of charged compounds from aqueous solutions is very challenging. Especially, for biotechnology applications efficient separation steps are crucial for the establishment of new processes.
The partition behavior of charged solutes between liposomes or micelles and aqueous solutions is not fully understood yet. Different theories exist that explain the transport mechanism of charged solutes. In theoretical studies it was shown that the charged solutes are hydrated in lipid bilayers, so called "water fingers" are formed. Another theory states that charged solutes are transported as ion-pairs. A third possibility is the protonation of charged solutes before transport into membranes.
In this work molecular dynamics (MD) simulations were used to understand the transport mechanism of charged solutes into lipid bilayers. Free energy profiles were calculated with different techniques. For the first time partition coefficients from calculated free energy profiles are rigorously compared to experimental results. The predictions are in good agreement with experimental results if the force filed is precise enough and the simulation set-up was appropriate. For instance, the simulation time to reach an equilibrated free energy profile was carefully investigated.
The MD simulations are used to investigate the pH dependence of partition coefficients and the salt concentration influence on partition of charged solutes. Experimentally it was shown that partition coefficients of charged solutes can increase with increasing salt concentration. However, the mechanism causing this dependency is not understood. We simulated the charged solutes in different configurations (protonated, ionized, ion-pair) at different salt concentrations. By comparing the predicted partition coefficients with experimental results, we could elucidate the partition mechanism. Especially, it was found that ion-pair formation does influence the result.
Additionally, the recently introduced extension of the thermodynamic model COSMO-RS to micelles and bilayers (termed COSMOmic) was investigated to show its comparability with MD results and its applicability to charged solutes.

(531b) Asymmetric Cross-Linked Polymer Membranes Constructed and Characterized By Molecular Dynamics Modeling and Simulations

Wang, J. C., Missouri University of Science and Technology
Liapis, A. I., University of Missouri-Rolla
Originally presented on: 11/6/2013 15:33:00 - 15:51:00

One of the most important properties in determining the performance and application of a membrane is the porous structure and more specifically the distributions in the membrane of the pore sizes and of the pore connectivity. In polymer membranes, the porous structure is determined mainly by the mutual interactions between the conformationally and configurationally flexible polymer chains, and the membrane structural and dynamic properties can be further modified by physical and/or chemical means such as cross-linking which covalently connects the polymer chains together. When the pore sizes are made to be characteristically dissimilar in different parts of a membrane, the result is an asymmetric membrane that, in contrast to conventional membranes with comparable pore sizes throughout, can provide additional advantages including fast and consistent molecular transport while requiring lower in magnitude gradients of externally applied forces. From both fundamental and practical points of view, asymmetric polymer membranes can be prepared by systematically utilizing cross-linkers differing in sizes and/or different extents of cross-linking in different parts of the membrane. While this appears to provide a plausible strategy for the molecular design of membranes, its practical feasibility still remains to be realized and requires detailed characterization. We present in this work a proof-of-concept study that employs the molecular dynamics (MD) modeling and simulation methodology to construct and characterize composite asymmetric polymer membranes formed by dextran chains and cross-linked by polyols. This selection of model membrane systems is based on the rationale that dextran and polyols provide desirable biocompatibility and good solubility in water, which have led to many bio-related applications. The pore structures formed by the dextran chains are characterized directly in terms of the distributions of the number of pore openings, pore radii, and pore volumes by dividing each of the polymer thin films into two levels of cubic lattices at the angstrom level and utilizing thin disks of varying sizes at each lattice point. The degree of pore connectivity can then be ascertained by the intersection between distribution curves that represent pores of different size ranges. The interaction energetics and effective mass transfer coefficients of water molecules and their distributions on the two different sides of the asymmetric membranes are evaluated and are used as an indication of the relative difficulty or ease associated with water sorption and desorption, which represent parts of the mechanisms underlying the functioning of polymer membranes. In general, the pores are larger in diameter and the polymer film is more flexible on the side of the asymmetric membrane having a lower dextran density. Within a distance corresponding to approximately three water molecular diameters from the dextran chains and the polyol cross-linkers, water molecules, in particular those adjacent to the dextran chains and cross-linkers, are found to have greater interaction energetics and, thus, are more strongly bound, and their mass transfer coefficients are significantly reduced in magnitude.

(531c) Application of Evapoporometry to Characterizing the Pore-Size Distribution of Hollow Fiber Membranes

Krantz, W. B., University of Colorado
Akhondi, E.,
Wicaksana, F.,
Originally presented on: 11/6/2013 15:51:00 - 16:09:00

Application of Evapoporometry to Characterizing the Pore-Size Distribution of Hollow Fiber Membranes

Ebrahim Akhondi[a], Filicia Wicaksana[a,b], William B. Krantz[a,c]

[a]Singapore Membrane Technology Center, Nanyang Technological University, Singapore

[b]Dept. of Chemical and Materials Engineering, University of Auckland, New Zealand

[c]Dept. of Chemical and Biological Engineering, University of Colorado, USA

Evapoporometry (EP) is a novel means for characterizing the pore-size distribution (PSD) in membranes. It is based on the principle that the vapor pressure is affected by the curvature of a volatile liquid contained in a porous material that is described by the Kelvin equation.Â If a wetting liquid is used to completely fill the pores, the vapor pressure will be depressed, whereas if a non-wetting liquid is used, the vapor pressure will be enhanced.Â If the membrane is pre-saturated with a wetting volatile liquid, the evaporation rate will monotonically decrease as a function of time because the liquid will evaporate progressively from the largest to the smallest pores. This is because the gas phase immediately adjacent to the membrane surface will be supersaturated in the vapor of the volatile liquid with respect to any pores having a diameter smaller than that of the pores that are draining.

EP involves first saturating the membrane with a volatile liquid that is not a solvent or swelling agent for the membrane material. The sample is then placed in a specially designed test cell on the pan of a microbalance that permits measuring the mass as a function of time. The slope of the mass versus time curve provides the evaporation rate, which can then be related to the vapor pressure at the interface between the liquid in the porous material and the ambient gas phase.Â The vapor pressure in turn can be related to the pore diameter via the Kelvin equation. The PSD then is then expressed as the mass fraction of pores as a function of the pore diameter; that is, EP gives a mass-based PSD and mass-based average pore diameter. In contrast to other techniques for determining the PSD, EP does not require assuming a pore geometry.

In prior studies EP was applied to characterizing the PSD of flat sheet membranes using only isopropyl alcohol (IPA) as the volatile wetting liquid. In this presentation the adaptation of EP to characterizing the PSD of polyvinylidene fluoride (PVDF) and polyacrylonitrile (PAN) hollow fiber ultrafiltration (UF) membranes using both IPA and water as the volatile wetting liquids will be discussed. The EP test cell and procedure had to be modified to properly seal the hollow fibers at the base of the test cell and to accommodate the free convection mass transfer associated with using water whose vapor, in contrast to IPA, is less dense than air.

EP characterization of the PSD is compared to that determined by liquid displacement porometry (LDP). This comparison indicates that the high pressures used in the LDP characterization (as high as 3.5 MPa) caused compaction of the UF membranes that resulted in the shift of the PSD to smaller pores. EP offers an attractive alternative to LDP since it does not require using any transmembrane pressure or specialized expensive equipment.

(531d) Sorption, Swelling, and Free Volume of Polybenzimidazole (PBI) and PBI/Zeolitic Imidazolate Framework (ZIF-8) Nano-Composite Membranes for Pervaporation

Shi, G. M., National University of Singapore
Chen, H. M., University of Missouri-Kansas City
Jean, Y. C., University of Missouri - Kansas City
Chung, T. S., National University of Singapore
Originally presented on: 11/6/2013 16:09:00 - 16:27:00

We have studied alcohol- and water-induced swelling and sorption of PBI and PBI/ZIF-8 nano-composite membranes and characterized both dry and wet membranes by positron annihilation lifetime spectroscopy (PALS). The degrees of membrane swelling after immersing in alcohols and water follow the order of ethanol > methanol > water > n-butanol. Addition of ZIF-8 particles into PBI membranes suppresses the ethanol-, methanol- and water-induced PBI swelling. The water-induced swelling is subdued most severely because of the hydrophobic nature and rigid structure of ZIF-8 particles, while the n-butanol-induced swelling is enhanced owing to a greater free volume in the PBI/ZIF-8 membrane. The PALS studies confirm that the high pervaporation permeability of PBI/ZIF-8 membranes is attributed to the high fractional free volume (FFV) created by large cavities of ZIF-8 particles. In addition, the alcohol saturated PBI/ZIF-8 membranes have significantly smaller free volume radii than alcohol saturated PBI membranes. A correlation among FFV, pervaporation permeability and sorption diffusion coefficient has been developed. This study also demonstrates the importance of using PALS to characterize membrane structure changes in solvent-saturated membranes at a molecular level and to correlate pervaporation performance with solvent-induced swelling.

(531e) Three-Dimensional Imaging of Polymeric Membranes for Determination of Critical Structural Metrics in Osmotic Membranes

McCutcheon, J., University of Connecticut
Manickam, S., University of Connecticut
Originally presented on: 11/6/2013 16:27:00 - 16:45:00

Engineered osmosis (EO) is a platform membrane separations technology with applications in water desalination (forward osmosis, FO), power production (pressure-retarded osmosis, PRO), and dewatering (direct osmotic concentration, DOC). A key obstacle in EO is the lack of an appropriately designed membrane that minimizes mass transfer resistance during osmosis. The most common resistance has been identified as internal concentration polarization (ICP), which is a film resistance that occurs in an asymmetric membrane's support structure. ICP combines the resistance to diffusion of solutes in contact with the membrane with the resistances imparted by the membrane support layer itself. The support layer properties that most impact ICP are the porosity, the thickness and the tortuosity of the overall structure. These are often combined into a single parameter, S, which represents a combined term (thickness X tortuosity/porosity). However this parameter has so far only been calculated from experimental flux measurements combined with model fitting, leading to erroneous calculations that can vary greatly from test to test because of inaccurate underlying assumptions. One could directly calculate these parameters if they were easily measured, but there are a lack of suitable characterization techniques for soft materials that enable these types of measurements. In this study the asymmetric structures of two commercially available thin film composite reverse osmosis (RO) membranes, the BW30 and SW30-XLE from Dow Water and Process Solutions, have been studied using an analytical (mercury intrusion porosimetry) and a 3D imaging (x-ray computed tomography) technique. A multi-length scale approach was used for the imaging technique in order to study both the bulk properties as well as visualize the microporous structure of the support layer. Structural metrics (porosity, tortuosity and thickness) of the membranes were then calculated from which, for the first time, structural parameters were directly measured. Large variations were observed between the "actual" structural parameter, from either technique, and the values obtained from osmotic flux tests. The results of the study indicate a need to revise the traditional approaches of characterizing membrane structures overall for any application.

(531f) Improved Organic Acid Purification Through Electrodeionization With Ionic Liquids

Lopez, A. M., University of Arkansas
Puente, A., University of Arkansas
Hestekin, J. A., University of Arkansas

Originally presented on: 11/6/2013 16:45:00 - 17:03:00

Organic acids and salts are used in a wide variety of industries, including petrochemical, cosmetic, and pharmaceutical industries. Research focused on the improvement of synthesis and purification techniques has been of great interest to these industries. The major limiting step in this process is the removal of wastewater and other impurities from the organic product. Therefore, many separation techniques have been employed to purify organic acids and salts for commercial use. Electrodialysis has shown significant promise due to the low energy costs associated with separation and the selectivity that membranes possess. However, this process does not remove water from the product solution, so further modification is still needed. Recently, our research has showed that the use of ionic liquids in electrodialysis has allowed organic salts separation from process waters. The addition of a bipolar membrane has allowed the conversion of organic salts into organic acids, adding versatility to the process and making the product recoverable through flash separations. However, the limitations of this process lie in the reduction of current efficiency and overall separation rates upon addition of ionic liquids, so further refinement is required. Through wafer-enhanced electrodeionization (WE-EDI), the current efficiency of ionic liquid assisted electrodialysis has improved along with the overall organic acid purification rate. Additionally, the energy requirement for this process has also been reduced. Through WE-EDI, ionic liquid assisted electro-separations can become a commercially viable process.

(531g) Batch and Continuous Drying of Lubricating Fluids With Novel Industrial Membrane Dehydrators

Nemser, S., Compact Membrane Systems, Inc
Majumdar, S., Compact Membrane Systems, Inc
Originally presented on: 11/6/2013 17:03:00 - 17:21:00

Ingression of ambient moisture over time into lubrication fluid systems in industrial gearboxes causes major operational and maintenance problems. Water can be present in lubrication systems as free, emulsified and dissolved water. The contamination level of water in lubrication oil adversely affects the service life of the associated components (e.g., industry studies show gear/bearing life increase from 5 years to 25 years). Accelerated metal surface fatigue can be promoted even if all the water present in the fluid is dissolved.

A compact novel membrane-based system to dewater lubricating oil in real time has been developed. With gear oil circulated on one side of the membrane module and a vacuum with ambient sweep air on the other side; the device can remove free, dispersed and dissolved water and extend the reliability and life of gearboxes, reduce maintenance costs, and increase uptime.

The paper will discuss the deleterious effects of water in lubricating oils. Literature data will be presented that documents the negative impact that water at various concentration levels can have on the service life of bearings. Data will also be presented illustrating the design and the real-time efficacy of a small footprint, novel membrane dehydrator to remove 100% of free and emulsified water and to reduce dissolved water to below 100 ppm. The paper will also provide quantitative insight into the consequences of batch versus continuous drying of lubricating fluids.

(531h) Up-Scaled Tubular High-Silica CHA-Type Zeolite Membranes for Industrial Purposes

Kyotani, T., Mitsubishi Chemical Corporation
Kikuchi, T., Mitsubishi Chemical Corporation
Sato, K., Mitsubishi Chemical Corporation
Originally presented on: 11/6/2013 17:21:00 - 17:39:00

Zeolite membranes have been a topic of much research in recent years, as a possible mean of performing efficient and low cost separations of organic molecules and gases. NaA (LTA)-type zeolite membrane has been commercialized for dehydration of neutral solvents such as alcohols. However, the utilization of the LTA-type zeolite membrane for dehydration has been limited to lower water concentration region (within 10wt.%), because the instability of LTA-zeolite under water rich condition. Zeolite membrane with high stability under various conditions is required for industrial purposes such as petrochemical plants. We developed 1m long tubular (16Φ) high-silica CHA-type zeolite membranes for industrial applications of dehydration under water-rich (over 10wt.%) and acidic feed conditions in which the LTA-type zeolite membranes cannot be utilized due to their instability.

The high-silica CHA-type zeolite membranes were synthesized hydrothermally using a template of adamantium hydroxide at 160 degrees centigrade for 12 - 48 hours by a secondary-growth method on the surface of tubular porous alumina supports up to 1m long. The membranes were calcined to remove the template [1].

Pervaporatiopn (PV) and vapor permeation (VP) performances of the high-silica CHA-type membranes were investigated by using hydrous feed solvents such as N-methylpyrrolidone (NMP), tetrahydrofuran (THF) and 2-propanol (IPA). Pure water flux of PV 120Â°C is 32 kg/m2h and shows long-term stability. As an example, permeation flux and separation factor in PV 130Â°C at feed composition of water (30wt.%)/THF (70wt.%) are 33.0 kg/m2h and 520, respectively. This result shows the higher water permeation flux and practical separation ability for the industrial purposes.

This study was financially supported in part by the New Energy and Industrial Technology Development Organization's (NEDO), Japan.

[1] K. Sato et al., J. Membr. Sci. **2012**, 409-410, 82-95.

*** Formatting Issue - Best Available Paper/Graphic ***

(531i) Deformation Based On Interfacial Tension During Microfiltration of Oil Drops On a Slotted Pore Membrane

Ullah, A.,

Originally presented on: 11/6/2013 17:39:00 - 17:57:00

Deformation based on interfacial tension during microfiltration of oil drops on a slotted pore membrane

A. Ullah

Department of Chemical Engineering Loughborough University, Leicestershire LE11 3TU, UK

Abstract

The effect of interfacial tension between two fluids, on the passage and rejection of oil droplets through slotted pore membranes is investigated. A mathematical model was developed in order to predict conditions for 100% cut-off of oil droplets through the membrane as a function of permeate flux rate. Good agreement of theoretical predictions with experimental data shows that the model can be applied to the filtration of deformable droplets through slotted pore membranes. At high interfacial tension (40 mN/m) with lower flux (200 l m^{-2} hr^{-1}) droplets of crude oil (27°API) were 100% rejected at droplet radius 4.3 µm using a 4 µm slotted pore membrane. At lower interfacial tension (5 mN/m), with the same flux rate, 100% rejection occurred at 10 µm droplet diameter using the same membrane. It was also found that the droplet rejection efficiency below the 100% cut-off was roughly linear with drop size, down to zero rejection at zero drop diameter. Hence, the model, coupled with this approximate correlation, can be used to predict dispersed oil drop concentration from a known feed drop size distribution.

Static force (F_{cx}) is the force responsible for the rejection of drops, given as follows [1]:

(1)

Where is interfacial tension between oil/water, is the radius of the drop, and is the half width of slot of the membrane and .

While drag force tries to pass the drops through the membrane and expressed as [2]:

(2)

where is a wall correction factor, is viscosity of the fluid, is the radius of the spherical droplet and is the velocity of the fluid. The droplet is under steady state conditions inside the pore, when becomes equal to and the droplets will be captured in this position. The droplet will deform and will pass through the membrane when and it will be rejected by the membrane in the case of [1].

A 4 µm slot width membrane was used for filtration experiments using the dead-end candle microfiltration system as illustrated in Fig.1.

Fig.1. Schematic view of dead-end microfiltration system.

Fig.2. illustrates the comparison of 100% cut-off (rejections) points of all the droplets obtained from experimental measurements and theoretical predications. The static force was obtained using eq (1), and drag force from eq (2). A satisfactory agreement between theoretical points and the experimental measured points shows that this model can be used to predict the 100% cut-off values of the oil types tested here.

Fig.2. Experimental measurements and predicted points of 100% cut-off during filtering vegetable oil droplets with various surfactants (Silica + gum Arabic, Gum Arabic, PVA, Tween20, Crude oil (31°API) and crude oil (27°API) with different flux rates (200, 400 and 600 l m^{-2} hr^{-1}).

References

[1] A. Ullah, R.G. Holdich, M. Naeem, V.M. Starov, Stability and deformation of oil drops during microfiltration on a slotted pore membrane, J Membrane Sci 401-402 (2012), 118-124.

[2] H.J. Keh, Y. Chen Po, slow motion of a droplet between two parallel walls, Chem Eng Sci 56 (2001) 6863-6871.

(535a) Modeling and Optimization of a MOF-Based Pressure/Vacuum Swing Adsorption Process Coupled With Material Selection

Gao, H., Northwestern University
You, F., Northwestern University
Snurr, R., Northwestern University
Originally presented on: 11/6/2013 15:15:00 - 15:37:00

The huge amount of carbon dioxide emitted every year has raised significant concern about global climate change. Extensive attention has been paid to Carbon Capture and Sequestration (CCS) to reduce CO_2 emission from large, centralized sources such as power plants. Pressure or vacuum swing adsorption (PSA/VSA) is a promising technology for carbon capture due to its relatively better separation performance, lower energy consumption and lower cost [1]. In order to achieve good performance with PSA/VSA, process optimization and material selection are two important considerations. Process variables, like adsorption and desorption pressures, and material properties, like the isotherm shape and heat of adsorption, can significantly influence process performance. Most previous works have considered these two aspects separately. They either optimize process variables with a fixed material [2, 3], or select materials independent of detailed process modeling [4, 5]. This separated approach has the drawbacks that the performance of a process configuration can be limited by improper choice of adsorbent and that material evaluation without a process model can lead to a non-optimal selection. In this work, we integrate material selection into process modeling by adding integer variables to model material selection. This will enable us to simultaneously choose materials as well as optimize process performance by solving a mixed integer dynamic optimization (MIDO) problem. Our objective is to maximize CO_2 purity and recovery at the same time.

A standard Skarstrom PSA/VSA cycle is modeled with the introduction of integer variables to model the selection of adsorbent materials from a number of promising candidates, such as zeolite 13X, zeolite 5A, ZIF-78, Ni-MOF-74, etc. In a Skarstrom cycle, there is a conflict between purity and recovery. In order to achieve good separation performance, we have two objectives: maximizing CO_2 recovery and purity at the same time. The resulting problem is a partial differential equations (PDEs) constrained multi-objective MIDO program. Multi-objective optimization can be solved using the ε-constraint method, which generates a Pareto-optimal curve by sequentially solving a set of subproblems with different ε values. The governing partial differential equations are discretized completely both in time and space. We apply a finite volume method to discretize the spatial domain and a Radau collocation scheme for temporal discretization. This transforms the MIDO problems to mixed-integer nonlinear programs (MINLP)which can be solved using general-purpose MINLP solvers.

Optimization is performed for several candidate materials used for PSA/VSA purposes. The accuracy of the optimal solutions are verified using MATLAB simulation in which the PDEs are only discretized in space and resulting ODEs are solved using ODE15s in MATLAB. Optimal solutions show the optimal operating conditions and best separation performance for the PSA/VSA cycle. Meanwhile, we identify the best material for the process. The Pareto-curve helps us understand the tradeoff between decision variables under different recovery requirements.

(535b) Energy and Cost Estimation for CO2 Capture From Dry Flue Gas Using a Vacuum Swing Adsorption Process

Susarla, N., National University of Singapore
Haghpanah, R., Stanford University
Karimi, I. A., National University of Singapore
Farooq, S., National University of Singapore
Rajendran, A., Nanyang Technological University
Originally presented on: 11/6/2013 15:37:00 - 15:59:00

There seems a consensus now that CO2 emissions to the atmosphere are a major cause for climate change. Nearly 40% of all the CO2 released annually to the atmosphere is from the fossil-fuel fired power plants. On an average, a 500 MW coal-fired power plant emits approximately 3 million tons of CO2 per annum [1]. Thus, the capture and concentration of CO2, especially from power-plant flue gas, is crucial. However, industrially, this remains a challenge because the energy and cost demanded by the capture process is prohibitive. Thus, it is imperative to develop an efficient and economical CO2 capture technology that can be readily retrofitted into existing power plants.

Recent studies [2-5] suggest that an adsorption-based CO2 capture can significantly reduce energy penalty as compared to amine-based absorption. Haghpanah et al. [4, 5] presented a comprehensive study of a variety of VSA configurations and their performance for CO2 capture. However, most studies [2-5] in the literature have used specific energy consumption (kWh per tonne of CO2 captured) as the sole indicator for comparing various capture methods, and neglected the issues of capital expenditure and plant footprint, which can be equally critical. Recently, Hassan et al. [6] proposed a simplified strategy to estimate the cost of a PSA/VSA process for CO2 capture using a single column.

In this work, we study and optimize in detail the best 4-step cycle identified by Haghpanah et al. [5] based on Zeochem zeolite 13X as the adsorbent for capturing CO2 from a dry flue gas. We use the comprehensive model of Haghpanah et al. [4] for simulating this process, which assumes dual-site Langmuir isotherm models and a linear driving force model for the gas-to-solid mass transfer. Owing to the high computational demand of the detailed model, we develop a surrogate model based on the Kriging approximation for the same and use that for optimization. The model based on kriging approximation enables us to do a constrained optimization. In addition, we develop an iterative algorithm that shuffles between the detailed and surrogate models for our single objective optimization. We present a detailed analysis of the various design and operational factors that affect the minimum energy required to capture CO2 with a purity of 90 mol% purity and a recovery of 90% from a dry flue gas using the proposed adsorption process. Also, we also determine the trade-off between energy, purity, recovery, and productivity for feeds from various emission sources. Furthermore, we examine the advantages of multi-stage separation. Finally, we present a procedure to design and operate a multi-column capture system for dry flue gas from a

500 MW coal-fired power plant and develop a mixed integer non-linear programming (MINLP) formulation based on the kriging approximation model to minimize the total annualized cost. We solve this MINLP model in GAMS using IPOPT and the detailed model in MATLAB to analyse the relative importance of energy consumption, land cost, adsorbent cost, plant footprint, operating expenditure (OPEX), and annualized capital expenditure (CAPEX) for such a capture plant.

References

[1] The future of coal: options for a carbon-constrained world. Massachusetts Institute of technology, Cambridge, MA, 2007.

[2] Dowling AW, Vetukuri SRR, Biegler LT. Large-scale optimization strategies for pressure swing adsorption cycle synthesis. AIChE Journal. 2012; 58.12: 3777-3791.

[3] Zhang J, Webley PA, Xiao P. Effect of process parameters on power requirements of vacuum swing adsorption technology for CO2 capture from flue gas. Energy Conversion and Management. 2008; 49(2), 346-356.

[4] Haghpanah R, Majumder A, Nilam R, Rajendran A, Farooq S, Karimi IA, Amanullah M. Multi-objective Optimization of a 4-step Adsorption Process for Post-combustion CO2 Capture Using Finite Volume Technique. Industrial & Engineering Chemistry Research. 2013; 52(11), 4249-4265.

[5] Haghpanah R, Nilam R, Rajendran A, Farooq S, Karimi IA. Cycle Synthesis and Optimization of a VSA Process for Post-combustion CO2 Capture. AIChE Journal. Under review, 2013.

[6] Hasan MMF, Baliban RC, Elia JA, Floudas CA. Modeling, Simulation, and Optimization of Postcombustion CO2 Capture for Variable Feed Concentration and Flow Rate. 2. Pressure Swing Adsorption and Vacuum Swing Adsorption Processes. Industrial & Engineering Chemistry Research. 2012; 51.48, 15665-15682.

(535c) CO2 Capture From Flue Gas By PSA Utilizing a Structured Adsorbent

Mohammadi, N., University of South Carolina
Abdollahi, A., University of South Carolina
Nicholson, M. A., University of South Carolina
Ebner, A. D., University of South Carolina
Holland, C. E., University of South Carolina
Ritter, J. A., University of South Carolina
Originally presented on: 11/6/2013 15:59:00 - 16:21:00

A cost effective pressure swing adsorption (PSA) process is under development for post-combustion CO_2 capture from coal fired power plants. Several research groups from around the world are showing that PSA is still a viable option, as no separation process has emerged as a clear winner for CO_2 capture. PSA is also attractive because it requires only electricity; it does not require any steam from the power plant. This project is being carried out with funding from DOE NETL and four collaborative partners:

The key challenge to the success of this project is two-fold: 1) an attrition resistant and low pressure drop structured adsorbent must be developed based on commercial zeolite that is compatible with the high velocities associated with shorter PSA cycle times, and 2) a PSA cycle configuration must be developed in concert with the structured adsorbent so that the resulting PSA process delivers an exceptional performance at reduced capital and operating costs. Since the rapid PSA concept is conceptually simple but difficult to implement, a select team has been assembled to carry out this ambitious project and demonstrate the feasibility of the PSA concept for CO_2 capture at the bench scale. This team includes researchers from USC, W. R. Grace, Catacel, and Battelle.

To date, a key development includes the development of a low energy, high feed throughput PSA cycle for CO_2 capture via PSA process simulation. This was an incredible achievement when considering that the bulk density was reduced from 710 kg/m^3 (typical for a packed bed of zeolite beads) to 400 kg/m^3 (entirely possible with a Catacel structured adsorbent core). It boasts around 3,000 L(STP)/hr/kg feed throughput and a separations energy of less than 18 kJ/mol CO_2 captured. This presentation will discuss the latest results obtained from Ritter and his group.

(535d) Post-Combustion CO2 Capture Using Structured Bed Loaded With Functionalized SBA-15 Sorbent

Mittal, N.,
Samanta, A., University of Alberta
Segura, J., Alberta Innovates - Technology Futures
Amiri, S.,
Sarkar, P., Alberta Innovates- Technology Futures
Gupta, R., University of Alberta
Originally presented on: 11/6/2013 16:21:00 - 16:43:00

Post-combustion CO_2 Capture Using Structured Bed Loaded with Functionalized SBA-15 Sorbent

Nikhil Mittal[1], Arunkumar Samanta[1], J. Segura[2], Saeid Amiri[3], Partha Sarkar[2], Rajender Gupta[1,*]

[1]Department of Chemical and Materials Engineering, University of Alberta

9107 - 116 St, ECERF, AB, T6G 2V4, Canada, Email: rajender@ualberta.ca, Tel: 780-492-6861, Fax: 780-49-2881

[2] Environment and Carbon Management Division, Alberta Innovates - Technology Futures, Edmonton, AB, Canada

[3]Wave Control Systems. Inc., Edmonton, Alberta, T6E 0C2, Canada

Abstract:

The conventional packed bed for post-combustion CO_2 capture suffers from significantly high pressure drop whereas the fluidized bed finds problems in sorbent attrition and loss. So, there is a need of more reliable and efficient bed configuration, which can offer a lower pressure drop, fast mass transfer kinetics, high working capacity and improved thermal management. An alternative configuration to packed bed is sorbents placed in structured bed reactor. The major advantage of using structured bed for post-combustion CO_2 capture is low pressure drop because of its straight gas flow channels. The hydrodynamics and external mass transfer can be optimized in this structured bed by adjusting channel size and channel shape, whereas the sorbent loading and/or internal mass transfer can be controlled by the thickness and structure of sorbent channel wall. The objectives of this investigation was to study the adsorption of CO_2 from a flue gas in a structured bed loaded with amine functionalized SBA-15 sorbent and to understand the effect of various design variables such as sorbent bed diameter, channel spacing, membrane layer thickness, and flue gas volumetric flow rate etc.

In the proposed structured bed, the solid sorbent is contained inside a porous alumina tubular membrane reactor (OD ~ 4mm, ID ~ 3mm and average pore diameter ~0.8 micron) and CO_2 laden bulk gas flows through the annular region. CO_2 diffuses across the barrier layer and adsorbs preferentially on the solid sorbent. The equipment was designed and built to host single and multi-tube reactors with different lengths with minimum modification. The concentration of CO_2 stream coming to and from the adsorption column was monitored as a function of time by a Pfeiffer mass spectroscopy (Model: OmniStar GSD 320). Each membrane tube was filled with approximately 0.2 g of 50 wt% PEI impregnated SBA-15 solid sorbent. Thermal swing adsorption cycles were initially tested at different temperatures for CO_2 adsorption using a simulated flue gas containing 10%CO_2 by vol in N_2. The breakthrough curves obtained for CO_2 adsorption from CO_2/N_2 mixture (10%CO_2 bal. N_2) on structured bed of 50 wt% PEI impregnated SBA-15 at 75 °C revealed that average CO_2 adsorption capacity was about 2.4 mmol.g^{-1} after ten consecutive cycles. Each adsorption/desorption cycle involved a 10 min of CO_2 adsorption at 75 °C followed by regeneration for 15 min at 105 °C. For this sorbent, equilibrium CO_2 adsorption capacity measured in TGA under similar condition was 3.07 mmol/g adsorbent. The thermogravimetric data also suggested that the CO_2 capture kinetics was also found to be fast and reached 90% of the total capacities within the first ten minutes. Finally, an optimum set of geometric parameters of the structured bed such as channel spacing and gas superficial velocity has been determined to improve adsorption kinetics and to reduce the pressure drop per unit length of bed.

(535e) A New Temperature Swing Adsorption Configuration for High Purity and Recovery CO2 Capture From Flue Gas With Increased Throughput

Maring, B. J., The University of Melbourne
Webley, P. A., Monash University
Originally presented on: 11/6/2013 16:43:00 - 17:05:00

Tethered amine materials have great potential for CO_2 capture by temperature swing adsorption because of their high CO_2/N_2 selectivity and large heats of adsorption (> 90KJ/mol) which correspond to highly temperature-dependent adsorption isotherms. Tethered amines also have a lower theoretical regeneration energy than aqueous amine solutions because the heat capacity of their solid substrates is lower than that of the water in an amine solution. The major drawback of TSA in industrial applications is the length of time required to heat and cool the large adsorbent beds which can be several hours or even days. We have developed a new TSA configuration which does not rely on external heat transfer and therefore can be run at high throughput, thereby reducing capital costs. Furthermore, the separation can be performed using a narrow temperature range at relatively low temperatures (90ºC to 150ºC), utilizing low quality steam for a reduced energy penalty. We performed numerical simulations of our configuration and achieved high purity CO_2 product with high recovery from flue gas.

(535f) Impact of Water Vapor and Oxygen On Amine Sorbents for Post-Combustion CO2 Capture

Hammache, S., URS-NETL
Hoffman, J. S., U.S. Department of Energy, National Energy Technology Laboratory
Egbebi, A., URS-NETL
Gray, M. L., United States Department of Energy
Originally presented on: 11/6/2013 17:05:00 - 17:27:00

Amine sorbents such as polyethyleneimine/SiO_2 are being considered for post-combustion CO_2 capture from coal-fired power plants. A successful sorbent must be stable under moist conditions present during adsorption and regeneration steps. Also, the presence of oxygen in flue gas may have an impact on the amine sorbents; thus understanding its impact on the amine in the presence of CO_2 is required. In this work, we are evaluating the stability of amine sorbents under steam during regeneration and the impact water and oxygen have on the sorbent during adsorption

The sorbent used in this work was synthesized by wet impregnation of SiO_2 (Fuji Silysia Cariact grade G-10 HPV) with polyethyleneimine (Aldrich PEI Mn-423). CO_2 adsorption-desorption cycling was conducted in a fixed bed reactor. The sorbent was tested prior and subsequent to O_2 pretreatment using dry cycles (10 vol. % CO_2/He at 60°C) to assess any change in its performance. Gas analysis of the reactor effluent was conducted using mass spectrometry.

The impact of 5% O_2/He on the sorbent was investigated at 80 °C. The sorbent was held at these conditions for 5 h, resulting in a decrease in CO_2 capture of 7 mol%. The impact of 5% O_2/He at 80 °C in the presence of water was also investigated. The sorbent was cycled ten times through 5% O_2-8% H_2O/He at 80 °C with the total time adding up to 5 h and regenerated under helium at 80°C. The sorbent was tested under dry conditions prior to and after exposure to O_2/H_2O to assess their impact on its performance. The CO_2 loading was found to decrease by 19 mol% compared to the fresh sorbent. However when the sorbent was exposed to 8% H_2O/He, the decrease in CO_2 loading was found to be only 9 mol %. Clearly the combination of H_2O and O_2 at 80 °C has the greatest negative impact on the sorbent.

The impact of O_2 in the presence of CO_2 on this sorbent and a silanated sorbent will be discussed as well. Characterization of the fresh and spent sorbents prior to and following O_2/H_2O/CO_2/He exposure will be presented.

(535g) Adsorption Isotherm Measurements of Gas Shales for Subsurface Temperature and Pressure Conditions

Wang, B.,
Haghpanah, R., Stanford University
Wilcox, J., Stanford University
Originally presented on: 11/6/2013 17:27:00 - 17:49:00

Carbon dioxide emissions from fossil fuels combustion have been significantly increasing compared to that in the pre-industrial era, which has caused significant increase in global average temperatures. To stabilize atmospheric CO2 emissions, one possible approach may be to inject and store CO_2 into gas shale, where significant amounts of methane are present and can be exploited and recovered. Experimental studies indicate that CO_2 has enhanced adsorption over CH_4, thus the injected CO2 may displace the adsorbed methane inside the gas shale, thereby potentially enhancing methane recovery efficiency. However, the adsorption properties of CO_2 and CH_4 on gas shale are not fully understood, and need to be investigated both experimentally and theoretically.

In our recent work, excess adsorption isotherms of CO_2 on gas shale samples have been measured under subsurface temperature and pressure conditions, using a Rubotherm magnetic suspension balance. The sample used in the study is from the Eagle Ford reservoir located in Texas, United States. Both core chip and powdered forms of sample have been investigated. According to our preliminary results, the shale sample in the powder form always gives higher gas capacity than the same sample in chip form. The ruduced macropore mass-transfer resistance in the powdered sample is a possible explanation for its higher gas capacity.

Since kerogen and clay are the major components contributing to the adsorption behavior in gas shale, adsorption isotherm measurements for carbon-based and clay-based model materials have been performed to determine the relative roles that each of these natural systems play on the overall shale adsorption mechanism and capacity estimates.

In addition, Grand Canonical Monte Carlo (GCMC) simulations have been used to predict the adsorption behavior of carbon-based material with different pore size distributions and chemical heterogeneity. Results from simulation and experiment are compared to further investigate the mechanisms of adsorption in gas shale.

(536a) Crystallization of L-Ascorbic Acid in a 18-L Airlift Crystallizer

Lakerveld, R., Delft University of Technology
Van Krochten, J., Delft University of Technology
Anisi, F., Delft University of Technology
Stankiewicz, A. I., Delft University of Technology
Kramer, H. J. M., Delft University of Technology
Originally presented on: 11/6/2013 15:15:00 - 15:40:00

Crystallization is applied extensively in the chemical and pharmaceutical industry to produce and purify solid particles. The performance of the process and final product are closely related to specific properties of the crystals such as purity, crystal size distribution (CSD), and shape. However, the design and control of industrial crystallization processes to manufacture a crystalline product with tailored properties remains a major challenge. Many interacting physical phenomena occur in solution crystallization processes, involving for example primary nucleation, growth and attrition of crystals. In current industrial equipment, those physical phenomena take place in the same environment and are, therefore, strongly entangled. Novel designs are needed to improve control over individual physical phenomena.

Attrition refers to the birth of new crystals caused by collisions of crystals, which is often undesirable as it reduces the crystal size and broadens the CSD. Recently, a new crystallization concept involving an airlift, which eliminates the need for a stirrer, has been tested for crystallization of ammonium sulphate from water[1]. The results demonstrate that attrition can be significantly suppressed compared to crystallization in a conventional stirred crystallizer.

Although these results are very promising for further development, the full potential of the concept remains largely unclear. An airlift crystallizer offers an exciting opportunity to bring the control of crystal product quality to a new level particularly for systems that are very sensitive to the mechanical impact of a stirrer. In particular, pharmaceutical compounds that are mostly produced batch-wise and in stirred-tank crystallizers often suffer from such sensitivity, which limits flexibility to produce a desired CSD due to uncontrolled nucleation. In order to apply an airlift system for crystallization successfully, the process conditions need to be optimized. Especially the control of the flow regime and the design of the sparger are essential for the performance of this new type of crystallizer.

The objective of the present study is to investigate experimentally the performance of an 18-L airlift crystallizer for crystallization of L-ascorbic acid from water in terms of hydrodynamic behavior and final product quality for seeded batch crystallization. First, the hydrodynamic behavior of the system has been investigated as function of the operating conditions and the design of the sparger via several in-line density measurements and in-line imaging. The results demonstrate that favorable hydrodynamic conditions in which the injected nitrogen is only present in the riser of the system can be achieved via a proper design of the sparger for a broad range of operating conditions. Secondly, the attainable product quality in terms of shape and CSD has been investigated via in-line imaging and off-line measurement of the supersaturation profile and CSD. A systematic comparison between the crystallization performance in an 18-L airlift crystallizer and crystallization in a conventional 5-L stirred crystallizer has been conducted. The results demonstrate a significant improvement of crystal shape and reduced nucleation in the airlift crystallizer. Furthermore, a clear distinction between grown seeds and new nuclei in both crystallizers allow us for the first time to quantitatively distinguish various forms of nucleation of the studied model system. Thirdly, the experimental data has been used to develop kinetic expressions for growth and nucleation, which allows for model-based optimization of the system to reveal the full potential of both a conventional crystallizer and airlift crystallizer. Finally, prospects for extending the system to continuous flow-mode will be discussed.

References

[1] A. Soare, R. Lakerveld, J.van Royen, G. Zocchi, A. I. Stankiewicz, H. J. M. Kramer. Minimization of Attrition and Breakage in an Airlift Crystallizer. *Ind.Eng.Chem.Res.* **2012**, 51, 10895-10909.

Acknowledgement

This work was supported by the European Commissions Framework 7 program through the OPTICO consortium

(536b) Mathematical Modeling and Analysis of Cooling Crystallization within Dual-Impinging-Jet Mixers

Jiang, M., Massachusetts Institute of Technology
Braatz, R. D., Massachusetts Institute of Technology
Originally presented on: 11/6/2013 15:40:00 - 16:05:00

In the pharmaceutical industry, good control of crystal size distribution (CSD) can improve process efficiency and formulation of drug product. An effective way to improve the control of CSD is through combining continuous seeding and growth with concentration control (Woo et al., 2011). To continuously produce small mono-disperse crystals, one well-studied approach is to combine solution and anti-solvent streams in a dual-impinging-jet (DIJ) mixer (e.g., Johnson and Prud'homme, 2003; Midler et al., 1994; Woo et al., 2011). The underlying principle is that, at appropriate flow rates, a DIJ mixer can generate high-intensity micro-mixing of fluids to quickly achieve a nearly homogeneous composition of high supersaturation before the onset of primary nucleation. Use of a DIJ mixer to generate crystals can sometimes remove the requirement of a milling step, which simplifies the overall manufacturing process and avoids the potential of generating an undesirable polymorphic transformation (Johnson and Prud'homme, 2003; Midler et al., 1994).

In past work, we experimentally demonstrated the use of a cooling DIJ mixer to combine hot and cold saturated solutions, rather than combining a solution with an anti-solvent, to generate seed crystals with a narrow size distribution for L-asparagine monohydrate (LAM) in aqueous solution (Jiang et al., 2012). The crystals were less than 10 microns in length, placing them in the correct size range for direct application in inhalers (Capstick and Clifton, 2012). Theoretically, if the mixing were perfect (that is, concentration and temperature were completely mixed at the molecular scale in the DIJ mixer), inspection of the phase diagram for the LAM-water system indicated that the supersaturation would not be high enough to nucleate crystals within a cooling DIJ mixer. The observed nucleation of crystals can be explained by the observation that the energy transfer rate at the interface between the impinging fluids is much faster than the mass transfer rate, so that the temperature of the hot solution near the interface can drop to the average temperature of the two solutions before its solution concentration can significantly change, resulting in the supersaturation in the hot fluid becoming sufficiently high to nucleate crystals.

While the ability to generate highly uniform-sized crystals of less than 10 microns using a cooling DIJ mixer would be a perfect technology for inhaler applications, this approach is not expected to be applicable to all combinations of pharmaceuticals and solvents. Currently, scientists/engineers have to assess whether any particular compound/solvent combination is suitable for use in a DIJ mixer by trial-and-error experimentation aided with experience (Tung et al., 2009). It would be useful at the early stage of research and development to be able to quickly identify compound-solvent combinations that cannot nucleate crystals within DIJ mixers, based on their physicochemical properties. This work answers this question for cooling jet DIJ mixers. We combine mathematical modeling of crystallization kinetics, thermodynamics, and transport with theoretical analysis to derive conditions in which a pharmaceutical/solvent combination is suitable for the nucleation of crystals within a cooling DIJ mixer. The mathematical models are in the form of partial differential equations that are simplified by exploiting symmetries and employing scaling analysis, to enable the derivation of analytical expressions that compare favorably with numerical simulations. The mathematical model and methodology are validated for the crystallization of LAM in aqueous solution.

References

Capstick, T. G. D., Clifton, I. J., 2012. Inhaler technique and training in people with chronic obstructive pulmonary disease and asthma, Expert Review of Respiratory Medicine 6(1), 91-103.

Jiang, M., Wong, M. H., Zhu, Z., Zhang, J., Zhou, L., Wang, K., Ford Versypt, A. N., Si, T., Hasenberg, L. M., Li, Y., Braatz, R. D., 2012. Towards achieving a flattop crystal size distribution by continuous seeding and controlled growth, Chemical Engineering Science 77, 2-9.

Johnson, B. K., Prud'homme, R. K., 2003. Chemical processing and micromixing in confined impinging jets, AIChE Journal 49(9), 2264-2282.

Midler, M., Jr., Paul, E. L., Whittington, E. F., Futran, M., Liu, P. D., Hsu, J., Pan, S. H., 1994. Crystallization method to improve crystal structure and size, U.S. Patent #5314506A, 11 pages.

Tung, H., Paul, E. L., Midler, M., McCauley, J. A., 2009. Crystallization of Pharmaceuticals: An Industrial Perspective. Hoboken, NJ: Wiley.

Woo, X. Y., Tan, R. B. H., Braatz, R. D., 2011. Precise tailoring of the crystal size distribution by controlled growth and continuous seeding from impinging jet crystallizers, CrystEngComm 13, 2006-2014.

(536c) Tailoring Desired Powder Properties in Pharmaceutical Development By Crystal Engineering

Wang, C., Bristol-Myers Squibb Company
Mahoney, M., Bristol-Myers Squibb Company
Shi, Z., Bristol-Myers Squibb Company
Wang, S., Bristol-Myers Squibb Company
Chan, S., Bristol-Myers Squibb
Remy, B., Bristol-Myers Squibb Company
Sfouggatakis, C., Bristol-Myers Squibb Company
Lai, C., Bristol-Myers Squibb Company
Originally presented on: 11/6/2013 16:05:00 - 16:30:00

In order to deliver pharmaceutical materials with powder properties suitable for formulation development, specific solid characteristics (among many evaluated) are determined to be critical for the improvement of process robustness in drug product processing. A model compound with a challenging crystal habit for solid dosage formulation is evaluated by multiple crystal engineering approaches to alter its physical properties. A number of material design opportunities were identified to promote the desired powder characteristics during the manufacturing process, including synthesis, crystallization, isolation, as well as post-isolation processing steps. In this study, the primary approach to enhance powder properties is driven by varying crystal shape and/or particle morphology of the subject material. Both crystal and particle morphologies can be tailored in (a) crystallization with optimized operating parameters to guide the size and morphology evolution, in (b) crystallization by an alternative means to facilitate controlled assembly formation, and in (c) direct powder processing to impact material properties. The various crystal engineering techniques encompassed in this study allow target powder properties to be met at different development stages while accommodating frequent process changes. The design strategy and the results from the applied crystal engineering techniques will be presented.

(536d) Morphology Control Through Cycles of Particle Breakage, Dissolution and Growth

Singh, M. R., Purdue University
Tung, H. H., AbbVie Inc.
Bordawekar, S., AbbVie
Ramkrishna, D., Purdue University
Originally presented on: 11/6/2013 16:30:00 - 16:55:00

Needles and plates are commonly observed morphologies in pharmaceutical crystallization which lead to difficulties in downstream processing causing higher cake resistance in filtration, resistance to free flow, and poor compressibility. The processibility of such ill-behaved crystals can be improved by reducing their aspect ratios through cycles of sequential processes involving particle breakage, dissolution and growth. These cycles are widely used in pharmaceutical industries to improve aspect ratio of crystals. The dissolution-growth cycles can improve the aspect ratios of 2D crystals only if the relative dissolution rate of the slowest face is greater than its corresponding relative growth rate. Particle breakage has the additional advantage of providing flexibility in attaining target morphologies within the feasible operating regimes. The particle breakage using either a Rotor-Stator or a Sonicator usually affects the longer dimension of the crystal, resulting in the production of fines and reduction in the aspect ratio. Successively, the dissolution of crystals will redistribute fines into the solution and may further improve the aspect ratio of crystals. Finally, the crystals are allowed to grow to increase their mass without much increase in their aspect ratio. In this work, we develop optimal cycles of particle breakage, dissolution and growth for shape control of Cholesterol crystals in a batch slurry which has the potential for considerable improvement in productivity.

(536e) Controls of Transformation Between Amorphous and Crystal Ceftiofur Free Acid

Gao, Y., National Engineering Research Center for Industrial Crystallization Technology, School of Chemical Engineering and Technology, Tianjin University
Wang, J., National Engineering Research Center for Industrial Crystallization Technology, School of Chemical Engineering and Technology, Tianjin University, Tianjin,China
Wang, Y., National Engineering Research Center for Industrial Crystallization Technology, School of Chemical Engineering and Technology, Tianjin University
Yin, Q., National Engineering Research Center for Industrial Crystallization Technology, School of Chemical Engineering and Technology, Tianjin University
Hao, H., National Engineering Research Center for Industrial Crystallization Technology, School of Chemical Engineering and Technology, Tianjin University
Originally presented on: 11/6/2013 16:55:00 - 17:20:00

Ceftiofur is a third generation cephalosporin antibiotic, which has a wide spectrum of activity against Gram-positive and Gram-negative bacteria including β-lactamase-producing strains. It can be used as veternary medicine as the sodium salt, the hydrochloride, and the free acid. Among them, the free acid was found to be the best for clinical therapy. Ceftiofur free acid could exist in either amorphous or crystalline state. Generally, amorphous ceftiofur free acid will be much easier to obtain in the separation or purification process of it .In this research, the crystallization process of ceftiofur was investigated in detail by using several PAT tools, such as ATR-FTIR, FBRM, PVM and Raman spectroscopy. The crystal morphology and the structure of the final product was also analyzed by traditional off line tools such as XRD, DSC and SNMR. The effect of operating parameter, such as solvent, temperature, dosing rate and agitation speed, on the solid state of the final product was evaluated. Optimum conditions for manufacturing both forms were successfully developed. It was also found from experiments that the amorphous state ceftiofur can transform into crystalline state in appropriate conditions although it may take a very long period. The mechanism of this transformation was investigated using the in situ tools mentioned above.

(536f) NIR-Based Process Understanding of Anti-Solvent Cocrystallization Between Carbamazepine and Saccharin

Choi, G. J.,

Originally presented on: 11/6/2013 17:20:00 - 17:45:00

Over the last decade, the cocrystal approach has been studied extensively as one of the most promising methods to improve the dissolution properties of insoluble drug substances. Cocrystal powders are typically screened or produced by mechanical grinding (wet or neat) or a solution method (evaporation or cooling). In this work, high-purity carbamazepine-saccharin (CBZ-SAC) cocrystals were produced by a novel method, anti-solvent addition. Among several solvents, methanol was found to perform well with water as the anti-solvent for the cocrystallization of CBZ and SAC. When water was added to the methanol solution of CBZ and SAC at room temperature under agitation, nucleation of CBZ-SAC cocrystals initiated within 2-3 min. Cocrystallization was complete after 30 min, giving a solid yield as high as 85% on a CBZ basis. The effects of initial concentrations, focusing on the SAC/CBZ ratio, were examined to establish optimal conditions. The whole anti-solvent co-crystallization process was monitored via in-line NIR measurements as well as at-line ATR-FTIR analysis of regularly sampled solutions. The nucleation and crystal growth of CBZ-SAC co crystals were detected by a significant increase in absorption in the range of 2400-2260 cm-1, associated with the formation of hydrogen bonds between the carbonyl group in CBZ and the N-H of SAC. When CBZ hydrates were formed as impurities during anti solvent co-crystallization, the hydrogen bonding between methanol and water was reduced greatly, primarily due to the incorporation of water molecules into the CBZ crystal lattice.
In conclusion, the whole anti-solvent cocrystallization process can be monitored for a comprehensive understanding with highly pure CBZ-SAC cocrystal powders as final products.

(537a) Modern Sieve Trays for Liquid-Liquid Extraction

Shiveler, G., Sulzer Chemtech USA, Inc.

Originally presented on: 11/6/2013 15:15:00 - 15:35:00

Meeting: Autumn AIChE National Meeting

San Francisco, CA ? November 3rd through November 7th, 2013

Topical 2: Separations

Title: ***"Modern Sieve Trays for Liquid-Liquid Extraction"***

Authors: Glenn Shiveler, Sulzer Chemtech USA Inc. ? Tulsa, OK

Abstract:

Sieve trays continue for find a role in modern liquid-liquid extractions where static devices are specified for various niches in the petroleum and chemical process industry.

There are several equipment features that have demonstrated that can improve the reliability and performance of sieve extractor trays. These include: extruded orifice holes for deck perforations, downcomer restrictors, structural truss orientation and using multiple downcomers for large diameter units. Properly designed inlet devices and liquid-liquid coalescers are also important considerations for achieving good performance in these applications.

This paper provides examples of recent industrial experience involving sieve trays equipped with special features.

(537b) Solvent Recovery After Extraction By Energy Efficient Membrane Separation Process

Nemser, S., Compact Membrane Systems, Inc
Kosaraju, P., Compact Membrane Systems, Inc
Originally presented on: 11/6/2013 15:35:00 - 15:55:00

Solvent extraction is often used to extract valuable solutes. After solvent extraction, solvents are often recovered from solute-solvent mixtures by solvent evaporation. Solvent evaporation is an energy intensive process as it involves solvent phase change. A perfluoropolymer-based nanofiltration membrane has been recently developed for recovering solvents from solute-solvent mixtures, generated by solvent extraction process. This membrane separation process is highly energy efficient as it does not involve solvent phase change. Because of the perfluoropolymer chemistry and unique permeation properties, the perfluoropolymer-based nanofiltration membrane offers superior chemical stability, thermal stability, anti-fouling tendency and separation performance. Additional benefits of the membrane separation process include lower solvent emissions and higher process safety. Nanofiltration process based on the unique perfluoropolymers can also be valuable for recovering solvents from solute-solvent mixtures that are thermally sensitive. This presentation discusses unique features of the perfluoropolymer-based nanofiltration membranes, and solute-solvent systems that are suitable for processing by the membrane separation process.

(537c) Continuous Downstream Processing Using Intermittent Counter-Current Extraction - Theory and Feasibility Studies

Hewitson, P., Brunel University
Sutherland, I. A., Brunel University
Ignatova, S., Brunel University
Originally presented on: 11/6/2013 15:55:00 - 16:15:00

The pharmaceutical industry requires high throughput, lower solvent usage extraction, separation and purification methods to minimise drug development times and reduce both costs and environmental impact.

Intermittent Counter-current extraction (ICcE) has been demonstrated as a potential liquid-liquid extraction method for high throughput, continuous separation of compounds into two product streams and for the isolation, concentration and enrichment of a low abundance target compound [1]. Scale up, using industrial type instruments supplied by Dynamic Extractions Limited, has also been successfully demonstrated [2, 3].

Both theory and methodology for determining optimal operating parameters will be presented. Experimental results for the scaling of ICcE separations from the semi preparative instrument (143 mL column volume) to preparative instrument (912 mL) and on to pilot scale instrument (4.6 L) using a model mixture of pharmaceutical compounds from across a polarity range with hexane/ethyl acetate/methanol/water (HEMWat) two phase systems. These results are compared to a theoretical model of ICcE to confirm the predicted order of compound elution. Throughput and solvent usage are compared for both ICcE and batch extractions.

[1] P. Hewitson, S. Ignatova, H. Ye, L. Chen I. Sutherland, J. Chromatogr. A 1216 (2009) 4187

[2] I. Sutherland, P. Hewitson, S. Ignatova, J Chromatogr. A. 1216 (2009) 8787

[3]P. Hewitson, S. Ignatova, I. Sutherland, J. Chromatogr. A. 1218 (2011) 6072

(537d) Extraction of Diesel From Drill Cuttings Using Compressed Carbon Dioxide

Tharanivasan, A. K., Schlumberger
Verma, S., Schlumberger Doll Research
Bingham, R., Schlumberger
Kapila, M., M-I SWACO
Originally presented on: 11/6/2013 16:15:00 - 16:35:00

Drilling for oil and gas wells produce large amounts of drill cuttings. The oil content of these drill cuttings, using oil-based muds, is usually above 15 wt% after undergoing physical separation using shaker table. Environmental regulations require oil or hydrocarbon content to be reduced to less than 1 wt% in the cuttings prior to disposal. Current disposal methods involve re-injecting the cuttings into the formation which poses environmental issues, or transporting the cuttings to land for incineration or burial, which is expensive. Therefore, there is a need for a cost-effective, energy-efficient, environmentally acceptable and safe technique to recover hydrocarbon from drill cuttings. Recently, a process based on compressed carbon dioxide is identified as a promising alternative for the oil recovery from drill cuttings. The advantages are better handling, easier recycling, recovering pristine oil and lesser energy requirements when compared to existing techniques. The recovery process is designed to take place in an extraction unit, where liquid CO_2 stream is continuously added and mixed with drill cuttings. At certain operating conditions, liquid CO_2 stream dissolves the oil and progressively decreases the oil content in drill cuttings. This stream is flashed at low pressure in a separator. CO_2 is recovered as vapor and recycled. Oil is collected as the liquid from the separator. The essential information required to determine the operating conditions are the phase behavior data for carbon dioxide-oil mixtures. In this presentation, thermodynamic data generating the CO_2-diesel phase diagram and its application in the optimization of the extraction cycle will be discussed.

Results, observations and conclusions

Carbon dioxide-diesel mixtures exhibit a complex phase behavior at certain pressure and temperature conditions. For example, liquid-liquid immiscibility region is observed at pressures between the vapor-liquid region and the single phase liquid region. In the liquid-liquid equilibrium region, the majority of the lighter diesel components are soluble into carbon dioxide-rich liquid phase. The partitioning of diesel components into the CO_2 liquid phase has been experimentally established. In this work, the phase behavior data is collected at 17 and 25°C for various carbon dioxide-diesel mixtures. Specifically, the liquid-liquid and the vapor-liquid-liquid phase transition pressures are measured for each mixture. When the mixture is depressurized at constant temperature, the phase transitions occur from a single liquid region to the liquid-liquid region and from the liquid-liquid region to the liquid-liquid-vapor region. Based on these measurements for mixtures of several compositions, a pressure-composition phase diagram has been established for the carbon dioxide-diesel system. Consequently, the operating pressures of the extraction cycle are determined for test temperatures.

(537e) Extraction, Characterization and Quantification of Naphthenic Acid From Crude Oils

Zou, F., DBR Technology Center
Painer, D., Graz University of Technology
Lux, S., Graz University of Technology
Andersen, S., Schlumberger
Zeng, H., Schlumberger, DBR Technololy Center
Siebenhofer, M., Graz University of Technology
Li, H., DBR Technology Center, Schlumberger
Zacharia, J.,
Originally presented on: 11/6/2013 16:35:00 - 16:55:00

Extraction, characterization and quantification of naphthenic acid from crude oils

Fenglou Zou, Simon Andersen, Huang Zeng, Huaping Li, Jose Zacharia

Schlumberger, DBR Technology Center, Edmonton, Canada, T6N 1M9

Carboxylic acids (RCOOH), also known as naphthenic acids in the oil industry, are natural surfactants in crude oils. It is assumed that the functional groups of naphthenic acid affects oil properties and behaviour, and further impacts emulsion stability and reservoir wettability. Carboxylic acids span the entire boiling point range from benzoic acid to asphaltenes, which makes acid separation and quantification challenging. Classical TAN (Total Acid Number) is not representative of the actual content of acids in crude oils because TAN determined using conventional KOH titration includes contributions from other compounds, such as hydrogen sulphide, thiols and alkylphenols. Naphthenates (salts of naphthenic acids) have industrial applications such as synthetic detergent, fuel and lubricating oil additives, wood preservatives and oil drying agents used in painting and wood surface treatment.

The present paper explores various methods applicable to acid determination including different extraction procedures and methods to calculate true acid content. Water extraction and IER (ionic exchange resin) extraction were found to have low repeatability, and contamination by non-acids was high. Acid extraction based on column chromatography using SAX (strong anionic exchange resin) as a stationary phase showed good repeatability and the repeatability was further improved by using instrumental solid phase extraction (SPE) based on commercial SAX stationary phase cartridges (Figure 1). 2D GC-TOFMS (two dimensional gas chromatography-time of flight mass spectrometry) was used for characterization and quantification of species within the detection range of the GC. The difference seen in distribution between IER and water extraction was clearly biased by the lack of solubility of certain acids in water.

Figure 1. Comparison of two individual acid extractions using SPE showing good repeatability for acid extraction.

(537f) Fractionation of An Oligomeric Pyrene Pitch Via Supercritical Extraction

Thies, M. C., Clemson University
Esguerra, D. F., Clemson University
Originally presented on: 11/6/2013 16:55:00 - 17:15:00

Semi-continuous, packed-column supercritical extraction was investigated for the fractionation of a polycyclic aromatic hydrocarbon (PAH) pitch into its constituent oligomers. The starting pitch had been produced by the catalytic polymerization of pure pyrene using $AlCl_3$. Although neat supercritical toluene (Tc = 318.6 °C; Pc = 41.1 bar) was shown to be an effective solvent for the recovery of both monomer and dimer cuts in high purities, higher-oligomer solubilities in toluene are notoriously low. However, the addition of N-methyl-2-pyrrolidone (NMP) as a co-solvent enhanced oligomer solubilities in the extractive, supercritical-solvent phase by a factor of 3, making possible the recovery of trimer and higher oligomers. An unexpected benefit of the use of NMP is that it was found to suppress the condensation reaction between toluene and the pyrene oligomers that can occur in the extraction column at temperatures above 350 °C. Apparently the NMP forms a Lewis acid-base complex with the $AlCl_3$ catalyst, preventing further reaction.

In this talk, we demonstrate how pyrene dimer and trimer cuts can each be recovered from the starting pyrene pitch in good yields at purities exceeding 99%. A supercritical co-solvent mixture consisting of 15 mol % NMP in supercritical toluene was found to be an effective solvent system for the desired separation. The dramatically different properties from the starting mixture that the dimer and trimer each exhibit once they are isolated and obtained in high purity will also be shown.

(537g) Molecular-Level Composition Model for Heavy Petroleum Resid Supercritical Fluid Extraction Fractions

Zhang, L., China University of Petroleum
Hou, Z., University of Delaware
Billa, T., University of Delaware
Horton, S. R., University of Delaware
Klein, M. T., University of Delaware
Xu, Z., China University of Petroleum
Shi, Q., China University of Petroleum
Zhao, S., China University of Petroleum
Xu, C., State Key Laboratory of Heavy Oil Processing,China University of Petroleum
Originally presented on: 11/6/2013 17:15:00 - 17:35:00

The utilization of an increasing content of heavy petroleum feedstocks forms a challenge for refineries around the world. The high content of aromatic and heteroatom species in resids leads to fouling and catalysts deactivation during catalytic processing. Prior separation will improve the ability to handle these feedstocks and permit obtaining maximum yields of the valuable liquid product by removing contaminates. Supercritical fluid extraction (SFE) technology, developed 20 years ago, separates heavy petroleum resid into several deep cuts by adjusting the solubility of supercritical solvent. Selectively processing different fraction in heavy petroleum resid based on their property can thus be achieved. Compositional variation during SFE is important for organizing the separation/reaction processes networks. However, due to the complexity of heavy petroleum molecules, this is not yet well understood.

In the present work, we developed a molecular-level composition model for heavy petroleum resid supercritical fluid extraction fractions. In the first stage, a structural attribute (core, inter-linkage and sidechain) library was built by the knowledge of heavy petroleum chemistry. Molecular identities in the fraction were determined by juxtaposition of the structural attributes from the library. The concentrations of constructed molecules were calculated by sampling probability density functions (PDFs) for the attributes. The properties of the fractions were then calculated from group contribution methods and structure-property correlations. The PDFs parameters were tuned by a global optimization algorithm to fit the experimental data points to obtain quantitative results. There were ~20,000 molecules in compositional model, including aromatic hydrocarbon and heteroatom species.

Molecular compositions for a series of heavy petroleum resid SFE fractions were thus calculated. The overall solubility parameters show good agreement with experimental measurement. By comparing the solubility parameters distribution, the separation efficiency of the extraction unit was obtained. Molecular distribution for different SFE fractions was also discussed and compared with mass spectrometry results. The present model will provide understanding of heavy petroleum processing and lead to optimization of its use.

(539a) Study of the Tcep-Zeolite System: Kinetics and Surface Chemistry Analysis Investigations

Grieco, S., O'Brien & Gere Engineers, Inc.
Ramarao, B. V., ESPRI, SUNY College of Environmental Science and Forestry
Originally presented on: 11/6/2013 15:15:00 - 15:35:00

Emerging contaminants (ECs) and micropollutants are becoming a global environmental problem as there is increasing information linking them to environmental and human health hazards. One micropollutant of interest is tris-2-chloroethyl phosphate (TCEP), an organo-phosphate ester used as a plasticizer and flame retardant. Recent studies by others have focused on the presence of TCEP in environmental and drinking waters, and several of these previous studies indicated that TCEP was the most frequently detected compound of the ECs investigated. Transport of TCEP from wastewater effluents to raw water sources is supported by research that identifies the presence of TCEP in surface waters similar to those detected in wastewater effluents. This supports the premise that the TCEP is not being removed from the water cycle.

In our previous work presented last year, adsorption of TCEP to zeolite was studied. Equilibrium adsorption isotherms and surface hydrophobicity/energy measurements were conducted. These results demonstrated excellent removal of TCEP using laboratory water and zeolite (>99% removal; Langmuir sorption capacities of 23 mg/g to 103mg/g). Adsorption was determined to be a function of both solution pH and adsorbent type. The observations of pH-dependant adsorption (given a non-ionizing adsorbate) and the additional lack of adsorption related to low zeolite concentrations and higher pH values suggested a complex function of pH and hydrophobic/hydrophilic interaction which required further investigation.

This paper presents the continued research efforts to better understand the adsorption mechanisms of the TCEP-zeolite system. Kinetic modeling and measurements at various TCEP-zeolite ratios were conducted. These data suggest a system with fast kinetics. The modeling was used to establish characteristics of possible reaction mechanisms. Also zeta potential and surface area measurements were conducted to support the kinetic data and better understand the effects of pH on adsorption. These results will be presented.

Thermo Gravimetric Analyses (TGA), Nuclear Magnetic Resonance (NMR), High Perforamnce Liquid Chromatography (HPLC), and Fourier Transform Infrared Technology (FTIR) were used to characterize the TCEP-zeolite adsorption system and postulate the potential adsorption mechanisms occurring. The results of these data will be presented and discussed.

(539b) Adsorption and Degradation of Perfluorooctanoic Acid

Almquist, C. B., Miami University
Berberich, J. A., Miami University
Originally presented on: 11/6/2013 15:35:00 - 15:55:00

PFCs are anthropogenic industrial compounds used in many applications, including surface coatings, surfactants, and flame retardants. They are, however, emerging contaminants due to their recalcitrant nature; they are persistent and very stable in air, water, and soil environments. As a result of their environmental stability, PFCs have been detected in water and soil samples all over the globe, and they have been reported as bioaccumulative, with detectable concentrations in humans and in animals. For reasons of their stability and bioaccumulative nature, and for possible health effects on humans, PFOA has been added to Annex B of the Stockholm Convention on Persistent Organic Contaminants.

In this research, the adsorption of PFOA has been studied using both commercially-available and lab-synthesized adsorbents in combination with both biocatalysts and inorganic catalysts. Immobilizing PFOA using adsorption combined with chemical or enzymatic degradation of PFOA was investigated and found to be superior to each technology alone.

(539c) Removal of Arsenic in Water With An Adsorbent Derived From Waste Concrete Sludge

Sasaki, T., Seikei University
Watanabe, M.,
Hongo, T., Seikei University
Iizuka, A., Tohoku University
Yamasaki, A., Seikei University
Originally presented on: 11/6/2013 15:55:00 - 16:15:00

Introduction:

Novel adsorbents for removal of arsenic in waster were prepared from concrete wastes including waste cement powder and concrete sludge. The waste cement powder is fine particle of which the main component is hydrated cement, which is generated as waste in the recycling process of aggregates from waste concrete. The concrete sludge is waste of raw concrete generated and disposed of from construction sites.

Experimental:

(1) Adsorbent from waste cement. The waste cement particle supplied from a waste concrete recycling plant was used as an adsorbent for the arsenic removal experiment after sieving. The diamter range was 53 to 106 μm.

(2) Adsorbent from concrete sludge. The concrete sludge, which is generated in a concrete product plant, is first diluted with water; this pretreatment could avoid the hardening reaction of concrete. Then solid was separated from aqueous phase, and dried under various temperatures. The dried solid was then pulverized and thieved. The adsorbent so prepared with the diameter range of 53 to 106 μm was used for the arsenic removal experiment.

The arsenic removal experiment was conducted in a batch reactor with stirring under room temperature. The concentration (As (V)) of the solution was changed in the range of 10 to 700 mg-As/L. The ratio of solution to the adsorbent was changed in the range of 67 to 400. The stirring rate was fixed at 350 rpm, which is considered to be high enough to neglect the mass transfer resistance in the aqueous phase. The solution was sampled with filtration, and the concentration change was measured with ICP-AES.

Results and discussion:

It was confirmed that the arsenic concentration decreased with time for all the adsorbents studied. When adsorbent derived from concrete sludge was used, the final concentration of arsenic after 24 h for the initial concentration at 100 mg/L was lower than 0.1 mg/L, which is below the effluent standard for the waste water in Japan. In addition, the removal rate was much higher with the adsorbents from concrete sludge than those from waste cement. The capacity of the arsenic removal with the adsorbent prepared from concrete sludge was about 170 mg-As/g-adsorbent based on the removal experiments of various initial concentrations. This capacity is comparable with the ones of the conventional adsorbents. The arsenic was removed mainly in the form of calcium salts according to XRD analysis of the solid phase after removal. By considering the preparation cost, which is derived from waste, the present adsorbents could be widely applicable to the water treatment, especially in developing countries.

(539d) Effect of Organic Anion On Selective Adsorption

Koong, L. F., The Hong Kong University of Science and Technology
Lam, K. F., The Hong Kong University of Science and Technology
Barford, J., The Hong Kong University of Science and Technology
Mckay, G., The Hong Kong University of Science and Technology
Originally presented on: 11/6/2013 16:15:00 - 16:35:00

Metal containing waste such as electronic waste is the fastest growing waste in the world and they pose a great challenge to waste management and disposal. Today we are not only interested in metal removal, but also recovery especially from waste, due to the dwindling metal reserves. Many researches have looked into leaching metals from batteries, printed circuit boards, magnet etc. Adsorption could make a good application in metal recovery from leachate due to its simple application and ability to operate at low concentration levels. Selective adsorption will be especially useful in recovering individual metal, especially in wastewater or leaching medium where multiple metal species can be present. The selectivity is strongly dependent on the interactions between the functional groups on the adsorbent and the metals. Another factor that could affect selectivity is the metal speciation. Organic anions are able to form complexes with metals, which can then affect the adsorption capacity, and most importantly, selectivity. Organic anions can be naturally found in water environment, the organic acids and derivatives also find wide applications in industry, such as textile, drug, paper, electroplating and so on, and therefore it is important to investigate the influence of anion on adsorption mechanism. This paper aims to demonstrate the effect of organic anions on selectivity of metals using the example of Copper and Cadmium in different anion system: nitrate, sulfate, acetate and formate salt. Iminodiacetic acid resin and chitosan are used in this study. It was found that when organic anions of COOH-based are present, the adsorption capacity is affected but not the selectivity, which Cu is still preferentially adsorbed. However, when amine based organic anions are present, the selectivity can be affected due to stronger complexation between the metal and anions. To demonstrate this EDTA is added to into the nitrate system of Cu-Cd and it is observed that Cd can be selectivity adsorbed at a right concentration of EDTA. EDTA is also removed by adsorption to return the selectivity back to Cu. Total Organic Carbon Analyzer is used to measure the concentration of EDTA in this case.

(539e) Optimization of Commercial Para-Xylene (PX) Simulated-Moving-Bed (SMB) Simulation Using Extended Langmuir Isotherm

Dang, T. T., Hankyong National University
Lim, Y. I., Hankyong National University
Originally presented on: 11/6/2013 16:35:00 - 16:55:00

In the petrochemical industry, Para-xylene is the high valuable compound, which exits in the mixture of ethylene benzene (EB), toluen and xylene isomers (para-xylene (PX), meta-xylene (MX) and ortho-xylene (OX)) after the catalytic reforming reaction. It is very difficult to separate PX from this mixture by the conventional distillation method because of components's closed boiling temperatures. SMB is the continuous counter-current multicolumn chromatographic process, which has advantages to achieve the high PX's purity. The objective of this study is to optimize the operating conditions of SMB unit using the design software, FAST-CHROM/SMB with the Extended Langmuir Isotherm (ELM). To get this objective, the dead volume and flushing scheme is developed and a novel Fowler isotherm adsorption is applied. The ELM parameters are obtained from single adsorption and binary adsorption experimental data. The optimization procedure comprises two levels to optimize bed-configuration parameters and flow rates. Optimal operating conditions of 7 and 8-zone are found by parametric study and maximizing recovery with a desired purity more than 99.7%.

(539f) Synthesis of Mesoporous Carbons: Oxidative Modification and Hydrothermal Carbonization for Efficient Chemical Activation

Jain, A., National University of Singapore
Srinivasan, M. P., National University of Singapore
Balasubramanian, R., National University of Singapore
Originally presented on: 11/6/2013 16:55:00 - 17:15:00

Preparation of mesoporous carbons and activated carbons in general has always been a resource intensive process. Generally, chemical activation is carried out by simple mixing and drying of the activating agent with the raw material, which might lead to ineffective utilization of the former. Improvised chemical activation using hydrothermal treatment and incorporation of oxygen functional groups can lead to more effective chemical utilization and therefore render the process less resource-intensive. In this work, the effect of oxygen functional groups present in the hydrothermally treated precursor on porosity and mesoporosity was investigated. The amount of oxygen containing functional group was varied by appropriate changes to the processing and activating conditions. Coconut shell precursors were subjected to hydrothermal treatment in the presence of hydrogen peroxide as the oxidizing agent at various temperatures. The hydrothermal pre-treatment was followed by chemical activation with $ZnCl_2$ accompanied by pyrolysis. The samples were characterized by surface area methods, titrations and microscopy. Hydrothermal treatment in the presence of H_2O_2 results in significant improvement in the mesoporosity of the final product compared to conventionally processed samples. Samples pretreated with H_2O_2 at 200^0C resulted in highest mesoporosity and this was consistent with the observation of the highest amount of oxygen functional groups in precursor. By the use of H_2O_2 and hydrothermal treatment there is an increase in mesoporous area by 144% and BET surface area by 35%. The hydrothermal pretreatment and surface oxygen functional groups in the precursor, coupled with the use of H_2O_2 as the oxidizing agent, together contribute to improved chemical activation, thereby increasing mesoporosity.

(539g) Biosorption of Heavy Metals Using An Acid Activated Walnut Shell

Andrade-Espinosa, G., Universidad Autonoma de San Luis Potosi
Berber-Mendoza, M. S.,
Bernal-Jacome, L. A., Universidad Autonoma de San Luis Potosi
Medellin-Castillo, N. A., Universidad Autonoma de San Luis Potosi
Originally presented on: 11/6/2013 17:15:00 - 17:35:00

In this research the removal of heavy metals [Pb (II), Cd (II), Zn (II)] ions from aqueous solutions by chemically modified walnut shell was studied. A response surface design analysis was performed to screen the variables affecting heavy metals removal efficiency. The effects of citric acid concentration, temperature and time of modification on the adsorption capacity of metals were examined in batch reactors. The walnut shell was characterized before and after modification by Scanning Electron Microscopy (SEM), Fourier Transform Infrared Spectroscopy (FTIR), Thermogravimetric Analysis (TGA) and acid/base titration. Also, charge distribution of the walnut shell was determined by equilibrium. Plus pH and temperature effects on the adsorption capacity of Pb, Cd and Zn were investigated by adsorption isotherms that were conducted in batch adsorbers. The results of the surface design showed that metal ions adsorption is highly dependent on the modification time and less dependent on the modification temperature and citric acid concentration, accordingly. On the other hand, SEM micrographs revealed no significant changes in the surface of the walnut shell after modification. This indicates that the degree of modification does not impair the physical properties of the shell. In addition, total acid sites content of modified walnut shell decreased from 2.6 to 2.4 $mmolg^{-1}$ which was reflected in a less amount of carboxylic sites (from 0.78 to 0.56 $mmolg^{-1}$). These last findings were supported by FTIR and charge distribution; treated material showed less accentuated energy absorption peaks at wavelengths between 1750 and 1700 cm^{-1} and the point of cero charge (PCC) changed 1 unit to more basic values of pH, which is attributed to lower concentration of carboxylic acids (COOH). TGA analysis showed a small shoulder at around 380 °C and an intensive peak at around 350 °C. These minor and major reactions can be attributed to the hemicelluloses and cellulose decomposition, respectively (Acikalin, 2011). In the other side, the sorption experiments reported, that, the heavy metals removal increased as pH rise from 3 to 7; resulting a maximum sorption capacity of 18, 12, 4 mgg^{-1} of Pb (II), Cd (II) and Zn (II), respectively, for untreated walnut shell. The sorption capacity of modified walnut shell was 2% higher than the untreated material. Finally, it was found a slight increase onto the sorption capacity as the temperature rise from 15 to 35 °C. These findings indicate that the untreated and modified walnut shell studied herein have a potential application in the removal of contaminants presents in aqueous phase.

(541a) Facile Anhydrous Proton Transport On Hydroxyl Functionalized Graphene

Choudhury, P., University of Pittsburgh
Saidi, W. A., University of Pittsburgh
Johnson, J. K., University of Pittsburgh
Originally presented on: 11/6/2013 15:18:00 - 15:42:00

Proton transport is a critical important property of proton exchange membrane (PEM) fuel cells. We propose a nano-structured material capable of fast water-free proton transport. These materials can be constructed from functionalized surfaces (e.g. graphene, SWNT, Si(001), MOFs, etc) with different functional groups (e.g. -OH, -COOH, -SO3H, -NH, etc.) that form hydrogen bonded networks. In this work, we predict that hydroxyl functionalized graphene will conduct protons in the complete absence of water due to a self-assembled hydrogen bond network of OH groups on the surface.Â Activation energy of hydroxyl functionalized graphene is similar to that of Nafion, a standard polyelectrolyte proton exchange membrane. The single-file proton diffusion and proton mobility on the surface of both 1-D and 2-D hydrogen bonded graphane membranes will be discussed.

Acknowledgement

We gratefully acknowledge support from DTRA under Contract No. HDTRA1-09-1-0008 and from NSF TeraGrid (XSEDE) resources under allocation numbers [TG-DMR100097], [TG-SEE090006], and [TG-DMR110091]. We thank K. D. Jordan , M. E. Tuckerman and J. T. Yates, Jr. for helpful discussions. We thank the Center for Simulation and Modeling at the University of Pittsburgh for providing computational support.

(541b) Polysulfone-Based Anion Exchange Membrane Fuel Cell: Molecular Dynamics Simulation Approach

Han, K. W., Georgia Institute of Technology
Ko, K. H., Georgia Institute of Technology
Choi, J. I., Georgia Institute of Technology
Jang, S. S., Georgia Institute of Technology
Originally presented on: 11/6/2013 15:42:00 - 16:06:00

In this study, the nanophase-segregated structures and transport properties of quaternary ammonium grafted polysulfone membranes are investigated using molecular dynamics (MD) simulation method. For this, we develop a new force field from a reference density functional theory modeling with B3LYP and 6-31G** in order to describe the hydroxide anion. The bond stretching force constant is determined to reproduce the quantum mechanical vibrational frequency. The atomic charges are determined by Mulliken population analysis. Through the annealing procedure, the nanophase-segregated structure is developed as a function of water contents such as 10 and 20 wt %. The extent of nanophase-segregation is evaluated by the structure factor analysis, which can be compared with the experimental small angle scattering data. Once the equilibrium structures are obtained, another MD simulation is run for data collection from which the diffusion of water and hydroxide are analyzed using the mean-square displacement analysis by assuming that the diffusion is a Gaussian diffusion. The nanophase-segregated structures and the transport properties are compared to the proton exchange membrane consisting of the same polymer backbone except for the acidic functional group.

(541c) Permethylcobaltocenium Based Hydroxide Exchange Membranes With High Stability

Gu, S., University of Delaware
Wang, J.,
Fang, Q., University of Delaware
Kaspar, R. B., University of Delaware
Zhang, B., University of Delaware
Yan, Y., University of Delaware
Originally presented on: 11/6/2013 16:06:00 - 16:30:00

Permethylcobaltocenium Based Hydroxide Exchange Membranes with High Stability

Shuang Gu, Junhua Wang, Qianrong Fang, Robert B. Kaspar, Bingzi Zhang and Yushan Yan*

Department of Chemical & Biomolecular Engineering, University of Delaware

5 Innovation Way, Newark, DE 19716, USA

yanys@udel.edu

Hydroxide (OH^-) exchange membranes (HEMs) have become the research focus in energy conversion electrochemical devices (fuel cells, electrolyzers, and solar hydrogen generators) during the past decade owing to their intrinsic ability to work with non-precious-metal catalysts as well as their desirable tolerance to CO_2 contamination[1]. However, unlike their counterpart—proton (H^+) exchange membranes (PEMs) that are based on the highly stable sulfonic-acid functional group ($-SO_3H$) and thus have excellent chemical and thermal stability, HEMs suffer from limited chemical stability due to their currently used ammonium hydroxide cationic functional group ($-NR_3^+OH^-$). The sensitive degradation pathways (such as nucleophilic substitution, Hoffman elimination, and ylide formation) of ammonium hydroxide not only limit the working temperatures for HEM applications but also raise the strong concern for long-term HEM stability. Alternative cationic functional groups based on the same nitrogen element (such as pyridinium, guadinium, and imidazolium) and other elements (phosphorus: phosphonium, sulfur: sulfonium, and ruthenium: bis(terpyridine)ruthenium complex[2]) have also been introduced, bringing new features (improved solubility, improved thermal stability, and/or increased charge number) for HEMs but the practice suggests their chemical stability still remain limited. The limited chemical stability of ammonium and other cations may be rationalized as the result of their flexible valence electron structures, allowing for many degradation pathways possible.

Alkali metal cations (typically, Li^+, Na^+, and K^+) have excellent chemical stability, because of their non-flexible closed-shell valence electron configurations ([$1s^2$], [$2s^22p^6$], [$3s^23p^6$], respectively) that are the same as those of noble gases, but they are very challenging to immobilize for HEM applications. Bearing a unit of positive charge, biscyclopentadienyl cobalt sandwich cation (e.g., $(C_5H_5)_2Co(III)^+$, or cobaltocenium) satisfies the 18-electron rule for transition metal complexes, in which Co(III) also has closed-shell valence electron configuration. The cobaltocenium hydroxide is considered as a strong base, and in fact the cobaltocenium is the most stable metallocenium in alkaline media[3]. Equally importantly, cobaltocenium can also be immobilized to polymers in both main-chain and side-chain.

Permethylcobaltocenium, i.e., decamethylated cobaltocenium, is expected to have even higher structure stability than the normal cobaltocenium, due to the charge delocalization caused by strong electron donation from the ten methyl groups[4]. This is evidenced by over 600 mV negative shift of formal potential observed [−1.24 of $(C_5Me_5)_2Co(III)^+/(C_5Me_5)_2Co(II)$ vs. −0.63 V of $(C_5H_5)_2Co(III)^+/(C_5H_5)_2Co(II)$ referring the SHE, the same solvent of CH_2Cl_2[5]]. Here we present, for the first time, the permethylcobaltocenium-functionalized polymer HEMs that show very high thermal and chemical stability, promising to applications in durable and robust electrochemical devices.

References

[1] J. R. Varcoe, R. C. T. Slade, *Fuel Cells* **2005**, *5*, 187-200.

[2] Y. P. Zha, M. L. Disabb-Miller, Z. D. Johnson, M. A. Hickner, G. N. Tew, *J Am Chem Soc* **2012**, *134*, 4493-4496.

[3] G. Wilkinson, F. A. Cotton, *Prog Inorg Chem* **1959**, *1*, 1-124.

[4] N. G. Connelly, W. E. Geiger, *Chem Rev* **1996**, *96*, 877-910.

[5] U. Koelle, F. Khouzami, *Angewandte Chemie-International Edition in English* **1980**, *19*, 640-641.

Key words: Cobaltocenium, Hydroxide exchange membrane, Stability, Fuel cells, Electrochemical devices.

(541d) Durability and Performance Of An Anion Exchange Membrane Separator For Electrochemical Energy Storage and Conversion Applications

Yun, S.,
Parrondo, J., Illinois Institute of Technology
Ramani, V., Illinois Institute of Technology
Originally presented on: 11/6/2013 16:30:00 - 16:54:00

All-vanadium RFBs (VRFB) are very interesting from an operational point of view because the intermixing of positive and negative solutions does not cause irreversible damage. In this work, we synthesized a novel AEM based on cardo-polyetherketone (PEK-C), and evaluated its stability under RFB working conditions. The membrane stability was evaluated by measuring the changes in the chemical structure and mechanical properties after immersion in vanadium (V) solutions (strongly oxidizing) for 1 month. Post-mortem analysis of the membrane using 2D NMR techniques confirmed that there was no degradation in chemical structure after such immersion. The vanadium (IV) membrane permeability of quaternized PEK-C/TMA^+ membrane ($8.2 \pm 0.2 \times 10^{-9}$ cm^2 s^{-1}) was 35 times lower than for Nafion® membranes ($2.9 \pm 0.2 \times 10^{-7}$ cm^2 s^{-1}) confirming the suitability of anion exchange membranes for RFB applications. Water absorption isotherms (VTI-SA, TA instruments) showed moderate water uptakes (30wt%) - this is a necessary requisite for adequate mechanical integrity during long term RFB operation. No changes were observed in the ultimate tensile strength of the membranes after 1 month immersion in the vanadium (V) solution (19 ± 2 and 21 ± 1 MPa respectively). Redox flow battery (5 cm^2 single cell) testing was carried out and the charge/discharge of the battery was performed at a current density of 100mA cm^{-2} over 20 cycles initially. The coulombic efficiency was close to 100%, and the voltage and energy efficiencies were around 80% for every cycle. The stability of the anion exchange membrane under RFB operating conditions is a promising result that will facilitate widespread use of inexpensive AEMs as separators in RFBs.

Likewise, the same membrane chemistries can also serve as separators in anion exchange membrane fuel cells. Studies on this front are currently ongoing; the presentation will also discuss the efficacy and stability of the above membranes in alkaline fuel cell applications.

(541e) Polymer/Carbon Nano-Tube Composites for Alkaline Fuel Cell Applications
Lue, J.,
Originally presented on: 11/6/2013 16:54:00 - 17:18:00

Polybenzimidazole (PBI)-functionalized multi-walled carbon nano-tubes (CNTs) are obtained via an ozone-mediation route. The PBI nano-composites containing small amounts (<1%) of CNTs are prepared using solution casting methods. These thermally stable PBI and PBI/CNT composites are doped with potassium hydroxide (KOH) solution to prepare hydroxide-conducting electrolytes for alkaline direct methanol fuel cell (ADMFC) applications. The CNTs-containing composites exhibit higher fractional free volumes and higher water diffusivities. The CNTs also promote ionic conductivity of electrolytes and improves the fuel cell performance. Gas diffusion electrodes (GDEs) without polytetrafluoroethene (PTFE) treatment give superior cell power density compared to commercial E-tek GDEs, which contain hydrophobic PTFE layers. When the fuel cell is fed with 2 M methanol in 6 M KOH (as the anode fuel) and humidified oxygen (as the cathode oxidant), the system achieves a maximum power density of 104.7 mW cm^{-2} at 90°C. These KOH-doped PBI/CNT composites have the potential to be used in high temperature alkaline fuel cell applications.

(541f) Di-Quaternized DABCO-Based PEEK Hydroxide Exchange Membranes With Improved Ionic Conductivity

Wu, X., Dalian University of Technology
He, G., Dalian University of Technology
Wang, J., Dalian University of Technology
Yan, X., Dalian University of Technology
Originally presented on: 11/6/2013 17:18:00 - 17:42:00

Di-quaternized DABCO-based PEEK hydroxide exchange membranes with improved ionic conductivity

Xuemei Wu, Gaohong He*, Junjun Wang, Xiaoming Yan

State Key Laboratory of Fine Chemicals, Research and Development Center of Membrane Science and Technology, Dalian University of Technology, Dalian 116024, PR China.

Abstract

Polymer Electrolyte Membrane Fuel cells are considered to be clean and efficient power sources. Among the several kinds, hydroxide exchange membrane fuel cells (HEMFCs) operate in alkaline media with faster kinetics of the oxygen reduction, lower methanol permeability and easier water management, therefore superior to proton exchange membrane fuel cells (PEMFCs) in some aspects such as cost and application on direct methanol fuel cells (DMFCs). Hydroxide exchange membrane (HEM) is a key component of HEMFC, which conducts hydroxide ions and prevents crossover of reactants. Significant research works have been done in recent years to develop new HEMs. However, there are still two main challenges facing with HEMs, i.e. low ionic conductivity and poor stability of basic groups. In this study, di-quaternized 1, 4-diazabicyclo [2,2,2] octane (DABCO)-based PEEK (di-QDPEEK) hydroxide exchange membranes are prepared via the two nitrogen atoms of DABCO to improve ionic conductivity. And the strain free conformation of DABCO provides good resistance to both Hoffman degradation and nucleophilic substitution by OH^-.

We design both mono- and di-quaternized DABCO-based PEEK membranes with similar ionic exchange capacity (IEC) to investigate the influence of distribution of quaternary ammonium functional groups on the microstructures and electrochemical properties of the HEMs. For low IEC around 0.8, di-QDPEEK membrane exhibits much higher ionic conductivity (2.9 times) and swelling ratio (1.7 times) as compared with the mono-QDPEEK membrane. However, for high IEC around 1.2, di-QDPEEK membrane exhibits much higher ionic conductivity (2.1 times) and much lower swelling ratio

(0.6 fold) as compared with the di-QDPEEK membranes. Based on our data, a schematic model with scattered or concentrated distributions of the quaternary ammonium groups in the mono- and di-quaternized membranes is proposed. In the di-quaternized membranes, quaternary ammonium functional groups concentrate in a local scale, which enhance the interactions between ionic groups, leading to well hydrophilic-hydrophobic micro-phase separation and better interconnected pathways for ions through the hydrophilic domains. As a result, di-quaternized membranes exhibit much higher ionic conductivity and special swelling behavior as compared with the mono-QDPEEK membranes of similar IEC. More details about the microstructures-properties correlations, such as TEM image, glass transition temperature and mechanical strength of the mono- and di-QDPEEK membranes will be present in this work. For di-QDPEEK of IEC about 1.69, high ionic conductivity of around 35.5 mS cm^{-1} at 20 ^{o}C is achieved, which is within the top values reported in the literatures.

Keywords: Separations, Hydroxide exchange membranes, Fuel cells, DABCO, PEEK, Ionic conductivity

Acknowledgements

The authors thank Program for National Science Fund for Distinguished Young Scholars of China (Grant no.21125628), NSFC (Grant no. 20976027 and 21176044) for supports of this work.

* **Corresponding author.** Tel: +86 411 84707892 E-mail address: hgaohong@dlut.edu.cn

(552a) Asymmetric Membranes With Internal Active Domains

Li, W., University of Kentucky
Walz, J. Y., Virginia Tech
Originally presented on: 11/6/2013 15:15:00 - 15:40:00

Despite the fast development of membranes, cost-effective, environmentally benign, and multi-functional design remains a challenge for next-generation materials. In this paper, we present a strategy combining extremely straightforward techniques for creating membranes with and highly adjustable properties (e.g. pore size, specific surface area, surface properties). One of the unique characteristics of these membranes is that the active layer is actually contained within the supporting layer rather than as an external layer (which is typical for asymmetric membranes). This structure is created by infiltrating a particle solution into the porous scaffold and then drying from one side of the membrane, such that evaporation pulls the particles to the open side. To create more complex structure, a two-step infiltration/drying process is used, in which microparticles and nanoparticles are infiltrated sequentially. The resulting active domains are of a thickness less than 1% of that of the membrane for micro-particles and even smaller for nanoparticles. This design is versatile as it works effectively with a large variety of particles, regardless of their size, shape, or material. Because of its compatibility with a large group of particles, this method is promising in building membranes with changing properties by infiltrating more than one type of particle.

(552b) Novel Polysiloxaneimide/Polyetherimide/Non-Woven Fabric Composite Membranes for Hydrophobic Pervaporation

Song, Z. W., Central South University
Jiang, L. Y., Central South University
Zhu, J. M., Central South University
Originally presented on: 11/6/2013 15:40:00 - 16:05:00

In this study, synthesis of polysiloxaneimide (PSI) was carried out and novel multi-layer composite membrane using PSI as selective top layer was prepared for pervaporation. The PSI was synthesized from 3,3,4,4-biphenyltetracarboxylic dianhydride (BPDA) and amino modified siloxane (dual-end type). Physiochemical characterization of the PSI with FTIR, NMR, XRD and GPC confirmed the success in synthesis. As for the composite membrane, non-woven fabric layer covered by crosslinked porous polyetherimide (PEI) layer was used as support. The PSI layer was formed by non-solvent induced phase inversion. Effects of preparation conditions in terms of support layer structure, PSI dope solution composition and post treatment on the pervaporation properties of the composite membranes were investigated. Pervaporation test for recovering ethanol from its aqueous solution (5wt.%) at 60°C showed that the highest separation factor obtained with composite membrane was ~3 with corresponding flux being ~0.5kg/m^2-hr. Based on the information accrued from pervaporation, FESEM, EDX etc., the problems existing with this new type of membrane and its fabrication methods were analyzed.

(552c) PIM-1/Matrimid Hollow Fiber Membranes for Natural Gas and Air Separation

Yong, W. F., National University of Singapore
Li, F. Y., National University of Singapore
Xiao, Y. C., Suzhou Faith and Hope Membrane Technology Co. Ltd.
Chung, T. S., National University of Singapore
Tong, Y. W., National University of Singapore
Originally presented on: 11/6/2013 16:05:00 - 16:30:00

Polymers of intrinsic microporosity (PIM-1) have received worldwide attention as new polymer type for gas separation but most PIM-1 researches have been conducted on dense flat membranes. For the first time, we have fabricated PIM-1/Matrimid membranes in a useful form of hollow fibers with promising separation performance. The newly developed hollow fibers comprising 5-15 wt% of highly permeable PIM-1 not only possess much higher gas-pair selectivity than PIM-1 but also have much greater permeance than pure Matrimid fibers. Data from Positron annihilation spectroscopy (PAS), field emission scanning electron microscopy (FESEM) and apparent dense layer thickness indicate that the blend membranes have an ultrathin dense layer thickness of less than 70 nm. High flux defect-free as-spun blend membranes with gas-pair selectivity more than 90% of the intrinsic values of dense films can be spun directly from dopes containing 5 wt% PIM-1 into Matrimid while post annealing and additional silicone rubber coating are needed for membranes containing 10 and 15 wt% PIM-1. Comparing to Matrimid, the CO_2 permeance of as-spun fibers containing 5 and 10 wt% PIM increases 78% and 146%, respectively (e.g., from original 86.3 GPU to 153.4 GPU and 212.4 GPU) without compromising CO_2/CH_4 selectivity. The CO_2 permeance of the fiber containing 15 wt% PIM displays a greater improvement of 2.8 folds to 243.2 GPU with a CO_2/CH_4 selectivity of 34.3 after silicon rubber coating. Under binary gas tests of 50/50 CO_2/CH_4, this fiber shows a CO_2 permeance of 188.9 GPU and a CO_2/CH_4 selectivity of 28.8. The same fiber also has an impressive O_2 permeance of 3.5 folds higher than the pristine Matrimid (e.g., from original 16.9 GPU to 59.9 GPU) with an O_2/N_2 selectivity of 6.1. The newly developed membranes demonstrate an exceptional gas separation performance and have a great potential to be used for natural gas purification, air separation and CO_2 capture.

(552d) Effect of Polymer Molecular Weight On CO2 Separation Performance of SAPO-34/6FDA-DAM/Additive Ternary Mixed Matrix Membranes

Avci, A. H., Istanbul Technical University
Tantekin-Ersolmaz, S. B., Istanbul Technical University
Originally presented on: 11/6/2013 16:30:00 - 16:55:00

Polyimides are of interest in separation of industrial gas streams due to their outstanding thermal and chemical stability. The polyimide prepared from 4,4-hexafluoro isopropylidene diphthalic anhydride (6FDA) and 2,4,6-trimethyl-m-phenylene diamine (DAM) is attractive due to its high permeability for many gases but its selectivity for commercially important gas pairs is much lower than the industrially relevant values. We have previously reported a proof-of-concept study demonstrating that ternary (polymer-zeolite-additive) mixed-matrix membranes may be a possible route to tailor permselectivity [1]. It has been shown that by using SAPO-34 as the zeolite and 2,4,6-triaminopyrimidine (TAP) as the compatibilizing additive, separation properties of highly permeable 6FDA-DAM can be tailored to commercially attractive levels for CO_2/CH_4 application. The use of the additive served two purposes: (1) the amine functional groups of TAP formed hydrogen bonds with the OH groups of the zeolite and the carbonyl groups of the polyimide and hence providing good adhesion at the zeolite-polymer interface, (2) reducing the permeability and increasing the selectivity of polymer due to its antiplasticization effect so that permeabilities of zeolite and polymer become compatible and zeolite incorporation would favor the selectivity. The results were surprisingly good. A CO_2 permeability (35°C and 4 bar) of 8.3 Barrer and an ideal CO_2/CH_4 selectivity of 259 were obtained at 2% TAP and 20% SAPO-34 addition and this performance was much above the upper-bound. In this work, we investigate the effect of polymer molecular weight on the material and separation properties of these ternary mixed-matrix membranes. Membranes are characterized by Fourier transform infrared spectroscopy (FTIR), differential scanning calorimetry (DSC), thermal gravimetric analysis (TGA), scanning electron microscopy (SEM), and single and binary-gas permeation measurements at 35°C for CO_2/CH_4separation. Results indicate that polymer molecular weight is a key parameter in determining the final membrane properties. The effect of the additive varies significantly when the polymer molecular weight is varied from 30,000 to 100,000. Results are interpreted in terms of additive-polymer interaction.

[1] H. Ishakoglu, E. Erpek, S. Halitoglu, M. G. Ahunbay, S. B. Tantekin-Ersolmaz, Polymer-zeolite-additive three-component mixed matrix composite membrane for CO_2/CH_4 separation, presented at Euromembrane 2009, Montpellier, France, Sept. 06-10, 2009.

(552e) Formation of Uniform CHA Zeolite Layers

Kim, E., Korea University
Cai, W.,
Baik, H.,
Choi, J., Korea University
Originally presented on: 11/6/2013 16:55:00 - 17:20:00

All-silica CHA zeolites[1] hold promise to separate CO_2 from other molecules (e.g., N_2 in the post-combustion process) through their molecular sieving. In particular, the pore size of CHA zeolites (0.370 nm × 0.417 nm)[2] can allow only CO_2 (0.33 nm) to pass through the pore channels, while excluding the transport of larger N_2 (0.364 nm).[3] However, conventionally synthesized near cubic Si-CHA particles have a wide size distribution (~1-20 μm).[4-6] The wide size distribution prevents forming a close-packed layer, though the layer is critical in manufacturing zeolite films as desired in the secondary growth methodology.[7,8]

In an attempt to overcome this, anisotropic plate-like Si-CHA particles, discovered as the minor product along with the dominant near cubic Si-CHA particles, were adopted to make a uniform Si-CHA layer. Selected area electron diffraction and X-ray diffraction characterizations verified the CHA phase of the plate-like particles. The plate-like CHA particles could be selectively deposited on a porous the α-Al_2O_3 disc despite their lower abundance in a mixture. The uniform CHA layer was formed with a high surface coverage and the uniformity of the plate-like CHA deposits allowed a preferred out-of-plane orientation. The subsequent secondary growth of the seed layer allowed for the continuous CHA membranes. The corresponding CO_2/N_2 separation performance will be discussed in detail.

References

[1] Camblor, M. A.; Corma, A.; Diaz-Cabanas, M. J.; Baerlocher, C. *Journal of Physical Chemistry B* **1998**, *102*, 44.

[2] Hedin, N.; DeMartin, G. J.; Roth, W. J.; Strohmaier, K. G.; Reyes, S. C. *Micropor. Mesopor. Mater.* **2008**, *109*, 327.

[3] Breck, D. W. *Zeolite molecular sieves: strcucture, chemistry, and use*; John Wiley & Sons, Inc., **1974**.

[4] Eilertsen, E. A.; Arstad, B.; Svelle, S.; Lillerud, K. P. *Micropor. Mesopor. Mater.* **2012**, *153*, 94.

[5] Olson, D. H.; Camblor, M. A.; Villaescusa, L. A.; Kuehl, G. H. *Microporous Mesoporous Mater.* **2004**, *67*, 27.

[6] Trzpit, M.; Rigolet, S.; Paillaud, J. L.; Marichal, C.; Soulard, M.; Patarin, J. *Journal of Physical Chemistry B* **2008**, *112*, 7257.

[7] Jiang, H. Y.; Zhang, B. Q.; Lin, Y. S.; Li, Y. D. *Chinese Science Bulletin* **2004**, *49*, 2547.

[8] Snyder, M. A.; Tsapatsis, M. *Angew. Chem., Int. Ed.* **2007**, *46*, 7560.

(552f) Preparation of Economic and Robust Macroporous Membranes From Natural Kaolin Clays

Alsyouri, H. M., The University of Jordan
Ghrair, A., Royal Scientific Society
Abu Daabes, M., German Jordanian University
Originally presented on: 11/6/2013 17:20:00 - 17:45:00

Inorganic materials have been widely used in membrane separation technologies due to their attractive thermal and strength characteristics unmatched by the polymeric counterparts. Permeation properties have also been a focus of attention that led to many reported ceramic membrane configurations and preparation strategies. The high cost is one of the limitations in ceramic membranes stemmed by elevated costs of inorganic precursors and high energy requirements. A cost-effective ceramic membrane combining attractive transport and strength properties is highly desirable in this area. One way of addressing this challenge is to utilize natural clays. Jordan is abundant of many types of clays (e.g., Kaolin, bentonite, zeolite, quartz, etc) that can be used for manufacturing of composite membranes. Our ultimate objective is to prepare zeolite membrane supported on kaolin support from natural clays for use in ultra-filtration of brackish layer. This abstract summarizes our work in the preparation and testing of an effective kaolin membrane support.

We used three types of raw kaolin clays (K1, K2 and K3) mined from different regions in Jordan named: Assalat, Batnil-ghoul, and Wadil-hafira respectively. Kaolin and mineral contents were evaluated by XRD and XRF techniques. All clays were grinded into particles less than 20 micron and used without any pre-refining. Membrane supports were prepared by pressing the wetted clay powder in a mold at 1000 psi followed by sintering at 1100C for four hours yielding three sets of supports: K1, K2 and K3 . The supports were tested by steady state helium gas permeation at room temperature for evaluation of flux permeance, and average pore size properties.

XRD patterns and XRF elemental analysis revealed that clays are composed of three major minerals (Kaolin, Quartz and Calcite). The proportions of minerals vary with the source with kaolin content increasing progressively from K1 to K3. This combination of minerals is very interesting because it contains the major constituents used in commercial supports. Kaolin ($Al_2O_3.2SiO_2.2H_2O$) is a plastic and moldable material, while quartz (SiO_2) is refractory and can enhance the strength of the composite. During sintering, the quartz transforms into a glassy phase and holds the particles together. Calcite or limestone ($CaCO_3$) may appear to be harmful because it yields CO_2 upon heating which creates extra voids in the support. If these voids can develop into cracks, the calcite can be removed by an acid washing pretreatment. All supports were physically strong having a compressive strength exceeding 200 MPa. Attempts to purify the kaolin significantly decreased the strength to 78 MPa.

Gas permeation properties shows that from K1 to K3, the permeance progressively decreases two orders of magnitude from 1E-5 to 1E-7 mol/s.m2.Pa emphasizing the influence of minerals on the inner structure. These values are close to permeance of values of commercial alpha alumina supports. The supports are all macroporous as revealed by the average pore size evaluated from the fitted equations. The pore size decreases progressively from K1 to K3 in the same order of increasing kaolin content. It is interesting to note that gas transport in these membranes is a combination of viscous and Knudsen flows with percentage contributions ranging between 65/35 to 30/70 based on the pore size. The Knudsen mechanism becomes more dominant for smaller pore supports.

(593a) Constant Flux Crossflow Filtration Evaluation of Bioinspired Fouling-Resistant Membranes

Miller, D. J., University of Texas at Austin
Kasemset, S., University of Texas at Austin
Wang, L., The University of Texas at Austin
Paul, D. R., The University of Texas at Austin
Freeman, B. D., The University of Texas at Austin
Originally presented on: 11/7/2013 8:30:00 - 9:00:00

Polydopamine, formed by the spontaneous polymerization of dopamine in buffered, aqueous solution, has recently found use as a hydrophilic surface modification agent. Polydopamine is a catecholamine polymer inspired by the adhesive proteins secreted by many sessile marine organisms, and has been found to form a robust coating on a wide variety of polymeric membranes. In addition to hydrophilizing the surface, polydopamine can also be used as a surface primer for the conjugation of other molecules, such as poly(ethylene glycol). The surfaces of polysulfone ultrafiltration membranes were modified with polydopamine and polydopamine-*g*-poly(ethylene glycol) hydrophilic coatings. Unmodified and modified membranes were challenged with a soybean oil emulsion feed at six different permeate fluxes in constant flux crossflow filtration fouling studies. The threshold flux was determined for each membrane. Above the threshold flux, modified membranes generally exhibited lower transmembrane pressures than unmodified membranes. However, below the threshold flux, modified membranes had higher transmembrane pressures than unmodified membranes, likely due to an increase in mass transfer resistance resulting from the surface modification. If the membranes are to be operated below the threshold flux to achieve sustainable operation, one might conclude that these surface modifications did not improve the overall performance of the membranes during fouling, since the resistance to mass transfer of the modified membranes was not lower than that of the unmodified membrane. However, the surface-modified membranes had a lower pure water permeance, and, therefore, higher initial resistance, than the unmodified membrane. To account for this difference, polydopamine-*g*-poly(ethylene glycol) modified membranes were compared to membranes with a thicker polydopamine coating and to an unmodified membrane with a smaller pore size than that used in the surface modification studies; in this way, all three membranes had similar initial resistances. In this case, the unmodified membrane exhibited a much higher transmembrane pressure during fouling than the modified membranes, so when membranes are compared at the same initial resistance, the surface modifications improved fouling resistance. Therefore, a potential strategy to achieve fouling resistance in a membrane of a desired flux and rejection is to modify the surface of a more permeable (and perhaps lower rejection) membrane, thereby making the resulting modified membrane fouling resistant but leaving it with the desired flux and rejection characteristics.

(593b) CO2 Separation Properties of [Bmim][Tf2N]-PVDF Blend Composite Membranes

Gao, S.,
Zhang, D.,
He, G., Dalian University of Technology
Originally presented on: 11/7/2013 9:00:00 - 9:30:00

[bmim][Tf_2N]-PVDF blend composite membranes were fabricated with polypropylene (PP) ultrafiltration membranes as support layers using the solvent evaporation technique that show a best CO_2 permeance of 97.8 GPU with maximum CO_2/N_2 and CO_2/CH_4 selectivity of 36.2 and 23.8, in which the ionic liquid 1-butyl-3-methylimidazolium bis(trifluoromethylsulfonyl)imide ([bmim][Tf_2N]) was dispersed in poly(vinylidene fluoride) (PVDF) matrix. [bmim][Tf_2N] with high CO_2 permeability plays a crucial role in CO_2 separation. Membrane permeances for CO_2, N_2, and CH_4 and the CO_2/N_2 and CO_2/CH_4 selectivity were evaluated with three kinds of pure permanent gases (CO_2, N_2, CH_4). The CO_2 and CH_4 permeabilities of the blend composite membranes are approximately predicted by Maxwell model. The gas separation performance for CO_2/N_2 and CO_2/CH_4 of most of these blend composite membranes surpass the Robeson's 2008 upper bound limit, and are stable at trans-membrane pressure up to 0.7 MPa. Therefore, they are promising materials for CO_2 capture from flue gas and natural gas.

(593c) Development of Biomimetic Reverse Osmosis Membranes Via Molecular Self-Organization of Phospholipid Bilayer

Saeki, D., Kobe University
Yamashita, T.,
Sako, F.,
Matsuyama, H., Kobe university, Center for Membrane and Film Technology
Originally presented on: 11/7/2013 9:30:00 - 10:00:00

Biomimetic separation membranes have been attracting attention as future water treatment membranes such as reverse osmosis and forward osmosis[1,2]. The biomimetic membrane can transport target molecules highly effectively and selectively. In this study, we demonstrate two approaches to prepare biomimetic water treatment membranes via molecular self-organization. One is using electrostatic interaction between a polyamide support membrane and cationic liposomes (Figure 1A). The polyamide membrane was immersed into the cationic liposome suspension, and the liposomes were adsorbed onto the membrane. By pressurized at given pressure, the adsorbed liposomes were transformed into a flat bilayer structure. The prepared membrane showed higher salt rejection enough to use of a RO membrane process. Furthermore, addition of gramicidin A as a water channel molecule into the liposomes improved the water permeability. The other is using hydrophobic interaction between an alumina support membrane and liposomes. The alumina membrane was hydrophobilized by a silane coupling agent and immersed into the liposome suspension, and lipid bilayers were spontaneously formed onto the alumina membrane. The bilayer formation was confirmed by a fluorescence recovery after photobleaching (FRAP) method. The prepared membrane via hydrophobic interaction also showed higher salt rejection. These results indicate that the two presented approaches could be applicable to fabrication of new structural water treatment membranes.

References

1. Y. Kaufman, *et al.*, *Langmuir*, 26, 7388 (2010).

2. H.L. Wang, *et al.*, *Small*, **8**, 1185 (2012).

(593d) Preparation and Gas Separationperformance of the Inorganic Nano-Particles Carbon Composite Membranes

Li, L., State Key Laboratory of Fine Chemicals, Carbon Research Laboratory, School of Chemical Engineering, Dalian University of Technology
Wang, T., State Key Laboratory of Fine Chemicals, Carbon Research Laboratory, School of Chemical Engineering, Dalian University of Technology
Wang, N.,
Zhao, X.,
Cao, Y., Dalian Institute of Chemical Physics, Chinese Academy of Sciences
Qiu, J., State Key Laboratory of Fine Chemicals, Carbon Research Laboratory, School of Chemical Engineering, Dalian University of Technology
Originally presented on: 11/7/2013 10:00:00 - 10:30:00

Carbon membranes, novel materials derived from the pyrolysis of the polymeric precursors, have a well-developed ultramicropore network and exhibit higher gas permeability and selectivity than polymeric membranes. Nevertheless, the gas permeability for pure carbon membranes available now cannot satisfy the requirement of commercial applications due to the disordered inter-connective nano-channels in the pore structure, which results in the high diffusion resistance of the gas molecules penetrating through the membrane.

Incorporating functional materials into membrane precursors to fabricate composite carbon membranes has been regarded as an effective way to tune the disordered pore structure of pure carbon membranes, and thus to greatly improve the gas permeability. In this paper, the inorganic nano-particles/carbon composite membranes with high gas sepraration performance are designed and perpared by incorporating the inorganic nano-particles (nano-oxides, zeolites and ordered mesoporous materials) into the polymeric precursor and pyrolyzing. The effects of species, sizes and pore structure morphology of embeded nano-particls on the structure and gas separation performance of obtained composite carbon membrane are investigated. The as-made composite membranes are characterized by high-resolution transmission electron microscopy (HRTEM), high-resolution scanning electron microscopy (HRSEM), X-ray diffraction (XRD) and N_2 adsorption, of which the gas separation performance in terms of gas permeability and selectivity are evaluated using the single gas (H_2, CO_2, O_2, N_2, CH_4) and gas mixtures (CO_2/N_2 and CO_2/CH_4, 50/50 mol. %). The results indicate that the gas permeability of inorganic nano-particles/carbon composite membranes are enhanced by 3-10 times compared with the pure carbon membranes, which is attributed to the additonal channels for promoting gas diffusion created by the dispersion of nano-particles, involving the micro-interfacial gaps (interfacial pores) and the inner ordered channels of the porous particles. The nano-particles with small particle sizes, large pore channels and three dimensional pore structure would be benificial to enhancing the gas permeability and maintaining the high gas selectivity of the obtained composite carbon membranes, wihich suggests that the gas separation performance is controlled by porous structure of the nano-particles/carbon composite membranes tuned by changing the properties of the embeded nano-particles. The composite carbon membranes have both the higher permeability and selectivity with porous particles (zeolites and ordered mesoporous materials). The gas diffusion mechanism and the relevant model of gas diffusion pathways in the nano-particles/carbon composite membranes is also discussed.

(593e) C2H2 Absorption Using 1-Methyl-2-Pyrrlione Solution in a Hollow Fiber Membrane Contactor
Ruili, G., Shihezi University
Xiaoxu, W.,
Xiaopeng, Z.,
Jianshu, Z.,
Originally presented on: 11/7/2013 10:30:00 - 11:00:00

Based on the principle of gas exchange in lung between environment and blood, microporous membrane as gas-liquid contactor was developed and successfully explored in gas absorption. The study is focused on the development of a membrane gas absorption process for C_2H_2 removal from coal pyrogenation by hydrogen plasma using 1-Methyl-2-Pyrrlione (NMP) solvent in combination with the more hydrophobic microporous polytetrafloroethylene (PTFE) hollow fiber membrane (pore diameter 0.22μm, pore volume 52%). The effects of C_2H_2 concentrations in the feed, gas feed flow rate, liquid flow rate and membrane module length on the mass transfer flux of C_2H_2 and absorption efficiency were investigated. The results showed that, with the increase in C_2H_2 concentration, gas feed flow rate and liquid flow rate, the mass transfer flux of C_2H_2 and absorption efficiency increased. The increase in module length resulted in a slight decrease in mass transfer flux of C_2H_2 and absorption efficiency. The highest C_2H_2 absorption efficiency was 99% for 8% of C_2H_2 in the feed at 30cm of module length. In addition, membrane absorption-desorption system composed of a membrane absorption unit and a hot regeneration unit was operated continuously for 114 h to investigate its duration performance. The results were that the absorption efficiency of C_2H_2 maintained at 90%, C_2H_2 concentration in desorption gas was 98.77% and PTFE membrane material presented excellent NMP absorbent stability.

(601a) Optimized Synthesis and Characterization of Silica-Cellulose 3,5-Dimethylphenylcarbamate Hybrid Spheres and Investigation On Its Enantioseparation Ability On Beta-Blocker Drugs

Weng, X.,
Bao, Z., Zhejiang University / New Mexico State University
Zhang, Z., Zhejiang University
Su, B., Zhejiang University
Xing, H., Zhejiang University
Yang, Y., National Laboratory of Secondary Resources Chemical Engineering, Zhejiang University
Ren, Q., Zhejiang University
Originally presented on: 11/7/2013 8:30:00 - 8:45:00

The current commercial chiral stationary phases mostly rely on the coating and immobilizing effort to be fixed on the macroporous silicon spheres, which leads to a low loading capacity and lack of enough mechanical strength, limiting the further industrial preparation in a large scale. Synthesis of silica-cellulose 3,5-dimethylphenylcarbamate hybrid spheres adopting sol-gel method could effectively enhance the loading capacity and mechanical strength of packing material as chromatography packing material, aiming at the ultimate industrial application of preparative column and Simulated Moving Bed etc. However, a detailed and comprehensive characterization is still in the need to optimize the preparation condition to obtain a more applicable hybrid material. During the investigation process of optimization, an end-capping treatment is adopted and proved to be one indispensable factor to improve the enantioseparation ability and applicability of packing material. Otherwise a serious absorbance aroused by residual hydroxyl groups such as peak tailing may occur, making it totally impossible to separate racemates with certain polarity. A careful characterization and discussion is conducted to investigate the effectiveness of end-capping treatment as well as the overall property of hybrid material. Beta-blocker drugs belong to the family of basic drugs with pharmaceutical and commercial value, effective enantioseparation of which is meaningful and important. By end-capping most of residual hydroxyl groups of packing material, we are able to obtain hybrid material with ideal chiral recognition ability and enough mechanical strength, the resolution of β-blocker drugs on the end-capped material is also effective and of potential value for industrial applicability.

(601b) Monolithic Columns in Thin Layer Format for TLC-MS Separations

Lamprou, A., ETH

Lin, Z., Lawrence Berkeley National Laboratory

Lu, Y., Lawrence Berkeley National Laboratory

Svec, F., University of California at Berkeley

Originally presented on: 11/7/2013 8:45:00 - 9:00:00

Porous polymer monoliths are applicable as stationary phases for a variety of chromatographic modes. Among them, thin layer chromatography (TLC) is a facile and inexpensive separation method, where the mobile phase is propagating through a porous layer by capillary actuation. Owing to its flat format, TLC may be coupled with advanced mass spectrometric (MS) detection methods, like matrix-assisted laser desorption/ionization (MALDI) and laser ablation electrospray ionization (LAESI). It also offers the advantage of increased throughput by permitting to run several samples in parallel lines, while the developed plates can be stored for archiving. Motivated by the increasing attention that TLC is receiving during the last decade, very thin polymeric monolith layers have been prepared on suitable supports and applied for the separation of biomolecules, i.e. peptides, proteins and nucleosides.

The performance of such monolithic thin layers is controlled by both their porous structure and surface chemistry. Thus polystyrene-based, as well as polymethacrylate-based layers have been prepared using UV- or thermally-initiated polymerizations, in the presence of different porogenic mixtures. Their pore surface was subsequently modified via photografting or attachment of functional moieties via chemical reactions.

Specifically, the styrene-based thin layers were hypercrosslinked by Friedel-Crafts alkylation, in order to create a multiplicity of additional mesopores (see Figure 1). Hence, excellent separations of small molecules were achieved (see Figure 2). The methacrylate-based layers were photografted with different functional monomers, whereby the exposure to the initiating UV light was spatially controlled in a fashion that enabled the formation of a diagonal hydrophobicity gradient. Through this adjustment, during two sequential developments in reversed phase and ion-exchange mode, the solutes were able to encounter the gradient twice. Thus truly 2-dimensional separations, with superior resolution and selectivity were realized.[1,2]

In another implementation, noble metal nanoparticles were attached on the methacrylate-based thin layers, followed by a final modification with polar branched polymer chains, enabling the separation of peptides and nucleosides in hydrophilic interaction (HILIC) mode. Moreover, a functional monomer bearing weakly acidic groups, together with conjugated double bonds may be directly copolymerized with a crosslinker and appropriate porogens. The resulting monolithic thin layers can be used

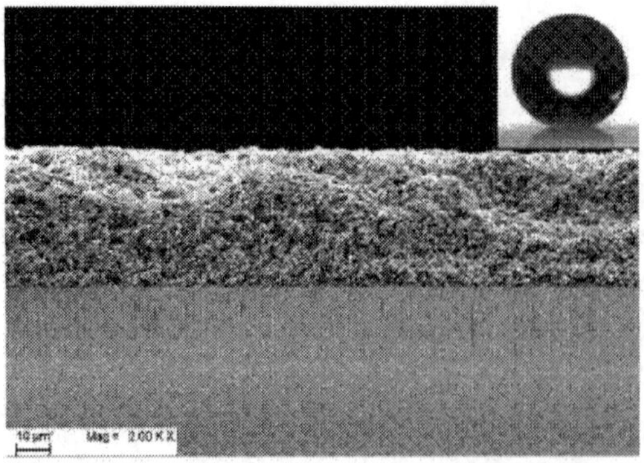

Figure 1. Hydrophobic hypercrosslinked polystyrene-based thin monolithic layer (Scanning Electron Microscopy image).

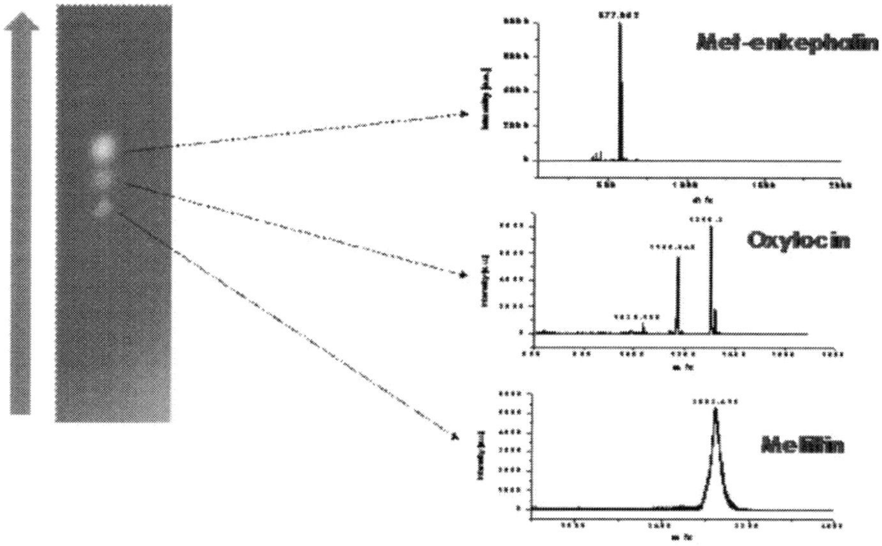

Figure 2. Separation of three peptides on a polystyrene-based thin monolithic layer (left) and subsequent analyte detection by MALDI (right). Mobile phase was 65% acetonitrile in water, with 0.1% trifluoroacetic acid, Development time was 15 min.

as weak-cation exchangers, as well as assist the energy transfer from a laser beam to the adsorbed analytes, thus being ideal substrates for MS detection immediately after the separation. The desorption/ionization of the separated compounds from such functional plates, as well as from those bearing noble metal nanoparticles was actually facilitated without the addition of any matrix medium, which is typically applied. Coupling TLC separation with direct MS detection provides the advantages of simplified sample preparation, as well as improved analytical sensitivity and reproducibility.

1. Han, Y.; Levkin, P.; Abarientos, I.; Liu, H.; Svec, F.; Frechet, J. M. J., Analytical Chemistry 2010, 82 (6), 2520-2528.

2. Urbanova, I.; Svec, F., J. Sep. Sci. 2011, 34, 2345-2351.

(601c) Elucidation of Time and Bed Height Dependent Stationary Phase Intra-Particle Concentration Profiles in a Chromatographic Column
Ozdural, A. R., Hacettepe University
Originally presented on: 11/7/2013 9:00:00 - 9:15:00

Elucidation of Time and Bed Height Dependent Stationary Phase Intra-Particle Concentration Profiles in a Chromatographic Column

Ahmet R. Özdural

Chemical Engineering Department, Hacettepe University, 06800 Ankara, Turkey

ABSTRACT

A simulation program has been developed for the prediction of intra-particle solute concentration profiles in chromatographic separations, where multicomponent analysis under nonlinear isotherm conditions is made. The validity of the assumption of local equilibrium (LE) conditions is uncertain in the dynamic state in a chromatographic column [1]. Therefore General Rate (GR) model is employed, where external mass transfer and intraparticle transport are included as separate identities. Thus the time taken for the solute to equilibrate between the stationary and mobile phases is taken into account for the prediction of the resulting band shape of a chromatographic peak.

The model presented in this work allows gaining an insight on the concentration gradient within a stationary phase particle. Moreover the model illustrates how the intra-particle concentration gradient is affected with column height, time, mobile phase flow rate, mobile phase concentration, sample volume, sample concentration, particle diameter, as well as isotherm properties. It is shown that the methodology presented in this work allows quantitative investigations of solute adsorption dynamics within a single stationary phase particle of a column. This adds a new tool to the available methods for characterizing and optimizing chromatographic processes.

Reference:

[1] Özdural A.R. (2011) Modeling Chromatographic Separation. In: Murray Moo-Young (ed.), Comprehensive Biotechnology, Second Edition, volume 2, pp. 681–695, Elsevier.

(601d) Relay Simulated Moving Bed Chromatography: Concept and Design Rules
Mota, J. P. B., Universidade Nova de Lisboa
Originally presented on: 11/7/2013 9:15:00 - 9:30:00

Relay Simulated Moving Bed Chromatography: Concept and Design Rules

Jose P. B. Mota

Requimte/CQFB, Departamento de Quimica, Faculdade de Ciencias e Tecnologia, Universidade Nova de Lisboa, 2829-516 Caparica, Portugal

Corresponding author email: pmota@fct.unl.pt

We present a new class of multicolumn chromatographic processes that change the classical way of handling the product outlets of simulated moving-bed (SMB) chromatography to avoid the use of flow controllers or an extra pump—the objective is to have just two- or three-way valves at a column outlet—while maintaining the analogy with the SMB in terms of displaced volumes of fluid per switch interval. In this class of processes the flow through a zone (or column) is always in one of three states: (i) frozen, (ii) completely directed to the next zone (or column), or (iii) entirely diverted to a product line. We use the term *relayed stream* to refer to this particular type of manipulation of the outflow from a column.

For this class of processes we derive a SMB analog—the R-SMB process—and demonstrate, under the framework of the equilibrium theory, that this process has the same separation region as the classical SMB for linear adsorption systems. In addition, the results from the equilibrium theory show that the R-SMB process consists of two distinct cycles that differ only in their intermediate sub-step (Fig. 1): one cycle (R-SMB$^-$) for values of the selectivity, α, smaller than $(3 + 5^{1/2})/2$ and another cycle (R-SMB$^+$) for larger values of α; in the former case no product stream is collected during the intermediate sub-step, whereas in the latter case both product streams are collected. The selectivity is defined as $\alpha = V_2/V_1$, where $V_i = [\epsilon + (1-\epsilon)K_i]V$, ϵ is the bed porosity, K_i is the Henry constant of component i and V is the column volume.

We also examine the R-SMB process under conditions of finite column efficiency and compare its performance against those of the classical open- and closed-loop SMBs. Our simulation results (Fig. 2) show that the R-SMB process requires less desorbent and is more productive than the standard SMB processes under conditions of finite column efficiency and that the comparison increasingly favors the R-SMB as the column efficiency decreases. We provide an experimental proof-of-concept for two linear separation problems.

(601e) Analysis of the Dynamic Phenomena in Liquid Chromatography Systems With Reactions in the Mobile Phase

Ling, L., Purdue University
Wang, N. H. L., Purdue University
Originally presented on: 11/7/2013 9:30:00 - 9:45:00

Reactions in liquid chromatography systems have many important applications. One can estimate the reaction rate constants for irreversible reactions, or the reaction stoichiometry and equilibrium constants for reversible reactions. For large-scale production using a reversible reaction, if the products are well separated from each other in a chromatography column, one can obtain purified products and achieve a higher conversion than the equilibrium conversion in a batch reactor. Additives (ligands or complexants) can be added to the mobile phase to enhance solute separation if a sorbent does not have sufficient selectivity for the solutes. Proteins and polymers can aggregate in chromatography systems and result in either multiple peaks or merged broad peaks in elution chromatography or multiple breakthroughs and plateaus in capture chromatography. The number of peaks, peak retention times, and the relative peak areas in elution and the number of plateaus in capture may depend on sample concentration, sample size, temperature, linear velocity, and column length. In this study, the adsorption/separation dynamic phenomena of six typical reactions are analyzed using dimensionless groups and rate model simulations.

The results provide overall guidelines for understanding, design, and optimization of chromatography systems with reactions. One can design experiments to find out whether reactions occur, whether the reactions are reversible or irreversible, and whether the reactions are first order or higher order. One can also determine the rate constants of an irreversible reaction from the peak areas as a function of the residence time. One can find the equilibrium constants and reaction stoichiometry from the peak ratios at different sample concentrations. For large scale production using a chromatography reactor, a reversible reaction can reach 100% conversion if the reaction rate is relatively large compared to the convection rate, and if the loading pulse size is sufficiently small to ensure the separation of the product bands. High purity products can be obtained if the diffusion rate is large and the axial dispersion rate is small relative to the convection rate. For reversibly aggregating systems, if the reactions reach equilibrium, one can separate the various aggregates by increasing the convection rate relative to the reaction rates. This separation can be achieved by increasing the sample concentration or the linear velocity of the mobile phase, or by reducing the temperature or the column length. One can merge multiple plateaus and increase the dynamic binding capacity in capture chromatography by increasing the reaction rate relative to the convection rate. For ligand-assisted separation, if the ligand concentration is sufficiently large and if the complexed solutes do not adsorb, then the overall selectivity is the ratio of the sorbent selectivity to the ligand selectivity, leading to a separation even if the sorbent has no selectivity. The sorbent selectivity should be opposite to that of the complexant selectivity to have synergistic effects on separation.

(601f) Development of Reactive Chromatography Systems for Equilibrium-Limited Reactions

Oh, J.,
Agrawal, G., Georgia Institute of Technology
Bommarius, A. S., Georgia Institute of Technology
Kawajiri, Y., Georgia Institute of Technology
Donaldson, M. E., The Dow Chemical Company
Frank, T. C., The Dow Chemical Company
Originally presented on: 11/7/2013 9:45:00 - 10:00:00

Reactive chromatography is a process that combines reaction and separation in a single unit that leads to a greater process performance and productivity [1]. This process is especially advantageous when the reaction is equilibrium limited, and the in-situ separation of product shifts the equilibrium in the direction of conversion increase [2].

In the present work, we study the application of reactive chromatography to the synthesis of an ester using AMBERLYST™ 15 as a catalyst and adsorbent. Among numerous esters, this study focuses on the production of propylene glycol methyl ether acetate (DOWANOL™ PMA glycol ether acetate), one of the most commonly used esters with a high industrial demand. DOWANOL™ PMA glycol ether acetate is the second-most used propylene glycol ether with nearly 90% of its use in surface coatings [3]. It is very efficient at dissolving resins used in paints, inks, lacquers, and other types of surface coatings such as in automotive, architectural, metal-coil, and industrial maintenance coatings [4]. Also, it is used in household products such as cleaners, paints (including spray paint), lacquers, varnishes, and pesticides. However, no study has been conducted on reactive chromatography for the formation of DOWANOL™ PMA glycol ether acetate, either through the esterification of 1-methoxy-2-propanol (DOWANOL™ PM glycol ether) with acetic acid or through the transesterification of DOWANOL™ PM glycol ether with ethyl acetate.

In this presentation, a case study of process development for an ester product is discussed. The dynamics of batch reaction and the fixed-bed adsorptive reaction are investigated by carrying out batch reaction experiments and chromatographic pulse tests. Stirred batch reactor experiments were conducted at various temperatures, stirring speeds, catalyst particle size and loading, and mole ratio of reactants. Reaction equilibrium and kinetic parameters together with their dependence on temperature were determined by fitting the model to the experimental data. Measurement of the adsorption equilibrium constant and reaction parameters were conducted by the pulse tests using a single chromatographic column. Since the resin acts as both adsorbent and catalyst, experiments were performed with nonreactive mixtures to obtain adsorption parameters first, and then reactive mixtures were injected to obtain reaction parameters. In addition to the development of model, the feasibility and efficiency of reactive chromatography where the conversion exceeds the reaction equilibrium of the batch reaction were demonstrated.

References

[1] V. Gyani, S. Mahajani, Reactive Chromatography for the Synthesis of 2-Ethylhexyl Acetate. *Separation Science and Technology*, 43: 2245-2268, **2008.**

[2] A. Rodigues, C. Pereira, J. Santos. Chromatographic Reactors. *Chem. Eng. Technol.* **2012,** 35, No7, 1171-1183

[3] Dow Chemical Company. *Product Safety Assessment* **2008**

[4] S.T. Cragg, R.J Boatman *Patty's Toxicology* **2001** (7)

(601g) An Alternative Method for Mono, Di and Tri Esters of Citric Acid Separation

Orjuela, A., Universidad Nacional de Colombia
Rivera, J. L.,
Suaza, A.,
de Leon, G. V.,
Trujillo, C. A., Universidad Nacional de Colombia
Rodriguez, G., Universidad Nacional de Colombia
Originally presented on: 11/7/2013 10:00:00 - 10:15:00

Citric acid triesters are well known green plasticizers that are preferred over phthalates on medical devices and children products applications. Its production is accomplished by batch esterification using large excess of alcohol and long reaction times. During esterification of citric acid a broad distribution of products is formed: mono, di and triesters together with dimmers and oligomers produced by self esterification of hydroxyl and acid moieties of the acid. This distribution is in general inevitable due to the reversible nature of the esterification reaction.

In order to enhance esterification of citric acid, a simultaneous reaction and separation process such as reactive distillation (RD) can be applied. By using this technology, chemical equilibrium limitations can be overcome and selectivity can be improved due to rapid separation of products. However there is poor or none characterization of intermediate products of reaction. In order to develop a RD process for citric acid esterification it is required to know phase equilibria an properties of intermediate compounds (mono and di esters). In this direction preparative chromatography is a highly efficient method for separation of chemicals with similar structures and moieties. This process is suitable for separation of thermolabile, heat sensitive, non volatile components.

In this work a semi-continuous preparative chromatography process was developed for separation of ethyl citrate esters. Initially, partial esterification reaction was carried out using a stirred glass reactor using Amberlyst 70 as catalyst. An especially developed polymeric packing was used as stationary phase and the reactive mixture was used as mobile phase with and without solvent.

A high purity mono ethyl, di ethyl and tri ethyl citrate fractions were obtained, and they were analyzed by High Performance Liquid Chromatography (HPLC) and Nuclear Magnetic Resonance (NMR) to confirm the chemical identity of obtained substances. Some physicochemical properties of pure intermediate esters useful for RD modeling are reported.

(601i) Multi-Stage Thermal Simulated Moving Bed Concentrator: An Adsorptive Process for Wastewater Treatment and Solute Recovery

Soepriatna, N., Purdue University
Wang, N. H. L., Purdue University
Wankat, P. C., Purdue University
Originally presented on: 11/7/2013 10:30:00 - 10:45:00

Wastewater generated from the effluent of various industrial activities is one of the main sources for water pollution. Many industries recycle and/or reuse this wastewater to both reduce water pollution and supply water for the plant. Many technologies have been developed and used for wastewater treatment. Adsorption provides an interesting and attractive alternative for wastewater treatment because of its ability to purify wastewater while concentrating the solutes contained in wastewater. Conventional adsorption systems often have major disadvantages. For example, regenerating the carbon in a kiln does not recover the solutes and eventually destroys the carbon. More gentle thermal adsorption processes and solvent regeneration of the adsorbent are typically operated as batch processes and have low efficiency of adsorbent use. Chromatography is very useful for separation of closely related compounds, but it has several limitations, such as low throughput, low efficiency (because the adsorbent packed in a column is not fully utilized), and solute dilution. The need for large amounts of a chemical desorbent can pose additional waste issues or high energy consumption for desorbent recovery. Although chromatography would not normally be used for water treatment, chromatography in a simulated moving bed (SMB) can be adapted to wastewater. The isothermal simulated moving bed (SMB) overcomes the throughput and efficiency limitations. However, solute dilution and high desorbent consumption still persist. The integration of SMB process with temperature swing adsorption (TSA), or thermal SMB, has been shown to overcome the inherent limitations of isothermal SMB[1-3]. In addition, a single 4-zone thermal SMB unit can simultaneously separate and concentrate binary dilute solutes as well as recover desorbent/solvent[4]. A different thermal SMB configuration with potential for water purification and concentration of one or more dilute solutes (thermal SMB concentrator) has been shown to have thermal energy consumption that is as low as 3% of an equivalent single-stage evaporator[5]. In the thermal SMB processes, adsorbent is regenerated at high temperature and solutes are adsorbed at cold temperature to regenerate the desorbent. This results in an extended ability of thermal SMB to separate solute and recover desorbent. In dilute systems separation is achieved without the use of additional desorbent. In this study, the application of two-zone, thermal SMB concentrator is extended to a multi-stage cascade of two-zone, thermal SMB concentrators for use in concentrating very dilute solutes and simultaneously producing purified water. The optimum operating conditions are determined by extending the standing wave design (SWD) method[4] to the two zone system and to a cascade of two zone, thermal SMB concentrators. The separation achieved at the optimum conditions is checked with detailed simulations using Aspen Chromatography.

References

1. Migliorini C, Wendlinger M, Mazzotti M. Temperature gradient operation of a simulated moving bed unit. *Ind Eng Chem Res*. 2001;40:2606–17.
2. Kim JK, Abunasser N, Wankat PC. Thermally Assisted Simulated Moving Bed Systems. *Adsorption*. 2005;11:579–584.
3. Jin W, Wankat PC. Thermal Operation of Four-Zone Simulated Moving Beds. *Ind Eng Chem Res*. 2007;46:7208–7220.
4. Soepriatna N, Wang NHL, Wankat PC. Standing Wave Design of a Four-Zone Thermal SMB Fractionator and Concentrator (4-zone TSMB-FC) for Linear Systems. *Adsorption*. 2013:*In Press*.
5. Lee JW, Wankat PC. Thermal Simulated Moving Bed Concentrator. *Chem Eng J*. 2011;166:511-522.

(601j) Applications of Supercritical Fluid Simulated Moving Bed On the Separation of Active Compound From Chinese Herbs

Liang, M. T., I-Shou University

Originally presented on: 11/7/2013 10:45:00 - 11:00:00

Supercritical carbon dioxide with various cosolvnet can be applied as desorbent for a SMB to reduce the use of organic solvents and to eliminate the dilution effect by chromatographic processing. A lab-scale SF-SMB with six columns was setup in I-Shou university, Kaohsiung, Taiwan, and several bioactive compounds in the crude extract of Chinese herbs, such as sesamin, sesamolin, resveratrol, emodin, Tashinone, and etc., were sucessfully purified or concentrated by the lab-scale SF-SMB. The design and operation of a SF-SMB will be illustrated in this presentation, and effects of cosolvent concentartion and pressure drop across the columns will be treated as the elution gradient to enhance the separation. The Triangle theory based on the volumetric flow rate was also applied to setup the operating conditions and to explain the experimental results. Compared with traditional liquid desorbent, the supercritical fluid provides fast mass transfer and clear cut of the separation from the crude extract of Chinese herbs, which contains bioactive compounds at low concentration and generally mixed with other multiple components. With the supercritical carbon dioxide as the major desorbent, the author found that traditional silica used for normal phase chromatography is generally suitable for the SF-SMB. Examples to purify the bioactive components from Chinese herbs will be presented and difficulties for dealing with the multicomponent system will also be discussed.

(606a) High-Throughput Membrane Protein Crystallization Condition Screening Using X-Ray Transparent In Meso Microfluidic Chips

Schieferstein, J. M., University of Illinois at Urbana-Champaign
Khvostichenko, D. S., University of Illinois at Urbana-Champaign
Pawate, A. S., University of Illinois at Urbana-Champaign
Kenis, P. J. A., University of Illinois at Urbana-Champaign
Originally presented on: 11/7/2013 8:30:00 - 8:55:00

The malfunction of membrane proteins has been tied to the initiation and progression of many diseases and, consequentially, membrane proteins account for ~60% of all available drug targets. High-resolution 3-D structures of proteins are essential to understand the structure-function relationship of these molecules and the changes therein that contribute to disease states. However, the amphipathic nature of membrane proteins, the difficulty of expressing and purifying them, and their low yield upon purification have all made them very difficult to successfully screen and crystallize. An emerging technique called the *in meso*method has increased success rates of membrane protein crystallization. This method suffers from limited adoption because of a variety of factors such as the requirement of specialized equipment, expert labor, or expensive robots. Upon successful crystallization, the next bottleneck is the manual handling of the fragile crystals while mounting them on a loop or a capillary, and subsequent cryo-cooling to collect X-ray diffraction data – a process where many crystals suffer damage resulting in poor diffraction data.

To address these challenges we have developed high-throughput X-ray transparent microfluidic chips for *in meso* crystallization screening. The chips are designed with 12-, 24-, and 48- well arrays requiring < 20 nL of protein solution per well and also simultaneously automate the formulation of each sample. X-ray transparency allows these microfluidic chips to circumvent the need for manual crystal harvesting, thus preserving the initial quality of crystals. These *in meso* chips enable the use of lipidic mesophases for crystallization screening on an X-ray transparent platform, a previously unachieved accomplishment. Our microfluidic platforms simplify *in meso* sample preparation and allow for crystallization condition screening, crystal growth and the acquisition of high-resolution diffraction data on chip for structure solution. We have validated our approach on bacterial photosynthetic reaction center and are currently working on challenging membrane proteins from the heme-copper oxidase superfamily.

(606b) Solvent Effect On Crystal Morphology: A Hybrid Approach of a Mechanistic Model and Molecular Dynamics

Kim, S. H., Iowa State University
Doherty, M. F., University of California, Santa Barbara
Originally presented on: 11/7/2013 8:55:00 - 9:20:00

The control of crystal morphology during crystallization processes has a great impact on the end-use properties of the materials (e.g., optical properties) and the downstream performance of the manufacturing process such as the separation characteristics. The crystal morphology depends on external factors such as solvent, supersaturation, temperature, additives and impurities. Various theoretical and computational approaches have been carried out to understand the relation between controlling the crystal morphology and the external growth factors.

Recently, a realistic crystal growth model was developed for non-centrosymmetric solute molecules[1] based on the spiral growth model proposed by Burton, Cabrera and Frank (BCF). This model well explained the non-isotropic behavior of complex drug molecules to predict the crystal shape. In solution growth, the interfacial energies between the solute and the solvent are applied to the spiral growth model. However, the interfacial energies are calculated at the macroscopic scale, so it is hard to relate the microscopic behavior of the interface between the solute and the solvent such as the free energy barriers of the solute molecules that control the rate of incorporation of solute into kink sites.

Molecular simulations have been used to understand the interfacial interactions between solute and solvent molecules at an atomistic level[2] . We applied the atomistic information from the molecular simulations into the mechanistic model of the crystal growth to consider the detailed solvent effect for organic molecules of realistic complexity such as active pharmaceutical ingredients (APIs).

[1] Kuvadia, Z. B.; Doherty, M. F. Spiral growth model for faceted crystals of non-centrosymmetric organic molecules grown from solution. *Crystal Growth & Design* 2011, **11**, 2780

[2] Piana, S.;Gale, J. D. Understanding the barriers to crystal growth: Dynamical simulation of the dissolution and growth of urea from aqueous solution. *Journal of the American Chemical Society* 2005, **127**, 1975

(606c) The Understanding of Nucleation Aspects of Tolbutamide Polymorphs in Solution

Yani, Y., Iowa State University
Chow, P. S., Institute of Chemical and Engineering Sciences, A*STAR (Agency for Science, Technology and Research)
Tan, R., Institute of Chemical and Engineering Sciences, A*STAR (Agency for Science, Technology and Research)
Originally presented on: 11/7/2013 9:20:00 - 9:45:00

The understanding of nucleation aspects of tolbutamide polymorphs in Solution

Yin Yani,*† Pui Shan Chow,† and Reginald B. H. Tan†,‡

† Institute of Chemical & Engineering Sciences, A*STAR (Agency for Science, Technology and Research), 1 Pesek Road, Jurong Island, Singapore 627833

‡ Department of Chemical & Biomolecular Engineering, National University of Singapore, 4 Engineering Drive 4, Singapore 117576

* Corresponding author. Tel: (65) 6796 3852. Email: yin_yani@ices.a-star.edu.sg

Abstract

Experimental works1, 2 have shown that tolbutamide (TB) Form I is preferentially nucleated from solution. Raman spectra collected for TB solute in saturated solution closely matches with that of TB Form I in suspension2. This observation shows that there could be some unique interactions of tolbutamide molecules or solvent molecules in solution that eventually lead to the nucleation of TB Form I. We use molecular dynamics simulation to observe the conformations and arrangements of tolbutamide molecules in saturated and supersaturated solutions. Various solvents were considered in the simulations, such as, ethanol, methanol, and acetonitrile. The simulation results show that, at pre-nucleation stage, TB molecules in solutions arranged close to the structure of TB Form I in vacuum. In addition to that, our simulation has determined that majority of the single TB molecules in solutions arrange more like the type of TB conformation found in Form I.

References

1. Keraliya, R. A.; Soni, T. G.; Thakkar, V. T.; Gandhi, T. R., Effect of Solvent on Crystal Habit and Dissolution behavior of Tolbutamide by Initial Solvent Screening. Dissolution Technologies 2010, 17 (1).

2. Thirunahari, S.; Chow, P. S.; Tan, R. B. H., Quality by Design (QbD)-Based Crystallization Process Development for the Polymorphic Drug Tolbutamide. Crystal Growth & Design 2011, 11 (7), 3027-3038.

(606d) Predicting Protein Solubility and Crystallization Behavior Based On the Second Osmotic Virial Coefficient

Brandenbusch, C., Technische Universitat Dortmund
Herhut, M., Technische Universitat Dortmund
Sadowski, G., Technische Universitat Dortmund
Originally presented on: 11/7/2013 9:45:00 - 10:10:00

The biocatalytic production of complex pharmaceutical proteins (e.g. monoclonal antibodies) has gained an increased industrial interest. Within these processes, the total production costs are often dominated (up to 80 %) by the downstream processing. In state-of-the-art engineering, protein purification is usually achieved by a series of cost-intensive chromatographic steps. One interesting alternative to the conventional processing is protein precipitation, as a first capture step, and crystallization for final product polishing.

In order to develop a precipitation or crystallization step, the solubility of the target protein in aqueous solutions and the type of solid formed (crystal or precipitate) are required as crucial information. In order to supply this data, within this work, an innovative and simple solubility model to measure and predict the protein solubility and crystallization behavior based on the second osmotic virial coefficient (B_{22}) is developed. By use of a potential-of-mean-force model (DLVO) with parameters fitted to B_{22} data at low precipitant concentration, B_{22} at high precipitant concentrations can be efficiently predicted, thus decreasing the experimental effort. By relating B_{22} to the asymmetric activity coefficient, the solid-liquid-phase equilibrium of a protein in solution, consisting of solvent, protein, and precipitant, can be calculated. This combination now enables for the prediction of the protein solubility over a broad concentration range of precipitant with a minimal set of experimental data.

The value of B_{22} herein serves as an ideal measure, as it combines information on the crystallization or precipitation behavior of the protein under the applied conditions, e.g. temperature, pH, type of precipitant used (salt or polymer e.g. PEG) and precipitant concentration.

In order to show the applicability of the model for varying protein sizes, experimental data of B_{22} for two proteins (14 and 144 kDa) in water at varying conditions were measured by static light scattering. The results were compared to experimental and literature data. The results point to the possibility of qualitative and quantitative prediction of the protein solubility in aqueous solution based on two B_{22} and two solubility data at varying precipitant concentrations. This modeling strategy supports a cost-efficient development of crystallization steps.

(606e) Conversion of FBRMÂ® CLD Data to Psd Data: Application to a Needle-Shaped Industrial Case, for the Estimation of Crystallization Kinetics

Mitchell, N., Process Systems Enterprise
Yang, B. S., Boehringer Ingelheim Pharmaceuticals, Inc.
Bermingham, S. K., Process Systems Enterprise
Kim, S., Boehringer Ingelheim Pharmaceuticals, Inc.
Mumtaz, H., Process Systems Enterprise
Originally presented on: 11/7/2013 10:10:00 - 10:35:00

Focused Beam Reflectance Measurement (FBRM®) is a commonly employed Process Analytical Technology (PAT) for tracking the evolution of crystallization processes. FBRM® data can utilised to indicate the onset of nucleation, and the relative rates of other crystallization mechanisms, such as secondary nucleation, crystal growth and/or agglomeration. However, the FBRM® technique does not provide a measure of the in-situ Particle Size Distribution (PSD), but merely the Chord Length Distribution (CLD) of the particles present in the measurement volume of the FBRM® probe. In order to utilise FBRM® data to quantitatively track crystallization processes, a conversion method between FBRM® measurements and PSD data, typically measured via laser diffraction, is necessary. Conversion of CLD to PSD data is not trivial, as numerous process variables, such slurry concentration, crystal shape, refractive indices of the solution and the crystal can significantly affect the measured FBRM®CLD data. This work concerns the application of two conversion methods to an industrial case, which exhibits needle-like crystals. This type of crystal shape represents the most difficult of the application cases for CLD to PSD conversion techniques. Indeed, to the authors' knowledge, all of the published literature on the topic concerns the application of these conversion methods to the more simplistic case of block-like or spherical shaped crystals.

In this work, two conversion methods were applied to the sample industrial case. The first method correlates certain statistics of the measured FBRM® CLD data to an attribute of the corresponding PSD, measured via laser diffraction. This method accounts for the average aspect ratio of the crystals present. In this way, this method accounts for the effect of different crystal shapes on the measured FBRM® data. The second approach utilised in this work, involved the use of the moments (zeroth to sixth), determined from the FBRM® CLD data. These FBRM® based moments were then normalised with respect to the solution concentration and the initial PSD for the seed crystals. This method was previously employed for the simultaneous estimation of the kinetic parameters for secondary nucleation, crystal growth and agglomeration for a paracetamol, methanol and water anti-solvent crystallization process[1]. Both of these techniques were found to provide reasonably accurate (±10% on laser diffraction measurements), indications of certain attributes of the PSD. Finally, the successful application of these conversion techniques allowed the FBRM® data to be quantitatively employed in the estimation of crystallization kinetics for the sample industrial case.

---

[1]Ó'Ciardhá, C.T., Hutton, K.W., Mitchell, N.A., Frawley, P.J. (2012) 'Simultaneous parameter estimation and optimization of a seeded antisolvent crystallization', Cryst. Growth & Des., 12(11), 5247-5261.

(606f) Problematic Crystal Shapes - Causality and Solvent Selection Guidance Based On Periodic Bond Chain Diagrams

Lovette, M. A., Eli Lilly

Originally presented on: 11/7/2013 10:35:00 - 11:00:00

Crystals with needle like or flat plate like shapes, as typified by aspect ratios ~ 1 : 1 : >100 and 1 : >100 : >100, are often the steady-state growth shapes in the crystallization of pharmaceutical compounds from solution. Crystals with such high aspect ratios may be problematic during subsequent processing steps; e.g., filtration, drying. Furthermore, the crystallization of active pharmaceutical ingredients (APIs) that result in such shapes present additional concerns during the additional drug product processing. Although mechanical techniques such as milling and non-mechanical techniques such as thermal cycling may be able to avoid the isolation of crystals with high aspect ratio shapes, it is often more desirable (especially during early development) to determine appropriate solvents that avoid such shapes altogether. In this presentation, a causality for the formation of needles will be discussed. This causality stems from the three dimensional network of anisotropic intermolecular interactions within a crystal; which are understood using periodic bond chain diagrams. This causality will be applied to different case studies that demonstrate the importance of the dispersive and associative interactions within the crystal and the set of corresponding solvents that are likely abate the formation of needle shaped crystals. From these case studies, a classification system for different types of needles will be provided. The foundations of a similar causality for flat plates and the role of solubility will also be described.

(609a) Evaluation of Various Hybrid Extraction-Chromatography Separation Process Options Using Model-Based Approach

Minceva, M., University of Erlangen-Nuremberg
Goll, J., Friedrich Alexander University Erlangen-Nuremberg
Arlt, W., University of Erlangen-Nuremberg
Originally presented on: 11/7/2013 8:30:00 - 8:50:00

Centrifugal partition chromatography (CPC), also known as counter-current chromatography (CCC), is a versatile and highly selective separation technology, which combines the principles of extraction and chromatography.

Like in extraction the separation is achieved as a result of the different partitioning of the solutes between the phases of a biphasic liquid system, and similar to chromatography one of the two liquid phases is kept stationary. However, there are some substantial differences, which make this technology unique. Namely, the preferred range of the partition coefficient of the solutes to be separated in CPC/CCC is between 0.5 and 2.5, considerably lower than in extraction. Unlike in liquid-solid chromatography a centrifugal force is used to keep one of two liquid phases stationary.

In CPC/CCC both liquid phases can be used as stationary phase and their role can be switched during the separation run. This opens up the possibility for several original batch and continuous operating modes, which are not possible or realizable in extraction and in liquid chromatography with solid stationary phases.

In this work an overview of the existing CPC/CCC processes is given, with a special focus on the continuous operating options. It is shown that the selection of the best operating mode for a given separation task is not always easy and straightforward. Hence, for making right selection the available process options should be systematically evaluated. This issue is addressed through a model-based approach, which combines experimentally derived thermodynamic, hydrodynamic and mass transfer data with a relatively simple mathematical model, i.e. the “cell model”. The proposed approach enables easy evaluation of the separation performances of different CPC/CCC operating modes.

(609b) Modeling of Liquid-Liquid Extraction for Uranyl Nitrate Separation

Rao, R. R., Sandia National Laboratories
Noll, H., Graz University of Technology
Fritz, M., Graz University of Technology
Jove-Colon, C. F., Sandia National Laboratories
Moffat, H. K., Sandia National Laboratories
Siebenhofer, M., Graz University of Technology
Originally presented on: 11/7/2013 8:50:00 - 9:10:00

Liquid-liquid extraction is a widely used separation method for an extensive range of metals including actinides. The Gibbs energy minimization (GEM) method is used to compute the complex chemical equilibria for the acidified aqueous uranyl nitrate solution by a mixture of tri-n-butyl phosphate (TBP) and organic at room temperature. The organic phase is treated as an ideal mixture defined by eight TBP complexes plus the inert diluent. The Pitzer method is used to capture non-idealities in the concentrated uranyl nitrate, which is the electrolyte phase. The generated extraction isotherms are in good agreement with reported experimental data for various TBP loadings and electrolyte concentrations demonstrating the adequacy of this approach to analyze complex multiphase multicomponent systems. The model is robust and yet flexible allowing for expansion to other LLE systems and coupling with computational tools for parameter analysis and optimization.

(609c) High Efficient Separation of Mono- and Di-Tocopherol Polyethylene Glycol Succinates Using Ionic Liquid Extraction

Kong, L.,

Originally presented on: 11/7/2013 9:10:00 - 9:30:00

High efficient separation of mono- and di-tocopherol polyethylene glycol succinates using ionic liquid extraction

Liyun Kong, Qiwei Yang, Huabin Xing, Zongbi Bao, Yiwen Yang*, Qilong Ren

(Key Laboratory of Biomass Chemical Engineering of Ministry of Education, Department of Chemical and Biological Engineering, Zhejiang University, Hangzhou 310027, China)

* Corresponding author. Tel.: +86-571-87952773 Fax: +86-571-87952773

E-mail address: ceywyang@zju.edu.cn

Tocopherol polyethylene glycol 1000 succinate (TPGS 1000), which is a water soluble derivative of natural vitamin E, is synthesized by the esterification of tocopherol acid succinate (TAS) with polyethylene glycol (PEG). Due to the equal activity of the two hydroxyl groups in PEG, two types of esters (mono- and di-TPGS) exist in commercial TPGS product. However, when we mention TPGS, we usually refer to mono-TPGS, and the typical characteristics are also on account of mono-TPGS. Being an excellent amphiphilic molecule with HLB number of 13, TPGS can be used as solubilizer, emulsifier, absorption enhancer and as a vehicle for lipid-based drug delivery formulations. Two hydrophobic TAS in the molecule of di-TPGS lead to the lower HLB number and almost insoluble in water, making di-TPGS have different uses.

Recommended methods for the separation of mono- and di-TPGS were chromatographic methods which had low treatment capacity, were solvent consuming, or required large equipment investment, so that they were not economically favorable for large-scale production. Here we proposed an extraction method which has the properties of high capacity, low solvent consumption, small investment. Therefore, it is suitable for continuous and industrial production.

Serious emulsification occurred when using water-ethyl acetate extraction system for the separation of mono- and di-TPGS, resulted from the amphiphilic structure of TPGS. Meanwhile the distribution coefficient of mono-TPGS was as low as 0.012. Addition of sodium chloride could eliminate emulsification, but resulted in no distribution of mono-TPGS in water. Organic solvents such as methanol, ethanol and *n*-

propanol can be used to reduce emulsification too. Extraction results of these solvents didn't come up to expectation with distribution coefficient of mono-TPGS less than 0.3 at 293 K.

Introduction of ionic liquids (ILs) got excellent result. Efficient demulsification and separation were achieved simultaneously with the ternary biphasic ILs/water-ethyl acetate extraction systems. This might be owing to the special structure and property of ILs which are liquid salts at room temperature, and are composed of asymmetric organic cations and organic or inorganic anions. ILs of [$C_n$mim][Cl] and [C_nPy][Br] with different alkyl chains (n=2 ~ 12) were investigated. The distribution coefficients of mono- and di-TPGS increased rapidly with the increment of the alkyl chain length in cations. For [C_nPy][Br] at the initial concentration of 5 mol% in water at 303 K, when n=6, 8 and 12, the distribution coefficient of mono-TPGS ($D_{mono\text{-}TPGS}$) were 1.62, 7.75 and 15.09, coupled with the selectivity (mono-TPGS to di-TPGS) as high as 188.37, 22.79 and 8.53, respectively. $D_{mono\text{-}TPGS}$ in [C_nPy][Br]- water/ethyl acetate extraction systems were hundreds of times higher than $D_{mono\text{-}TPGS}$ in water/ethyl acetate system.

When the initial concentration of ILs ([C_8Py][Br] and [C_{12}Py][Br]) in water varied from 0.4 to 20 mol%, $D_{mono\text{-}TPGS}$ reached a maximum at 4 mol%. Maximum existed in the distribution coefficient with the increase of the initial molar percent of ILs in water revealing the diluent's swing effect in this extraction process. Inverse extraction-temperature relationship was observed due to the inverse solubility-temperature behavior of the typical nonionic surfactant structure of mono- and di-TPGS. Finally, this method can be successfully expanded to the separation of other mono- and di- polyethylene glycol alkyl esters, such as mono- and di-polyoxyethylene laurates, mono- and di-polyoxyethylene palmitates, mono- and di-polyoxyethylene oleates and mono- and di-polyoxyethylene stearates, etc.

(609d) Investigations of Use of Ionic Liquid in Extraction of Picolinic Acid From Aqueous Solutions

Kumar, S., Motilal Nehru National Institute of Technology (MNNIT)
Chomal, N.,
Kumar, A.,
Originally presented on: 11/7/2013 9:30:00 - 9:50:00

Abstract

Picolinic acid ($C_6H_5NO_2$) is an isomer of nicotinic acid, a pyridine compound with a carboxyl side chain at the 2-position. It is involved in the manufacturing of amino acids like tryptophan and phenylalanine etc. To improve the biological production of picolinic acid and its derivatives by fermentation, it is necessary to develop new environment friendly and efficient separation methods that improve the productivity of the acid (1, 2). The reactive extraction by a suitable extractant has been found to be a most promising alternative to the conventional separation processes to recover this acid from the fermentation broth and aqueous solution (3-5). Phosphorus bonded oxygen bearing extractants such as tri-*n*-butyl phosphate (TBP) and tri-*n*-octyl phosphine oxide (TOPO), and long chain aliphatic amines such as tri-*n*-octylamine (TOA), tri-octyl methyl ammonium chloride (Aliquat 336) and Amberlite LA-2 are effectively used for the reactive extraction of different carboxylic acids (3-8). The increasing emphasis on cleaner and environmentally benign extraction procedures has led to the systematic investigation of systems containing ionic liquids (ILs) - a new class of non-volatile alternative solvents. The unique physicochemical properties of ILs and the relative ease with which these properties can be fine-tuned by altering the cationic or anionic moieties comprising the IL have led to intense interest in their use as alternatives to conventional organic solvents in a wide range of synthetic, catalytic, and electrochemical applications. In this study we have investigated the use green solvents, ionic liquids with for the extraction of picolinic acid along with the amine based extractant, tri-*n*-octylamine (TOA).

The extraction equilibrium experiments are carried out with aqueous (acid + water) and organic (IL + TOA) solutions in conical flasks of 25 ml and shaken at 100 rpm for 12 hours in a temperature controlled magnetic stirrer water bath at constant temperature (298.15 K). The aqueous solutions of picolinic acid are prepared in the concentration range of 0.01 to 0.18 M using doubled distilled water. The organic solutions are prepared by dissolving ionic liquid and TOA. After attaining equilibrium, the mixture of aqueous and organic phases is kept for separation in a separating funnel (60 ml) for 2 hrs at 298.15 K. After the separation of both phases, the aqueous phase acid concentration is analyzed with titration using NaOH solution of 0.008 N and phenolphthalein as an indicator. The acid concentration in the organic phase is calculated by mass balance. The experimental data are analyzed by calculating distribution coefficient ($K_D = C_{org}/C_{aq}$) and degrees of extraction [$E = K_D / (1 + K_D)$].

The equilibrium experiments are carried out for the extraction of picolinic acid using two ionic liquids (1-butyl,3-methylimidazolium hexafluorophosphate, C4MIMPF6 and 1-octyl,3-methylimidazolium hexafluorophosphate, C8MIMPF6) with and without TOA. The effect of different parameters such as type of IL, volume phase ratio, composition of IL in organic phase and concentration of acid in aqueous phase on the extraction is also determined. The extraction efficiency is found to be increased by increasing the TOA concentration in IL upto a maximum value and after that it is showing the decrease in degree of extraction. Due to high viscocity and very low quantity of IL, pure IL is not found better extractant for the extraction. 1-octyl,3-methylimidazolium hexafluorophosphate, C8MIMPF6 containing higher alkyl chain showed less extractability as compared to C4MIMPF6. The aqueous phase concentration is also affecting the extraction efficiency of IL and TOA. It is found to increase the degree of extraction with an increase in acid concentration at constant organic phase concentration. The highest degree of extraction is achieved 94% at acid concentration of 0.18 mol/L and organic phase composition (TOA:IL = 2:3 v/v).

It can be seen from data presented by Datta et al, 2012, Very high amount of TOA (equal molar solution and 1:1 volume phase ratio) is required to achieve the degree of extraction more than 90%. The present investigations on the use of IL for extraction of picolinic acid show that small amount of IL can replace the huge amount of TOA to give same or better degree of extraction.

Keywords: Green Solvent; Ionic liquid; Equilibria; Reactive extraction; Tri octyl amine.

Refernces:

1. Wang, L.R.; Fang, Y. (2005) UV-Raman study and theoretical analogue of picolinic acid in aqueous solution. J. Molec. Spect., 234: 137.
2. Gonzalez, J.M.; Fernandez, M.A.; Pizarro, C. (1997) Application of weakly basic copolymer polyacrylamide (acrylamide-co- N,Ndimethylaminoethyl methacrylate) gels in the recovery of citric acid. J. Euro. Poly., 33: 475.
3. Kumar, S.; Babu B.V. (2009) Process intensification of nicotinic acid production via enzymatic conversion using reactive extraction. *Chem. Biochem. Eng. Q.* 23: 367.
4. Kertes, A. S.; King, C. (1986) Extraction chemistry of fermentation product carboxylic acids. *J. Biotechnol. Bioeng.*, 28: 269.
5. Kumar, S.; Babu, B.V. (2008) Process intensification for separation of carboxylic acids from fermentation broths using reactive extraction. *J. Fut. Eng. Technol.*, 3 (3): 19.
6. Kumar, S.; Wasewar, K. L.; Babu, B. V. (2008) Intensification of nicotinic acid separation using organophosphorous solvating extractants by reactive extraction. *Chem. Eng. Tech.* 31 (11): 1584.
7. Datta, D., **Kumar, S.** Reactive Extraction of Pyridine Carboxylic Acids with N, N-Dioctyloctan-1-amine Dissolved in Five Different Diluents. *Sep.Sci.Technol.* 2013, 48(6), 898-908.
8. Datta, D., **Kumar, S.**, Wasewar K.L., Babu B.V. Comparative Study on Reactive Extraction of Picolinic Acid with Six Different Extractants (Phosphoric and Aminic) Dissolved in Two Different Diluents (Benzene and Decane-1-ol). *Sep.Sci.Technol.* 2012, 47(7), 997-1005.

(609e) Solvent Extraction and Separation of Nickel and Vanadium From Sulfate Leach Liquor of Power Plant Fly Ash Using D2EHPA and Cyanex 272

Vahidi, E., University of South Florida
Noori, M., University of Tehran
Rashchi, F., University of Tehran
Babakhani, A., University of Tehran
Originally presented on: 11/7/2013 9:50:00 - 10:10:00

Fly ash generated by power plants is categorized as a special waste by US EPA. These residues from fossil fuel combustion are captured by pollution control equipment and disposed in special landfills. In landfills, leaching of heavy metals such as nickel and vanadium into ground water, soil, and/or surface water is the main potential environmental concern. Furthermore, due to the limited storage capacity of landfills, safe disposal becomes more and more costly. Therefore, it is imperative to develop means of both diverting these types of wastes from the solid waste stream, and also recovering nickel and vanadium as valuable metals from fly ashes from environmental and economical viewpoints. In this investigation, recovery and separation of nickel and vanadium from leach liquors of power plant fly ash were investigated by solvent extraction using D2EHPA, Cyanex 272, and their mixtures in various proportions. The mentioned fly ash contains about 2.6% nickel and 3.3% vanadium and according to the leach liquor after the sulfuric acid leaching, synthetic sulfate solutions of nickel and vanadium were prepared in 2gr/L of each metal. After dilution of the organic solvents in kerosene with the ratio of 1 to 4, experiments were carried out in the pH range of 1.0–5.0 in steps of 0.5 at ambient temperature. The investigation of solvent extraction of nickel and vanadium by sole D2EHPA was performed and the extraction percentages determined to be 95% and 80% for nickel and vanadium, respectively. However, the co-extraction of nickel and vanadium by D2EHPA can be increased with increasing equilibrium pH and temperature. It was shown that using sole D2EHPA, pH_{50} (the pH at 50% metal extraction) values for nickel and vanadium were 3.5 and 2, respectively; which is not appropriate for the efficient separation of nickel and vanadium simultaneously. Adding Cyanex 272 to D2EHPA in the organic phase, leads to a right shifting of extraction isotherm of nickel and a slight left shifting of the extraction isotherm of vanadium and finally improves the separation of nickel over vanadium. To optimize the recovery and separation process of nickel and vanadium from the sulfate leach liquor, the influence of different D2EHPA to Cyanex 272 ratios, various temperatures such as 25, 40, and 60°C and two different diluents including kerosene and carbon tetrachloride were studied. As a result, optimum separation of vanadium over nickel was achieved with a Cyanex 272 to D2EHPA ratio of 3. Based on the optimum results, pH_{50} values for nickel and vanadium were changed from 3.5 to 4.5 and from 2 to 1.75, respectively.

(609f) Solventless Extraction of Metals and Organics Using Trioctylphosphine Oxide Impregnated Hollow Fiber Membranes

Praveen, P., National University of Singapore
Loh, K. C., National University of Singapore
Originally presented on: 11/7/2013 10:10:00 - 10:30:00

Liquid-liquid extraction using organic solvents is a very effective technique for removal of organic and inorganic contaminants from wastewater. However, phase dispersion and consequent downstream separation problems are major limitations. Membrane facilitated extraction, as in supported liquid membranes (SLM), can alleviate these challenges and provide a low-cost, low-energy alternative to conventional extraction. However, the instability of SLM during prolonged operation makes this technique unsustainable.

To alleviate the challenges in solvent extraction, we have immobilized a solid extractant, trioctylphosphine oxide (TOPO), into polypropylene hollow fiber membranes (HFM). The strategy was to dissolve TOPO in a volatile carrier solvent and soak the HFM in this solution. The carrier solvent was then evaporated from the HFMs using dry air, causing a phase change resulting in impregnation of TOPO within the HFM walls. The resulting extractant impregnated hollow fiber membranes (EIHFM) can facilitate solventless extraction of various organic and inorganic compounds and has the advantages of high solute retention capacity of extraction, non-dispersive mass transfer of adsorption and the large specific interfacial area of the HFM.

The EIHFMs were characterized and used as the organic phase in the extraction of organic compounds such as phenols and carboxylic acids, and inorganic metals such as chromium. The EIHFMs exhibited high sorption capacity and rates, and equilibrium was achieved within 30 min in most of the experiments. The effects of EIHFM concentration, contact time, feed solute concentration and pH on extraction was evaluated, and a kinetics model was developed to elucidate the mass transfer of solutes from the aqueous phase to the EIHFMs. Finally, the stability of the EIHFMs was demonstrated during repeated extraction/stripping cycles. These results indicate that the use of EIHFMs can be quite promising in the extraction of organic and inorganic chemicals from aqueous solutions.

(609g) Large Phase Ratio Extraction and Enrichment of Rare-Earth Ions By a Novel Gas-Assistant Column Extractor

Huang, K., Institute of Process Engineering, Chinese Academy of Sciences
Liu, H.,
Originally presented on: 11/7/2013 10:30:00 - 10:50:00

We proposed a novel gas-assistant column extractor to perform extraction and enrichment of rare-earth ions from low content aqueous leach solutions with large aqueous-to-oil phase ratios. Batch and continuous experiments were conducted to test enrichment and separation efficiency of rare-earth ions. Experimental investigations on the feed-in solutions containing La, Pr, Nd and Lu about 10 to 100 mg/L were carried out in order to determine the extraction performance of the gas-assistant column extractor. The results demonstrate that enrichment and separation efficiency of light and heavy rare-earth ions is dependent on air bubbling rate, solvent flow rate and aqueous-to-oil phase ratio. The operational aqueous-to-oil phase ratio can be reached even more than 300:1, and percent extraction of rare-earth ions were above 95% after 5min. of air bubbling. The present work highlights a novel strategy to achieve extraction processes requiring higher aqueous-to-oil phase ratios, so that difficulties in enrichment and recovery of rare-earth ions with low concentrations (<100 mg/L) by traditional mixer-settler can be solved.

(619a) Nanoscale Interfacial Design Concept Towards High Flux CO2 Separation Maintaining High Selectivity and Longevity

Kocherlakota, L. S., University of Washington
Pham, T. H., University of Washington
Overney, R. M., University of Washington
Originally presented on: 11/7/2013 8:30:00 - 8:55:00

Nanoscale Interfacial Design Concept Towards High Flux CO_2 Separation Maintaining High Selectivity and Longevity

Lakshmi S. Kocherlakota, Tiep Pham, and RenÃ© M. Overney*

Contemporary environmental and energy production areas are driving current research for pursuing green and sustainable technologies for an economically viable and energy efficient solutions. Membrane based separation technology offers a less energy intensive and environmentally friendly alternative for gas purification processes. Particularly novel organic-inorganic nanocomposite membranes have been shown to push the gas permeability and selectivity capabilities of polymeric membrane materials into the attractive regions based on commercial separation requirements. One of the most notable permeation materials in regards of its high permeation values of up to several thousand Barrer, and its reverse-selectivity towards the lighter component in H_2/CO_2 and N_2/CO_2 is poly(l-trimethylsilyl-1-propyne) (PTMSP), a high free volume glassy polymer. Problematic in nanocomposites are particle dispersion and environmental health concerns due to the small particle size. These two shortcomings are resolved potentially by switching to "inverse composites", in which the polymer matrix is replaced by a porous membrane that provides the interfacial/dimensional constraints, while the polymer itself is injected into the pores. Porous membranes also provide mechanical and temporal stability. In this paper, we discuss how beneficial interfacial constraints, concerning both permeation and selectivity in PTMSP (found in ultrathin film membranes[1]), can be imposed controllably in macroscopic inverse composites, thereby maintaining the enhanced transport properties of thin films, and provide relative improved longevity. PTMSP embedded nanoporous anodic aluminum oxide (AAO) hybrid membranes reveal gas permeabilities for helium, nitrogen and CO_2 that exceeded 5 to7 times the values of pristine bulk PMTSP. Most notably, the gas selectivities of CO_2/He and CO_2/N_2 could also be enhanced by ~40%, and physical aging caused by free-volume collapse, could be reduced 20- fold compared to ultrathin film systems due to pore wall stabilization. Our current studies on polyethylene oxide (PEO) based polymeric materials which are being extensively studied for CO_2 separation are also revealing improved separation capabilities in AAO-confined hybrid polymeric structures. We found that AAO-confined PEO exhibits CO_2/He separation of ~10, which represents a 40% enhanced selectivity over the bulk membranes.

1. Kocherlakota LS, Knorr DB, Foster L, and Overney RM. Polymer 2012;53(12):2394-2401.

(619b) Investigation of Water Quantification in Ionic Liquids Utilizing Near Infrared Absorption of Water

Fagnant, D. Jr.,
Goff, G. S., Los Alamos National Laboratory
Runde, W., Los Alamos National Laboratory
Brennecke, J. F., University of Notre Dame
Originally presented on: 11/7/2013 8:55:00 - 9:20:00

Ionic Liquids (ILs) are potential candidates to replace traditional organic solvents for a variety of separation processes. Some ILs are capable of forming aqueous biphasic systems which allow for solvent extraction of various solutes from aqueous solutions. The ability to quantify water content in an IL phase is important to implementing an ionic liquid as a solvent in a real world process. The water content of an IL affects physical properties such as density, viscosity and conductivity and also significantly affects energy costs for heating and cooling.

Here we investigate and demonstrate the use of near infrared absorption spectroscopy to quantify water content in an imidazolium-based IL. Near infrared absorption offers an accurate, rapid and in situ method for water quantification in the ionic liquid phase, which provides a non-destructive alternative to Karl-Fischer titration/coulometry. The change in water absorption at different water concentrations in the IL is investigated.

(619c) Mixed Gas Sorption of CO2/CH4 Mixtures in PIM-1 and PTMSP Membranes: Experiments and Modeling

Vopicka, O.,
De Angelis, M. G., University of Bologna
Sarti, G. C., Alma Mater Studiorum - UniversitÃ di Bologna
Guiver, M. D., National Research Council Canada
Originally presented on: 11/7/2013 9:20:00 - 9:45:00

Sorption of pure methane, carbon dioxide and their binary mixtures in two glassy polymers, poly(1-trimethylsilyl-1-propyne) (PTMSP), and the first polymer of intrinsic microporosity (PIM-1), has been studied experimentally and theoretically, at 35.0 °C.

Measurements were obtained on a newly designed pressure decay sorption apparatus for mixtures of gases, having the basic construction according to Sanders et al. [1], but with a more versatile procedure than that used in [1], which allowed to measure sorption isotherms at constant partial pressure of one component of the gas mixture. Indeed this novel method allows one to measure sorption isotherms *i*) at constant composition of the gaseous phase, *ii*) at constant fugacity of one component or *iii*) at constant equilibrium pressure. The first protocol, in particular, allows to mimic better the real constraints faced when dealing with a membrane separation process, where one has a gas stream of fixed composition, in which only the total pressure can be varied, by compression. The pressure decay apparatus is coupled to a gas chromatograph Varian CP-4900 Micro-GC equipped with a capillary column and with a thermal conductivity detector for analysis of the gas phase composition.

In the case of PTMSP, the mixture n-C_4/CH_4 was initially considered, to provide a direct comparison with literature data [2] and validation of the method. Indeed, the sorption of n-C_4/CH_4mixtures showed a reasonable agreement with the existing mixed gas sorption data [2].

On the other hand, the CO_2/CH_4mixed sorption data in PTMSP are completely new, and were measured in the range from 0 to 33 atm of total equilibrium pressure, and from 5 to 90 mol.% of carbon dioxide in the gaseous phase.

Furthermore, the same characterization of CO_2/CH_4 mixed sorption was performed in PIM-1: the pressure range inspected was the same as in PTMSP, while the composition of CO_2 ranged from 10 to 50 mol.%.

PTMSP membrane was cast from a solution of toluene, immersed in methanol and then dried under vacuum before characterization; its density was 0.77±0.01 g/cm^3.

The PIM-1 membrane was prepared from a filtered ca. 2.0 wt.% chloroform solution of PIM-1 and heated in vacuum at 70 °C, then submerged in methanol and dried under vacuum at 70 °C. The density of pure PIM-1 was (1.143±0.008) g/cm^{3}at 25°C.

For both PTMSP and PIM-1, the mixed gas solubility differs significantly from the pure gas value, and, in particular, the solubility of both components is depressed by the presence of the second one, as it often happens in glassy polymers.[3] The solubility selectivity ranges between 2 and 6 for PTMSP and between 5 and 10 for PIM-1. The methane solubility, however, is more significantly depressed by CO_2 than that of CO_2 is decreased by CH_4, therefore the real solubility selectivity (CO_2/CH_4) for PTMSP and PIM-1 is higher than the ideal solubility selectivity. Such effect becomes more significant with increasing the mole fraction of CO_2 in the gaseous phase and with pressure, and is more significant for PIM-1 than for PTMSP. Indeed, the real solubility selectivity becomes 3 times higher than the ideal one in PTMSP for a fraction of 70 mol.% of CO_2 in the gas phase, while for PIM-1 such point is reached with a lower concentration of CO_2(50 mol.%).

Both results indicate the presence of a competition for available polymer matrix sites, which is not surprising due to the nature of physical sorption in glassy matrices, and possibly also of different interactions between polymer and penetrants.

To investigate that behavior, the Non-Equilibrium Lattice Fluid model (NELF) was used [3], while the widely used Dual Mode Sorption (DMS) model was also considered as a reference tool. The NELF model, as well as the DMS, does not require additional parameters for the prediction of the mixed gas behavior, and is fully predictive provided a few pure gas sorption data in the polymer matrix. Indeed, binary interaction parameters are the same as in the pure gas case, and the swelling induced by the mixture is estimated from pure gas swelling. Remarkably, in the DMS model, only competition (depression) effects are accounted for, because the mixed gas additional term (positive) appears only in the denominator of the expression for solubility.

The NELF model provided quantitative predictions of the mixed gas sorption of CO_2 and CH_4under pure- and mixed-gas conditions in PTMSP and in PIM-1. The solubility selectivity is also predicted, although with less accuracy, by the NELF model. The DMS model works fairly well in the case of PTMSP, but provides poorer predictions than the NELF model of the mixed gas solubility in PIM-1.

Sorption of mixtures of CO_2 and CH_4 in PTMSP and in PIM-1 was predictable with the NELF model with an accuracy that is comparable to the experimental one, which could reduce the need for the laborious measurements of mixed gas sorption in polymers. Better insights and interpretation of the mixed gas sorption mechanism can also be obtained by using the NELF model.

References:

[1] E.S. Sanders, W.J. Koros, H.B. Hopfenberg, V.T. Stannett J. Membr. Sci. 18 (1984) 53-74.

[2] R.D. Raharjo, B.D. Freeman, E.S. Sanders, Polymer 48 (2007) 6097-6114.

[3] M. Minelli, S. Campagnoli, M.G. De Angelis, F. Doghieri, G.C. Sarti, Macromolecules 44 (2011) 4852-4862.

(619d) Diffusive Separation of Propylene/Propane With ZIF-8 Membranes

Hara, N., The University of Tokyo
Yoshimune, M., National Institute of Advanced Industrial Science and Technology (AIST)
Negishi, H., National Institute of Advanced Industrial Science and Technology (AIST)
Haraya, K., National Institute of Advanced Industrial Science and Technology (AIST)
Hara, S., National Institute of Advanced Industrial Science and Technology (AIST)
Yamaguchi, T., Tokyo Institute of Technology
Originally presented on: 11/7/2013 9:45:00 - 10:10:00

Metal-organic frameworks (MOFs) have been recognized as a new candidate for gas and vapor separation. Among several varieties of MOFs, ZIF-8 is one of the most common materials for separation membranes. ZIF-8 is composed of zinc atoms and 2-methyl imidazole as ligands, forming the sodalite structure with large cavities, 1.16 nm in diameter, which are connected with a small pore aperture of 0.34 nm diameter [1]. Recently, several reports have focused on the potential of ZIF-8 for propylene/propane separation on the basis of their difference of diffusivity [2].

The most common preparation methods of ZIF-8 membranes are the in situ growth method and the secondary growth method. Pan et al. reported the separation of propylene/propane mixture, in which the ZIF-8 membrane is prepared by the secondary growth method with a separation factor of up to 50 for propylene/propane binary mixture [3]. Kwon et al. reported the preparation of ZIF-8 membranes using microwave-assisted seeding and secondary growth with the separation factor of up to 40 for propylene/propane binary mixture [4].

To improve the propylene/propane separation properties, diffusivity and solubility separation factors should be separately investigated from the single-component permeation measurements. In this study, we clarified the contribution of diffusive separation for propylene/propane with ZIF-8 membrane based on the analysis of diffusivity and solubility obtained from the membrane permeation properties. Permeation properties were studied using the single-component gas permeation properties, and superior separation properties for propylene/propane were achieved for ZIF-8 membranes with uniform selective layer. Permeability, diffusivity, solubility, and their dependencies on temperature were studied, and the contribution of diffusivity on propylene/propane separation is thoroughly discussed.

(295 words)

[1] K. S. Park, Z. Ni, A. P. Cote, J. Y. Choi, R. D. Huang, F. J. Uribe-Romo, H. K. Chae, M. O'Keeffe and O. M. Yaghi, Exceptional chemical and thermal stability of zeolitic imidazolate frameworks, Proc. Natl. Acad. Sci. USA 103 (2006) 10186-10191.

[2] K. H. Li, D. H. Olson, J. Seidel, T. J. Emge, H. W. Gong, H. P. Zeng and J. Li, Zeolitic Imidazolate Frameworks for Kinetic Separation of Propane and Propene, J. Am. Chem. Soc. 131 (2009) 10368-10369., C. Zhang, R. P. Lively, K. Zhang, J. R. Johnson, O. Karvan and W. J. Koros, Unexpected Molecular Sieving Properties of Zeolitic Imidazolate Framework-8, J. Phys. Chem. Lett. 3 (2012) 2130-2134.

[3] Y. C. Pan, T. Li, G. Lestari and Z. P. Lai, Effective separation of propylene/propane binary mixtures by ZIF-8 membranes, J. Membr. Sci. 390 (2012) 93-98.

[4] H. T. Kwon and H. K. Jeong, Highly propylene-selective supported zeolite-imidazolate framework (ZIF-8) membranes synthesized by rapid microwave-assisted seeding and secondary growth, Chem. Commun. 49 (2013) 3854-3856.

(619e) Simulation of Propylene/ Propane Plasticization in 6FDA-Based Polyimide Membranes

Halitoglu-Velioglu, S., Istanbul Technical University
Ahunbay, M. G., Istanbul Technical University
Tantekin-Ersolmaz, S. B., Istanbul Technical University
Originally presented on: 11/7/2013 10:10:00 - 10:35:00

The permeation properties of hydrocarbon gases through polymeric membranes are of interest for the development of membrane processes for separation of olefin/paraffin mixtures, such as propylene (C_3H_6)/propane (C_3H_8), in the petrochemical industry. Propylene is largely a byproduct of ethylene production, which is the largest volume organic produced in the petrochemical industry. Furthermore, propylene is produced in propane dehydrogenation, and natural gas to olefins and other olefin conversion processes. While rubbery polymers have very low selectivity for propylene/propane, glassy polyimides (PIs) and polyimide-co-polypyrrolones based on 6FDA (4,4'-(hexafluorisopropylidene)dipthalic anhydride) have been reported to exhibit the best performance [1]. However, a major issue for these polyimides is the plasticization of the membrane during the permeation of these condensable hydrocarbons.

Propylene and propane, both being condensable, tend to plasticize polymeric membranes, even at partial pressures as low as 2 bars. There are several experimental studies investigating plasticization effect of propane and propylene in polyimides reported in literature [1-4]. The observed plasticization pressure among these studies was between 2 and 5 atm. In order to suppress plasticization, cross-linking and thermal treatment have been used. On the other hand, there are no molecular simulation studies reported in the literature that focus on propane and propylene-induced plasticization in polyimides.

In this study, molecular simulation techniques are used to estimate the degree of plasticization of 4,4-hexafluoroisopropylidene-diphthalic anhydride (6FDA)-based PI membranes induced by sorption of propane (C_3H_8) and propylene (C_3H_6). The diamines used to build the polyimide structure are 2,4,6-trimethyl-m-phenylene diamine (DAM), 2,5-dimethyl-p-phenylenediamine (DPX), and 4,4-oxydianiline (ODA). To the best of our knowledge, this is the first simulation study examining propane/propylene plasticization in 6FDA based PIs. The sorption simulations are carried out in the Grand Canonical Monte Carlo (GCMC) ensemble. To reproduce the hydrocarbon-induced plasticization effect, sorption-relaxation cycles are applied until C_3H_6 or C_3H_8 concentration converges. The increase in the FFV of the resulting polymer structure is considered as the extent of plasticization. The change in the sorption selectivity is determined for both pure gases and binary C_3H_6/C_3H_8 mixtures for % 50/50 feed composition. The glass transition temperatures (Tg) of the polymers are estimated by observing the volume change of the simulation cells during cooling down from 760 K to 380 K. Rotational time autocorrelation functions of the unit vectors defining to the distance between the diamine and dianhydride moities are also calculated in order to analyze changes in the local dynamics of the polymers due to hydrocarbon sorption. Our recent work [5] on the CO_2-sorption induced plasticization of these polymers have shown that the CO_2 sorption capacity at 35°C and 10 bars depend on the FFV of the swollen PI, yielding similar CO_2 concentrations for similar FFV values. Thus, particular interactions between CO_2 and sorption sites have limited differential contribution on sorption capacity. However, no such relationship was obtained for propane and propylene at 25°C and 1.13 bar due to changing site preferences, indicating the strength of sorbate-site interactions is significant on sorption capacity, as well as the FFV.

[1] Das, M., Koros, W.J., *J. Memb. Sci.,* 365, 399-408, 2010.

[2] Staudt-Bickel, C., Koros, W. J., *J. Memb. Sci.,* 170, 205, 2000.

[3] Tanaka, K., Taguchi, A., Hao, J. Q., Kita, H., Okamoto, K., *J. Memb. Sci.,* 121, 197, 1996.

[4] Jiang, L. Y., Chung, T. S., Rajagopalan, R., *Chem. Eng. Sci.,*63, 204-216, 2008.

[5] Velioglu, S., Ahunbay, M. G., Tantekin-Ersolmaz, S.B., *J. Memb. Sci.,* 417, 217-227, 2012.

(625a) Membrane Modification and Synthesis for Fouling Control

Andrews, R., University of Cincinnati
Co, C. C., University of Cincinnati
Ho, C. C., University of Cincinnati
Originally presented on: 11/7/2013 8:30:00 - 8:55:00

Membrane processes are widely used in the biotechnology and pharmaceutical industries for sterile filtration of fermentation media, initial harvesting of therapeutic products, and purification of buffers and proteins. Protein fouling is one of the critical factors governing the membrane performance in various filtration processes. In this study, we utilize the protein resistant property of polyethylene oxide (PEO) to modify the membranes to investigate the potential of utilizing PEO for fouling reduction. Three surface modification techniques have been investigated: covalent, ionic, and self assembled monolayers. In the first system, polyethylene oxide (PEO) containing pluronic surfactant was UV grafted to 0.2 micron mixed cellulose esters membranes. Upon coating the membrane with a monolayer of allyldimethylchlorosilane (ADCS). The silanized membrane was covalently linked to Pluronic F127, a triblock copolymer of poly ethylene oxide and poly propylene oxide (PEO-PPO-PEO) by UV irradiation at wavelength above 215 nm. In the second system, polyelectrolytesterminated with polyethylene glycol, poly (oligoethylene glycol methacrylate co methacrylic acid), (OEGMA/MA), was applied to modify positively charged PVDF membranes. In the third system, self-assembled monolayer with different terminal functional groups was formed on silver membranes. Membranes modified covalent or ionic approach showed minimum effect on flux decline while membranes modified with alkanthiols with a terminal tri(ethylene glycol) functional groups showed significant reduction in flux decline during bovine serum albumin (BSA) filtration. This shows that the density of the PEO group on the surface is critical in fouling control. To increase the PEO density, we developed a new approach to fabricate membranes with PEO as the side chains in the polymer that forms the membrane using Glassy Self Assembly Templating. This approach can lead to membranes with highly uniform and finely tunable nanometer-size pores whose dimensions dictated by the quasi-equilibrium thermodynamics of the glassy sugar/surfactant template. In addition PEO can be incorporated throughout the membrane to reduce fouling potential.

(625b) Responsive Ultrafiltration Cellulose Acetate Membranes Composed of N-Isopropylacrylamide (NIPAAm) and Superparamagnetic Iron Oxide (SPIO) Nanoparticles

Chede, S., The University of Toledo
Bothun, G., University of Rhode Island
Escobar, I. C., The University of Toledo
Originally presented on: 11/7/2013 8:55:00 - 9:20:00

Membrane fouling occurs when there is reversible or irreversible attachment of macrosolutes present in the water to the membrane surface. Membrane replacements due to fouling can add significantly to the operating costs of the membrane system. Reversible fouling can be minimized by crossflow operation, backflushing during filtration and/or by chemical pretreatment, but it may take time and energy and as well as have to meet safety guidelines with respect to chemicals used. On the other hand, irreversible fouling cannot be corrected by backflushing or crossflow filtration, and it results in permanent flux decline. Among irreversible foulants, natural organic matter (NOM) is considered to be a major contributor. NOM is composed of wide range of hydrophilic and hydrophobic components; hence, any static hydrophobic or hydrophilic membrane may be fouled. Therefore, a dynamic membrane able to alternate between being hydrophobic and hydrophilic is hypothesized to decrease fouling. Stimuli-sensitive polymers have been gaining attention in recent years due to their unique property that the polymer changes its conformation from a coiled and hydrophobic state to an uncoiled (or globular) and hydrophilic state in the presence of a stimulus. By alternating between the phases, the membrane surface can be made dynamic.

The purpose of this study was to develop stimuli responsive membranes to control fouling using membrane casting dopes made of cellulose acetate with N-isopropylacrylamide (NIPAAm). NIPAAm is a stimuli-responsive polymer, which offers the potential to collapse or expand the membrane as a function of changes in temperature and it is a reversible transition. As a response to a temperature decrease, this polymer expands into a hydrophilic state, while a temperature increase causes it collapse into a hydrophobic state. The phase change arises from the existence of a lower critical solution temperature (LCST) such that the polymer precipitates from aqueous solution as the temperature is increased. By continuously activating the film, it is hypothesized that a dynamic stimuli-responsive surface can prevent foulants from attaching to the membrane surface. However, alternating between hot and cold temperatures to stimulate the polymer is energy intensive, and during heating and cooling, the response time within the membrane is hindered by heat transfer resistances. These problems are hypothesized to be eliminated if the heating is localized and intensified within the membrane matrix by embedding small superparamagnetic iron oxide (SPIO) nanoparticles within the temperature sensitive membranes. SPIO nanoparticle heating is caused by Neel and Brownian relaxations. Thus, temperature activation can be triggered through alternating electromagnetic heating of the embedded SPIO nanoparticles.

(625c) Surface Analysis and Biofouling Resistance of Nanostructured Reverse Osmosis Membranes

Moses (Varin), K. J., University of California, Los Angeles
Cohen, Y., University of California, Los Angeles

Originally presented on: 11/7/2013 9:20:00 - 9:45:00

Water re-use could represent a significant sustainable water resource if upgraded to potable water quality. In this regard, reverse osmosis (RO) desalination is a promising technology for water desalination; however, despite modern advances in RO membrane technology, membrane mineral scaling and biofouling hinders the widespread use of RO membranes in brackish desalination and water re-use applications. This study focuses on the synthesis of a class of RO membranes with a dense hydrophilic polymeric surface brush layer that suppresses surface adhesion of organic, bacteria, and mineral foulants. The polymeric brush lower consists of covalently tethered and terminally-anchored water soluble chains whose segments are capable of a significant degree of Brownian motion. In order to assess the impact of the brush layer on nano-scale changes in surface topography, polyamide-silicon (PA-Si) surrogate surfaces were first developed to optimize the graft polymerization process. The UCLA atmospheric pressure plasma-induced graft polymerization (APPIGP) method was then utilized to generate a high surface density of active sites on a polyamide membrane support for subsequent graft polymerization onto the polyamide surface using suitable water soluble vinyl monomers. The resulting surface nano-structured (SNS)-PA-Si surfaces were characterized with respect to surface energy, roughness, and brush layer thickness. Performance evaluation of the SNS-thin-film composite (TFC)-PA RO membranes was subsequently carried out by examining their biofouling propensities, assessed via flux decline studies with secondary treated municipal waste water effluent, biofilm imaging with confocal laser scanning microscopy (CLSM), and extracellular polymeric substances (EPS) analysis. Additionally, the membrane effectiveness of permeability recovery was evaluated for cleaning regiments using DI water and chemical (Na_2·EDTA) cleaning. The present study demonstrated that it is feasible to tailor-synthesize the grafted brush layer to attain the desired tethered polymer layer topography and the level of surface free energy of hydration to promote biofouling resistance.

(625d) Novel Copper-Charged Anti-Microbial Membranes for Desalination

Asapu, S., University of Toledo
Seedorf, M., The University of Toledo
Escobar, I. C., The University of Toledo
Originally presented on: 11/7/2013 9:45:00 - 10:10:00

The need to control membrane fouling by microorganisms, also known as biofouling, is crucial in water separations applications. The purpose of this project was to investigate the charging of membrane surfaces with copper (Cu^{+2}) to control biofouling by developing anti-microbial membranes. The use of anti-microbial membranes would reduce the cost associated with chemical pretreatment and cleaning as well as chemical storage. To this end, glycidyl methacrylate (GMA), an epoxy, was used to attach a chelating agent, iminodiacetic acid (IDA) to facilitate the charging of copper to the membrane surface. Both cellulose acetate (CA) and CA-GMA membranes were cast using the phase-inversion method. The CA-GMA membranes were then charged with copper ions to make them resistant to microbial growth. The pore size distribution analysis of CA and copper charged membranes were conducted using various molecular weights of polyethylene glycol (PEG). CA and copper charged membranes were characterized chemically using Fourier Transform Infrared (FTIR), and with respect to copper dispersion on the membrane surface using scanning electron microscope (SEM), hydrophilicity changes using contact angle. The permeation experiments were conducted with DI water and then subjected to protein rejection measurements. The permeation of the copper-charged membranes was initially lower than the cellulose acetate membrane during the filtration of DI water. The membranes were then subjected to bovine serum albumin (BSA) and lipase filtration. The copper charged membrane showed higher permeation values for of both proteins as compared to CA membranes. This resulted from increased resistance from the IDA and copper on the surface during DI water filtration, and later, from reduced fouling from both proteins. The rejection of BSA and lipase was the same for both the copper charged and CA membranes. Therefore, the modified membranes have a potential to be used as low-biofouling membranes in the future.

(625e) Changes in Physiochemical Properties and Performance of a Polyamide-Based Nanofiltration Membrane Due to Chlorine Exposure

Do, V. T., Nanyang Technological University
Tang, C. Y., Nanyang Technological University
Originally presented on: 11/7/2013 10:10:00 - 10:35:00

Owing to their high selectivity and water permeability, polyamide-based (PA) thin film composite (TFC) membranes are widely employed in water and wastewater treatment using nanofiltration (NF) and reverse osmosis (RO) processes. However, the performance and life -span of these membranes are susceptible to attack of chlorine, which is widely used to control membrane bio-fouling. In this study, the effects of chlorine exposure conditions on the performance of a commercial PA-TFC membrane, namely NF90 were investigated and the conditions that resulted in an optimum membrane performance were identified. The membrane samples were exposed to sodium hypochlorite (NaOCl) of different concentrations: 10, 100, 1000 ppm Cl at pH 5, 7 and 9 for 100 h. The membrane performance was evaluated based on the water permeability values and the rejection of NaCl and inorganic contaminants of significant environmental and health concern (boron (H_3BO_3) and arsenic ($H_2AsO_4^-$)). At pH 5, the water permeability of the chlorinated membranes decreased when NaOCl concentrations increased. However, at the pH 7 and 9, the water permeability of the chlorinated membranes increased with NaOCl concentrations. The rejection of charged NaCl and arsenic improved in most chlorinating conditions except for 1000 ppm Cl at pH 7 and 9. Meanwhile, the rejection of neutral boric acid was lower in most cases, except for 10 ppm Cl and pH 5. The disparate changes in membrane performance due to different chlorine treatments were explained in terms of competing mechanisms: membrane tightening, bond cleavage by N-chlorination and chlorination promoted polyamide hydrolysis. X-ray Photoelectron Spectroscopy (XPS), attenuated total reflection-Fourier transform infrared (ATR-FTIR) spectroscopy, zeta potential and contact angle measurement were the analytical methods to provide evidences for the mechanisms.

(625f) Ultrathin Coatings Tethered to Commercial Membranes Via Vapor Phase Treatment for Fouling and Chlorine-Resistance

Yang, R., Massachusetts Institute of Technology

Jang, H.,

Stocker, R.,

Gleason, K. K., Massachusetts Insitute of Technology

Originally presented on: 11/7/2013 10:35:00 - 11:00:00

Division: membrane based separation?membrane surface engineering

Ultrathin coatings tethered to commercial membranes via vapor phase treatment for fouling and chlorine-resistance

Rong Yang, Hongchul Jang, Roman Stocker, Karen K. Gleason

Fouling refers to the unintended accumulation of organic matters, biomolecules or organisms on wetted structures, and the subsequent formation of biofilms. It is very prevalent in membrane-based separation processes and is extremely undesirable as fouling leads to impaired membrane performance and shortened membrane lifetime. It has become the bottle-neck to improve the efficiency of membrane-based separation processes, especially for reverse osmosis (RO) operations.[1] The current antifouling methods in industrial practice are insufficient, costly and time consuming; methods such as periodic cleaning also shortens the membrane lifetime. Membrane surface engineering such as physical adsorption and chemical bond formation are potential solutions to the membrane fouling problem.[1] However, surface modification is no longer effective in preventing fouling once the first layer of foulants has adsorbed to the membrane surface. Therefore, stable coatings that impart ultralow-fouling characteristics to a surface are highly desired for membrane-based separations. Zwitterionic coatings are promising candidates to reduce fouling during membrane separations given the ultra-low absorption of protein and high resistance to bacteria adhesion and biofilm formation in long-term tests.[2]

Most membranes are made of polymeric materials for the low cost and ease of fabrication. However, such membranes usually lack solvent- or thermal-resistance, which makes the existing coating methods for zwitterionic materials inapplicable. A vapor phase deposition method at low temperature avoids the potential for such damages. Here, we report on the synthesis of zwitterionic polymers through vapor phase treatment. A copolymer containing sulfobetaine zwitterionic groups was synthesized through an initiated chemical vapor deposition (iCVD) technique and was grown directly on various membrane substrates including commercial membranes as well as electrospun nanofiber mats. Thin film composite (TFC) RO membranes were chosen because of the significant impact of fouling on the membrane performance. This also demonstrates the benignness of the iCVD conditions; although TFC RO membranes are extremely sensitivity to heat and solvent,[1] full membrane function was retained after iCVD without compromising the delicate polyamide layer. It is worth noting that the coating procedure is substrate-independent and no re-optimization is required for other membrane substrates.

The iCVD zwitterionic coating is unique because the antifouling zwitterionic groups are highly concentrated on the surface, precisely where the antifouling properties of the zwitterionic groups can be exploited. Angle-resolved photoelectron spectroscopy (ARXPS) measurements revealed that 97% of the total nitrogen was in the zwitterionic form within the top 3 nm, whereas when we probed the top 9 nm, only 30% was zwitterion (**Fig. 1**). It is important to note that the currently observed high concentration of zwitterionic groups at the surface is exactly opposite to that for films formed by solution-phase methods.[3] Coated surfaces exhibited 93% and 98% reduction in surface absorption of bovine serum albumin (BSA) and sodium alginate (SA), as measured with the quartz crystal microbalance with dissipation monitoring (QCM-D). Cell adhesion tests were carried out using *Escherichia coli* and *Vibrio cyclitrophicus*, and the coated RO membranes showed greatly reduced bacteria attachment while on the bare RO membrane surface, significant bacteria adhesion and formation of a confluent biofilm were observed.

The *in situ* monitor of polymer coating growth inside an iCVD chamber enabled the precise control over coating thickness; ultrathin and pin-hole free coating with the thickness of ~30 nm was deposited on commercial RO membranes. The inherent crosslinking in the iCVD copolymer made the coating insoluble in water. A strategy to covalently attach the coating to RO membranes was developed to increase the stability of the coating and adapt the coated membrane for long-term usage in water processing. This is the first time that the library of iCVD functional groups has been extended to charged zwitterionic moieties, and the zwitterionic coatings have been applied on delicate substrates, such as RO membranes.

The iCVD zwitterionic coatings were then improved to possess chlorine-resistance. Chlorine-resistance is of interest for membrane-based separation because chlorine bleach is the most economical and prevalent disinfection technique in water industry. To be compatible with the disinfection procedure and to further reduce surface fouling by using chlorine in addition to surface engineering, the resistance to oxidizing agents is highly desirable for antifouling coatings. However, currently available zwitterionic coatings are fabricated almost exclusively with acrylate monomers containing ester and/or amide linkages and do not stand up to oxidizing agents.[4, 5] For example, continuous exposure to water containing even a few parts per billion (ppb) chlorine degrades the amide chemistry significantly.[6] The highly desirable chlorine-resistant zwitterionic coatings have not been fabricated to the best of our knowledge. Here, we report for the first time, a novel zwitterionic chemistry that is also chlorine-resistant. The coating derives from a polymer that contains no labile linkages susceptible to oxidative damages by chlorine.

The scale-up of the iCVD process has been demonstrated in a roll-to-roll system,[7] which is compatible with the existing infrastructures in membrane industries. Combining the significant antifouling performance with high stability in strong oxidizing agent, as well as the scalability of the iCVD process, this surface modification scheme can reduce the energy usage, improve the reliability and lower the environmental impact of RO and potentially other membrane-based separation processes.

Reference

1. Elimelech, M.; Phillip, W. A. *Science* **2011,** 333, (6043), 712-717.

2. Jiang, S.; Cao, Z. *Adv. Mat.* **2010,** 22, (9), 920-932.

3. Yang, S.; Zhang, S.-P.; Winnik, F. M.; Mwale, F.; Gong, Y.-K. *J. Biomed. Mater. Res., Part A* **2008,** 84A, (3), 837-841.

4. Chiantore, O.; Trossarelli, L.; Lazzari, M. *Polymer* **2000,** 41, (5), 1657-1668.

5. Butt, C. M.; Young, C. J.; Mabury, S. A.; Hurley, M. D.; Wallington, T. J. *The Journal of Physical Chemistry A* **2009,** 113, (13), 3155-3161.

6. Park, H. B.; Freeman, B. D.; Zhang, Z.-B.; Sankir, M.; McGrath, J. E. *Angewandte Chemie International Edition* **2008,** 47, (32), 6019-6024.

7. Gupta, M.; Gleason, K. K. *Thin Solid Films* **2006,** 515, (4), 1579-1584.

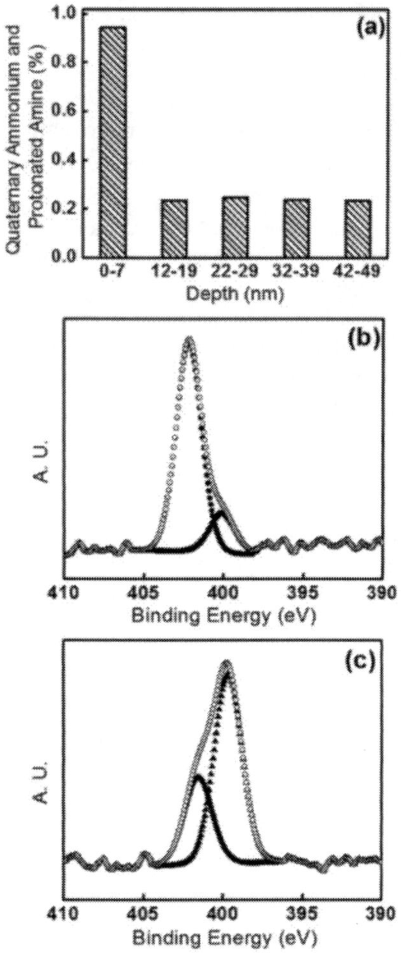

Figure 1. The concentrations of quaternary ammonium and protonated amine along the coating cross-section and ARXPS spectra. a) Composition of the zwitterionic component (quaternary ammonium) at different depth in the film, obtained by depth-profiling XPS. b) High resolution ARXPS N (1s) scans with photoelectron takeoff angles of 19.5^{o} and c) 90°.

(629a) Material Science Meets Rigidity Theory Meets Mechanical Engineering: What Can We Learn About Flexibility of MOFs From the Reduced Topological Representations?

Sarkisov, L., University of Edinburgh
Smit, B., University of Amsterdam
Originally presented on: 11/7/2013 8:30:00 - 8:46:00

Flexibility of Metal-Organic Frameworks (MOFs) has become one of the most fascinating topics in material science and crystal engineering. Indeed, over the years a number of MOF structures have been discovered that exhibit a broad and diverse range of conformational effects, from local rotational freedom of linkers to swelling and negative thermal expansion to large amplitude deformations of the framework, such as breathing effects. Given virtually infinite number of possible MOFs, this behaviour can be exploited to develop smart, stimuli responsive, programmable MOFs for controlled drug release, detection and energy storage. Current development of these applications is hindered by poor understanding of the mechanisms of the deformation processes on molecular level. Here we present a systematic way to detect potential flexibility modes in various classes of MOFs. The approach builds on the reduced topological representations of MOF structures, combined with the rigidity theory analysis. As case studies we consider flexibility modes in IRMOF-1, MIL-53 and several other materials.

(629b) Interrogating Simple Pore Models Using Flat-Histogram Sampling Methods to Understand the Effect of Sorbent Flexibility On Fluid Adsorption

Shen, V. K., National Institute of Standards and Technology
Siderius, D. W., National Institute of Standards and Technology
Originally presented on: 11/7/2013 8:46:00 - 9:02:00

While it is well known that sorbent flexibility can affect observed adsorption behavior, rigid sorbents are still commonly used in the study of adsorption via molecular simulation for both simplicity and practicality. Past attempts at incorporating sorbent flexibility in simulation have largely relied on computationally demanding atomistic descriptions of the sorbent framework. In this work, we present two simple pore models for flexible sorbents. The first model accounts for internal sorbent flexibility in a grand canonical ensemble formulation. The second model accounts for volume fluctuations via an osmotic ensemble formulation. In both cases, we use flat-histogram sampling methods to calculate precisely the relevant multi-dimensional probability distributions, which we will show are directly related to specific thermodynamic potentials and the deformation properties of the sorbent. Using a square-well fluid as the adsorbate, we are able to reproduce experimentally observed adsorption behavior, such as adsorption-induced deformation and breathing phase transitions. More importantly, we are able to identify the physically relevant features (or properties) of the sorbent that lead to these interesting and exploitable sorbent properties.

(629c) First Principles Derived, Transferable Force Fields for CO2 Adsorption in Cation-Exchanged Zeolites

Fang, H., Georgia Institute of Technology
Kamakoti, P., ExxonMobil Research and Engineering
Cundy, S., ExxonMobil Research and Engineering
Li, Q., ExxonMobil Research and Engineering
Ravikovitch, P. I., ExxonMobil Research and Engineering
Paur, C., ExxonMobil Research and Engineering
Aronson, M., Univ. of California, Santa Barbara
Sholl, D. S., Georgia Institute of Technology
Originally presented on: 11/7/2013 9:02:00 - 9:18:00

The development of accurate force fields is vital for predicting adsorption in porous materials. Previously, we introduced a first principles-based transferable force field for CO_2 adsorption in siliceous zeolites (Fang et al. *J. Phys. Chem. C* **2012**, *116*, 10692). In this study, we extend our approach to CO_2 adsorption in cationic zeolites which possess more complex structures. Na-exchanged zeolites are first chosen for demonstrating the approach. These methods account for several structural complexities including Al distribution, cation positions and cation mobility, all of which are important for predicting adsorption. The simulation results are validated with high-resolution experimental measurements of isotherms and microcalorimetric heats of adsorption on well-characterized materials. The choice of first-principles method has a significant influence on the ability of force fields to accurately describe CO_2-zeolite interactions. The PBE-D2 derived force field, which performed well for CO_2 adsorption in siliceous zeolites, does not do so for Na-exchanged zeolites; the PBE-D2 method overestimates CO_2 adsorption energies on multi-cation sites that are common in cationic zeolites with low Si/Al ratios. In contrast, a force field derived from the DFT/CC method performed well. Agreement was obtained between simulation and experiment not only for LTA-4A on which the force field fitting is based, but for other two common adsorbents, NaX and NaY. Recently, we have expanded this approach to other cations including divalent Ca-exchanged zeolites as well as Ca/Na mixed zeolites. The effect of residual H_2O on CO_2 adsorption will also be considered.

(629d) Ab Initio Carbon Capture in Open-Site Metal Organic Frameworks - Force Field Development and Its Applications

Lin, L. C., University of California-Berkeley
Lee, K., Lawrence Berkeley National Laboratory
Kim, J., Lawrence Berkeley National Laboratory
Smit, B., University of Amsterdam
Originally presented on: 11/7/2013 9:18:00 - 9:34:00

Carbon capture and sequestration (CCS) is a viable strategy proposed to mitigate the negative environmental impacts derived from burning fossil fuels. Recently, nano-porous adsorbent materials such as metal-organic frameworks (MOFs) have shown to potentially provide a more energy-efficient way to separate CO_2 out of flue gases. In particular, MOFs that possess unsaturated open metal sites are known to interact strongly with CO_2 molecules, and thus emerging as promising materials for the separation applications. However, from the computational point of view, the strong interaction is poorly modeled by common force fields, resulting in underestimation of the adsorption properties of CO_2 in Mg-MOF-74 by as much as an order of magnitude. Accordingly, a systematic methodology has been proposed to generate accurate force fields using high-level quantum chemical calculations[1]. In this method, interaction energies of a few selected CO_2 configurations computed by quantum calculations are used to properly map into the force field parameters. In our previous study[1], Møller-Plesset second-order perturbation theory (MP2) with representative small framework fragments was used in the force field development. To facilitate the automation of this methodology as well as resolve the intrinsic discrepancies between cluster and periodic calculations, a methodology based on the periodic density functional theory (DFT) calculations is introduced in this work. The calculated binding geometry and adsorption properties from the DFT-derived force fields show excellent agreements with the ones from DFT calculations and experiments, respectively. Finally, we have also utilized the new DFT-derived force field to study the CO_2 dynamics in Mg-MOF-74 and made connections to our previous study[2].

Reference

[1] A. Dzubak, L.-C. Lin, J. Kim, J.A. Swisher, R. Poloni, S.N. Maximoff, B. Smit, L. Gagliardi, Ab initio carbon capture in open-site metal–organic frameworks, *Nat. Chem.* 2012 (4), 810-816.

[2] L.-C. Lin, J. Kim, X. Kong, E. Scott, T.M. McDonald, J.R. Long, J.A. Reimer, B. Smit, Understanding CO_2 Dynamics in Metal-Organic Frameworks with Open Metal Sites, *Angew. Chem. Int. Ed.* 2013 (52), 4410-4413.

(629e) Virial Coefficient Theory for Adsorption of Gas Mixtures

Yang, J. H., University at Buffalo, The State University of New York
Schultz, A. J., University at Buffalo, The State University of New York
Errington, J. R., University at Buffalo, The State University of New York
Kofke, D. A., University at Buffalo, The State University of New York
Originally presented on: 11/7/2013 9:34:00 - 9:50:00

Reliable prediction of the properties of mixtures in the presence of adsorbents is a key capability for the design of adsorption processes. Molecular-based approaches are needed to include the key physical considerations to perform predictive modeling, and several such approaches are available. These include classical density functional theory, integral equation theories, and molecular simulation. Each of these methods has advantages and disadvantages relating to the complexity of physical features that can included in the treatment, as well as computational cost and suitability for analysis. Among this field of considerations a special place is held by methods based on a virial treatment of the properties. Such methods can incorporate almost any level of detail in the molecular model, including molecular flexibility, quantum effects, multibody interactions, *etc.*, and (unlike molecular simulation) yield an analytic equation for the behavior. On the other hand, the range of application of the treatment is limited to low-density phases, which nevertheless are still of practical interest.

Virial-based methods are well established for bulk fluids, where the approach has classically provided capabilities somewhere between thermodynamic models and molecular simulations. This powerful, systematic method can be applied to inhomogeneous systems as well. Even though the treatment of the general imperfect gas can easily extended to the present problem, the study of gas-surface system is far less well developed. The theory was developed by the 1962 papers by Bellemans, who then formulated a Mayer-cluster theory of an inhomogeneous system and showed a unique way to separate the surface terms proportional to the surface area

In the present work, the virial coefficient theory is applied to the gas-surface system. We report a study of gas mixtures adsorbed on solid surface using inhomogeneous virial coefficients. The relevant surface variables for gas-surface system are directly expressed in terms of interactions between molecules themselves, and these molecules with the surface. The Mayer-sampling Monte Carlo (MSMC) method is applied to calculate the cluster integrals which contain surface contributions. We first report the results of calculation of "surface" virial coefficients of a mixture evaluated by using MSMC. Thermodynamic properties such as pressure, isosteric heat of adsorption are calculated by a virial expansion. Also we examine the selectivity at the limit of zero pressure which is directly related to the intermolecular forces. Comparison with molecular simulation demonstrates the range of applicability of the virial treatment.

(629f) An Improved Classical Density Functional Theory for the Description of Adsorption Thermodynamics

Hlushak, S., Vanderbilt University
McCabe, C., Vanderbilt University
Cummings, P. T., Oak Ridge National Laboratory
Originally presented on: 11/7/2013 9:50:00 - 10:06:00

An improved classical density functional theory [1] (CDFT) based on the accurate bulk thermodynamics predicted by the first-order mean-spherical approximation (FMSA) [2] is applied to study the thermodynamics of adsorption of simple gases into porous materials.
The CDTF theory is implemented in multiple dimensions, which allows studies of various confinement geometries, and makes use of Fast Fourier Transform to integrate the convolution integrals appearing during the calculation of the weighted densities. Contributions from the attractive interactions are taken into account using accurate structural properties [3,4,5] and energy-route thermodynamics obtained using the FMSA theory enhanced with the simplified exponential approximation [6,7,8]. The introduced improvements allow us to quickly and accurately calculate isotherms and heats of adsorption.

[1] S. Hlushak, C. McCabe, P. Cummings. J. Chem. Phys., 137, 104104 (2012).
[2] Y. Tang, and B. C.-Y. Lu, J. Chem. Phys., 99, 9828 (1993).
[3] Y. Tang, J Chem. Phys. 127, 164504 (2007).
[4] S. Hlushak, A. Trokhymchuk, S. Sokolowski, J. Chem. Phys. 130, 234511 (2009).
[5] S. Hlushak, A. Trokhymchuk, S. Sokolowski. J. Chem. Phys. 134, 114101 (2011).
[6] Y. Tang, B. C-Y. Lu. AIChE journal, 43, 2215-2226 (1997).
[7] S. Hlushak, A. Trokhymchuk, Cond. Matt. Phys., 15, 23003 (2012).
[8] S. Hlushak, P. Hlushak, A. Trokhymchuk, J. Chem. Phys., 138, 164107 (2013).

(629g) Computational Screening of Microporous Organic Solids: Development of Design Principles and Analysis of BET Theory

Abbott, L. J., Pennsylvania State University
Colina, C. M., Pennsylvania State University
Originally presented on: 11/7/2013 10:06:00 - 10:22:00

Organic molecules and dendrimers of intrinsic microporous (OMIMs and DIMs) are amorphous, organic microporous materials, which are of interest for a variety of applications, including gas separation and storage. They incorporate rigid and awkward structures to promote inefficient packing, similar to polymers of intrinsic microporosity (PIMs). Molecular simulations have been utilized to screen an array of >30 OMIMs and DIMs with a wide variety of structures in order to guide future synthetic efforts. Analysis of the simulations has revealed the role of three structural design aspects for increasing porosity in these types of materials. First, molecular rigidity has been observed to have a profound effect on pore formation, and simulations of ideal rigid-body OMIMs have resulted in materials with surface areas larger by an order of magnitude. Second, efficient packing and molecular interpenetration is hindered by the inclusion of bulky terminal units, such as *tert*-butyl and adamantyl functional groups. Third, molecular structures that are more three-dimensional, instead of planar, introduce greater "internal molecular free volume" and help promote more inefficient packing to increase porosity.

In addition to providing insight into important design principles, our computational screening has enabled a closer look into the applicability of BET theory for deriving apparent surface areas from adsorption data in these types of microporous materials. Specifically, we have compared geometric surface areas measured from the simulations with BET surface areas calculated from simulated nitrogen adsorption isotherms at 77 K, as is done experimentally. Our results have shown a systematic over-prediction by the BET surface areas, especially for the ultramicroporous materials. We utilize the array of intrinsically microporous materials we have simulated, which have surface areas covering the range of 100 to >1500 m^2 g^{-1}, to improve understanding of this relationship.

(629h) Computer-Assisted Screening of Ordered Crystalline Nanoporous Adsorbents for Separation of Alkane Isomers

Dubbeldam, D., University of Amsterdam
Krishna, R., University of Amsterdam
Calero, S., University Pablo de Olavide
Yazaydin, O., University of Surrey
Originally presented on: 11/7/2013 10:22:00 - 10:38:00

The separation of linear, mono-branched, and di-branched isomers of alkanes is of significant importance in the petrochemical industry, in particular for producing high-octane gasoline. Using computational techniques to screen a large variety of adsorbent materials, ZIF-77 emerges as the best adsorbent. It has the unique ability to fractionate isomer mixtures according to the degree of branching.

We have rationalized the adsorption behavior of hexane (and heptane) isomers and our screening study provides a thorough understanding. First, it is an illadvised endeavor to use large-pore MOFs for alkane separations. Above a certain threshold (depending on the alkane chain-length) the excess pore volume just goes to waste. Below the threshold for selectivity, both branching hierarchies are possible, but the one corresponding to the largest pore volume (the normal hierarchy) is probably most efficient, also because adsorption and diffusion selectivity go hand-in-hand. To properly evaluate a structure at least a 5-component mixture is needed, with both the 2,2-di-branched and the 2,3-di-branched included, as these have usually very different behaviors. The simulated breakthrough curves provide the best way of assessing the efficiency of a structure. The efficiency depends on a) the selectivity, and b) the accessible pore volume. CoBDP provides a very good compromise between the two, but the selectivity of ZIF-77 is so high that this sieve fractionates to yield individual pure components. The screening strategy we present[1] can be applied to other problems, such as, for example, separating alkane mixtures with C10–C18 hydrocarbons and xylene isomers.

(1) Dubbeldam, D., Krishna, R., Calero, S., Yazaydin, A.O., "Computer-Assisted Screening of Ordered Crystalline Nanoporous
Adsorbents for Separation of Alkane Isomers", *Angewandte Chemie International Edition,* 2012, 51, 11867-11871.

(629i) Branched Versus Linear Alkane Adsorption in Carbonaceous Slit Pores

Harrison, A.,
Krueger-Venus, J.,
Cracknell, R., Shell Global Solutions
Sarkisov, L., University of Edinburgh
Originally presented on: 11/7/2013 10:38:00 - 10:54:00

The presence of carbonaceous deposits on the internal surfaces of a spark ignition engine has been linked in the literature to impaired vehicle performance, as manifested by increased knocking, higher fuel consumption, higher emissions and other adverse effects. One of the proposed mechanisms is the adsorption and desorption of fuel components in the pores within the deposit. In this presentation we investigate this mechanism in more detail by considering single component adsorption of normal and branched alkanes in selected model slit pores representing the structure of the deposits. We further extend these studies to the binary mixture of normal heptane and isooctane, corresponding to a primary reference fuel blend. In particular, we show that in larger pores adsorption selectivity towards isooctane is about 1.2 on average throughout the pressure range. However, in the smaller 10Å pore selectivity towards isooctane can be in excess of three as a result of packing effects. These results are then placed in the context of engine performance issues.

(640a) The Influence of Plasticization and Dual-Mode Competetive Effects On CO2 Transport in Thermally Rearranged (TR) Polymers

Gleason, K. L., The University of Texas at Austin
Smith, Z. P., The University of Texas at Austin
Liu, Q., The University of Texas at Austin
Paul, D. R., The University of Texas at Austin
Freeman, B. D., The University of Texas at Austin
Originally presented on: 11/7/2013 8:30:00 - 8:55:00

In this work, results of mixed-gas permeation tests involving CO_2 permeation in TR polymers derived from 3,3'-dihydroxy-4,4'-diamino-biphenyl (HAB) and 2,2'-bis-(3,4-dicarboxyphenyl) Hexafluoropropane Dianhydride (6FDA) are presented. Polyimide HAB-6FDA films were converted to polybenzoxazole (PBO) to varying degrees by following different thermal treatment protocols. Permeation tests with these films were conducted using CO_2-containing mixtures at temperatures from -15 to 70°C, and pressures up to 50 atm. The bulk of the results are for 50% CO_2/CH_4 mixtures. Preliminary results for CO_2/N_2 mixtures are presented as well, and exhibit trends similar to the CO_2/CH_4 results. Mixed-gas results are compared against pure-gas measurements. CO_2/CH_4 mixed-gas selectivity at 35°C was 20-30% higher than the simple ratio of pure-gas permeabilities. Strong dual-mode competitive effects were responsible for a decrease in CH_4 and N_2 permeability in the presence of CO_2, with very little change to CO_2 permeability in the mixture. Plasticization is not evident in the pure-gas CO_2 permeation data. However, in the mixed-gas experiments, CH_4 permeability was found to increase gradually with increasing CO_2 pressure. The degree of plasticization was found to decrease as the degree of thermal conversion to PBO increased.

(640b) Facilitated Transport in Hydroxide-Exchange Membranes for Post-Combustion CO2 Separation

Xiong, L., University of Delaware
Gu, S., University of Delaware
Jensen, K., University of Delaware
Yan, Y., University of Delaware
Originally presented on: 11/7/2013 8:55:00 - 9:20:00

Facilitated transport, hydroxide-exchange membranes have been fabricated for the separation of CO_2 from post-combustion flue gas feed. First, a correlation between expected basicity of the fixed-site functional group and CO_2 separation performance is highlighted. Later, this observation is used to suggest that phosphonium may be a good candidate for achieving high separation performance. Separation tests show that quaternary phosphonium-based membranes achieve separation performance above Robeson's Upper Bound, an empirical bound describing the inverse relationship between permeability and selectivity. Specifically, permeabilities as high as 1090 and selectivities as high as 275 have been achieved. As expected, increases in CO_2 permeability are observed when increasing the degree of functionalization. In contrast, there is an small decrease in CO_2/N_2 selectivity when increasing the degree of functionalization, attributed to the increasingly large water uptakes at higher degree of functionalization. The high performance found in the membrane can be attributed to the combination of the quaternary phosphonium moiety and efficient transport channels.

(640c) CO2 Removal From High Pressure Natural Gas With Hybrid Fixed-Site-Carrier Membranes: Experimental, Process Design and Feasibility Analysis

He, X., Norwegian University of Science and Technology

Kim, T. J., SINTEF

Hägg, M. B., NTNU

Originally presented on: 11/7/2013 9:20:00 - 9:45:00

Natural gas (NG) is becoming one of the most attractive growing fuels for world primary energy consumption. However, the raw natural gas usually contains considerable amount of CO_2 which should be removed to meet the natural gas network grid specifications. Membrane systems possess of small footprint, low capital and operating costs, being environmentally friendly, and exhibiting process flexibility, which show a great potential for CO_2 removal from natural gas even though it has only 5% of the market today. Commercial gas separation membranes for natural gas sweetening are mostly made from cellulose acetate (CA), cellulose triacetate (CTA) and polyimide (PI). The challenges of relatively low separation performance (both CO_2 permeance and CO_2/CH_4 selectivity) and membrane plasticization for these commercial polymeric membranes direct to the development of novel, high performance composite membranes. Our previous work has already documented that the carbon nanotubes (CNTs) reinforced PVAm / polyvinyl alcohol (PVA) blend FSC membranes could improve the membrane performance, especially at high pressure operation [1]. Thus, the high performance hybrid FSC membranes were prepared using an optimized preparation condition in this work.

Carbon nanotubes reinforced PVAm/PVA blend FSC membranes were prepared to improve the mechanical strength and reduce the compaction influence at high pressure operation. A special casting procedure was employed to coat an CO_2-selective layer with a pre-determined thickness as reported by Kim et al. [2]. An optimized heat treatment procedure is conducted to enhance the membrane mechanical strength, separation performance and stability. The size of the prepared large flat-sheet hybrid FSC membrane is 30 cm ×30 cm.

Membrane separation performance was characterized by high pressure gas permeation testing with the synthetic mixed gases of 10% CO_2 - 90% CH_4 and 50% CO_2 - 50% CH_4. The high pressure pilot-scale module involves 3 sheets (the membrane area of each sheet is 110 cm^2) that can be operated in parallel or individually. A constant feed flow of 3000 Nml/min was humidified (RH 100%) and fed into the membrane module at 30 °C and different pressures without sweep gas, and the developed FSC membranes showed a high CO_2 permeance of 0.084 - 0.218 m^3 (STP) / (m^2.h.bar) with a CO_2/ CH_4 selectivity of 17.9 - 34.7 at different feed pressures from 40 to 10 bar. Moreover, the membrane capacity was also investigated by testing the whole module (area: 330 cm^2) at 30 °C and 30 bar using different feed flow rates. It was found that the CH_4 purity in retentate can achieve the requirement (i.e., >96%) using a single stage membrane unit at a low feed flow rate (high stage-cut), but the CH_4 losses are quite high which indicated that part of methane should be recovered using a 2nd?stage membrane unit- this could be used to guide the process simulation.

Based on the experimental data, process simulation was conducted by HYSYS integrated with an in-house membrane program (ChemBrane) to evaluate the process feasibility of membrane systems for CO_2 removal from high pressure natural gas. A two-stage membrane system was designed for process feasibility analysis. Process simulation was conducted on the basis of a natural gas feed flow of 5×10^5 Nm^3/h contains 10% CO_2 / 90% CH_4. The minimization of specific power consumption and natural gas sweetening cost (economic cost estimation using CAPCOST_2008 [3]) were applied to process optimization, and the results are shown in Table 1. It can be found that membrane process with the developed FSC membranes was feasible for CO_2 removal, even at a relatively low CO_2 feed concentration (~10 %) in natural gas, and it is possible to achieve > 96 % CH_4 purity in the retentate stream and a low CH_4 loss < 2%. A minimum cost of 5.91e-3 \$/ Nm^3 sweet natural gas produced was found at a feed pressure of 10bar in the 2nd stage to achieve the given separation requirements (i.e., CH_4 purity > 96%, CH_4 losses < 2% and captured CO_2 purity > 95%), which is lower compared to 6.40e-3 \$/ Nm^3 product with amine absorption [4]. Thus, the developed CNTs reinforced hybrid FSC membranes show great potentials for CO_2 removal from high pressure natural gas.

Table 1 Simulation results

Parameters	Simulation results
Feed pressure: 1st / 2nd stage , bar	40/10

Sweet NG productivity, Nm^3/h	4.67E+05
CH_4 purity in sweet NG, %	96.00
CH_4 losses, %	0.35
CO_2 purity, %	95.19
CO_2 recovery, %	62.50
Total membrane area, m^2	2.62E+05
Specific power consumption, kWh/Nm^3 sweet NG	4.11E-02
NG sweetening cost, $\$/Nm^3$ sweet NG	5.91E-03

Acknowledgements

The authors acknowledge the NaGaMa project (partners: Norwegian Research Council, Statoil and Petrobras) for financing this work. The high pressure pilot-scale module designed by PHILOS in South Korea is also acknowledged. The authors also thank SHOWA DENKO K. K. company in Japan to provide the carbon nanotubes for this study.

References

[1] He
X, Hägg MB. Hybrid Fixed-site-carrier Membranes for CO_2/CH_4
Separation. Euromembrane 2012. London, UK.

[2] Kim
T-J, Vrålstad H, Sandru M, Hägg M-B. Separation performance of PVAm composite membrane for CO2 capture at various pH levels. J Membr Sci. 2013;428(0):218-24.

[3] Turton
R, Bailie RC, Whiting WB, Shaeiwitz JA. Analysis, Synthesis, and Design of Chemical Processes, Third Edition. Prentice Hall, New Jersey. 2008.

[4] Peters
L, Hussain A, Follmann M, Melin T, Hägg MB. CO2 removal from natural gas by employing amine absorption and membrane technology?A technical and economical analysis. Chem Eng J. 2011;172(2?3):952-60.

ADDIN EN.REFLIST

(640d) Synthesis, Characterization, and Gas Transport Properties of Two Isomeric Thermally Rearranged (TR) Polymers for Gas Separation Membranes

Liu, Q., The University of Texas at Austin
Borjigin, H., Virginia Tech
Paul, D. R., The University of Texas at Austin
McGrath, J. E., Virginia Tech
Freeman, B. D., The University of Texas at Austin
Originally presented on: 11/7/2013 9:45:00 - 10:10:00

Thermally rearranged (TR) polymers, which are typically polybenzoxazoles with a rigid-rod structure, have recently gained interest in gas separation application due to their excellent gas transport and separation performance, particularly for CO_2/CH_4 mixtures. In fact, for these separations, TR polymers have permeabilities and selectivities that are above the Robeson upper bound. Another attractive feature is their apparent strong resistance to plasticization. In this study, the permeabilities of H_2, N_2, O_2, CH_4, and CO_2 were measured over pressure ranging from 3 atm to 17 atm at 35°C for aromatic polyimides based on 3,3'-diamino-4,4'-dihydroxy biphenyl and 2,2'-bis-(3,4-decarboxyphenyl) hexafluoropropane dianhydride (AHB-6FDA). Thermally rearranged (TR) polybenzoxazoles (PBO) were obtained by thermally treating the polyimides between 350 and 450°C. The the gas transport properties of AHB-6FDA are compared with those of its isomer based on 3,3-dihydroxy-4,4'diamino-biphenyl and 2,2'-bis-(3,4-decarboxyphenyl) hexafluoropropane dianhydride (HAB-6FDA). Both polyimides and the corresponding TR polymers showed similar gas transport properties.

The effects of polyimide precursor ortho-positioned functional groups on gas transport properties were also investigated. Acetate-containing AHB-6FDA displayed much higher gas permeability and comparable selectivity relative to hydroxyl-containing AHB-6FDA for both the polyimide and its corresponding TR polymers. Thus adding bulkier groups at the ortho-position improves gas transport performance in these polymers.

(640e) Process Modeling and Optimization of Post-Combustion Carbon Capture With N2-Selective Membranes

Narakornpijit, K., Stanford University
Yuan, M., Stanford University
Haghpanah, R., Stanford University
Wilcox, J., Stanford University
Originally presented on: 11/7/2013 10:10:00 - 10:35:00

Membrane separation has been proposed as a post-combustion carbon capture technology for its potential advantages such as a small footprint, environmental friendliness (i.e., no solvent involved in the process), and a lower energy requirement. Current modeling and optimization work has focused on post-combustion capture using polymeric, CO_2-selective membranes, where the CO_2 in flue gas is selectively permeated through the membrane and is concentrated in the permeate stream. One of the limitations of CO_2-selective membranes is that the low CO_2 concentration in flue gas (12–15 vol% from coal-fired power plants, 5–8 vol % from natural gas–fired power plants) provides a low driving force for separation, which requires that multistage separation and/or stream recycling be employed to achieve DOE's capture target of 90% capture (removal efficiency) and 95% product purity.

The concept of a N_2-selective membrane has been explored as a post-combustion carbon capture technology with the goal of further reducing the land and energy requirements of membrane-based capture processes. Two factors may contribute to the potential energy savings. First, the predominantly higher N_2 concentration in flue gas may provide a higher driving force for membrane separation, eliminating part of the feed compression requirements. Second, the concentrated CO_2 stream exits from the high-pressure retentate stream, which may displace part of the compression energy required for pipeline transport. The concept of N_2-selective membranes has been experimentally proven with pure vanadium (V) and its alloys with ruthenium (Ru) as membrane materials. In our modeling work, flue gas from a coal-fired power plant containing ~15 vol% CO_2 is treated with an ideal N_2-selective membrane (assuming minimal defects present in the material, and only N_2 passes across the membrane) in the countercurrent flow scheme. The membrane optimization literature mostly deals with single-objective optimization of costs or energy use. However, a capture plant should satisfy two criteria that have direct implications for operating and fixed costs, namely low parasitic energy consumption and a small footprint. In the present study, we performed multi-objective optimization using Genetic Algorithm (GA) to minimize energy consumption and membrane area simultaneously. We also conducted parametric analysis to investigate the potential of hybrid N_2- and CO_2-selecitve membrane configurations to further lower the energy consumption of the capture processes. The results of these studies will be presented.

(640f) Benchmarking Of Hydrogen Selective Membranes Through Experiments and Modelling

Pieterse, J. A. Z., Energy Research Centre of the Netherlands
Boon, J.,
Dijkstra, J. W., Energy research Centre of the Netherlands
Originally presented on: 11/7/2013 10:35:00 - 11:00:00

Palladium (-alloy) membranes are a promising option for hydrogen separation in industrial hydrogen production and in pre-combustion CO_2 capture [1,2]. Given the state of the art, benchmarking of membranes from different vendors is a crucial step in the demonstration of the maturity and performance under industrially relevant conditions. A multitude of membrane manufacturers and developers exists, yet at the moment, performance figures have been published at different conditions, and including the impact of externals, e.g. different membrane module designs. The present contribution provides an approach to assess the performance of H_2-selective membranes under industrially relevant conditions, independent from the impact of module design. The strategy presented here, consisting of experiments and model development, will be a valuable tool in benchmarking membrane performance. In order to evaluate membrane performance without artifacts induced by the experimental module, a membrane model has been devised. Systematic experimental tests were performed (extended for up to 1500 hours on stream) with membranes that were delivered by a number of different suppliers, allowing for model development and experimental validation. The staged experimental approach allows for investigating systematically the effects of intrinsic permeation, hydrodynamics, and concentration polarization. Membranes are tested at a scale of 300 to 800 cm^2 in a pilot setup [3] capable of accommodating up to 6 membranes of 100 cm length. Experiments have been performed that started with pure H_2 feed, without sweep, subsequently followed by introducing N_2 on the feed side, and N_2 sweep gas. Using a phenomenological description for the palladium layer and the dusty gas model for the membrane support, coupled to a 2D Navier–Stokes solver with a convection-diffusion equation to account for possible concentration polarisation, all relevant mass transfer resistances are adequately modeled [4]. For the conditions investigated, the main resistances to mass transfer are concentration polarisation in the retentate, hydrogen permeation through the metallic palladium layer, and a diffusional resistance in the support layer. Finally, the effect of inhibition by syngas components was measured for representative gas compositions and incorporated in the permeation equation. The 2D model was then used for evaluating and comparing the intrinsic membrane performance and for predicting membrane performance at commercial scale. In parallel, membrane and membrane-module robustness data extracted from the long term stability experimental testing program provided valuable insight in the technological status.

References

[1]. N.W. Ockwig and T.M. Nenoff, Chemical Reviews 107 (2007) 4078-4110.

[2]. G.Q. Lu, J.C Diniz da Costa, M. Duke, S. Giessler, R. Socolow, R.H. Williams, T. Kreutz, Journal of Colloid and Interface Science 314 (2007) 589.

[3]. D. Jansen, J.W. Dijkstra, Energy Procedia, 1 (2009) 253.

[4]. J. Boon, J.A.Z. Pieterse, J.W. Dijkstra, M. van Sint Annaland, Int. J. of Greenhouse Gas Control 11 (2012) 122.

(647a) The Novel Cal-C Process for Single-Stage Removal of CO2, SOx and NOx - Experimental Studies and Process Analysis and Integration

Deshpande, N., The Ohio State University
Phalak, N., The Ohio State University
Wang, W., The Ohio State University
Fan, L. S., The Ohio State University
Originally presented on: 11/7/2013 8:30:00 - 8:55:00

A multiple-gas cleanup process is proposed for the removal of CO_2, SO_x and NO_x from coal combustion flue gas. This cleanup occurs at high temperatures (500-650 Â°C) in a single-stage gas-solid fluidized-bed reactor. This process ? named Cal-C process ? utilizes limestone-derived calcium sorbent and inexpensive coal char, and consists of the three-step calcium looping developed at OSU with the addition of char to the carbonator. The calcium sorbent is re-circulated through the system in a closed loop after regeneration and reactivation, while fresh addition of char is warranted due to complete consumption.

The process has been verified experimentally using a lab-scale fluidized-bed reactor setup and the simultaneous removal of all the species of interest has been established. Based on previous studies, lignite coal chars are known to be suitable for this application. In this work, experiments are performed to quantify the consumption of char on the basis of selectivity of char toward NO. Different parameters such as calcium to char loading, O_2 concentration and NO concentration are studied for their effect on this selectivity. Further, the fate of gaseous species such as CO and H_2, which are likely to be produced in-situ, is investigated and possible process conditions are identified for minimizing unwanted byproducts. Different process configurations with respect to the entry of char in the process loop are also analyzed experimentally. Thermogravimetric analysis of the hydration (reactivation) reaction is also conducted to study the effect of presence of char on the hydration kinetics. Using ASPEN simulations, the process analysis is conducted and different process configurations are evaluated with respect to the opportunities for heat recovery, minimization of excess char requirement, etc.

(647b) Carbon Dioxide Absorption Into Aqueous Solution of Hydrophilic Amine-Functionalized Ionic Liquids

Lu, B., Zhejiang University
Wang, X., Zhejiang University
Jin, J.,
Li, W.,
Originally presented on: 11/7/2013 8:55:00 - 9:15:00

Carbon dioxide (CO_2), as one of the greenhouse gases (GHG), is currently responsible for over 60% of the enhanced greenhouse effect. Since global warming was aggravated, carbon capture and sequestration to reduce the release of $CO_2$to atmosphere is of critical importance. Chemical solvents (such as amines) have been extensive used, but still suffer some disadvantages, which restricting the promotion of the technology.

In recent years, Ionic liquids (ILs) are frequently used as solvents for gas absorption operations due to their immeasurably low vapor pressure, which can improve the process economy and general efficiency of gas separations. However, gas dissolution in ILs is a physical phenomenon with no chemical reaction and the absorption rate is quite slow. In order to introduce chemical reaction, a variety of ILs were integrated with ions incorporating functional groups. In the open literature, functional ILs were mostly dissolved in conventional ILs and owns high viscosity, which is unbeneficial to industrial applications.

Herein, we designed ether functionalized imidazolium based ILs with glycine anions for the absorption of CO_2. And a Hydroxy group was also introduced to this ILs in order to improve the solubility of the ILs in aqueous solution. The absorption of CO_2 by hydrophilic amine-functionalized ILs ([C_2OHmim]Gly) and Sodium glycinate (SG) were investigated respectively in a double stirred-cell absorber with a planar gas–liquid interface at 15 kPa CO_2 partial pressure. The absorption rate and capacity of [C_2OHmim]Gly into aqueous solution were higher than SG solution. Moreover, the IR spectra of [C_2OHmim]Gly was also proved that the functionalized ILs containing hydroxyl and glycine anions.

It has been widely confirmed that, vaporization and decomposition during regeneration is one of most non-negligible dissipation of aqueous amine, which increases the capital cost of the process. The regenerate efficiency of solvent is important for industrial application. Therefore, the regenerate experiments were investigated in an oil bath at 375.15 ± 0.5 K, and sustained approximately 90 minutes. The multi-cycle regenerate efficiency of [C_2OHmim]Gly and SG were compared with each other. The regenerate efficiency were all decreased as the regeneration cycle was increased but [C_2OHmim]Gly were decreased quite slow. After the fourth regeneration, the efficiency of [C_2OHmim]Gly just decreased 13%, SG almost decreased

25%. The result indirectly proved that the [C_2OHmim]Gly were well recycled and has a higher stability than SG by thermal regeneration even after multi-cycle experiments

As oxygen is one of the most typical components in the flue gas and may give rise to undesired side reactions, effect of O_2 was investigated and the O_2 concentration gradually increased in a range of 0-8 % (V/V). The absorption rate and capacity of CO_2 into [C_2OHmim]Gly and SG were all decreased in the presence of oxygen, but the decline trend of [C_2OHmim]Gly was relatively small. The capacity of [C_2OHmim]Gly only reduced 5.9 %, but the capacity of SG almost reduced 17 % in 4%O_2 compared with the capacity in absence of O_2. Oxygen is susceptible to induce oxidative degradation of alkanolamines. This undesirable breakdown not only reduces their CO_2 absorption capacity but also introduces unwanted degradation products, thus forcing the solution to be eventually discarded. Therefore, the multi-cycle regenerate experiments in present of oxygen (4%O_2) were also investigated. And the differences of regenerate efficiency of these two solutions were significant.

The above experiment results indicated that the hydrophilic amine-functionalized ILs ([C_2OHmim]Gly) we synthesized not only owns a high CO_2 absorption rate and capacity but also can enhance the antioxidant capacity of the solution, and can be well recycled after multi-cycle regeneration, possesses great potential for making up the deficiencies of alkanolamine, which is beneficial to industrial applications.

(647c) The Effect of Tethering Strategies On the Surface Structure of Amine-Functionalized Ionic Liquids: Inspiration On the CO2 Capture

Xing, H., Zhejiang University
Yan, Y., Zhejiang University
Yang, Q., Key Laboratory of Biomass Chemical Engineering of Ministry of Education, Department of Chemical and Biological Engineering
Bao, Z., Zhejiang University
Su, B., Zhejiang University
Yang, Y., Zhejiang University
Ren, Q., Zhejiang University
Originally presented on: 11/7/2013 9:15:00 - 9:35:00

Amine-functionalized ionic liquids (ILs) are one of the most attractive and promising absorbents for CO_2 capture, because of its high performance and notable advantages over traditional organic amine absorbents. For ILs absorbing CO_2, it is of primary importance to study the structure and orientation of cations and anions at the interface, which presents a great impact on the surface properties and the transportation of CO_2 across the liquid-gas interface. However, despite a variety of cation/anion-tethered amine-functionalized ILs being reported, the choice of the tethering location of the amine group is often somewhat arbitrary and lacks enough consideration in the view of interface chemistry.

Therefore, this work focused on this problem and illustrated the significance of the tethering strategy of the amine functional group on the interfacial structure of ILs for the first time, by investigating the interfacial composition and orientation of cation-/anion-tethered amine-functionalized ILs through molecular dynamic simulation. The results showed that when the amine group is tethered to the cation, strong interaction between the amine group and anions results in a compact interfacial structure without the cation/anion enrichment. However, for amino acid-based ILs with an anion-tethered strategy, the amine group tends to have a higher preference in the outer layer of interface. Therefore, from the viewpoint of interfacial adsorption and transfer, the amine-functionalized ILs with an anion-tethering strategy is considered of great benefit for the CO_2 capture, especially when the amine group is on the terminal side of a relatively long alkyl chain.

(647d) A Carbonate Absorption Process With Crystallization-Enabled High Pressure Stripping for Post-Combustion CO2 Capture: A Kinetic Study and Process Development for Bicarbonate Crystallization

Ye, Q., University of Illinois at Urbana-Champaign
Lu, Y., University of Illinois at Urbana-Champaign
Originally presented on: 11/7/2013 9:35:00 - 9:55:00

A novel carbonate absorption process with high pressure stripping is currently being developed for post-combustion CO_2 capture at the University of Illinois. The process employs an aqueous solution of carbonate, such as K_2CO_3, as a solvent. Bicarbonate in the CO_2-rich absorber effluent is crystallized by cooling the solution from about 70 °C to 30-35 °C in a crystallizer. The resulting bicarbonate slurry is used for CO_2 stripping at a pressure of at least 10 bars. A preliminary techno-economic analysis indicates that the process has an overall energy penalty 30 – 40% lower than that of the bench mark monoethanolamine process, primarily due to lower stripping heat and CO_2 compression work requirements.

The phase transition of bicarbonate during cooling of the CO_2-rich solution is a key feature of the process. A kinetic study of bicarbonate crystallization has been carried out using a laboratory-scale mixed-suspension, mixed-product-removal (MSMPR) reactor. Parametric tests with respect to agitation rate, mean residence time, temperature, and super-saturation level were performed to evaluate their effects on kinetic rates such as crystal nucleation and growth rates. The results were fit to Mydlarz and Jones (1993) model. Empirical regression formulae were then developed to correlate kinetic rates and crystal dimension with process operating conditions. Based on model predictions, mass and energy balance calculations were conducted to predict the performance of a new design configuration employing multiple continuous stirred tank reactors (CSTR) for crystallizing bicarbonate in the proposed CO_2 capture process. This presentation will provide a summary of the kinetic study of $KHCO_3$ crystallization and its implications for the related process design.

(647e) Carbon Capture and Utilization By Three-Step Aqueous Phase Mineralization

Hemmati, A., Sharif University of Technoligy
Bu, J., Institute of Chemical and Engineering Sciences (ICES - A*Star)
Shayegan, J.,
Sharratt, P., A*Star. Institute of Chemical and Engineering Sciences
Originally presented on: 11/7/2013 9:55:00 - 10:15:00

The increase in greenhouse gasses concentration in the atmosphere has contributed to global warming in recent decades. Carbon sequestration is one of the solutions to control the CO_2 concentration in atmosphere. Â A pH-swing mineral carbonation is the process of extracting magnesium or calcium from mineral rock, followed by precipitation of extracted minerals in separate reactors. The process was found to be feasible method for potential industrial application to reduce CO_2 emission by carbon sequestration. This study investigates production of value-added materials for a three-series mineralization reactor in aqueous phase. Magnesium, iron and small portions of some other minerals were extracted from magnesium silicate rock, were precipitated in two steps by increasing the pH from 0.5 up to 10. Product analysis showed that in each step, high purity stable solids such as silica, iron hydroxide II and III, hydrated magnesium carbonate were produced. The CO_2 source was Na_2CO_3 which was the product generated from the scrubbing reaction between a power plant's flue gas and NaOH to serve as a potential carbon capture method. Mass balance for the mineral sequestration process was investigated for future scale up. CO_2 uptake was determined by magnesium carbonate elemental analysis.Â It was calculated that for each ton of sequestrated CO_2 approximately 3.22 ton of serpentine rock (mainly $(Mg_3Si_2O_5(OH)_4)$ was required and also 25 m^3 of 1M NaOH, 35 m^3 of 1M Na_2CO_3 and 80 m^3 of 1M HCl were utilized. The resultant value-added products from the process would be 3.1 ton of 99.74% pure nesquehonite $(Mg_2CO_3{\cdot}3H_2O)$ and 0.82 ton of iron oxide and hydroxide with 90% purity. Crystalline structure, particle size distribution and morphology of hydrated magnesium carbonate produced, were strongly dependent on the last precipitation reactor conditions including temperature, feeding flow rate, mixing speed and base concentration. Additionally, different composition and crystalline structure of magnesium carbonate produced in the last precipitation step in different temperature ranging 0 to 60 Â°C, shows that the process is capable of producing value-added products including nesquehonite, hydromagnesite and dypingite under a wide range of ambient temperature. Â In future investigations, carbon footprint of the process chain will be studied.Â

(647f) Wood Ash Suspensions in Water As a Solvent in CO2 Absorption Processes

Petho, S. L., Escola PolitÃ©cnica da Universidade de SÃ£o Paulo
Salvagnini, W. M., Polytechnic School of the University of Sao Paulo
Taqueda, M. E. S., Polytechnic School of the University of Sao Paulo
Originally presented on: 11/7/2013 10:15:00 - 10:35:00

Wood ash is a solid waste with organic and inorganic components, obtained from combustion of wood. Its chemical composition may vary according the tree species, soil type, climate and combustion characteristics. The resulting suspension of ash in water has typical properties of high concentrations of carbonates and high pH values. Together with these qualities, the suspension is an interesting solvent in the study of a CO_2 absorption process with low demand of purity and energy.

There are studies about the methodology of determining the calcium carbonate equivalence in ashes because in its composition there other substances like alumino-silicates, silicates and iron oxides that can lead to inaccurate results. So the aim of this study is to develop a simple technique to determine the absorption capacity of unknown ash-aqueous suspensions with carbonates

The experiments were made with wood fuel ashes from a restaurant furnace and validation tests with water and NaOH. The apparatus have a simple characteristics of dismantle all its parts, including the gas tubes and the electrical wirings, forming a called absorption cell, which could be weighted before and after the absorption process.

A simple device was successfully developed that could predict the absorption potential for CO_2 of a solvent with unknown composition in ambient conditions of temperature and pressure. The soluble extracted from *Pterodon emarginatus* Vogel. wood, popularly known in Brazil as Faveiro, had 5.33% of calcium carbonate equivalence. The absorption process with an aqueous solution of this solubles, at aprox. 700 mmHg and 20°C, absorbed 0.77g CO_2 / 100g ashes. Ashes from Eucalyptus has 89.74% of calcium carbonate equivalence and its solubles, 11.01%. The absorption process with the ash-water suspension, at aprox. 700 mmHg and 20°C, absorbed 12.6g CO_2 / 100g ashes.

Analyzing the results of pH in function of time suggested that the absorption of CO_2 occurred because of the total transformation of the carbonates in bicarbonates. This study suggests that wood fuel ashes are a potential solvent for CO_2 absorption.

(647g) CO2 Chemical Transformation to Valuable Commodities: Toward a Utopia for Fossil-Fuel-Based Economies

Taheri Najafabadi, A.,

Originally presented on: 11/7/2013 10:35:00 - 11:00:00

Energy and environment have been always classified as the top human challenges by mid-century. In the contemporary fossil-fuel-based civilizations up to 90% of the energy needs are supplied by the unsustainable resources. Therefore, striking a delicate balance between the versatilities of the fossil fuels and their negative environmental impacts (e.g. global warming and climate change) is crucial. Implementing robust strategies to positively impact global carbon balance ultimately leads to the utopia for the fossil-fuel-based economies; with taking full advantage of carbonaceous fuels while minimizing their undesirable effects. This is where carbon capture and sequestration technologies play a critical role to mitigate CO_2 from the large scale emitters. Current recipes for the CO_2 reduction from large-scale energy consumers (e.g. power stations and cement works) mainly rely on carbon capture and storage (CCS), having three proposed generic solutions: post-combustion capture, pre-combustion capture, and oxy fuel combustion; where the separated carbon dioxide is sent for the physical storage options afterwards. However, researchers believe CCS strategy does not solely resolve the problem permanently, just shifting that from the atmosphere to elsewhere. In fact, long-term ecological impacts of using the earth as a gigantic reservoir for CO_2 are off the beaten track. Adding the potential hazards of CO_2 leakage to the earth surface noticeably exacerbates the prospect of this approach. Considering all the facts, the ultimate solution of realistically facing with CO_2 sequestration concerns is the chemical transformation of CO_2 to valuable commodities such as fuels (through, for instance, Fischer–Tropsch chemistry) or polymers (through successive copolymerization and chain growth). This sustainably reduces carbon emissions, remarkably benefiting from CO_2-derived chemical commodities, so-called carbon capture and conversion (CCC). Nevertheless, CO_2 reduction chemistry is challenging due to the presence of large energy barriers, requiring noticeable catalysis. This work aims to review the most recent advances in the CCC research area selectively (i.e. CO_2 conversion to fuels and CO_2 copolymerization) from chemical engineering viewpoint in terms of both materials and process design. Some of the most promising studies will be presented in detail with the emphasis on subsidizing more research on CO_2 conversion technologies considering the global carbon management urgencies.

(650a) Fundamental Investigation of Protein-Ligand Interactions in Multimodal Chromatography Using Nuclear Magnetic Resonance Spectroscopy and Isothermal Titration Calorimetry

Srinivasan, K., Rensselaer Polytechnic Institute
McCallum, S. A., Rensselaer Polytechnic Institute
Parimal, S., Rensselaer Polytechnic Institute
Barquera, B., Rensselaer Polytechnic Institute
Lopez, M.,
Cramer, S. M., Rensselaer Polytechnic Institute
Originally presented on: 11/7/2013 12:30:00 - 12:48:00

Recently, the use of multimodal (MM) chromatography, nuclear magnetic resonance (NMR) and molecular dynamics simulations (MD) in our group has revealed the presence of a preferred binding face on the surface of ubiquitin. Although trends in chromatographic retention data were consistent with that of NMR, there is a need to quantify effects of synergy and avidity involved in these interactions. A novel approach with systems of varying ligand density was employed to extract quantitative information about the co-operative effects of proteins binding to self-assembled monolayers (SAM) of MM ligands. Gold nanoparticles (AuNPs) were functionalized with a SAM of MMC ligands and titrations were performed on ^{15}N labeled ubiquitin. Chemical shift perturbations obtained from the ^{15}N-^{1}H HSQC spectra were used to determine the residue specific dissociation constants, which provided quantitative information about variation in strength of binding across the protein surface. In the present study two different MM cation exchange ligand systems, namely the 'Nuvia cPrime' like ligand and the 'Capto MMC' like ligand were employed to delve into the differences in protein binding at a molecular level. Although a similar binding face was observed on Ubiquitin for both these ligand systems, a quantitative comparison revealed a localized binding region on Ubiquitin for the 'Capto MMC' system and a more diffuse binding region for 'Nuvia cPrime'.

In order to investigate the thermodynamic processes governing these interactions, Isothermal Titration Calorimetry (ITC) was employed. Titrations were performed for these protein-nanoparticle systems to obtain quantitative information regarding the overall binding constant, the stoichiometry and thermodynamic parameters which define the driving forces of binding in these systems. Binding in both ligand systems was found to be primarily entropically driven. Titrations were also carried out at different temperatures to further investigate the relative enthalpic and entropic contributions. Finally, the overall binding constants obtained from ITC were compared to residue specific binding constants obtained from NMR in order to study effects of synergy involved in the binding of these systems.

To investigate effects of co-operativity and avidity involved in these systems, nanoparticles were synthesized with different ligand densities of these MM ligands. NMR and ITC were then employed to characterize protein binding to SAMs of varying ligand density. These molecular level investigations could potentially reveal the importance of ligand density on protein selectivity in multimodal systems which could have a significant impact on the development of more selective bioprocesses.

(650b) Adsorption and Diffusion of Bioalcohols in Zeolite SAPO-34 and MOF ZIF-8

Cousin Saint Remi, J.,
Denayer, J., Vrije Universiteit Brussel
Baron, G., Vrije Universiteit Brussel
Originally presented on: 11/7/2013 12:48:00 - 13:06:00

Bioalcohols, like biobutanol, form a promising alternative to petroleum-based chemicals and fuels. A major challenge in the economical production of biobutanol is its separation from the aqueous medium in which it is produced by ABE (Acetone-Butanol-Ethanol) fermentation. Given the low concentration of biobutanol in the fermentation broth (typically 20 g/l), the recovery via distillation would be energy- and cost-intensive. Among the alternative methods, adsorption was identified as the most energy-efficient recovery technique.

Adsorptive separation of butanol from aqueous ABE fermentation broth requires adsorbents with high stability, capacity and selectivity towards alcohols and low affinity for water. Additionally, smaller and larger alcohols or other components should be excluded!

From screening and experimental work two suitable materials emerged: SAPO-34 zeolite and ZIF-8 MOF. From the vapor/liquid phase adsorption equilibrium and breakthrough experimentation it could be concluded that, water is not adsorbed at all by ZIF-8, ethanol and acetone (major side products) are adsorbed to a much lower extent as compared to butanol (capacities up to 30 wt%), making ZIF-8 a highly attractive adsorbent for bio-butanol recovery.[1] Liquid phase adsorption and diffusion measurments showed that, while most zeolites with tubular pores have higher adsorption affinity for longer molecules, SAPO-34 (small cages connected via even smaller windows) exhibits an inverse chain length selectivity[2] and a nonuniform chain length dependent diffusion,[3] allowing removal of short chain alcohols from mixtures with longer ones and hence purification of butanol. A two step process combining ZIF-8 and SAPO-34 offers perspectives for the economical and low energy intensive production of butanol.[4]

[1] J. Cousin Saint Remi, T. Remy, D. De Vos, G.V. Baron, J.F.M. Denayer, ChemSusChem (2011) 4, 1074-1077.

[2] T. Remy, J. Cousin Saint Remi, R. Singh, P.A. Webley, G.V. Baron, J.F.M. Denayer, J. Phys Chem C (2011) 115, 8117-8125.

[3] J. Cousin Saint Remi, G.V. Baron, J.F.M. Denayer, J. Phys Chem C (2013), DOI: 10.1021/jp312287k

[4] J. Cousin Saint Remi, G.V. Baron, J.F.M. Denayer, Adsorption (2012), 18, 367–373.

(650c) Quaternized Chitosan Electrospun Membranes for Virus Removal

Mi, X., Michigan Technological University
Heldt, C. L., Michigan Technological University
Originally presented on: 11/7/2013 13:06:00 - 13:24:00

For remote or personal water systems, small pore size membranes can be added to reject bacteria and viruses from water purely based on size; but, the small pore size required to remove these pathogens often increases membrane fouling and leads to large membrane areas, high transmembrane pressures, low water flux, and frequent changing of membranes. This increases the cost of water filtration. In order to make clean water more accessible to everyone, there is an urgent need for improved water disinfection technology. Our goal is to create functionalized, large pore-sized membranes that have the ability to absorb virus and bacteria. Membranes with pores on the order of microns have the advantage of increased flow rate and reduced membrane fouling. As a first step in our goal to create water filtration membranes, we have created membranes made of biocompatible, inexpensive, and environmentally friendly chitosan that have been functionalized to contain a quaternary amine. This quaternary amine polymer is electrospun with polyvinyl alcohol to create high surface area adsorbants. These membranes have an increased surface area as compared to microfibers and are on the same size scale as viruses, creating a curved surface that has the potential to improve virus adsorption. The cationic quaternary amine polymer has a strong positive charge that facilitates virus adsorption. Surfaces have been created that remove >99.9% of a model, non-enveloped virus. The relationship between fiber diameter and pore size will allow us to characterize the binding of viruses to electrospun nanofibers. This study will improve our understanding of virus binding to nanofibers and allow us to engineer improved membranes for future virus removal systems.

(650d) Competitive Protein Adsorption On Silica

Darwish, A. M., University of Cincinnati
Robie, T.,
Thiel, S. W., University of Cincinnati
Originally presented on: 11/7/2013 13:24:00 - 13:42:00

Protein separations are critical in the pharmaceutical industry; unfortunately, the behavior of complex protein mixtures can be difficult to predict. To investigate this problem, the competitive adsorption of myoglobin (17.6 kDa, pI=7.2) and lysozyme (14.3 kDa, pI=11.35) on were used as test proteins for competitive adsorption on acid-washed 1000Å porous silica. Adsorption isotherms were obtained at room temperate using 0.02 M potassium phosphate buffer at pH 7; initial liquid-phase protein concentrations ranged from 0 – 570 μM for myoglobin and from 0 – 2800 μM for lysozyme. These proteins were chosen due to their similar molecular weights and high isoelectric points. In the absence of myoglobin, lysozyme follows a Type II isotherm, while in the absence of lysozyme, myoglobin follows a Type I isotherm. In competitive adsorption experiments, lysozyme decreases myoglobin adsorption as expected. In the presence of myoglobin, lysozyme follows a Type I isotherm; at low concentrations, myoglobin decreases lysozyme adsorption. As additional myoglobin is added, lysozyme adsorption increases. The observed influence of myoglobin on lysozyme adsorption is not consistent with simple competition for surface area or functional groups. A thermodynamically-based mathematical model is used to explain the observed adsorption behavior.

(650e) Development of Fiber Enhanced Dual-Modal Cellulose Monolith for Influenza Virus Separation

Pan, X., Texas A&M University
Jinka, S., 2. Texas Tech University, Institute Environmental & Human Health
Ramkumar, S. S., The Institute of Environmental and Human Health, Texas Tech University
Karim, M. N., Texas Tech University
Singh, V.,
Originally presented on: 11/7/2013 13:42:00 - 14:00:00

Development of fiber enhanced dual-modal cellulose monolith for influenza virus separation

Xinghua Pan[1], Sudheer Jinka[2], **Seshadri Ramkumar[2],** M. Nazmul.Karim[1]*, Vinitkumar Singh[2]

1. Artie McFerrin Department of Chemical Engineering, Texas A&M University
2. Institute Environmental & Human Health, Texas Tech University

*Corresponding author, Tel: (979) 845-9806, E-mail: naz.karim@che.tamu.edu

Influenza virus imposes a serve threat to humans and animals, and its frequent mutation challenges the vaccine immunoprophylaxis, which still remains the most efficient method for flu epidemic prevention [1]. Recent widely applied, virus-producing hosts such as mammalian cells and *E. coli*have been reported for faster vaccine production compared to conventional chicken egg hosts [2]. The purification of virus or virus-like particles (VLP) from host cells is still a time consuming procedure and has been undergoing extensive study [2-4].

A monolith is a single-piece, interconnected, and porous stationary phase material which is widely applicable for separation technologies such as gas and liquid chromatography, high performance liquid chromatography, and capillary electrochromatography. The separation can target small molecules and large complex biomolecules. The advantages of monoliths compared to traditional porous beads are their mechanical robustness, ease of in-situ preparation at low cost, no void volumes, pore size controllability, and easy of functionalization. The advantages of the monolith also include easy shaping into various modes such as capillaries, columns, micropipette tips, and microfluidic channels. Besides, a highly interconnected porous structure allows high hydraulic permeability and convective flow dynamics in the monolith, which allows for higher flux rates.

Cellulose is among the most abundant bioproduct in plants. Its excellent biodegradability, sustainability, and biocompatibility make it as one of the ideal candidates for medical and chemical applications. The organic solution used for processing cellulose for membranes could be recycled, making the whole process an environmentally friendly and petroleum-independent green process. Numerous hydroxyl groups on cellulose are easy to modify for different separation purposes. Cellulose-based materials have been commonly applied in separation technologies such as centrifugation and filtration membranes, ion-exchange chromatography, and affinity chromatography.

The preparation of cellulose monolith regarding its preparation method, mechanical properties, and pore size control has not been exclusively reported yet. To advance the virus separation technology faster and more efficiently, we propose a novel method to prepare dual-modal cellulose fiber enhanced monoliths for the separation of the influenza virus directly from mammalian cell culture media.

The interconnected structure of the monolith was contributed both by the porogen generating pores and fiber interactions. To address the variance in expression of surface glycoproteins among different strains, as well as maintaining the specific separation, a dual-mode chromatography utilizing pseudo-affinity and anion-exchange immobilization was prepared for use with influenza virus.

The cellulose monolith structure was characterized by Scanning Electron Microscopy. The two different modifications of anion-exchange and pseudo-affinity were quantified by element analysis. An H1N1 influenza virus strain (A/WSN/33) was replicated and harvested in MDCK cells and used directly for separation efficiency evaluation, after freeze-thaw operation. The results suggested the novel cellulose monolith could be a promising candidate for influenza virus separation.

Reference:

[1] Finelli L, Fiore A, Dhara R, Brammer L, Shay DK, Kamimoto L, et al. Influenza-associated pediatric mortality in the United States: increase of Staphylococcus aureus coinfection. Pediatrics. 2008;122:805-11.

[2] Wolff MW, Reichl U. Downstream Processing: From Egg to Cell Culture-derived Influenza Virus Particles. Chemical engineering & technology. 2008;31:846-57.

[3] Vicente T, Roldao A, Peixoto C, Carrondo MJT, Alves PM. Large-scale production and purification of VLP-based vaccines. Journal of invertebrate pathology. 2011;107:S42-S8.

[4] Opitz L, Hohlweg J, Reichl U, Wolff MW. Purification of cell culture-derived influenza virus A/Puerto Rico/8/34 by membrane-based immobilized metal affinity chromatography. Journal of virological methods. 2009;161:312-6.

(650f) Monolith Technology In Isolation Of Human Immunoglobulin G Subclasses

Leblebici, P., Department of Chemical Engineering, Laboratory of Separation and Reaction Engineering, Associate Laboratory LSRE/LCM,University of Porto
Pais, L. S., Laboratory of Separation and Reaction Engineering, Associate Laboratory LSRE/LCM
Rodrigues, A. E., University of Porto, Faculty of Engineering
Originally presented on: 11/7/2013 14:00:00 - 14:18:00

Human immunoglobulin G (hIgG) constitutes an important therapeutic antibody for a number of diseases. There is an increasing demand for highly purified hIgG and its subclasses due to the wide application in *in vitro* diagnostics and immunotherapy. [(1)] hIgG consists of four subclasses (IgG1, IgG2, IgG3, and IgG4) that show small differences in some of their physicochemical characteristics and biological properties. Quantitatively, the relative serum concentrations of the human IgG subclasses are as follows: IgG1 > IgG2 > IgG3 ≈ IgG4. The four subclasses show more than 95% homology in the amino acid sequences of the constant domains of the γ -heavy chains. [(2)] Protein A affinity chromatography is a common method for antibody purification due to its well-characterized structure and high affinity to IgG. [(3)] To our knowledge, until now Protein A **monolith** column has not been used for the isolation of individual hIgG subclasses. Monolithic columns have attracted significant attention for the purification of large biomolecules due to its flow independent binding capacity that enables rapid separations at high flow rates. [(4)]

In the present study, a rapid separation method was developed to obtain pure hIgG subclass fractions with monolith technology. pH elution was applied using Protein A monolith column in a stepwise method and conditions were optimized. Since IgG3 has very low affinity to Protein A, elution was occurred immediately upon injection of the sample. IgG1 and IgG4 were eluted simultaneously under the applied conditions. Furthermore, IgG2 was separated with significant purity. Results indicated that the separation between IgG2 and IgG1- 4 could be achieved within 20 min run using a Protein A monolith column. The separation process time could be reduced more by increasing the flow rate without any loss in selectivity. Furthermore optimization of the mobile phase composition is under investigation for the efficient separation of IgG1 and IgG4. According to the present results, monolithic Protein A column could be proposed as a time efficient separation media for the enrichment of IgG2 which is the most often directed antibody against bacterial pathogens. [(5)]

1. K. Huse, H.-J Böhme, and G. H. Scholz, *J. Biochem. Biophys. Methods*, 51, 217-231, (2002).
2. R. G. Hamilton, *Clin. Chem.*, 33/10, 1707-1725, (1987).
3. S. Hober, K. Nord, and M. Linhult, *J Chromatogr B*, 848, 40-47, (2007).
4. A. Podgornik, M. Barut, A. Strancar, D. Josić, and T. Koloini, *Anal. Chem.* 72, 5693- 5699, (2000).
5. L. Hammarstrom and C.I.E. Smith, *Monogr. Allergy*, 19, 122-133, (1986).

(650g) Design and Characterization of Affinity Peptides Based On the Interactions With Human Immunoglobulin

Wei, Y.,
Li, C., Institute of Process Engineering, Chinese Academy of Sciences
Zhang, L., Institute of Process Engineering, Chinese Academy of Sciences
Xu, J.,
Xu, X., Institute of Process Engineering, Chinese Academy of Sciences
Originally presented on: 11/7/2013 14:18:00 - 14:36:00

The increase in huge demand of antibodies, the most promising biopharmaceuticals for the treatment of a number of diseases in recent years, has resulted in a need for the more efficient, less costly, more specific downstream production process to separate and purify antibodies from serum and mammalian cell culture than the current processes accounting for 50-80% of the overall production costs. In this study, a series of polypeptides were designed based on the interactions between the peptides and IgG or HSA using the molecular simulations. The binding energy between the lead peptide ligands and Fc fragment of antibody was evaluated using the AutoDock. The adsorption behaviour between Fc fragments and lead peptide ligands in the aqueous environment was further investigated using the Gromacs. The associated constant and dissociated constant of ligand to antibody were measured using the SPR technology. The dynamic binding capacity, purity and recovery were measured when the agarose was used as media. The SPR results indicated that the interactions between the lead affinity peptides and IgG or HSA varied with pH, ionic strength and buffer. The selectivity of the lead peptides to IgG was optimized at NaCl concentration of 150 mM at pH 7.0 when HEPES was used as the buffer. These results demonstrated the potential for rational design and synthesis of affinity peptides which can be used in the downstream production processes to separate and purify antibodies with high efficient and low cost.

(650h) Protein Adsorption Equilibria and Uptake Kinetics to Poly(ethylenimine)-Based Ion Exchangers

Yu, L., Tianjin University
Dong, X., Tianjin University
Sun, Y., Tianjin University
Originally presented on: 11/7/2013 14:36:00 - 14:54:00

Sepharose FF was modified with poly(ethylenimine) (PEI) to prepare anion exchangers of 10 different ionic capacities (ICs, 100–1220 mmol/L). Adsorption equilibria and kinetics of bovine serum albumin (BSA) were systematically studied. We found a critical IC (cIC, 600 mmol/L), above which both protein capacity and uptake rate increased drastically. Then, five PEI-Sepharose FF resins of typical ICs were selected to explore the effect of ionic strength (IS) on the adsorption equilibria and uptake kinetics of BSA. Commercially available DEAE (IC=160 mmol/L) and Q Sepharose FF (IC=269 mmo/L) resins were used for comparisons. It is found that at similar ionic capacities, protein adsorption capacities on both the PEI-Sepharose FF resins and the commercial resins decreased with increasing IS, but on the capacity sensitivity to salt concentration, the former was lower than the latter. In addition, the effective diffusivities (D_e) of the former were smaller than the latter in the entire IS range studied. The low IS sensitivity of adsorption capacity of the PEI-Sepharose FF resins could be interpreted by the increase of pore accessibility with increasing IS; the smaller D_e values in the PEI-Sepharose FF resins were considered due to the lack of surface diffusion in the PEI-Sepharose FF resins of low PEI densities. For the PEI-Sepharose FF resins of high ICs (520, 740 and 1220 mmol/L), both protein capacity and $D_e$values increased first and then decreased with increasing IS. The increasing trend of protein capacity in the low IS range was considered due to the increase of accessible pores for BSA. The rise-fall trend of D_e was attributed to the dependencies of the "chain delivery" effect on protein capacity and binding strength, both of which are related to IS. Moreover, the IS sensitivity of the D_e for the resins of ICs>cIC (740 and 1220 mmol/L) was much higher than those of ICs<cIC, further proving that the "chain delivery" effect in PEI layer did contribute significantly to the overall mass transfer at IC>cIC. Furthermore, the two PEI-Sepharose FF resins of ICs>cIC kept high adsorption capacities and D_e values up to 200-300 mmol/L NaCl. Therefore, the operating IS ranges for these two PEI-Sepharose FF resins can be much broader than traditional ion-exchange media.

(657a) Porous Membranes With Tailored Protein Selectivity By Grafting Or Pore-Filling With Multifunctional Polymers

Ulbricht, M., Universitat Duisburg-Essen
Adrus, N., Universitat Duisburg-Essen
Schwark, S.,
Yang, Q.,
Yin, D.,
Originally presented on: 11/7/2013 12:30:00 - 12:55:00

Porous membranes with tailored protein selectivity by grafting or pore-filling with multifunctional polymers

Mathias Ulbricht, Nadia Adrus, Sebastian Schwark, Qian Yang, Dongxu Yin

Lehrstuhl für Technische Chemie II, Universität Duisburg-Essen, 45117 Essen, Germany, mathias.ulbricht@uni-essen.de

There is strong interest in using the highly selective recognition phenomena known from biological systems for technical applications. Synthetic membranes are an attractive platform because they can be used in various formats and configurations, as separator or contactor, to facilitate separation processes [1]. Here we focus on porous membranes with tailored biomimetic polymeric hydrogel layers with their primary use in protein separation [2]. However such membranes could also be interesting as combined recognition/transducer element in sensor systems [2].

Separations using macroporous membrane adsorbers combine fast convective transport through pores with selective and reversible binding of the target substances on functionalized pore walls. This is one example for the use of membranes as contactor. There is large interest in improving the performance for established materials and developing novel materials. We will discuss recent own work toward higher affinity and selectivity for specific proteins based on grafting of functional polymeric layers to various porous polymeric membranes. A high binding capacity is obtained by well-defined three-dimensional layer architectures; using controlled surface-initiated polymerization methods is crucial to achieve that goal. Examples include synthetic glycopolymers [3], zwitterionic polymers [4] and polymers presenting mixtures of different binding groups for epitops on protein surfaces [5]. In all these cases, cooperative multivalent binding is a key to achieve high selectivity. As an alternative we present the preparation of protein-imprinted polymeric layers obtained by a novel two-step grafting method where protein is used as template to create specific binding sites in a cross-linked functional hydrogel layer [6]. The transfer of preparations from model isoporous track-etched membranes to commercial polymeric base membranes with higher capacity, and the resulting protein binding selectivities and capacities under dynamic conditions will also be discussed.

Finally, the combination of reactive pore-filling with functional hydrogel via *in situ* cross-linking polymerization toward a tunable and switchable ultrafiltration membrane [7] with the preparation of protein-imprinted hydrogels [8] will be explored with the ultimate aim to obtain membranes where the barrier permeability is regulated by the presence of a specific protein.

[1] E. Drioli, L. Giorno (Ed.), *Membrane operations: Innovative separations and transformations*, Wiley–VCH, Weinheim, **2009**.

[2] Q. Yang, N. Adrus, F. Tomicki, M. Ulbricht, *J. Mater. Chem.* **2011**, *21*, 2783.

[3] Q. Yang, M. Ulbricht, *Macromolecules* **2011**, *44*, 1303-1310.

[4] Q. Yang, M. Ulbricht, *Chem. Mater.* **2012**, *24*, 2943-2951.

[5] S. Schwark, M. Ulbricht, *Eur. Polym. J.* **2012**, *48*, 1914-1922.

[6] D. Yin, M. Ulbricht, *J. Mater. Chem.* B **2013**, *accepted*.

[7] N. Adrus, M. Ulbricht, *J. Mater. Chem.* **2012**, *22*, 3088-3098.

[8] N. Adrus, M. Ulbricht, *Polymer* **2012**, *53*, 4359-4366.

(657b) Calculation of Phase Diagrams of Polyetherimide/N, N-Dimethylacetamide/Nonsolvent Casting Systems and Prediction of Performances of Ultrafiltration Membranes

He, G., Dalian University of Technology

Originally presented on: 11/7/2013 12:55:00 - 13:20:00

Calculation of phase diagrams of polyetherimide/N, N-dimethylacetamide/nonsolvent casting systems and prediction of performances of ultrafiltration membranes

Ming Tan, Gaohong He[*], Yan Dai, Rujie Wang, Junli Xiao

State Key Laboratory of Fine Chemicals, R&D Center of Membrane Science and Technology, School of Chemical Engineering, Dalian University of Technology, Dalian, 116024

Abstract: Polyetherimide (PEI)/N, N-dimethylacetamide (DMAc) ternary phase diagrams with nonsolvent (NS) of H_2O, polyethylene glycol (PEG400) and 1-butanol (BuOH) were calculated by using Hildebrand and Hansen solubility parameter and Flory-Huggins theory. The phase diagrams indicate that it should use Hansen solubility parameter and correct constant Â¦Ã □ of 1 to calculate interaction parameters of NS/DMAc/PEI casting system, and then the interaction parameter between BuOH solution and PEI was computed. Besides, PEI/DMAc/H_2O-BuOH quaternary phase diagram was simulated, which is consistent with the experimental cloud points. The phase diagrams show that when volume ratio of BuOH to H_2O is 9.327, 5.5% BuOH solution consisted of 5% BuOH and 0.5% H_2O leads to phase separation of 15-20wt.% PEI casting solution. Simultaneously, when volume ratio of BuOH to H_2O is 22.86, 6.5% BuOH solution including 6.2% BuOH and 0.3% H_2O brings on phase separation of the casting solution. Therefore, additive BuOH can enhance the phase separation of the PEI casting solution and hereby induce higher porosity of the membrane. Moreover, molecular size of BuOH is bigger than DMAc. Hence, compared with DMAc, diffusion of BuOH into the water coagulation bath causes larger pore size. However, when additive BuOH increases to a certain value, the greatly increased viscosity of the casting solution hinders the outlet of the solvent DMAc, additive and inlet of the nonsolvent water and thereby causes a thicker skin layer. As a result, it can be predicted that membrane PWF first increases then decreases, and RR decreases with the increases of BuOH concentration. The prepared membranes validated our predictions.

Key words: phase diagram; solubility parameter; interaction parameter; ultrafiltration; Flory-Huggins

[*] Corresponding author Tel.: +86-411-84707892; Fax: +86-411-84707700; E-mail: hgaohong@dlut.edu.cn

(657c) Hierarchically Structured Membranes With Oriented Pores for Bio-Inspired Separations

Meoto, S. L., Rensselaer Polytechnic Institute
Coppens, M. O., Rensselaer Polytechnic Institute
Originally presented on: 11/7/2013 13:20:00 - 13:45:00

Hierarchically Structured Membranes with Oriented Pores for Bio-Inspired Separations

Authors: Silo Meoto (Presenter), Marc-Olivier Coppens

Protein channels in cell walls are remarkably efficient for separations, showing high permeation of, and selectivity for a desired species. Examples include AQP1 (water channel) and KcsA (potassium channel). Knowledge of the physico-chemical principles that are responsible for the superior performance of such biological nanopores could be applied to the design of artificial membranes with similar properties. We discuss the development of an inorganic membrane with ordered arrays of oriented nanochannels, whose diameter and surface chemistry can be tailored to enhance and control separation selectivity.

A hierarchically structured membrane has been synthesized, with nanochannels of silica grown inside a macroporous support. Using a soft-templating method, the silica grows around cylindrical micelles, formed from a surfactant, which act as structure-directing agents. Silica with cylindrical nanopores is obtained upon removal of the surfactant. A systematic design was used to ensure that the pore channels of the macroporous support are filled tightly with silica nanostructures, guaranteeing uniformity over macroscopic distances across the support. Based on the mechanism underlying the performance of biological nanopores, desired functional groups have been tethered to the inner surface of the confined silica nanochannels to create similar steric and electrostatic interactions. Membrane permeability and selectivity were investigated as a function of pore size and charge effect, created by different chemical functional groups.

The fabricated membrane was characterized by small-angle X-ray scattering (SAXS), scanning electron microscopy (SEM), transmission electron microscopy (TEM) and nitrogen adsorption/desorption measurements after the removal of the surfactant. Membrane performance was tested by representative bio-separations.

(657e) Bioinspired Fabrication of Polydopamine-Based Membranes for Gas Separation

Wu, H., Tianjin University
Li, Y., Tianjin University
Jiang, Z., Tianjin University
Originally presented on: 11/7/2013 14:10:00 - 14:35:00

Abstract: Facile fabrication of composite membranes with ultrathin, defect-free and stable active layer is one of the key issues in scaling up numerous energy-related dense membrane processes such as gas separation, gasoline desulfurization and alcohol dehydration, etc. Inspired by marine mussel's bioadhesion phenomenon and principle, a facile and versatile approach for fabricating composite membranes with ultrathin and stable active layer by coating dopamine on microporous substrates followed by polymerization and post treatment was presented. The thickness and the free volume of the polydopamine layer could be tailored by varying dipping conditions including the dopamine concentration, dipping time and pH value of the solution. Then, polymer–inorganic hybrid membranes were prepared by using polymeric materials as the membrane bulk matrix and metal ions-reinforced polydopamine nanoaggregates as the filler. The structure of the hybrid membrane could be tuned by varying the composition of the organometallic nanoaggregates, resulting in the well-controlled size-dependent transport of gas molecules. Alternatively, an additional layer of polyethyleneimine was coated on the polydopamine membrane. The resultant polyethyleneimine-coated polydopamine membrane exhibited simultaneously high CO_2 permeance, high CO_2/CH_4 selectivity and high stability. We hope these findings open an effective approach to large-scale fabrication of high-performance composite membranes for gas and other energy-related separations.

(662a) Application of Dynamic Mean Field Theory to Linear Pores With Inkbottle and Side Stream Pore Structures

Schneider, D., University of Leipzig
Valiullin, R., University of Leipzig
Monson, P. A., University of Massachusetts
Originally presented on: 11/7/2013 12:30:00 - 12:50:00

This work is motivated by recent results for adsorption of nitrogen in porous silicon (PSi) that show surprising behavior featuring, for instance, desorption from mesopores at unexpectedly high gas pressure. Much of the behavior is apparently inconsistent with the known adsorption thermodynamics for fluids confined in pores which have the geometry that has been assumed for PSi. In order to investigate a possible role for dynamics in the observed behavior we have carried out a study of several kinds of linear pore geometries using dynamic mean field theory. The geometries include inkbottle pores and sidestream pore structures as well as pores closed at one end. Several new dynamic mechanisms of capillary condensation and evaporation emerge in this work, including a surprising nucleation phenomenon for pores closed at one end.

(662b) Experimental and Theoretical Comparison of Gas Desorption Energies On Metallic and Semiconducting Single Walled Carbon Nanotubes

Chen, D. L., University of Pittsburgh
Mandeltort, L., University of Virginia
Al-Saidi, W. A., University of Pittsburgh
Cole, M. W., The Pennsylvania State University
Yates, J. T. Jr., University of Virginia
Johnson, J. K., University of Pittsburgh
Originally presented on: 11/7/2013 12:50:00 - 13:10:00

Single walled carbon nanotubes (SWNTs) have been widely studied as adsorbents from both simulations and experiments due to their high surface areas and precisely defined pores. It is widely known that SWNTs can be either metallic or semiconducting based on differences in the chiral indices, (n,m), which determine how the carbon hexagons are arranged on the surface of the SWNT. All simulations of adsorption on SWNTs using classical potentials have implicitly assumed that the adsorption potential between a molecule and a SWNT is independent of whether the SWNTs are metallic or semiconducting. However, it is reasonable to assume that metallic SWNTs, which have extremely high longitudinal polarizabilities, would exhibit stronger van der Waals forces with adsorbate molecules than semiconducting SWNTs. We use precise temperature programmed desorption experiments and van der Waals-corrected density functional theory to probe the differences in binding energies for gas molecules adsorbed on the surfaces of metallic and semiconducting SWNTs. We have measured and calculated the binding energies for Xe, a spherical adsorbate, and the highly anisotropic n-heptane molecule on metallic and semiconducting-SWNTs having about the same diameter. Binding energies computed from van der Waals corrected density functional theory are in good agreement with experiments.

(662c) Cage Effects in the Adsorption of Hydrocarbons On the Metal-Organic Framework Uio-66

Duerinck, T., Vrije Universiteit Brussel

Denayer, J., Vrije Universiteit Brussel

Baron, G., Vrije Universiteit Brussel

Originally presented on: 11/7/2013 13:10:00 - 13:30:00

Cage effects in the adsorption of hydrocarbons on the Metal-Organic Framework UiO-66

T. Duerinck[1,2], J.F.M. Denayer[1]

[1] *Vrije Universiteit Brussel, Department of Chemical Engineering, Pleinlaan 2, 1050 Brussel, Belgium.*
[2] *Georgia Institute of Technology, School of Chemical & Biomolecular Engineering, Atlanta, GA 30332 USA.*

The metal-organic framework UiO-66 (Zr- terephthalate) has a complex structure of octahedral (~11Å) and pyramidal pores (~8Å). UiO-66 shows excellent chemical and thermal stability.[1] Analogue materials using modified BDC-linkers are fairly straightforward synthesizable (e.g. Me, NO_2, NH_2, halides...).[2] Several reports in the adsorption properties of this framework for permanent gasses or water exist but only a few on hydrocarbons exist. The separation potential for hexane and xylene isomers was studied for several UiO-66 variants.[3-5]

In this work, we tried to get a deeper insight in the adsorption mechanisms at pore level by studying a wide range of molecules: n-, single and double branched alkanes, cycloalkanes and ?alkenes, aromatics and hetero atom functionalized derivatives. A pulse gas chromatographic study, using columns packed with pellets of UiO-66, UiO-66-Me, UiO-66-NO_2 and UiO-66-Me_2 was performed in the range of 200-300°C. Using the van ?t Hoff equation, adsorption properties (enthalpy and entropy values, pre-exponential factors, Henry constants) were determined.

Remarkably, the adsorption enthalpy of n-alkanes does not linearly increase with carbon number. Non-linearity effects are observed around C_5-C_7 chain lengths. Significant entropy effects define the overall favourable Gibbs free energy. The UiO-66 framework shows significant adsorption preference for cyclic and branched alkanes with maxima around cyclo-octane for cycloalkanes and 3,3-dimethylpentane compared to the linear counterparts. Remarkably the adsorption enthalpy and entropy factors show a drop in absolute value at C_6-C_7 on UiO-66-Me and C_7-C_8 UiO-NO_2.

Further analysis by investigating the relative confinement factors[6] between C_n and C_{n+1} isomers shows changes that can be related to change in relative closeness to the pore wall. The functionalized UiO-66 materials have a relatively higher confinement, indicating that the presence of an additional group leads to a reduction in effective pore size. This change in effective pore size was further investigated by comparing cage sizes and average molecular dimensions of the pore size. In addition, Monte Carlo simulations were performed to investigate the molecular position of the adsorbates in the structure. It was shown that most of the selectivity effects arise from the interplay between the molecular dimension of the adsorbate and the adsorbent's cage size. Adsorbates preferentially adsorb in the smallest, tetrahedral cage whenever possible. Larger alkanes (typically C_8 or bigger) do not optimally fit in the tetrahedral cage and preferentially sit in the octahedral cage. Not only the largest dimension but also overall shape of the adsorbate plays an important role. More spherical molecules interact stronger with the cage walls than linear molecules due to the relatively small cage dimension of smallest cage. In some cases, entropy effects play a major role in the separation of stereoisomers. This is especially true for cyclic adsorbates (e.g. dimethylcyclohexanes).

Acknowledgements
T. Duerinck and J.F.M. are grateful to FWO Flanders for financial support.

References

1 Valenzano, L. *et al.*
Disclosing the Complex Structure of UiO-66 Metal Organic Framework: A Synergic Combination of Experiment and Theory. *Chemistry of Materials* **23**, 1700-1718 (2011).

2 Kandiah, M. *et al.*
Synthesis and Stability of Tagged UiO-66 Zr-MOFs. *Chemistry of Materials* **22**, 6632-6640 (2010).

3 Barcia, P. S. *et al.*
Reverse shape selectivity in the adsorption of hexane and xylene isomers in MOF
UiO-66. *Microporous and Mesoporous*
Materials **139**, 67-73 (2011).

4 Moreira, M. A. *et al.*
Reverse Shape Sellectivity in the Liquid-Phase Adsorption of Xylene Isomers in
Zirconium Terephthalate MOF UiO-66. *Langmuir*
28, 5715-5723 (2012).

5 Mendes, P. A. P. *et*
al. Hexane isomers sorption on a functionalized metal-organic framework. *Microporous and Mesoporous Materials* **170**,
251-258,
doi:10.1016/j.micromeso.2012.12.017 (2013).

6 Denayer, J. F. M. & Baron, G. V. The confinement
factor: A thermodynamic parameter to characterize microporous adsorbents. *Adsorption-Journal of the International Adsorption Society* **11**, 85-90
(2005).

(662d) Rapid Characterization Of Porous Materials By Thermal Response Measurements

Leistner, M., Fraunhofer IWS
Grahlert, W., Fraunhofer IWS
Dreisbach, F., Rubotherm GmbH
Kaskel, S., Technische Universitat Dresden
Originally presented on: 11/7/2013 13:30:00 - 13:50:00

Nitrogen physisorption at 77 K is a common and versatile method for the characterization of porous materials. But a drawback is the usually long measurement time that is needed for a single experiment. For a rapid characterization of porous materials, the thermal response method is presented. Thereby a porous sample is exposed to a test gas and its temperature change caused by the release of the heat of adsorption is measured in real-time using an optical temperature sensor [1-3].
The measured thermal response curve is analyzed and information on the materials adsorption properties can be obtained. In this way, with the recently introduced infraSORP high-throughput screening tool, up to 12 samples can be simultaneously tested for their adsorption capacity and specific surface area. Further, a novel infraSORP test setup enables access to adsorption kinetics, pore structure and the isotherm characteristics of a sample within some minutes of test time [3]. As an example for the various possibilities of the method, several experimental results using different porous materials and test gases are presented but also the limitations of the method will be discussed.

[1] P. Wollmann, M. Leistner, U. Stoeck, R. Grünker, K. Gedrich et al.., Chem Commun 2011;47:5151-5153.
[2] P. Wollmann, M. Leistner, W. Grählert, O. Throl, F. Dreisbach et al. Microporous and Mesoporous Materials 2012;149:86 - 94.
[3] M. Leistner, W. Grählert, S. Kaskel, Chemie Ingenieur Technik, DOI: 10.1002/cite.201200119.

(662e) Diffusion of N2 and CO2 in 13X Zeolite From Volumetric Frequency Response Measurements Up to 10 Hz

Hossain, M. I., University of South Carolina
Ebner, A. D., University of South Carolina
Holland, C. E., University of South Carolina
Ritter, J. A., University of South Carolina
Originally presented on: 11/7/2013 13:50:00 - 14:10:00

Frequency response methods have been developed as an alternative technique for the determination of mass transfer mechanisms controlling the transport of gases into and out of microporous adsorbent materials. Recent results in the literature are showing that this technique seems to be the best for discriminating the controlling mechanism. It is also possible to reveal two or more different controlling mechanisms that may dominate in different frequency ranges.

The present talk will focus on the use of the volumetric frequency response (VFR) approach for the identification on the main mechanism governing the transport of N_2 and CO_2 in zeolite 13X. In a recently commissioned VFR apparatus, experiments were carried out with pure gases at pressures of 200, 400 and 750 Torr, temperatures of 25, 40 and 55 °C and frequencies between 0.7×10^{-5} and 10 Hz. The results show that the transport of both species is purely diffusive (non Darcinian) and that for the case of CO_2, the diffusive mechanism is consistent with macropore diffusion. The results also show that these mechanisms can be unequivocally identified under the unavoidable thermal affects that appear at lower frequencies. The results were used to show the dependency of the linear driving force (LDF) mass transfer coefficient on frequency in the regions where both the Glueckauf and Nakao/Suzuki approximations dominate.

(662f) High Capacity Ammonia Adsorption On Functionalized Carbon Materials

Qajar, A., PennState University
Peer, M., Pennsylvania State University
Rajagopalan, R., Penn State University
Foley, H. C., Pennsylvania State University
Originally presented on: 11/7/2013 14:10:00 - 14:30:00

Porous carbon materials are functionalized using acid treatment. Ammonia adsorption capacity and adsorption dynamics are characterized using high pressure ammonia adsorption at room temperature. The oxygen content and type of functional groups were characterized using XPS, FITR spectroscopy and EELs pattern. Our study showed microporous carbons with little mesoporosity and hierarchical structure were ideal for acid treatment and then ammonia adsorption. The highest uptake was achieved at on a microporous carbon with more than 1 cc/g pore volume and acid treatment for 1 hr at 90 °C. The ammonia uptake was 17 mmol/g at 1 bar and 25 °C with 14 mmol/g reversible adsorption capacity. The synergy between high adsorption affinity of the oxidized surface and strong attractive forces between the adsorbed ammonia molecules increased ammonia adsorption on the carbons. Here we observed having a portion of ammonia irreversibly adsorbed to the adsorbent surface is actually helpful to rise the reversible capacity as well.

(662g) Characterization of Ag/TiO2 Adsorbents; Comparing and Contrasting Crystallite Size Using Oxygen Chemisorption and XPS Techniques

Davis, Z. D., Auburn University
Tatarchuk, B. J., Auburn University
Originally presented on: 11/7/2013 14:30:00 - 14:50:00

In the past several years, silver based sorbents have been investigated due to the desirable properties that these sorbents possess with respect to the desulfurization of logistic fuels. Particle size and dispersion are two crucial parameters determined from the chemistry occurring at the topmost atomic layers of the adsorbent. The surface characterization techniques: oxygen chemisorption and x-ray photoelectron spectroscopy (XPS) have proven to be useful in defining the Ag/TiO_2 adsorbent system. The use of oxygen chemisorption has detected nanometer scale occurrence while XPS showed sub-nanometer scale occurrence of Ag supported on titania. The discrepancy in the results from these two techniques can be attributed to the underlying perspectives and inherent assumptions intrinsic to each technique. In this study, the introduction of the multi-speciation behavior of Ag was proposed. This will account for both the reducible and irreducible behavior of Ag. This characterization will improve the accuracy of oxygen chemisorption to give a better representation of the real behavior occurring on the surface of the adsorbent.

(679a) Research On Technology for Recovering Ethanol, n-Propanol and Isobutanol From Alcohols-Water System

Li, C., Hebei University of Technology
Hu, Y., Hebei University of Technology
Fang, J., Hebei University of Technology
Wang, B.,
Originally presented on: 11/7/2013 12:30:00 - 12:55:00

As alcohols-water azeotropes and mixtures of ethanol and n-propanol which have similar structure and chemical properties are difficult to separate, the technology for recovering ethanol, n-propanol and isobutanol from alcohols-water system becomes one of industrial difficult problems. Take industrial alcohols-water system as an example, it is composed by water (20%, mass), ethanol (40%), n-propanol (30%) and isobutanol (10%). After this mixture is treated by extractive distillation with salt and azeotropic distillation, their existing problems are introduced respectively: in distillates, high pure ethanol product and n-propanol product can't be obtained, and water content is not less than 8%; large dosage of azeotropic agent is required, causing high energy consumption.

This paper proposes a new technology of saline extraction combining with azeotropic distillation and vacuum distillation to separate alcohols-water system. In saline extraction process, Adding extracting agent and inorganic salt alters distribution coefficients of alcohols and water in solution and makes solution layered. Extraction phase containing less water content than the mixture inflows into subsequent azeotropic distillation column, which makes the energy consumption of technology reduced greatly. In azeotropic distillation process, extracting agent of saline extraction process shows the same effect as azeotropic agent, and formation of ternary azeotrope of extracting agent, ethanol and water can overcome the difficult that ethanol and n-propanol are not easy to separate. Separation of n-propanol and isobutanol by vacuum distillation can increase their relative volatility and product purities. The product purity of ethanol, n-propanol and isobutanol could reach 95%, 99% and 97% correspondingly by using this technology. Compared with other separation techniques, this technology is applicable to separating alcohols-water system with high water content. This paper also studies the effect of kinds of inorganic salts and extracting agents, different mass ratios of inorganic salt to extracting agent and PH on distribution coefficients of alcohols and water in saline extraction process, energy consumption of technology and product purities through a series of contrast experiments. The optimal technology parameters are determined, which provides basis for industrialized production of alcohols-water system separation.

(679b) Laboratory Tools Available to Solid-Liquid Separations Consultants in Industry
Wolff, S.,
Originally presented on: 11/7/2013 12:55:00 - 13:20:00

Solid-liquid separations have a long history in the chemical industry from catalyst or product recovery to impurity removal to wastewater treatment. Although the two primary separation methods are filtration or centrifugation, the variety of devices available can seem endless. The critical first step is a good understanding of the process and separation goals. From there, laboratory testing further guides and narrows the selection process. A thorough laboratory slurry characterization can include several analytical measurements as well as preliminary filtration and/or centrifugation evaluations. Standard testing devices are available and briefly described. Scale-up to the commercial process should then include pilot operation of all solid-liquid separation steps with representative feed stocks. This is of particular importance for processes utilizing bio-derived feed stocks where impurity profiles are significant and more varied than traditional feed stocks. It is well known that commercial processes involving particles are slower to achieve nameplate production rates; proactive laboratory and pilot testing of separation steps is one strategy to address this commercialization phenomenon.

(679c) Peptide Nanofiltration By Quality By Design

Marchetti, P., Imperial College London
Butte, A., Lonza AG
Livingston, A. G., Imperial College London
Originally presented on: 11/7/2013 13:20:00 - 13:45:00

Peptide nanofiltration by *Quality by Design*

Recently, nanofiltration (NF) techniques have been introduced by the pharmaceutical industries as part of the downstream processes for peptides, to perform concentration, purification and salt/solvent exchange, and they have been demonstrated as being suitable for integration with conventional purification techniques, providing savings in terms of time and costs. Application of NF has been supported by the development of new membrane materials with high stability in organic solvents (commonly used in the peptide industry), and efficient membrane modules [1,2,3].

During process development, process parameters and quality attributes are investigated, with the aim of establishing a relationship among them. Critical Quality Attributes (CQAs) are physical or chemical characteristics, which must be controlled to ensure the quality of the product. Critical Process Parameters (CPPs) are process inputs, which have a direct and significant effect on CQAs, when they are varied within the experimental range [4]. Typical CQAs for a mixture after NF are specifications regarding concentration of main product, impurities, ionic and solvent composition. To support new initiatives and provide guidance for pharmaceutical process development, the International Conference on Harmonisation (ICH) of Technical Requirements for Registration of Pharmaceuticals for Human Use introduced the Quality by Design (QbD) concept [5]. This concept was defined as "a systematic approach to development that begins with predefined objectives and emphasizes product and process understanding and process control, based on sound science and quality risk management" [5], and it has become an important tool in assisting the industry to move towards a more scientific approach to pharmaceutical development. As a result of all this knowledge, the company can: (i) select the best process conditions with a limited number of experimental data; (ii) continually monitor the process to assure consistent product quality; and (iii) update the process without requiring further experimental effort.

Understanding and, possibly, modelling all the possible effects and sources of variations on the quality of the final product is often difficult. Consequently, Design of Experiments (DoE) methods are extensively applied in process design [6]. DoE is characterized by reduction or minimization of the total number of trials by the simultaneous variation of all potential influencing factors. Application of DoE provides scientific understanding of the effects of multiple process parameters on product CQAs, allows the identification of interactions among process parameters (impossible for the 1-factor-at-a-time approach) and favours the selection of optimal working conditions. DoE methods have been largely employed for the development of chromatographic techniques, to identify the important factors affecting the retention performance and optimize the separation [7,8]. Studies of NF by DoE are, on the other hand, few [9,10] and rarely show how information from DoE can be used to support process modelling for process selection [11].

In this paper, it is shown how empirical DoE models can (i) provide phenomenological understanding of the transport mechanism through nanofiltration membranes for a specific solute of interest, and (ii) successfully support the process modelling for concentration and diafiltration, providing a methodology to select the most appropriate filtration technique for a given separation problem. In the first part of this work, the permeation of three model peptides, produced by Lonza (Visp, Switzerland) in acetonitrile/water mixtures through two ceramic NF membranes (InoporÂ® Nano 450 and InoporÂ® Nano 750, both composed of TiO_2/Al_2O_3) is studied by DoE and the results are analysed by statistical Analysis of Variance (ANOVA). Typical peptide mixtures are composed of peptides and salts (as counter-ions or buffers) dissolved in solvent mixtures. Two peptides, one with preferential solvation for acetonitrile over water (PEP_1) and the other with preferential solvation for water over acetonitrile (PEP_2), are studied under conditions largely below the isoelectric point; a third peptide (PEP_3) is studied under conditions close to its isoelectric point. For each peptide, peptide rejection, flux and trifluoroacetic acid (TFA-H) rejection are modelled as function of five operating parameters (peptide concentration, %v TFA-H, pressure, cross-flow velocity, and %v acetonitrile/water). The effects of the operating parameters and their interactions are discussed in terms of positive or negative effects on the responses and the role of preferential solvation and effect of salt on peptide flux and rejection is highlighted. For the peptide under conditions close to its isoelectric point (PEP_3), reversible formation of micelles is observed and its effect on the permeation performance identified.

In the second part of this work, DoE analysis is used to support process selection for the first model peptide, PEP_1. The peptide mixture exits the preparative chromatography with a composition of 0.06 to 0.1%v TFA-H / 30%v ACN/water; NF can be successfully used to perform concentration and diafiltration to (i) increase the peptide concentration, (ii) reduce the operating volumes and (iii) reach the required composition of 0.02%v TFA / 0.003%v ACN/water, before entering the lyophilization.

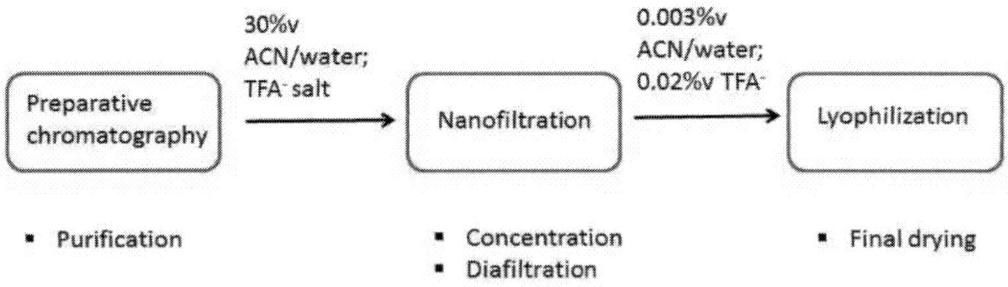

Figure 1. Downstream process (DSP) for PEP_1.

The best operating conditions for concentration are found by numerical optimization of the statistical models obtained by DoE. The statistical models from DoE are afterwards included in the mathematical framework of the diafiltration process, to calculate the evolution of peptide, counter-ion and solvent concentrations over time, compare constant volume vs. variable volume diafiltration modes, and select the best process in terms of operating time and solvent consumption. Preconcentration followed by constant volume diafiltration is compared to variable volume diafiltration, in terms of operating time and solvent consumption. For the case study in this work, the former strategy (i.e. preconcentration + constant volume diafiltration) is identified as the best one.

Bibliography

[1] Yang, X. J., Livingston, A. G., Freitas dos Santos, L., *Experimental observations of nanofiltration with*

organic solvents, Journal of Membrane Science 190, (2001) 45-55.

[2] Vandezande, P., Gevers L. E. M., Vankelecom I. F. J., *Solvent resistant nanofiltration: separating on a molecular level,* Chemical Review Society 37 (2008) 365-405.

[3] Tsuru, T., Miyawaki, M., Kondo, H., Yoshioka, T., Asaeda, M., *Inorganic porous membranes for nanofiltration of nonaqueous solutions*, Separation and Purification Technology 32 (2003) 105.

[4] Yu, L. X., *Pharmaceutical Quality by Design: product and process development, understanding and control*, Pharmaceutical Research 25 (2007) 781-791.

[5] International Conference On Harmonisation of Technical Requirements for Registration of Pharmaceuticals for Human Use, August 2009, *ICH Harmonised Tripartite Guideline: Pharmaceutical Development Q8 (R2)*

Â http://www.ich.org/products/guidelines/quality/article/quality-guidelines.html.

[6] Shivhare, M., McCreath, G., *Practical considerations for DoE implementation in Quality by Design,* BioProcess International 8(6) (2010) 22-30.

[7] Hibbert, D. B., *Experimental design in chromatography: a tutorial review*, Journal of Chromatography B 910 (2012) 2-13.

[8] Atkinson, A. C., Tobias, R. D., *Optimal experimental design in chromatography*, Journal of Chromatography A 1177 (2008) 1-11.

[9] Ahmad, A. L., Leo, C. P., Shukor, S. R. A., *Statistical design of experiments for dye-salt-water separation study using bimodal porous silica/gamma-alumina membrane*, Desalination and Water Treatment 5 (2009) 80-90.

[10] Polom, E., Szaniawska, D., *Rejection of lactic acid solutions by dynamically formed nanofiltration membranes using a statistical design method*, Desalination 198 (2006) 208-214.

[11] RomÃ¡n, A., Vatai, G., IttzÃ©s, A., KovÃ¡cs, Z., Czermak, P., *Modeling of diafiltration processes for demineralization of acid whey: an empirical approach*, Journal of Food and Process Engineering 35 (2012) 708-714.

(679d) Synthesis of Optimal Separation System for Benzene Removal From Gasoline

Ahmadi Pouya, R., Islamic Azad University, North Branch of Tehran

Originally presented on: 11/7/2013 13:45:00 - 14:10:00

Generally a gasoline production from catalytic reformer sends to hydrogenation for the separating of benzene and its precursors from gasoline. Indeed for access to desirable products, we need to inject hydrogen at hydrogenation unit. Whereas whole gasoline is sending to hydrogenation unit then the consumption of hydrogen will increase. A suggestion for decrease hydrogen demand, we can design a separation system as Three Product System (TPS) for catalytic reformer. By designed system, initially we can separate C6 fraction from output gasoline and then solely sending C6 fraction to hydrogenation unit result in reduction of hydrogen's requirement. In this article has been discussed different form of TPS and estimated the performance of theirs with design costs and energy consumption .Ultimately with comparing of their performance, optimum separation system has been suggested for case study.

(679e) Operation of Energy Efficient Divided Wall Column

Shamsuzzoha, M., King Fahd University of Petroleum and Minerals(KFUPM)

Originally presented on: 11/7/2013 14:10:00 - 14:35:00

Based upon the known design of dividing wall column the required energy usage to run the column is V_0 and during operation, actually the column is run at the vapor split ratio R_v that is found to be non- optimal. Here the value of V_0 is actually lower than what is required for the separation of three products. This situation is likely to happen more frequently in industrial operation and a condition could be that the available energy to run the column is even lower than the minimum energy required for optimal operation of the column. This study emphasize on analyzing the effect on product compositions in case of the column is operated with the energy lower than the optimal energy required for running the column. For this analysis various simulations has been conducted and the effect on product compositions is examined.

(685a) Influence of Polydopamine Deposition Conditions On Oil/Water Fouling Resistance and Properties of Polysulfone Ultrafiltration Membrane

Kasemset, S., University of Texas at Austin
Miller, D. J., University of Texas at Austin
Sharma, M. M., University of Texas at Austin
Freeman, B. D., The University of Texas at Austin
Originally presented on: 11/7/2013 12:30:00 - 12:55:00

One of the major challenges in water purification membrane application is fouling which leads to permeate flux decline and/or increase in transmembrane pressure difference. Membrane surface modification can be used to develop anti-fouling membrane via several approaches such as increasing surface hydrophilicity, decreasing surface roughness, or decreasing surface charge. A facile method of membrane surface coating by using "polydopamine" has proved to enhance fouling resistance of several types of membrane to oil/water fouling due to its high hydrophilicity. Polydopamine reacts from dopamine and coats on virtually any surfaces where the reaction conditions are major key parameters to control the deposition of the coating. These reaction conditions include dopamine concentration, reaction time, pH of buffered reaction solution, oxygen content, etc.

In this study, the influence of polydopamine deposition conditions on fouling resistance of flat-sheet polysulfone ultrafiltration is evaluated via crossflow fouling tests. In addition, the membrane properties such as pure water permeance, surface hydrophilicity, molecular weight cut off are determined to gain a fundamental understanding of this coating at various reaction conditions.

(685b) High Performance Magnetically Responsive Membranes for Ultrafiltration

Dong, R., University of Arkansas
Qian, X., University of Arkansas
Wickramasinghe, S. R., University of Arkansas
Originally presented on: 11/7/2013 12:55:00 - 13:20:00

Membrane fouling is a major problem in numerous membrane based separation processes. During ultrafiltration, membrane performance is compromised by the formation of a concentration polarization boundary layer consisting of rejected species at the membrane surface. This boundary layer provides an additional resistance to permeate flow, resulting in a decrease in flux. Further, since the concentration of rejected species is higher in the boundary layer than the bulk feed, the apparent rejection coefficient of retained species is often lower. In addition, concentration polarization can lead to deposition of rejected species on the membrane surface leading to reversible as well as irreversible fouling.

While many surface modification methods have been described that suppress adsorption of rejected species on the membrane surface, they do not suppress concentration polarization. Here we modify the surface of commercially available nanofiltration membranes in order to impart a magnetically responsive nanostructure. In an oscillating magnetic field, movement of the grafted polymer chains leads to break up of the concentration polarization boundary layer. Results presented for BSA filtration indicate higher permeate fluxes in the presence of an oscillating magnetic field. In addition, use of an oscillating magnetic field leads to greater flux recovery.

(685c) A Variety of Surface-Engineered Membranes

Sirkar, K. K., New Jersey Institute of Technology

Originally presented on: 11/7/2013 13:20:00 - 13:45:00

We have developed a variety of novel surface-engineered porous/microporous membranes for different membrane separation techniques over the years. The separation techniques involved more often than not principles involved in immobilizing gas-liquid, liquid-liquid interfaces at membrane pore mouths. Conventional membrane separation techniques have also been used. We will provide a brief overview of these surface-engineered membranes and the associated separation and other applications. The processes/techniques where these have been applied are: direct contact membrane distillation for desalination; pervaporation for biofuel separation from dilute broths; membrane solvent back extraction using porous hydrophilic membranes; supported liquid membranes and membrane contactors for CO_2 separation; polypropylene (PP) substrate-based interfacially polymerized membranes for nanofiltration; integrated filtration-chromatography process; thermo-sensitive membrane for controlled release. The membrane surface engineering utilized plasma polymerization, interfacial polymerization, PEI insolubilization and crosslinking, PP surface hydrophilization etc. The structures created superhydrophobic surfaces, smaller surface pore sizes, composite layers having different hydrophobicities, hydrophilic-hydrophobic membranes, thermosensitive surface grafts etc.

(685d) Novel Membranes for Hydrophobic Interaction Chromatography

Vu, A., University of Arkansas
Qian, X., University of Arkansas
Wickramasinghe, S. R., University of Arkansas
Originally presented on: 11/7/2013 13:45:00 - 14:10:00

Responsive membranes have been developed for hydrophobic interaction chromatography. Poly N-vinylcaprolactam (PVCL) chains were grown from the surface of regenerated cellulose membranes using atom transfer radical polymerization (ATRP). Modified membranes were characterized using scanning electron microscopy, infrared spectroscopy, X-ray photoelectron spectroscopy contact angle. ATRP allows us to independently control chain density and molecular weight. The grafting degree of PVCL increases with increasing ATRP initiator immobilization time.

PVCL is a thermo-responsive polymer with a lower critical solution temperature (LCST) that depends on the concentration of salt ions present in solution. The LCST decreases below room temperature in buffer containing 1.8 M $(NH_3)_2SO_4$ where the polymer adopts a more hydrophobic/collapsed conformation. At low ionic strength, the LCST remains above room temperature. Contact angles for PVCL in 20 mM phosphate buffer with and without 1.8 M $(NH_3)_2SO_4$ were determined. Contact angles in the buffer with high ionic strength were higher than those in low ionic strength.

Adsorption and desorption of bovine serum albumin (BSA) and immunoglobulin G (IgG) have been investigated. Loading was conducted in high ionic strength buffer. Elution was conducted in low ionic strength buffer. By using a responsive ligand that changes its conformation during loading and elution, high protein recoveries were obtained. In addition efficient fractionation of proteins is possible. Ligands that respond to solution ionic strength show promise for high performance hydrophobic interaction chromatography.

(685e) Responsive Functionalized Membranes for Environmental Catalysis and Tunable Separations

Xiao, L., University of Kentucky
Bhattacharyya, D. B., University of Kentucky
Originally presented on: 11/7/2013 14:10:00 - 14:35:00

Membrane based separation process has been widely used in water detoxification, water purification and waste water treatment. Stimuli-responsive polymers have been gained much attention due to their unique and excellent physical and chemical responses to changes in physiological conditions and have applied in different areas: such as, drug delivery, bioseparation and tissue engineering, etc. Various types of responsive membranes for regulated permeation have been devised, such as pH, temperature and light, etc. The porous membranes with responsive polymer chain have advantages over hydrogel membrane for mechanical strength and quick response to external signal. In our work, temperature (poly-N-isopropylacrylamide, NIPAAm) and pH (polyacrylic acid, PAA) responsive membrane has been designed and synthesized, in which pores open or close in response to temperature and pH values. Hydrophylized polyvinylidene fluoride (PVDF) membrane was used as support due to its chemical resistance properties and good thermal stability. In a narrow temperature or pH range, hydration-dehydration changes lead to the membrane conformation change, resulting in the stimuli-responsive permeability of gating membrane. The hydrophilic /hydrophobic transition of the membrane around response temperature and pH was investigated. The effects of temperature and pH on solute permeability and rejection were also determined. Besides pore-filling method, these temperature and pH responsive polymer were also covalently bonded with PVDF membrane with basified reaction followed with free-radical polymerization to enhance the stability of responsive polymers within the membrane. In addition, we also extended our work into catalytic area with immobilized nanoparticles in membrane platform for water detoxification.

This research was supported by NIEHS-SRP and by DOE-KRCEE.

(685f) Sulfonated Polybenzimidazole Membranes for Pervaporation Dehydration of Acidic Solvent

Wang, Y., Huazhong University of Science & Technology
Gruender, M., PBI Performance Products, Inc.,
Originally presented on: 11/7/2013 14:35:00 - 15:00:00

In this study, sulfonated polybenzimidazole (SPBI) membranes are developed and investigated for pervaporation dehydration of acetic acid, via a two-step modification on membrane surface with the sulfonation by sulfuric acid aqueous solution first followed by a thermal treatment at 450 °C. Both steps are found indispensible in order to produce a stable SPBI membrane with enhanced acid resistance and superior separation performance. Effects of the sulfuric acid concentration and thermal treatment duration have been investigated and found to have significant impact on the pervaporation performance of the resultant SPBI membranes. Various characterizations (FTIR, XPS, TGA and XRD) are employed to elucidate the physicochemical changes of membranes as a function of chemical and thermal modifications. In addition, effects of pervaporation temperature and feed composition are studied not only in terms of flux and separation factor, but also of membrane intrinsic permeance and selectivity. The best pervaporation performance of the SPBI membrane has a flux of 207 g/m^2h and a separation factor higher than 5000 for dehydration of a 50/50 wt.% acetic acid/water feed solution at 60 °C, which not only outperforms the conventional distillation process, but also surpasses most other polymeric pervaporation membranes reported in literature. It is therefore believed that the novel developed SPBI membrane may have great potential for pervaporation dehydration of acidic organics, as well as other applications that demand acid-proof materials.

(687a) Controllable Phase Transition of CuBTC Metal-Organic Framework and Its Application in CO2 Capture and Storage

Huang, L., North Carolina State University
Zhang, S.,
Gubbins, K. E., North Carolina State University
Lu, X., Nanjing university of technology
Originally presented on: 11/7/2013 12:30:00 - 12:47:00

As one of the most studied metal-organic frameworks, CuBTC has been considered promising in the fields of adsorption, separation and catalytic reaction. However, the expensive commercial price and the thermal/hydrostatic stability are two limiting factors for industrial utilization of CuBTC. Here we report a controllable and reversible 1D/3D phase transition of CuBTC structures, and demonstrate its application in CO_2 capture and storage. By *ab initio* density functional theory (DFT) calculation and reactive molecular dynamics simulation, we study at the electron and molecular level the stability of 1D and 3D CuBTC structures, the potential reversible phase transition, and the CO_2/CuBTC interactions. Together with the experimental evidence, we discuss the mechanism of the controllable 1D/3D phase transition, the role of hydrogen bond network, the ratio of 1D/3D strucutures and its experimental implementation. The fast and controllable characteristics of the 1D/3D phase transition make CuBTC material a promising candidate in CO_2 separation and storage.

(687b) Thermodynamic Stability of Hypothetical ZIF-8 Polymorphs Using Molecular Simulations

Gee, J., University of California Santa Barbara
Sholl, D. S., Georgia Institute of Technology
Originally presented on: 11/7/2013 12:47:00 - 13:04:00

Zeolitic imidazolate frameworks (ZIFs) are a class of metal-organic frameworks (MOFs) that are known to exhibit structural polymorphism. Despite the large number of reported ZIF polymorphs, the factors governing the stability of these materials remain poorly understood. In this work, we present a novel approach to investigate the thermodynamic stability of MOF polymorphs using molecular simulations. Our work employs a combination of thermodynamic integration (TI) and OFAST simulations to predict the free energy of solvation, ΔG_{solv}, and vibrational density of states (VDOS) calculations to determine the configurational energy of the framework. This approach is applied to a series of hypothetical ZIF-8 polymorphs under solvothermal conditions. Our simulations predict that the framework stability is governed by the configurational energy rather than the ΔG_{solv} of the framework. We find good agreement between OFAST and the more precise TI method for the prediction of the ΔG_{solv} of the framework. This methodology can be applied to study the effect of structural topology, solvent, and temperature on the stability of the framework for a wide range of nanoporous materials.

(687c) Adsorption of 5-Hydroxymethyl-Furfural (HMF) From Liquid Water/Dimethyl Sulfoxide (DMSO) Solutions Onto Hydrophobic Zeolites: An Experimental and Simulation Study

Xiong, R., University of Delaware
Leon, M., University of Delaware
Nikolakis, V., University of Delaware
Sandler, S. I., University of Delaware
Vlachos, D. G., University of Delaware
Originally presented on: 11/7/2013 13:04:00 - 13:21:00

The adsorption of 5-hydroxymethylfurfural (HMF), dimethyl sulfoxide (DMSO), and water from binary and ternary mixtures in hydrophobic silicalite-1 and dealuminated Y (DAY) zeolites at ambient conditions was studied via experiments and molecular modeling. HMF and DMSO adsorption isotherms were measured and compared to those calculated using a combination of grand canonical Monte Carlo and expanded ensemble (GCMC-EE) simulations. A method based on GCMC-EE simulation for dilute solutions combined with the Redlich-Kister (RK) expansion (GCMC-EE-RK) is introduced to calculate the isotherms over a wide range of concentrations. The simulations, using literature force fields, are in reasonable agreement with the experimental data. In HMF/water binary mixtures, large pore hydrophobic zeolites are much more effective for HMF adsorption but less selective because large pores allow water adsorption due to H2O-HMF attraction. In HMF/DMSO/water ternary mixtures, the HMF loading decreases with increasing DMSO fraction, rendering separation of HMF from water/DMSO mixtures by adsorption difficult. We show that the ratio of the energetic interaction in the zeolite to the solvation free energy is a key factor in controlling separation from liquid mixtures. Overall, our findings could have an impact on the separation and catalytic conversion of HMF and the rational design of nanoporous adsorbents for liquid-phase separations in biomass processing.

(687d) Preferential Adsorption From Liquid Ethanol-Water Mixtures in Alumina Pores

Phan, A. T., The University of Oklahoma
Striolo, A., University of Oklahoma
Cole, D. R., Oak Ridge National Laboratory
Originally presented on: 11/7/2013 13:21:00 - 13:38:00

Equilibrium MD simulations were employed to investigate interfacial liquid ethanol-water mixtures on several alumina slit-shaped pores. The results were quantified in terms of preferential adsorption from the bulk, density profiles, in-plane radial distribution functions, residence times at contact with the substrate, self-diffusion coefficients, orientational distribution functions, hydrogen bonds distributions, and reorientation autocorrelation functions. When pure ethanol is simulated, the results show that ethanol molecules in the first monolayer near a solid surface have a well-organized arrangement, while those in the second layer show bulk-like features. The predicted orientational distribution for ethanol molecules within the first layer is consistent with experimental results. When water is added to the system, our results show that water quickly replaces ethanol molecules from the first adsorbed layer, and preferentially partitions within narrow slit-shaped alumina pores. We will discuss how such results could be useful for the design of separation processes. The results could also be important for better understanding the behavior of fluids in sub-surface environments.

(687e) Computer Simulation of Water Sorption On Flexible Protein Matrices

Palmer, J. C., Princeton University
Debenedetti, P. G., Princeton University
Originally presented on: 11/7/2013 13:38:00 - 13:55:00

In the last two decades, protein therapeutics have emerged as the fastest-growing segment of the pharmaceutical industry, resulting in more than 200 new treatments for diseases such as AIDS, leukemia, melanoma, cystic fibrosis, and multiple sclerosis. Despite these advances, however, there remain a number of technical challenges in formulating and manufacturing new protein-based therapeutics. One of the most difficult tasks is to develop therapeutic formulations with long-term stability against the chemical degradation and protein aggregation processes that compromise the safety and efficacy of drugs that are stored in solution over long periods of time.

The standard approach for increasing the stability of protein therapeutics is to remove the biomolecules from their aqueous environment by incorporating them into vitrified solid materials using a process known as lyophilization or freeze-drying. In some cases, however, proteins often exhibit diminished therapeutic activity when they are reconstituted with water following storage in the glass state. To understand this behavior, we use computer simulation to explore how proteins are affected by dehydration and subsequent rehydration. We demonstrate a new computational approach we have developed to study water sorption on protein crystals and powders, and show that this technique allows us, for the first time, to make rigorous contact with experimental measurements designed to probe the protein hydration process. We also show that we are able to go beyond the current capabilities of experimental techniques and investigate how changes in hydration level can give rise to microscopic mechanical stresses that may alter the structure and biological activity of some proteins.

(687f) Adsorption of Polyols and Furfurals in All-Silica Zeolites

Bai, P., University of Minnesota
Tsapatsis, M., University of Minnesota
Siepmann, J. I., University of Minnesota
Originally presented on: 11/7/2013 13:55:00 - 14:11:00

Polyols and furfurals are important commodity compounds that can be produced through biomass conversion. To increase the efficiency and lower the energy cost of the separation of these molecules from reactor effluents, adsorptive separation using nanoporous materials has been proposed as a promising technique. In order to better understand the capacity and the selectivity of all-silica zeolites to polyols and furfurals, configurational-bias Monte Carlo simulations in the Gibbs ensemble were performed to compute adsorption isotherms from solution phases. The TraPPE force field is extended to furfural and its derivatives and intramolecular potentials are determined for some polyols; good agreement with experimental liquid density and vapor pressure data is found for the unary vapor-liquid equilibria of these compounds. The trends in the adsorption isotherms, the free energies of transfer, and selectivities of several selected polyol and furfural molecules are explained in terms of hydrogen-bond speciation and confinement effects in zeolite structures.

(687g) Self-Assembled Monolayer of Carboxylate On Calcite $(10\bar{1}4)$ Surface: Molecular Dynamics Simulation Approach

Chun, B., Georgia Institute of Technology

Lee, S. G., Georgia Institute of Technology

Brunello, G. F., Georgia Institute of Technology

Choi, J. I., Georgia Institute of Technology

Jang, S. S., Georgia Institute of Technology

Originally presented on: 11/7/2013 14:11:00 - 14:28:00

Self-Assembled Monolayer of Carboxylate on Calcite $(10\bar{1}4)$ Surface:

Molecular Dynamics Simulation Approach

Byeong Jae Chun, Seung Geol Lee, Giuseppe F. Brunello, Ji Il Choi and Seung Soon Jang

Computational NanoBio Technology Laboratory, School of Materials Science and Engineering, Georgia Institute of Technology, 771 Ferst Drive NW, Atlanta, GA 30332-0245

To study the wettability alteration of calcite surface in oil reservoir, we investigate the molecular adsorption of carboxylate on the calcite $(CaCO_3)$ $(10\bar{1}4)$ surface using full-atomistic molecular dynamics (MD) simulation as well as density functional theory (DFT). There are three types of carboxylate adsorbate used for this study; benzoate, dodecanoate (C12), and octadecanoate (C18). Using DFT calculation, the molecular details of carboxylates are determined including the quantum charge and optimal configuration. Once the fundamental information of the adsorption models is attained, MD simulations are performed to obtain the equilibrium structures and evaluate the thermodynamic stability of the chemisorbed carboxylate and their self-assembled monolayer (SAM) on calcite. Our force field contains the newly developed parameters for the off-diagonal van der Waals interactions of carboxylate functional group with calcite surface. The given original force fields are employed for van der Waals interactions between the same type of atoms or molecules. Based on the thermodynamic stability of each adsorption case achieved by our MD simulations, we determine the optimal packing density of carboxylate SAM on the calcite surface for each case and characterize the corresponding structures. In order to determine the effect of solvation on the packing and structures of SAM on calcite surface, we simulate the same systems in the presence of solvent phase such as water and octane. Once equilibrated structures of carboxylate SAM are obtained, the free energy barrier for detachment of carboxylate from the surface is evaluated by applying potential of mean force (PMF) in order to assess the kinetics of wettability alteration as well as thermodynamic stability.

(687h) Elucidating the Details of Protein Transport in Polymer-Grafted Ion Exchangers Through Multi-Scale Modeling

Basconi, J., University of Virginia
Carta, G., University of Virginia
Shirts, M. R., University of Virginia
Originally presented on: 11/7/2013 14:28:00 - 14:45:00

In protein separations using ion-exchange chromatography, matrices with pores containing charged polymer grafts have exhibited increased binding capacity, faster adsorption kinetics, and a different diffusion mechanism as opposed to traditional macroporous matrices, though these enhancements depend strongly on the protein properties and concentration, as well as the buffer salt concentration. We use multi-scale modeling to study the factors which affect protein transport in polymer-grafted and macroporous materials, which in general depends on both the equilibrium adsorption and diffusion behavior of proteins in the different phases.

We have developed coarse-grained molecular models suitable for molecular dynamics (MD) simulations of proteins diffusing through representative pores of both polymer-grafted and macroporous materials. In simulations of lysozyme interacting with agarose-based matrices with charged dextran polymers, we obtain good agreement with experimental results for equilibrium binding capacities and effective transport rates, as well as their dependence on system parameters such as polymer graft density and surface topology.

The kinetic and equilibrium information from these MD simulations is also employed in a simple rate model to study the macroscopic protein concentration profiles in the different materials over time. With this model we observe faster adsorption kinetics and a different adsorption front for the polymer-grafted material as compared to the macroporous material, which is qualitatively consistent with experiments. We apply this multi-scale modeling approach to different types of proteins and adsorbed phases, as understanding how protein transport rates are affected by various protein and material properties should provide insight into the general mechanism by which charged polymers can enhance chromatographic performance.

(687i) Investigation of the Critical Point of Adsorption of Polymer Chains On Porous Adsorbents

Cimino, R. T., Rutgers, The State University of New Jersey
Neimark, A. V., Rutgers, The State University of New Jersey
Originally presented on: 11/7/2013 14:45:00 - 15:01:00

Liquid phase chromatography at critical conditions (LCCC) is a widely used technique for the separation of polymeric mixtures of similar molecular weight but different chemical properties. The critical point of adsorption (CPA) is defined as the point at which adsorption is independent of molecular weight, and is observed experimentally. While theory has been developed to evaluate the CPA for ideal chains, an adequate theoretical description of the CPA of real chains is still lacking. This work provides a theoretical basis for defining the CPA of a realistic polymer chain based on the free-energy and chemical potential of the chain, and provides model calculations of the CPA on a porous substrate using Gauge Cell Monte Carlo. In developing the model, several modes of adsorption are considered: adsorption approaching the external surface of an adsorbent particle, complete adsorption within a pore, adsorption of chains tethered to the external surface of the particle, and the adsorption of chains which are partially translocated – called "flower" configurations. The summation of the Boltzmann-weighted free energies of these configurations are used as the partition function for calculating system variables – notably the partition coefficient and chemical potential of the polymer chain.

(694a) Homogeneous Nucleation of Ice in Droplets and Free-Standing Films of Supercooled Water

Haji-Akbari, A., Princeton University
Debenedetti, P. G., Princeton University
Originally presented on: 11/7/2013 12:30:00 - 12:52:00

The problem of homogeneous ice nucleation is of immense importance to fields as diverse as biology and chemistry, to geology and atmospheric sciences. Of particular interest is the formation of ice in confined geometries. One such example is the nucleation of ice in droplets and free-standing thin films, systems that are especially important in geological and atmospheric sciences. In this work, we use molecular simulations to compute the rates and unravel the mechanism of ice formation in droplets and thin films of supercooled water. The water model used in this study is the monoatomic potential proposed by Molinero et al. [1]. This model has no explicit hydrogen atoms, and the local tetrahedrality of water molecules that arises from the formation of hydrogen bonds is enforced by including a three-body interaction term in the inter-molecular potential. The lack of electrostatic interactions in the Molinero water leads to faster dynamics and increased computational efficiency.

Since the formation of critical nuclei within a supercooled liquid is a rare event, it cannot be properly sampled from a single long Molecular Dynamics (MD) trajectory of the system even at relatively large supercoolings. This makes conventional MD unsuitable for studying nucleation. We therefore combine conventional molecular dynamics with Forward-Flux Sampling (FFS) which is an advanced sampling technique developed for computing the rates of rare events such as protein folding, evaporation and crystallization [2]. We calculate crystallization rates for temperatures 220-235 K, which are then used for computing activation energies and critical nucleus sizes. We also compare our findings with nucleation rates in the bulk to understand how geometric confinement affects the crystallization process. Our findings will help us understand how a free interface affects ice formation in water, which will be an invaluable knowledge for both statistical physicists and atmospheric scientists.

1- Molinero V, Moore EB, *J. Phys. Chem. B* **113(13)**: 4008-16 (**2008**).

2- Allen RJ, Frenkel D, ten Wolde PR, *J. Chem. Phys.* **124**: 194111 (**2006**).

(694b) Growth of Tetrahydrofuran (THF) Clathrate Hydrates: A Molecular Dynamics Study

Wu, J. Y., National Taiwan University
Lin, S. T., National Taiwan University
Originally presented on: 11/7/2013 12:52:00 - 13:14:00

The growth of tetrahydrofuran (THF) clathrate hydrates and methane-THF clathrate hydrate were calculated by Molecular dynamic (MD) simulations. The melting points of THF hydrate determined from MD simulations were 277.5K at 1atm and 270K at 1000atm, in good agreement with the experimental data (277.6K at 1atm and 274K at 1000atm). The melting point of THF hydrate was found to correlate strongly with the solubility of THF in water. The solubility of THF in water may decrease from completely miscible to less than 0.01 (mol THF/mol solution) at 298K and 1atm by changing the interaction parameters between THF and water. In the latter case, the melting point of THF clathrate hydrate decreased to less than 260K at 1atm. The occupancy of THF in the $5^{12}6^4$ cages strongly stabilized the structure II hydrate. In the growth process, the all $5^{12}6^4$ cages are filled with THF (occupancy is 1). The promoting effect of THF on CH_4 clathrate hydrates was successfully simulated. With the adding THF in the system, the melting temperature of CH_4 clathrate hydrate increased about 50 K.

(694c) Molecular Dynamics Simulations Of Vapor Bubble Growth and Detachment On A Heating Surface- Investigation Of The Contact LINE Dynamics

Bao, J., City College of New York ,City University of New York
Koplik, J., City College of the City University of New York
Rumschitzki, D. S., The City College of The City University of New York
Originally presented on: 11/7/2013 13:14:00 - 13:36:00

Based on our previous simulations of conductively-driven quasi-static vapor bubble growth in an axisymmetric, cylindrical cell comprised of solid and liquid phases of finite thicknesses with fluid motion and heat transfer obeying small Reynolds, Peclet, Capillary and Bond numbers, we found that the appearance and the motion of the three-phase contact line (CL) is a critical element, central to physics needed to explain the large heat enhancement in nucleate boiling. However, our simulations also suggested that different ad hoc models for the poorly-understood motion of the CL with phase change occurring at the CL as the vapor bubble grows do have a significant effect on vapor bubble deformation and detachment when the Bond number is no longer small. Therefore, we have used molecular dynamics (MD) to simulate a nano-scale version of our three-phase system.

Our MD simulation domain of interest is a cuboid, which is composed of the fluid region sandwiched between two parallel solid walls. The fluid phase is made of argon atoms interacting via a 12-6 Lennard-Jones (LJ) potential. The atoms of the two solid walls are tethered via springs to their lattice positions and also interact with both the fluid and each other via LJ potentials. We determine the critical interaction parameters between the solid and fluid atoms in LJ form by considering the balance between the wettability of the solid surface, which is key to observe nucleation and the heat transfer rate through the solid to the fluid. Periodic boundary conditions are applied at the 4 vertical boundaries of the domain. After achieving thermal equilibrium of all the phases in the simulation domain (solid walls and liquid argon) at the uniform reduced temperature of T=0.75 (the fluid saturation temperature), we start to expend the top wall gradually at constant reduced temperature T=0.75 to some specified pressure at which no vapor bubble appears via cavitation. Next, we instantaneously increase the temperature at the bottom layer of the bottom wall while keeping the original lower temperature at the top wall, both maintained by thermostating. This maintains a temperature gradient through the fluid phase. Meanwhile, the pressure at the top wall (which matches that in the bulk fluid) is kept fixed by allowing the top surface to freely slide up and down.

Without applying a uniform body force (that in some way might mimic gravity) to the fluid atoms, we adjust the control pressure and the heating temperature and find the region of the pressure vs temperature plot where a vapor bubble nucleates and eventually grows to cover the entire solid surface, where a bubble nucleates but does not grow to cover the entire solid surface and where no bubble nucleates in a reasonable amount of time. By then artificially applying a uniform body force that mimics a relevant Bond number, we then successfully nucleate and grow a vapor bubble to a moderate size and observe bubble necking and detachment. We find that,, despite the existence of temperature slip between solid and liquid, the vapor bubble volume still grows as $t^{3/2}$, t being time, as in our earlier conduction-only continuum calculations. This scaling is also in agreement with numerous experimental observations. Furthermore, our temperature profiles show that the vapor bubble surface at the three-phase CL appears to be at the liquid saturation temperature, which confirms the assumptions made in our continuum calculations' boundary condition at liquid-vapor interface and where it contacts the solid surface. By further increasing this artificial uniform body force (i.e., Bond number) and then keeping it constant, we find the body force indeed deforms the bubble and causes it to detach from the surface. This provides critical information as to how system parameters affect the size of the vapor residue left on the surface and on whether cooler fluid flowing downward to replace the rising detached bubble succeeds in quenching this residue. During this process, we reset/raise the temperature at the top wall to avoid vapor condensation at the top of vapor bubble as the bubble grows, detaches and rises into regions of lower temperature, without changing the magnitude of the uniform body force. These simulations result in, among other things, the time evolution of the CL motion. The radius of CL initially expands with the growth of the bubble at low Bond number. Then, as the bubble grows to a size where the body force becomes significant, the CL contracts sharply but continuously as the bubble deforms until detachment. Finally, we incorporate this physical CL motion into our continuum calculation to simulate the vapor bubble growth until detachment in a continuum model.

(694d) Molecular Simulation of Homogeneous Crystal Nucleation and Growth in Polyethylene Melts

Yi, P., Massachusetts Institute of Technology
Rutledge, G. C., Massachusetts Institute of Technology
Originally presented on: 11/7/2013 13:36:00 - 13:58:00

The crystallization of polymers, like that of small molecules, can be described by a two-stage process of homogeneous nucleation and crystal growth. Yet it is also different from that of small molecules in having chain folding phenomenon and the formation of lamellae structure due to the interplay between long, flexible chain molecules. Chain folding and the lamellar crystalline morphology is essential to the structure and properties of the majority of commodity and high performance plastics, but the understand of their development on the microscopic level remains incomplete.

To study this process, using a realistic united-atom force field, molecular dynamics simulations were performed at about 30% supercooling from the melts of *n*-pentacontahectane (C150) and a linear polyethylene (C1000), both of which are long enough to exhibit the chain folding that is characteristic of polymer crystallization. The nucleation rate was calculated and the critical nuclei were identified using a mean first-passage time analysis. The nucleation rate was found to be insensitive to the chain length in this range of molecular weight. The critical nucleus contains about 150 carbons on average and is significantly smaller than the radius of gyration of the chains, at this supercooling. Using the survival probability method, the critical nucleus size was determined as a function of temperature. A cylinder model was used to characterize the shape of the crystal nuclei and to calculate the interfacial free energies. The interfacial free energies were found to be temperature-dependent. The nucleation free energy barrier and nucleation rate as functions of temperature were also calculated and compare favorably with experiments. At the end of the crystal growth stage, stable crystal lamella forms. A chain segment analysis was performed to characterize the topology of the lamella-melt interface in terms of loops, bridges and tails. The length distribution of loops is broad, supporting the "switchboard model" for the crystals formed at deep supercooling.

References: [1] P. Yi, C.R. Locker and G.C. Rutledge, Macromolecules 2013 (in press).

(694e) Pore-Scale Study of the Effect of Precipitate On Dissolution Using Lattice Boltzmann Method

Kang, Q., Los Alamos National Laboratory
Chen, L., Xi'an Jiaotong University
Originally presented on: 11/7/2013 13:58:00 - 14:20:00

A mesoscopic model based on the lattice Boltzmann method (LBM) is developed to investigate the effects of secondary precipitate on the dissolution of primary solid phase. The model can predict coupled multiple physicochemical processes including fluid flow, mass transport, chemical reaction, dissolution, precipitation consisting of nucleation and crystal growth, as well as dynamical evolutions of pore geometries. Effects of dissolution and precipitation reaction kinetics, molar volumes of primary and secondary solid phases, initial powder size and surface roughness of the primary solid phase, as well as nucleation and crystal growth mechanisms on the dissolution and precipitation processes are investigated in terms of rate and amount of dissolution and precipitation. Several types of dissolution and precipitation processes are identified based on the morphology and structures of the precipitates and the extent the precipitates affect the dissolution of the primary solid phase. Simulation results are also compared with existing experimental results. Depending on the conditions, the precipitates are found to have trivial, favorable or detrimental effects on the dissolution of the primary phase.

(694f) Prediction of Crystal Habits of Urea Using Molecular Dynamics Simulations and Comparision to Experimental Results

Vetter, T., ETH Zurich
Salvalalglio, M., ETH Zurich
Mazzotti, M., ETH Zurich
Parrinello, M., ETH Zurich
Originally presented on: 11/7/2013 14:20:00 - 14:42:00

Crystals grown at moderate and low levels of supersaturation typically exhibit a faceted shape that reflects the ordered arrangement of molecules within the crystal lattice and the environment in which the crystal was grown. In solution crystallization, solvents and additives (or impurities) can significantly change crystal shape by selectivley discouraging (or promoting) the growth of individual crystal faces. While these effects manifest in habit changes on the scale of whole crystals, their underlying reason is the interaction between solvent and additive molecules with the crystal on a molecular scale (Salvalaglio et al., 2012).

Investigating crystal growth at the molecular level is therefore of great importance in order to enhance our current understanding of crystallization processes. Moving towards this goal is a multi-scale (and multi-disciplinary) challenge that can greatly benefit from the interplay between theory, simulations and experimental techniques. In this work we therefore combine advanced molecular simulation methods (Laio and Parrinello, 2002, Barducci et al., 2008, Bonomi et al., 2009) with a theoretical crystal growth model that describes rough crystal growth and growth by the birth and spread (2D nucleation) mechanism (Ohara and Reid, 1973) on a molecular scale. Using molecular simulations that explicitly include solute, solvent and additive molecules allows us to obtain the physical parameters in our theoretical crystal growth model, which allows calculating relative growth rates of different crystal faces. The relative growth rates can then be used to predict crystal habits that originate from the simulated conditions.

Here we examine the paradigmatic case of urea, which grows from vapor exposing three well defined crystal faces: {001}, {110} and {111} (Docherty et al., 1993). Experimental evidence (Vetter, 2012) highlighted that the solvent and additives in solution caused a wide variety of crystal habits that all consisted of combinations (or subsets) of these faces. In order to rationalize this finding we have developed a representation tool, the Shape Diagram (SD), in which all possible steady state morphologies of urea crystals for different growth rate ratios can be assessed.

The rational display of the accessible crystal habits allowed comparing experimental results with computational predictions of the surprisingly wide ensemble of urea growth shapes obtained from different solvents and additives. The experimentally found and computationally predicted crystal habits were found to be in good agreement, thus demonstrating that the developed simulation methodology and theoretical model are able to predict crystal habits from a variety of supersaturations, solvents and additives (Salvalaglio et al.,

2013).

References:

A. Barducci, G. Bussi, and M. Parrinello (2008). Well-Tempered Metadynamics: A Smoothly Converging and Tunable Free-Energy Method. Phys. Rev. Lett., 100:020603

M. Bonomi, A. Barducci, and M. Parrinello (2009). Reconstructing the equilibrium Boltzmann distribution from well-tempered metadynamics. J. Comput. Chem., 30:1615

K. Chow, H. H. Y. Tong, S. Lum, and A. H. L. Chow (2008). Engineering of pharmaceutical materials: An industrial perspective. J. Pharm. Sci., 97:2855-2877.

R. Docherty, K. J. Roberts, V. Saunders, S. Black and R. Davey (1993). Theoretical analysis of the polar morphology and absolute polarity of crystalline urea. Faraday Discuss., 95:11-25.

M. Ohara and R. C. Reid (1973). Modeling crystal growth rates from solution, Prentice Hall Inc., Englewood Cliffs.

A. Laio and M. Parrinello (2002). Escaping free-energy minima. Proc. Natl. Acad. Sci., 99:12562{12566, 2002.

M. Salvalaglio, T. Vetter, M. Mazzotti, and M. Parrinello (2012). Uncovering Molecular Details of Urea Crystal Growth in the Presence of Additives. J. Am. Chem. Soc., 134:17221-17233.

M. Salvalaglio, T. Vetter, M. Mazzotti, and M. Parrinello (2013). Controlling and Predicting Crystal Shapes: The Case of Urea. Submitted to Nature Materials.

T. Vetter (2012). PhD thesis, ETH Zurich, 2012. (http://e-collection.library.ethz.ch/view/eth:6639)

(694g) Predicting Dispersions in Crystallization Process

Singh, M. R., Purdue University
Ramkrishna, D., Purdue University
Originally presented on: 11/7/2013 14:42:00 - 15:04:00

Crystallization is one of the most commonly used and relatively well explored techniques in industries. Yet there are beleaguering issues towards fundamental understanding of this process. One of the curious observations is the appearance of unavoidable dispersions in nucleation and growth rates of crystalline materials grown under homogenous bulk conditions. These dispersions can contribute significantly towards polydispersity in crystalline products. The classical models such as Gibbs critical nucleation theory and Burton-Cabrera-Frank growth models primarily relate nucleation and growth to supersaturation. Therefore, the dispersions in nucleation and growth rates can be related to the local fluctuations in supersaturation. These local fluctuations are driven by the Brownian motion of solute molecules which affects their local concentration and temperature. This article presents a stochastic model to describe fluctuations in supersaturation and thereby dispersions in crystalline materials. The stochastic model is derived from the framework of density and temperature fluctuations. The resulting dispersions in size and shape of nuclei are obtained by multi-dimensional maximization of Gibbs free energy in Morphology Domain of Potassium Acid Phthalate. The model predictions are validated with the experimental measurements of growth rate dispersions in Potassium Acid Phthalate crystals. The proposed framework provides the first rational and quantitative basis for observed dispersions in crystallization processes.

(700a) Sifel-Based Microfluidic Platforms for Liquid-Liquid Extraction of Radioactive Metals

Goyal, S., University of Illinois at Urbana Champaign
Desai, A. V., University of Illinois at Urbana Champaign
Li, H., Washington University School of Medicine
Riechert, D., Washington University in St. Louis
Kenis, P. J. A., University of Illinois at Urbana-Champaign
Originally presented on: 11/7/2013 12:30:00 - 12:55:00

Radioisotopes of metals (*e.g.*, Cu-64 and Y-86) produced employing the cyclotron are crucial in nuclear medicine for Positron Emission Tomography (PET) imaging. The extraction of radioisotopes produced from the cyclotron involves dissolution of the target material (containing impurities) in an aqueous acid followed by separation of the isotope of interest from the bulk solution using liquid-liquid extraction (LLE). Conventionally, the UNEX process has been developed to isolate radionuclides from an acidic waste [1-3]. However, the UNEX process is designed for large-scale separation of radionuclides and thereby requires considerable dilution of the radionuclide and requires large quantities of organic solvents. In addition, the process requires large-scale radiation shielding resulting in large size of the equipment, as well as long times for interface stabilization and reduced kinetics resulting from dilution.

These issues can be overcome by using microfluidic platforms that require low sample volumes (eliminating the need for excessive dilution), allow rapid interface stabilization, minimize the wastage of organic solvents, and enhance the extraction efficiency of radioactive metal resulting from high interfacial area and small diffusional lengths. Specifically, we are interested in reactive extraction of isotope of interest (Cu-64) from the mixture in aqueous acid phase with the chelating agents into the organic phase [4, 5, 6]. In this work, we present on the design, development, and optimization of solvent resistant polymer-based microfluidic platform to enable LLE for separation of radioactive metals (*e.g.*, Cu-64) of interest from the mixture obtained from the cyclotron. We developed polymer based microfluidic platforms over glass or silicon-based microfluidic platforms to enable simple and less involved fabrication procedure. We systematically optimized several factors affecting the interface stability with the aid of an analytical model, including geometry of the micro-channel, surface functionalization, and the flow rates of the two immiscible streams as well as predicted extraction efficiencies as a function of different paramters.

In this work, we demonstrate the application of the developed microfluidic platform to extract radioactive copper (Cu-64) from an aqueous solution into a solution of a chelating agent in an organic solvent (*e.g.*, 2-Hydroxy-4-*n*-octyloxybenzophenone Oxime in toluene) followed by back extraction into the aqueous phase [7]. We studied the effect of pH, contact time between the phases (length of the microchannel), concentration and the type of chelating agents, concentration of the isotope of interest, as well as of the impurities introduced in the solution on the extraction efficiency. Radioactive dosimeter and ICP-OES or ICP-MS was used to quantify extraction efficiencies. The developed microfluidic platforms provided quantitative information on the extraction process compared to conventional techniques, due to the superior control over the spatial and temporal properties of the interface and enabled rapid extraction times and high extraction efficiencies owing to minimal diffusion lengths.

[1] J. D. Law, K. N. Brewer, R. S. Herbst, T. A. Todd and D. J. Wood, Waste Management 19 (1999) 27-37.

[2] J. D. Law, R. S. Herbst, T. A. Todd, V. N. Romanovskiy, V. A. Babain, V. M. Esimantovskiy, I. V. Smirnov and B. N. Zaitsev, , Solvent Extraction and Ion Exchange 19 (2001) 23-36.

[3] V. N. Romanovskiy, I. V. Smirnov, V. A. Babain, T. A. Todd, R. S. Herbst, J. D. Law and K. N. Brewer, Solvent Extraction and Ion Exchange 19 (2001) 1-21.

[4] Y. Baba, M. Iwakuma and H. Nagami, Industrial & Engineering Chemistry Research 41 (2002) 5835-5841.

[5] T. Tasaki, T. Oshima and Y. Baba, Talanta 73 (2007) 387-393.

[6] C. Priest, J. Zhou, R. Sedev, J. Ralston, A. Aota, K. Mawatari and T. Kitamori, International Journal of Mineral Processing 98 (2011) 168-173.

[7] S. Goyal, A.V. Desai, R. Lewis, D. Ranganathan, H. Li, D.E. Riechert, and P.J.A. Kenis, in preparation.

(700b) Modeling and Experimental Validation of Free Fatty Acids (FFAs) Removal From Crude Vegetable Oils Using Fiber Reactor Technology

Moran, C., Texas A&M University-Kingsville

Mills, P. L., Texas A&M University-Kingsville

Rivas, R. C., Texas A&M University- Kingsville

Originally presented on: 11/7/2013 12:55:00 - 13:20:00

Modeling and experimental validation of Free Fatty Acids (FFAs) removal from crude vegetable oils using fiber reactor technology

A combination of experimental measurements, reactor modeling, and model validation is used to develop a highly efficient approach for removal of free fatty acids (FFA) from vegetable oils. The process revolves around a new reactor concept called the Fiber Reactor?, which is a novel contactor for conducting reactions involving two partially miscible liquid phases that offers from one to two orders-of-magnitude improvement in liquid-liquid mass transfer rates when compared to conventional reactors, such as stirred tanks. The primary objective of this presentation is to describe recent research results that provide further understanding of Fiber Reactor? performance by investigating intricacies of the engineering technology for application to FFA extraction. Emphasis is placed on comparing the effectiveness of reactive and non-reactive extraction technologies for removal of FFA's from crude vegetable oil by comparing observed mass transfer coefficients, acid removal efficiency and overall cost of operation. Fiber Reactor? parameters of interest include: fluid pressure drop, fluid flow patterns, fluid flow regimes, liquid-fiber interfacial areas, mass transfer coefficients, heat transfer coefficients, and fiber particulars (material of construction, layout, size, geometry, and surface/wetting properties). Validation of the model-predicted results is also provided using experimental data generated using fiber reactor having diameters of ¼-in and ½-in, respectively.

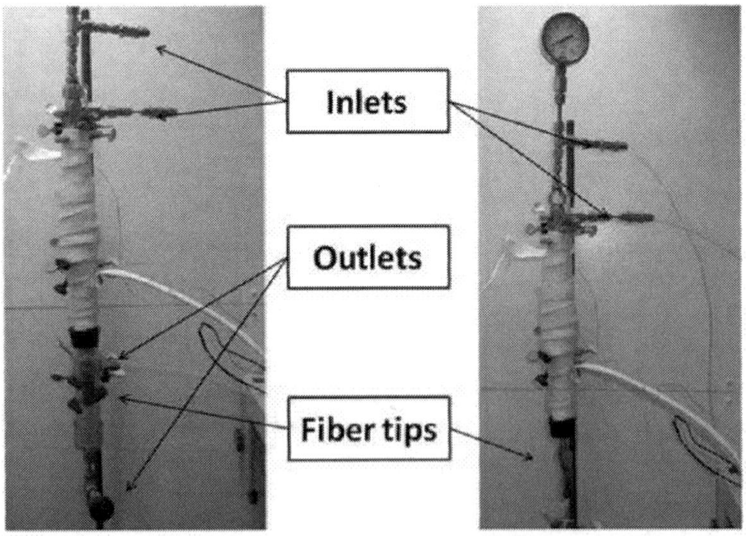

Figure 1. Schematic of the Fiber reactor

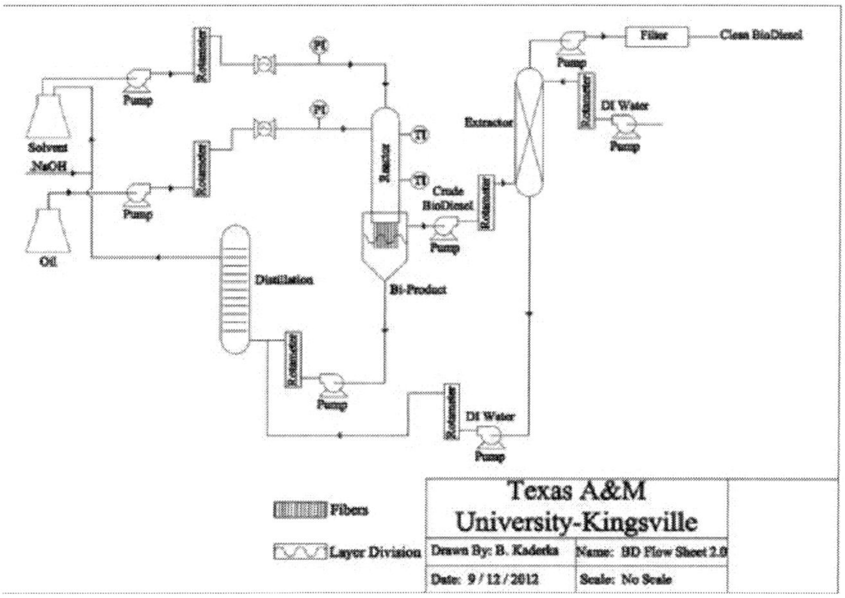

Figure 2. Process flow diagram of a typical experimental setup for transesterification of lipids.

(700d) Environmentally Benign Vegetable Oil (coconut oil) - Based Supported Liquid Membrane for Simultaneous Extraction and Recovery of Heavy Metals (Cd & Pb) From Its Aqueous Solutions

Bhatluri, K. K., IIT Guwahati
Manna, M., IIT Guwahati
Saha, P., Indian Institute of Technology Guwahati
Ghoshal, A. K., Indian Institute of Technology Guwahati
Originally presented on: 11/7/2013 13:45:00 - 14:10:00

The growing environmental concern all over the world obligates the chemical industries to minimize the discharge of effluents (wastewater) that may contain various pollutants, especially heavy metals (Hg, Cd, Pb, Ni, As and Cr etc.), that are hazardous to environment and living organisms [1-3]. Heavy metals are dangerous and not biodegradable contaminants that tend to be accumulated in living organisms, causing various diseases and disorders as it leads to illness such as brain damage, memory loss, chromosome breakage, dysfunction of liver, kidney and central nervous system, blindness, deafness and hence it is important that they be removed before effluent disposal. The major sources of effluent containing heavy metals (*viz.*Cd and Pb) are industries like Chloro-alkali, pulp & paper, battery, fertilizer, mining & combustion and rubber processing [4-7]. Major bottleneck of the separation of heavy metals from industrial effluent is the fact that they are found in trace quantities. The separation and purification of trace amount of solute from their dilute solutions has been an interesting problem for the scientists and engineers for decades.

Some established [8] techniques of heavy metal removal from wastewater include chromatography, adsorption and ion-exchange, crystallization, precipitation and coagulation, solvent extraction, *etc.* However the applications of such processes are having their own merits and demerits. Hence, it is necessary to develop some cost effective yet environment friendly alternatives for recovery of trace elements from their dilute solutions. In recent times, membrane based technology has emerged as an efficient process for concentration and purification of pollutants in aqueous solutions. Liquid membrane based separation technique is one such extraction/recovery process where transport of solute (metal component) occurs through two extraction processes in series – from solute rich source phase to solvent phase and thereby from solvent phase to strip phase. The liquid membrane based separation can be implemented through various configurations such as Bulk Liquid Membrane (BLM), Supported Liquid Membrane (SLM) and Emulsion Liquid Membrane (ELM) [2].

This work is aimed at investigation of simultaneous extraction and recovery of heavy bi-metals (cadmium and lead) from aqueous solutions through a coconut oil-based supported liquid membrane (SLM). The selection of extractant (carrier) and diluent were carried out by two phase equilibrium studies from various environmentally benign solvents, extractants and their combinations. Aliquat 336 (carrier agent) in coconut oil was found as the best combination for the extraction purpose. Ethylenediaminetetraacetate (EDTA) was selected as stripping agent keeping in view of its chelating capacity to metal ions. The performance of LM based separation process was investigated by studying the effect of various operating conditions *viz.* carrier concentration in membrane phase, stripping agent concentration, feed phase pH and molar feed ratio of Cd (II) and Pb(II). The optimum process conditions for the heavy metals transportation was found as follows: feed phase pH of 6.5, membrane phase as 0.5 % (v/v) aliquat 336 in coconut oil, polymeric PVDF membrane of pore size 0.2μm as solid support and 0.015 *M* EDTA as stripping agent. The initial feed concentration was 5 mg L^{-1} and total operating time was 10 h. The optimum extraction and recovery of cadmium were found as 79% and 67% respectively and that for lead were found as76% and 66% respectively.

Literature Cited

[1] M. Mulder, Basic principle of membrane technology, Kluwer Academic Publishers, Dordrecht, 1991.
[2] V.S. Kislik, Liquid membranes Principles & Applications in Chemical Separation & Wastewater Treatment, British Library, Elsevier, 2010.
[3] K. Chakrabarty, Liquid membrane based technology for removal of pollutants from wastewater, Doctor of philosophy thesis from IIT Guwahati, 2010.
[4] K.P. Lisha, S.M. Maliyekkal, T. Pradeep, Manganese dioxide nanowhiskers: A potential adsorbent for the removal of Hg (II) from water. Chemical Engineering Journal, 160 (2010) 432–439.
[5] F. Zahir, S.J. Rizwi, S.K. Haq, R.H. Khan, Low dose mercury toxicity and human health. Environmental Toxicology and Pharmacology, 20 (2005) 351-360.
[6] R.N. Yadava, Waste water treatment and waste management: proceedings of the international conference on water and environment, Allied Publishers Private Limited, 2003.
[7] K.A. Krishnan, T.S. Anirudhan, Removal of mercury (II) from aqueous solutions and chlor-alkali industry effluent by steam activated and sulphurised activated carbons prepared from bagasse pith: kinetics and equilibrium studies. Journal of Hazardous Materials B, 92 (2002) 161–183.
[8] W. Kaminski, W. Kwapinski, Applicability of liquid membranes in environmental protection. Polish Journal of Environmental Studies, 9 (2000) 37-43.

(700e) Central Composite Design for the Modeling of the Glycolic Acid Reactive Extraction

Datta, D., Birla Institute of Technology & Science, Pilani
Kumar, S., Motilal Nehru National Institute of Technology (MNNIT)
Originally presented on: 11/7/2013 14:10:00 - 14:35:00

The development of an industrial process requires study of the effect of various operating parameters which can be achieved by exhaustive experimentation. The experiments are carried out by varying numerous experimental units to evaluate the performance of the system in terms of single/many output variable (s). These experimental data are useful to draw many valuable results and inferences about the system and the process. Therefore, an experimenter needs to plan and design the experiments, and analyze the results. Central composite design is applied for the construction of empirical models based on experimental design and data. The developed model is then used to estimate the optimum of the process variables (parameters) to maximize or minimize the dependent variable. This method is widely applied for the modeling and optimization of separation processes for the extraction of various compounds from different liquid and solid phases. The equilibrium reactive extraction of glycolic acid from aqueous solution is studied using tri-*n*-octylamine (TOA) as extractant dissolved in organic solvents (MIBK). Experimental studies are designed considering central composite design method to investigate the main and interaction effects of initial acid concentration in the aqueous phase (C_{in}, mol/L), initial amine composition in the organic phase (, %v/v) and temperature (^{o}C) on the degree of extraction, Y (%). The developed model is then used to find out optimum process parameters of reactive extraction.

(700f) Extraction Properties of Novel Polyethylene Oxide Diphosphate-Based Chelating Agents Towards Lanthanides and Their Solubilities in Supercritical Carbon Dioxide

Duan, D.,
Su, B., Zhejiang University
Zhang, Z., Zhejiang University
Bao, Z., Zhejiang University
Yang, Y., Zhejiang University
Ren, Q., Zhejiang University
Originally presented on: 11/7/2013 14:35:00 - 15:00:00

These CO_2-philic open-chain organophosphorous chelating ligands, i.e. diethylene glycol bis (2, 4, 4-trimethyl-1-pentyl) dimethyl diphosphate (EG2TP), diethylene glycol bis (3, 5, 5-trimethyl-1-hexyl) dimethyl diphosphate (EG2TH), and diethylene glycol dioctyl dimethyl diphosphate (EG2Oct), ethylene glycol bis(2-isopropoxyethyl) dimethyl diphosphate (EG2IPE), triethylene glycol bis(2-isopropoxyethyl) dimethyl diphosphate (EG3IPE), tetraethylene glycol bis(2-isopropoxyethyl) dimethyl diphosphate (EG4IPE) were synthesized and characterized. Solubilities of these ligands in $scCO_2$ were determined at different combinations of temperature (313.15-333.15 K) and pressure (9-20 Mpa), which increase in the order of EG4IPE (MW= 554.50) < EG3IPE (MW= 510.45) < EG2Oct (MW= 518.56) < EG2IPE (MW= 466.40) < EG2TH (MW= 546.61) < EG2TP (MW= 518.56). Solubility parameter and connectivity index of these compounds were calculated and discussed, which indicated that branched side chains of the ligands play an important part in increasing solubility in $scCO_2$. These experimental data are in agreement with computed data via a semi empirical model, in which the average absolute relative deviations lie in the range of 2.3% to 5.0%. Also the quantum chemical calculation and COSMO-RS method were introduced to estimate the structure effect on the interaction with CO_2 by computing the charge density distribution on the surface of these compounds. As expected, better masking effect of the side chains results in higher solubility. The effect of these ligands on supercritical fluid extraction of selected rare earth metals (La^{3+}, Ce^{3+}, Pr^{3+}, Nd^{3+}, Sm^{3+}, Gd^{3+}, Er^{3+}, and Yb^{3+}) was investigated at 313.15 K and 20 Mpa. The extraction efficiency of the system was found to increase in the order EG4IPE < EG3IPE < EG2IPE < EG2Oct < EG2TP < EG2TH with a range from 55% to 92%. The rationale behind different selectivities towards these metals was also discussed in comparison to other traditional organophosphorous agents.

(702a) Low-Pressure Sweep Membrane Modules for Carbon Dioxide Capture

Kniep, J., MTR
Hao, P., MTR
Chan, K., MTR
Fulton, D.,
Freeman, B., Electric Power Research Institute
Merkel, T., MTR
Baker, R., MTR
Originally presented on: 11/7/2013 12:30:00 - 12:55:00

To mitigate the harmful effects of global climate change, CO_2 in power plant flue gas must be captured and either sequestered or utilized in processes such as enhanced oil recovery. Membrane technology is an attractive CO_2 capture option because of advantages such as energy-efficient passive operation, tolerance to acid gases and oxygen, no hazardous chemical emissions or handling issues, minimal water requirements, and no steam use requiring modifications to the existing boiler.

Working with DOE, MTR has previously demonstrated the feasibility of a two-step membrane process design to recover CO_2 from power plant flue gas. The second step of this process is a selective sweep membrane module that strips CO_2 from the flue gas and recycles it back to the boiler with the incoming combustion air. MTR is developing a membrane module specifically designed for high gas flow, low pressure, sweep operation. These membrane modules contain up to 500 m^2 of membrane area, which is 10 to 20 times the membrane area of modules currently being used for flue gas CO_2 capture. These large area modules will reduce the manifolding complexity, footprint, and cost of the very large membrane plants required for flue gas treatment. By lowering the pressure drop through the modules, less compression/blower energy is needed to move gases through the module, which reduces the parasitic energy loss of the capture process. In this presentation, results of the development work to date and future plans, including slip stream field testing at the National Carbon Capture Center (NCCC) operated by Southern Company in Birmingham, AL, will be discussed.

The development of large, efficient, low cost, sweep membrane modules is an enabling technology for membrane CO_2 capture and sequestration systems. MTR is investigating several alternative process designs to remove CO_2 from coal power plant flue gas, and while all of the designs use a membrane contactor to recycle CO_2 to the power plant boiler, the second CO_2 separation unit need not be a membrane unit. The selection of different hybrid separation technologies, hybrid process designs, and process economics will be presented.

(702b) Major Interest of Hollow Fiber Membrane Contactors for An Ammonia-Based CO2 Capture Process: Detailed Analysis of An Innovative Process

Makhloufi, C.,
Roizard, D., Nancy Universite
Favre, E., Laboratoire Reaction et Genie des Procedes LRGP- CNRS, Universite de Lorraine
Originally presented on: 11/7/2013 12:55:00 - 13:20:00

Up to now, the reference technology for CO_2 capture in postcombustion makes use of monoethanolamine (MEA) as chemical solvent in a packed column. Despite high capture efficiencies, MEA induces too high energy penalties. Consequently, an abundant literature is discussing the use of aqueous ammonia for flue gas scrubbing. Indeed ammonia is cheaper than MEA, shows a higher absorption capacity and, most of all, its energy regeneration demand is much lower. Nevertheless, the high ammonia volatility is a considerable drawback which causes a high NH_3 slip within the flue gas. Specific operatic conditions are required to limit NH_3 losses and a post absorber scrubbing step is therefore needed.

Amongst growing technologies, membrane contactors appear as a promising technology able to allow significant process intensification when compared to conventional gas/liquid contactors. So far, no study dealing with the use of membrane contactors for the CO_2 capture in aqueous ammonia has been disclosed. For the first time, a detailed analysis of an ammonia based CO_2 capture process using hollow fiber membrane contactors will be given during this presentation. Particularly, membrane contactors ability to both improve CO_2 mass transfer performances and mitigate NH_3 volatilization compare to conventional gas/liquid contactors will be discussed.

First of all, it will be shown that CO_2 absorption in ammonia can not be performed using conventional microporous hollow fiber membrane contactors and composite membranes made of a thin polymeric layer facing the liquid phase must be used. In that sense, composite hollow fiber membrane contactors have been specifically designed, prepared and tested to favor CO_2 absorption and, in the same time, to limit the ammonia slip in the flue gas.

This involved precisely selecting the dense layer and it will be shown how an inadequate material selection could lead to its significant degradation. The chosen dense skin must be thin, highly permeable to CO_2 and selective for CO_2 over NH_3. This should allow reaching high CO_2 mass transfer coefficient while reducing drastically the ammonia slip in the flue gas. This was undoubtedly one of the main challenges. Indeed, it has to be recalled that a faster permeation of CO_2 compared to NH_3 in dense polymers was totally unexpected and to our knowledge unexplored. For the first time, 4 chemically resistant polymers showing a reverse selectivity CO_2/NH_3 were identified. These meaningful results obtained thanks to time lag experiments will be presented in this presentation.

In a second step, CO_2 absorption experiments in aqueous ammonia using lab scale composite hollow fiber membrane contactors (CHFMC) have been successfully carried out. The effect of the main operating conditions namely ammonia concentration, temperature, gas and liquid flow rates, nature of the dense layer, on the CO_2 absorption performances and the ammonia loss has been studied. The overall volumetric CO_2 absorption capacity in CHFMC has been compared to the one obtained in packed columns. It has been concluded that membrane contactors strongly enhance the CO_2 absorption performances in comparison to conventional contactor. Moreover, a sharp decrease of the ammonia slip has been observed. The above results will be detailed in the presentation and the performances and advantages of this new CO_2 absorption process will be discussed further.

(702c) Membrane Assisted Liquid Absorbent Regeneration (MALAR):Innovative Use of Membranes in Liquid Absorbent Based CO2 Capture

Feron, P.,
Zhao, S.,
Originally presented on: 11/7/2013 13:20:00 - 13:45:00

CO_2 capture and geological storage (CCS) is a necessary option to limit CO_2 emissions from existing power stations. In the CCS chain the capture and compression of CO_2 to pipeline or injection pressure represents around 70% of the costs and all of the energy requirements of the process. Typical output reductions for 90% capture of CO_2 at a power station are around 30% and these are very large indeed. The leading capture technology is based on a liquid absorbent which chemically reacts with CO_2. The liquid absorbent is regenerated thermally and it will be re-used during many cycles. The CO_2produced from this process is also of high purity and can easily be stored in the underground after compression.

The energy required for the liquid absorbent regeneration is extracted as steam from the power station. This steam is then not available for power generation. Hence the output of the power station is reduced, which is undesirable. It is therefore necessary to develop absorbent regeneration processes which have much lower energy requirement. The starting point is to analyse the thermal energy requirement in more detail. It has the following contributions:

- The energy required to break the bond between CO_2 and the active component in the solvent as determined by the chemical formulation of the liquid absorbent.
- The heat required for bringing the solvent up to the regeneration temperature, which is a function of the CO_2 loading of the liquid absorbent.
- The evaporation enthalpy for the stripping steam which leaves the stripper together with the CO_2, which is determined by the temperature of the steam-saturated CO_2-product leaving the stripping column.

The overall energy penalty also incorporates the electricity use of the capture and compression process. This penalty is however to a large extent determined by the heat required to release the CO_2 from the liquid absorbent, typically around 4 MJ/kg CO_2. Dependent on the fuel type, this represents between 20 and 40% of the energy contained in the incoming fuel.

The ongoing development of new liquid absorbents will very likely reduce the first two contributions, i.e.

new amines with lower binding energy and higher CO_2-loadings have been identified. However the energy required for the production of stripping steam is still needed. In fact all the energy needed for the regeneration is brought in via the stripping steam and is transferred to the liquid absorbent via condensation. Hence efficient control of the steam balance is important for the energy performance of the capture process.

The MALAR process builds on this foundation, aiming to reduce the energy requirement to less than 2 MJ/kg CO_2 using a **M**embrane **A**ssisted **L**iquid **A**bsorbent **R**egeneration process. This process concept lends itself for application with any type of liquid absorbent.

The proposed innovation aims to efficiently integrate heat recovery in the liquid absorbent regeneration process. This is achieved by the use of membranes for transfer of heat using condensation and evaporation processes. The process concept is an entirely new one utilising membranes as gas-liquid contactors in such a way that the (common) occurrence of liquid leakage through the membrane is not an issue.

In the MALAR process energy is recovered via transfer of steam through a membrane. In the membrane condenser latent heat is recovered from the wet CO_2 product into a portion of the cold CO_2-rich liquid absorbent; in the membrane evaporator latent heat is recovered from the hot CO_2-lean absorbent into the stripping column. The use of membranes for direct heat transfer via evaporation or condensation enables quite small approach temperatures which would be unachievable and/or uneconomical with conventional heat exchangers. The compactness of the membrane equipment further adds to the achievement of small temperature differences and thereby a very energy efficient process. Two porous membrane configurations using either hyrodophobic or hydrophilic are proposed for the exchange of latent heat.

This application of membranes is a so-called membrane contactor application, i.e. the membrane is used to contact a gas flow with a liquid flow. Such operations often suffer from minor amounts of liquid permeating through the membrane to the gas phase. This liquid is subsequently carried away by the gas flow, which is generally not desired, and particularly if the gas enters the atmosphere. It will furthermore hamper the efficient transfer of components between the two phases. In the MALAR application envisaged here, liquid leakage is not critical as the heat transfer function is not affected and the liquid will be contained in the stripper column and mixed with the liquid absorbent already present.

A 50% reduction in thermal energy required for the liquid absorbent regeneration can be achieved by using the MALAR concept in combination with improved liquid absorbents, resulting in a thermal energy requirement below 2 MJ/kg CO_2. The improvements are coming from:

- A lower gas temperature exiting the stripper column reducing the evaporation loss
- A closer temperature approach through the use of the membrane heat exchangers
- The liquid absorbent having a double CO_2-loading and lower reaction enthalpy, which a feature of new liquid absorbents developed over the last 2 decades.

In summary: the benefits of the MALAR are:

- Energy efficient capture process through novel heat recovery method
- Smaller size equipment through use of high specific surface area membrane equipment
- Applicable to a wide range of liquid absorbents
- Different membrane types (hydrophobic and hydrophilic) are feasible
- Use of a membrane contactor in a non-critical application with respect to liquid leakage

(702d) Pressure Swing Membrane Absorption for Pre-Combustion CO2 Capture From Shifted Syngas

Sirkar, K. K., New Jersey Institute of Technology
Chau, J., New Jersey Institute of Technology
Jie, X., New Jersey Institute of Technology
Obuskovic, G., New Jersey Institute of Technology
Originally presented on: 11/7/2013 13:45:00 - 14:10:00

An advanced pressure swing membrane absorption (PSMAB) technology utilizing a hollow fiber membrane contactor device and a cyclic absorption-desorption process is being developed to produce purified hydrogen at a high pressure for IGCC-CCS plant's combustion turbine from low temperature (L-T) post-shift reactor synthesis gas and simultaneously obtain a purified CO_2 stream suitable for sequestration. The low temperature water gas shift reactor product stream is likely to be available at ~20 atm and around 150-200^{0}C. The experimental studies are employing He as a surrogate for H_2. Studies have been carried out with a 40%CO_2-60% He feed gas mixture coming in at temperatures up to 100^{0}C and pressures varying from 100 to 250 psig (689-1723 kPag). The nonvolatile absorbent liquid consists of an ionic liquid 1-butyl-3-methylimidazolium dicyanamide ([bmim][DCA]) (among others) containing nonvolatile amines. Two different cyclic processes have been studied; most of the performance will focus on a 5-valve cycle. Process performance studies have focused on the duration of various parts of the cycle, the effects of pressure and temperature variation, and the effect of hollow fiber membrane module design on the quality of the two product streams namely, the He-rich stream and the CO_2-rich stream. A numerical model has been developed to predict the process performance of the simpler 3-valve cycle for pure ionic liquid as the absorbent. Solubility and diffusion coefficient data for the gases are needed in this model; these were determined for the pure ionic liquid absorbent using a pressure-decay-dual transducer apparatus at temperatures of 323, 353, 363, and 373 K and at pressures up to 1.38 MPa (~200 psig).

(702e) Application of Composite Polymeric Membranes for Efficient Regeneration of Physical Solvents

Pesaran, A.,
Tande, B., University of North Dakota
Originally presented on: 11/7/2013 14:10:00 - 14:35:00

The most dominant CO_2 capture technology used for pre-combustion capture involves the application of physical solvents. Unlike chemical solvents, physical solvents do not react with the solute and they physically dissolve the acid gases, which are then stripped without the need to be heated by means of pressure swing techniques or combination of heat and pressure letdown. However, despite low energy required to regenerate the physical solvents and their high capacity to capture and separate acid gases from the syngas produced in a gasification plant, physical solvents have some disadvantages including CO_2 pressure loss and the energy required to pump the solvent to the high pressure absorber.

The primary objective of this work is to evaluate the use of composite polymeric membranes, for the recovery of CO_2 from CO_2-rich solvent streams. To achieve this purpose, an experimental bench-scale setup was built to investigate and quantify CO_2 removal capacity from the rich solvent across different types of membranes. This setup primarily consists of a high pressure absorber where solvent is loaded and saturated with CO_2 and the membrane module where CO_2 diffuses through the membrane. In order to increase the driving force across the membrane, sweep gas is used (Nitrogen). To evaluate the absorption process, absorber temperature and pressure is measured and recorded continuously using a NI data logger and the LabView software. The absorber is equipped with a home-made cooling water coil and a relief valve to adjust the absorber temperature and pressure respectively. Two 400 W inline pencil heaters are used and controlled with a temperature controller to adjust the solvent temperature downstream of the membrane module. The solvent line temperature and pressure are also recorded continuously. CO_2 concentration in the sweep gas stream is measured using the Non-Dispersive Infrared CO_2 analyzer and Agilent 7850A GC- FID depending on the concentration of CO_2 in the sweep gas. Dimethyl ether of polyethylene glycol (Selexol) is used as the solvent since it is reputed to be one of the major physical solvents for CO_2 removal. To evaluate the effectiveness of different types of membranes, CO_2 permeation rate and membrane selectivity were measured for different membranes. To better understand the effect of system pressure and solvent flow rate on different experiment responses such as CO_2 flux, selectivity and % recovery, a two-factor two -level full factorial design with two replicates an three center points were performed with the membrane that showed the most promising results in the screening study.

The results of the screening study indicated that PDMS based membranes (PERVATECH and PERVAP

4060) shows higher CO_2 permeability compared to PVAOH based membranes (PERVAP 1211 and PERVAP 1201). The best membrane for further analysis and experiments to find the optimum operational conditions was chosen as PEVAP 4060 from SULZER due to its high CO_2 flux and selectivity compared to other membranes.

A statistical analysis was performed to identify the significant factors for each individual response such as permeation rate, leak rate and selectivity. For CO_2 flux, pressure appeared to be strongly significant. However, solvent flow rate had no significant effect on the rate of CO_2 permeation. The immediate conclusion from this observation is that the main mass transfer is controlled by the membrane. With respect to the solvent leak, the analysis of Pareto charts suggested pressure to be significant and solvent flow rate to be insignificant. As the system pressure increases, the liquid on the upper chamber of the membrane module forces itself more into the membrane and hence, solvent leak increases. However, increasing the solvent flow rate leads to higher liquid velocity on top of the membrane and shows no significant changes on the rate of solvent leak. Neither system pressure nor solvent flow rate found to be significant considering the selectivity as the experiment's response. Finally, regarding the percent recovery, both the system pressure and solvent flow rate appeared to be significant. This is mainly because at elevated pressures, mole fraction of CO_2 in solvent increases and thus a higher driving force for CO_2 permeation exists. As a result of this, percent of recovery increases by pressure as confirmed by the Main Effects plots. On the other hand, by increasing the solvent flow rate, more amount of CO_2 enters the upper membrane chamber but the mass transfer resistant through the membrane prevents more CO_2 from being transported. Thus, there is the condition of introducing more CO_2 to the upper chamber but still a slow mass transport process being controlled by the membrane which eventually decreases the separation ability of the membrane (percent recovery).

In order to examine the chemical stability and structural integrity of the membranes after being exposed to the high pressure solvent, a series of post-experiment characteristic tests such as FTIR and DSC were performed. The results of these studies revealed no major changes.

(702f) Efficient Capture of CO2 From Pulverized Coal Fired Plants

Dada, E. A., ChemProcess Technologies (CPT), LLC
Olanrewaju, K., University of Iowa
Drayton, R. A., THS, Inc
Originally presented on: 11/7/2013 14:35:00 - 15:00:00

As demand for fossil fuel based energy increases, the need to sustainably manage and mitigate carbon dioxide (CO_2) emissions, the largest volume contributor to global warming, is unequivocally necessary. Various studies discussing different methods of capturing/sequestrating $CO_{2,}$ either by separation from a stream of effluent flue gases after a combustion process or through fixation, have been thoroughly reviewed. These methods range from chemical and physical absorption and adsorption, to forestation, Ocean fertilization, photosynthesis process, mineral carbonation and other *micro*-scale sequestration schemes. As this study will focus more on the technical options of separating CO_2 from pulverized coal-fired plant (a *post* combustion process), the current method of chemically absorbing CO_2 from stream of flue gases is fraught with challenges such as: the energy intensity of the process, limited loading capacity, solvent degradation and the corrosive effect of the solvent upon the process units. While physical absorption and adsorption address some of the limitations within chemical sorption process, the shortcomings still persist in their inability to handle large volume of flue gases with low CO_2 partial pressures. This work discusses two novel solutions that address the issues concerning the current technical approaches of sorbing carbon from effluent flue gases at coal-fired plants. It is proposed to design an energy efficient process that is cost effective and novel solvent formulation that an enhance CO_2 loading capacity due to their higher chemical stability and lower corrosive characteristics.

(707b) CO2 Capture and Conversion Based On ILs
Zhang, S., Institute of Process Engineering, Chinese Academy of Sciences
Originally presented on: 11/7/2013 12:55:00 - 13:20:00

CO_2 Capture and Conversion Based on ILs

Suojiang Zhang*, Jian Sun

Institute of Process Engineering, Chinese Academy of Sciences,100190, Beijing, PR China

**Corresponding author: sjzhang@home.ipe.ac.cn*

An ionic liquid based CO_2 capture technology has been developed. To enhance the absorption efficiency, a series of ILs were synthesized and screened including amino functionalized IL.[1] The performance of IL mixture for CO_2 capture was also carried out.[2]The system integration result indicated that the new process resulted in about 27% reducing of energy consumption comparing to the traditional amine process.

As a way out for CO_2 capture, the synthesis of cyclic carbonates via cycloaddition of CO_2 to epoxides is studied, where the synthesized ILs are found to be very efficient catalysts for the cycloaddition reaction. Our previous work indicated that the hydroxyl group in solvent, IL or carrier all could play a promotion role on the reaction. [3] Based on the fundamental understanding, hydroxyl or carboxyl-functionalized ILs were developed and found to be much more active.[4] Furthermore, these ILs are chemically immobilized on solid supports, including silica, resin, zeolites, and chitosan, which show high stability and efficiency.[5]

Refferences

[1] a) X. P. Zhang, X. C. Zhang, H. F. Dong, Z. J. Zhao, S. J. Zhang, Y. Huang. ***Energy Environ. Sci***., 2012, 5: 6668-6681. b) J. M. Zhang, J. Sun, X. C. Zhang, Y. S. Zhao, S. J. Zhang. ***Greenhouse Gas Sci Technol***., 2011, 2:142-159.

[2] a) Y. S. Zhao, X. P. Zhang, Y. P. Zhen, H. F. Dong, G. Y. Zhao, S. J. Zeng, X.Tian, S. J. Zhang. ***Int. J. Greenh. Gas Con***., 2011, 5:367-373; b) Y. S. Zhao, X. P. Zhang, H. F. Dong, Y. P. Zhen, G. H. Li, S. J. Zeng, S. J. Zhang. ***Fluid Phase Equilib***., 2011,302: 60-64.

[3] a) J.Q. Wang, J. Sun, W.G Cheng, S.J. Zhang. ***Phys. Chem. Chem. Phys.***, 2012, 14, 11021-11026; b) J.

Sun, J. Y. Ren, S. J. Zhang, W. G. Cheng. ***Tetrahedron Lett.***, 2009, 50(4): 423-426.

[4] a) X. Chen, J. Sun, J. Q. Wang, W. G. Cheng. ***Tetrahedron Lett.***, 2012, 53: 2684-2688; b) J. Sun, L. J. Han, W. G. Cheng , J. Q. Wang, X. P. Zhang, S. J. Zhang. ***ChemSusChem.***, 2011, 4(4): 502-507; c) J. Sun, S. J. Zhang, W. G. Cheng, J. Y. Ren. ***Tetrahedron Lett.***, 2008, 49(22): 3588-3591.

[5] a) W. G. Cheng, X. Cheng, J. Sun, J.Q. Wang, S. J. Zhang. ***Catal. Today***, 2013, 200:117-124; b) J. Sun, J.Q. Wang, W. G. Cheng, J. X. Zhang, X. H. Li, S. J. Zhang, Y. B. She. ***Green Chem.***, 2012, 14: 654-660. c) J. Sun, W. G. Cheng, W. Fan, Y. H. Wang, Z. Y. Meng, S. J. Zhang. ***Catal. Today***, 2009, 148: 361-367.

Acknowledgments

This work was supported by National Basic Research Program of China (2009CB219901), and National Science Fund of China (21006117).

(707c) Modeling of CO2 Capture Using Microalgae

Andrade, L. M., University of Sao Paulo
Mendes, M. A., University of Sao Paulo
Silveira, C. A.,
Oller Nascimento, C. A., University of Sao Paulo
Originally presented on: 11/7/2013 13:20:00 - 13:40:00

In order to model the growth of microalgae in a open-pond reactor, a 1000L system has been built and operated under realistic outdoor conditions. Environmental data (light intensity, temperature, pH) were captured with a datalogger and the growth was measured. A phenomenological mathematical model was built for the open reactor. The comparison of the simulated data to the experimental data shows good agreement.

(707d) Biogas Sweetening Via a Carbamate Crystallization Approach

Zheng, S., ZhejiangUniversity
Liu, Q.,
He, Y.,
Shi, Y.,
Originally presented on: 11/7/2013 13:40:00 - 14:00:00

The CO_2 in biogas significantly limited its potential applications. The regeneration process of aqueous alkanolamine based CO_2 capture often requires a large amount of energy, which makes the process costly. In order to enhance the energy efficiency of the capture, a carbamate crystallization approach was proposed in this study where water was replaced with non-aqueous solvents. It was observed that triethylenetetramine (TETA) carbamate crystallize out as supernatant solid from ethanol solution upon the adsorption of CO_2. A TETA/non-aqueous solvent solution shows higher absorption capacity and absorption rate than alkanolamine-water solution. The mole ratio of CO_2/TETA in carbamate was determined to be around 2:1 by elemental analyzer. The regeneration capabilities of solid carbamates was also investigated. Thermal behavior of the carbamate was examined with differential scanning calorimetry and thermogravimetric analysis. It was shown that the decomposition of TETA-carbamate commenced at about 90°C. The separation of carbamate as solid phase can offer great advantage by letting less volume to regenerate and make this approach a promising alternative for biogas sweetening.

(707e) Preparation and Absorption Behaviors Study of Doped Li4SiO4 At Low CO2 Concentration
Qi, Z., East China University of Science and Technology
Sai, Z., East China University of Science and Technology
Originally presented on: 11/7/2013 14:00:00 - 14:20:00

Preparation and Absorption Behaviors Study of Doped Li_4SiO_4 at Low CO_2 Concentration

Sai Zhang, Qi Zhang*,Qian Ye, Haiyan Wang, Zibin Zhu

Dept. of Chemical Engineering, East China University of Science and Technology, Shanghai 200237, China

Abstract

Different K-doped Li_4SiO_4 samples were prepared to improve the pure Li_4SiO_4 adsorbent's sorption/desorption properties in the critical SESMR practical conditions with low CO_2 partial pressure and steam atmosphere. The optimized sample was analyzed by sorption experiments and simulation with FactSage5.5 Software. The results show that the K-doped Li_4SiO_4 has a better CO_2 absorption ability than pure Li_4SiO_4 as the K can form a eutectic compound with Li_4SiO_4 which enhances the Li^+ and O^{2-} ions diffusion and thus facilitating the CO_2 absorption ability of Li_4SiO_4. It is also found that steam not only enhances the Li^+ mobility but also reacts with Li_4SiO_4, therefore increase the mass uptake of the adsorbent. In order to make the adsorbent more practical, sorption-desorption cycle processes were also carried out. The results show that the optimized Li_4SiO_4 has a lower desorption temperature and stable absorption ability.

Furthermore, the sorbents, Li_4SiO_4 and different K-doped Li_4SiO_4, were evaluated using a bench scale unit, of which the results can provide reference for the subsequent industrial design. Li_4SiO_4 and different K-doped Li_4SiO_4 were prepared by solid-state reaction method. The results was obtained from the studies of different Li_4SiO_4 samples in a fixed-bed reactor with 15%CO_2 and 85%N_2 at 575Â°Ã¦. The pure Li_4SiO_4 showed a short breakthrough time, approximately 20 min, while modified Li_4SiO_4 samples presented a relative 20~40 min longer breakthrough time under same conditions. The results of the experiment indicate that doping K into Li_4SiO_4 can improve the adsorption capacity of the sorbents. Comparing with MgO-based K_2CO_3[1] and hydrotalcites[2] sorbents, which breakthrough time are no more than 40min and 13min, respectively, the relatively long breakthrough time of K-doped Li_4SiO_4 exhibits huge potential for industrial applications. K-Li_4SiO_4 with good performance at low CO_2 partial pressure can be applied into sorption enhanced steam methane reforming system, in which the CO_2 concentration can be also as low as 15%. Meanwhile, the regeneration period can be prolonged by doping appropriate amount K into Li4SiO4, which means cost for recycling can be reduced as a result. K-doped Li4SiO4 has a certain value in industrial application, both practicability and economically.

Key words: CO_2 acceptors, Lithium orthosilicate, on-line absorption behaviors

References

[1]Soo Chool Lee, Ho Jin Chae, et al. Development of Regenerable MgO-Based Sorbent Promoted with K2CO3 for CO2 Capture at Low Temperatures[J]. Environ.Sci. Technol, 2008, 42: 2736-2741

[2]Hendricus Th. J. Reijers, Sacha E. A.Valster-schiermeier, et al. Hydrotalcite as CO2 Sorbent for Sorption-Enhanced Steam Reforming of Methane[J]. Ind. Eng. Chem. Res, 2006, 45, 2522-2530

(707f) Zr Doping Effect On CeO2 As the Catalyst in the CO2 - Ethylbenzene Reaction

Wang, Q.,
Li, W., Taiyuan University of Technology
Originally presented on: 11/7/2013 14:20:00 - 14:40:00

Under the thermodynamic limitation , dehydrogenation of ethylbenzene (EB) to styrene (ST) with CO_2 is widely investigated. The lattice oxygen mobile capacity of catalyst is the important influencing factors in the CO_2-EB reaction. CeO_2 of the CaF_2 structure with the excellent oxygen storage/release capacity (OSC) is preferred as the catalyst. The addition of Zr^{4+} to the cubic structure of CeO_2 can greatly increase the OSC and improve its surface area, thermal stability, mechanical strength and surface acidity. The Zr content of $Ce_xZr_{1-x}O_2$ solid solutions strongly influence the crystal structure, the mobility of lattice oxygen and the catalytic performance. $Ce_xZr_{1-x}O_2$ solid solutions replacing the pure CeO_2 as the catalyst could be used in oxidative dehydrogenation of EB to ST with CO_2 for changing the lattice structure of CeO_2 and improving the transmission capacity of lattice oxygen.

In this work, $Ce_xZr_{1-x}O_2$ solid solutions with different Zr proportions were synthesized using hydrothermal method. The synthesized catalysts were characterized by X-ray power diffraction pattern, Raman spectroscopy and N_2-adsorption. It was found that the catalysts were solid solution mixture having specific surface area more than 100 m^2/g. By the means of the H_2 temperature-programmed reduction, the lattice oxygen mobile capacity was calculated, corresponding to the order is $Ce_{0.5}Zr_{0.5}O_2 > Ce_{0.7}Zr_{0.3}O_2 > Ce_{0.3}Zr_{0.7}O_2 > CeO_2$. The performance of the catalysts for dehydrogenation of EB to ST was evaluated in a fixed-bed reactor under the conditions 823 K, 1 atm, and 20 of the CO_2/EB molar ratio. Results showed that the lattice oxygen mobile capacity of $Ce_xZr_{1-x}O_2$ influenced the catalytic activity and promoted activity of CO_2. The $Ce_{0.5}Zr_{0.5}O_2$ showed the highest activity with the EB conversion of 50% and the ST selectivity above 96%. The mechanistic model of EB dehydrogenation with CO_2 over the $Ce_xZr_{1-x}O_2$ catalysts was proposed.

(707g) Combined Natural Gas Liquefaction and Power Generation Plant With Carbon Dioxide Sequestration

Verma, S., Schlumberger Doll Research
Ramakrishnan, T. S.,
Originally presented on: 11/7/2013 14:40:00 - 15:00:00

It is generally well accepted that an increase in anthropogenic emissions is a leading cause of the increase in levels of carbon dioxide (CO_2), one of the greenhouse gases, in the atmosphere. Generation of electrical power is responsible for a roughly 40% of these emissions. Design of new zero emission power plants is critical in any effort to reduce emissions of CO_2.

We discuss the design of a combined natural gas liquefaction and power generation plant with no atmospheric emissions of greenhouse gases. More specifically, this design relates to (i) a zero emission process for power generation as the CO_2produced during the power generation cycle is sequestered into a terrestrial formation and (ii) a process by which the high pressure natural gas stream is expanded to provide refrigeration and power. The recovered energy may be used to liquefy part of the natural gas stream to provide pure liquefied natural gas (LNG).

This plant uses natural gas from a well as a fuel. The high-pressure stream is liquefied in a reverse-Brayton cycle and the remainder of the natural gas stream is fed to a solid oxide fuel cell to produce power. The effluent of the solid oxide fuel cell is completely combusted to reduce it to CO_2 and water. This stream is subsequently cooled to remove the condensed water.

Energy recovered in expansion of natural gas is used for CO2 stream compression, reducing the energy required to by 13.7%. Alternatively, if LNG is not desired as a product, then as much as 30% of the compression energy may be supplied from expansion of natural gas. Overall power plant efficiencies of 50 – 60% are possible. About 27.5% wt/wt of the incoming natural gas stream may be liquefied in this cycle.

(708a) Control of Metal-Organic Framework Crystal Structure By Ligand Functionalization: Functionalized HKUST-1 Derivatives

Kulkarni, A. R., Georgia Institute of Technology
Cai, Y., Georgia Institute of Techonlogy
Walton, K. S., Georgia Institute of Technology
Sholl, D. S., Georgia Institute of Technology
Originally presented on: 11/7/2013 15:15:00 - 15:35:00

Control of Metal-Organic Framework Crystal Structure by Ligand Functionalization: Functionalized HKUST-1 Derivatives

*Ambarish R. Kulkarni, Yang Cai, Krista S. Walton and David S. Sholl**

School of Chemical & Biomolecular Engineering, Georgia Institute of Technology.

Metal-Organic Frameworks (MOFs) are a novel class of materials consisting of metal centers bonded to organic ligands to form a 3-dimensional porous network. Recent computational and experimental approaches have focused on functionalization of the organic linker to alter the local chemical environment of the framework, which in turn can be used to tune macroscopic properties such as diffusion and adsorption. These studies inherently assume that functionalization of the linker will not affect the topology of the parent MOF.

Previously, we have reported the synthesis of a new topology for CuBTC MOF using methyl and ethyl functionalized BTC (BTC = 1,3,5-benzenetricarbozylate) precursors [*Cryst. Growth Des.*, **2012**, 12 (7), pp 3709–3713]. In this work, we study a set of six functionalized BTC precursors (methyl, ethyl, methoxy, nitro, acetamide and bromo) to experimentally show that the use of functionalized ligands within the CuBTC family leads to multiple distinct crystal structures. Further, we perform periodic Density Functional Theory (DFT) calculations to explain the experimental observations. Our DFT calculations show that show that the preference of a particular topology for a given functionalized CuBTC polymorph can be explained and rationalized using computational methods. Specifically, using a cluster model to study the secondary building unit, we assess the thermodynamic stability of the polymorphs in terms of energy penalties associated with the distortion of both the BTC linker and the framework. This molecular level insight of the interactions between the functional group and the topology allows the prediction of topology for a given functionalized ligand and opens the possibility of controlling the topology of a MOF by an appropriate choice of the functionalized precursor.

(708b) Effect of Isophtalate Moieties On The Adsorption Properties of Light Hydrocarbons in Tbo-MOF. Application to Methane Purification and Storage

Belmabkhout, Y., King Abdullah University of Science and Technology

Originally presented on: 11/7/2013 15:35:00 - 15:55:00

Effect of Isophtalate Moieties on The Adsorption Properties of Light Hydrocarbons in tbo-MOF. Application to Methane Purification and Storage

Youssef Belmabkhout[1], Jarrod Eubank[2], Vincent Guillerm[1], Hasnaa Mouttaki[2], and Mohamed Eddaoudi[1,2].

[1]*KAUST Advanced Membranes & Porous Materials Center, Functional Material Design, Discovery and Development (FMD[3])King Abdullah University of Science and Technology*
4700 King Abdullah University of Science and Technology Thuwal 23955-6900 (Kingdom of Saudi Arabia)

[2]*Department of Chemistry, University of South Florida, 4202 East Fowler Avenue, Tampa, Florida 33620, USA.*

**Corresponding author email: youssef.belmabkhout@kaust.edu.sa*

Equilibrium adsorption of methane (CH_4), C_{2+} gases (i.e., ethane (C_2H_6), ethylene (C_2H_4), propane (C_3H_8) and propylene (C_3H_6)) and carbon dioxide (CO_2) were measured on **tbo**-MOF-1[1] and its isoreticular, functionalized analogue, **tbo**-MOF-$\{CH_2O[Ph(CO_2H)_2]\}_2$ (or **tbo**-MOF-2) [2] characterized by pendant isophthalate moiety freely pointing into the cavities [*J. Am. Chem. Soc.*, 2011, 133 (36), 14204–14207]. Both **tbo**-MOFs were evaluated in the spectra of methane storage and C_{2+}/CH_4 separation. The parent **tbo**-MOF-1 showed high gravimetric and volumetric CH_4, close to the United State Department of Energy (DOE) target for methane storage at 35 bar and room temperature, while the presence of pendant isophthalate moiety in the analogue compound **tbo**-MOF-2 led to a decrease in CH_4 uptake. Nevertheless, the additional isophthalate moiety increased the affinity for propane, propene, ethane, and ethylene at low pressures vs CH_4, due mainly to the enhanced interactions with the isophthalate moiety and the reduced size of the cavities. Using Ideal Adsorption Solution Theory (IAST), the predicted mixture adsorption equilibria for C_3H_8/CH_4, C_3H_6/CH_4, C_2H_6/CH_4, C_2H_4/CH_4 and C_3H_8/CO_2 systems showed high adsorption selectivity for C_{2+} over methane for **tbo**-MOF-2 as compared to the parent **tbo**-MOF-1. The relatively high affinity of **tbo**-MOF-2 to C_{2+} over CO_2 make it suitable for it use in subsequent or/and simultaneous removal of C_{2+} from natural gas and refinery off-gases.

References

[1] Eubank, J. F.; Mouttaki, H.; Cairns, A.J.; Belmabkhout, Y.; Wojtas, L.; Luebke, R.; Alkordi, M.; Eddaoudi, M. *The quest for modular nanocages: Tbo -MOF as an archetype for mutual substitution, functionalization, and expansion of quadrangular pillar building blocks*. *J. Am. Chem. Soc.* 2011, 133, 14204-14207.

(708c) Post-Synthetic Functionalization of Magnesium Metal-Organic Framework With Tetraethylene Pentamine for CO2 Adsorption

Su, X., Massachusetts Institute of Technology
Simeon, F., Massachusetts Institute of Technology
Huq, A., Oak Ridge National Laboratory
Speakman, S. A., Massachusetts Institute of Technology
Hatton, T. A., Massachusetts Institute of Technology
Originally presented on: 11/7/2013 15:55:00 - 16:15:00

Metal-organic frameworks (MOFs) are a class of microporous and mesoporous materials with great potential in gas adsorption and separation. MOFs are composed of organic linkers coordinated to metal centers, and produce open-framework structures that rank among the materials with the highest surface areas currently available. Amongst the metal-organic frameworks used for carbon dioxide adsorption, Mg-MOF-74 is an interesting option to study due to its very high equilibrium and dynamic uptake, the economic feasibility of its reagents as well as its ease of synthesis.[1-2] Its cylindrical pores and unsaturated metal centers can adsorb between 21-26% CO_2, and represents one of the solid-state adsorbents with the highest capacity for CO_2.

Post-synthetic functionalization of MOFs is a technique to impart properties inaccessible by direct synthesis. For gas adsorption applications, functionalization with amines plays an interesting role due to their ability to improve adsorption behavior of MOFs under different conditions.[3] We grafted the Mg-MOF-74 with tetra-ethylene pentamine (TEPA, a multi-unit amine), in order to study its interaction with the framework and its effect on the CO_2 adsorption. Multi-unit amines (similar to poly-ethylene-imines) have been shown recently to provide high carbon dioxide uptake capacity as well as regenerability over multiple cycles.[4] We expect that through functionalization of longer-chain amines, to improve adsorption over current ethylene diamine studies, since one amine and one Mg-site are lost during the grafting process, whereas for TEPA, we gain one primary amine and 3 secondary amines.

Gas adsorption testing on our MOF/TEPA materials suggests that CO_2 total uptake capacity is dependent on amine loading. CO_2 uptake is reduced over the parent MOF when amines are put in excess, but improved when limiting amine is added. The crystal structure from XRD seems to be identical pre- and post-functionalization, as well as pre-post-capture. After functionalization in excess amine, the surface elemental analysis through XPS gives $C_{55.98\%}N_{16.82\%}O_{22.5\%}Mg_{4.71\%}$ and $C_{54.03\%}N_{19.38\%}Mg_{4.52\%}O_{21.7\%}$. The ratio of N/Mg is around 5, which indicates that each surface Mg-site has a tetra-ethylene pentamine grafted onto it. We speculate that the surface Mg-sites are all grafted, but the large amount of amines induces pore-blocking which reduces bulk CO_2 mass transfer. We grafted the equivalent of 1% of the excess amount of amines, and the surface composition became $C_{54.03\%}N_{10.77\%}Mg_{7.15\%}O_{28.05\%}$, with improved CO_2 uptake over Mg-MOF-74.

Neutron powder diffraction studies are currently being carried out to investigate the crystal structure of the post-synthetically functionalized MOF. We attempt to identify the location within the unit cell of the amines and their conformation, as well as potential sites for CO_2 adsorption. We expect detailed diffraction studies will allow us to draw structure-performance relationships, and gain understanding on the adsorption mechanisms behind adsorption of CO_2 on amine-functionalized MOFs.

References

1. Caskey, S. R.; Wong-Foy, A. G.; Matzger, A. J., Dramatic tuning of carbon dioxide uptake via metal substitution in a coordination polymer with cylindrical pores. *Journal of the American Chemical Society* **2008,** *130* (33), 10870-1.

2. Britt, D.; Furukawa, H.; Wang, B.; Glover, T. G.; Yaghi, O. M., Highly efficient separation of carbon dioxide by a metal-organic framework replete with open metal sites. *Proceedings of the National Academy of Sciences of the United States of America* **2009,** *106* (49), 20637-40.

3. Choi, S.; Watanabe, T.; Bae T.; Sholl, D.; Jones, C. W. Modification of Mg / DOBDC MOF with Amines to Enhance CO 2 Adsorption from Ultra-Dilute Gases. *Journal of Physical Chemistry Letters* **2012,** *3*, 1136-1141.

4. Zhao, J.; Simeon, F.; Wang, Y.; Luo, G.; Hatton, T. A., Polyethylenimine-impregnated siliceous mesocellular foam particles as high capacity CO2 adsorbents. *RSC Advances* **2012,** *2* (16), 6509-6509.

(708d) Adsorption Characteristics of Coordinatively Unsaturated Metal Sites Containing Dhtp Series of Metal Organic Frameworks

Mishra, P., Indian Institute of Technology Guwahati
Edubilli, S., Indian Institute of Technology Guwahati
Mandal, B., Indian Institute of Technology Guwahati
Gumma, S., Indian Institute of Technology Guwahati
Originally presented on: 11/7/2013 16:15:00 - 16:35:00

During last decade, a new class of crystalline materials commonly known as metal organic frameworks (MOFs) has been widely investigated for their potential use in gas storage, gas separation and catalysis. 2,5-dihydroxyterephthalate (dhtp) series of MOFs is one of the most interesting MOF series due to presence of large number of coordinatively unsaturated metal sites.

In this study, we report high pressure adsorption isotherms of four industrially relevant gases (CO_2, CO, CH_4 and N_2) on Mdhtp (M = Mg, Mn, Co, and Ni) MOFs at three temperature over a wide pressure range. These gases were purposefully chosen to cover a wide range of adsorpbate polarity and polarizability and observe their effects on the adsorption characteristics in unsaturated metal sites containing MOFs. For gases like CO_2 and CO which have significant polarity, these MOFs exhibit two different adsorption sites and a DSL model was necessary to describe the adsorption behavior; however for other relatively non polar gases a variant of Langmuir formulation was sufficient.

The MOF containing, Ni has shown the strongest lateral interactions (Henry's constant) for CO; this can be readily attributed to the small ionic radius of Ni^{2+}. On the other hand, the effect of metal atoms on adsorption characteristics of relatively non-polar gases like CH_4 and N_2 was found to be negligible, indicating the role of polarity for specific interaction with unsaturated metal sites. In contrast to CO, CO_2 has the highest Henry's Constant for Mg^{2+}. Mg^{2+} has a fixed coordination number of 6 with an exceptional preference for oxygen-containing ligands (and hence exhibits high Henry constants); other metal atoms studied in this work have coordination number ranging from 4 to 6. In addition, it was observed that the metals in the framework have negligible effect on the saturation capacity of the gases. Any slight differences observed in saturation loadings can be readily attributed to difference in pore volume (due to presence of different metal atoms) of these samples.

(708e) Metal-Organic Nanotubes With Molecular Valves

McNutt, N., University of Tennessee
Murdock, C.,
Jenkins, D.,
Keffer, D., University of Tennessee
Originally presented on: 11/7/2013 16:35:00 - 16:55:00

Novel metal-organic frameworks with one-dimensional channels – metal-organic nanotubes (MONTs) – show potential for extremely selective separations based on their nanoscale structure. A class of MONTs has been synthesized in which phenyl groups provide flexible rotors that act as molecular-scale valves, either allowing transport down the axial dimension of the nanotube when in the "open" position, or preventing transport in the "closed" position. The ability to control the position of the valve remains a subject of investigation. In this work, we present a comparison between experimentally determined crystal structures and simulated molecular models of a suite of MONTs. The experimental work indicates that the choice of solvent (DMA, DMF, NMP, water) can be used to control the orientation of the valve. This choice, and the relative influences of steric hindrance and electronic effects are assessed. The simulation work provides an understanding of the molecular-level mechanism. Structural and energetic properties are reported, which yield insight into a broader application of molecular valve control in MONTs.

(708f) Unusual Patterns in the Adsorption and Diffusion of C1-C6 Alcohols and Other Organic Compounds in the ZIF-68 Metal-Organic Framework

Van Der Perre, S., Vrije Universiteit Brussel

Lannoeye, J.,

De Vos, D., Katholieke Universiteit Leuven

Baron, G., Vrije Universiteit Brussel

Denayer, J., Vrije Universiteit Brussel

Originally presented on: 11/7/2013 16:55:00 - 17:15:00

Zeolitic Imidazolate Frameworks (ZIFs) are crystalline microporous materials, consisting of transition metals (M = Co, Cu, Zn, etc.) linked by imidazolate (Im) or functionalized Im ligands. This subfamily of the Metal-Organic Frameworks (MOFs) displays good chemical and structural stability, which make them attractive as separation media in liquid phase separations. ZIF-68 is an interesting member of this family, with a complex pore structure consisting of two one-dimensional channels formed by three different types of cages. The small channel is composed of alternating small and medium cages, while the large channel comprises the large cages [1].

In the present work, the adsorption behavior of ZIF-68 is studied via vapor phase kinetic and equilibrium experiments with C1-C6 alcohols and other organic compounds (acetone, acetonitrile, C6-isomers, tri-isopropylbenzene) and kinetic batch experiments of alcohol mixtures in water. S-shape isotherms are obtained for polar adsorbates (fig.1), which is related to (1) the presence of two chemically different channels in the ZIF-68 framework and (2) the rather apolar nature of the material, which results in unfavorable adsorption of polar compounds at low concentration. Depending on the component, the adsorption capacity varies between 0.25 and 0.30 g/g. A comparison to the Argon isotherm indicated that both types of channels of ZIF-68 are densely filled with the different organic compounds.

Uptake of the different alcohols occurs slowly on the large crystals (around 70 μm) used in this study. Moreover, two different diffusion regimes can be observed, in which molecular transport in the two-pore systems occurs at different rates. Besides, it was observed that bulky molecules, with a kinetic diameter significantly larger than the pore windows, are adsorbed in large amounts, which gave reasons to think that also this ZIF material, just as ZIF-7 [2] and ZIF-8 [3-5], has a certain degree of structural flexibility to enlarge the free aperture of the channels, enhancing the diffusion process. Liquid phase experiments indicate the potential of ZIF-68 in the selective uptake and separation of alcohol/water mixtures.

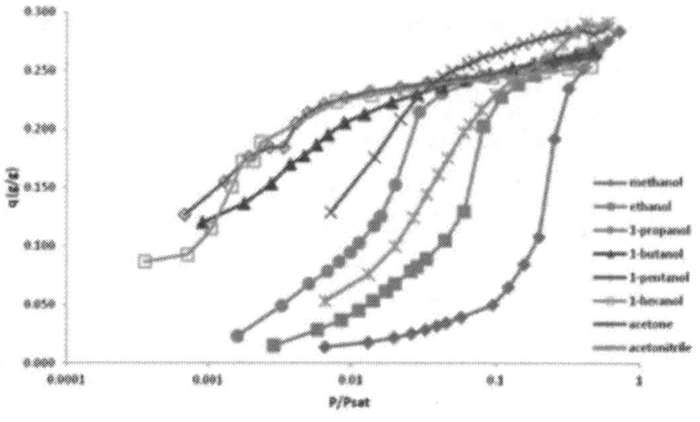

Fig.1: Adsorption isotherms of C1-C6 alcohols, acetone and acetonitrile at 50 °C.

References

[1] R. Babarao, S. Dai and D. Jiang, J. Phys. Chem. C 115 (2011) 8126?8135

[2] C. Gücüyener, J. van den Bergh, J. Gascon and F. Kapteijn, J. Am. Chem. Soc. 132 (2010) 17704-17706

[3] S.A. Moggach, T.D. Bennett and A.K. Cheetham, Angew. Chem. 121 (2009) 7221-7223

[4] D. Fairen-Jimenez, S.A. Moggach, M.T. Wharmby, P.A. Wright, S. Parsons and T. Düren, J. Am. Chem. Soc. 133 (2011) 8900-8902

[5] J.C. Saint Remi, T. Rémy, V. Van Hunskerken, S. Van der Perre, T. Duerinck, M. Maes, D.E. De Vos, E. Gobechiya, C.E.A. Kirschhock, G.V. Baron and J.F.M. Denayer, ChemSusChem 4 (2011) 1074-1077.

Acknowledgements

The authors are grateful to the Agency for Innovation by Science and Technology in Flanders (IWT) for support in the SBO MOFshape project.

(708g) Enhanced Growth of Metal Organic Frameworks On Polymer Fiber Mats Using Atomic Layer Deposition

Zhao, J.,
Losego, M. D.,
Lemaire, P. C. Sr., North Carolina State University
Gong, B.,
Blevins, T. M.,
Oldham, C. J.,
Williams, P. S.,
Walls, H. J.,
Shepherd, S. D.,
Peterson, G. W., U.S. Army Research, Development and Engineering Command, Edgewood Chemical Biological Center
Parsons, G. N., North Carolina State University
Originally presented on: 11/7/2013 17:15:00 - 17:35:00

Immobilization of metal organic frameworks (MOFs) promotes the application to hazardous gas adsorption, membrane separation, catalytic coating and sensors. However, conventional MOF synthesis methods are not easily adapted to MOF-polymer composite materials. Previous reports of MOFs in or on polymer fibers led to composites with low MOF mass fraction or limited accessible MOF surface area. By depositing an Al_2O_3 thin film via atomic layer deposition (ALD), we modified the surface of non-woven polypropylene (PP) fiber mats to facilitate the nucleation of HKUST-1 [$Cu_3(BTC)_2$]. Using solvothermal MOF synthesis, ALD-coated PP substrates showed good uniform MOF growth with high coverage (as observed by SEM), whereas similar fiber substrates without ALD led to poor growth and non-uniform coverage. The MOF growth on ALD-coated fibers produced a mass gain of 388% and BET surface area of MOF-PP samples was up to 798 m^2/g. Considering the mass of the MOF only, this corresponded to 1046 m^2/g for the MOF surface area on the fibers, representing good quality of HKUST-1 without activation. The MOF crystals were found to be strongly attached to the fibers, with ~90% of the MOF mass withstanding a high-rate gas-flow adhesion test. Breakthrough tests showed dynamic loading of ammonia was up to 4.9 mol/kg on MOF-PP samples with ALD. We will discuss how ALD thickness affects MOF growth and the application of this technique to other polymer materials.

(708h) Systematic Investigation of the Influence of Metal Sites On Water Stability and Carbon Monoxide Adsorption

Stults, K. A., Georgia Institute of Technology
Flemming, C., Georgia Institute of Technology
Schoenecker, P. M., Georgia Institute of Technology
Cmarik, G., Georgia Institute of Technology
Walton, K. S., Georgia Institute of Technology
Originally presented on: 11/7/2013 17:35:00 - 17:55:00

MOF-74 is a well-known metal-organic framework consisting of one-dimensional channels with coordinately unsaturated metal centers desirable for adsorption, providing discrete, regularly-spaced active sites. The variety of metal centers that can be included with relative ease creates a favorable system for investigating the role of open metal sites in adsorption in metal-organic frameworks. Specifically, these materials are promising for breathing air purification, removing toxic chemicals such as carbon monoxide. A series of MOF-74 isostructures were synthesized and investigated for water stability and carbon monoxide capacity. Activation conditions were evaluated to minimize the impact of varying amounts of unevacuated solvent between samples to isolate the influence of the open metal site. The water capacity of the examined materials is greater than or equal to that previously reported, with Co-MOF-74 being the least sensitive to pre-experimental conditions. Like Mg-MOF-74 reported previously, Co-MOF-74 and Zn-MOF-74 retain some crystallinity as determined via PXRD, but cannot be fully regenerated, as determined by porosity characterization after water exposure. Co-MOF-74 and Ni-MOF-74 were found to interact strongly with carbon monoxide, and Co-MOF-74 was found to have a higher CO capacity than any previously reported material at 25°C and 1 bar, making this a promising material for further investigation into air purification. Heats of adsorption for CO were calculated for each analog and found to be comparable to previous predictions or measurements.

(711a) Thin Film Composite Membranes for Reverse Osmosis Desalination

Perera, D. H. N., University of Cambridge
Nataraj, S. K., University of Cambridge
Thomson, N., University of Cambridge
Qiblawey, H., Qatar University
Sivaniah, E., University of Cambridge
Originally presented on: 11/7/2013 15:15:00 - 15:37:00

Desalination by reverse osmosis (RO) has become an important process to produce fresh water for domestic and industrial uses[1]. We develop RO membranes using different polymeric materials as the selective layer in thin film composite (TFC) membrane structures. TFC structure is selected due to its low resistance to the water flow and ability to tailor the selective layer without affecting the porosity of the support[2]. Cellulose acetate thin film composite (CA TFC) membranes were successfully made on microporous polymeric substrate by spin coating a dilute cellulose acetate solution. This membrane initially had low salt rejection (~ 60 %) and was later developed up to the level of a RO membrane (~ 90 % salt rejection) by treating in room temperature water baths. The pure water flux through these membranes showed a strong relationship with the time in the water bath. A significant decrease in the surface roughness of the cellulose acetate selective layer was also observed with the time in the water bath. Here the CA TFC membrane was successfully used as a model to investigate the effect of absorbed water on the evolution of the CA selective layer which is not directly observable with asymmetric CA membranes due to the complexity of the process. The surface of the CA TFC membrane was further treated with silver nanoparticles to incorporate anti-fouling properties. This treatment was found to reduce the bacterial adhesion by around 4 orders of magnitude.

We also develop novel TFC RO membranes with polymers of intrinsic microporosity (PIMs) and polyamide for the selective layers. While having excellent salt rejection (> 95 %), these membranes open new pathways for developing chemical-resistant RO membranes. The effects of treatment methods in improving the flux of these RO membranes are also investigated.

References:

1. L. Greenlee, D. Lawler, B. Freeman, B. Marrot, and P. Moulin, *Water res.*, 2009, **43**, 2317-2348.

2. K. P. Lee, T. C. Arnot, and D. Mattia, *J. Membrane Sci.*, 2011, **370**, 1 – 22.

(711b) Adsorptive Psf/Metal-Based Nanoparticle Hollow Fiber Membranes for Fluoride Removal From Aqueous Solution

Jinsong, H., National University of Singapore
J.Paul, C.,
Originally presented on: 11/7/2013 15:37:00 - 15:59:00

An innovative PSF/ metal-based nano-particles(NPs) blend hollow fiber membrane was successfully developed for remove fluoride from aqueous solution. The removal efficiency of fluoride onto the blend membranes by both batch adsorption experiments and continuous filtration were investigated. The batch adsorption results showed that the membrane could effectively remove fluoride in a wide pH ranging from 3 to 10. The adsorption equilibrium could be reached in 24 h, and the maximum adsorption capacity was 61.31mg/g, which is much better than many other sorbents. The co-existing anions such as HCO_3^-, SiO_3^{2-} and PO_3^{2-} had different effect of the fluoride removal. Furthermore, the membrane could remove fluoride efficiently though the continuous filtration, even in the presence of humic acid. The regenerated membrane exhibited good performance for fluoride removal. The potential mechanism for the removal of fluoride was proposed based on XPS analysis. In addition, the cytotoxicity analysis revealed that the blend membrane was free from any toxic leachate during the application.

(711c) Novel Reverse Osmosis Membranes for Seawater Desalination

Zhao, L., The Ohio State University
Ho, W. S. W., The Ohio State University
Originally presented on: 11/7/2013 15:59:00 - 16:21:00

Reverse Osmosis (RO) is one of the most promising technologies for water desalination, which serves as an important solution to the global water shortage. However, the hydrophilicity of the aromatic polyamide (PA) thin-film-composite (TFC) RO membrane is insufficient, which results in a considerably limited water flux. In addition, traditional RO membrane is very vulnerable to fouling from large molecules.

A hydrophilic additive is defined as a chemical compound which contains at least one hydrophilic portion and one reactive portion that can react with either MPD or TMC during the interfacial polymerization. Therefore, the additive is chemically bound to the cross-linking network formed by MPD and TMC. It is believed that the hydrophilic additive incorporated in the TFC membrane can provide an additional pathway for the molecular transport of water. As a result, the incorporation of hydrophilic additive is able to improve the water flux of the RO membranes, thereby increasing the water productivity.

In this work, *o*-aminobenzoic acid-triethylamine salt was selected as the specific hydrophilic additive to investigate its effects on membrane separation performance under seawater desalination conditions for the first time. In addition, different membrane preparation conditions were investigated and optimized to further improve the water flux of the synthesized membranes. After optimization, the membrane samples were also tested by using the seawater from Port Hueneme, CA to determine their desalination performance for the real application. Furthermore, the fouling resistance of the high-flux membranes was compared with that of the standard TFC membranes.

The high-flux RO membranes were synthesized through advanced interfacial polymerization where *o*-aminobenzoic acid-triethylamine salt was added in the aqueous MPD solution. The membrane synthesized with the hydrophilic additive exhibited significantly improved desalination performance, in a side-by-side comparison with the membrane synthesized without the additive. After optimizing the concentration of the hydrophilic additive in the MPD solution, the water flux of the synthesized membrane was enhanced by about 62%, and the salt rejection remained the same. For instance, the membrane synthesized with 1.0 wt% *o*-aminobenzoic acid/triethylamine salt showed 39.2 gfd (gallons/ft^2/day) water flux and 99.46% salt rejection. The incorporation concentration of *o*-aminobenzoic acid-triethylamine salt was maintained at 1.0 wt% for the optimization of other membrane synthesis conditions to enhance water flux without compromising salt rejection. With 40 wt% isopropanol (IPA) in the aqueous MPD solution, the water flux increased to 44.4 gfd, and salt rejection was still above 99.4%. The effects of additional drying time after soaking polysulfone support in the amine solution and hydrocarbon removal time after interfacial polymerization were also investigated.

The high-flux membranes synthesized under the optimal conditions were tested by using the seawater from Port Hueneme, CA with an average turbidity of 0.032 Nephelometric Turbidity Units (NTU) and a silt density index (SDI) (15 min) of 2.2. The average water flux was about 41.8 gfd and the average salt rejection was above 99.4%. The lower water flux was found to result from the higher solid concentration in the real seawater compared to the synthetic seawater which contains 3.28% NaCl. In addition, the synthesized high-flux RO membrane exhibited an excellent stability in terms of water flux (average 40 gfd) and salt rejection (average 99.4%) during a 30-day test with the seawater from Port Hueneme, CA at 25°C and 800 psi applied pressure across the membrane. This good stability was verified by comparing FTIR spectra of the same membrane sample before and after the 30-day test. These results are important for the commercialization of this high-flux RO membrane for seawater desalination.

Sodium alginate was selected as the foulant to investigate membrane fouling, since it is a common compound in seawater from seaweeds. The membrane fouling resistance was significantly improved with the incorporation of the hydrophilic additive. For example, in the presence of 50 ppm sodium alginate in the feed solution, the water flux reduction of the membrane synthesized with the hydrophilic additive was 19.6% compared to 32.0% for the membrane synthesized without the additive.

(711d) Fabrication of Newly Developed Nanofiltration (NF) Composite Membranes Via IP for Aqueous Dye Removal

Shao, L., Harbin Institute of Technology
Cheng, X., Harbin Institute of Technology
Liu, Y., Harbin Insitute of Technology
Wang, Z.,
Zhang, Y.,
Originally presented on: 11/7/2013 16:21:00 - 16:43:00

Taking into account of energy consumption and environment issues, nanofiltration emerges as a process substituting for the conventional separation processes and purification technique such as distillation for their high energy consumption in separating mixtures on molecular level. Among all of the configuration of nanofiltration modules, hollow fiber nanofitration module has higher the obvious advantages of higher packing density, less pre-treatment and maintenance, which has motivate many researchers' interests. In this study, thin-film-composite (TFC) nanofiltration (NF) membranes were newly developed through interfacial polymerization (IP) on the lumen side of hollow fiber support membranes by exploring novel combination of reacting monomers and the optimum conditions of fabrications and further utilized to remove typical dyes in aqueous solutions at different conditions. Based on the different combinations of reacting monomers, the molecular weight cut off (MWCO) of NF membranes in the hollow fiber module can be adjusted to be around 300.0 g mol^{-1} with the high pure water permeability of 3.1 L m^{-2} h^{-1} bar^{-1} at 3.0 bar. The hollow fiber NF membranes demonstrated different rejections for different salts in the order of $R(Na_2SO_4)>R(MgSO_4)>R(NaCl)>R(MgCl_2)$ and high pressure resistance. For both of Safranin O and Aniline blue dyes, the rejection can be higher than 90% under testing conditions, and the excellent performance of NF membranes to remove dyes from waste water can be maintained in a wide range of pH values although the best performance was obtained at the pH value of 11. Therefore, the novel NF membranes are promising for removing crucial dyes as the next-generation technology for industrial wastewater treatment.

(711e) Magnetically Responsive Self-Cleaning Nanofiltration Membranes

Song, G.,
Qian, X., University of Arkansas
Wickramasinghe, S. R., University of Arkansas
Originally presented on: 11/7/2013 16:43:00 - 17:05:00

Magnetically responsive nanofiltration (NF) membranes with superparamagnetic nanoparticles attached to the chain ends of the grafted polymer nanolayers have been shown to be effective in breaking concentration polarization at the membrane-liquid interface under an appropriate external oscillating magnetic field. NF270 is one of the most commonly used NF membranes for removing low molecular weight species and divalent or trivalent salt ions from the feed solutions. Earlier studies for magnetically responsive NF membranes have focused on 500 ppm NaCl and 2000 ppm $MgSO_4$ model salt solutions only. Here the effects of the frequency of the oscillating field, salt ion type and its concentration on the performance of NF270 are systematically investigated. Salt solutions investigated include NaCl, $CaCl_2$, $MgCl_2$, $MgSO_4$ and Na_3PO_4 as well as their mixtures with concentration ranging from 500 to 5000 ppm. In addition, the effects of grafting degree including both chain length and chain density are also studied. Our studies show that the higher the salt concentration, the higher the valence of the salt ions and the higher the grafting degree, the stronger the effects of the magnetic field on permeate flux and salt rejection.

(711f) Synthesis of Biomimetic Polybenzimidazole Nanofiltration Membranes

Wagh, P., The University of Toledo
Escobar, I. C., The University of Toledo
Originally presented on: 11/7/2013 17:05:00 - 17:27:00

The search for new technologies for water treatment is an increasingly important research topic. Depending on the application, water is desired in varying degrees of purity. For some applications, such as desalination, a very high degree of purification is desired. This project relies on the idea of combining the ultra-efficient functioning of biological molecules with the productivity of synthetic membranes. A successful formulation should lead to the production of water purification membranes with high selectivity along with high water flux. Aquaporin, a highly selective water channel protein, has received worldwide attention because of its potential to form biomimetic membranes with high flux and rejection for water reuse and desalination. The high water permeability and selectivity of aquaporins makes aquaporin-based biomimetic membranes potentially attractive for water purification and "green-energy" production. In nature, aquaporins move water across membranes from areas of low dissolved solids to an area of higher dissolved solids by osmosis. However, if pressure is applied to the side of the membrane containing higher concentrations of dissolved material, pure water can flow through the aquaporins back into the low concentration side of the system. In this sense, pressure can be used to induce reverse osmosis across a membrane embedded with aquaporins and effectively purify water.

The overarching objective of this project is to form a biomimetic membrane made of unaltered aquaporin dispersed in a membrane selective layer and capable of operation under high hydraulic pressure. A challenge of the current study is to design and prepare an assembly that allows artificial biomimetic membranes with embedded aquaporin proteins to sustain hydraulic water pressure gradients without losing their integrity and performance. To this end, aquaporins will be dispersed in hydrophilized polybenzimidizole (PBI) membranes that are chemically and mechanically strong so that the resulting biomimetic membrane will be able to withstand high hydraulic pressures. Membranes with pores or channels specifically engineered to maximize water transport or ion selectivity would offer the potential for dramatically more efficient operation compared to conventional polymer membranes.

(711g) Nanoclay Based TFC Membranes for RO Water Desalination

Zaidi, S. M. J., The university of Queensland
Fadillah, F.,
Matsuura, T.,
Khan, Z.,
Vinu, A.,
Originally presented on: 11/7/2013 17:27:00 - 17:49:00

Nanoclay based TFC membranes for RO water Desalination

S. M. Javaid Zaidi[a]*, F. Fadhillah[b], Z.U. Khan[c], T. Matsuura[d], A. Vinu[a]

[a]The University of Queensland, Australia, [b]Imam Mohammed University, Saudi Arabia, [c]King Fahd University of Petroleum & Minerals, Saudi Arabia, [d]Ottawa University, Canada

Abstract

Nowadays water shortage is a common problem in almost all countries around the world. Population growth and industrial development have significantly increased the demand for water. In this context Desalination has been identified as a strategic technology for conversion of seawater into drinking water. Although reverse osmosis (RO) desalination is poised to become a critical clean water technology and promises to greatly increase the supply of clean water through seawater desalination and purification of brackish water, the currently used RO technology based on polyamide thin film composite membranes remains energy-intensive, non-selective, and fouling-prone. Bio-fouling remains a major concern which leads to higher pumping pressures, shorter membrane life and lower permeates quality. These issues create an urgent need for significant innovations in RO membrane technology to overcome these problems. One of such innovation relates to the development of a new class of nanoparticle filled polymer composite materials that can be used to fabricate RO membranes. In this research we report the development of new type of the RO membranes by interfacial polymerization dispersing nanoclay particles into the polymer matrix.

Thin film composite polyamide (PA) membrane was fabricated using interfacial polymerization technique and tested for their performance for RO desalination. Nanoclays were incorporated into PA membrane to enhance membrane performance. The performance of these membranes carried out for RO desalination show high salt rejection and better permeation characteristics. These membranes have potential to be used in commercial RO desalination.

*j.zaidi@uq.edu.au

Tel: 61-7-33464167

(720a) Unified Pressure and Electroosmotic Flow Enhancement Mechanism in Carbon Nanotube Membranes

Hinds, B. J. III, University of Kentucky
Originally presented on: 11/7/2013 15:15:00 - 15:40:00

Carbon nanotubes have three key attributes that make them of great interest for novel membrane applications 1) atomically flat graphite surface allows for ideal fluid slip boundary conditions and extremely fast flow rates 2) the cutting process to open CNTs inherently places functional chemistry at CNT core entrance for chemical selectivity and 3) CNT are electrically conductive allowing for electrochemical reactions and application of electric fields gradients at CNT tips. In general, the transport mechanisms through CNT membrane are a) ionic diffusion is near bulk expectation with no enhancement from CNT b) gas flow is enhanced by ~1-2 order of magnitude due to specular reflection off of flat graphitic surface c) and pressure driven flux of a variety of solvents (H2O, hexane, decane ethanol, methanol) are 4-5 orders of magnitude higher than conventional Newtonian flow [*1*] due to atomically flat graphite planes inducing nearly ideal slip conditions. Nearly all applications require chemical selectivity in what is allowed to pass across the membrane. However the act of placing selective functional chemistry at pore entrance or along the core of CNTs, dramatically/completely eliminates the enhanced flow effects by eliminating the near perfect slip boundary condition[*2*]. Needed is a mechanism to pump chemicals through the pore where selective chemistry is. This is routinely achieved in protein channels where permeates are accelerated through regions of precise functionality. The CNT membrane, with tips functionalized with charged molecules, is a nearly ideal platform to induce electro-osmotic flow with high charge density at pore entrance and a nearly frictionless surface for the propagation of plug flow. Through diazonium electrochemical modification we have successfully bound anionic surface charge to CNT tips and along CNT cores. High electro-osmotic flows of 0.16 cm/s-V at are seen by the pumping of neutral caffeine molecules. Improvements in electroosmotic power efficiency of 25-112 fold are seen in CNTs compared to conventional nanoporous materials with atomically rough interfaces [*3*]. Use of the electro-osmotic phenomenon for responsive/programmed transdermal drug delivery devices is discussed with the voltage gated delivery of clonidine and nicotine across CNT membrane at therapeutically useful fluxes [*4*]. In small diameter SWCNTs ion mobilities are seen to be ~6 fold enhanced due to induced electroosmotic flow. Electroosmotic flow enhancements of 10,000 fold are seen [*5*] and are consistent with pressure driven flow enhancements.

1 "Nanoscale hydrodynamics: Enhanced flow in carbon nanotubes" Majumder, M.; Chopra, N.; Andrews, R; Hinds, B.J *Nature* **2005**, 438, 44.

2 'Mass Transport through Carbon Nanotube Membranes in three different regimes: ionic diffusion, gas, and liquid flow' Mainak Majumder, Nitin Chopra, B.J. Hinds *ACS Nano* **2011** 5(5) 3867-3877

3 'Highly Efficient Electro-osmotic Flow through Functionalized Carbon Nanotubes Membrane' Ji Wu, Karen Gerstandt, Mainak Majunder, B.J. Hinds, *RCS Nanoscale* **2011** 3(8) 3321-28

4 " Programmable transdermal drug delivery of nicotine using carbon nanotube membranes" J. Wu, K.S. Paudel, C.L. Strasinger, D. Hamell, Audra L. Stinchcomb, B. J. Hinds *Proc. Nat. Acad. Sci.* **2010** 107(26) 11698-11702.

5 "Electrophoretically Induced Aqueous Flow through sub-Nanometer Single Walled Carbon Nanotube Membranes" Ji Wu, Karen Gerstandt, Hongbo Zhang, Jie Liu, and Bruce. J. Hinds *Nature Nano* **2012** DOI: 10.1038/NNANO.2011.240

(720b) Improving the Separation Performance of Nanocomposite Membrane By Aligning Carbon Nanotubes Into Polymeric Matrix

Zhao, H.,
Zhou, Z., Zhejiang University
Zhang, L., Zhejiang University
Chen, H. L., Zhejiang University
Originally presented on: 11/7/2013 15:40:00 - 16:00:00

The nanocomposite membranes containing carbon nanotubes (CNTs) have shown better separation performance than the bare membranes. For example, several studies have demonstrated that a higher permeation flux can be obtained by incorporating functionalized CNTs in the polymeric membrane. However, the improvement in separation performance is limited due to the disordered state of CNTs in the nanocomposite membranes.

In this work, a facile method, hot-press combined with peel-off (HPPO), is established, through which CNTs can be successfully vertically aligned on the polymeric membrane substrate. Shear force and mechanical stretch are proposed to be the main forces to align the tubes perpendicular to the substrate surface during the peel-off process. The TEM images show that the alignment of CNTs keeps their orientation in a thin hybrid membrane by coating polymer casting solution. The stable alignment of CNTs by HPPO greatly contributes to the improved separation performance of the nanocomposite membranes. The CNTs facilitate the permeation of components due to their nature molecular channel and the microvoids between the polymer matrix and tubs.

(720c) Translating Carbon Molecular Sieves (CMS) Dense Film Gas Transport Performance to Asymmetric Hollow Fiber Membranes

Bhuwania, N., Georgia Institute of Technology
Koros, W. J., Georgia Institute of Technology
Williams, P. J., Shell International Exploration and Production (US) Inc
Originally presented on: 11/7/2013 16:00:00 - 16:20:00

Fundamental studies on Carbon Molecular Sieves (CMS) dense film membranes have shown enhanced gas separation ability based on its bimodal pore distribution. The simplified structure of CMS membrane is generally assumed to compose a 'slit-like' dual-pore morphology consisting of micropore (~6-20Å) and ultra-micropore (<6Å). The nano-size pore distribution is obtained by the thermal degradation of polymer precursors under controlled atmospheres. For industrial application, it is important to translate this high separation performance in hollow fiber membrane configuration. In our talk, we will discuss an approach to overcome the challenge in maintaining the asymmetric morphology of CMS hollow fiber membranes. To achieve the superior gas separation performance and asymmetric morphology in final CMS hollow fibers, we use the sol-gel reaction between organic-alkoxy silane and moisture (V-Treatment). CMS V-Treated hollow fibers membranes are evaluated for both pure and mixed gas feeds.

(720d) Slip and Spin-Coupling: Foundations of the Flow Enhancement of Fluids in Graphene Slit Channels and Carbon Nanotubes With Important Applications in Nanoscale Fluid Pumping

Todd, B. D., Swinburne University of Technology
Kannam, S.,
De Luca, S.,
Hansen, J.,
Daivis, P. J., RMIT University
Originally presented on: 11/7/2013 16:20:00 - 16:45:00

We present both theoretical and simulation studies that clearly demonstrate the importance of several non-classical phenomena that are fundamental features of fluid flow under extreme confinement. These are: (1) the prevalence of slip and (2) the strong coupling of molecular spin to linear translational momentum for molecularly structured fluids. In the first of these, we utilize a newly developed equilibrium based model[1] to accurately predict the slip velocity and slip lengths of systems such as water or methane flowing in graphene nanochannels and carbon nanotubes[2-4]. We demonstrate that traditional nonequilibrium molecular dynamics simulations of such systems are far less efficient and accurate than the easily implemented model we propose. Our results confirm large flow enhancements that are accurately quantified and compared to existing experimental, theoretical and simulation data in the literature. Next, we show that ignoring the coupling of spin angular momentum to linear translational motion of a highly confined fluid can lead to significant over-estimation of the predicted flow rates using conventional Navier-Stokes treatments. By including spin-coupling into the extended Navier-Stokes equations, hydrodynamic prediction is seen to be very accurate down to length scales of a few atomic diameters[5]. We also demonstrate how this knowledge, coupled with our knowledge of slip, can be used to pump molecular fluids such as water via non-intrusive application of a rotating electric field[6-8].

[1]J.S. Hansen, B.D. Todd and P.J. Daivis. Phys. Rev. E **84**, 016313 (2011).

[2]S. K. Kannam, B.D. Todd, J.S. Hansen and P.J. Daivis. J. Chem. Phys. **135**, 144701 (2011).

[3]S. K. Kannam, B.D. Todd, J.S. Hansen and P.J. Daivis. J. Chem. Phys. **136**, 024705 (2012).

[4]S. K. Kannam, B.D. Todd, J.S. Hansen and P.J. Daivis. J. Chem. Phys. **138**, 094701 (2013).

[5]J.S. Hansen, J.C. Dyre, P.J. Daivis, B.D. Todd and H. Bruus. Phys. Rev. E **84**, 036311 (2011).

[6]J. D. Bonthuis, D. Horinek, L. Bocquet, and R. R. Netz, Phys. Rev. Lett. **103**, 144503

(2009).

[7]J.S. Hansen, H. Bruus, B.D. Todd, and P.J. Daivis. J. Chem. Phys. **133**, 144906 (2010).

[8]S. De Luca, B.D. Todd, J.S. Hansen and P.J. Daivis, J. Chem. Phys. **138**, 154712 (2013)

(720e) Coupling Electrodialysis With a Novel Carbon Nanotube Topology Structure for Desalination: An Idea Demonstration

Chen, Q., Tsinghua University
Kong, X., Tsinghua University
Lu, D., Tsinghua University
Liu, Z., Tsinghua University
Originally presented on: 11/7/2013 16:45:00 - 17:05:00

Carbon nanotube (CNT) has been one of the research spotlights in recent years. By virtue of the several orders of magnitude faster water flow than macroscopic hydrodynamics and the impressive anti-fouling properties, aligned CNT membrane has become a super-star candidate for the next-generation reverse-osmosis (RO) desalination membrane materials. However, the shadow casting over the promising application of CNTs lies in its lack of intrinsic water/ion selectivity. Despite many significant research attempts by using steric hindrance or electrostatic interactions, researchers failed to achieve high-flux and high-selectivity at the same time, either because of the difficulty to synthesize small-diameter CNTs or the reduction of water flux once entrance of CNTs is functionalized. To obtain both high selectivity and flux, we design a novel membrane, which is filled with honeycomb-shaped CNT, and demonstrate the feasibility of desalination with high flux and selectivity by using molecular dynamics simulation. In operation, an external electric field vertical to the flow direction is applied to serve as the driving force for the spatial segregation of ions and water. In our honeycomb-CNT model, the diameters of all the CNTs are identical, ranging from 1.3 nm to 2.0 nm in different simulations. Each separation stage contains two Y-junctions, with a unit length as long as 5 nm. There are four CNT entrances and outlets on the graphene layer which functions as the membrane surface, resulting in the pore density around 6.7×10^{12} pores/cm^2. Concentrated streams, consisting of Na^+ and Cl^-, flow through the two outlets near anode and cathode, respectively, while the dilute streams flow through the two outlet streams in the middle. The ion concentration of the incoming stream is set to be 1-2 mol/L and the hydrostatic pressure at different side of the membrane is 400 MPa to enable more permeation events observed in 60 ns simulation. Under the electric field of 0.2-0.6 V/nm, desalination is successfully realized: the ion permeation ratio, defined as the ratio between ion concentrations of the dilute and the incoming streams, is close to 0%, indicating pure water is obtained in the middle outlet. In addition, the selectivity, defined as the ratio between ion concentrations of the concentrate and the dilute streams, can be as high as 5.5 ± 1.1. For better understanding of separation mechanism, further analysis was carried out. It is demonstrated that, first of all, low ion flux in the dilute streams is attributed to both steric hindrance and electric-field driven migration, which is consistent with our initial guess. Secondly, free energy barrier for ions to move upwards or downwards at the Y-junction is approximately 5 kcal/mol, which is the major driving force for the spatial distribution of ions. Thirdly, the importance of vertical ion transport is further demonstrated in the aligned CNT model, in which the lack of vertical connectivity leads to poor ion rejection performance even if the vertical electric field is applied. Finally, faster water transport in the concentration streams is subjected to examination, which probably resulted from the electro-osmosis effect. In conclusion, this hybrid approach can solve problems caused by both traditional electrodialysis (ED) and CNT membrane. On the one hand, CNT membrane is equipped with additional driving force for high selectivity. On the other hand, the lower ion concentration inside CNT membrane will significantly reduce the energy consumption of ED desalination. Notably, our model is not far from experimental practices. For example, techniques to synthesize sub-2-nm or Y-shaped CNTs are currently available, and applying electric field is much easier than charge modification. Therefore, our simulation provides an insight into the design of a more complex CNT membrane to achieve desalination.

(720f) Simulations and Experiments of Carbon Nanotube-Polyamide Nanocomposite Membranes for Water Desalination

Chen, H. Y., University of Pittsburgh
Chan, W. F., Virginia Polytechnic Institute and State University
Marand, E., Virginia Tech
Johnson, J. K., University of Pittsburgh
Originally presented on: 11/7/2013 17:05:00 - 17:30:00

We present a combined simulation and experimental study of zwitterion functionalized carbon nanotube (CNT) membranes for water desalination. We constructed model CNT membranes for the simulation studies by embedding four CNTs between graphene sheets to prevent fluid flow around the outside of the CNTs. Each end of each CNT was functionalized with one or more zwitterion groups. The model membrane was placed a periodically repeated saltwater box in order to study the flux of water and salt ions through the membrane. A pressure drop of about 200 MPa was applied across the membrane in order to generate a high water flux over the course of the simulation. The ion rejection ratio was found to be 100% when two zwitterions were affixed to the ends of (20,0) nanotubes, which are 1.56 nm in diameter. We found that five zwitterions per tube end was required to completely block ion transport in a 2.03 nm diameter (26,0) CNT. In contrast, the rejection ratio for non-functionalized CNTs and functionalized CNTs with five COOH groups at each tube end was found to be 0% and 20%, respectively. We have synthesized asymmetric carbon nanotube/polyamide membranes with varying loadings of CNTs and assessed these membranes for water desalination in experiments. The flux of water was found to be increased from 6.8 to 28.7 gallons per square foot per day as the weight fraction of zwitterion functionalized CNTs in the polyamide membrane was increased from 0 to 20 wt%. Importantly, the ion rejection ratio also increased from 97.6% to 98.6% with increasing weight fraction of functionalized CNTs. Additional simulation tests revealed the ion blocking mechanism is almost exclusively due to a steric hindrance, with a minor contribution due to electrostatic repulsion. Theoretical predictions indicate that an ideal CNT/polymer membrane having a loading of 20 wt % CNTs has a maximum flux of about 20000 gallons per square foot per day at the conditions of our experiments.

(720g) Explaining Ultra-High Water Flow Rates Observed in Carbon Nanotube Membranes
Mattia, D.,
Calabrò, F.,
Originally presented on: 11/7/2013 17:30:00 - 17:50:00

Explaining ultra-high water flow rates observed in carbon nanotube membranes

Davide Mattia,1* Francesco Calabr?2

1Department of Chemical Engineering, University of Bath, UK

2 DIEI, Universit? di Cassino, Italy

d.mattia@bath.ac.uk, www.bath.ac.uk/nanotech

Liquid flow through nanoporous media has been studied in disciplines as diverse as membrane science, soil permeability and cell physiology. In all these fields, though, the emphasis has been on the macroscopic outcome, often neglecting the effects of intermolecular forces and liquid-pore wall physicochemical interactions on liquid behaviour. The availability of carbon nanotubes (CNTs) as a platform for mass transport studies at the nanoscale has generated a large number of experimental and molecular dynamics studies which have shown unexpected fluid behaviours compared to the macroscale. The most striking one has been the observation of several orders of magnitude higher liquid flow velocities compared to the no-slip Haagen-Poiseuille equation. ADDIN EN.CITE ADDIN EN.CITE.DATA
3C456E644E6F74653E3C436974653E3C417574686F723E4D61747469613C2F417574686F723E3C596561
723E323030383C2F596561723E3C5265634E756D3E3932353C2F5265634E756D3E3C446973706C617954
6578743E3C7374796C652066616365D22737570657273637269707422E312C20323C2F7374796C653E
3C2F446973706C6179546578743E3C7265636F72643E3C7265632D6E756D6265723E3932353C2F726563
2D6E756D6265723E3C666F726569676E2D6B6579733E3C6B657920617070D22454E222064622D69643D
227A737461323272736E64357A706665737A326F706464777774746632737866736678303922 3E393235
3C2F6B65793E3C2F666F726569676E2D6B6579733E3C7265662D74797065206E616D653D224A6F75726E
616C2041727469636C65223E31373C2F7265662D747970653E3C636F6E7472696275746F72733E3C6175
74686F72733E3C617574686F723E4D61747469612C20446176696465 3C2F617574686F723E3C61757468
6F723E476F676F7473692C20597572793C2F617574686F723E3C2F617574686F72733E3C2F636F6E7472
696275746F72733E3C7469746C65733E3C7469746C653E5265766965773A207374617469632616E6420
64796E616D69632062656861766696F72206F66206C6971756964732696E736964652063617262 6F6E20
6E616E6F7475626573 3C2F7469746C653E3C7365636F6E646172792D7469746C653E4D6963726F666C75
69646963732616E64204E616E6F666C75696469633733C2F7365636F6E646172792D7469746C653E3C2F
7469746C65733E3C706572696F646963616C3E3C66756C6C2D7469746C653E4D6963726F666C75696469
637320616E64204E616E6F666C7569646963733C2F66756C6C2D7469746C653E3C616262722D313E4D69
63726F666C7569642E204E616E6F666C7569642E3C2F616262722D313E3C2F706572696F646963616C3E
3C70616765733E3238392D3330353C2F70616765733E3C766F6C756D653E353C2F766F6C756D653E3C6E
756D6265723E333C2F6E756D6265723E3C65646974696F6E3E31372F30342F323030383C2F6564697469
6F6E3E3C6B6579776F7264733E3C6B6579776F72643E6E616E6F666C75696469633733C2F6B6579776F72
643E3C6B6579776F72643E636172626F6E206E616E6F7475626573 3C2F6B6579776F72643E3C6B657977
6F72643E66696C6C696E673C2F6B6579776F72643E3C2F6B6579776F7264733E3C64617465733E3C7965
61723E323030383C2F796561723E3C2F64617465733E3C63616C6C2D6E756D3E303932353C2F63616C6C
2D6E756D3E3C75726C733E3C72656C617465642D75726C733E3C75726C3E687474703A2F2F64782E646F
692E6F72672F31302E313030372F7331303430342D3030382D303239332D353C2F75726C3E3C2F72656C
617465642D75726C733E3C2F75726C733E3C656C656374726F6E69632D7265736F757263652D6E756D3E
31302E313030372F7331303430342D3030382D303239332D353C2F656C656374726F6E69632D7265736F
757263652D6E756D3E3C2F7265636F72643E3C2F436974653E3C436974653E3C417574686F723E576869
7462793C2F417574686F723E3C596561723E323030373C2F596561723E3C5265634E756D3E3737333C2F
5265634E756D3E3C7265636F72643E3C7265632D6E756D6265723E3737333C2F7265632D6E756D626572
3E3C666F726569676E2D6B6579733E3C6B657920617070D22454E222064622D69643D227A7374613232
72736E64357A706665737A326F706464777774746632737866736678303922 3E3737333C2F6B65793E3C
2F666F726569676E2D6B6579733E3C7265662D74797065206E616D653D224A6F75726E616C2041727469
636C65223E31373C2F7265662D747970653E3C636F6E7472696275746F72733E3C617574686F72733E3C
617574686F723E576869746279 2C204D2E3C2F617574686F723E3C617574686F723E517569726B652C20
4E2E3C2F617574686F723E3C2F617574686F72733E3C2F636F6E7472696275746F72733E3C7469746C65
733E3C7469746C653E466C75696420466C6F7720696E20436172626F6E204E616E6F7475626573 20616E
64204E616E6F70697065733C2F7469746C653E3C7365636F6E646172792D7469746C653E4E6174757265

***** Formatting Issue - Best Available Paper/Graphic *****

204E616E6F746563686E6F6C6F67793C2F7365636F6E646172792D7469746C653E3C2F7469746C65733E
3C706572696F646963616C3E3C66756C6C2D7469746C653E4E6174757265204E616E6F746563686E6F6C
6F67793C2F66756C6C2D7469746C653E3C616262722D313E4E61742E204E616E6F2E3C2F616262722D31
3E3C2F706572696F646963616C3E3C70616765733E38372D39343C2F70616765733E3C766F6C756D653E
323C2F766F6C756D653E3C6B6579776F7264733E3C6B6579776F72643E4E616E6F666C75696469633C2F
6B6579776F72643E3C6B6579776F72643E6C697175696420666C6F773C2F6B6579776F72643E3C6B6579
776F72643E666C6F773C2F6B6579776F72643E3C6B6579776F72643E6E616E6F747562653C2F6B657977
6F72643E3C6B6579776F72643E4E616E6F706970653C2F6B6579776F72643E3C6B6579776F72643E7761
73686275726E206571756174696F6E3C2F6B6579776F72643E3C6B6579776F72643E426F73616E717565
7420657175174696F6E3C2F6B6579776F72643E3C2F6B6579776F7264733E3C64617465733E3C796561
723E323030373C2F796561723E3C2F64617465733E3C63616C6C2D6E756D3E303737333C2F63616C6C2D
6E756D3E3C776F726B2D747970653E7265766965773C2F776F726B2D747970653E3C75726C733E3C2F75
726C733E3C2F7265636F72643E3C2F436974653E3C2F456E644E6F74653EFF00 1, 2
3C456E644E6F74653E3C436974653E3C417574686F723E4D61747469613C2F417574686F723E3C596561
723E323030383C2F596561723E3C5265634E756D3E3932353C2F5265634E756D3E3C446973706C617954
6578743E3C7374796C6520666163653D22737570657273637269707422E312C20323C2F7374796C653E
3C2F446973706C6179546578743E3C7265636F72643E3C7265632D6E756D6265723E3932353C2F726563
2D6E756D6265723E3C666F726569676E2D6B6579733E3C6B657920617070D22454E222064622D69643D
227A737461323272736E64357A706665737A326F706464777777474663273786673667830392239323 5
3C2F6B65793E3C2F666F726569676E2D6B6579733E3C7265662D74797065206E616D653D224A6F75726E
616C2041727469636C65223E31373C2F7265662D747970653E3C636F6E7472696275746F72733E3C6175
74686F72733E3C617574686F723E4D61747469612C20446176696453C2F617574686F723E3C61757468
6F723E476F676F7473692C20597572793C2F617574686F723E3C2F617574686F72733E3C2F636F6E7472
696275746F72733E3C7469746C65733E3C7469746C653E5265766965773A207374617469632616E6420
64796E616D6963206265686176696F72206F66206C697175696473206E736964652063617262F6E20
6E616E6F7475626573C2F7469746C653E3C7365636F6E646172792D7469746C653E4D6963726F666C75
696469637320616E64204E616E6F666C7569646963733C2F7365636F6E646172792D7469746C653E3C2F
7469746C65733E3C706572696F646963616C3E3C66756C6C2D7469746C653E4D6963726F666C75696469
637320616E64204E616E6F666C7569646963733C2F66756C6C2D7469746C653E3C616262722D313E4D69
63726F666C7569642E204E616E6F666C7569642E3C2F616262722D313E3C2F706572696F646963616C3E
3C70616765733E3238392D3330353C2F70616765733E3C766F6C756D653E353C2F766F6C756D653E3C6E
756D6265723E333C2F6E756D6265723E3C6564697469F6E3E31372F30342F323030383C2F6564697469
6F6E3E3C6B6579776F7264733E3C6B6579776F72643E6E616E6F666C75696469633733C2F6B6579776F72
643E3C6B6579776F72643E636172626F6E206E616E6F7475626573C2F6B6579776F72643E3C6B657977
6F72643E66696C6C696E673C2F6B6579776F72643E3C2F6B6579776F7264733E3C64617465733E3C7965
61723E323030383C2F796561723E3C2F64617465733E3C63616C6C2D6E756D3E303932353C2F63616C6C
2D6E756D3E3C75726C733E3C72656C617465642D75726C733E3C75726C3E687474703A2F2F64782E646F
692E6F72672F31302E313030372F7331303430342D3030382D303239332D353C2F75726C3E3C2F72656C
617465642D75726C733E3C2F75726C733E3C656C656374726F6E69632D7265736F757263652D6E756D3E
31302E313030372F7331303430342D3030382D303239332D353C2F656C656374726F6E69632D7265736F
757263652D6E756D3E3C2F7265636F72643E3C2F436974653E3C436974653E3C417574686F723E576869
7462793C2F417574686F723E3C596561723E323030373C2F596561723E3C5265634E756D3E3737333C2F
5265634E756D3E3C7265636F72643E3C7265632D6E756D6265723E3737333C2F7265632D6E756D626572
3E3C666F726569676E2D6B6579733E3C6B657920617070D22454E222064622D69643D227A7374613232
72736E64357A706665737A326F70646477777474663273786673667830392233E3737333C2F6B65793E3C
2F666F726569676E2D6B6579733E3C7265662D74797065206E616D653D224A6F75726E616C2041727469
636C65223E31373C2F7265662D747970653E3C636F6E7472696275746F72733E3C617574686F72733E3C
617574686F723E5768697462792C204D2E3C2F617574686F723E3C617574686F723E517569726B652C20
4E2E3C2F617574686F723E3C2F617574686F72733E3C2F636F6E7472696275746F72733E3C7469746C65
733E3C7469746C653E466C75696420466C6F7720696E20436172626F6E204E616E6F747562657320616E
64204E616E6F70697065733C2F7469746C653E3C7365636F6E646172792D7469746C653E4E6174757265
204E616E6F746563686E6F6C6F67793C2F7365636F6E646172792D7469746C653E3C2F7469746C65733E
3C706572696F646963616C3E3C66756C6C2D7469746C653E4E6174757265204E616E6F746563686E6F6C
6F67793C2F66756C6C2D7469746C653E3C616262722D313E4E61742E204E616E6F2E3C2F616262722D31
3E3C2F706572696F646963616C3E3C70616765733E38372D39343C2F70616765733E3C766F6C756D653E
323C2F766F6C756D653E3C6B6579776F7264733E3C6B6579776F72643E4E616E6F666C75696469633C2F
6B6579776F72643E3C6B6579776F72643E6C697175696420666C6F773C2F6B6579776F72643E3C6B6579
776F72643E666C6F773C2F6B6579776F72643E3C6B6579776F72643E6E616E6F747562653C2F6B657977
6F72643E3C6B6579776F72643E4E616E6F706970653C2F6B6579776F72643E3C6B6579776F72643E7761
73686275726E206571756174696F6E3C2F6B6579776F72643E3C6B6579776F72643E426F73616E717565
7420657175174696F6E3C2F6B6579776F72643E3C2F6B6579776F7264733E3C64617465733E3C796561
723E323030373C2F796561723E3C2F64617465733E3C63616C6C2D6E756D3E303737333C2F63616C6C2D
6E756D3E3C776F726B2D747970653E7265766965773C2F776F726B2D747970653E3C75726C733E3C2F75

*** Formatting Issue - Best Available Paper/Graphic ***

726C733E3C2F7265636F72643E3C2F436974653E3C2F456E644E6F74653E0000

These experimental results have generated an intense debate about the underlying physical phenomena and whether such flow can be explained within the continuum model on which traditional fluid mechanics is based or whether there is a size threshold below which it no longer applies (Figure 1). These results also hold great promise for filtration applications in general and water desalination in particular. ADDIN EN.CITE

<EndNote><Cite><Author>Lee</Author><Year>2011</Year><RecNum>1294</RecNum><DisplayText><style Article">17</ref-type><contributors><authors><author>Lee, Kah Peng</author><author>Arnot, Tom C.</author><author>Mattia, D.</author></authors></contributors><titles><title>A review of reverse osmosis membrane materials for desalination--Development to date and future potential</title><secondary-title>Journal of Membrane Science</secondary-title></titles><periodical><full-title>Journal of Membrane Science</full-title><abbr-1>J. Membr. Sci.</abbr-1><abbr-2>J Membr Sci</abbr-2></periodical><pages>1-22</pages><volume>370</volume><number>1-2</number><keywords><keyword>Desalination</keyword><keyword>Reverse osmosis (RO)</keyword><keyword>Membrane material</keyword><keyword>Membrane performance</keyword><keyword>Nano-materials</keyword></keywords><dates><year>2011</year></dates><isbn>0376-7388</isbn><call-num>1294</call-num><urls><related-urls><url>http://www.sciencedirect.com/science/article/B6TGK-51SPS08-2/2/6f8473a0951f30842ad2aace7de13864</url></related-urls></urls><electronic-resource-num>DOI: 10.1016/j.memsci.2010.12.036</electronic-resource-num></record></Cite></EndNote>3

Figure 1. Water plugs in 40-60 nm and 2-5 nm diameter carbon nanotubes, respectively, showing a transition from a continuum to a sub-continuum behaviour. ADDIN EN.CITE ADDIN EN.CITE.DATA

3C456E644E6F74653E3C436974653E3C417574686F723E4D61747469613C2F417574686F723E3C596561
723E323030363C2F596561723E3C5265634E756D3E3634393C2F5265634E756D3E3C446973706C617954
6578743E3C7374796C652066616365D2273757065727363726970742223E342C20353C2F7374796C653E
3C2F446973706C6179546578743E3C7265636F72643E3C7265632D6E756D6265723E3634393C2F726563
2D6E756D6265723E3C666F726569676E2D6B6579733E3C6B657920617070D22454E222064622D69643D
227A737461323272736E64357A706665737A326F706464777777474663273786673667830392 23E363439
3C2F6B65793E3C2F666F726569676E2D6B6579733E3C7265662D74797065206E616D653D224A6F75726E
616C2041727469636C65223E31373C2F7265662D747970653E3C636F6E7472696275746F72733E3C6175
74686F72733E3C617574686F723E4D61747469612C20442E3C2F617574686F723E3C617574686F723E52
6F7373692C204D2E20502E3C2F617574686F723E3C617574686F723E4B696D2C20422E4D2E3C2F617574
686F723E3C617574686F723E4B6F726E6576612C20472E3C2F617574686F723E3C617574686F723E4261
752C20482E482E3C2F617574686F723E3C617574686F723E476F676F7473692C20592E3C2F617574686F
723E3C2F617574686F72733E3C2F636F6E7472696275746F72733E3C7469746C65733E3C7469746C653E
456666656374206F662047726170686974697A6174696F6E206F6E207468652057657474616269C6974
7920616E6420456C65637472696963616C20436F6E647563746976697479206F6620435644204361726
6E204E616E6F747562657320616E642046696C6D733C2F7469746C653E3C7365636F6E646172792D7469
746C653E4A6F75726E616C206F66205068797369636616C204368656D69737472792042C2F7365636F6E
646172792D7469746C653E3C2F7469746C65733E3C706572696F646963616C3E3C66756C6C2D7469746C
653E4A6F75726E616C206F66205068797369636616C204368656D69737472792042C2F66756C6C2D7469
746C653E3C616262722D313E4A2E20506879732E204368656D2E20423C2F616262722D313E3C2F706572
696F646963616C3E3C70616765733E39383530202D393835353C2F70616765733E3C766F6C756D653E31
31303C2F766F6C756D653E3C6E756D6265723E32303C2F6E756D6265723E3C6B6579776F7264733E3C6B
6579776F72643E616E6E65616C696E673C2F6B6579776F72643E3C6B6579776F72643E7765747474696E67
3C2F6B6579776F72643E3C6B6579776F72643E4356443C2F6B6579776F72643E3C6B6579776F72643E6E
616E6F747562653C2F6B6579776F72643E3C6B6579776F72643E636F6E647563746976697479 3C2F6B65
79776F72643E3C6B6579776F72643E77617465723C2F6B6579776F72643E3C6B6579776F72643E677261
70686974697A6174696F6E3C2F6B6579776F72643E3C2F6B6579776F7264733E3C64617465733E3C7965
61723E323030363C2F796561723E3C2F64617465733E3C63616C6C2D6E756D3E303634393C2F63616C6C
2D6E756D3E3C75726C733E3C2F75726C733E3C2F7265636F72643E3C2F436974653E3C436974653E3C41
7574686F723E4D61747469613C2F417574686F723E3C596561723E323030373C2F596561723E3C526563
4E756D3E3836393C2F5265634E756D3E3C7265636F72643E3C7265632D6E756D6265723E3836393C2F72
65632D6E756D6265723E3C666F726569676E2D6B6579733E3C6B657920617070D22454E222064622D69
643D227A737461323272736E64357A706665737A326F706464777777474663273786673667830392 23E38

*** Formatting Issue - Best Available Paper/Graphic ***

36393C2F6B65793E3C2F666F726569676E2D6B6579733E3C7265662D74797065206E616D653D22436F6E
666572656E63652050726F63656564696E6773223E31303C2F7265662D747970653E3C636F6E74726962
75746F72733E3C617574686F72733E3C617574686F723E4D61747469612C20442E3C2F617574686F723E
3C617574686F723E526F7373692C204D2E20502E3C2F617574686F723E3C617574686F723E59652C2048
2E3C2F617574686F723E3C617574686F723E476F676F7473692C20592E3C2F617574686F723E3C2F6175
74686F72733E3C2F636F6E7472696275746F72733E3C7469746C65733E3C7469746C653E3C7374796C65
20666163653D226974616C69632220666F6E743D2264656661756C74222073697A653D223130302522E
496E207369747520466C7569643C2F7374796C653E3C7374796C6520666163653D226E6F726D616C2220
666F6E743D2264656661756C74222073697A653D223130302522E2053747564696573206E20436172
626F6E204E616E6F747562657320776974682044696D657465727320616E67696E672066726F6D20
3120746F20353030206E6D3C2F7374796C653E3C2F7469746C653E3C7365636F6E646172792D7469746C
653E3574682041534D452F575345415320496E7465726E6174696F6E616C20436F6E666572656E6365
206F6E20466C756964204D656368616E69637320616E64204165726F64796E616D6963733C2F7365636F
6E646172792D7469746C653E3C74657274696172792D7469746C653E4D617468656D617469637320616E
6420436F6D707574657273206E20536369656E636520616E6420456E67696E656572696E673C2F7465
7274696172792D7469746C653E3C2F7469746C65733E3C70616765733E3239372D3239393C2F70616765
733E3C6B6579776F7264733E3C6B6579776F72643E436172626F6E206E616E6F747562653C2F6B657977
6F72643E3C6B6579776F72643E6E616E6F666C7569646963733C2F6B6579776F72643E3C6B6579776F72
643E666C756964207472616E73706F72743C2F6B6579776F72643E3C6B6579776F72643E7765747469E
673C2F6B6579776F72643E3C6B6579776F72643E636F6E7461637420616E676C653C2F6B6579776F7264
3E3C6B6579776F72643E73757266616365206368656D69737472793C2F6B6579776F72643E3C2F6B6579
776F7264733E3C64617465733E3C796561723E323030373C2F796561723E3C2F64617465733E3C707562
2D6C6F636174696F6E3E417468656E732C20477265656563653C2F7075622D6C6F636174696F6E3E3C6973
626E3E3937382D3936302D383435372D39392D37203C2F6973626E3E3C63616C6C2D6E756D3E30383639
3C2F63616C6C2D6E756D3E3C75726C733E3C2F75726C733E3C2F7265636F72643E3C2F436974653E3C2F
456E644E6F74653E0000 4, 5
3C456E644E6F74653E3C436974653E3C417574686F723E4D61747469613C2F417574686F723E3C596561
723E323030363C2F596561723E3C5265634E756D3E3634393C2F5265634E756D3E3C446973706C617954
6578743E3C7374796C6520666163653D22737570657273637269707422E342C20353C2F7374796C653E
3C2F446973706C6179546578743E3C7265636F72643E3C7265632D6E756D6265723E3634393C2F726563
2D6E756D6265723E3C666F726569676E2D6B6579733E3C6B657920617070D22454E222064622D69643D
227A737461323272736E64357A706665737A326F706464777774746632737866736678303922E363439
3C2F6B65793E3C2F666F726569676E2D6B6579733E3C7265662D74797065206E616D653D224A6F75726E
616C2041727469636C65223E31373C2F7265662D747970653E3C636F6E7472696275746F72733E3C6175
74686F72733E3C617574686F723E4D61747469612C20442E3C2F617574686F723E3C617574686F723E52
6F7373692C204D2E20502E3C2F617574686F723E3C617574686F723E4B696D2C20422E4D2E3C2F617574
686F723E3C617574686F723E4B6F726E6576612C20472E3C2F617574686F723E3C617574686F723E4261
752C20482E482E3C2F617574686F723E3C617574686F723E476F676F7473692C20592E3C2F617574686F
723E3C2F617574686F72733E3C2F636F6E7472696275746F72733E3C7469746C65733E3C7469746C653E
456666656374206F66204772617068697469A6174696F6E206F6E2074686520576574746162696C6974
7920616E6420456C656374726963616C20436F6E647563746976697479206F6620435644204361726 26F
6E204E616E6F747562657320616E642046696C6D733C2F7469746C653E3C7365636F6E646172792D7469
746C653E4A6F75726E616C206F66205068797369636616C204368656D697374727920423C2F7365636F6E
646172792D7469746C653E3C2F7469746C65733E3C706572696F646963616C3E3C66756C6C2D7469746C
653E4A6F75726E616C206F66205068797369636616C204368656D697374727920423C2F66756C6C2D7469
746C653E3C616262722D313E4A2E20506879732E204368656D2E20423C2F616262722D313E3C2F706572
696F646963616C3E3C70616765733E39383530202D393835353C2F70616765733E3C766F6C756D653E31
31303C2F766F6C756D653E3C6E756D6265723E32303C2F6E756D6265723E3C6B6579776F7264733E3C6B
6579776F72643E616E6E65616C696E673C2F6B6579776F72643E3C6B6579776F72643E7765747469E67
3C2F6B6579776F72643E3C6B6579776F72643E4356443C2F6B6579776F72643E3C6B6579776F72643E6E
616E6F747562653C2F6B6579776F72643E3C6B6579776F72643E636F6E647563746976697479793C2F6B65
79776F72643E3C6B6579776F72643E77617465723C2F6B6579776F72643E3C6B6579776F72643E677261
70686974697A6174696F6E3C2F6B6579776F72643E3C2F6B6579776F7264733E3C64617465733E3C7965
61723E323030363C2F796561723E3C2F64617465733E3C63616C6C2D6E756D3E303634393C2F63616C6C
2D6E756D3E3C75726C733E3C2F75726C733E3C2F7265636F72643E3C2F436974653E3C436974653E3C41
7574686F723E4D61747469613C2F417574686F723E3C596561723E323030373C2F596561723E3C526563
4E756D3E3836393C2F5265634E756D3E3C7265636F72643E3C7265632D6E756D6265723E3836393C2F72
65632D6E756D6265723E3C666F726569676E2D6B6579733E3C6B657920617070D22454E222064622D69
643D227A737461323272736E64357A706665737A326F7064647777747466327378667366783039223E38
36393C2F6B65793E3C2F666F726569676E2D6B6579733E3C7265662D74797065206E616D653D22436F6E
666572656E63652050726F63656564696E6773223E31303C2F7265662D747970653E3C636F6E74726962
75746F72733E3C617574686F72733E3C617574686F723E4D61747469612C20442E3C2F617574686F723E
3C617574686F723E526F7373692C204D2E20502E3C2F617574686F723E3C617574686F723E59652C2048

2E3C2F617574686F723E3C617574686F723E476F676F7473692C20592E3C2F617574686F723E3C2F6175
74686F72733E3C2F636F6E7472696275746F72733E3C7469746C65733E3C7469746C653E3C7374796C65
20666163653D226974616C69632220666F6E743D2264656661756C74222073697A653D223130302522 3E
496E207369747520466C7569643C2F7374796C653E3C7374796C6520666163653D226E6F726D616C2220
666F6E743D2264656661756C74222073697A653D223130302522 3E2053747564696573206 96E20436172
626F6E204E616E6F747562657320776974682044696 16D657465727320616E67696E672066726F6D20
3120746F20353030206E6D3C2F7374796C653E3C2F7469746C653E3C7365636F6E646172792D7469746C
653E3574682049415354452F57534541532049 6E7465726E6174696F6E616C20436F6E666572656E6365
206F6E20466C756964204D656368616E69637320616E64204165726F64796E616D6963733C2F7365636F
6E646172792D7469746C653E3C7465727469617279 2D7469746C653E4D617468656D61746963732 0616E
6420436F6D7075746572732 0696E20536369656E636520616E6420456E67696E656572696E673C2F7465
7274696172792D7469746C653E3C2F7469746C65733E3C70616765733E3239372D3239393C2F70616765
733E3C6B6579776F7264733E3C6B6579776F72643E436172626F6E206E616E6F747562653C2F6B657977
6F72643E3C6B6579776F72643E6E616E6F666C75696469637 33C2F6B6579776F72643E3C6B6579776F72
643E666C7569642074726 16E73706F72743C2F6B6579776F72643E3C6B6579776F72643E77657474696E
673C2F6B6579776F72643E3C6B6579776F72643E636F6E7461637420616E676C653C2F6B6579776F7264
3E3C6B6579776F72643E737572666163652063686 56D69737472793C2F6B6579776F72643E3C2F6B6579
776F7264733E3C64617465733E3C796561723E323030373C2F796561723E3C2F64617465733E3C707562
2D6C6F636174696F6E3E417468656E732C20477265656365 3C2F7075622D6C6F636174696F6E3E3C6973
626E3E393738 2D39363 02D383435372D39392D37203C2F6973626E3E3C63616C6C2D6E756D3E30383639
3C2F63616C6C2D6E756D3E3C75726C733E3C2F75726C733E3C2F7265636F72643E3C2F436974653E3C2F
456E644E6F74653E0000 The micrograph on the left has been obtained using an environmental SEM by preferential condensation of water inside a CNT. The micrograph on the right was obtained using high resolution TEM and shows the coexistence of water liquid and vapor in a capped CNT.

The attempts made so far to explain these phenomena have focused on the slip length at the wall and its possible dependencies on a wide range of parameters, from contact angle to roughness, viscosity and shear rate. All these solutions are somewhat unsatisfying when slip values in the order of tens of micrometers for membranes with CNTs with diameters of only few nanometers are obtained. Moreover, no model has yet been capable of explicitly relating the flow enhancement with solid-liquid molecular interactions occurring at the channel walls.

Here a model capable of explaining the observed dependence of the flow enhancement on the membrane channels' characteristic dimensions and surface wettability is proposed. The model can accurately predict the effect of each component for membrane channels and nanotubes of different sizes, wall surface chemistry and structure. Comparison with the experimental and modelling data available shows good agreement (Figure 2). ADDIN EN.CITE <EndNote><Cite><Author>Mattia</Author><Year>2012</Year><RecNum>1472</RecNum><DisplayText><style db-id="zsta22rsnd5zpfesz2opddwwttf2sxfsfx09">1472</key></foreign-keys><ref-type name="Journal Article">17</ref-type><contributors><authors><author>Mattia, D.</author><author>Calabr?, F.</author></authors></contributors><titles><title>Explaining high flow rate of water in carbon nanotubes via solid-liquid molecular interactions</title><secondary-title>Microfluidics and Nanofluidics</secondary-title></titles><periodical><full-title>Microfluidics and Nanofluidics</full-title><abbr-1>Microfluid. Nanofluid.</abbr-1></periodical><pages>1-6</pages><keywords><keyword>work of adhesion</keyword><keyword>surface diffusion</keyword><keyword>slip</keyword><keyword>solid-liquid interaction</keyword></keywords><dates><year>2012</year></dates><publisher>Springer Berlin / Heidelberg</publisher><isbn>1613-4982</isbn><call-num>1472</call-num><urls><related-urls><url>http://dx.doi.org/10.1007/s10404-012-0949-z</url></related-urls></urls><electronic-resource-num>10.1007/s10404-012-0949-z</electronic-resource-num></record></Cite></EndNote>6 The proposed model is based on a two-phase model for water flow in cylindrical channels with reduced viscosity at the wall, , and a non-zero velocity at the wall which is solely a function of solid-liquid interactions (ex. graphitic carbon and water) and the applied pressure. By solving the Haagen-Poiseuille equation with this velocity at the wall term, the following expression for the fluid flow enhancement, , at the wall is obtained:

where and are the tube/channel/pore length and radius, respectively; is the thickness of the annular region and is the liquid bulk density; is the surface diffusion term arising from the chemical potential generated by the pressure gradient, ADDIN EN.CITE
<EndNote><Cite><Author>Ruckenstein</Author><Year>1983</Year><RecNum>1314</RecNum><Di
splayText><style no-slip boundary condition of
hydrodynamics</title><secondary-title>Journal of Colloid and Interface
Science</secondary-title></titles><periodical><full-title>Journal of Colloid and
Interface Science</full-title><abbr-1>J. Colloid Interface Sci.</abbr-1><abbr-2>J
Colloid Interface
Sci</abbr-2></periodical><pages>488-491</pages><volume>96</volume><number>2</number>
<keywords><keyword>slip</keyword><keyword>surface
diffusion</keyword></keywords><dates><year>1983</year></dates><isbn>0021-9797</isbn>
<call-num>1314</call-num><urls><related-urls><url>http://www.sciencedirect.com/scien
ce/article/B6WHR-4CT7DM2-NX/2/a2dbcea1f67474c1626147d2f0f2fb1c</url></related-urls><
/urls><electronic-resource-num>10.1016/0021-9797(83)90050-4</electronic-resource-num
></record></Cite></EndNote>7 and is the work of adhesion which represents the strength of the solid-liquid interaction at the tube/channel/pore wall. The two latter terms can be measured independently and at the macroscale for any solid-liquid couple.

Two recent publications have offered further validation of the proposed model: First, a molecular dynamics paper of water flow in CNTs of different lengths has found a linear dependence of the flow enhancement with the former. ADDIN EN.CITE
<EndNote><Cite><Author>Nicholls</Author><Year>2011</Year><RecNum>1358</RecNum><Displ
ayText><style Jason</author></authors></contributors><titles><title>Water transport
through (7,7) carbon nanotubes of different lengths using molecular
dynamics</title><secondary-title>Microfluidics and
Nanofluidics</secondary-title></titles><periodical><full-title>Microfluidics and
Nanofluidics</full-title><abbr-1>Microfluid.
Nanofluid.</abbr-1></periodical><pages>257-264</pages><volume>12</volume><number>1-4
</number><keywords><keyword>Physics and
Astronomy</keyword></keywords><dates><year>2011</year></dates><publisher>Springer
Berlin /
Heidelberg</publisher><isbn>1613-4982</isbn><call-num>1358</call-num><urls><related-
urls><url>http://dx.doi.org/10.1007/s10404-011-0869-3</url></related-urls></urls><el
ectronic-resource-num>10.1007/s10404-011-0869-3</electronic-resource-num></record></
Cite></EndNote>8

Second, water flow measurements in hydrophilic alumina nanochannels with diameters ranging from 20 nm to 100 nm have shown to be well represented by the above equation. For the smallest channels tested, the water permeability is more than double than the theoretical prediction using the Haagen-Poiseuille equation. Though such an enhancement is significantly smaller than what has been measured in hydrophobic carbon nanotubes, it clearly shows that flow enhancement and water slippage occur on hydrophilic surfaces as well, contrary to existing theoretical models (Figure 2, red empty circles ○). ADDIN EN.CITE
<EndNote><Cite><Author>Lee</Author><Year>2012</Year><RecNum>1486</RecNum><DisplayTex
t><style Peng</author><author>Leese, Hannah</author><author>Mattia,
Davide</author></authors></contributors><titles><title>Water flow enhancement in
hydrophilic
nanochannels</title><secondary-title>Nanoscale</secondary-title></titles><periodical
><full-title>Nanoscale</full-title></periodical><keywords><keyword>slip</keyword><ke
yword>flow
enhancement</keyword><keyword>Hagen-poisuille</keyword><keyword>AAO</keyword></keywo
rds><dates><year>2012</year></dates><publisher>The Royal Society of
Chemistry</publisher><isbn>2040-3364</isbn><call-num>1486</call-num><urls><related-u
rls><url>http://dx.doi.org/10.1039/C2NR30098B</url></related-urls></urls></record></
Cite></EndNote>9

*** Formatting Issue - Best Available Paper/Graphic ***

Finally, the above equation has been successfully used to model molecular dynamics simulations of water flow in 2nm diameter silicon carbide (SiCNTs, yellow diamonds ♦) and boron nitride (BNNTs, blue squares ■) nanotubes.10

THE GRAPHIC ON THIS PAGE WAS NOT AVAILABLE ON THE E-MEDIA VERSION OF THIS PAPER.

Figure 2. Normalized flow enhancement as a function of nanotube radius (dashed curve). Literature: ○ - (Thomas and McGaughey, Nano Lett, 2009); □ - (Holt et al., Science, 2006); × - (Thomas et al., Int J Therm Sci, 2010); ▲ - (Majumder et al., Nature, 2005); ♦ - (Du et al., Langmuir, 2011); ■ - (Whitby and Quirke, Nano Lett, 2007). Literature values (open symbols for MD and full symbols for experimental) are reported as a range between minimum and maximum possible values. ○ Enhancement values for water flow in alumina nanopores; ♦ SiCNTs; ■ BNNTs; Adapted from ref.s ADDIN EN.CITE ADDIN EN.CITE.DATA
3C456E644E6F74653E3C436974653E3C417574686F723E4D61747469613C2F417574686F723E3C596561
723E323031323C2F596561723E3C5265634E756D3E313437323C2F5265634E756D3E3C446973706C6179
546578743E3C7374796C652066616365D22737570657273637269707422E362C20393C2F7374796C65
3E3C2F446973706C6179546578743E3C7265636F72643E3C7265632D6E756D6265723E313437323C2F72
65632D6E756D6265723E3C666F726569676E2D6B6579733E3C6B657920617070D22454E222064622D69
643D227A737461323272736E64357A706665737A326F70646477777474663273786673667830922 3E31
3437323C2F6B65793E3C2F666F726569676E2D6B6579733E3C7265662D74797065206E616D653D224A6F
75726E616C204172746963656223E31373C2F7265662D747970653E3C636F6E7472696275746F72733E
3C617574686F72733E3C617574686F723E4D61747469612C20442E3C2F617574686F723E3C617574686F
723E43616C616272C3B22C20462E3C2F617574686F723E3C2F617574686F72733E3C2F636F6E74726962
75746F72733E3C7469746C65733E3C7469746C653E4578706C61696E696E6720686967682066C6F7720
72617465206F6620776174657220696E20636172626F6E206E616E6F74756265732076696120736F6C69
64E280936C6971756964206D6F6C6563756C617220696E746572616374696F6E733C2F7469746C653E3C
7365636F6E646172792D7469746C653E4D6963726F666C7569646963737320616E64204E616E6F666C7569
646963733C2F7365636F6E646172792D7469746C653E3C2F7469746C65733E3C706572696F646963616C
3E3C66756C6C2D7469746C653E4D6963726F666C7569646963737320616E64204E616E6F666C7569646963
733C2F66756C6C2D7469746C653E3C616262722D313E4D6963726F666C7569642E204E616E6F666C7569
642E3C2F616262722D313E3C2F706572696F646963616C3E3C70616765733E312D363C2F70616765733E
3C6B6579776F7264733E3C6B6579776F72643E776F726B206F6620616468657369F6E3C2F6B6579776F
72643E3C6B6579776F72643E737572666163652064696666757369F6E3C2F6B6579776F72643E3C6B65
79776F72643E736C69703C2F6B6579776F72643E3C6B6579776F72643E736F6C69642D6C697175696420
696E746572616374696F6E3C2F6B6579776F72643E3C2F6B6579776F7264733E3C6461746573733E3C7965
61723E323031323C2F796561723E3C2F64617465733E3C7075626C69736865723E537072696E67657220
4265726C696E202F2048656964656C626572673C2F7075626C69736865723E3C6973626E3E313631332D
343938323C2F6973626E3E3C63616C6C2D6E756D3E313437323C2F63616C6C2D6E756D3E3C75726C733E
3C72656C617465642D75726C733E3C75726C3E687474703A2F2F64782E646F692E6F72672F31302E3130
30372F7331303430342D3031322D303934392D7A3C2F75726C3E3C2F72656C617465642D75726C733E3C
2F75726C733E3C656C656374726F6E69632D7265736F757263652D6E756D3E31302E313030372F733130
3430342D3031322D303934392D7A3C2F656C656374726F6E69632D7265736F757263652D6E756D3E3C2F
7265636F72643E3C2F436974653E3C436974653E3C417574686F723E4C65653C2F417574686F723E3C59
6561723E323031323C2F596561723E3C5265634E756D3E313438363C2F5265634E756D3E3C7265636F72
643E3C7265632D6E756D6265723E313438363C2F7265632D6E756D6265723E3C666F726569676E2D6B65
79733E3C6B657920617070D22454E222064622D69643D227A737461323272736E64357A706665737A32
6F706464777777474663273786673667830922 3E313438363C2F6B65793E3C2F666F726569676E2D6B65
79733E3C7265662D74797065206E616D653D224A6F75726E616C204172746963656C223E31373C2F7265
662D747970653E3C636F6E7472696275746F72733E3C617574686F72733E3C617574686F723E4C65652C
204B6168205065656E673C2F617574686F723E3C617574686F723E4C656573652C2048616E6E61683C2F61
7574686F723E3C617574686F723E4D61747469612C20446176696465D3C2F617574686F723E3C2F617574
686F72733E3C2F636F6E7472696275746F72733E3C7469746C65733E3C7469746C653E5761746572206
6C6F7720656E68616E63656D656E7420696E2068796472F7068696C6963206E616E6F6368616E6E656C
733C2F7469746C653E3C7365636F6E646172792D7469746C653E4E616E6F7363616C653C2F7365636F6E
646172792D7469746C653E3C2F7469746C65733E3C706572696F646963616C3E3C66756C6C2D7469746C
653E4E616E6F7363616C653C2F66756C6C2D7469746C653E3C2F706572696F646963616C3E3C6B657977
6F7264733E3C6B6579776F72643E736C69703C2F6B6579776F72643E3C6B6579776F72643E666C6F7720
656E68616E63656D656E743C2F6B6579776F72643E3C6B6579776F72643E486167656E2D706F69737569
6C6C653C2F6B6579776F72643E3C6B6579776F72643E41414F3C2F6B6579776F72643E3C2F6B6579776F
7264733E3C64617465733E3C796561723E323031323C2F796561723E3C2F64617465733E3C7075626C69
736865723E54686520526F79616C20536F6369657479206F66204368656D69737472793C2F7075626C69
736865723E3C6973626E3E323034302D333336343C2F6973626E3E3C63616C6C2D6E756D3E313438363C
2F63616C6C2D6E756D3E3C75726C733E3C72656C617465642D75726C733E3C75726C3E687474703A2F2F

64782E646F692E6F72672F31302E313033392F43324E523330303938423C2F75726C3E3C2F72656C6174
65642D75726C733E3C2F75726C733E3C2F7265636F72643E3C2F436974653E3C2F456E644E6F74653E00
5200 6, 9
3C456E644E6F74653E3C436974653E3C417574686F723E4D61747469613C2F417574686F723E3C596561
723E323031323C2F596561723E3C5265634E756D3E313437323C2F5265634E756D3E3C446973706C6179
546578743E3C7374796C6520666163653D227375706572736372697074223E362C20393C2F7374796C65
3E3C2F446973706C6179546578743E3C7265636F72643E3C7265632D6E756D6265723E313437323C2F72
65632D6E756D6265723E3C666F726569676E2D6B6579733E3C6B657920617070003D22454E222064622D69
643D227A737461323272736E64357A706665737A326F70646477777474663273786673667830392223E31
3437323C2F6B65793E3C2F666F726569676E2D6B6579733E3C7265662D74797065206E616D653D224A6F
75726E616C2041727469636C65223E31373C2F7265662D747970653E3C636F6E7472696275746F72733E
3C617574686F72733E3C617574686F723E4D61747469612C20442E3C2F617574686F723E3C617574686F
723E43616C616272C3B22C20462E3C2F617574686F723E3C2F617574686F72733E3C2F636F6E74726962
75746F72733E3C7469746C65733E3C7469746C653E4578706C6F696E696E6720686967682066C6F7720
72617465206F662077617465722069E20636172626F6E206E616E6F7475626573207669612736F6C69
64E280936C6971756964206D6F6C6563756C617220696E746572616374696F6E733C2F7469746C653E3C
7365636F6E646172792D7469746C653E4D6963726F666C75696469637320616E64204E616E6F666C7569
646963733C2F7365636F6E646172792D7469746C653E3C2F7469746C65733E3C706572696F646963616C
3E3C66756C6C2D7469746C653E4D6963726F666C756964696373320616E64204E616E6F666C7569646963
733C2F66756C6C2D7469746C653E3C616262722D313E4D6963726F666C7569642E204E616E6F666C7569
642E3C2F616262722D313E3C2F706572696F646963616C3E3C70616765733E312D363C2F70616765733E
3C6B6579776F7264733E3C6B6579776F72643E776F726B206F6620616468657369F6E3C2F6B6579776F
72643E3C6B6579776F72643E7375726661636520646966667573696F6E3C2F6B6579776F72643E3C6B65
79776F72643E736C69703C2F6B6579776F72643E3C6B6579776F72643E736F6C69642D6C697175696420
696E746572616374696F6E3C2F6B6579776F72643E3C2F6B6579776F7264733E3C6461746573E3C7965
61723E323031323C2F796561723E3C2F64617465733E3C7075626C69736865723E537072696E67657220
4265726C696E202F2048656964656C626572673C2F7075626C69736865723E3C6973626E3E313631332D
343938323C2F6973626E3E3C63616C6C2D6E756D3E313437323C2F63616C6C2D6E756D3E3C75726C733E
3C72656C617465642D75726C733E3C75726C3E687474703A2F2F64782E646F692E6F72672F31302E3130
30372F7331303430342D3031322D303934392D7A3C2F75726C3E3C2F72656C617465642D75726C733E3C
2F75726C733E3C656C656374726F6E69632D7265736F757263652D6E756D3E31302E313030372F733130
3430342D3031322D303934392D7A3C2F656C656374726F6E69632D7265736F757263652D6E756D3E3C2F
7265636F72643E3C2F436974653E3C436974653E3C417574686F723E4C65653C2F417574686F723E3C59
6561723E323031323C2F596561723E3C5265634E756D3E313438363C2F5265634E756D3E3C7265636F72
643E3C7265632D6E756D6265723E313438363C2F7265632D6E756D6265723E3C666F726569676E2D6B65
79733E3C6B657920617070003D22454E222064622D69643D227A737461323272736E64357A706665737A32
6F70646477777474663273786673667830392223E313438363C2F6B65793E3C2F666F726569676E2D6B65
79733E3C7265662D74797065206E616D653D224A6F75726E616C2041727469636C65223E31373C2F7265
662D747970653E3C636F6E7472696275746F72733E3C617574686F72733E3C617574686F723E4C65652C
204B616820506E673C2F617574686F723E3C617574686F723E4C656573652C2048616E6E61683C2F61
7574686F723E3C617574686F723E4D61747469612C20446176696465653C2F617574686F723E3C2F617574
686F72733E3C2F636F6E7472696275746F72733E3C7469746C65733E3C7469746C653E57617465722066
6C6F7720656E68616E63656D656E7420696E20687964726F7068696C6963206E616E6F68616E6E656C
733C2F7469746C653E3C7365636F6E646172792D7469746C653E4E616E6F7363616C653C2F7365636F6E
646172792D7469746C653E3C2F7469746C65733E3C706572696F646963616C3E3C66756C6C2D7469746C
653E4E616E6F7363616C653C2F66756C6C2D7469746C653E3C2F706572696F646963616C3E3C6B657977
6F7264733E3C6B6579776F72643E736C69703C2F6B6579776F72643E3C6B6579776F72643E666C6F7720
656E68616E63656D656E743C2F6B6579776F72643E3C6B6579776F72643E486167656E2D706F69737569
6C6C653C2F6B6579776F72643E3C6B6579776F72643E41414F3C2F6B6579776F72643E3C2F6B6579776F
7264733E3C64617465733E3C796561723E323031323C2F796561723E3C2F64617465733E3C7075626C69
736865723E54686520526F79616C20536F6369657479206F66204368656D69737472793C2F7075626C69
736865723E3C6973626E3E323034302D333336343C2F6973626E3E3C63616C6C2D6E756D3E313438363C
2F63616C6C2D6E756D3E3C75726C733E3C72656C617465642D75726C733E3C75726C3E687474703A2F2F
64782E646F692E6F72672F31302E313033392F43324E523330303938423C2F75726C3E3C2F72656C6174
65642D75726C733E3C2F75726C733E3C2F7265636F72643E3C2F436974653E3C2F456E644E6F74653E00
0000

References:

ADDIN EN.REFLIST 1. D. Mattia and Y. Gogotsi, Microfluid. Nanofluid., 2008, 5, 289-305.

2. M. Whitby and N. Quirke, Nat. Nano., 2007, 2, 87-94.

3. K. P. Lee, T. C. Arnot and D. Mattia, J. Membr. Sci., 2011, 370, 1-22.

4. D. Mattia, M. P. Rossi, B. M. Kim, G. Korneva, H. H. Bau and Y. Gogotsi, J. Phys. Chem. B, 2006, 110, 9850 -9855.

5. D. Mattia, M. P. Rossi, H. Ye and Y. Gogotsi, in 5th IASME/WSEAS International Conference on Fluid Mechanics and Aerodynamics, Athens, Greece, 2007, pp. 297-299.

6. D. Mattia and F. Calabr?, Microfluid. Nanofluid., 2012, 1-6.

7. E. Ruckenstein and P. Rajora, J. Colloid Interface Sci., 1983, 96, 488-491.

8. W. Nicholls, M. Borg, D. Lockerby and J. Reese, Microfluid. Nanofluid., 2011, 12, 257-264.

9. K. P. Lee, H. Leese and D. Mattia, Adv Math Lett, 2012.

10. K. Ritos, D. Mattia, F. Calabr? and J. Reese, 2013.

(725a) Characterization of Hydorophilicity/Hydrophobicity of Zeolites By High-Resolution Gas and Water Vapor Adsorption

Endo, A., National Institute of Advanced Industrial Science and Technology
Shimomura, M., National Institute of Advanced Industrial Science and Technology
Kamimura, Y., National Institute of Advanced Industrial Science and Technology
Originally presented on: 11/7/2013 15:15:00 - 15:35:00

Zeolites are crystalline microporous materials and widely used as catalysts in chemical industries. It is very important to establish a methodology for evaluating porous structure and surface hydrophobicity/hydrophilicity of these materials in order to eclucidate their catalitic activity and adsorption behavior.

In this study, we report the characterization of meso- and micro- porous structure of zeolite with different Si/Al ratio using high resolution gas and water vapor adosption. The gas adsorption isotherms were measured using Belsorp-max in the pressure range of $p/p_o = 10^{-8} - 0.995$ at 77.4K for N_2 , 87.3K for Ar. The water vapor adsorption isotherms were measured in the pressure range of $p/p_o = 10^{-5} - 0.95$ and 298.15K for water vapor.

To our best knowledge, this is the first report of the adsorption isotherms of water vapor in the above-mentioned pressure range. At very low pressure regions, the uptake of nitrogen and water vapor was quite sensitive to the Si/Al ratio of zeolites. The relationship between the amount and location of Al site in zeolites and water uptake at a very low relative poressre region will be discussed.

(725b) Equilibrium Adsorption Of Nitrogen, Oxygen and Their Binary Mixtures On Pelletized LiX Zeolite

Wu, C. W.,
Kothare, M. V., Lehigh University
Sircar, S., Lehigh University
Originally presented on: 11/7/2013 15:35:00 - 15:55:00

New pure gas and binary gas mixture adsorption isotherms of nitrogen and oxygen were measured on a pelletized sample of LiX zeolite in the pressure range of 0 – 6 atmospheres at 273.1, 303.1 and 338.1°K by employing an isothermal column dynamic method. Measurements included low pressure isotherm data in order to establish the Henry's law regions. The data show that the zeolite exhibits significant energetic heterogeneity for adsorption of these gases. A pure and mixed gas heterogeneous isotherm model was successfully used to describe the experimental data. Constraints imposed on the model parameters by adsorption thermodynamics and Henry's law were obeyed in course of the data fit.

It is shown that adsorbent heterogeneity plays a significant role in determining the mixed gas adsorption characteristics for this system. A homogeneous isotherm model may empirically fit the pure gas isotherms fairly well but such a model will be very misleading in estimating mixed gas isotherm data.

(725c) Multicomponent Adsorption Characteristics of Different Solid Sorbent Materials

Rother, J., Ruhr-University Bochum
Fieback, T. M., Ruhr-University Bochum
Dreisbach, F., Rubotherm GmbH
Originally presented on: 11/7/2013 15:55:00 - 16:15:00

With the increasing importance of biogas as an energy source it is necessary to purify the corresponding gas mixtures. To the end, the importance of sorptive gas purification has increased significantly over the last years. Main aim for development and improvement of such processes is the evaluation of new and conventional sorbent materials. For this purpose a large amount of potential sorbent materials, different activated carbons , carbon molecular sieves and zeolite molecular sieves as well as the new substance class of metal-organic frameworks (MOFs) were analyzed. In addition to commercial available MOFs, a special type of functionalized MOFs was considered. These substances promise high selective sorption behavior.

Presented will be a short introduction to a new developed volumetric multi-sample sorption measuring apparatus, which allows the fast measurement of the ad- and desorption of gases in a technically relevant pressure and temperature range, as well as the gravimetric magnetic suspension balance for sorption measurements with highest accuracy.
After that different comparable measurements with pure gas sorption isotherms for the characterization of common sorbent materials will be presented.

For cleaning and purification processes not only the pure gas sorption measurement but also the determination of selective sorption of multicomponent gas mixtures is essential. Therefore a combination of a volumetric gas dosing system, for gas mixture generation, a magnetic suspension balance, for gravimetric sorption determination, and a gas analysis system, for determining the selective sorption, was used. Herewith the sorbent materials with best pure gas isotherms were analyzed with multicomponent gas mixtures and the results of this measurements will also be presented.

(725d) Estimating CO2 Permeability From Atomistic Models of Pims for Screening of Gas Separation Performance

Hart, K. E., Pennsylvania State University
Colina, C. M., Pennsylvania State University
Originally presented on: 11/7/2013 16:15:00 - 16:35:00

There is tremendous demand for a porous material that has the ability to selectively separate the individual components from a gaseous mixture. Several materials are being pursued, where polymeric membranes offer several distinct advantages as potential adsorbent candidates. Polymers of intrinsic microporosity (PIMs) are a class of glassy polymers that create large free volumes as a result of bulky, rigid, and contorted backbones. However, a fundamental trade-off in membrane performance exists, where there is a balance between high permeability and high selectivity for any given gas pair and both are necessary for industrial viability. It has been shown, both experimentally and through molecular simulations, that modest modifications in the functionality of adsorbent materials may result in profound changes in the gas adsorption properties. However, to date, a systematic computational analysis to develop the design principles necessary to understand and exploit the adsorbing potential of amorphous polymers has been elusive due to the difficulty in generating a physically accurate molecular model and computationally expensive calculations required for performance evaluation.

Through predictive molecular simulations, several structural functionalities on the PIM backbone are examined in this work, and it is found that the pore volume can be substantially increased with certain functional groups. More specifically, large bulky groups present on the spirocenter site of contortion were most effective at prohibiting efficient packing of the polymer chain, thus increasing surface area. Typically an increase in total pore volume of adsorbent improves permeability at the expense of gas pair selectivity. However, it is also shown, by means of an estimated permeability calculation using free volume theory in combination with grand canonical Monte Carlo adsorption simulations, that the permeability of the hypothetical functionalized PIMs may be increased without subsequent decreasing of mixed-gas selective performance for CO_2/N_2 and CO_2/CH_4.

(725e) Investigation of Surface Heterogeneity Behavior of Porous Silica Towards CO2 Adsorption: A Theoretical Approach

Loganathan, S., Indian Institute of Technology Guwahati
Tikmani, M., Indian Institute of Technology Guwahati
Ghoshal, A. K., Indian Institute of Technology Guwahati
Originally presented on: 11/7/2013 16:35:00 - 16:55:00

In recent years, mesoporous silica materials are profusely used as CO_2 capture adsorbents, because, these materials exhibit a unique combination of high surface area (>1000 m^2/g), large pore sizes (3-30 nm) and high pore volume (2.6 cm^3/g). In the present work, we address the physical and chemical features characterizing the interaction of CO_2 with the silica surface. For this study, we have considered the silica surface of our recently developed pore-expanded material (MCM-41) (**A**) (surface area: 1045.21 m^2/g, pore volume: 2.58 cc/g and average pore size: 30 nm) through sol-gel method for CO_2 capture. Gravimetric adsorption study of CO_2 on **A** was conducted at four different temperatures (30, 45, 60, 75 °C) and pressures ranging from 0-25 bar using Rubotherm magnetic suspension balance. Langmuir multi-site model was used to quantitatively determine the individual contribution of different surface groups of silica to the overall CO_2 adsorption capacity. Heat of adsorption values for different active sites was estimated using the temperature dependent form of Langmuir multisite parameter (*b*j). The two different values of heat of adsorption (7.75 kJ/mol and 22.19 kJ/mol) clearly indicated that CO_2 interacts with site 1 (siloxane bridges) through dispersive forces and site 2 (silanol groups) through hydrogen bonding. Interestingly, it was observed that the surface apparently becomes homogeneous in nature as the temperature increases. This is because, at low temperature all the sites don't have the same energy, and rather there is a distribution of energies. At higher temperature (~75 °C), the adsorbing sites of **A** became energetically uniform resulting in an almost homogenous surface. To find out the fraction of surface covered by the two functionalities at different temperatures, we developed a methodology relating the maximum adsorption capacity obtained by Langmuir and Langmuir multi-site model. The shift of α value from 0.82 to 0.97 with increase in temperature indicated the less effectiveness of the H-bond and existence of the dispersive force only for CO_2 adsorption.

(725f) Comparison of Narrow Pore Silicate and Phosphates Based CO2 Sorbents: The Good and The Bad

Cheung, O., Stockholm University
Liu, Q., Stockholm University
Bacsik, Z., Stockholm Univerisity
Hedin, N., Stockholm Univerisity
Originally presented on: 11/7/2013 16:55:00 - 17:15:00

Effective carbon capture from flue gas requires sorbents with desirable properties, these properties include: high CO_2 capacity, high CO_2 selectivity over N_2, fast adsorption kinetics, and many others. Inorganic crystalline sorbents with narrow pore window apertures have significant advantages; the uptake of CO_2 on these materials is comparably very high at the partial pressures of CO_2 in flue gas, and that they potentially can be tuned into CO_2-selective molecular sieves. Hence, we study such sorbents and here we compared properties of the silicate and phosphate based sorbents we studied. The sorbents include Zeolite LTA (and ZK-4), silico- and aluminophosphates (SAPOs and AlPOs) and titanium silicates (ETS-4 and CTS-1). We showed that zeolite NaKA (17% K^+) and NaK-ZK-4 (25% K^+) were both highly selective (with ideal CO_2-N_2 selectivity > 200). Silicoaluminosphoshpates (SAPO) had very high CO_2 capacity (SAPO-56 - 5.6 mmol/g, 273K, 101kPa) and retained over 95% of the original CO_2 capacity after cycling. Aluminophosphates (AlPO) and titanium silicates (CTS-1) were selective towards CO_2 sorption and offered reasonable CO_2 capacity. AlPOs also showed good cyclic capacity. AlPOs and Zeolite ZK-4 (with high Si/Al ratio) were relatively hydrophobic, which could be important for CO_2 capture from non-dried flue gas. In this work we rationalized and highlighted the advantages and the drawback of each of these narrow pore sorbents. This could be important for the research and development of good CO_2 sorbents.

(725g) Effect of Confinement On the Phase Behavior of Wetting and Nonwetting Adsorbates: Combining High Resolution Gas Adsorption and Mercury Porosimetry for An Advanced Textural Characterization of Mesoporous Molecular Sieves

Thommes, M., Quantachrome Instruments
Guillet-Nicolas, R.,
Ahmad, R., Quantachrome Instruments
Cychosz, K., Quantachrome Instruments
Kleitz, F., Laval University
Originally presented on: 11/7/2013 17:15:00 - 17:35:00

In order to investigate the effects of confinement on the phase behavior of nonwetting fluid (Hg) and wetting fluids (Nitrogen and argon at their respective boiling temperatures) we performed a systematic study of the sorption behavior of N_2 at 77 K and Hg porosimetry at 298 K in both 3-D and 2-D ordered large pore systems, *i.e.*, cubic *Ia3d* KIT-6 and hexagonal *p6mm*SBA-15 silica materials with pore diameters ranging from ca. 7 nm up to 11 nm

Sructural details of ordered mesoporous silica, i.e. KIT-6 and SBA-15 were obtained by performing a systematic study of nitrogen adsorption at 77 K including so-called scanning experiments [1] of the adsorption hysteresis loops (which allows, for instance to obtain details about pore network characteristics). Applying proper density functional theory methods allowed us to obtain accurate pore size data of a series of KIT-6 and SBA-15 silica materials with tailored mean pore diameters. The effect of pore network characteristics on mechanical stability were probed by performing careful mercury intrusion/extrusion experiments up to pressures of ca. 4000 atm. No collapse of the pore structure was observed when Hg intrusion/extrusion cycles were performed on fully 3-D mesopore systems, *i.e.*KIT-6, even for materials with pore diameters up to 13 nm.

To the best of our knowledge this is the first successful example of Hg porosimetry on ordered mesoporous molecular sieves such as KIT-6 silica. Depending on details of the pore networks characteristics reversible mercury intrusion/extrusion cycles or collapse of the pore network was found for SBA-15 silica samples.

Our results using these ordered mesopore structures confirmed that the underlying mechanism of mercury intrusion/extrusion is thermodynamically equivalent to the mechanism of capillary evaporation/condensation of a wetting fluid. From the analysis of the mercury intrusion/extrusion data we derive mercury vapor adsorption/desorption isotherms which demonstrate the effect of pore diameter on the vapor-liquid phase transition which occurs for the non-wetting fluid mercury at pressures which are larger than the saturation vapor pressure $p_0$of the bulk fluid, the smaller the pore diameter the larger the relative pressure where the condensation of mercury occurs.

Combing advanced physical adsorption and mercury porosimetry results provides new insights into the correlation between the pore structure/pore network characteristics and the properties (including mechanical stability) of ordered mesoporous silica materials.

[1] R.Cimino, K.A Cychosz, M. Thommes, Neimark, Colloids and Surfaces A, in press (2013)

[2] a) F. Porcheron, P.A. Monson, M. Thommes, *Langmuir*, **2004**, *20*, 6482; b) F. Porcheron, M. Thommes, R. Ahmad, P.A. Monson, *Langmuir*, **2007**, *23*, 3372; c) M. Thommes, R. Skudas, K.K. Unger, D. Lubda, *J. Chromatogr A*, **2008**, *1191*, 57.

(731a) Binary Adsorption Measurements With Net Adsorption Framework

Talu, D. O., Cleveland State University
Otto, K., Cleveland State University
Farran, A. M., Cleveland State University
Originally presented on: 11/7/2013 15:15:00 - 15:35:00

The ***net*** adsorption thermodynamic framework was introduced by Gumma and Talu (*Langmuir* 2010, 26(22), 17013-17023). Net adsorption fundamentally differs from ***excess*** and ***absolute*** adsorption by the placement of Gibbs dividing surface which is necessary to quantify any adsorption process. Net adsorption is a natural choice for microporous solids. It removes any ambiguity in adsorption phenomena by circumventing questionable measurements and quantities such as; 1) solid volume necessary for excess adsorption (typically based on non-adsorbing helium assumption), and 2) pore volume necessary for absolute adsorption (typically based on liquid density assumption in micropores).

This paper presents the first set of net adsorption data collected for pure and binary mixtures of methane, carbon dioxide, ethylene and propylene in silicalite at several temperatures. The measurement techniques are discussed and contrasted with primarily excess adsorption. Advantages and disadvantages are discussed. Data is examined for thermodynamic consistency and compared to ideal adsorbed solution predictions for both net and excess adsorption.

(731b) Hydrogen Adsorption On Ordered Mesoporous Carbons At Pressures Up to 1500 Bar

Deng, S., New Mexico State University
Klyamkin, S. N.,
Chuvikov, S. V.,
Krechetov, I. S.,
Astakhov, M. V.,
Saha, D., Oak Ridge National Laboratory
Yuan, B., New Mexico State University
Originally presented on: 11/7/2013 15:35:00 - 15:55:00

Hydrogen adsorption isotherms on four ordered mesoporous carbon (OMC) samples were measured volumetrically at 81, 273 and 298 K and hydrogen pressures up to 1500 bar. The OMC samples were synthesized with both soft-template and hybrid soft-hard template methods and characterized for their pore textural properties (BET, pore size distribution and pore volume), topology with TEM, and other physical and chemical properties. The four porous carbon samples have quite different pore textural properties (BET specific surface area varying from 570 to 1912 m^2/g). Hydrogen adsorption and desorption isotherms were measured on 1 g of sample by a volumetric method on a Sieverts type experimental unit specifically designed for studies of solid gas reactions under gas pressures up to 2000 bar. A metal hydride storage system was used as a source of high-purity hydrogen (99.9999%) for this work. An equation of state suitable for highly compressed hydrogen was employed for calculating the adsorption amounts. To evaluate the excess adsorption capacity of the carbon adsorbents the adsorption system was calibrated with helium prior to the hydrogen adsorption measurements. The hydrogen adsorption and desorption isotherms on different OMC adsorbents will be compared and discussed. An excess adsorption amount as high as 4.2 wt.% was observed on an OMC sample with BET of 1912 m^2/g at 81 K and 30 bar. The effects of pore textural properties (BET, pore size and pore volume) on hydrogen adsorption will be discussed in detail.

(731c) Measuring Methods for Non-Conventional Natural Gas Recovery

Paschke, T.,

Dreisbach, F., Rubotherm GmbH

Originally presented on: 11/7/2013 15:55:00 - 16:15:00

Introduction

Research for sustainable energy supply from conventional and non-conventional sources, energy conversion and storage as well as CO_2 sequestration is of increasing importance in context of problems related with global warming and high energy prices. These energy related processes do usually involve high pressures, high temperatures, corrosive and/or toxic substances.

As a result of that measuring instruments for investigating and improving these processes are required to provide reliable results even under these severe conditions. Gravimetric analyzers equipped with magnetic suspension balances (MSB) can be used for different kinds of measurements ? for example thermogravimetric, ad- or absorption, solubility, phase equilibrium ? in an extremely wide temperature and pressure range, even with corrosive, toxic and explosive reaction gases and vapors.

Applications

In the presentation an overview about recent research activities of our international customers in the field of energy supply using MSB instruments will be given. The following applications will be discussed particularly:

Coal bed methane recovery and CO_2 sequestration

During the coalification process, significant amounts of CH_4 are generated and retained permanently. The chemistry and the micro-structural features inherited from the original biomass confer to coal a large internal surface area, suggesting that sorption is indeed the main storage mechanism of gas in coal seams. Thus, the ad- and desorption properties of CH_4 on coal samples under geological pressure and temperature conditions are important measures for the development of coal bed methane recovery processes and can be measured with MSB instruments. As an example CH_4 adsorption data at two temperatures on a coal sample is provided in the following diagram [1].

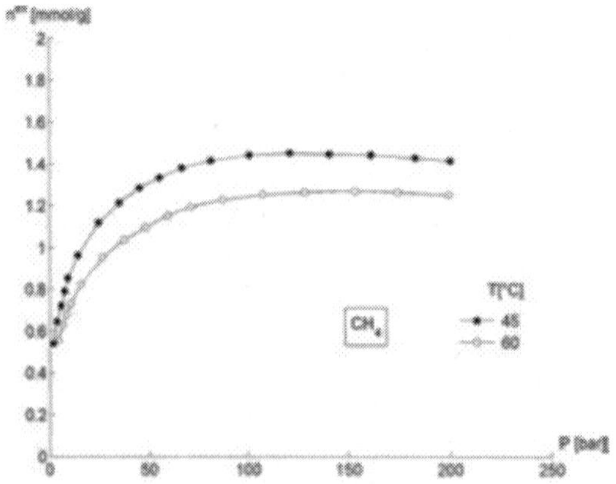

Diagram 1: CH_4 adsorption data at two temperatures on a coal sample as function of the pressure [1].

Coal bed methane recovery combined with CO_2 sequestration:

It has been suggested, that the recovery of coal bed methane might be enhanced by injecting CO_2 in the coal seam. Through an in situ adsorption/desorption process the displaced methane is produced and the adsorbed CO_2 is permanently stored. This process is called enhanced coal bed methane (ECBM) recovery. For proving the process and determining the necessary data gas mixture adsorption data have to be measured on the coal samples. MSB instrument equipped with gas mixing and dosing systems can perform such measurements and provide reliable gas mixture adsorption data. As an example the excess adsorption of the gas mixture CH_4/CO_2 with two different compositions is shown in the next diagram [2].

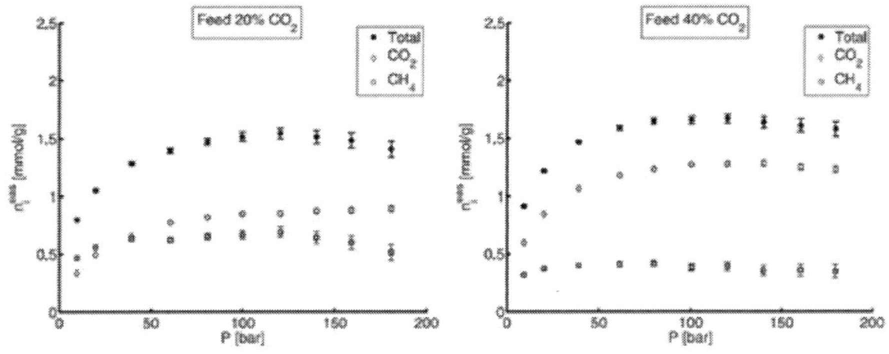

Diagram 2: Adsorption data of CH_4 and CO_2 mixtures with two different feed compositions a coal sample [2].

Shale gas recovery

Although the recovery of non-conventional resources of natural gas ? so called shale gas ? is controversially discussed in Germany, this application is intensively researched in other parts of the world. Critical for the efficiency and the environmental effects are the agents used currently for the hydraulic fracturing. During this process a liquid mixture is pumped into the geological structure and generates cracks and porosity which leads to improved diffusivity for the natural gas. Additives in the mixture are used to maintain the cracks and porosity even when the liquid is displaced by the gas. These additives are potentially toxic and may migrate into ground water reservoirs. A research topic currently is to generate the required improved diffusivity by using other fracturing agents. One of the potential substances is highly pressurized CO_2. We have developed especially for this research a highest pressure MSB which allows studying the mass change of complete geological bore cores under pressures up to 700 bar. A number of these instruments were installed recently in China. First results for the temperature dependent adsorption of CH_4 on a rock sample are given in the figure below.

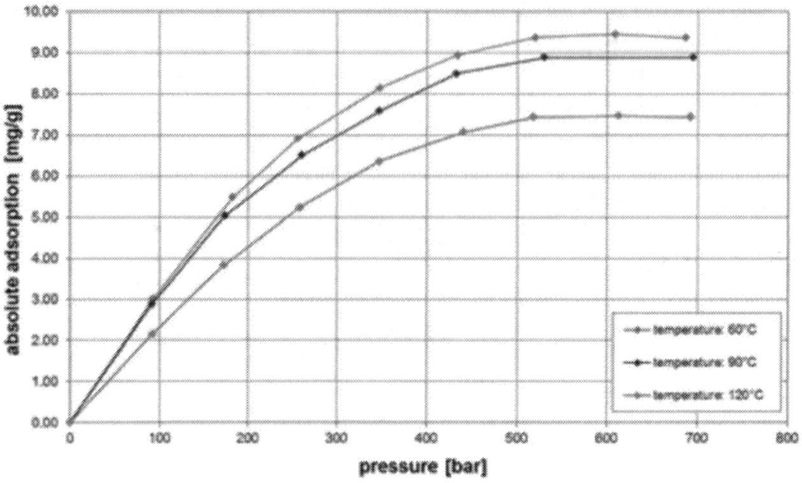

Diagram 3: CH_4 adsorption data on a Chinese shale sample up to highest pressures of 700 bar at three temperatures.

Summary

In the presentation several examples of measurements in the above application fields will be presented. The typical experimental setup, procedure of measurements and results ? mainly generated by our customer's in their laboratories - will be presented and discussed.

Literature

[1] S. Ottiger, R. Pini, G. Storti, M. Mazzotti, R. Bencini, F. Quattrocchi, G. Sardu and G. Deriu, Environmental Progress, Vol.25 (2006), No.4

[2] S. Ottiger, R. Pini, G. Storti and M. Mazzotti, Adsorption, Vol. 14 (2008)

(731d) On the Variability and Reproducibility of Equilibrium Adsorption Isotherm Measurements From Different Laboratories

Erden, L., University of South Carolina
Ebner, A. D., University of South Carolina
Nicholson, M. A., University of South Carolina
Holland, C. E., University of South Carolina
Ritter, J. A., University of South Carolina
Trinh, D., Marshall Space Flight Center
Shapiro, A., Marshall Space Flight Center
Knox, J. C., Marshall Space Flight Center
Mitchell, L. A., Vanderbilt University
LeVan, M. D., Vanderbilt University
Originally presented on: 11/7/2013 16:15:00 - 16:35:00

The accurate and precise measurement of equilibrium adsorption isotherms of pure gases on commercial adsorbents is perhaps the most important piece of information required in the development of new cyclic adsorption processes based on pressure and/or temperature swing adsorption. In principle, the measurement appears to be straightforward via commercially available or in-house developed volumetric and gravimetric adsorption apparatuses. In practice, the measurement tends to be difficult, time consuming and error prone, with accuracy and precision always being questioned.

The precision of an instrument can be checked very easily by making the same measurements several times to verify reproducibility. It is much more difficult to check the accuracy of the instrument. Usually an independent measurement must be made with a different instrument.

Teams from the University of South Carolina, Vanderbilt University and the Marshall Space Flight Center set out to measure the equilibrium adsorption isotherm of CO_2 on Grace Davison 13X zeolite at 25 ^oC. This adsorbent is being used by NASA to remove metabolic CO_2 from spacecraft cabins; it has many other commercial and developmental applications. It also represents a stringent adsorbate-adsorbent pair to test different apparatuses because the quadrupole in the CO_2 molecule interacts very strongly with the cations present in the 13X zeolite crystals.

This seemed like a fairly simple and straightforward task for each team. However, it became quite a challenge to try to explain the differences that appeared in the equilibrium adsorption isotherms obtained from the different laboratories. During this presentation details about this round-robin exercise between these three laboratories will be provided, along with compelling and revealing results about the variability and reproducibility of equilibrium adsorption isotherms of common gases on commercial adsorbents.

(731e) High Throughput Synthesis and Screening for Discovery of Improved Sorbent Materials for Gas Separations

Kaye, S. S., Wildcat Discovery Technologies Inc.
Originally presented on: 11/7/2013 16:35:00 - 16:55:00

Title: High Throughput Synthesis and Screening for Discovery of Improved Sorbent Materials for Gas Separations

Authors: Justin Dutton, Steven Kaye*

Wildcat Discovery Technologies Inc.

The separation of industrial gases via adsorption processes is often a key component of the cost and energy required for their manufacture. For example, 24% of the energy required to produce ethylene is used to separate it from other alkenes and alkanes. As such, the development of more efficient and selective adsorbents is of paramount importance. Research is often slowed by the large number of potential sorbents and the need to simultaneously optimize multiple performance metrics (e.g. working capacity, kinetics, impurity tolerance, etc.).

Wildcat Discovery Technologies has developed a modular, scalable platform for the high throughput synthesis and screening of sorbent materials that enables evaluation of >1000 samples per week. Central to this platform is our parallel gas sorption assay. Using an integrated mass spectrometer, this instrument is capable of rapidly evaluating materials for pressure and temperature swing adsorption by measuring a wide variety of gas sorption properties, including gas uptake, rate of sorption, and isotherms at variable temperatures using either pure gases or mixtures. Coupled with Wildcat's high throughput synthesis platform, this allows parallel evaluation of virtually any sorbent under realistic conditions for a target application.

In this talk, I will discuss the development and validation of Wildcat's high throughput sorbent discovery platform as well as our initial results from our discovery efforts.

(731f) Automated Analysis of ZLC Experiments: Fitting of Kinetic and Isotherm Parameters

Friedrich, D., University of Edinburgh
Mangano, E., University of Edinburgh
Hu, X., University of Edinburgh
Ferrari, M. C., University of Edinburgh
Brandani, S., University of Edinburgh
Originally presented on: 11/7/2013 16:55:00 - 17:15:00

Here we present the combination of our adsorption process simulator with a hybrid optimisation strategy for the automated analysis of a series of ZLC experiments.
The Zero Length Column (ZLC) method is an established technique for the measurement of the kinetic and equilibrium properties of adsorption systems. Briefly, in the ZLC method the desorption curve of a small amount of adsorbent previously equilibrated with a known gas mixture is measured. By changing the experimental conditions, i.e. flow rate of the purge gas, the system can be operated under equilibrium or kinetic control; thus equilibrium and kinetic properties can be investigated in the same series of experiments. The reliable measurements for fast and slow kinetics combined with the small required sample quantity (less than 10mg) make the ZLC method a very useful and flexible tool for kinetic and equilibrium measurements for a wide range of materials.
In order to automate and improve the analysis of the experimental data we have applied our in-house adsorption simulator to the simulation of the ZLC system. The process simulation is built from units representing the separate parts of the system: feed, purge, zero length column and detector. This representation allows the explicit description of the ZLC and the surrounding parts of the system. For example, the dead volume of the system and the detector response are included through fitting of the blank experiments. The used full-scale column model facilitates the direct application of the estimated parameters in complex, large-scale PSA cycles. Furthermore, the column unit implements different adsorption isotherms, different mass transfer models as well as isothermal and non-isothermal models. Thus the simulator provides an accurate description of the dynamics in the ZLC system.
By linking the ZLC simulation to a hybrid optimisation strategy (combining a global Multi-Objective Genetic Algorithm with a local optimisation routine) we generate an automated tool for the estimation of the kinetic and equilibrium parameters from several experimental runs. The optimisation algorithm tries to minimise the least square error between the experimental data and the simulation output. In the first step the parameters of the empty system, i.e. no adsorbent in the ZLC, are fitted with empty system experimental runs. Then in the second step only the kinetic and equilibrium parameters in the ZLC are fitted. The combination of the adsorption simulator and the hybrid optimisation allows us to fit the parameters with greater certainty compared to the previously used simplified analytical models. Specifically the simultaneous fitting of several experimental curves at different flow rates and temperatures provides a robust estimation for both the equilibrium and kinetic parameters of the adsorbent.
The flexibility of the adsorption simulator, i.e. different kinetic and equilibrium models, allows the investigation of the dominant physical mechanisms. For example, the error in the parameter estimation from kinetically controlled zeolite 13X experiments using the single-site Langmuir isotherm is much larger than the error obtained using the dual-site Langmuir isotherm. Subsequent comparison of the fitted isotherms to the isotherm obtained from equilibrium controlled experiments shows that the large error for the single-site Langmuir fit corresponds to a large error in the fitted adsorption isotherm.

(731g) Vacuum-Assisted Ion Exchange of Zeolite Membranes

Kim, S. J., University of Cincinnati
Sholl, D. S., Georgia Institute of Technology
Jones, C. W., Georgia Institute of Technology
Liu, Y., The Dow Chemical Company
Moore, J. S., Massachusetts Institute of Technology
Dixit, R., The Dow Chemical Company
Pendergast, J., The Dow Chemical Company
Nair, S., Georgia Institute of Technology
Originally presented on: 11/7/2013 17:15:00 - 17:35:00

Zeolite membranes with a variety of crystal structures and pore sizes can effectively perform molecular separations due to preferential adsorption and differences in diffusion rates. There have been significant efforts to modify the zeolite membranes in order to improve separation/catalytic performance using methods such as chemical vapor deposition, coking treatments, and catalytic hydrothermal cracking. One way to modify the membrane properties is through ion exchange, which is known to control the adsorption, diffusion, and catalytic properties of zeolite materials. However, the conventional ion-exchange methods (in which one immerses the zeolite membrane into the ion exchange solution) often lead to slow and limited exchange of ions. This is because of the large hydrated form of the ions in solutions, small zeolitic pore size, and the thick zeolite membrane layer, thereby causing a low driving force for ion exchange.

We will discuss a new vacuum-assisted 'flow-through' technique to effectively ion exchange the zeolite membranes. Na-containing ZSM-5 (MFI) zeolite membranes with different Si/Al ratios were synthesized on a porous α-alumina disk by in situ crystallization method and by secondary growth. In this vacuum-driven pressure differential method, the membrane surface is gently immersed into the ion exchange solution bath. We show how the large increases in the exchange of Ga^{3+}, Zn^{2+}, and Pt^{2+} ions can be achieved with this method. We find that, in contrast to conventional ion exchange involving counter-diffusion of the ions being exchanged in and out, the vacuum-driven method creates a unidirectional displacement of one type of ion with the other. The Na^{+} ions are transported into the vapor phase on the vacuum side in the form of hydrated clusters, and replaced with e.g., Ga^{3+} ions from the liquid side. Finally, we find that the creation of a low-pressure vapor phase is critical for this process, above and beyond the existence of a pressure differential. A similar liquid-phase process with the same pressure differential as the vacuum-driven method, but using a pressurized liquid feed and an atmospheric liquid permeate, leads to a very low degree of ion exchange. The talk will further discuss the mechanisms responsible for the large enhancement in ion exchange by the vacuum-driven process over conventional methods.

(732a) Virus Purification By Aqueous Two Phase Extraction System

Vijayaragavan, K. S., Michigan Technological University
Heldt, C. L., Michigan Technological University
Originally presented on: 11/7/2013 15:15:00 - 15:35:00

The prevention of diseases by vaccination is one of the most significant medical achievements and currently there is considerable interest in development of viral vaccines for human therapy. However, modest financial returns for vaccine businesses have been causing severe obstacles in new vaccine technology development. Additionally, urgencies and short timelines in downstream processing to establish scaled up clinical trials integrated with upstream processes demand a growing need for simple, easy scale up and cost effective technique. One such technique that is rarely applied to virus purification is aqueous two phase systems (ATPS). ATPS, consisting of a liquid-liquid biphasic system widely recognized for separation and purification of biomolecules like proteins and cell organelles, was first developed in Sweden by Huddleston in 1950's. Partitioning of protein molecules using ATPS is a highly complex phenomenon due to hydrogen bonding, charge interactions, Van der Waal's forces and surface hydrophobicity. We are currently using porcine parvovirus (PPV), a non-enveloped DNA virus of 18-26 nm diameter, as a model virus to develop vaccine purification techniques. The parvovirus particles are produced in porcine kidney cells (PK-13) containing fetal calf serum. Therefore, they contain serum proteins that can be immunogenic. We propose applying ATPS to effectively concentrate and purify PPV, primarily on the basis of hydrophobicity. Based on experimental studies, PPV has been recently validated by our laboratory to be more hydrophobic than typical blood and host cell proteins. For our ATPS experiments we have focused on polymer-citrate salt systems. We have successfully identified our ideal ATPS system after screening three PEG molecular weights and bulk salt concentrations ranging between 23- 41 w/w%. We report 98% virus recovery including both the top and interphase of the ideal ATPS system and removal of the serum proteins. The study presents a new technique for vaccine production and advance scientific understanding of viral purification strategies from extraneous contaminant proteins.

(732b) Subcritical Water Extraction of Protein From Sunn Hemp
Nyankson, E., Auburn University
Originally presented on: 11/7/2013 15:35:00 - 15:55:00

2013 American Institute of Chemical Engineers Annual Meeting

San Francisco, CA

November 3-8, 2013

Subcritical Water Extraction of Protein from Sunn

Emmanuel Nyankson*, Hema Ramsurn*, Alex Roman*, Ram B. Gupta* and Jorge Mosjidis[+]

*Department of Chemical Engineering, Auburn University, Auburn, AL 36849

[+]Department of Agronomy and Soils, Auburn University, Auburn, AL 36849-5127

Sunn Hemp, which originated in India, is a tropical or sub- tropical plant that when grown in the continental United States performs like a summer annual. This tropical plant is reported to have added 134 to 147 pounds of nitrogen per acre after 60 days of growth. Studies have shown that Sunn Hemp contains between 20-30 wt% crude protein. This high protein content has made Sunn Hemp a good source of animal feed. The protein present in Sunn Hemp can also be extracted and used as a nutritional source for humans. Traditionally in an industrial set-up, organic solvents are commonly used for the extraction of food components. The choice of organic solvent is dictated by the polarity, solubility and mass transfer characteristics of the compounds to be extracted. However, long extraction time and toxic waste generation has resulted in the quest for new extraction methods.

Water as reaction medium is safe, non-toxic, inexpensive and readily available. Subcritical water is liquid under pressure at temperatures between 100°C and 374°C, the critical temperature of water. The unique properties of subcritical water such as low dielectric constant and polarity make it a better replacement for organic solvents. Water can act as an acid/base catalyst precursor below the critical point and at high pressures since the ionic product is up to three orders of magnitude higher than under ambient conditions. In the subcritical region, the relatively low dielectric constant enhances ionic reactions which promote reactions like hydrolysis. Furthermore, subcritical water enhances the solubility of the organic compounds, hence improving extraction.

In this study, subcritical water will be used to extract protein from a particular species of Sunn Hemp, developed in the Department of Agronomy and Soils at Auburn University. The effects of temperature (160-240 °C), reaction time (30 mins and 60 mins), different biomass feeding methods (dry powder versus slurry feeding), different catalysts (NaOH and Na_2CO_3) and different catalyst amount (2.5-28 wt %) on the protein yield were investigated. The maximum protein extraction yield of 74% was obtained at 240°C with 28 wt% of Na_2CO_3. Reaction time does not have major effect on the protein yield: protein yields of 69 and 66% were obtained at 220 °C with 15 wt% Na_2CO_3 using 30 and 60 minutes reaction time, respectively. The use of 6 wt% of NaOH at 220°C for reaction time of 30 minutes resulted in a protein yield of 60% compared to 48% obtained when using the same amount of Na_2CO_3. A comparison between the Lowry and Bradford assay in the protein quantification was performed. Protein yields obtained using the Bradford assay were about 12 to 20% below those obtained by the Lowry assay, indicating detection limitations of the assays with respect to the type of amino acids in the peptide sequence. Amino acid analysis of the extract showed the presence of mainly polar amino acids.

(732c) Extraction of Chemicals From Fermentation Broths Using a KARR Column

Glatz, D. J., Koch Modular Process Systems, LLC
Originally presented on: 11/7/2013 15:55:00 - 16:15:00

Fermentation processes have become more important in recent years for the creation of valuable chemical products. Liquid-liquid extraction (LLE) is a logical first step in the recovery process since many of the chemicals produced are organic chemicals with higher boiling point than water. These processes generally require a high number of theoretical stages to achieve that required separation. As such, agitated LLE columns are necessary instead of static columns, including packed columns and columns with sieve trays. However, most fermentation broths have a high tendency towards emulsification when contacted with solvents for extraction. Agitated columns using rotating internals only intensify the tendency towards emulsification. The KARR® Column however, uses a reciprocating plate stack with open area greater than 55% for providing the agitation during counter current flow of the two phases. This type of agitation has proven to work effectively when columns with rotating internals have failed. This paper will present several case studies where the KARR® Column was used to provide solutions for several such processes.

(732d) EFFECT OF PRESSURE ON THE EXTRACTION OF BIO OIL = FROM OIL PALM EMPTY FRUIT BUNCH USING SUPERCRITICAL CO2

Mohamed, M., Universiti Teknologi PETRONAS
Yusup, S., Universiti Teknologi PETRONAS
Goto, M., Nagoya University
Wahyu, D., Kumamoto University
Originally presented on: 11/7/2013 16:15:00 - 16:35:00

This paper highlights the conversion study of oil palm empty fruit bunch to bio-oil using supercritical carbon dioxide (SC-CO_2). While maintaining the amount of feed which is 10g and particle size of less than 150μm, effects of process conditions such as temperature within the range of 60-90°C, pressure within 25-50 MPa and flowrate of CO_2 within 3-6ml/min are investigated within two hours of extraction time. The product is diluted in 10ml dichloromethane (DCM) for the analysis using gas chromatography-mass spectrometry (GC-MS). The composition of fatty acid, ester, alcohol and other long chain hydrocarbon varies depending on the process condition. Hexadecanoic acid known as palmitic acid and dodecanoic acid 1, 2, 3-propanetriyl ester are identified as the major components of the bio oil. Approximately about 53% percent of hexadecanoic acid is produced when the temperature and pressure is at the highest value which is 90°C and 50 MPa respectively, while the flowrate of CO_2 is maintained at 5ml/min. When the extraction pressure and flowrate of CO_2 are kept constant at 40MPa and 5ml/min respectively, the amount of hexadecanoic acid obtained reduced with temperature. In addition, the ester composition is increased from 17% to 24% when the temperature is decreased from 80°C to 60°C. At constant temperature and pressure of 70°C and 40 MPa, both amount of ester and hexadecanoic acid reduced when the flow rate of CO_2 is reduced from 6ml/min to 3 ml/min.

(732e) Tunable Aqueous Polymer Phase Impregnated Resins: A New Approach for Protein Recovery

van Winssen, F. A., TU Dortmund
Merz, J., TU Dortmund
Schembecker, G., TU Dortmund
Originally presented on: 11/7/2013 16:35:00 - 16:55:00

Tunable Aqueous Polymer Phase Impregnated Resins:

A new approach for protein recovery

Nowadays, a variety of specialized biotechnological products is needed in food, cosmetic and pharmaceutical industry and the demand increases steadily. As a consequence, the Downstream Processing has to face complex and diverse challenges like high product sensitivities towards temperatures, organic solvents, or pH values. Additionally, comparatively low product concentrations, a high number of unspecified impurities, and strict demands for final product quality complicate the Downstream Processing.

Current Downstream Processing techniques suffer from several drawbacks such as high costs, irreversible product deactivation, and possible product contamination due to leaching of ligands, for example. Thus, Downstream Processing often represents the bottleneck of the whole production process and its costs can make up to 80% of the whole production costs.

Hence, there is a constant need for efficient alternatives to conventional Downstream Processing methods for the purification of biotechnological products. One of these alternative techniques is Aqueous Two-Phase Extraction (ATPE). For ATPE, two partially immiscible aqueous phases are used as extraction system. Aqueous two phase systems (ATPS) can consist of two aqueous polymer or salt phases, or the combination of one aqueous polymer and one aqueous salt phase. The main advantage of ATPE as a Downstream Processing technique for biotechnological molecules is the biocompatibility due to the high water content of approximately 80 wt.-%.The low interfacial tension between the two aqueous phases allows fast mass transfer rates and emulsification is reached at lower agitation speed. Thus, proteins that are sensitive to mechanical stresses are less damaged. Beside the advantages, an industrial application of ATPE is still limited because of some major drawbacks: The low interfacial tension and low density difference between the two aqueous phases lead to intensive and stable emulsifications and therefore, long phase separation times. Thus, the footprint for the required mixer/ settler equipment is high, or additional equipment like centrifuges are needed. Various attempts have been made to decrease the demixing time of ATPS, like differently designed extraction columns or acoustic and temperature driven phase coalescers. Another approach to overcome the long settling times of classical ATPE is the so called 'Tunable Aqueous Polymer-

Phase Impregnated Resins' (TAPPIR®)-Technology (Schembecker et al., 1 April 2011, AZ 10 2011 001 743.). This technology eliminates the need of phase emulsification and separation by immobilizing one aqueous phase inside porous solids prior to extraction. The phase contact is achieved by the suspension of the impregnated solids in the second aqueous phase. Thus, emulsification and phase separation is realized in one single step. The separation principle of the TAPPIR®-Technology is based on the technique of solvent impregnated resins (SIR) introduced by Warshawsky. In contrast to SIR, the TAPPIR®-Technology is meant to be used as novel approach for Downstream Processing of biotechnological products. The challenges in downstream processing of this product division are different from those present in the SIR applications where no special attention has to be paid on product sensitivities. For example, the use of organic solvents should be avoided because of possible denaturation effects of the proteins. Thus, for the TAPPIR®-Technology ATPS are used to create a more biocompatible environment and gentle extraction conditions.

Regarding the expensive phase forming polymers like PEG to form ATPS, the TAPPIR®-Technology offers the opportunity to reuse the impregnated particles for multiple separation runs. Thus, the recycle of the impregnated particles might lead to an overall cost reduction and could increase the industrial acceptance to ATPE. In terms of process integration, a potential advantage might be the combination of ATPE in the TAPPIR®-Technology with size exclusion separation effects due to different pore sizes of the used particles.

However, the TAPPIR®-Technology is new and thus, many factors need to be investigated in regard to their impact on the separation system. The selection of the ATPS determines the separation efficiency and thus, the partitioning of product and contaminants between immobilized and bulk aqueous phase. The type of the ATPS in combination with the concentration of the phase forming components and other properties, such as the molecular weight of the polymer, determine its physicochemical properties, such as the density difference between the phases, the phases' viscosities and the interfacial tension. Additionally, neutral additives such as sodium chloride offer the possibility to individually influence the partitioning behaviour of product and contaminants without having significant impact on the miscibility. The form, size and especially porosity and pore size of the solid particles influence transport processes, hydrodynamics, and impregnation behaviour. Regarding the number of influencing factors, the set-up or optimization of the TAPPIR®-Technology can be complex. On the other hand it offers the possibility to react flexible and to tailor a selective and customized purification strategy for biotechnological products.

To show the applicability of the TAPPIR®-Technology the partitioning of proteins in an aqueous polymer-salt system is investigated. The biodegradable and sustainable aqueous polyethylene glycol (PEG) /sodiumcitrate system and macroporous particles, which had no adsorptive capacities towards the proteins investigated, are used. The effects of particle material and size and the addition of the neutral additive sodium chloride are determined in regard to separation efficiency. The whole process of target protein extraction into the immobilized aqueous PEG phase and the back-extraction into a second aqueous sodiumcitrate phase is investigated. To evaluate the performance of the TAPPIR®-Technology, the results are compared to classical ATPE experiments. It could be shown, that the TAPPIR®-Technology can efficiently separate the target protein out of the mixture. Additionally, it could be demonstrated, that the target protein is extracted back into a second aqueous sodiumcitrate phase. Comparing the overall recovery of target protein the results achieved via TAPPIR®-Technology are comparable to the one yielded by classical ATPE. Thus, these new technology can be an efficient alternative to classical purification strategies and can offer new alternatives for the use of ATPE.

(732f) Salting-Out Extraction of Bio-Based Chemicals
Xiu, Z.,
Originally presented on: 11/7/2013 16:55:00 - 17:15:00

Salting-out Extraction of Bio-based Chemicals

Jian-Ying Dai, Ya-Qin Sun and Zhi-Long Xiu*

School of Life Science and Biotechnology, Dalian University of Technology, Dalian116024, China

*E-mail: zhlxiu@dlut.edu.cn, Tel: +86-411-84706369

Abstract

The bio-refinery from cheap biomass attracts much attention and endeavor, which is mainly focused on the improvement of strains and fermentation. Down-stream processing for bio-based chemicals becomes bottle-neck in industrial production because so many impurities are introduced into the fermentation broths. Here we summerized salting-out extraction (SOE) based on the partition difference of chemicals in two phases consisting of salts and polymer or hydrophilic solvent, hydrophobic solvent, amphipathic chemicals. The effects of solvents and salts on the formation of two phases were introduced as well as its application in the recovery of bio-based chemicals, such as 1,3-propanediol, 2,3-butanediol, acetoin, lactic acid, succinic acid. The diols such as 1,3-propanediol and 2,3-butanediol could be recovered at a yield above 95%, and the recovery of lactic acid also reached 90%, especially from fermentation broths without pretreatment. At the same time, most of the impurities (cells and proteins) were removed in the same step extraction.

For example, in the recovery process of 1,3-propanediol methanol (35%, v/v) and phosphate (to saturation) were directly added into the fermentation broth (pH 10.7). After mixing and standing 98.1% 1,3-propanediol was partitioned to the top phase, meanwhile 99.85% cells, 92.4% proteins, 90.1% formic acid, 76.7% lactic acid, 87% lactic acid, 87% acetic acid, 88.6% succinic acid and 81% fumaric acid were removed from the top phase. If the fermentation broth was treated by a solvent with weaker polarity, the partition coefficient (K) would also decrease. In a SOE system consisted of $K_2HPO_4$and alcohol, the K value was decreased from 7.87 to 2.4 when the used alcohol was switched from methanol to 2-butanol. 2,3-Butanediol and acetoin demonstrated similar behavior as 1,3-Propanediol. At suitable conditions more than 95% of target product was recovered.

Due to the high hydrophilicity of organic acids, the recovery of organic acids was a little lower compared with diols and acetoin. In a SOE systems composed of 26% methanol and 25% K_2HPO_4 (w/w) 86% lactic acid was recovered, while in a system consisted of 30% ethanol and 14% K_2HPO_4 (w/w) the recovery was 90.6%.

References

[1] Z. Li, B. Jiang, Z. Xiu. *Sep. Purif. Technol.* **2009**, 66, 472.

[2] Z. Li, H. Teng, Z. Xiu. *Process Biochem.* **2011**, 46, 586.

[3] J. Dai, Y. Zhang, Z. Xiu. *Chinese J. Chem. Eng.* **2011**, 19, 682.

[4] G. Liu, B. Jiang, H. Wang, J. Dai, Z. Xiu. *CIESC Journal* **2009**, 60, 2798.

[5] L.-H. Sun, B. Jiang, Z.-L. Xiu. *Biotechnol. Lett.* **2009**, 31, 371.

[6] B.-C. Wei, Z.-Y. Song, Y.-Q. Sun, Z.-L. Xiu. *The Chinese Journal of Engineering* **2012**, 12(1), 44.

(732g) Recovery of Bioactive Catechins From Aqueous Extract of Tea Leaves (Camellia sinensis) By Liquid Membrane (LM) Technique Leading to the Formation of Metal-Catechin Complex

Manna, M., IIT Guwahati
Bhatluri, K. K., IIT Guwahati
Saha, P., Indian Institute of Technology Guwahati
Ghoshal, A. K., Indian Institute of Technology Guwahati
Originally presented on: 11/7/2013 17:15:00 - 17:35:00

Catechin compounds aid great pharmacological benefits to human health such as reduction of cholesterol and hypertension, anti-oxidation, anti-microbial, protection against cardiovascular disease and cancer (Ghoshal et al, 2012). They are found in leaves and flowers of tea plants (*Camellia sinensis*). Recovery of these bioactive components from hot (60Â¢ÂºC) water extract of tea leaves using liquid membrane based separation has been studied in the present work. Liquid membrane technology is an advancement of the conventional solvent extraction process with the best advantage of simultaneous extraction and stripping of the target solute along with other advantages such as very low consumption of extractant, lower energy consumption, low installation cost *etc.* Liquid membrane, in general is a thin liquid film interposed between two other phases of different composition, *viz.* feed phase and strip phase (both immiscible with membrane phase). The target solute is extracted in the feed side interfaces and diffuses through the liquid membrane due to concentration gradient and as soon as it reaches to the strip side interface, it is stripped by the stripping agent due to its higher affinity to the solute. Tributyl phosphate (TBP), a neutral carrier in *n*-decane is identified as the most efficient solvent-carrier combination (membrane phase) in two phase equilibrium study (Manna et al., 2012). On the other hand, metals play an important role in medical science with their biological activity as complexing agent, in addition to the fact that they are essential in human diet chart. Metal-catechin complex (Mhatre et al., 1993) is important species for some biological functions like gene expression, apoptosis, enzyme regulation and neurotransmission. Various transition metals such as Cu^{2+}, $Zn^{2+,}$ $Co^{2+,}$ $Fe^{2+,}$ Mg^{2+}, Ca^{2+} as well as Al^{2+} and Al^{3+} (in their respective aqueous salts) were tried as striping agent and also for formation of complexing agent. $AlCl_3$ was found as the best stripping as well as catechins complexing agent due its larger standard reduction potential. The metal (Al^{+3}) complexation with various catechin compounds were confirmed through the change of UV-vis. spectrum of the bioactive compounds. Al-catechin complex was found as 2:1 in elementary analysis in excess (molar basis with respect to catechins in stripping phase) $AlCl_3$ in stripping phase. The complex formation was confirmed also by FT-IR and X-ray diffraction analyses. 60% extraction and 48% recovery of the catechin have been achieved at the optimum conditions of the operating parameters *viz.*12% (v/v) TBP in the membrane phase, 100 mg L^{-1} (0.00034 M) feed phase concentration, pH 4.0, 0.02 M $AlCl_3$ as stripping agent and stirring speed of 400 rpm of both the feed and receiving phases.

Literature cited

1. Manna M.S., Bhatluri K.K., P. Saha and A.K. Ghoshal (2013). “Transportation of catechin (±C) using physiologically benign vegetable oil as liquid membrane”, Industrial & Engineering Chemistry Research, in press, DOI: 10.1021/ie3017863.

2. Ghoshal, A. K.; P. Saha; Manna, M. S.; Bhatluri, K. K. (2012) “Environmentally Benign Liquid membrane Technology for Chemical, Pharmaceutical and Medicinal Applications”, presented in 2nd Indo-German workshop on “Advances in Reaction and Separation Process” , 20-22 Feb 2012, Bad Herrenalb, Germany.

3. Mhatre, S. N., Iyer, R. K., Moorthy, P. N., 1993. Characterization of aluminium complexes in tea extract. Part I-Al NMR studies, Magnetic Resonance in Chemistry, 31(2), 169-175

(740a) Countercurrent Centrifugal Extraction: A New and Highly Efficient Extraction Apparatus Based On Centrifugal Partition Chromatography

Schwienheer, C.,
Merz, J., TU Dortmund
Schembecker, G., TU Dortmund
Originally presented on: 11/7/2013 15:15:00 - 15:40:00

Centrifugal partition chromatography (CPC) is a kind of liquid-liquid partition chromatography in which the separation mechanism is based on different distribution of the components between two immiscible liquid phases. As an advantage compared to other chromatographic devices no supporting solid phase is used in CPC. Thus, CPC provides gentle separation conditions (e.g. no irreversible adsorption) and a high amount of stationary phase (60-80% of the total column volume) which is accessible to the sample solutes and thus, providing a high capacity.

To immobilize one liquid phase without a supporting solid phase, the column of a CPC consists of a cascade of chambers connected by ducts in serial and aligned around a central axis of rotation. Caused by the rotation of the rotor, a centrifugal force is generated which is used to immobilize the liquid (stationary) phase in each chamber, while the mobile phase is pumped through the stationary one along the whole chamber cascade. Due to different distribution behavior between the two phases, the components of a mixture injected at the beginning of the cascade will elute at different times. E.g., a component that distributes more to the stationary phase will be retained for a longer time in the rotor.

Depending on the density ratio of the two liquid phases two different operation modes can be distinguished, namely "ascending mode" ($\rho_{stat}/\rho_{mob} > 1$) and "descending mode" ($\rho_{stat}/\rho_{mob} < 1$). For both modes, the rotor is initially filled with the phase, which is chosen to be the stationary one. Then, the mobile phase is pumped into the rotor and some part of stationary phase is replaced until equilibrium is reached and no further stationary phase is eluted (name: retention of stationary phase). In general, a high retention of stationary phase in the chambers is desired for CPC operations. The retention of stationary phase is mainly influenced by the selection of the phase system with its physical properties (densities, viscosities and interfacial tension) and the operating parameters of the CPC (mode of operation, volume flow of mobile phase and rotational speed of the rotor).

For the operation in CPC basically all liquid systems which form two immiscible liquid phases can be chosen. Aqueous organic systems are common for the separation of natural products and there are guidelines for the selection of suitable phase and operating conditions. Aqueous two phase systems (ATPS), however, are relatively new as phase system for CPC operation. Here, both phases are mainly composed of water enriched with phase forming components (two incompatible polymers, one polymer and one salt or two different salts) above a critical amount in each phase. ATPS provide gentle conditions for the separation of biomolecules (e.g. proteins) and thus, allow the CPC technique to deal with another application field. However, due to their physical properties (higher viscosities, lower density difference and lower interfacial tension when compared to common aqueous organic systems) the efficient operation in CPC is difficult and up to now it has not been well investigated.

The separation of a substance mixture with CPC is generally a discontinuous process and its set-up and application is similar to preparative chromatography. To increase the capacity for industrial applications other CPC operations were developed. A new CPC concept, the sequential CPC (sCPC), derives from the fact, that each of the liquid phases used can be the mobile phase and that the mode of operation (ascending or descending) can be switched during a separation. By this technique the sample is feed continuously between two conventional CPC rotors, which are connected in serial. The mode of operation in sCPC is sequentially switched between ascending and descending mode for a defined period of time. The mobile phase in each sequence is pumped in the inlet of rotor one and a product stream is collected at the outlet of rotor two, or vice versa. Thus, comparable to other cyclic chromatographic technics (e.g. simulated moving bed chromatography), two product streams can be collected discontinuously, each stream enriched with some components of the mixture to be separated. By this purification technology a higher capacity of separation, compared to classical CPC can be achieved.

To further increase the capacity for industrial applications the principle of countercurrent centrifugal extraction (CCE) was developed and patented (PCT/EP2012/068999) by the chair of plant and process design at TU Dortmund in Germany. With the CCE technology it is possible to pump both liquid phases simultaneously in counter flow direction over the whole chamber cascade allowing a continuous separation. The continuous working principle is achieved by connecting each chamber with two channels, one for each liquid phase. Additionally, rotary joints providing five channels in total are necessary (in CPC only two channels) to maintain a continuous flow of both phases as well as the sample feed. The whole set-up consists of one rotor only, which simplifies the set-up compared to sCPC. The substance mixture to be separated can be fed continuously at the middle of the chamber cascade. According to the distribution coefficients between the two liquid phases the separation of the target substances occurs in direction of the heavy or light phase flow. The principle of CCE is similar to liquid-liquid extraction processes. However, a particularly advantage of CCE technology is the design of combined "mixer-settler-units" in small scale. By use of the centrifugal field it is possible to use instable phase systems, which form more stable emulsions with long settling times. This is mainly important for the use of ATPS, where the process efficiency is often limited by the settling time of the phase system. Additionally, a high number of separation stages can be realized in CCE and it is possible to adjust the number of stages by the number of CCE discs used. Another advantage is the easy scale-up by variation of the chamber size.

In the presentation, the construction and operating principle of the CCE technology will be explained in detail. Additionally, experimental results for the separation of substances using organic-aqueous systems will demonstrate its efficiency and thus, will show the applicability of this new technology.

(740b) MINLP Optimization of a Membrane-Absorption-Hybrid System for the Removal of CO2 From OCM Product Gas

Esche, E., Berlin University of Technology

Müller, D., Berlin University of Technology

Song, S., Berlin Technical Institute

Wozny, G., Berlin University of Technology

Originally presented on: 11/7/2013 15:40:00 - 16:05:00

The Oxidative Coupling of Methane (OCM) presents a possibility for catalytically turning methane into longer hydrocarbons, especially ethylene and ethane. Hence, it could be an opportunity for replacing oil with methane from natural or biogas for producing base chemicals such as polymers, ethylene oxide etc.

As part of the Cluster of Excellence ?Unifying Concepts in Catalysis?, funded by the German Research Foundation, a mini-plant has been built at Berlin University of Technology (Technische Universität Berlin). The mini-plant features several types of reactors, e.g. fixed-bed, membrane, and fluidized-bed reactors. For the subsequent product gas separation an absorption-desorption process, two gas separation membranes, and an adsorption-desorption unit for the removal of carbon dioxide (CO_2) are installed. An initial introduction of a single membrane module upstream to the absorption unit showed a decrease of the energy required for the removal of 90% of the CO_2 contained in the feed gas from 5.02 MJ to 2.76 MJ per captured kilogram of CO_2 [1].

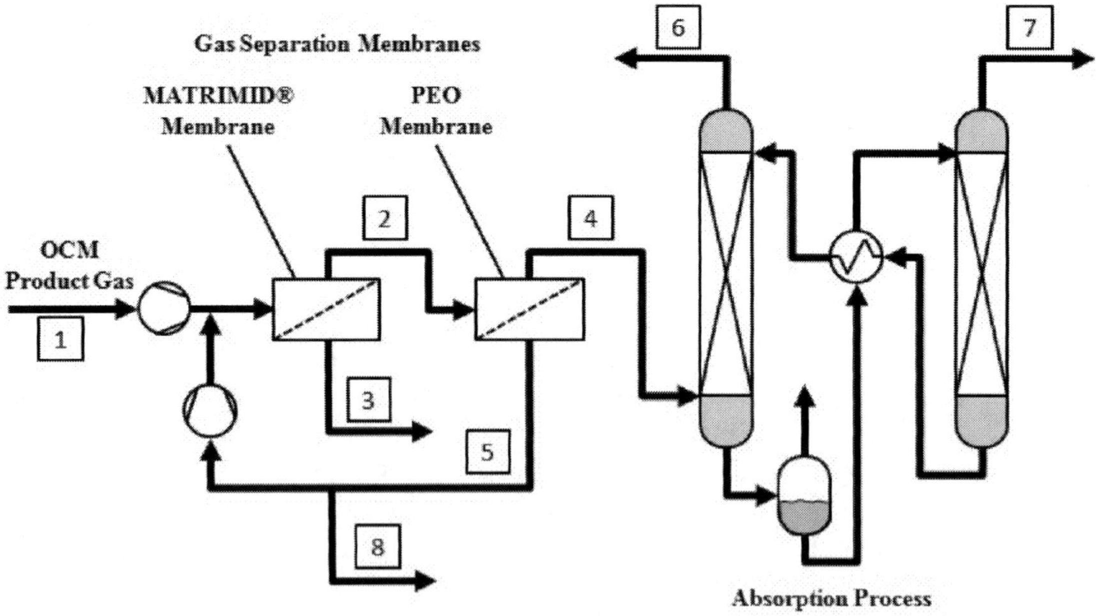

Figure 1: Simplified process flow sheet for the initial membrane-absorption-hybrid system for the removal of CO_2 from OCM product gas [2].

Hereby, the determination of the optimal combination and operation conditions of these reaction and separation units only through experimental investigations has proven to be challenging. On top of that, there are fluctuations in all mass and energy flows, which have to be taken into regard to ensure a safe and reliable operation. Consequently, simulative optimization under uncertainty of the entire superstructure is proposed. As a first step, the stand-alone amine-based absorption desorption process was modeled and fitted to mimic the actual behavior of the installed unit in the mini-plant. As a second step the two-stage membrane system was added upstream to the absorption unit to achieve a decrease of the required energy duty for the CO_2 removal. The OCM product gas is fed to a first polyimide (MATRIMID®) membrane module, which shows a high selectivity for CO_2 compared to ethylene, ethane, and methane. Additional CO_2 is removed in the second module, which contains a polyethylene oxide (PEO) membrane. Given a slightly lower CO_2/C_2H_4 selectivity a part of its permeate stream is recycled and added to the OCM product gas at the inlet. The retentate flow of the second membrane module is then fed to the absorption desorption unit to achieve the required CO_2 removal of 90%.

Initial NLP optimization studies of the proposed structure depicted in Figure 1 [2] showed a further decrease of the energy required per captured kilogram of CO_2 from 2.76 down to 2.57 MJ, while keeping the ethylene loss in the hybrid system below 5%. This is mostly achieved by a considerably lower feed gas pressure of 21.2 instead of 32.0 bar and an increase of the CO_2 concentration after the recycle before the first membrane. In order to optimize the system validated process models have been developed both for the membrane models as well as the absorption process [2], which will be adapted here.

In this contribution, the complexity of the membrane section of the hybrid system is increased. Apart from the two-stage combination of Matrimid® and PEO membrane modules several other combinations with even more membrane types are of interest. A possible superstructure with a maximum of six membranes is given in Figure 2.

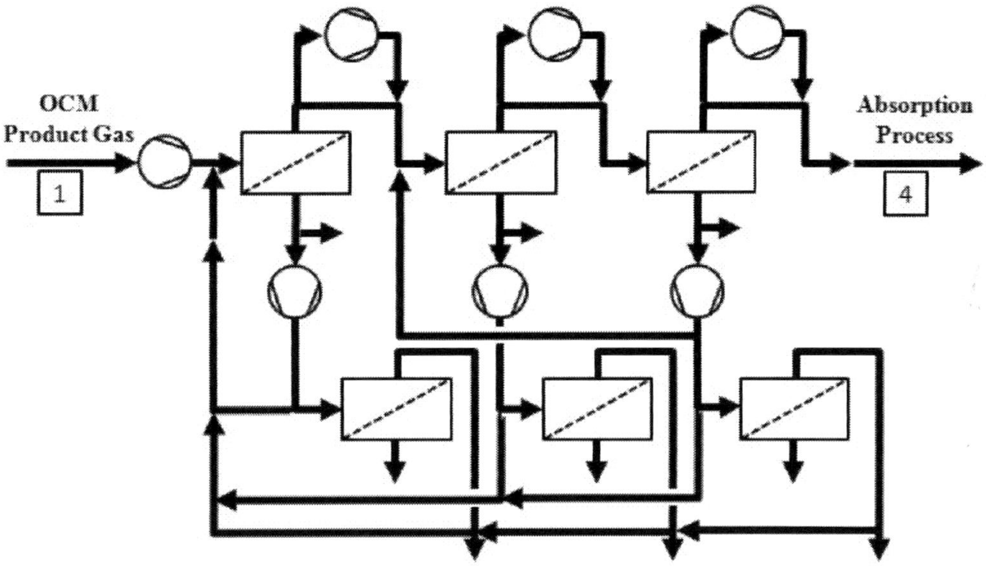

Figure 2: Superstructure showing a variety of possible connections to combine up to four membrane modules before feeding the gas to the absorption process. The membranes can contain either Matrimid® or PEO as permeable polymers.

Based on this superstructure alone, 4096 different structures for the membrane section by itself are possible. Seeing as the two-stage membrane together with the absorption system showed a nice behavior in terms of computation time and complexity, an MINLP formulation of the new system is done. The superstructure shown above is added to the absorption process. The whole system is mathematically described with the models already derived in previous studies [2] using the web-based modeling environment MOSAIC [3] from the Berlin University of Technology to formulate and to export the optimization problem to AMPL. Therein, various solvers such as Couenne [4] and Bonmin [5] are employed to solve the MINLP. To increase the numerical stability of the MINLP optimization a method is implemented to update the starting values of inactive parts of the superstructure.

The objective of the MINLP optimization is the minimization of the specific energy required per kilogram of CO_2 removed, while keeping the CO_2 removal rate at 90% and the ethylene loss below 5%. Apart from the structural binary variables, the decision variables consist of the membrane types, the length and operation pressure of each membrane module, recycle flow rates as well as the main operational variables of the absorption process, i.e., feed gas pressure, absorbent flow rate, as well as heating duty in the desorption.

In this contribution details on the modeling and optimization of the MINLP are presented to discuss the benefit of more complex gas separation process structures to reduce product loss as well as the required energy. Additionally, the experimental validation in the actual mini-plant will be discussed.

Acknowledgements: This work is part of the Cluster of Excellence ?Unifying Concepts in Catalysis? coordinated by the Berlin University of Technology (TechnischeUniversität Berlin). Financial support by the German Research Foundation (Deutsche Forschungsgemeinschaft, DFG) within the framework of the German Initiative for Excellence is gratefully acknowledged.

References

[1] Stünkel, S., Bittig, K., Godini, H., Jaso, S., Martini, W., Arellano-Garcia, H., and Wozny, G. (2012) Process Development in a Miniplant Scale ? A Multilevel - Multiscale PSE Approach for Developing an Improved Oxidative Coupling of Methane Process, 11th International Conference on Process Systems Engineering, Singapore.

[2] Esche, E., Müller, D., Song, S., Wozny, G. (2013) Optimization of a Membrane-Absorption System for the Removal of CO2 from OCM Product Gas, 6th International Conference on Process Systems Engineering (PSE ASIA 2013), *accepted for publication.*

[3] Kuntsche, S., Arellano-Garcia, H., and Wozny, G. (2011) MOSAIC, an environment for web-based modeling in the documentation level, Computer Aided Chemical Engineering 29, 1140-1144 ISBN 978-0-444-54-298-4.

[4] Belotti, P., Lee, J., Liberti, L., Margot, F., Wächter, A. (2009) Branching and bounds tighteningtechniques for non-convex MINLP, Optimization Methods and Software 24 (4-5), 597 ? 634.

[5] Bonami, P., Biegler, L.T., Conn, A.R., Cornuéjols, Grossmann, I.E., Laird, C.D., Lee, J., Lodi, a., Margo, F., Sawaya, N., Wächter, A. (2005) An algorithmic framework for convex mixed integer nonlinear programs. Tech. rep., IBM Research Report RC23771. To appear in *Discrete Optimization.*

(740c) Stereo Graphical Method for Feasibility Analysis of Reactive Distillation

Gao, X., Tianjin University
Li, H., Tianjin University
Sun, X.,
Li, X., Tianjin University
Originally presented on: 11/7/2013 16:05:00 - 16:30:00

Stereo Graphical method for feasibility analysis of reactive distillation

Xin Gao[a, b], Hong Li[a, b, [*]], Xiaoyu Sun[a], Xingang Li[a, b]

[a] *School of Chemical Engineering and technology, Tianjin University, Tianjin 300072, China*

[b] *National Engineering Research Center of Distillation Technology, Tianjin 300072, China*

Abstract

The Stereo Graphical (SG) method is a novel convenient, reliable, and systematic approach for feasibility studies of reactive distillation. This method was developed by innovatively taking temperature into account. Reactive distillation studies using SG can be established based on traditional methods, which visualize the temperature. The exact reactive distillation curve of a given system can be acquired when the given operation pressure kept constant, such that this curve describes the reaction zone of the reactive distillation process. The feasibility of using reactive distillation in a reaction system can be assessed using SG (feasibility study), thereby providing useful information such as the feasible range of feed composition and the need for a non-reaction zone. In this paper, three cases were used to elaborate the application of the proposed SG method. Reactive distillation was proven feasible for the methyl tert-butyl ether, ethyl tert-butyl ether, and tert-amyl methyl ether systems. Given the large differences in their reaction equilibrium constants, these systems differ in terms of the details of their conceptual designs. This method is expected to have accurate results when it is conveniently applied in the study of other reaction systems.

---

[*] Corresponding author. Tel: +86-022-27404701-8503; E-mail: lihongtju@tju.edu.cn

(740d) Novel Process of NH3-N Removal and Recovery in Non-Ferrous Industry Wastewater: Fundamental, Simulation and Commercial Projects

Lin, X.,
Liu, C. M.,
Cao, H. B., Institute of Process Engineering, Chinese Academy of Sciences
Originally presented on: 11/7/2013 16:30:00 - 16:55:00

Over 2 million tons of NH_3-N was discharged in China per year recently. Non-ferrous industry wastewater containing NH_3-N and heavy metals is a major contributor of high concentration NH_3-N. Combining distillation and air-stripping technology, NH_3-N is removed and recovered based on relative volatility difference of NH_3 and H_2O in this study. Phase equilibrium, heat equilibrium are calculated to determine process parameters. Hydrodynamic performance like flow field distribution is also calculated. Then, design parameters for distillation trays or structured packing are determined based on calculations and simulations. Besides, the phase diagram of $(NH_4)_2SO_4$-NH_4Cl-H_2O in 60℃-100℃, the typical salts composition of nonferrous wastewater, is studied to ensure commercial plants' continuous operation. NH_3-N removal tower with different high efficient column internals are designed to fit different wastewater treatment requirement. The novel process have been used in treatment of NH_3-N wastewater of vanadium, chromium, zirconium, nickel, cobalt, molybdenum, niobium, tantalum and rare earth industry. The projects operation results shows the novel process reached a nearly zero-emission effect with >99.5% NH_3-N was removed and then recovered as ammonia or ammonia salts.

(740e) Purification Process Technology of Coal Bed Methane (CBM)

Ko, D. H., GS E&C
Row, S. W., GS e&C
Ahn, E. S.,
Won, W., Sogang Univ.
Originally presented on: 11/7/2013 16:55:00 - 17:20:00

Natural gas adsorbed on coal beds underground is called Coal Bed Methane (CBM). The CBM compositions are mainly methane and other impurities such as carbon dioxide, ethane, and so on. The compositions may vary according to the CBM site.

To use the CBM as an energy source, gas treating processes to remove sour gas may be required depending on the purposes of the natural gas utilization. This study focuses on the development of the energy saving and flexible separation / purification processes of CBM.

At first, we compared three main methods for separation processes, i.e., absorptions, adsorptions, and membranes through simulations and experiments.

Our simulation of absorption process using Pro-II software shows that the sour gas mole fractions of CO2 and H2S in the purified gas stream are less than 10 and 5 ppm respectively when the feed gas (CH4 rich gas) stream consists of CO2 (~10%) and H2S (~5%). The absorption processes using amine as a solvent have an excellent purification performance and is generally adopted in the large scale gas treatment plants requiring a highly purified gas product.

The simulation (gPROMS software) and experiments of a pressure swing adsorption (PSA) process were performed. When the adsorbent is carbon molecular sieve (CMS) and the feed gas consists of 10% CO2 and 90% CH4, the purified product purity is more than 98% and the recovery is about 80%. The purity can be enhanced up to more than 99% CH4. The adsorption process using CMS has a very good purification performance and is generally used in the small to medium large scale gas treatment processes.

We checked membrane performances through the simulation (gPROMS software) and experiments. When the feed gas consists of 10% CO2 and 90% CH4, the purified product gas (CH4) has more than 97% purity. The membrane process has a relatively low purification performance but a simple and compact configuration compared with the other processes such as the absorption and adsorption processes. The membrane process is mostly used in a small scale process but recently a somewhat large scale membrane process is under the development and commercialized.

This study focuses on the two processes – adsorption and membrane processes – as candidates for the conceptual design work, considering the utilization purposes and amount of the purified CBM gas as well as transportation problems of the pilot plant facilities to the corresponding site in our R&D program. Here, the scale of CBM feed gas is expected as a 10~2500 Nm3/hr.

With the two processes of adsorption and membrane processes, we performed a conceptual design study for the gas treatment by comparing eight types of candidate process configurations. The candidate processes involve the single processes (PSA only or membrane only) and hybrid processes consisting of PSA and membrane systems. The power consumptions per the corresponding product (purified gas) flow rate are compared, and a suitable process configuration is adopted in view of power consumption, purity, and recovery of the product gas. A hybrid process with a recycle stream is better than single process in view of the power consumption, the purity, and the recovery of product gas (CH_4).

As a conclusion, the conceptual design is performed for the CBM purification processes adopting PSA and membrane processes, resulting in the development of the energy efficient hybrid system of PSA and membrane, which can achieve more than 97% purity and 95.5% recovery of CH_4. Also we checked the pressure vacuum swing adsorption process that can acquire more than 97% purity and 90% recovery of CH_4. As a result, a basic design package for the CH_4 gas purification process was developed to construct a pilot plant to purify CBM feed gas.

(744a) Treatment of High-Strength Saline Phenolic Wastewater in a Chemostat Coupled With a Forward Osmotic Membrane Bioreactor

Praveen, P., National University of Singapore
Nguyen, D. T., National University of Singapore
Loh, K. C., National University of Singapore
Originally presented on: 11/7/2013 15:15:00 - 15:40:00

Industrial wastewater from many sources, such as leather and petrochemical industry, contain high quantities of salts and toxic organics. Desalination of such effluents with prior removal of the organic contaminants results in the release of these pollutants in the environment at high levels. On the other hand, biodegradation of high concentration of toxic organics in saline wastewater is difficult due to the inhibitory effects these chemicals exert on the biodegrading microorganisms. A simple solution to alleviate these inhibitions is the dilution of the wastewater but it results in two problems: (1) increase in the process volume, and (2) large water requirement.

In this research, we have developed a novel bioreactor system by integrating a chemostat with a forward osmosis (FO) membrane system for treatment of phenolic wastewater at high salinity. The strategy was to dilute the wastewater to achieve non-inhibitory concentration of salts and phenol, phenol was metabolized in the chemostat and the resulting effluent was concentration and desalinated in the FO module. Water recovered through FO using was used for the dilution of the feed solution. Wastewater at phenol concentration of 1000-2000 mg/L and sodium chloride concentration of 0.5-1 M was diluted 3-5 times and pumped into the bioreactor at the dilution rate of 30-90 mL/h. Phenol was completely degraded in the chemostat within 10 h at low dilution rates, but a longer time was required to reach steady state at higher dilution rates. The specific growth rate decreased with increasing phenol concentration and at low dilution factor. The cell growth rate did not change significantly below salt concentration of 0.2 M, however, the growth rate decreased at higher concentration. In the absence of salt, the recovery of water from the chemostat effluent was simple. As the salt content of the effluent increased, a higher draw solute concentration was required for effective desalination.

It was also observed that the water flux through the FO membranes decreased over a period of 24 h which could be due to the presence of biofilms on the membranes, and also due to concentration polarization. The biofilms could be removed from the membranes by osmotic backwash at the end of operation, which alleviating concentration polarization required thorough washing of the membranes with deionized water. These results show that the chemostat-FO system can treat saline phenolic wastewater effectively.

(744b) A Fundamental Investigation of Single Bubble and Bubble Swarm Behavior With Respect to Fouling Prevention in Membrane Systems

Böhm, L., Technische Universität Berlin

Kraume, M., Technische Universität Berlin

Originally presented on: 11/7/2013 15:40:00 - 16:05:00

A fundamental investigation of single bubble and bubble swarm behavior with respect to fouling prevention in membrane systems

L Böhm[1], M Kraume[1]

[1]Chair of Chemical and Process Engineering, Technische Universität Berlin, Germany

FH6-1, Fraunhoferstrasse 33-36, 10587 Berlin, Germany. Tel: +493031472791, email: lutz.boehm@tu-berlin.de

Aim

Aeration is a common measure to detach deposition layers from membranes. A detailed view at commercially available flat sheet membrane modules reveals that although aeration is a main cost factor in the operation of membrane bioreactors (MBR), no optimal design and operation with respect to the cleaning efficiency was found yet. Different gap distances between the membrane cushions, different bubble sizes, additional mechanical cleaning with the help of granulates, pulsating aeration etc. are applied. This investigation's focus is gaining a fundamental understanding about the behavior of single bubbles and bubble swarms in such systems. The complexity of the behavior of a bubble swarm in an MBR is rather high. The fundamental approach of investigating the single bubble and bubble swarm behavior in one rectangular channel will help getting a deeper insight into the bubble dynamics in such systems. Especially the investigations with model solutions imitating the rheological behavior of activated sludge can deepen the understanding of the cleaning process.

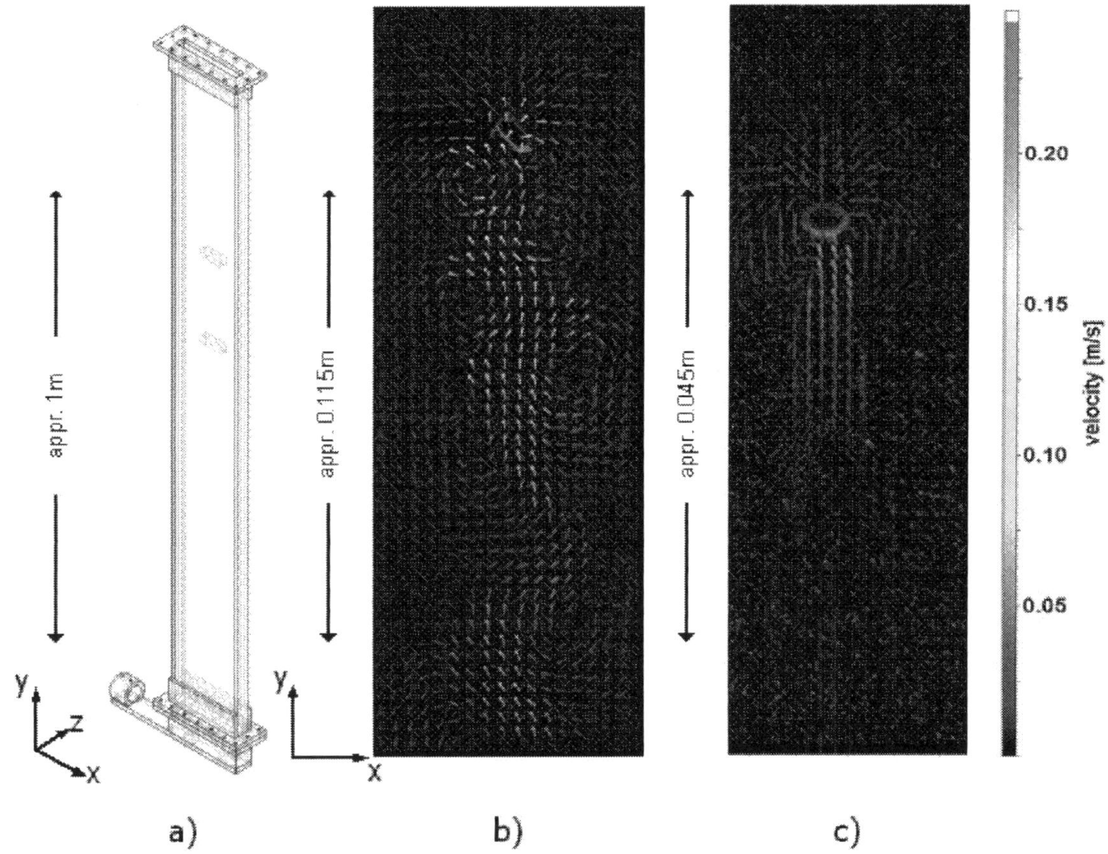

Figure 1. Schematic of the rectangular channel (a), front view of the flow field near a 5 mm bubble in a 5 mm channel in stagnant water (b) and in non-Newtonian Xanthan solution

Methods

For this purpose, an experimental apparatus was built which represents one of the channels with a rectangular cross section area between two flat sheet membrane cushions [1]. Acrylic glass is used to make optical measurements possible (Fig.1a). The fully automated rig is used to investigate single bubbles rising in the channel with the help of the electrodiffusion method for the determination of the generated shear stress on the walls, with a high speed camera for the determination of the bubble dynamics and with particle image velocimetry for the determination of flow field parameters in the liquid surrounding the bubble. In these investigations, the gap distance between the plates (3-7 mm), the single bubble size (3-9 mm), the superimposed liquid velocity (0-0.2 m/s) and the rheology of the liquid (Newtonian, non-Newtonian) were varied. For the investigation of the bubble swarm, in this list of parameters, the single bubble size is replaced by the aeration rate and aerator type.

Results

Exemplary flow fields near a single bubble are shown in Fig.1b) for the rise in water and c) for the rise in Xanthan solution. The concentration of Xanthan is chosen in a way, so that the shear thinning behavior of the solution is according to the one of activated sludge in MBRs [3]. The flow fields show the strong influence of the rheology of the continuous phase on the rise behavior. In case of the rise in water, the typical vortices which detach periodically from both sides of the bubble are visible. This leads to an oscillating movement of the bubble. Due to the confinement by the walls, especially for the bubbles equal or larger in size in comparison to the gap distance, only a zigzag-movement in the x-y-plane is found. For bubbles smaller than the gap distance, a zigzag-movement in the y-z-plane is appearing as well. Depending on the bubble size, in unconfined geometries the bubble in the size range investigated here would rise in a zigzag or helical motion. In case of the rise in the Xanthan solution, an almost symmetric flow field with respect to the vertical center line is found. Therefore, if appearing at all, only minor oscillations were found in any direction. The very particular flow field in the non-Newtonian liquid was similarly found for bubbles rising in unconfined geometries [2]. Near the bubble, the liquid is pushed around the bubble's edges on both sides. In the wake of the bubble, a triangular area with an upward flow on the edges of the triangular and a downward flow in the center of the triangular was found. This is due to the lower viscosity of the liquid in the high shear zone induced by the bubble movement. This effect will be visible for bubble swarms as well. In the investigation of the bubble swarm with the high speed camera, the bubble dynamics, bubble rise velocities und rising path are compared to the single bubble results and literature values e.g. from Yamanoi and Kageyama [4].

Conclusion

This fundamental investigation illustrates the influence of the liquid rheology in MBRs on the bubble dynamics and therefore the cleaning effect. With shear thinning liquids, e.g. the oscillation parallel to the membrane will be dampened and therefore the membrane area that is influenced by the bubble will be reduced. Without oscillation normal to membrane, the generated shear stress will be reduced as the liquid film thickness between bubble and wall increases in comparison to the film thickness in water. The investigation of the bubble swarm will show if the effects found for single bubbles can be confirmed.

References

[1] L. Böhm, A. Drews, and M. Kraume. Bubble induced shear stress in flat sheet membrane systems - serial examination of single bubble experiments with the electrodiffusion method. *J. Membr. Sci.*, 437:131?140, 2013.

[2] D. Funfschilling and H. Z. Li. Flow of non-Newtonian fluids around bubbles: PIV measurements and birefringence visualisation. *Chem. Eng. Sci.*, 56(3):1137 ? 1141, 2001.

[3] S. Rosenberger, K. Kubin, and M. Kraume. Rheology of activated sludge in membrane bioreactors. *Eng. Life Sci.*, 2(9):269?275, 2002.

[4] I. Yamanoi and K. Kageyama. Evaluation of bubble flow properties between flat sheet membranes in membrane bioreactor. *J. Membr. Sci.*, 360(1-2):102?108, 2010.

(744c) Oscillatory Ethanol Extraction By Pervaporation From a Membrane Bioreactor

Noriega, M. A., National University of Colombia
Fontalvo, J., Universidad Nacional de Colombia Sede Manizales
Gomez Garcia, M. A., National University of Colombia
Originally presented on: 11/7/2013 16:05:00 - 16:30:00

Energy consumption for ethanol production is intensive due to low concentration of the product in the fermentation broth which is purifies in a series of distillations columns. And also, ethanol concentration can not be easily increased from 9%v/v in conventional fermentation due to product inhibition of the yeast. A hybrid fermentation and pervaporation process promises to reduce inhibition while a high concentration of ethanol is obtained by a selective membrane. Based on pervaporation experimental data, this work analyses several continuous hybrid configurations of fermentation and pervaporation systems. A conventional fermentation system (CFS), a hybrid fermentation pervaporation system (FPS) and a hybrid oscillatory fermentation-pervaporation system (OFPS) are compared in terms of ethanol productivity, product concentration and energy required for purification by distillation of the product stream.

(744d) Modeling and Simulation of Membrane Reactor for Biodiesel Production

Noriega, M. A., National University of Colombia
Narvaez, P. C., Universidad Nacional de Colombia
Cadavid, J. G., National University of Colombia
Originally presented on: 11/7/2013 16:30:00 - 16:55:00

Biodiesel is an important renewable fuel whose production has been limited mainly by high residence times in the reaction and separation sections due to the presence of soaps and high viscosity gels. This work studies the possibility of reducing this problem by implementing a falling film reactor integrated whit hollow fiber membranes. This reactor takes advantage of both systems: while production of biodiesel using a falling film reactor is characterized by high reaction rates and productivity and a positive effect on separation stages because interfacial mass transfer area is achieved without mixing, the integration of a membrane system allows it increase the yield to fatty acid methyl esters as well as the simultaneous purification of the biodiesel obtained. In this work a mathematical model to predict the behavior of falling film reactor assisted by membranes was developed. The model describes a PFR reactor and the flux through the membrane by the transport equations of UF. A sensitivity analysis was performed to understand the operation of the membrane reactor and identify the main variables of the system. The model predicts the appropriate operating conditions range for experimental evaluation. Simulation of the process allows us to conclude that there is a strong drag effect exerted by methanol on the glycerol and that the conversion and yield in the reactor increase when the biodiesel and glycerol flux through the membrane rise, but they decreases when the flux of methanol increases, indicating that the amount of methanol fed to the reaction system is a main operating variable because it has strong influence on the transport through the membrane.

(744e) Oxidative Dehydrogenation of Propane By Metal-Loaded Mesoporous Membranes

Arafat, A., King Abdulaziz University
Bamufleh, H., King AbdulAziz University
Alhamed, Y., King Abdulaziz University
Originally presented on: 11/7/2013 16:55:00 - 17:20:00

Oxidative Dehydrogenation of Propane By Metal-Loaded Mesoporous Membranes

Abstract:

The catalytic performance of V and/or Pd-containing high-surface mesoporous siliceous materials (TUD-1) membranes in oxidative dehydrogenation of propane is investigated. Catalysts were prepared by anchoring of mesoporous TUD-1 mesoporous silica on alumina tubes by hydrothermal synthesis procedures. Catalytic reactions were performed on automatic gas phase reactor purchased from PID, Spain. The reactions were followed by online GC analysis. The effect of TUD-1 texture and the nature of metal loaded on the catalytic performance of the membrane catalyst are explored. Prepared catalysts were characterized by XRF for determination of metal content, XRD, SEM and N_2-adsorption for study of morphology and texture, and H_2-TPR and DR UV?vis spectroscopy. All prepared materials were tested in propane oxidative dehydrogenation (ODH) at 550 ◦C and the obtained catalytic results were correlated with their structural and surface characteristics. Results obtained revealed that the structure of mesoporous silica support plays crucial role in propane ODH reaction. The catalytic performance of investigated materials in C3-ODH can be correlated with population of all tetrahedrally coordinated metals and only isolated monomeric MOx complexes, respectively.

(744f) A Dual-Membrane Airlift Reactor for Cyclohexanone Ammoximation Reaction Over TS-1

Xiangrong, Z., Nanjing University of Technology

Honglin, M., Nanjing University of Technology

Weihong*, X., Nanjing University of Technology

Rizhi, C., Nanjing University of Technology

Yiqun, F., Nanjing University of Technology

Originally presented on: 11/7/2013 17:20:00 - 17:45:00

Cyclohexanone oxime is one of the most important intermediate products of caprolactam which was used to produce Nylon-6[1,2]. The coupling technology of reaction and ceramic membrane separation had been employed to realize in situ product removal[3]. Previous researches adopted cyclohexanone, ammonia gas and hydrogen peroxide to be interactants, tert-butanol as a solvent, to run one-step synthesis of oxime along with TS-1[4] in the split-membrane reactor. A large amount of tert-butanol was needed to recycle in the current process, which resulted in an increase of subsequent separation energy consumption. Without any solvents, it upsets the mass transfer of reactants to the pore system leading to low activity. In addition, concentration polarization and membrane pollution could affect stability of membrane separation process in long-term simultaneous reaction and separation processes.

In order to cope with difficulties mentioned above, a noval dual-membrane airlift reactor with a ceramic outer-membrane as a reactant ammonia distributor and another ceramic inner-membrane as a TS-1 separator was developed for the cyclohexanone ammoximation reaction over TS-1 without any solvents, as shown in Figure 1. Ammonia gas was dispersed into the reaction system through a ceramic outer-membrane. Another tubular porous ceramic membrane was employed as a membrane separator for in situ separation of catalysts from the products. Aeration was introduced to form two-phase flow on membrane surface, overcoming concentration polarization, increasing the shear force and local mixing velocity and controlling of membrane fouling caused by long-term consecutive reaction. To investigate the feasibility of the consecutive reaction in this new proposed dual-membrane reactor, the effect of membrane distributor, influence of aeration on membrane fouling, and optimal operation conditions were further studied.

It was interested to find that the introduction of membrane distribution could increase the selectivity by 34.6%, because many tiny ammonia bubbles could be produced with a membrane distributor to strengthen the mix of gas phase and organic phase. Membrane for separation could effectively prevent TS-1 passing micropores to realize the coupling reaction and separation. Aeration had little effect on conversion and selectivity, but had positive on controlling resistant. The conversion, selectivity, and membrane resistance are all closely related to the operation conditions. The optimum operating conditions were that residence time, pressure, temperature and catalyst concentration were 100 min, 0.3 MPa, 343 K and 30-35 g/L respectively, the molar ratio of cyclohexanone, ammonia and hydrogen peroxide was 1:1.59:1.2. On these conditions, the convertion of 87% and selectivity of 76% were obtained in continuous reaction process of 25h. Due to serious membrane fouling,the catalyst concentration was reduced to 10 g/L to inspect continuous cultivation stability of 100h. The reaction stability decreased with time while the membrane resistance increased, as shown in Figure 2. The experiments gave the further direction of investigation on the separation process stability.

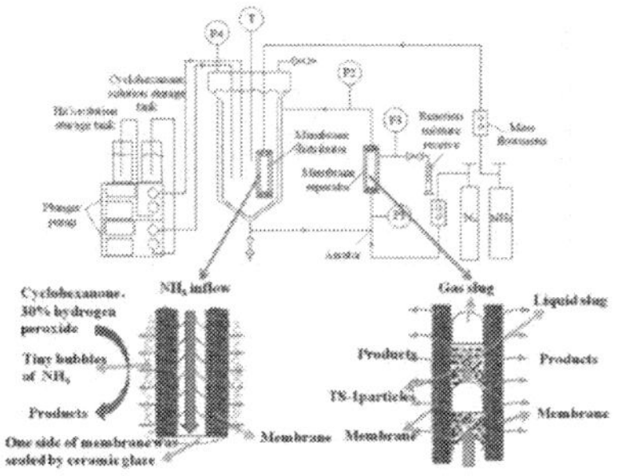

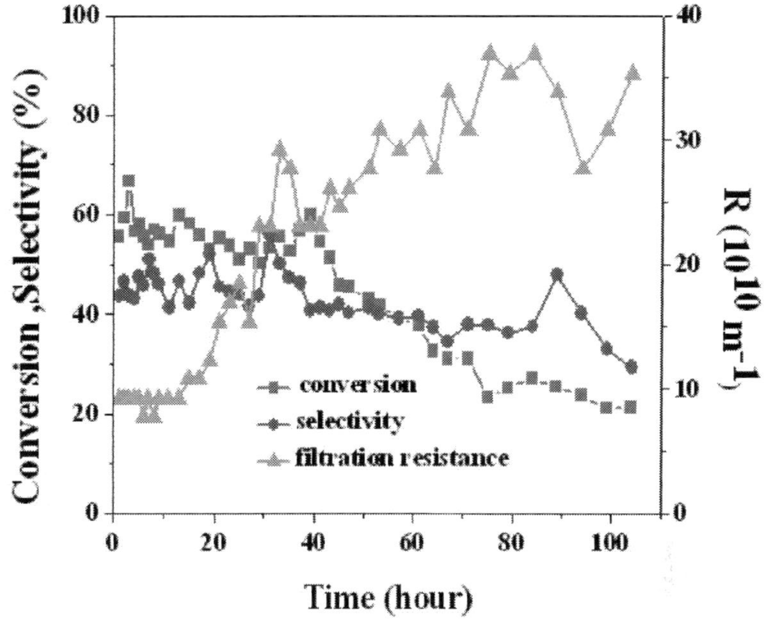

Figure1. Diagram of a novel dual-membrane airlift reactor system

Figure 2. Stability of the continuous reaction and membrane separation process.

References

1. Thomas JM, Raja R. Design of a "green" one-step catalytic production of epsilon-caprolactam (precursor of nylon-6). *Proceedings of the National Academy of Sciences of the United States of America.* Sep 2005;102(39):13732-13736.

2. Mokaya R, Poliakoff M. Chemistry - A cleaner way to nylon? *Nature.* 2005;437(7063):1243-1244.

3. Lu C, Chen R, Xing W, Jin W, Xu N. A submerged membrane reactor for continuous phenol hydroxylation over TS-1. *Aiche J.* 2008;54(7):1842-1849.

4. Yip ACK, Hu X. Formulation of Reaction Kinetics for Cyclohexanone Ammoximation Catalyzed by a Clay-Based Titanium Silicalite-1 Composite in a Semibatch Process. *Ind. Eng. Chem. Res.* 2011;50(24):13703-13710.

Acknowledgments

Financial supports from the National Natural Science Foundation of China (No. 21125629), the National Basic Research Program (No.2009CB623406), the National High-Tech Research and Development Program of China (2012AA03A606). ADDIN EN.REFLIST

(745a) Incorporation of Hydrophobic Silica Nanoparticles On Polyetherimide (PEI) Hollow Fiber Membranes for CO2 Absorption in a Membrane Contactor

Wang, R.,
Zhang, Y.,
Originally presented on: 11/7/2013 15:15:00 - 15:40:00

The essential properties of the membrane used for contactor application for CO_2 capture by chemical absorbent include high hydrophobicity, high surface porosity with small pore size, low mass transfer resistance and high chemical resistance to the liquid absorbents. Asymmetric hollow fiber membrane prepared from the non-solvent induced phase separation (NIPS) process usually exhibits a relatively dense skin layer supported by a porous substrate. The most mass-transfer resistance in the membrane comes from the dense skin layer, while the porous substrate only provides mechanical support and fractional resistance. The pore size and pore size distribution of the skin layer significantly affect the CO_2 absorption flux and are associated with membrane wetting problem since the skin layer is in direct contact with the liquid absorbents.

In this presentation, we are going to report the fabrication of a highly porous polyetherimide (PEI) membrane with large pore sizes by using a triple-orifice spinneret in the hollow fiber spinning process, followed by novel incorporation of fluorinated silica ($fSiO_2$) nanoparticles into the top layer of the membrane to make the membrane surface highly hydrophobic and chemical resistant to prevent the membrane from wetting caused by the large pore size or/and less hydrophobic surface during the CO_2 absorption process.

Experiments revealed that the highly porous and cellular surface structure of the PEI membrane is favorable for nanoparticle embedment. The $fSiO_2$ nanoparticles were incorporated on the PEI hollow fiber membrane, leading to a significant improvement in surface hydrophobicity, as evidenced by the advancing contact angle value of 123.2°, receding contact angle value of 107.2°, and contact angle hysteresis of only 15.9°. A stable CO_2 absorption performance by using sodium taurinate as the liquid absorbent was attained over a 60-day long-term test, indicating the great potential of PEI-$fSiO_2$ membranes in practical membrane contact application for CO_2 capture.

(745b) Mesoporous Fluorocarbon-Modified Silica Aerogel Membranes Enabling Long-Term Continuous CO_2 Capture With Large Absorption Flux Enhancements

Tung, K. L., National Taiwan University
Lin, Y. F., National Tsing-Hua University
Originally presented on: 11/7/2013 15:40:00 - 16:05:00

The use of a membrane contactor combined with a hydrophobic porous membrane and an amine absorbent has attracted considerable attention for the capture of CO_2 because of its extensive use, low operational costs, and low energy consumption. The hydrophobic porous membrane interface prevents the passage of the amine absorbent but allows the penetration of CO_2 molecules that are captured by the amine absorbent. Herein, highly porous SiO_2 aerogels modified with hydrophobic fluorocarbon functional groups (CF_3) were successfully coated onto a macroporous Al_2O_3 membrane; their performance in a membrane contactor for CO_2 absorption is discussed. The SiO_2 aerogel membrane modified with CF_3 functional groups exhibits the highest CO_2 absorption flux and can be continuously operated for CO_2 absorption for extended periods of time. This study suggests that a SiO_2 aerogel membrane modified with CF_3 functional groups could potentially be used in a membrane contactor for CO_2 absorption. Also, the resulting hydrophobic SiO_2 aerogel membrane contactor is a promising technology for large-scale CO_2 absorption during the post-combustion process in power plants.

(745c) Spray-Assembled Composite Polymer-Clay Thin Films As Selective Layers in Reverse Osmosis Membranes

Kovacs, J., Massachusetts Institute of Technology
Hammond, P. T., Massachusetts Institute of Technology
Originally presented on: 11/7/2013 16:05:00 - 16:30:00

The development of novel membranes for use in continuous reverse osmosis (RO) processes can reduce the operating costs for modern desalination operations through increasing selectivity and reducing the cost of manufacturing. Composite, cross-linkable polyelectrolyte and clay thin films assembled via the layer-by-layer (LbL) deposition process can serve as a substitute for the polyamide selective layer in commercially-available membrane systems. The flexibility of the spray layer-by-layer (spray-LbL) assembly process enables the deposition of large asymmetric thin films orders of magnitude faster than possible through traditional dip-LbL processing. Composite thin films were assembled on polysulfone nanofiltration membranes, with the composition and physical properties fine-tuned by manipulating the film assembly conditions such as spray time of the film components as well as the number of layers deposited. Using mathematical modeling on data collected from dead-end permeation experiments, permeability coefficients for water and salt through selective layers of varying film composition were calculated. These experimental results confirm the viability of using composite polyelectrolyte-clay thin films as selective layers, with significant increases in water permeability observed when compared against commercially-available polyamide membranes.

(745d) Development of Multimodal Membrane Adsorbers for Protein Separations

Wang, J., Clemson University
Zhou, J., Clemson University
Anderson, L., Clemson University
Sproul, R., Clemson University
Husson, S. M., Clemson University
Originally presented on: 11/7/2013 16:30:00 - 16:55:00

This presentation will describe our latest research efforts using surface engineering to develop multimodal membranes as a new class of high-productivity chromatographic materials for applications in the purification of therapeutic proteins. Multimodal membranes incorporate ligands that can bind proteins through a combination of Coulombic interactions, hydrophobic interactions, and/or hydrogen bonding. This advancement has the potential to widen the operation ranges for chromatographic separations. In addition, the high-productivity nature of this membrane platform will help to alleviate the tradeoff that manufacturers must now make between load processing time and dynamic binding capacity.

Membranes were prepared using surface-initiated ATRP (atom transfer radical polymerization) of poly(glycidyl methacrylate) 'tentacles' from regenerated cellulose membranes. In a second step, 4-mercaptobenzoic acid was coupled to the membranes by an epoxide ring-opening reaction. ATR-FTIR measurements were carried out to support successful ligand incorporation. Protein binding experiments with bovine immunoglobulin G showed that binding capacities of the multimodal membranes were dependent on pH and ionic strength. A maximum binding capacity above 60 mg IgG/mL was found near the isoelectric point of the protein. The capacity remained above 30 mg IgG/mL even at 300 mM ionic strength.

(745e) Surface Modification Synthesis of Defect-Free Vertically Oriented Ordered Mesoporous Silica Membranes

Lin, J., Arizona State University
Seshadri, S.,
Originally presented on: 11/7/2013 16:55:00 - 17:20:00

Ordered mesoporous materials have highly ordered pore structure, well defined pore diameter in the mesopore range, high surface area and unique pore connectivities. These materials in membrane morphology such that the pore channels are aligned perpendicular to the membrane surface are desirable as it will allow for better pore accessibility for use in a number of applications. To make defect free ordered mesoporous membranes with vertically aligned pores has been a major challenge for researchers working in this area. In this paper, we will show that vertically oriented ordered mesoporous silica membranes can be synthesized as silica plugs filling the macron-sized straight pores of hydrophobic track-etched polycarbonate membrane support by an counter-diffusion self-assembly method. However, these membranes have gaps between the mesoporous silica plugs and support pore wall. We will report results of various methods studied to fill these gaps. We found that these defects can be eliminated by filling the membrane gaps with an alkoxysilane followed by exposure to humid air to allow controlled hydrolysis and condensation resulting in formation of microporous silica within the gaps. Molecular probing gas permeation and helium/nitrogen (or oxygen) binary separation tests, coupled with surface characterization methods, show that final membranes contain ordered mesopores of about 2.7 nm pore diameter, running through the membrane, with gaps sealed by microporous silica having a pore size less than 0.55 nm. The results show that the liquid deposition technique is an effective way for defect sealing in counter-diffusion self-assembly method membranes and this approach could possibly be used in other systems with equal success.

(745f) Imparting New Functions to Zeolitic Imidazolate Framework ZIF-8 Membranes Via Ligand Exchange and Post Synthetic Modification

Kwon, H. T., texas A & M university
Jeong, H. K., Texas A&M University
Originally presented on: 11/7/2013 17:20:00 - 17:45:00

Imparting new functions to Zeolitic-imidazolate Framework ZIF-8 Membranes via Ligand Exchange and Post Synthetic Modification

Hyuk Taek Kwon[1], and Hae-Kwon Jeong*[1,2]

[1]Artie McFerrin Department of Chemical Engineering and [2]Materials Science and Engineering Program

Texas A&M University, College Station, TX 77843-3122

* Corresponding author: hjeong7@mail.che.tamu.edu

Metal-organic frameworks (MOFs), crystalline organic-inorganic hybrid materials, have revolutionized the research fields of porous materials since their discoveries due to their attractive features such as high surface area, regular and sub-molecular size of pore structure, and recognition and binding ability of specific molecules.[1] The diversity of MOFs, originated from theoretically unlimited combinations of organic (ligand) and inorganic (metal ion) constituents, enlarges tremendously their applications.[2]

Material properties of MOFs can be fine-tuned via a couple of strategies. One is through the direct construction of MOFs with pre-designed ligands and metal ions having customized functions.[3] The other is post synthetic treatments, for instance, ligand/metal exchange (PSE)[4,5] and/or post synthetic modification (PSM) on organic ligands.[3] The former approach has less freedom in terms of property modification because even if the synthesis starts with designed ingredients, often times it results in unintended structure or is even challenging to crystallize the MOFs due to the fact that the crystallization of MOFs is very specific to synthesis conditions such as temperature and solvent properties where ligands and synthesis solvents play a role of structure directing agent. On the other hand, the post synthetic treatments, which are performed in rather mild conditions compared to direct synthesis, can impart new functions to parent MOFs in a rather facile manner, while retaining the integrity of parent crystal structures as proven on a series of robust MOFs.[4,6]

Here, we would like to present the modification of chemically inert ZIF-8 membranes, synthesized by counter diffusion concept, via ligand exchange and subsequent post synthetic modification. The effect of newly introduced functional groups on membrane properties such as hydrophobicity and gas separation performance will be systematically discussed.

References and Notes

1. Eddaoudi M, Kim J, Rosi N, et al. Systematic Design of Pore Size and Functionality in Isoreticular MOFs and Their Application in Methane Storage. *Science.* January 18, 2002 2002;295(5554):469-472.

2. Kuppler RJ, Timmons DJ, Fang Q-R, et al. Potential applications of metal-organic frameworks. *Coordination Chemistry Reviews.* 2009;253(23–24):3042-3066.

3. Tanabe KK, Cohen SM. Postsynthetic modification of metal-organic frameworks-a progress report. *Chemical Society Reviews.* 2011;40(2):498-519.

4. Kim M, Cahill JF, Fei H, Prather KA, Cohen SM. Postsynthetic Ligand and Cation Exchange in Robust Metal–Organic Frameworks. *Journal of the American Chemical Society.* 2012/10/31 2012;134(43):18082-18088.

5. Kim M, Cahill JF, Su Y, Prather KA, Cohen SM. Postsynthetic ligand exchange as a route to functionalization of 'inert' metal-organic frameworks. *Chemical Science.* 2012;3(1):126-130.

6. Fei H, Cahill JF, Prather KA, Cohen SM. Tandem Postsynthetic Metal Ion and Ligand Exchange in Zeolitic Imidazolate Frameworks. *Inorganic Chemistry.* 2013/04/01 2013;52(7):4011-4016.

(754a) Nucleation in the Highly Supersaturated Boundary Layer During Solvent Freeze-Out

Poon, G.,
Peters, B., University of California Santa Barbara
Originally presented on: 11/7/2013 15:15:00 - 15:40:00

Nucleation kinetics, induction times, and metastable zone widths are often modeled in quiescent, quasi-steady, and/or spatially uniform concentration fields. However, freezing an aqueous solvent can concentrate the solute and effectively increase the supersaturation. During the freezing process, a boundary layer of giant supersaturation develops ahead of the moving ice front. We solve for the time-dependent concentration in the boundary layer under conditions where the growing ice perfectly excludes the solute. We then develop stochastic models of nucleation in the boundary layer. Whether heterogeneous on the ice surface, or homogeneous in the boundary layer, nucleation is dramatically accelerated by the growing ice. For methane hydrates, which form at similar conditions to ice, induction times for nucleation can be reduced by as much as 10^{80} times because of the moving supersaturation zone.

(754b) On Limits of Metastability in Solution Crystallization: Induction Supersaturation and Spinodal

Bhamidi, V., Abbott Laboratories
Kenis, P. J. A., University of Illinois at Urbana-Champaign
Zukoski, C. F., University of Illinois at Urbana Champaign
Originally presented on: 11/7/2013 15:40:00 - 16:05:00

Our previous research on slowly supersaturated systems suggested that a limit of metastability in solution crystallization can be determined based solely on energetic considerations of nucleation. For such systems, we define an induction supersaturation (a 'working metastability limit') at a supersaturation at which the energetic barrier to nucleation drops to the order of thermal energy of the system. In this work, we examine this limit of metastability in light of recent experimental results by several researchers. We present density functional calculations of energy barrier to nucleation on our experimental data, and estimate the probability of nucleation in a slowly-supersaturated system from this energy barrier. These results are compared to the predictions based on classical nucleation theory. A general agreement between the two methods of calculations is observed. The induction supersaturation is contrasted against the spinodal decomposition limit. Modern concepts of nucleation and metastability limits are revisited in the context of our results, and the implications of our experimental findings are discussed.

(754c) Observing Ice Nucleation From Water Microdroplet Freezing Below "Homogenous Nucleation Temperature Limit"

Laksmono, H., SLAC National Accelerator Laboratory
McQueen, T.,
Sellberg, J.,
Huang, C.,
Loh, D.,
Sierra, R., SLAC National Accelerator Laboratory
Starodub, D.,
Norlund, D.,
Beye, M.,
DePonte, D.,
Martin, A.,
Barty, A.,
Feldkamp, J.,
Boutet, S.,
Williams, G.,
Bogan, M., SLAC National Accelerator Laboratory
Nilsson, A., SLAC National Accelerator Laboratory
Originally presented on: 11/7/2013 16:05:00 - 16:30:00

Water is one of the most fascinating and arguably the most important substance in our life, environment, and industry, due to many anomalous physical properties, and many phases of solid it forms. Although there have been many experiments performed to understand water and its phase transition, there is still a region below "homogenous nucleation temperature limit", ~232K, where water is not well understood due to rapid crystallization complicating experimental studies. Understanding phase transition within this region is also important to predict cloud properties and phases of water in the interstellar space. To date, the limited number of experiments performed limits our knowledge of water freezing within this region. A recent development in an ultrafast and ultrabright X-ray Free Electron Laser at the Linac Coherent Light Source (LCLS) enable researchers to probe matter in high resolution by taking snapshots using femtoseconds long X-ray pulses. Here we will present our ongoing effort to understand phase transition below the "homogenous nucleation temperature limit" from our recent wide angle X-ray scattering experiment using a specialized setup to take advantage of the X-ray laser at LCLS.

(754d) Populations and Size Distributions of Solute-Rich Mesoscale Structures in Aqueous Amino Acid Solutions and Their Role in Crystal Nucleation

Jawor-Baczynska, A., University of Strathclyde
Lee, H., University of Minnesota
McCormick, A., University of Minnesota
Moore, B. D., University of Strathclyde
Sefcik, J., EPSRC Centre for Innovative Manufacturing in Continuous Manufacturing and Crystallisation University of Strathclyde
Originally presented on: 11/7/2013 16:30:00 - 16:55:00

Aqueous solutions of well soluble molecules such as small amino acids are usually assumed to be essentially homogenous systems with some degree of local structuring due to specific interactions on the sub-nanometre scale (e.g. molecular clusters, hydration shells) usually not exceeding several solute molecules. However, recent theoretical and experimental studies have indicated the presence of much larger molecular assemblies or mesospecies in solutions of small organic and inorganic molecules as well as proteins[1].

We investigated both supersaturated and undersaturated aqueous solutions of two simple amino acids (glycine and DL-alanine) using Dynamic Light Scattering (DLS), Brownian Microscopy/Nanoparticles Tracking Analysis (NTA) and Cryogenic Transmission Electron Microscopy (Cryo-TEM). Colloidal scale mesospecies (nanodroplets), were previously reported in supersaturated solutions of these amino acids and were implicated as intermediates species on a non-classical crystallization pathway[2, 3].

Surprisingly, we found that these mesostructures were also present in undersaturated solutions at solute concentrations well below the solid-liquid equilibrium (saturation) concentration at a given temperature, with mean sizes ranging from 100 to 300 nm and size distributions widening towards larger sizes with increasing solute concentrations. While DL-alanine mesostructures appeared to be solid-like and could be filtered out by 100 nm syringe PTFE filters, glycine mesostructures were comfortably passing through these filters and are thought to be liquid-like.

Using experimentally measured size distributions of these mesostructures from NTA, SAXS intensity data were fitted using various structural models and an excellent agreement was obtained using a mass fractal model with the fractal dimension d_f=2.7 suggesting fairly compact clusters with highly irregular interfaces. This is not surprising, since it is expected that the surface tension between the mesostructures and surrounding bulk solution would be very low as they only differ in glycine concentration. The result may be an undulating interfacial topology similar to that found in complex surfactant or block-copolymer systems.

We note that the mesospecies are not a separate phase and the system is better described as a thermodynamically stable mesostructured liquid containing solute-rich domains dispersed within bulk solute solution. Hence, at a given temperature, if excess crystalline solid is present, the complete mesostructured liquid will be in equilibrium with the solid and solute molecules will be stably distributed between solute-rich mesospecies and the surrounding bulk solution. The local concentration of glycine in mesostructures is higher than that in the surrounding bulk solution so when the system is supersaturated nucleation is more likely to occur in them. However, if the nuclei produced are smaller than the critical nucleus size, they will redissolve when they are exposed to the surrounding bulk solution.

1. P. G. Vekilov, *Crystal Growth & Design*, 2004, **4**, 671-685.

2. Y. Ma, H. Cölfen and M. Antonietti, *The Journal of Physical Chemistry B*, 2006, **110**, 10822-10828.

3. A. Jawor-Baczynska, J. Sefcik and B. D. Moore, *Crystal Growth & Design*, 2013, **13**, 470-478.

(754e) Crystallization of Non-Needlelike β-Glyine From Water Solutions Via Droplet Evaporation

Trauffer, D., Bucknell University
Snyder, R. C., Bucknell University
Originally presented on: 11/7/2013 16:55:00 - 17:20:00

Glycine is among the most well studied organic molecular systems with regards to its nucleated crystalline polymorphs and resulting growth morphologies. Because of this existing work, it has become a common case study for crystallization science and technology such as those performed with laser induced (Zaccaro, et. al. 2001) or templated (Kang, et. al. 2000) nucleation as well as morphology prediction (Snyder and Doherty, 2009) and additive behavior (Lovette, et. al, 2008). Additionally, it serves as a basis for understanding the molecular foundation of crystallization phenomena though studies ranging from atomic force microscopy and solution based studies to molecular simulation (Chen and Trout, 2010). Despite this body of study, the basic drivers towards nucleating the various polymorphs readily attainable at atmospheric pressure (e.g. the stable γ form, the commonly found α form, or the metastable β form) are not fully understood (Chen and Trout, 2010). Recently, monodisperse droplet evaporation has been shown to be an effective tool at forming metastable polymorphs and unique particle morphologies for another small organic molecular system, succinic acid (Carver and Snyder, 2012). In that case the polymorphism and morphology was strikingly different in comparison with those readily obtained through more traditional solution crystallizations.

In this talk, we will highlight our work on forming crystalline glycine particles through the evaporation of monodisperse droplets produced using a vibrating orifice aerosol generator (VOAG). The metastable β polymorph of glycine is often obtained from water-alcohol solvent mixtures or other more complex solutions. In this work, the β polymorph is formed as expected from water-ethanol mixtures. However, we also show for the first time that the β polymorph can be obtained from pure liquid water solutions at ambient temperature and pressure. Thus, this work provides evidence that solvent is not the only dominant factor in determining glycine polymorph formation, which had long been the proposed determining factor. Rather, through droplet evaporation we demonstrate that supersaturation likely plays a dominant role under some conditions in glycine polymorph determination. The β polymorph of glycine also usually forms a long needlelike habit; however, we show that our crystalline particles are nearly spherical with some ridging features. Thus, this method changes the aspect ratio from more than 10:1 to 1:1. Such a habit modification is usually a daunting task from a perspective of modifying the chemistry; however, in this case only the process of crystal formation needed to be changed to have a dramatic impact (e.g. not just shortening the needles) with a single pure solvent. Finally, we will discuss the broader potential for the use of monodisperse droplet evaporation as a tool for future crystallization studies with regards to crystal nucleation, polymorphism, growth morphologies and amorphous particle formation.

References

Zaccaro, J., Matic, J., Myerson, A. S., Garetz, B. A.. *Cryst. Growth. Des.* **2001**, *1*, 5-8.

Kang, J. F., Zaccaro, J., Ulman A., Myerson, A. S. *Langmuir.* **2000**, *16*, 3791-3796.

Snyder R. C., Doherty M. F., *Proc. Royal. Soc. A.* **2009**, *465*, 1145-1171.

Lovette, M. A., Browning, A. R., Griffin, D. W, Sizemore, J. P., Snyder, R. C. Doherty M. F., *Ind. Eng. Chem. Res.* **2008**, *47*, 9812-9833.

Chen, J., Trout, B. L ., *J. Phys. Chem. B*, **2010,** *114*, 13764-13772.

Carver, K. M., Snyder, R. C. *Ind. Eng. Chem. Res.* **2012**, *51*, 15720-15728.

(754f) A Mechanistic Model for Crystal Growth of Calcite

Dandekar, P., UC Santa Barbara
Doherty, M. F., University of California, Santa Barbara
Originally presented on: 11/7/2013 17:20:00 - 17:45:00

Calcite ($CaCO_3$) is one of the most abundant simple salts in the geological environment. It is also used in paints, paper making and the cement industry. Crystal growth of calcite and the effect of impurities, both inorganic (e.g., Mg, Sr[1]) and organic (e.g., aspartate, citrate[2]), on the growth spirals has been extensively studied via experiments in the biomineralization literature. We present a high-fidelity growth model that starts from the crystal structure data and is able to calculate the relative growth rates on all the faces of the crystal and can thus predict the morphology. The model accounts for the long-range solid-solid interactions; it also captures the effect of the solvent on the crystal growth of calcite surfaces. The intermolecular interactions in the solid have been modeled using the Periodic Bond Chain (PBC) theory[3]. The concept of a building unit of the PBC has been applied to account for the stoichiometry and dipole moment properties of the PBCs. The model also allows for the presence of multiple growth units along the step edges by using the kink rate model developed by Kuvadia & Doherty[4]. Our model correctly predicts the shape of the growth spirals on the (10-14) surface of calcite grown from aqueous solution, including the presence of an asymmetric spiral, as reported in the literature from AFM experiments[5]. The effect of Ca^{2+} to CO_3^{2-} ion ratio in the solution on the growth process, specifically on the step velocities of the asymmetric edges of the growth spiral, has been captured by this model. The model also explains the effect of inorganic impurities such as Mg^{2+} ions on the growth spirals of calcite.

This high-fidelity crystal growth model is capable of being generalized to a wide class of ionic crystals, including such important materials as titanium dioxide, potassium dihydrogen phosphate (KDP), etc. Ultimately, the model will provide a tool for predicting crystal morphology so that suitable growth media can be designed (e.g., solvent(s), additives, supersaturation) to deliver inorganic crystals of desired shapes depending on their applications.

References:

1. De Yoreo, J. J. and Dove, P. M. Science 306(5700), 1301–1302 (2004).

2. Davis, K. J., Dove, P. M., and De Yoreo, J. J. Science 290(5494), 1134–1137 (2000).

3. Hartman, P. and Perdok, W. G. Acta Crystallogr. 8(1), 49–52 Jan (1955).

4. Kuvadia, Z. B. and Doherty, M. F. Cryst. Growth Des. 11, 2780–2802 (2011).

5. Larsen, K., Bechgaard, K., and Stipp, S. L. S. Geochom. Cosmochim. Acta 74, 2099–2109 (2010).

(774a) A Novel Fluidized Bed Hydrator for Steam Reactivation of Calcium-Based CO2 Sorbent

Wang, A.,
Wang, D., The Ohio State University
Deshpande, N., The Ohio State University
Phalak, N., The Ohio State University
Wang, W., The Ohio State University
Fan, L. S., The Ohio State University
Originally presented on: 11/8/2013 8:30:00 - 8:55:00

Calcium-based solid sorbents has been widely applied for the removal of acid gases, such as sulfur dioxide (SO_2), sulfur trioxide (SO_3), and hydrochloric acid (HCl), from coal-fired flue gas. Amid possible carbon dioxide (CO_2) regulations, scientists have explored ways to extend this technology for CO_2 removal especially from sizable point sources at low cost. Here at the Ohio State University (OSU), researchers have developed a 3-step CaO-based carbonation-calcination reaction (CCR) process that operates in the range of 450 – 950 °C for the simultaneous removal of both CO_2 and SO_2 from the post-combustion flue gas. The decay in sorbent reactivity for CO_2 capture over multiple reaction cycles poses a significant challenge to large-scale cyclic carbonation-calcination process. Thermal sintering at elevated temperatures and the subsequent loss in pore volume both contribute to decreasing CO_2 capture capacity. Intermediate hydration has been suggested to be one of the most promising approaches to overcome this limitation by regenerating the particle pore volume and surface area. However, most existing hydration studies have been performed at laboratory scale using thermogravimetric techniques, often investigating conditions not amenable to the 3-step CCR process. In addition, a reactor design specifically suited to high temperature steam hydration is warranted. Experimental data from such a reactor can offer insights to the mass transfer limitations associated with the gas-solid reaction of the extremely cohesive calcium hydroxide ($Ca(OH)_2$) particles under fluidized bed conditions. For the first time, this study reports the design of a bench-scale high-temperature steam hydrator for reactivation of CaO sorbent. The hydrator, consisting of a fluidized-bed reactor with additional internals such as an agitator and a cooling coil, was evaluated using cold-flow tests. Based on the results obtained, a high temperature reactor unit was constructed. The high-temperature steam hydration tests were performed at isothermal conditions in the range of 300-500 °C and 0.6-1.0 atm steam partial pressure. In particular, at an average temperature of 473 °C and P_{H2O} = 1 atm, the 30 min hydration conversion exceeded 70%. Sorbent reactivity toward CO_2 was substantially recovered by steam hydration and the correlation follows a linear trend. Lastly, the cooling coil design demonstrated a proof-of-concept for extracting the heat from this highly exothermic high temperature (>450 °C) reaction. This further expands the application of such reactor design beyond CO_2 capture, for other applications such as chemical heat pumps.

(774b) Dynamic Modeling of a Solid-Sorbent CO2 Capture System

Modekurti, S., West Virginia University
Bhattacharyya, D., West Virginia University
Zitney, S. E., National Energy Technology Laboratory
Miller, D. C., U.S. Department Of Energy, National Energy Technology Laboratory
Originally presented on: 11/8/2013 8:55:00 - 9:20:00

A one-dimensional (1D), two-phase, non-isothermal, dynamic model of a solid sorbent-based CO_2 capture system has been developed in Aspen Custom Modeler (ACM). The system consists of a two-stage bubbling fluidized bed (BFB) adsorber-reactor for CO_2 capture, a single stage moving bed regenerator for the sorbent regeneration, a multi-stage integral-gear centrifugal compressor system for CO_2compression, a pneumatic transport system for solids transport between the adsorber and regenerator, and the balance of the plant. In addition, a heat recovery system has been modeled for the regenerator which includes pre- and post-heat exchangers using steam/boiler feed water as the heating/cooling medium. The solid sorbent is NETL 32D developed at National Energy Technology Laboratory (NETL). The system of partial differential algebraic equations (PDAEs) has been generated using conservation equations, hydrodynamics, and reaction kinetics along with the appropriate boundary and initial conditions.

Flue gas from a supercritical pulverized coal (SCPC) plant enters at the bottom of the lower stage adsorber and regenerated sorbent enters the upper stage adsorber. Embedded heat exchangers have been considered for cooling of the bed using water as the cooling medium. The CO_2-rich sorbent is pneumatically transported to the regenerator through a pre-heater. An embedded heat exchanger in the regenerator maintains the unit at near-isothermal conditions. The regenerated sorbent passes through a low pressure steam generator that helps to recover heat for improving the overall efficiency. The medium selected for transporting the solid sorbent between the adsorber and regenerator is the one that minimizes the auxiliary power requirement without violating the overall target for CO_2 capture. The CO_2 rich gas stream from the regenerator is compressed to pipeline pressure by a multi-stage integral-gear compressor. A glycol absorption system is used to satisfy the pipeline specification for water in the CO_2stream. For off-design operation of the compressors, surge detection and control algorithms are developed.

The PDAE system with the appropriate boundary conditions is solved in ACM using the well-known method of lines which involves discretization of the spatial domain and dynamic integration of the time domain. The transient responses of various process variables such as CO_2 capture, flue gas and sorbent temperatures, and loading of the sorbent are studied by simulating typical disturbances such as change in the flue gas temperature, flowrate, and composition. A lower level control system is designed to maintain the overall CO_2 capture at the desired value in the face of such disturbances.

(774c) Comparative Environmental Life Cycle Assessment of CO2 Capture Systems Based On Ionic Liquid and Monoethanolamine Solvents

Farahipour, R.,
Karunanithi, A. T., University of Colorado Denver
Originally presented on: 11/8/2013 9:20:00 - 9:45:00

Carbon capture, storage and utilization (CCUS) is increasingly being viewed as a global climate mitigation option. Central to a viable CCUS system is the development of cost-effective methods for removal of CO_2 from flue-gas streams of power plants and other stationary CO_2 sources. In addition to economic viability it is important to consider the life cycle impacts of CCUS methods in order to evaluate the net benefits. Moreover, in addition to climate mitigation we would need to consider other issues such as increased life cycle impacts on human and/or ecosystem health. Amin-based (e.g., monoethanolamine) absorption systems for CO_2 capture is a fairly mature technology. However, they can be very energy intensive and less attractive from an environmental stand point due to emissions of the used volatile solvent components. In recent years, because of their low volatility, ionic liquids show significant promise as a CO_2 absorption solvent for post-combustion CO_2 capture. In this work, we present process simulation results of post-combustion CO_2 capture with monoethanolamine (MEA) and ionic liquid bmim [PF6]. The heat and material balances derived from these simulations are utilized to construct the life cycle inventory for the capture process. In addition data related to other upstream processes such as solvent (MEA and ionic liquid) production and downstream processes such as CO_2 compression, transportation, storage and/or utilization is used to perform a life cycle assessment with an aim to compare the environmental impacts of absorption systems based on these two different solvents.

(774d) Whole-Chain CCS System Modelling: Enabling Technology to Help Accelerate Commercialisation and Manage Technology Risk

Matzopoulos, M., Process Systems Enterprise Ltd.
Ramos, A., Process Systems Enterprise, Inc.
Originally presented on: 11/8/2013 9:45:00 - 10:10:00

The commercial implementation of CCS still faces significant challenges. Many of these arise from the fact that the whole chain – and, eventually, whole CO_2 transportation network – needs to be considered as a single system in order to make design and operation decisions that satisfactorily address the commercial imperatives and risk requirements of the various stakeholders along the chain. Even a quick analysis shows that design and operating decisions at the power plant can have a significant effect on storage providers at the other end of the chain, and vice versa. A systems modelling approach is essential, but there are currently no tools that can satisfactorily provide this capability over the whole CCS chain.

In order to address this, a £3m project was commissioned and co-funded by the ETI and project participants, who comprise E.ON, EDF, Rolls-Royce, Petrofac (via its subsidiary CO2DeepStore), Process Systems Enterprise (PSE) and E4tech. The project is aimed at delivering a robust, fully integrated tool-kit that can be used by CCS stakeholders across the whole CCS chain that will be released as a commercially-supported software product at the end of the project.

Because of the requirement for many stakeholders to be able to model areas of the chain beyond their own specific processes (for example, a key requirement for power generators is the ability to investigate the effects on their operation of amine and compression systems attached to their plants), the tool-kit will provide process models for the whole chain. This will also allow it to be used by groups who require a whole-chain view, such as engineering companies or government departments who need to quantify policy decisions.

The tool-kit includes models for conventional generation (pulverised coal and combined cycle gas turbine), new generation (gasification and oxyfuel), solvent-based carbon capture, compression, transmission and injection. The individual process models are mostly 'medium-to-high fidelity': multicomponent streams; equilibrium methods for vapour-liquid separation; comprehensive compressor models allowing multi-stage, multi-section compressors with manufacturers' curves; and distributed pipeline models for construction of pipeline networks that can take elevation into account. The capture models use rate-based techniques for accurate quantification of chemical and physical capture in order to quantify energy penalties accurately and allow meaningful analysis of transient operations.

Using the system modelling tool-kit it will be possible to look at single areas such as amine plants in detail; investigate partial-chain operations – for example, power generation, capture and compression; or analyse interactions across the whole chain and eventually the whole CO2 transmission network with multiple sources and multiple sinks.

The tool-kit also caters for interfacing of external software packages such as E.ON's PROATES, in order to simplify interoperability analysis while allowing companies to preserve existing workflows where necessary. In addition an advanced custom modelling capability means that it is easy to add models of new processes and combine them with existing flowsheeting components.

This presentation describes the status of developments to date in whole chain CCS system analysis and plans to roll the tool-kit software out to the wider 'CCS community' following the current testing phase.

(774e) Optimization of CO2 Capture, Utilization and Sequestration (CCUS) Supply Chain Networks

Hasan, M. M. F., Princeton University
Boukouvala, F., Princeton University
Floudas, C. A., Princeton University
Originally presented on: 11/8/2013 10:10:00 - 10:35:00

CO_2 capture, utilization and storage (CCUS) is an enabling technology toward reducing CO_2 emissions from stationary sources which include power plants, cement production plants, iron and steel plants, refineries, petrochemicals and gas processing plants. More than 60% of the total CO_2 emissions in the United States are attributed to stationary sources [1]. CCUS connects CO_2 sources and utilization/storage sites through three sequential operations: capture, compression, and transportation. CO_2 is first captured and compressed at the source plant, and then transported via pipeline to a site where it is utilized or injected for sequestration for geological storage.

It is critical to integrate capture, compression, transportation, utilization and sequestration in an optimal fashion. However, standalone development of CCUS components have made the design of large-scale CCUS networks a complex and challenging task. In this work, we present a mixed-integer optimization (MILP) model for the synthesis of CCUS supply chain networks for nationwide and regional CO_2 management. We design the CCUS supply chain network by considering simultaneous selection of source plants, capture technologies, capture materials, and locations of utilization and sequestration sites for the first time. We consider real geographic locations and reliable storage estimates from the available database. Furthermore, our cost model includes the costs of capture, compression, transportation and injection for sequestration, and also revenues from CO_2 utilization. The capture and compression costs are derived using advanced modeling, simulation and optimization techniques for a wide range of flue gas compositions and flow rates [2-3], which are crucial to incorporate stationary sources of different types in a large CCUS network. We apply our model to determine the CCUS supply chain network with minimum net expenditure for a given CO_2 reduction level.

Our model selects the optimal nationwide CCUS supply chain network from 3317 stationary sources, 491 saline formations, 1837 unmineable coal areas, and 161 utilization locations (oil and gas reservoirs). We also allow four leading capture technologies (absorption, membrane, pressure swing adsorption (PSA) and vacuum swing adsorption (VSA)), and 9 top materials (2 solvents, 4 zeolites [4-5], and 3 membranes) to choose from for CO_2 capture. Results suggest that it is possible to capture 50% of the current CO_2 emissions from the stationary sources in the United States at a cost less than $30 for each ton of CO_2. Results for the CCUS in South Central United States, and Texas are also favorable. We demonstrate that it is crucial to consider both material and technology selections as integral parts of a large-scale CCUS network design.

References:

[1]. DOE/NETL. Carbon Sequestration Atlas of the United States and Canada, 2012.
[2]. Hasan, M. M. F.; Baliban, R. C.; Elia, J. A.; Floudas, C. A. modeling, simulation, and optimization of postcombustion CO_2 capture for variable feed concentration and flow rate. 1. Chemical absorption and membrane processes. Ind. Eng. Chem. Res. 2012a, 51, 15642-15664.
[3]. Hasan, M. M. F.; Baliban, R. C.; Elia, J. A.; Floudas, C. A. modeling, simulation, and optimization of postcombustion CO_2 capture for variable feed concentration and flow rate. 2. Pressure swing adsorption and vacuum swing adsorption processes. Ind. Eng. Chem. Res. 2012b, 51,15665-15682.
[4]. Hasan, M. M. F.; First, E. L.; Floudas, C. A. Cost-effective CO_2 capture based on in silico screening of zeolites and process optimization. Submitted for publication, 2013.
[5]. U.S. Provisional Patent Application #61/761,436 and #61/765,284.

(776a) Influence of Packing On Gas Diffusion in Samaria Aerogel Catalyst By Diffusion NMR

Mueller, R., University of Florida
Zhang, S., University of Florida
Neumann, B., University of Bremen
Baumer, M., University of Bremen
Vasenkov, S., University of Florida
Originally presented on: 11/8/2013 8:30:00 - 8:45:00

Recent investigations have shown that different macroscopic catalyst packings made from the same material can lead to very different catalytic performance for catalytic processes involving small gas molecules. These observations can be attributed to an influence of diffusion of reactant and product species on the catalytic performance. In this work, a pulsed field gradient (PFG) NMR technique employing a high magnetic field of 17.6 T was used to study self-diffusion of carbon dioxide in alumina stabilized samaria aerogel, a promising porous catalyst for gas-phase reactions. Such rare-earth aerogels exhibit high porosity and surface area with active sites directly integrated into the pore framework. PFG NMR diffusion studies were performed with the catalyst formed into the following two types of samples: macroscopic monoliths and beds of particles with sizes around 200 μm. The behaviour of CO_2 self-diffusion will be presented and discussed for both types of packings at different CO_2 loading pressures.

(776b) Efficient and Accurate Methods for Characterizing Effects of Framework Flexibility On Molecular Diffusion of CH4 in 8MR Zeolites

Awati, R., Georgia Institute of Technology
Ravikovitch, P. I., ExxonMobil Research and Engineering
Sholl, D., Georgia Institute of Technology
Originally presented on: 11/8/2013 8:45:00 - 9:00:00

Molecular dynamics (MD) and transition state theory (TST) methods are becoming efficient tools for predicting diffusion of molecules in nanoporous materials. The accuracy of predictions, however, often depends upon a major assumption that the framework of the material is rigid. This saves considerable amount of computational time, and is often the only method applicable to materials for which accurate force fields to model framework flexibility are not available. In this study, we systematically characterize the effect of framework flexibility on diffusion in four model zeolites with typical cage-window structures (LTA, CHA, ERI, and BIK) that exhibit different patterns of window flexibility. In this study it is shown that the molecules with kinetic diameters comparable to (or larger than) the size of the window, the rigid framework approximation can produce order(s) of magnitude difference in diffusivities as compared to the simulations performed with a fully flexible framework. It is also shown that the simple recipes to include the effect of framework flexibility are not generally accurate.

To account for framework flexibility effects efficiently and reliably, two new methods are introduced in which the flexible structure is approximated as a set of discrete rigid snapshots obtained from simulations of dynamics of an empty framework, using either classical or, in principle, *ab-initio* methods. In the first method, MD simulations of diffusion are performed in a usual manner but by replacing the rigid structure with a new random snapshot at a certain characteristic frequency corresponding to the breathing motion of the window, while keeping positions of adsorbate molecules constant. In the second method, cage to cage hopping rates in each rigid snapshot are computed using TST and net diffusivity was obtained by taking the average of hopping rates over a distribution of snapshots. Excellent agreement is obtained between diffusivities predicted with these two new methods and direct MD simulations using fully flexible structures. Both methods are orders of magnitude more efficient than the simulations with the fully flexible structure. The new methods are broadly applicable for fast and accurate predictions of both infinite dilution and finite loading diffusivities of simple molecules in zeolites and other nanoporous materials, generally without the need for an accurate flexible force field.

(776c) Linear Driving Force for Shaped Adsorbents

Harriott, G., Air Products and Chemicals
Jin, B., Air Products and Chemicals
Originally presented on: 11/8/2013 9:00:00 - 9:15:00

The well-known Linear Driving Force mass transfer model represents a quasi-steady approximation of transient diffusion in which mass flux is proportional to the departure from saturation with transfer coefficient K. In the original derivation, K was determined for a spherical particle from an integral of the transient response to a step change in external concentration [1]. This elaborate approach can only be used to derive formulae for K in other shapes if an explicit representation of the transient exists. Although the value of K in a particular particle may be determined by numerical simulation and subsequent integration of the transient, repeated computations are necessary to indicate the effects of shape parameters. Consequently, ad-hoc shortcut methods have been applied to generate approximate formulae for K in terms of shape parameters based on assuming a spatial concentration profile within the particle (often quadratic).

Here we present a new route to the LDF model based on asymptotic analysis of the diffusion equation that does not require construction of the full transient solution and thereby facilitates mass transfer analysis and computation. We correct errors in mass transfer coefficients derived by shortcut methods and present exact solutions for a variety of simple shapes such as ellipsoids, cuboids, hollow cylinders and filled spheres as well as fast computations of K for complex shapes that are not amenable to analytical solution. The new formulation readily incorporates external mass transfer resistance, thermal coupling and multi-component sorption.

[1] Glueckauf, E., Theory of chromatography Part 10 – formulae for diffusion into spheres and their application to chromatography, Trans. Farad. Soc. **51**, 1540-1551 (1955).

(776d) Toward Efficient Prediction of Gas Self-Diffusivity in Porous Materials

Liu, Y., University of California, Riverside
Fu, J., University of California, Riverside
Wu, J., UC Riverside
Originally presented on: 11/8/2013 9:15:00 - 9:30:00

We present an efficient computation procedure for rapid prediction of gas self-diffusivity in nanoporous materials by combination of the Knudsen theory, Rosenfeld's entropy scaling method, and a classical density functional theory (DFT). The self-diffusivity conforms to the Knudsen diffusion at low molecular density and the effects of intermolecular interactions at higher density are accounted for with Rosenfeld's entropy scaling method. The excess entropy required in the scaling analysis is calculated from the classical DFT that is able to predict the adsorption isotherms quantitatively. The hybrid computational procedure has been calibrated with MD simulation for the adsorption of simple fluids such as H_2, He, Ne and CH_4 in metal organic frameworks. It predicts different types of diffusion curves in excellent agreement with simulation observations. The calculation time cost is within one minute, which is appropriate for practical application. The theoretical procedure is expected extendable to more complicated fluids in porous materials including polymers.

(776e) Predicting Mixture Diffusion in Zeolites: Accounting for Strong Correlations With a Simplified Theory and Examining the Role of Mixture Adsorption Thermodynamics

Rao, S. M., UOP LLC
Coppens, M. O., Rensselaer Polytechnic Institute
Originally presented on: 11/8/2013 9:30:00 - 9:45:00

The Maxwell–Stefan (MS) equations have been extensively used to describe mixture diffusion in zeolites. Correlation effects can arise due to molecules unsuccessfully attempting to pass each other in the very narrow pores of the zeolite, e.g., at high loadings of adsorbed molecules. In an extreme situation, diffusion may even become anomalous, giving rise to a single-file phenomenon. These correlation effects are captured in the so called "exchange terms" within the framework of the MS equations. The exchange terms are often calculated from pure component data using empirical approaches with different fits for different combinations of zeolite hosts and molecules. These computations, in turn, can be fairly tedious, especially for large multicomponent systems. It is therefore advantageous to be able to use a simplified theory to capture the effect of strong correlations. To this end, the MS equations are simplified to predict mixture diffusion behavior in the case of strong correlation effects. Intermolecular exchange coefficients are not required. No empirical relations are employed. This simplified model is used to predict the diffusion behavior in the separation of a CH_4-C_3H_8 mixture through a silicalite membrane.[1]

The calculation of mixture thermodynamic correction factors in the MS equations can also pose a computational hurdle especially for large systems, and if theories such as the Ideal Adsorbed Solution Theory (IAST) have to be employed. It is expected that, if a vacant adsorption site is available, the adsorption of a species inside the zeolite pore should be dictated by the pure component adsorption isotherm alone, and that mixture adsorption thermodynamics should not play a role, or else correlation effects should be present in adsorption as well, in which case theories such as the IAST cannot be used. To this end, the MS equations are simplified to employ only the pure component thermodynamic correction factors. This simplified model was used to predict mixture diffusion for two cases of practical interest: [1] (i) the separation of a N_2-CH_4 mixture through a SAPO-34 membrane, and (ii) the alkylation of C_6H_6 with C_2H_4 over H-ZSM-5.

In all cases, predictions from the simplified models are favorably compared to results from the more complex MS equations or experiments, indicating that the simplified models can serve as good starting approximations, and can also speed up mixture diffusion calculations in zeolites.

References:

1. Rao, S. M.; Coppens, M.-O. J. Phys. Chem. C 2012, 116 (51), 26816-26821

(776f) Direct Simulations of Molecular Transport and Permeation of Gases in Polymeric Membrane Materials for Separation Processes

Frentrup, H., Imperial College London
Hart, K. E., Pennsylvania State University
Colina, C. M., Pennsylvania State University
Muller, E. A., Imperial College London
Originally presented on: 11/8/2013 9:45:00 - 10:00:00

The microscopic characteristics of highly permeable polymers depend strongly on material processing conditions. Membranes made with these materials suffer from physical ageing, which causes their suitability for separation processes to rapidly deteriorate over time. In contrast, Polymers of Intrinsic Microporosity (PIM) owe their microporosity to a rigid and contorted molecular structure which provide them with increased and stable free volumes and high surface areas making them ideally suitable as membrane polymers [1, 2]. The permeability and selectivity of these polymer membrane is further influenced by the polymer's pore size distribution and pore connectivity [3]. Molecular models of PIMs allow a detailed investigation of these properties by providing a direct insight into the polymer's microporous structure [4, 5]. A molecular understanding of the transport phenomena in polymeric membrane materials at the nanoscale is required for further advances in membrane separations.

Molecular dynamics studies of gas permeation for three industrially relevant gases, namely helium, methane and carbon dioxide, through PIMs are presented. Helium, being an inert and light penetrant, exhibits a mobility which is almost two orders of magnitude higher than that of large penetrants such as methane and strongly-adsorbing carbon dioxide. As the diffusion of large penetrant molecules is very slow [6], obtaining transport properties from equilibrium molecular dynamics simulations for these molecules is extremely difficult since the necessary time scales are hardly accessible for traditional approaches.

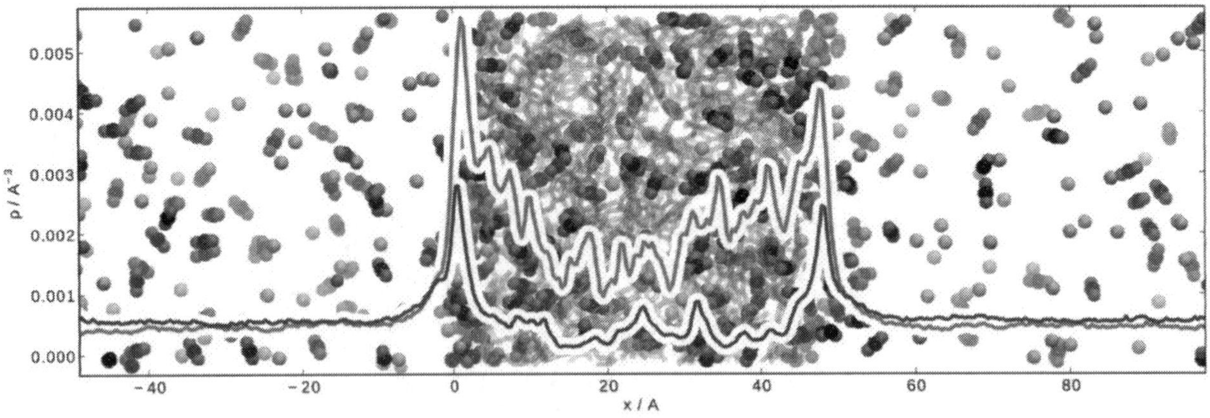

Figure 1: Snapshot of a simulation of a gas mixture methane (blue) and carbon dioxide (red-black) at T=293K, bulk pressure p=10bar, and a bulk composition x=0.5, in direct contact with an ultra-thin polymer layer. The enhanced adsorption of CO_2 in the polymer membrane is evident in the overlayed density profile. The flux (not shown) of CO_2 is however an order of magnitude smaller.

In response to these issues, the direct simulation of gas permeation through an atomistically-detailed ultra-thin PIM-1 membrane was performed with a highly efficient boundary-driven non-equilibrium molecular dynamics approach [7], mimicking the experimental procedure for determining gas diffusion in polymers. The methodology is not limited by model complexity, system size or other computational difficulties. The method employed here avoids these issues by applying an external perturbation to the system which accelerates the diffusion process, imposing a concentration gradient developing a steady-state response. Process relevant properties, namely permeation and solubility are obtained directly from the measured gas flux and density distribution. It uniquely allows for a direct comparison of permeation and solubility selectivity for pure gases with experimental studies on PIM-1 polymer membranes, as well as the simulation of mixed gas streams [8].

The steady-state nature of the methodology allows for an accumulation of simulation statistics, reducing the error in the results. The NEMD methodology ensures that the properties computed in non-equilibrium simulations are in the linear response regime and thus amenable to extrapolation in order to extract the underlying transport coefficients. Furthermore, the methodology is not limited to homogeneous systems. Discontinuities in the microscopic topology can be treated in a straight-forward manner. The results show that differences in the microscopic properties have a remarkable influence on separation characteristics of PIMs. This has decisive implications for gas separations, considering that, while high permeation is desirable, it often goes at the expense of reduced selectivity.

References

[1] P. Budd et al., *J. Mem. Sci.*, **251**, 263-269 (2005).
[2] G. S. Larsen, P. Lin, K. E. Hart, and C. M. Colina, *Macromolecules*, **44**, 6944-6951 (2011).
[3] P. Budd et al., *J. Mem. Sci.*, **325**, 851-860 (2008).
[4] K. E. Hart, L. J. Abbott, and C. M. Colina, *Mol. Sim.*, **39**, 397-404 (2013).
[5] L. Sarkisov and A. Harrison, *Mol. Sim.*,**37**, 1248-1257 (2011).
[6] S. Neyertz and D. Brown, *Macromolecules*, **41**, 2711-2721 (2008).
[7] H. Frentrup, E. A. Müller et al., *Mol. Sim.*, **38**, 540-553 (2012).
[8] S. Thomas et al., *J. Mem. Sci.*, **333**, 125-131 (2009)

(776g) CO2-Propane Mixtures in Silica Pores: Relation Between Structure and Dynamics

Le, T.,
Phan, A. T., The University of Oklahoma
Striolo, A., University of Oklahoma
Cole, D. R., Oak Ridge National Laboratory
Originally presented on: 11/8/2013 10:00:00 - 10:15:00

We report here equilibrium molecular dynamics simulation results obtained for mixtures of carbon dioxide and propane confined in narrow slit-shaped silica pores. A wide range of thermodynamic conditions (from sub-critical to super-critical) has been considered, as well as a wide range of fluid compositions. The pore width was changed to quantify the effect of confinement on the results, and the pore surface properties were changed to modulate the preferential pore-fluid interactions. We observed evidence for preferential adsorption of carbon dioxide near the protonated –OH groups on the surface, which yields slow self-diffusion coefficients at low loadings. Non-monotonic changes in the self-diffusion coefficients as a function of pore loading are observed, and explained, tentatively, based on the competitive adsorption of the various fluids near the pore surfaces. We will discuss how these predictions compare with experimental NMR results. The results could be important for designing separation devices, and also for better understanding the behavior of fluids in sub-surface environments.

(776h) Assessing the Performance of Microfibrous Entrapped Sorbent Enhanced Gas Phase Filter in a Full Scale Test Rig

Xu, G., Auburn University
Tatarchuk, B., Auburn University
Originally presented on: 11/8/2013 10:15:00 - 10:30:00

A full scale test rig was designed and constructed based on the ASHREA Standard 52.2 to systematically assess the performance of the filters in terms of pressure drop, gas life, single pass removal efficiency, etc. Pre-qualification exams including the air leakage, face velocity uniformity, challenge gas uniformity were conducted to verify the reliability of the test rig. The filters were subjected to a 5ppm hexane gas and operated at 200fpm face velocity. For a commercial carbon filter (24"x24"x1/2", 2kg carbon), the initial removal efficiency was only 60%, the 50% breakthrough time was 150min, and the pressure drop was as high as 0.8" W.G. The slow mass transport of large particle size and the severe channeling and bypass flow were considered as the main reason of the low efficiency of filter. To improve the performance of the filters, a novel multi-element structured array (MESA) and composite bed design was developed. The MESA can significantly increase the carbon load and available media area. For a W shape MESA, the carbon load increased to 8kg per unit, and the initial removal efficiency increase to about 80%, while the 50% breakthrough time increased to about 1000min under the same operation condition. Meanwhile, the composited bed applied microfibrous entrapped sorbent (MFES) as a polishing layer to remove the penetrated or by-passed gas from the preceding bed, increasing the initial remove efficiency from 60% to more than 99%. The combination of MESA and composite bed provides a solution to deliver clean air with long protection period. This is very important to protect the power system such as fuel cells stack in the naval ships and stationary power plant.

(776i) Kinetic Modeling of CO2 Adsorption On Pore-Expanded MCM-41

Loganathan, S., Indian Institute of Technology Guwahati
Tikmani, M., Indian Institute of Technology Guwahati
Ghoshal, A. K., Indian Institute of Technology Guwahati
Originally presented on: 11/8/2013 10:30:00 - 10:45:00

Adsorption equilibrium and kinetics are critical parameters to assess the performance of an adsorbent. Equilibrium sorption studies are important in determining the efficiency of an adsorbent. The kinetic analysis allows the determination of the residence time required for completion of adsorption process. In a dynamic process such as adsorption in a fixed bed column, efficiency of an adsorbent and its capacity to withstand large adsorbate flows are associated with its rate of adsorption. Therefore, from a practical standpoint, it is important to characterize the rate-controlling steps for an adsorption process. More recently, our research group fabricated a pore-expanded MCM-41 (**A**) for CO_2 capture application. In the current work, we investigated the kinetics of CO_2 adsorption on **A**. Kinetic measurements of pure CO_2 in **A** sample at four different temperatures (30, 45, 60, and 75 ºC) were performed in the pressure range of 1-11 bar, using a Rubotherm magnetic suspension balance. To analyze the adsorption rate behavior of CO_2 on **A** under different pressure and temperature condition, we considered both pseudo first-order kinetic model and the pseudo second-order kinetic model to obtain the overall mass transfer coefficient. The pseudo-second order model failed to accurately fit the experimental data whereas pseudo-first order model provided a pretty good agreement between the measured and predicted data throughout the entire range of pressure and temperature analyzed. To further investigate the mechanism of CO_2 adsorption on **A**, intra particle diffusion and Boyd's models were considered. A conclusion regarding mass transfer mechanism is derived from the information provided by both the models. It was found that external mass transport mainly governed the rate-limiting process of adsorption of CO_2 on **A** at low pressure conditions. Interestingly, it was also interpreted that pore diffusion becomes more important at higher pressure conditions.

(776j) Adsorption Equilibrium and Mass Transfer Rates for Argon, Oxygen, and Nitrogen in a Carbon Molecular Sieve

Mitchell, L. A., Vanderbilt University
Giesy, T. J., Vanderbilt University
LeVan, M. D., Vanderbilt University
Originally presented on: 11/8/2013 10:45:00 - 11:00:00

Separation of oxygen from air using zeolite-based PSA processes usually results in a product with a 5% argon impurity caused by the similarity of oxygen and argon isotherms on most zeolites. It has been suggested that carbon molecular sieve (CMS) materials could be used to further purify this oxygen product, as uptake rates of oxygen and argon in CMS differ by more than an order of magnitude. Furthermore, CMS materials have already proven to be effective adsorbents for production of nitrogen from air by PSA. Thus, there exists sufficient motivation to study adsorption and transport of atmospheric gases on these materials.

In this work, we present our findings from adsorption equilibrium and adsorption rate studies of argon, oxygen, and nitrogen on Shirasagi MSC-3R type 172, a CMS material that is relatively uninvestigated in the literature. Pure component excess adsorption isotherms have been measured using a volumetric method at various temperatures and at pressures as high as 100 bar. Due to very slow uptake rates in these systems, a transport model was derived to determine final equilibrium loadings. Pure excess isotherms were modeled using the Toth equation. Adsorption rate experiments were performed using various frequency response techniques. Pure component transport mechanisms and parameters were determined at pressures up to 1 bar. We show the effect of pressure on the transport coefficients and discuss the possible implications of this dependence. The effect of a second gas on pure component transport parameters is investigated by performing frequency response experiments on binary gas systems.

(777a) Ion Transport in Anion Exchange Membranes for Water Purification and Power Generation Applications

Geise, G. M., The University of Texas at Austin
Hickner, M. A., The Pennsylvania State University
Logan, B. E., Pennsylvania State University
Originally presented on: 11/8/2013 8:30:00 - 8:55:00

The global need for sustainable water and energy sources presents an opportunity for polymer science as many of the membrane-based technologies used to purify water [e.g., reverse osmosis (RO), forward osmosis (FO), and electrodialysis (ED)] and emerging membrane-based power generation technologies [e.g., pressure-retarded osmosis (PRO) and reverse electrodialysis (RED)] rely on polymeric membranes to control rates of water and ion transport. In ED and RED processes, ion exchange membranes are used to restrict the transport of either cations or anions across the membrane while simultaneously allowing passage of oppositely charged species. Currently, the performance and understanding of anion exchange membranes (AEMs), which contain positively charged functional groups such as quaternary ammonium moieties, lags behind that of cation exchange membranes (CEMs), which contain tethered negative charges such as sulfonate moieties. Fundamental structure-transport property relations for AEMs are therefore needed to optimize and extend technologies such as ED and RED. The electrical potential-driven ionic resistance and permselectivity of AEMs based on quaternary ammonium-functionalized poly(sulfone), poly(phenylene oxide), and poly(styrene) backbones were measured to study the influence of degree of functionalization, water sorption, polymer backbone polarity, and morphology on ion transport properties. The polymer films were characterized during immersion in two different aqueous solutions containing either sodium chloride or ammonium bicarbonate, as this thermolytic salt is of interest for RED, and these polymers were compared to commercially available membranes. Ion transport properties measured using each of these electrolytes were analyzed with regard to ion hydration and dilute solution mobilities of the charge carrying ion. For many polymers, the relative rates of ion transport for different electrolytes can be described in terms of the dilute solution mobilities of the charge-carrying ions. For some highly functionalized polymers, however, dilute solution mobilities do not provide a reliable estimate of the intrinsic resistance of a polymer for a given electrolyte based on the intrinsic resistance measured using a different electrolyte. This phenomenon is rationalized by considering the nature of the charge carrying ion, water sorption of the polymer, and polymer morphology.

(777b) Surface Modification of Forward Osmosis Membranes for Improved Fouling Resistance

Romero-Vargas Castrillon, S., Yale University
Lu, X., Yale University
Shaffer, D., Yale University
Elimelech, M., Yale University
Originally presented on: 11/8/2013 8:55:00 - 9:20:00

Forward Osmosis (FO) utilizes the osmotic pressure difference between two solutions of different concentration to drive water permeation across a semipermeable membrane. As an emerging membrane technology, FO holds promise in a variety of applications, including seawater desalination, wastewater reclamation and osmotic membrane bioreactors. Despite significant advances in FO membrane development, membrane fouling, due to adsorption of organic molecules on the membrane surface, remains a major technical problem, decreasing process performance and membrane life. Here we report the fabrication, characterization and testing of thin film composite (TFC) FO membranes containing polyethylene glycol (PEG) surface grafts for improved fouling resistance. The membranes comprise a microporous polysulfone support over which a polyamide selective layer is synthesized by interfacial polymerization. To facilitate PEG grafting, a second interfacial polymerization is carried out between ethylene diamine and acyl chloride groups on the nascent polyamide layer, yielding a selective layer rich in primary amine groups. The resulting amine-rich active layer is functionalized with PEG diglycidyl ether, whose epoxide groups readily react with primary amines. Surface characterization by ATR-FTIR spectroscopy and zeta-potential measurements shows the presence of PEG and amine groups on the membrane surface, and contact angle measurements demonstrate that the PEGylated active layer is more hydrophilic than that of control polyamide membrane surfaces (whose active layer does not contain a grafted PEG layer). We report dynamic fouling experiments using alginate as model organic foulant, showing that membrane fouling is significantly reduced in the case of PEGylated membranes, due to the presence of a PEG barrier that hinders foulant adsorption. Finally, we use AFM adhesion force measurements to demonstrate that the interaction energy between a carboxylated colloidal probe and the PEGylated membrane is significantly weakened as compared to the control polyamide surface.

(777c) Polydopamine Modified Commercial Thin Film Composite Membranes for Forward Osmosis: Flux and Desalination Performance

Arena, J. T., University of Connecticut

Manickam, S. S., University of Connecticut

Reimund, K. K., University of Connecticut

Freeman, B. D., The University of Texas at Austin

McCutcheon, J. R., University of Connecticut

Originally presented on: 11/8/2013 9:20:00 - 9:45:00

Forward osmosis is a water treatment technology seen a tool to aid in the treatment, desalination, and purification of feed waters often unsuitable for other pressure driven membrane separation processes[1-3]. Until recently there was a single commercially available membrane for use in forward osmosis processes; the cellulose triacetate manufactured and sold by Hydration Technology Innovations (HTI^{TM}). Being made from cellulose acetate this membrane is vulnerable to hydrolysis under alkaline conditions, reducing membrane selectivity. In recent years commercial and academic interest has focused on developing more chemically resilient forward osmosis membranes. Thin film composite (TFC) membranes have garnered great interest as an alternative platform for forward osmosis.

TFC comprise the vast majority of commercial reverse osmosis and nanofiltration membranes being used for such purposes over the last 30 years. TFC membranes are built upon a three tiered structure comprising of an ultra-thin aromatic polyamide layer supported by a polysulfone (PSu) or polyethersulfone (PES) layer that is cast onto a polyester (PET) nonwoven. Early work using TFC membranes in FO by McCutcheon found the performance of TFC RO membranes to be inferior to that of CA FO membranes[4]. Later work found, lack of support layer wetting, leading to severe internal concentration polarization (ICP) and hindered osmotic flux due to a reduced effective porosity[5]. A possible solution to this problem is to make a membrane with an intrinsically hydrophilic support; however, this would require a retuning of the delicate interfacial polymerization process, which can be impacted by the support layer properties. An alternative is to take an existing membrane that has been optimized for superior permselectivity and modify the support layer to increase its hydrophilicity.

Later work by Arena utilized polydopamine (PDA) for TFC membrane *support* modification for improved hydrophilicity observing a significant improvement in the water flux of PDA modified TFC RO membranes in the pressure retarded osmosis (PRO) orientation with no transmembrane pressure[6]. The work presented here is a follow up on that work finding similar performance improvements with these same PDA modified membranes in the forward osmosis (FO) orientation. Additionally these membranes were tested under desalination conditions using the ammonia-carbon dioxide (NH_3-CO_2). Water and ion fluxes were measured from these tests and found unequal rates of cation and anion transport through these membranes.

Additionally water and ion fluxes for NH_3-CO_2 draw solution with varying ammonia to carbon dioxide ratios were studied using commercial membranes, seeking to gain better understanding on the effects of varying ammonia to carbon dioxide ratios on membrane water and ion fluxes.

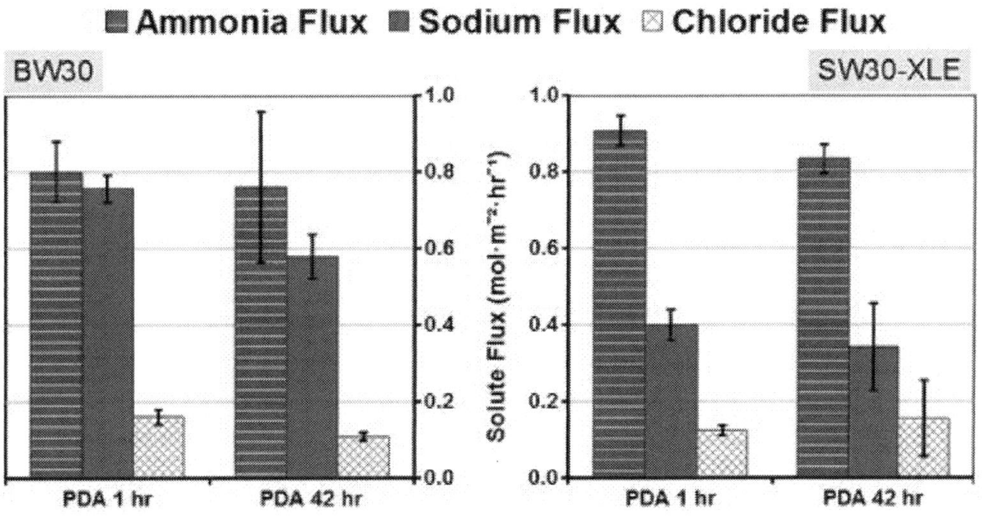

Figure 1: Solute fluxes for osmotically driven sodium chloride rejection. The lined bar represents ammonia species reverse solute fluxes, the solid bar represents sodium ion forward solute fluxes, and the cross-hatched bard represents chloride ion forward solute fluxes at 23±1 °C, 0.25 m/s draw and feed cross-flow velocity.

Works Cited

[1] McGinnis RL, Hancock NT, Nowosielski-Slepowron MS & McGurgan GD. Pilot demonstration of the NH_3/CO_2 forward osmosis desalination process on high salinity brines. *Desalination* (2013) **312**: pp. 67-74.

[2] Cath TY, Childress AE & Elimelech M. Forward osmosis: principles, applications, and recent developments. *Journal of Membrane Science* (2006) **281**: pp. 70-87.

[3] Hancock NT, Black ND & Cath TY. A comparative life cycle assessment of hybrid osmotic dilution desalination and established seawater desalination and wastewater reclamation processes. *Water research* (2012) **46**: pp. 1145-1154.

[4] McCutcheon JR, McGinnis RL & Elimelech M. A novel ammonia-carbon dioxide forward (direct) osmosis desalination process. *Desalination* (2005) **174**: pp. 1-11.

[5] McCutcheon JR & Elimelech M. Influence of membrane support layer hydrophobicity on water flux in osmotically driven membrane processes. *Journal of Membrane Science* (2008) **318**: pp. 458-466.

[6] Arena JT, McCloskey BD, Freeman BD & McCutcheon JR. Surface modification of thin film composite membrane support layers with polydopamine: enabling use of reverse osmosis membranes in pressure retarded osmosis. *Journal of Membrane Science* (2011) **375**: pp. 55-62.

(777d) Novel Hydrophilic Nylon 6,6 Supported Thin Film Composite Membranes for Engineered Osmosis

McCutcheon, J. R., University of Connecticut
Huang, L., University of Connecticut
Originally presented on: 11/8/2013 9:45:00 - 10:10:00

Clean water and energy are essential for public health and economic prosperity. Recent studies show that Engineered Osmosis (EO) may be a viable technology for desalination, water and wastewater treatment, and power generation. However, the absence of a membrane designed for such processes is a significant obstacle hindering further advancements of this technology. Conventional thin film composite (TFC) reverse osmosis membranes yield poor water flux in EO due to their thick and hydrophobic polysulfone (PSu) support layers resulting in internal concentration polarization (ICP). In this study, a new class of thin film composite membranes based on hydrophilic nylon 6,6 supports has been designed for the first time. Two different support structures of nylon 6,6 have been employed: one is conventional phase-inversion casting microfiltration support provided by 3M®, another is lab-made electrospun nanofiber mat. The ultra-thin polyamide (PA) selective layer has been built via *in situ* interfacial polymerization.

Results show that defect-free polyamide selective layer was able to form on both support structures and its permselectivity can be carefully controlled by manipulating the monomer concentration, varying the support pore size or adding co-solvent into the organic phase (hexane). Osmotic flux tests revealed that both the nylon 6,6 cast film and the electrospun nanofiber supported TFC membranes exceeded the performance of commercial forward osmosis membrane from Hydration Technology Innovations® (HTI). The cast-film supported TFC showed matched water flux with 10X lower reverse salt flux than HTI-CTA membrane largely due to the high permselectivity of the TFC. The flux performance was further improved by nanofiber supported TFC membrane, which showed 1.5X water flux than HTI-CTA membrane in PRO mode and 2X water flux in FO mode. The high osmotic flux of the nanofiber supported TFC, as well as the small flux difference between PRO mode and FO mode, can be attributed to its favorable support structure (low thickness, high porosity, low tortuosity and hydrophilicity) that helps to mitigate ICP, as indicated by its low structural parameter of only 200 microns. This is one of the lowest structural parameter that has been ever reported.

In general, this study demonstrate the new concept of using hydrophilic nylon 6,6 to substitute conventional PSu as the support for TFC EO membranes. This new class of nylon 6,6 based TFC membranes supported by conventional cast film or electrospun nanofibers would enable application of EO in seawater desalination and power generation.

(780a) Continuous Enzymatic Hydrolysis of Biomass in a Membrane-Assisted Reactor

Malmali, M., University of Arkansas
Stickel, J., National Renewable Energy Laboratory
Wickramasinghe, S. R., University of Arkansas
Originally presented on: 11/8/2013 8:30:00 - 8:55:00

Although enzymatic hydrolysis of lignocellulosic biomass is usually conducted in batch mode, continuous operation is desirable for a number of reasons. Higher overall investment costs due to larger reactor volumes, higher running costs due to frequent startup/shut down, one-time use of enzymes as well as enzyme separation costs are disadvantages of batch operation. Recovery and reuse of the enzymes during enzymatic hydrolysis of biomass is attractive as the enzyme cost is a major economic constraint. In addition, high sugar concentrations are desirable for subsequent fermentation steps, but obtaining high sugar concentrations directly from batch enzymatic hydrolysis is limited by product inhibition of the enzymes and material handling of slurries at high-solids concentrations. Development of a continuous saccharification process with integrated membrane separations overcomes the limitations of batch operation. Membrane separation steps integrated with the enzymatic hydrolysis reaction could enable recovery and reuse of enzymes, as well as concentration of product sugars. Further capital cost will be lower due to smaller process stream volumes as a result of higher concentrations.

In this study we have designed and tested an integrated continuous membrane reactor. The integrated membrane reactor includes a microfiltration/ultrafiltration membrane for recovery and recycle of the enzyme as well as a nanofiltration membrane for concentration of the sugars. Here, we present results for our continuous integrated membrane reactor. These results are compared to a batch reactor. We show that glucose production rates and product sugar concentrations are higher for the integrated continuous membrane reactor.

(780b) Small-Scale Production of Ammonia From a Combined Reactor-Membrane-Adsorber

Himstedt, H., University of Minnesota
McCormick, A. V., University of Minnesota
Schmidt, L. D., University of Minnesota
Cussler, E. L., University of Minnesota
Originally presented on: 11/8/2013 8:55:00 - 9:20:00

Ammonia is one of the world's most important chemicals; over 130 million metric tons are produced annually. Without modern fertilizers derived from ammonia two billion people would starve. Ammonia is currently produced via the Haber-Bosch process in large capital-intensive factories. Despite a century of optimization, single-pass ammonia production is only roughly 20% and requires large amounts of sacrificial fossil fuels to obtain raw materials.

We present work on a hybrid membrane/adsorption reactor system capable of producing ammonia from wind power. Hydrogen and nitrogen are obtained from electrolysis of water and membrane separation of air, respectively. The ammonia produced can be separated by either a selective membrane or pressure-swing adsorption. We believe membrane separation is more effective for larger scale operation, but the simpler absorption is better for smaller scale. By combining reaction and absorption into a single unit operation, ammonia conversion can be increased. The system performs as a mixed-media membrane reactor without the potential physical limitations of a membrane structure. Such a system may make sustainable distributed production of ammonia possible.

(780d) Membrane Processes for Conversion of Solar Energy to Electricity

Lipscomb, G., University of Toledo
Patil, R., University of Toledo
Originally presented on: 11/8/2013 9:45:00 - 10:10:00

The world faces increasing demand for energy as the population grows and becomes more affluent. This demand poses significant challenges as the consumption of fossil fuel energy sources ultimately is not sustainable and has the potential to effect climate change.

The conversion of solar energy to electricity offers one alternative to fossil fuel use. Currently, solar energy is converted to electricity using both thermal and photovoltaic processes. We propose an alternative route that uses membrane processes.

The proposed process utilizes a combination of either reverse electro-dialysis or pressure retarded osmosis to mix two working fluids of different salinity to produce electricity. The salinity difference is restored by demixing using either membrane distillation or membrane crystallization. This demixing is driven by solar thermal energy. The overall process leads to a conversion of solar energy to electricity.

The capabilities of the proposed process are explored through simulation. The relationships between membrane transport properties, area requirements, and working fluid properties are determined from which electricity generation costs can be estimated.

(780e) Industrial Membrane Hybrid Driers

Campos, D.,
Nemser, S., Compact Membrane Systems, Inc
Pennisi, K. J.,
Majumdar, S., Compact Membrane Systems, Inc
Originally presented on: 11/8/2013 10:10:00 - 10:35:00

This paper describes a chemically and thermally resistant membrane system for the dehydration of organic solvents and chemicals. The hydrated chemical can be fed to the system either in the liquid or the vapor phase. The feed to the membrane is separated into an organic rich fraction in the retentate and a water rich fraction in the permeate. The solvent recovery depends on the permeability of the membrane to the solvent and the purity required in the solvent-rich fraction. Applications of this technology include (but are not limited to) drying organics that are difficult to dry by conventional technologies (e.g., azeotropes or solvents that have boiling points close to water), heat sensitive chemicals, reacting mixtures in which the chemical reaction is inhibited by the presence and/or generation of water. This technology is been used commercially for the successful dehydration of lubricating oils. Pilot scale test data are presented on the application of this technology for drying ethanol, IPA, THF, toluene and phosphate esters. The membrane system can be combined with conventional distillation to create a hybrid process for removing water from organics resulting in cost and energy savings. Examples of this hybrid drying process are presented along with the economic advantages derived from using the hybrid technology.

(780f) A Generic Approach of Hybrid Membrane / Condensation Process : Potentialities and Limitations

Belaissaoui, B., Laboratoire Reaction et Genie des Procedes LRGP- CNRS, Universite de Lorraine
Favre, E., Laboratoire Reaction et Genie des Procedes LRGP- CNRS, Universite de Lorraine
Originally presented on: 11/8/2013 10:35:00 - 11:00:00

In numerous industrial applications, gaseous streams containing an inert gas and a condensable compound have to be treated by a separation process. A direct condensation strategy is often applied for that purpose, but it requires in certain cases a cryogenic operation (e.g. a liquid nitrogen cold box), typically when the target condensor temperature is lower than –30 C. This situation occurs for low boiling compounds and/or when a low residual concentration of the condensable compound is necessary. In that event, an alternative approach based on hybrid membrane condensation process can be proposed: the principle consists to achieve a pre-concentration of the condensable compound by membrane separation (vapor permeation), in order to enable condensation to occur on the permeate side of the membrane at a higher temperature. This concept has been already succesfully applied for Volatile Organic Compounds (VOC's) recovery from air [1-4], where increasingly stringent regulatory requirements make recovery of VOC's of utmost importance. Nevertheless, the hybrid membrane condensation concept could possibly be of interest for other industrial applications such as drying of natural gas, hydrocarbon removal from hydrogen purification, ammonia recovery or carbon post combustion capture [5].

In this study, a generic approach of a hybrid membrane/condensation process has been developped and will be compared to a direct condensation strategy, taken as the baseline. It will be shown that, based on only two key parameters, namely the condensable compound solubility parameter and the membrane selectivity, the potential interest of the hybrid approach can be rapidly evaluated. The Hildebrand solubility parameter includes indeed the cohesive energy of the condensable compound and its molar volume; this property can be related to the condensation temperature, the heat of condensation or the solubility of a compound in an elastomeric membrane.

A large range of solubility parameter data, covering gases (CO_2, NH_3), VOC's (hexane, dicholoroethane, acetone or vinyle chloride) and water has been investigated through a series of simulations of the hybrid process. The influence of different membrane process parameters, e.g., recovery ratio, pressure ratio, stage cut has been carried out. The condensation temperature has been evaluated for each set of conditions and each compound, and, from this, the corresponding overall energy requirement of the process is estimated.

As a major outcome of this study, a generic process selection map will be presented. Based on the solubility parameter and membrane selectivity, the potentiality of the hybrid membrane/condensation process to achieve attractive performances comparing to the standalone condensation process is analysed. Heuristic rules and recommandations leading to an increased energy efficiency of the hybrid membrane condensation process covering diffrents solvent/inert/membrane systems will finally be proposed.

References

[1] Richard W. Baker, Noriaki Yoshioka, Judith M. Mohr, Amy J. Khan, Separation of organic vapors from air, Journal of Membrane Science, Volume 31, Issues 2–3, May 1987, Pages 259-271.

[2] H. Paul, C. Philipsen, F.J. Gerner, H. Strathmann, Removal of organic vapors from air by selective membrane permeation, Journal of Membrane Science, Volume 36, 1988, Pages 363-372.

[3] K. Kimmerle, C.M. Bell, W. Gudernatsch, H. Chmiel, Solvent recovery from air, Journal of Membrane Science, Volume 36, 1988, Pages 477-488.

[4] M. Leemann, G. Eigenberger, H. Strathmann, Vapour permeation for the recovery of organic solvents from waste air streams: separation capacities and process optimization, Journal of Membrane Science, Volume 113, Issue 2, 15 May 1996, Pages 313-322.

[5] B. Belaissaoui, D. Willson, E. Favre, Membrane gas separations and post-combustion carbon dioxide capture: Parametric sensitivity and process integration strategies, Chemical Engineering Journal 211–212 (2012) 122–132.

(784a) Characterizing the Gas Transport and Mechanical Properties of Thermally Rearranged (TR) Polymers

Liu, Q., The University of Texas at Austin
Paul, D. R., The University of Texas at Austin
Freeman, B. D., The University of Texas at Austin
Originally presented on: 11/8/2013 8:30:00 - 8:55:00

Polyimides containing ortho-positional functional groups, such as hydroxyl and acetate groups, undergo thermal rearrangement to form polybenzoxazole structures upon heating between 350 and 450°C. These thermally rearranged (TR) polymers show excellent gas transport properties and have strong resistance to plasticization. In this study, a series of TR homopolymer and copolymer were prepared with polyimide precursors based on HAB (3,3'-dihydroxy-4,4'-diaminobiphenyl), APAF (2,2'-bis (3-amino-4-hydroxyphenyl)-hexafluoropropane), and 6FDA (2,2'-bis-(3,4-dicarboxy-phenyl) hexafluoropropane dianhydride). Pure gas permeabilities of H_2, N_2, O_2, CH_4, and CO_2 were measured over pressure ranging from 3 atm to 17 atm at 35°C. Upon thermal rearrangement, APAF-6FDA has the best gas transport properties, presumably due to an enlarged cavity size and distribution of the TR polybenxazole structure. In addition, this study systematically documents the mechanical properties of the polyimides and TR polymers. HAB-6FDA polyimide has 10% elongation-at-break, whereas APAF-6FDA polyimide has only 4% elongation-at-break.

(784b) Vapor Permeation of Ethanol and Water in Thermally Rearranged (TR) Polymers

Gleason, K. L., The University of Texas at Austin
Paul, D. R., The University of Texas at Austin
Freeman, B. D., The University of Texas at Austin

Originally presented on: 11/8/2013 8:55:00 - 9:20:00

In this work, we present a high temperature permeation study of water and ethanol vapor in a thermally rearranged polyimide synthesized from 3,3'-dihydroxy-4,4'-diamino-biphenyl (HAB) and 2,2'-bis-(3,4-dicarboxyphenyl) Hexafluoropropane Dianhydride (6FDA). Polyimide films were converted to polybenzoxazole (PBO) by heating at 450 °C for 30 minutes. Under these conditions, nearly 100% conversion to PBO occurred. Chemical resistance of the PBO was investigated by exposing film samples to $50\%_{wt}$ ethanol/water at 120 °C and 3 atm for two weeks. Light gas permeability and selectivity was measured before and after exposure, and FTIR-ATR was used to probe changes to the chemical structure. Samples of the polyimide precursor exposed in this way were severely degraded, but fully converted PBO samples were little affected. Permeation tests involving water and ethanol vapors were performed at temperatures ranging from 110-190 °C and at pressures up to 6.4 atm. At 110 °C, water vapor permeability exceeded 3000 Barrer, and water/ethanol selectivity exceeded 60. Water vapor permeability was found to decrease with increasing temperature, indicating that water sorption was affected more strongly by the changing temperature than was diffusivity. On the other hand, ethanol permeability changed very little with temperature, indicating that the changes to sorption and diffusivity were more balanced for ethanol. For elevated pressures, mild plasticization effects were apparent, causing permeability to increase and selectivity to decrease with pressure.

(784c) Synthesis and Gas Transport of Thermally Rearranged Polyimides for Applications in Gas Separations

Smith, Z. P., The University of Texas at Austin
Sanders, D. F., The University of Texas at Austin
Guo, R., University of Notre Dame
McGrath, J. E., Virginia Tech
Paul, D. R., The University of Texas at Austin
Freeman, B. D., The University of Texas at Austin
Originally presented on: 11/8/2013 9:20:00 - 9:45:00

Over the past several years, research on thermally rearranged (TR) polyimides has shown interesting separation properties for a number of gas separations such as CO_2/CH_4 and hydrogen separations. TR polymers are traditionally synthesized via a solid state reaction of polyimides that contain reactive groups *ortho*-functional to the polyimide diamine. The precursor polyimides can be synthesized via a number of pathways, including chemical and thermal imidization. Furthermore, polybenzoxazole structures similar to those formed from polyimides can be synthesized by polyamides. The scope of this work is address the differences in gas transport properties for H_2, N_2, O_2, CH_4, and CO_2 as they relate to polybenzoxazole backbone structure, *ortho*-position reactive group, and synthesis of polybenzoxazoles from either polyimides or from polyamides.

Three *ortho*-hydroxy polyimides were synthesized via thermal imidization. The diamines for synthesizing these polyimides were 2,2-bis (3-amino-4-hydroxyphenyl)-hexafluoropropane (APAF), 3,3'-dihydroxy-4,4'-diamino-biphenyl (HAB), and the dianhydrides were 2,2'-bis-(3,4-dicarboxyphenyl) hexafluoropropane dianhydride (6FDA), and 4,4'-oxydiphthalic anhydride (ODPA). From these monomers, HAB-6FDA, APAF-6FDA, and APAF-ODPA were synthesized and converted to TR polymers by heating at 450°C for 30min. Conversion under these conditions resulted in nearly complete thermal rearrangement as determined by sample mass loss. The permeation properties of the precursor polyimides and TR polymers were determined for H_2, N_2, O_2, CH_4, and CO_2. Out of all of the samples considered, the APAF-6FDA polyimide and the APAF-6FDA TR polymer had the most promising transport properties out of all of the polyimides and all of the TR polymers, respectively. These samples had permeabilities near or above the 2008 Robeson upper bound for CO_2/CH_4 and H_2/CH_4separation.

A critical analysis on the effect of transport properties as they relate to the *ortho*-position leaving group was evaluated for HAB-6FDA. An *ortho*-hydroxy HAB-6FDA sample was modified via an esterification reaction post-imidization. This esterification reaction added different reactive functional groups to the *ortho*-position of the polyimide. In addition to the *ortho*-hydroxy group, an acetate, propanoate, and pivalate group were added to the polyimide backbone structure. These samples were thermally rearranged to different extents of conversion, and gas permeation was tracked as these samples were converted. Samples with bulkier leaving groups showed more dramatic improvements to gas permeability than *hydroxyl*-functional polyimides.

Finally, a polyamide sample of APAF-6FDA was prepared for gas transport properties. This sample can be converted into a polybenzoxazole via a condensation reaction at 350°C for 60min. The APAF-6FDA polyamide had consistently lower permeability than the APAF-6FDA polyimide. Furthermore, the polybenzoxazole synthesized from the polyamide had consistently lower permeability than the polybenzoxazole synthesized from the polyimide. It is believed that these differences in transport properties relate to the non-equilibrium nature of polybenzoxazoles prepared from different synthetic pathways.

(784d) Composite Hollow Fiber Membranes for Post Combustion CO2 Capture

Rowznaghi, A. A., Georgia Institute of Technology
Koros, W. J., Georgia Institute of Technology
Bhandari, D.,
Originally presented on: 11/8/2013 9:45:00 - 10:10:00

Composite Hollow Fiber Membranes for Post Combustion CO_2 Capture

Ali A. Rownaghi[1], William J. Koros[1], Dhaval Bhandari[2]

[1]School of Chemical and Biomolecular Engineering, 311 Ferst Dr. NW, Atlanta, GA 30332

[2]GE Global Research Center, 1 Research Cir, Niskayuna, NY 12309

Fabrication of composite hollow fiber membranes with a thin polymeric permselective layer coated on a highly porous substrate for gas separation applications is attractive to provide both high flux and selectivity. The selective layer and the highly porous substrate described in this work are made from a combination of organic/inorganic materials. High porosity polyamide-imide (PAI) hollow fiber substrates were fabricated and characterized by scanning electron microscopy. The porous substrates were then coated with poly[bis-(2-(2-methoxy) ethoxy)ethoxyphosphazene] (MEEP) to create a permselective skin layer. The fabricated membrane performance was verified by permeation tests with He, N_2 and CO_2 at 35 Â°C and 30 psi. The membranes were found to have CO_2/N_2 selectivity up to 37. The increase in CO_2/N_2 selectivity for some membranes was accompanied by an increase in CO_2 permeance.Â

Acknowledgment

The material described in the abstract is based upon work supported by the Department of Energy National energy Technology Laboratory (DOE-NETL) under award number DE-FE0007514.

Disclaimer

(784e) Exploring Light Gas Transport in Carbon Molecular Sieve Membranes Using Diffusion NMR

Mueller, R., University of Florida
Kanungo, R., University of Florida
Kiyono-Shimobe, M., Georgia Institute of Technology
Koros, W. J., Georgia Institute of Technology
Vasenkov, S., University of Florida
Originally presented on: 11/8/2013 10:10:00 - 10:35:00

Carbon molecular sieve (CMS) membranes, microporous membranes fabricated from pyrolysis of polymeric precursors, are promising materials for energy-efficient separations of light gases. The understanding and prediction of transport properties of CMS membranes from only macroscopic observations is a very challenging task due to the complex amorphous pore structure of CMS membranes. Here we report results of an application of a pulsed field gradient (PFG) NMR method for microscopic diffusion studies in CMS membranes. In our multinuclear PFG NMR studies we combine the advantages of using large magnetic field gradients and high magnetic field. This enables obtaining reliable diffusion data even under the conditions of small sorbate diffusivities and short NMR relaxation times. As a result, PFG NMR measurements could be performed for diffusion inside CMS membranes on microscopic length scales of displacements under well-defined, uniform sorption loading and temperature conditions. The results of these measurements will be presented and compared with the previously reported macroscopic diffusion data.

(784f) ZIF/Polyimide Asymmetric Mixed-Matrix Hollow Fiber Membranes With Significantly Enhanced Hydrocarbon Separation Performance

Zhang, C., Georgia Institute of Technology
Koros, W. J., Georgia Institute of Technology
Originally presented on: 11/8/2013 10:35:00 - 11:00:00

ZIF/Polyimide Asymmetric Mixed-Matrix Hollow Fiber Membranes with Significantly Enhanced Hydrocarbon Separation Performance

Chen Zhang, William J. Koros

School of Chemical and Biomolecular Engineering, Georgia Institute of Technology

Asymmetric hollow fibers are the most industrially practical geometry for gas separation membranes in terms of productivity and membrane packing efficiency. However, studies on polymer-based mixed-matrix membranes in the past two decades have been mostly limited to the geometry of films with only a few extended to the geometry of asymmetric hollow fibers due to many technical challenges. For those few examples[1-3] in which selectivity enhancements were observed in mixed-matrix hollow fiber membranes, defect-sealing post-treatments (e.g. PDMS coating) were required in most cases to realize the enhanced selectivity.

This study reports successful formation of high performance asymmetric mixed-matrix hollow fiber membranes using a zeolitic imidazolate framework (ZIF) and a high T_g (glass transition temperature) polyimide, in which a significant propylene/propane selectivity enhancement was realized without any defect-sealing post-treatments. Spinning dope compositions and spinning parameters were optimized to form mixed-matrix hollow fiber membranes with desirable properties using the dry-jet/wet-quench technique. Propylene/propane gas mixture permeation showed that adding ZIF molecular sieve particles (~17 wt%) in the hollow fiber sheath layer significantly enhanced propylene/propane selectivity over pure polyimide dense film membrane by over 70% while maintaining attractive propylene permeance. The intrinsic compatibility of hydrophobic ZIF particles with the polyimide matrix as well as the optimized spinning dope compositions were keys for success of this study.

To our best knowledge, this is the first study in the open literature that reports successful spinning of asymmetric mixed-matrix hollow fiber membranes showing remarkably enhanced gas separation selectivity without any defect-sealing post-treatments. The current study, therefore, has significant impacts on the development of polymer-based gas separation membranes. An application of ZIF/polyimide mixed-matrix hollow fiber membrane described in this study is to pre-purify the feed stream of C_3 splitters in olefin plants to reduce the energy cost for producing polymer-grade propylene.

References:

(1) Husain, S.; Koros, W. J.: Mixed matrix hollow fiber membranes made with modified HSSZ-13 zeolite in polyetherimide polymer matrix for gas separation. Journal of Membrane Science 2007, 288, 195-207.

(2) Dai, Y.; Johnson, J. R.; Karvan, O.; Sholl, D. S.; Koros, W. J.: Ultem®/ZIF-8 mixed matrix hollow fiber membranes for CO2/N2 separations. J. Membr. Sci. 2012, 401Â¨C402, 76-82.

(3) Ekiner, O. M.; Kulkarni, S. S.: Process for Making Hollow Fiber Mixed Matrix Membranes. US Patent 6663805, 2003.

(785a) Quantitative Computational Screening of Pd-Based Intermetallic Membranes for Hydrogen Separation

Chandrasekhar, N., Georgia Institute of Technology
Sholl, D. S., Georgia Institute of Technology
Originally presented on: 11/8/2013 8:30:00 - 8:55:00

Intermetallics are stoichiometric compounds of two or more metals, which are known to exhibit structural and physical properties often different from their constituent elements. In this work, 78 Pd-based binary intermetallic membranes were screened for hydrogen separation from gas mixtures. A systematic computational screening approach based on calculation of key properties like solubility, diffusivity and permeability of hydrogen through these materials was developed. No Pd-based intermetallic was found to have favorable hydrogen permeabilities for separation applications. The hydrogen solubility of different metastable structures of the same intermetallic was studied to understand the impact of structure on the hydrogen binding energy. Pd_3Mn was chosen since it is structure, hydrogen binding energy and activation barrier for diffusion was found to be similar to pure Pd. Different structures of the same intermetallic were found to have different hydrogen binding energy. Hydrogen stabilizes two Pd_3Mn metastable phases. These metastable phases are predicted to have favorable solubilities at very low hydrogen pressures. Our calculations provide a useful method for analyzing a large set of materials from first principles and shortlisting materials that show properties useful for applications such as hydrogen separation.

(785b) Characterization of New Pd, Pd-Au and Pd-Au-Pt Large Scale Membranes in Actual Coal Derived Syngas Atmospheres

Ma, Y. H., Worcester Polytechnique Institute

Guazzone, F., Worcester Polytechnic Institute

Catalano, J., Worcester Polytechnic Institute

Mardilovich, I., Worcester Polytechnic Institute

Kazantzis, N., WPI

Originally presented on: 11/8/2013 8:55:00 - 9:20:00

Characterization of new Pd, Pd-Au and Pd-Au-Pt large scale membranes in actual coal derived syngas atmospheres

Yi Hua Ma, Federico Guazzone, Ivan Mardilovich, Jacopo Catalano and Nikolas K. Kazantzis

Center for Inorganic Membrane Studies

Department of Chemical Engineering

Worcester Polytechnic Institute

Worcester, MA 0109, USA

Abstract

Composite Pd alloy membranes are highly suitable for the production of H_2 in membrane reactors due to their high stability at elevated temperatures and the infinite theoretical selectivity of dense Pd alloys towards H_2. The methane steam reforming (MSR) and water gas shift (WGS) reactions have been carried out in Pd-alloy membrane reactors by several researchers. In the effort to develop a WGS catalytic membrane reactor, our group at Worcester Polytechnic Institute has been studying the characteristics and stability of several composite Pd based membranes in actual coal derived syngas atmospheres at the National Carbon Capture Center (NCCC) in Wilsonville, Alabama. The H_2 flux, H_2 permeance and H_2 purity of pure Pd and Pd-Au alloy membranes have been studied at 450°C, 12.6 bar and an average H_2 feed concentration of 35%. It was demonstrated that Pd-Au membranes were characterized by an outstanding long term permeance and selectivity stability at 450°C in coal derived syngas for at least 473 h and H_2 purities as high as 99.9%. However, all tested membranes at NCCC showed a sharp initial decrease in H_2 permeance (in the order of 30-60% of the initial permeance) indicating the possible adsorption of contaminants on the Pd surface. XPS analysis of a Pd film exposed to the same syngas as that of the membranes showed the presence of Hg, Na, Mg, O, S and C.

The study of three membranes (Pd-Au, Pd-Au-Pt and pure Pd) at NCCC led to the following observations: (1) extended exposure (>150 h) of pure Pd membranes in coal derived syngas might have led to the appearance of cracks (seen in SEM micrographs) and subsequent drastic loss in selectivity, (2) Pd-Au membranes with high Au content at the surface did not show signs of brittleness/cracking, however, their permeance showed to be negatively affected by H_2S concentration (up to 10.6 ppm), but it was unexpectedly recovered up to a 100% after only 30 min in syngas without H_2S, and (3) the initial H_2 permeance loss of the Pd-Au-Pt membrane was only 25% indicating that Pt might be a possible alloy element for H_2S resistance.

Preliminary experiments in our laboratories with the Pd-Au and the Pd-Au-Pt membranes showed that at the operating conditions (450°C and 12.6 bar) and at the maximum ethylene and propylene concentrations in the syngas (< 1000 ppm) at NCCC, ethylene and propylene had no or negligible effect on the H_2 permeance of Pd-Au and Pd-Au-Pt membranes. Since the pure Pd membrane showed a 80% decrease in initial permeance when tested at NCCC, it appeared that the main reason for the initial H_2 permeance loss were sub-ppm levels of H_2S not possible to detect with the on site GC.

The mechanical stability of composite Pd membranes was also studied with a fourth membrane which was subjected to 150 temperature cycles between 250 and 450°C (shown in Figure 1(a)) and 100 additional temperature cycles between 250 and 400°C. During the first 150 cycles the H_2/He ideal selectivity declined from approximately 10,000 to around 2,000. However, during the additional 100 temperature cycles the selectivity (or N_2 leak permeance as seen in Figure 1(b)) remained unchanged indicating that the decrease in selectivity was mainly associated with the short dwell times at 450°C and not caused by the temperature cycles. The temperature cycling test showed the outstanding robustness of the prepared composite Pd membranes.

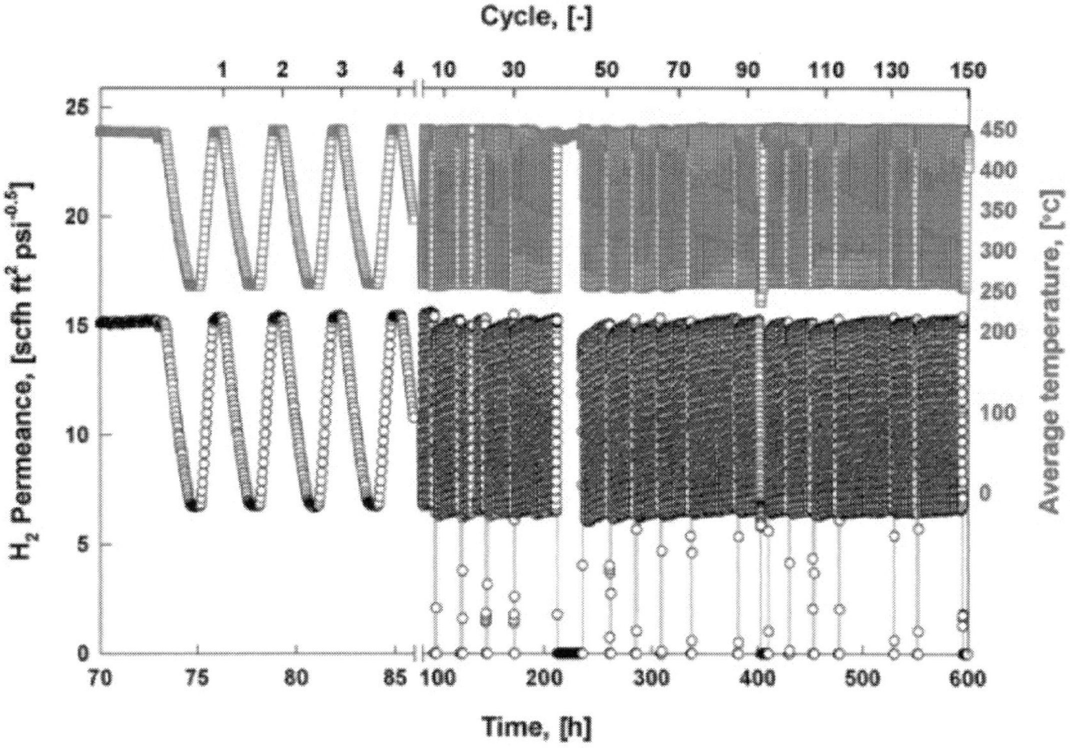

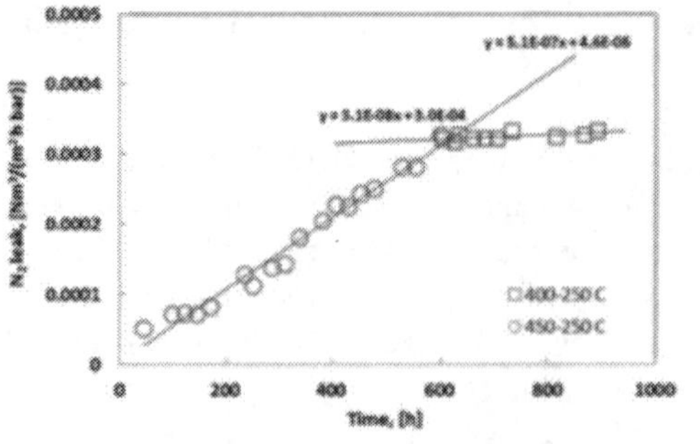

Figure 1. (a) H_2 permeance and temperature profile during the 150 cycles between 250-450°C. (b) N_2 leak permeation as a function of time during the first 150 cycles between 250-450°C and the additional 100 cycles between 250-400°C

(785c) Microstructured Palladium Alloys for Purification of Hydrogen

Dunbar, Z. W., Army Research Laboratory

Originally presented on: 11/8/2013 9:20:00 - 9:45:00

Highly selective metal membranes for hydrogen gas separation possess tremendous potential, with a variety of important applications. Large stationary purification systems have seen the most interest, for use in ammonia synthesis or in the US Department of Energy's Coal to Hydrogen initiative. Additionally, there is interest from the military for smaller systems, to support mobile hydrocarbon reforming for fuel cell applications. Unfortunately, the widespread adoption of metal membranes has thus far been stymied by their high cost and chemical sensitivity, particularly to sulfur.

In this work, these challenges are addressed using microfabrication technology. This reduces the precious metal requirements of the membrane to sub micron levels, while maintaining high selectivity. The thin palladium membrane is mechanically supported with a nickel honeycomb structure. As a result, the completed membrane is completely metallic, allowing for superior sealability and thermal cycling durability relative to metallic/ceramic membranes. Membrane flux, chemical stability and mechanical durability will be presented and discussed.

(785d) Thermal Stability of Palladium-Composite Membrane Fabricated By Surfactant Induced Electroless Plating (SIEP)

Ilias, S., North Carolina A&T State University
Islam, S.,
Originally presented on: 11/8/2013 9:45:00 - 10:10:00

Thermal stability of the palladium (Pd) membrane is a major challenge in hydrogen separation from mixture of gases at high temperature. Dense Pd-composite membranes on microporous stainless steel substrate (MPSS) were prepared by surfactant induced electroless plating (SIEP). Dodecyl trimethyl ammonium bromide (DTAB), a cationic surfactant was used in the Pd EP-bath. The pre- and post-annealing characterizations of the membranes were carried out by SEM, XRD, EDX, and AFM analysis. SEM images showed significant improvement of the membrane surface morphology, in terms of metal grain structures compared to the membranes fabricated by conventional EP process. Permeability tests were carried out with hydrogen and nitrogen at temperatures and pressures in the range of 523 - 823 K and 20 - 100 psi, respectively. The membranes showed excellent H_2-flux and selectivity. H_2 Flux through the membranes fabricated by SIEP shows large improvement compared to those by conventional EP with comparable perm-selectivity. Pd-MPSS membranes fabricated by SIEP and conventional EP methods were tested for thermal stability under thermal cycling (573-723-573 K) at 15 psi pressure drop for about 1200 hours. Some of these results will be discussed in this presentation.

(785e) Supported Molten Metal Membranes for Hydrogen Separation: Promise and Pitfalls

Yen, P. S., Worcester Polytechnic Institute
Ma, Y. H., Worcester Polytechnique Institute
Datta, R., Worcester Polytechnic Institute
Originally presented on: 11/8/2013 10:10:00 - 10:35:00

Palladium hydrogen separation membrane are being developed for reformers, as they are more compact, energy efficient, and selective as compared to the traditional technologies, i.e., absorption, pressure-swing adsorption, and cryogenic distillation. However, they are expensive, sensitive to poisons, and lack adequate durability. Therefore, the search for novel and cheaper materials to replace palladium is ongoing. We are, thus, developing entirely novel supported molten metal membranes (SMMMs) based on low melting metals and their alloys because of their higher diffusivity and greater potential robustness to poisons, and for their inherent ability to avoid sintering, hydrogen embrittlement, and thermal mismatch between the membrane and the support. In this presentation, we will describe the promise of SMMM and the pitfalls we have so far encountered in this quest.

Development of thin and stable SMMMs requires a basic understanding of the wettability and reactivity between porous support and the molten metals. This is a fine balance since some chemical interaction between the liquid metal and the support is necessary for adequate wetting, while too much of it dooms the membrane. Therefore, a goal of our work was to find a porous support material with both good wettability and stability. A thermodynamic assessment of molten Ga with various ceramic materials was performed to estimate the stability and wettability. Coupon test of selected material was also performed for comparison. It was found that materials with good wetting tend to be unstable and materials with good stability tend to have bad wetting with few exceptions among tested materials, most notable silicon carbide (SiC). A further complication is that the wetting changes with the gas composition. Such issues are incompletely understood and are being investigated. Hydrogen permeance experiments with dense gallium molten metal membrane showed that the measured hydrogen permeance of Ga is comparable to Pd membrane at 500-550 , which makes molten Ga a potential material as hydrogen separation membrane.

(785f) High Throughput Study of Ternary Pd-Alloys for Applications in Hydrogen Separation

Miller, J. B., Carnegie Mellon University
Kondratyuk, P., Carnegie Mellon University
Gumuslu, G., Carnegie Mellon University
Tarditi, A. M., Instituto de Investigaciones en Catalisis y Petroquimica (FIQ, UNL-CONICET)
Cornaglia, L., INCAPE
Gellman, A. J., Carnegie Mellon University
Originally presented on: 11/8/2013 10:35:00 - 11:00:00

Separation of hydrogen from mixed gas streams is an important unit operation in generation of carbon-neutral fuels and electricity from fossil- and bio-derived feedstocks. Pd-based membranes have received significant attention for the separation application in advanced coal gasification processes because of their near-perfect selectivity for hydrogen. In practice, pure Pd suffers from several limitations, including high cost, poor mechanical strength, and deactivation by minor components, most notably H_2S. H_2S can compromise membrane performance by reacting with Pd to form a low-permeability Pd_4S scale on the membrane surface. Pd has been alloyed with minor components to improve structural properties, improve permeability and impart resistance to H_2S. To achieve both high permeability and H_2S tolerance, we are applying high-throughput methodologies to development of $Pd_xCu_yAu_{1-x-y}$ ternary alloys across continuous composition space (x, y= 0 to 1).

Using e-beam PVD sources and a masking system of our own design, we deposited $Pd_xCu_yAu_{1-x-y}$ composition spread alloy films (CSAFs), thin (~100 nm) alloy films spanning broad, continuous composition space (x, y = 0 to 1), onto compact (14 x 14 mm) substrates. Structural characterization of an annealed (530 °C) CSAF, by electron backscatter techniques, reveals that addition of more than ~ 20 atom % Au prevents formation of the BCC form that displays high H-atom permeability in the PdCu binary. H_2-D_2 exchange activity measurements, performed using a unique multichannel reactor array of our own design, show that the alloy is most active for hydrogen dissociation at high Pd contents. Finally, surface composition characterization of an annealed CSAF that had been exposed to 1000 ppm H_2S in H_2 at 350 °C shows that sulfur uptake by the alloy is lowest at compositions in the vicinity of $Pd_{0.40}Cu_{0.20}Au_{0.40}$, suggesting that the CuAu combination is more effective than either Au alone or Cu alone for improving Pd's resistance to sulfur corrosion.

Results of permeability tests of prototype membranes made by sequential electroless deposition methods, with compositions suggested by the high throughput studies, will also be presented.

(789a) Kinetic Studies On Binary Clathrate Hydrates

Cai, L., Princeton University
Pethica, B. A., Princeton University
Debenedetti, P. G., Princeton University
Sundaresan, S., Princeton University
Originally presented on: 11/8/2013 8:30:00 - 8:52:00

While early clathrate hydrate studies aimed at the suppression of hydrate formation in oil pipelines, several applications have emerged where the formation of clathrate hydrates could be useful, including natural gas storage, carbon dioxide sequestration and desalination. In these settings, it is valuable to explore ways to produce hydrate at lower pressures, higher temperatures and increased formation rates than are presently available.

One innovation to facilitate hydrate formation is to employ a binary hydrate, which can boost the thermodynamic driving force significantly. In hydrates, the ice-like structures of hydrogen-bonded water networks form cavities of different sizes. While the occupation of each type of cavity by guest molecules promotes hydrate formation, no single guest species effectively fits all cage sizes at the same time. By introducing two guest species simultaneously -- one liquid primary former (such as cyclopentane) fitting the large cavities and one helper gas (such as methane or CO_2) fitting the small cavities – the cage stabilization effect is strengthened, allowing hydrate to form and grow under conditions unavailable to corresponding single-component hydrates.

Though thermodynamically favorable, the formation kinetics of binary clathrate hydrates are not well understood. Prior kinetic studies mainly deal with single component hydrates, but the growth of binary hydrates is more complicated. It entails the co-presence of hydrate, water, liquid former and gas helper, comprising a complex coexistence of four-phases. The hydrate growth process may take place primarily at the water - liquid hydrate former interface, consuming the helper gas dissolved in both liquid phases. The solubility of the helper gas in the liquid primary former is usually far below the hydrate stoichiometric level. For example, cyclopentane-methane hydrate has the sII structure with a 2:1 ratio of small:large cavities, and its formation consumes methane and cyclopentane roughly in the same ratio, yet a liquid phase composed of 67% methane and 33% cyclopentane does not exist in the temperature/pressure range of interest. As a result, transport issues arise additional to the factors involved in single-component hydrate growth since the hydrate growth is potentially limited by the transfer of the helper gas from the vapor phase to and between the two liquid phases.

This experimental study investigates the kinetics of the binary hydrate growth, with emphasis on the interphase transport of the hydrate components. The hydrates of the former pairs, cyclopentane-methane and cyclopentane-CO_2, are generated in a batch reactor as a function of temperature, pressure, stirring rate and level of saturation. The growth rate is tracked by the pressure and temperature trajectories, and comparative studies are performed to identify the roles of various factors involved in the hydrate growth, such as the degree of subcooling, agitation intensity, interfacial areas of the liquid phases and solubility of the helper gas in the primary hydrate former. These findings extend our fundamental understanding of the formation and growth kinetics of binary clathrate hydrates. They also guide the selection of promising binary hydrate formers, as well as the optimal operating conditions to realize economic clathrate hydrate production.

(789b) Nucleation and Growth Kinetics of Calcium Oxalate Crystallization in the Presence of Anionic Macromolecules

China, M., University at Buffalo - The State University of New York (SUNY)
Tsianou, M., University at Buffalo - The State University of New York (SUNY)
Originally presented on: 11/8/2013 8:52:00 - 9:14:00

We investigate the factors that control the nucleation and growth of important calcium biominerals such as calcium carbonate, calcium oxalate, and calcium phosphate. We establish that macromolecules (polyelectrolytes, polypeptides, surfactants) can control the final morphology and polymorph by i) face-selective adsorption on crystals (thermodynamic regime) or by ii) temporarily stabilizing the primary nanoparticles formed thus leading to mesoscale crystallization (kinetic regime). A combination of techniques which include Scanning Electron Microscopy (SEM), X-ray Diffraction (XRD), Fourier Transform Infrared Spectroscopy (FTIR), Thermogravimetric Analysis (TGA), and Calcium Ion Selective Electrode have been used to evaluate the effect of macromolecules on the nucleation, growth, and the morphology of obtained crystals. For example, sodium polystyrene sulfonate (PSS) was found to affect the morphology of the produced crystals and to preferentially promote the crystallization of calcium oxalate dihydrate (COD) over the thermodynamically stable calcium oxalate monohydrate (COM). At sufficiently high PSS concentrations, all calcium ions participate in complex formation with PSS and thus, the crystallization of calcium oxalate was completely inhibited. The effects of PSS on the nucleation and growth of calcium oxalate were remarkably reduced or completely eliminated when a cationic surfactant molecule was introduced in the system.

(789c) Estimation of Kinetic Parameters From in-Situ Measurements During Batch Cooling Crystallization

Li, H., Georgia Institute of Technology
Kawajiri, Y., Georgia Institute of Technology
Grover, M., Georgia Institute of Technology
Rousseau, R. W., Georgia Institute of Technology
Originally presented on: 11/8/2013 9:14:00 - 9:36:00

Recently many studies have focused on using in-situ measurements to monitor and control crystal size distribution (CSD) [1, 2, 3]. Our objective in this study is to model crystallization kinetics and design a control strategy to produce a desired CSD using information from in-situ measurements to estimate crystallization kinetics. The measurements include focused beam reflectance measurement (FBRM) for the solid phase and attenuated total reflectance-Fourier transform infrared spectroscopy (ATR-FTIR) for the liquid phase.

Using an empirical approach we proposed previously [4], CSD can be estimated from the chord length distribution (CLD) measured by FBRM. In the approach, crystals of different size ranges are prepared by sieving and their characteristic CLDs were measured. The matrix formulated from the characteristic CLDs is based on the linear FBRM model, which associated CSD with CLD. For the ATR-FTIR spectrum, the ratio of peak heights produced by solute and solvent was used to estimate the solute concentration.

Different temperature profiles and initial concentrations were used in batch cooling crystallizations of paracetamol from ethanol, in order to determine the influence of these variables on crystallization kinetics. Key findings from these experiments were (1) the occurrence of nucleation during cooling, as opposed to nucleation under isothermal conditions at the same supersaturation, increased the number of crystals produced; and (2) growth kinetics were highly sensitive to temperature.

A population-balance model was constructed to find the kinetic parameters that fit the experimental data. The method of moments was applied to the data to obtain ranges of the model parameters, which were then used as starting points for the optimization of the full population-balance model. With the resulting model and the fitted parameters, an optimal operation profile was determined and confirmed by subsequent experiments.

References:

[1] N. Kail, W. Marquardt, and H. Briesen. Estimation of particle size distributions from focused beam reflectance measurements based on an optical model. Chemical Engineering Science, 64(5): 984-1000, 2009.

[2] J. Worlitschek and M. Mazzotti. Model-based optimization of particle size distribution in batch-cooling crystallization of paracetamol. Crystal Growth & Design, 4(5):891{903, 2004.

[3] A. N. Saleemi, C. D. Rielly, and Z. K.Nagy. Comparative investigation of supersaturation and automated direct nucleation control of crystal size distributions using ATR-UV/vis spectroscopy and FBRM. Crystal Growth & Design, 12(4): 1972-1807

[4] H. Li, M. A. Grover, Y. Kawajiri, and R. W. Rousseau. Development of an empirical method relating crystal size distributions and FBRM measurements. Chemical Engineering Science, 89:142-151, 2013.

(789d) Secondary Nucleation of Lactose By Fluid Shear Forces

Paterson, A. H. J., Massey University
Wu, D.,
Agrawal, S.,
Originally presented on: 11/8/2013 9:36:00 - 9:58:00

Lactose has many uses in the pharmaceutical and food industries. It is produced by crystallisation from whey or whey permeates produced by the dairy industry. Understanding the mechanisms of lactose crystallization is critical to control crystal size, and thus maximise profitability. To do this the nucleation rate must be controlled both at the start of the crystallisation process and during the growth phase. Nucleation can occur as a primary event or by secondary mechanisms. If an evaporative crystallisation process is being used, then secondary nucleation is the dominant mechanism. If a batch cooling crystalliser is being used without the addition of any seeds, then there will be a primary nucleation event followed by the growth of those crystals. In this case, in order to maximise the size of the crystals produced, secondary nucleation is to be avoided. Secondary nucleation mechanisms involve initial breeding, polycrystalline breeding, macroabrasion, dendritic, fluid shear and contact.

This project focused on the effect of fluid shear forces on secondary nucleation under two different levels of absolute supersaturation (8.4 g and 10.7g lactose/100g water). The experimental methodology used circulated supersaturated solutions through a glass tube with immobilized crystals inside for 15 minutes at constant temperature (30°C). The pressure drop across the tube was measured by a differential pressure transducer and this was used to estimate the shear rate on the crystals. The sheared solution was incubated for two hours, allowing the nuclei to grow. A Malvern Mastersizer 2000 was used to estimate the number of nuclei.

Blank runs, using an agitator in a beaker when growing the nuclei showed that agitation induced both primary and secondary nucleation. This lead to the modification of using rolling tubes for the growth phase. Repeated blank runs showed that some primary nucleation caused by the action of pumping the supersaturated solution through clean glass tubes at similar fluid shear rates did cause some nucleation but at rates significantly below those found when crystals were present.

It was found that the fluid shear forces did induce secondary nucleation. The nucleation rate increased proportionally with absolute supersaturation solution level and shear rate from 200 to 4000 with the supersaturated solution of 8.4g lactose/100g water. There was no increase in shearing effect on nucleation rate detected when the shear rate was over 4000 . At the supersaturated solution of 10.7g lactose/100g water, there was no increase in shearing effect at shear rates above 3000 .

(789e) A Simplified Microseed Matrix Screening Method Using X-Ray Transparent High-Throughput Microfluidic Chips

Pawate, A. S., University of Illinois at Urbana-Champaign
Schieferstein, J. M., University of Illinois at Urbana-Champaign
Guha, S., University of Illinois at Urbana-Champaign
Varel, M. J., University of Illinois at Urbana-Champaign
Kenis, P. J. A., University of Illinois at Urbana-Champaign
Originally presented on: 11/8/2013 9:58:00 - 10:20:00

Membrane proteins (MPs) are important targets for crystallization because they are transducers of material and signals across cell membranes and intracellular organelles, and their malfunction accounts for many diseases in humans. One of the bottlenecks to obtaining diffraction quality crystals of membrane proteins is the need to screen thousands of crystallization conditions while negotiating their limited availability, instability in non-native lipids and detergents, and the long incubation times before the appearance of initial crystals.

A successful method to speed up the chances of finding the right crystallization conditions involves microseeding while setting up a matrix screening of the membrane protein. However, microseed matrix screening involves disturbing the crystallization drop to introduce the microseeds by either a skill intensive and manually laborious procedure, or by microseeding robots.

Here, we demonstrate a high-throughput X-ray compatible microfluidic platform for microseed matrix screening of crystallization conditions that enables rapid detection of crystallization conditions and optimization of microseeding protocols. The microfluidic chips, in 24-well and 96-well formats, allow for formulation of multiple microseed dilutions while consuming ~6 μL protein, precipitant and microseed solutions. Our X-ray transparent chips enable room-temperature data collection while minimizing radiation damage by a strategy of collecting small wedges of data from multiple crystals and merging them to create a complete dataset. We validated our approach by crystallizing photoactive yellow protein and optimized the microseeding protocol to obtain large, single crystals. Further, we applied our method to investigate microseeding of bacterial photosynthetic reaction center, cytochrome *c* oxidase, and qNOR.

(789f) Effect of Ultrasonic Irradiation On the Behavior of Primary Nucleation and Growth of Cloxacillin Benzathine

Li, J.,

Originally presented on: 11/8/2013 10:20:00 - 10:42:00

The effect of ultrasonic irradiation on the behavior of primary nucleation of cloxacillin benzathine in supersaturation solutions with different levels were studied in the present work. The solubility of cloxacillin benzathine in binary solvent mixtures of ethanol and water were measured by a dynamical method, using a medium-throughput multiple reactor (Crystal 16™, Avantium, Netherlands). Based on the solubility data, the induction periods of this system were measured under different experiment conditions. Compared to the induction time without ultrasonic irradiation, the induction period for different supersaturations decreased with increasing ultrasonic irradiation energy and subsequently increased to a certain degree. Above this level of ultrasonic irradiation, the induction period decreased dramatically. Besides that, the composition of solution and feeding rate of the reactant were investigated to figure out the influential factors to the primary nucleation behaviors for this system. Measurements of crystal size and crystal habits for irradiated and silent experiments showed that the particle size and aggregates were both affected by ultrasonic irradiation and is highly correlated with the supersaturation. The products with desired crystal size could be obtained by optimally controlling the conditions of supersaturation and ultrasounic irradiation.

(789g) Modeling of Dissolution and Re-Crystallization of Paracetamol in Ethanol

Pandit, A.,
Ranade, V. V., National Chemical Laboratory
Originally presented on: 11/8/2013 10:42:00 - 11:04:00

Dissolution and crystallization are important phenomena occurring in a variety of reactions and separations. Several factors like reactor configuration & operating parameters, temperature history, solute-solvent properties etc. influence these phenomena significantly. Hence, the modeling of dissolution and re-crystallization is very challenging. In the present study we wish to develop a modelling framework to quantitatively capture dissolution and crystallization. Experimentation was done using a paracetamol-ethanol system.

In our previous work[1], we carried out experiments where a saturated paracetamol-ethanol solution was cooled at a fixed cooling rate until after crystallization and then reheated at the same rate until after complete dissolution. The experiments we carried out in a Mettler OptiMAX reactor setup. The Mettler FBRM probe was used for monitoring the evolution of particle count and particle chord length distribution with time. These experiments were carried out for different heating/cooling rates (0.3, 0.5, 0.7 & 1 K/min). When the particle counts were plotted versus reactor temperature for different heating/cooling rates, a peculiar hysteresis was observed (shown in figure). As the heating/cooling rate is reduced, the hysteresis loop ?twists' as the particle count starts falling even after crystallization. This peculiarity was hypothized to occur because of Ostwald's ripening. Ostwald's ripening means the growth of larger particles at the expense of smaller particles. The particle count falls as many small particles would be consumed for the growth of a few large particles. Also, Ostwald's ripening is reported to occur at near equilibrium conditions. When the heating/cooling rate is slow, the solution is more ?near-equilibrium', which would promote the incidence of Ostwald ripening and would explain why the peculiarity is pronounced at lower rates. The hypothesis regarding the occurrence of Ostwald's ripening is supported by experimental evidence as the chord length distribution is seen to reflect the growth in larger particles although very slightly.

The molecular phenomena occurring in such systems are typically crystal growth, primary/secondary nucleation, Ostwald's ripening, aggregation, breakage and dissolution. All of these phenomena occur simultaneously in the system and also compete with each other depending on the thermodynamic state of the system. It is clear that the peculiar ?twisting' of the hysteresis loop arises because of the interplay between these phenomena. In the present study, we have attempted to quantitatively capture the hysteresis as also the ?twisting' of the hysteresis loop. A population balance equation was used as the basic canvas for this modelling framework. The population balance approach is already widely used to model dissolution[2]. The model proposed by Kubota[3] was used to determine the occurrence of first nucleation events as well as for obtaining the (initial) primary nucleation rate. Similarly, the various other phenomena are appropriately addressed by suitably modifying or adding certain elements to the basic population balance equation. The developed model was then used to qualitatively predict the ?twisting' effect of the hysteresis loop. Then, after appropriately calibrating the model, model predictions of the hysteresis loop were compared to experimental predictions. We hope that focusing our effort to quantitatively capture the observed peculiarity will help in developing a robust modelling framework for crystallization systems as well as serve for its rigorous validation.

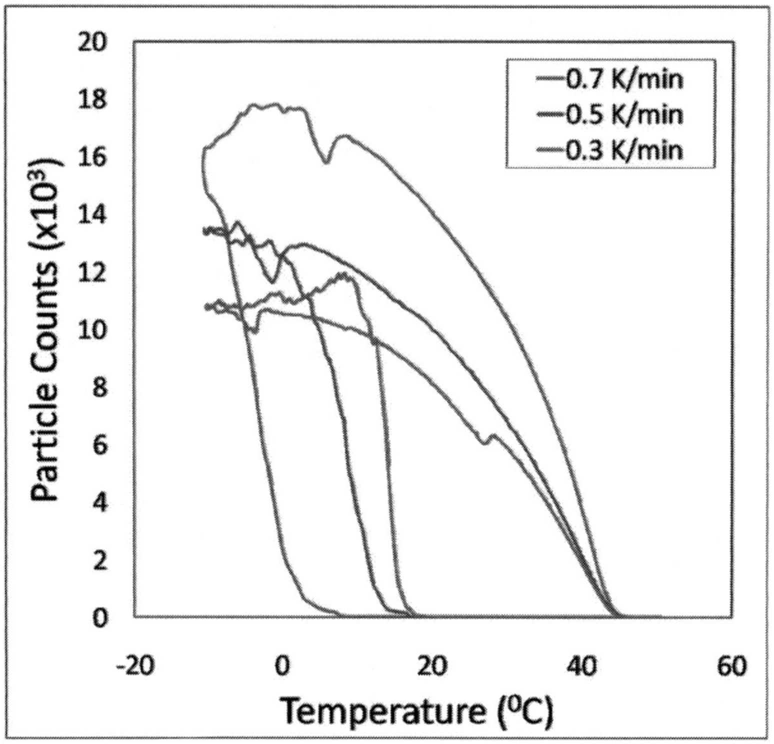

Comparison between the hysteresis observed during the dissolution and

Re-crystallization of Paracetamol in Ethanol at different heating/cooling rates

References:

[1] A. V. Pandit and V. V. Ranade, ?Hysteresis during Dissolution and Re-Crystallization?, Poster presentation at European Congress of Chemical Engineering (ECCE9), 2013.

[2] D. Ramakrishna, ?Population Balances: Theory and Applications to Particulate Systems in Engineering?, Academic Press, 2000.

[3] N. Kubota, ?A new interpretation of metastable zone widths measured for un-seeded solutions?, Journal of Crystal Growth, Volume 310, Issue 3, 2008, 629?634.

(802a) Aqueous Surfactant Two-Phase Systems for the Batch and Continuous Extraction Processes: Experimental Study and Modeling of the Solute Partitioning

Smirnova, I., Hamburg University of Technology

Originally presented on: 11/8/2013 12:30:00 - 12:50:00

In this work applications of the surfactant-based aqueous two phase systems for batch and continuous extraction processes are discussed. Special attention is given to the use of such systems for in situ product removal in biotechnological processes.

Based on the state of the art in the literature and the own work it is demonstrated that surfactant/water systems represent a suitable alternative to conventional solvents and can effectively be processed in batch processes and in continuous extraction columns. For this purpose, mainly aqueous nonionic surfactant solutions, which split into two phases if the temperature is increased, are applied. Presently many different types of nonionic surfactants are produced commercially, out of these numerous have been considered as potential solvent for the cloud point extraction. In this work the crucial thermophysical properties of nonionic surfactants are investigated to determine the potential of surfactant systems for extraction processes. One of the most important parameters for these applications is the partition coefficient, which quantifies the distribution of a specific solute within an aqueous micellar system. Besides this, information about the location of a solute in the system and possible energy barriers have to be retrieved either from experiments or from simulations. Since the experimental effort for these properties is known to be rather high, predictions based on thermodynamic models are favourable. In this work the suitability of the models COSMO-RS and COSMOmic for this purpose is discussed. Both models are applied for the prediction of the partition equilibria in a number of the systems with pure and mixed surfactants. Influences of the additives, pH and temperature on the partition equilibiria are presented. The results of the predictions are compared with own experimental data, obtained by the Micellar Liquid Chromatography (MLC) and Micellar Enhanced Unltrafiltration (MEUF).

Further, the implementation of different surfactant systems in continuous extraction processes is presented based on the own work with a stirred extraction column. The influence of different process parameter and surfactant types and ratios on the separation efficiency is shown.

Finally, potential of different thermodynamic models to describe the liquid-liquid equlibria and solute partition in surfactant-based aqueous solution is discussed.

In this work applications of the surfactant-based aqueous two phase systems for batch and continuous extraction processes are discussed. Special attention is given to the use of such systems for in situ product removal in biotechnological processes.

Based on the state of the art in the literature and the own work it is demonstrated that surfactant/water systems represent a suitable alternative to conventional solvents and can effectively be processed in batch processes and in continuous extraction columns. For this purpose, mainly aqueous nonionic surfactant solutions, which split into two phases if the temperature is increased, are applied. Presently many different types of nonionic surfactants are produced commercially, out of these numerous have been considered as potential solvent for the cloud point extraction. In this work the crucial thermophysical properties of nonionic surfactants are investigated to determine the potential of surfactant systems for extraction processes. One of the most important parameters for these applications is the partition coefficient, which quantifies the distribution of a specific solute within an aqueous micellar system. Besides this, information about the location of a solute in the system and possible energy barriers have to be retrieved either from experiments or from simulations. Since the experimental effort for these properties is known to be rather high, predictions based on thermodynamic models are favourable. In this work the suitability of the models COSMO-RS and COSMOmic for this purpose is discussed. Both models are applied for the prediction of the partition equilibria in a number of the systems with pure and mixed surfactants. Influences of the additives, pH and temperature on the partition equilibiria are presented. The results of the predictions are compared with own experimental data, obtained by the Micellar Liquid Chromatography (MLC) and Micellar Enhanced Unltrafiltration (MEUF).

Further, the implementation of different surfactant systems in continuous extraction processes is presented based on the own work with a stirred extraction column. The influence of different process parameter and surfactant types and ratios on the separation efficiency is shown.

Finally, potential of different thermodynamic models to describe the liquid-liquid equlibria and solute partition in surfactant-based aqueous solution is discussed.

(802b) Multi-Stage Aqueous Two-Phase Extraction of a Monoclonal Antibody in a Mixer-Settler Battery

Muendges, J., TU Dortmund University
Zalesko, A.,
Zeiner, T.,
Originally presented on: 11/8/2013 12:50:00 - 13:10:00

Monoclonal Antibodies (mAbs) are biologically active proteins produced by plasma cells in response to the presence of foreign substances. During the past years, the application of mAbs for therapeutic treatment has been increasing. Several diseases like cancer, asthma or autoimmune diseases are already treated by mAbs. Due to the fact, that mAb-based pharmaceuticals are mainly administered for chronic diseases at high doses, a large demand of these pharmaceuticals has to be satisfied. Through an intensified research on the field of upstream-processing, the mAb concentration in the fermentation broth has significantly been increased. Still purification of mAbs is a challenging step in the manufacturing process accounting for a major part of the total production costs. Since conventional unit operations like preparative chromatography are often limited by their capacity, buffer consumption and ligand cost, growing interest exists in finding alternative downstream-processing methods.

Aqueous two-phase systems (ATPS) are a promising option for the improvement of mAb purification. By mixing two different phase forming components (PFCs), e.g. two hydrophilic polymers or a hydrophilic polymer and a salt in water at high concentration, two liquid phases can form. One of the phases is rich in PFC 1, the other is rich in PFC 2. Both phases consist of 70-80 % of water and thus have an aqueous character. Purification of the target product is achieved because of a different partitioning of target component and impurities between both phases. Especially for sensitive products ATPS are a promising alternative as they offer mild separation conditions at high, continuous capacity.

In this work, an ATPS consisting of polyethylene glycol, phosphate salt and water is chosen for the purification of mAbs from a Chinese Hamster Ovary cell supernatant (CHO-CS). Essential for the design of an aqueous two-phase extraction (ATPE) is a fundamental knowledge of substance properties, liquid-liquid equilibrium data as well as partitioning of target component and impurities. Several influences on the purification of mAbs like polyethylene glycol molecular weight, pH, initial concentration of CHO-CS and concentration of additional sodium chloride were analyzed in single-stage extraction experiments prior to this work. To further enhance the efficiency of mAb extraction, ATPE is performed in multi-stage experiments in a miniplant mixer-settler battery. The feasibility of multistage ATPE at miniplant scale will be discussed with regard to operational challenges and different operation modes. Those include the position and phase of CS inlet. Furthermore the extraction of mAb from CHO-CS will be evaluated, with regard to purity and recovery of mAb. The influences of stage number, throughput and operation mode will be presented.

(802c) Aqueous Two-Phase Extraction of Trametes Versicolor Laccase Within Microfluidic Device: Experiments and Modeling

Novak, U.,
Nujic, M.,
Tisma, M.,
Znidarsic-Plazl, P.,
Plazl, I.,

Originally presented on: 11/8/2013 13:10:00 - 13:30:00

ATPS offers several advantages compared to traditional extraction with organic solvents, particularly regarding to the non-denaturing conditions and high biocompatibility The polyethylene glycol (PEG)/phosphate aqueous two-phase system (ATPS) was used for extraction of commercial *Trametes versicolor* laccase within a Ψ-branched inflow and outflow glass microchannel. The flow rates of both phases entering Ψ-branched were selected in order to enable the separation of phases at the exit of microchannels in microfluidic device. Due to decreased scales in Ψ-branched microchannel, the extraction of laccase from phosphate-rich phase to PEG-rich phase reached equilibrium with the highest obtained extracted fraction of laccase of 65 % (w/w). The diffusion coefficients for laccase in both phases were determined based on the experimental data obtained in a short Y-junction microfluidic chip and the corresponding mathematical model. Furthermore, a 3D mathematical model in dimensionless form was developed comprising convection and diffusion terms at steady state conditions and it was validated by experimental results in order to forecast the performance for a wide range of flow rates.

(802d) ATPS From Natural Products to Large Biomolecules

Aires-Barros, R.,
Conde, J. P.,
Chu, V.,
Fernandes, P.,
Azevedo, A.,
Silva, D.,
Soares, R.,
Originally presented on: 11/8/2013 13:30:00 - 13:50:00

ATPS from Natural Products to Large Biomolecules

R. Soares[a,b], D.F.C. Silva[a,b], A.M. Azevedo[a], P. Fernandes[a], V. Chu[b], J.P.Conde[b], M.R. Aires-Barros[a]

[a]IBB – Institute for Biotechnology and Bioengineering, Centre for Biological and Chemical Engineering, Department of Bioengineering, Instituto Superior T?cnico, Technical University of Lisbon, Lisbon, Portugal

[b]INESC Microsistemas e Nanotecnologias and IN-Institute of Nanoscience and Nanotechnology, Lisbon, Portugal

Aqueous two-phase systems (ATPS) have shown to be a valuable option for the downstream processing of biopharmaceuticals, combining a high biocompatibility and selectivity with an easy and reliable scale up and capability of continuous operation. A continuous operation process, based in ATPS, for the capture of human immunoglobulin G (IgG) from different mammalian cells culture supernatants has been developed and validated in a pump mixer settler battery [1].

The limited predictive design of this process, has restrict its applicability to current downstream procesing. ATPS in a microfluidic platform was designed and tested for mAbs extraction, as an effective tool to accelerate bioprocess design and optimization [2]. Furthermore, this platform has the potential of combining the process efficiency of ATPS with the reduced times and volumes associated with microfluidics, as well as the possibility to multiplex and parallel process in real downstream processes.

ATPS both in macro or microsystems are currently a trendy subject in macromolecule purification and concentration, however, it is largely unexplored for small molecules. ATPS was applied for extraction of a mycotoxin from wine, ochratoxin A (OTA), a small organic molecule produced mainly by *Aspergillus fungi*, which might accumulate in grapes before harvest and during storage. ATPS in microfluidic chips presents a very tempting possibility to integrate such purification and miniaturization in a chip for fast and on site detection of mycotoxins in red and white wine.

[1] P.A.J. Rosa, A.M. Azeved, S. Sommerfeld, M. Mutter, W. B?cker, M.R. Aires-Barros "Continuous purification of antibodies from cell culture supernatant: from concept to process", Biotechnol.J. (2013) (in press)

[2] D.F.C. Silva, A.M. Azevedo, P. Fernandes V. Chua, J.P. Conde, M.R. Aires-Barros (2012) "Design of a microfluidic platform for monoclonal antibody extraction using an aqueous two-phase systemJ. Chromatogr. A 1249, 1-7.

(802e) Interfacing Aqueous Two-Phase Systems With Countercurrent Chromatography: Track Record and Future Potential for Bioseparations

Ignatova, S., Brunel University
Sutherland, I. A., Brunel University
Fisher, D.,
Originally presented on: 11/8/2013 13:50:00 - 14:10:00

Countercurrent Chromatography (CCC) is a continuous form of liquid-liquid extraction, which the Brunel Institute for Bioengineering (BIB) team has developed from academic know-how to industrial competitive process scale. CCC has been used extensively for the separation and isolation of molecules across a wide range of polarity, typically natural products using aqueous-organic two-phase systems.

To extend CCC to bioseparations, biocompatible solvent systems, such as Aqueous Two-Phase Systems (ATPS) containing polyethylene glycol (PEG) and salt, are required. These differ from the conventional aqueous-organic systems as they are more viscous, have lower density difference and lower interfacial tension resulting in slower phase settling times.

Successful separations of biomolecules have been achieved using both hydrostatic and hydrodynamic CCC technologies at various scales.

Further protein separation studies to the purification of monoclonal antibody (mAb) products from various cell cultures, using PEG-inorganic salts ATPS opened up a new area of research. Clear separation of the mAb from impurities has been achieved with retention of biological activity using new invention allowing independent variation of mixing and settling. The need to optimise the CCC operational mode to preserve bioactivity will be discussed.

Additionally, CCC instruments have the potential to perform as a reactor/separator. In this mode a catalyst is immobilised by partitioning to the liquid stationary phase and substrate passed through the column dissolved in the liquid mobile phase. Reaction occurs in the zones of mixing, together with separation of product from substrate by the chromatographic action of the CCC. This approach has been successfully applied to enzymatic biotransformation of a chiral mixture of amino acids when the high value product of the bioreactor was continuously separated from the product and biocatalyst as it is formed, using a PEG-phosphate ATPS.

Results of our current research with industry will be presented in more details.

(802f) Separation of Monoclonal Antibodies From Chinese Hamster Ovary Cell Supernatant With Aqueous Two-Phase System

Wu, Q., Zhejiang University
Lin, D. Q., Zhejiang University
Yao, S. J., Zhejiang University
Originally presented on: 11/8/2013 14:10:00 - 14:30:00

The pharmaceutical and biotechnological industries are investing substantial resources in the development of antibody-based therapeutic products, especially for monoclonal antibodies (mAbs). Currently, the antibody expression of mammalian cells with titer of 5 g/L becomes common and the bottleneck of antibody production has shifted from antibody expression to downstream processing. Protein A affinity chromatography has been the important platform for the purification of antibodies in large-scale production, but it still has some limitations. Some alternative techniques have been developed to improve the downstream processes, including the extraction with aqueous two-phase system (ATPS).

In the present work, the ATPS with polyethylene glycol (PEG) and hydroxypropyl starch (HPS) was used to separate mAbs from Chinese hamster ovary (CHO) cell supernatant. Firstly, the phase diagrams of PEG/HPS ATPS were investigated and the effects of NaCl addition on the phase diagrams were determined. The results indicated that the addition of NaCl could lead to a shift on the binodal curve and the phase separation would occur at higher PEG and HPS concentrations. The effects of pH, NaCl addition and the load of CHO cell supernatant on the partitioning of mAbs were investigated. It was found that with 6% CHO cell supernatant and 15% NaCl addition at pH 6.0 the yield of mAbs in the top phase was 97.6% with high purity of 96.0%. The back extractions of mAbs into a phosphate-rich bottom phase with the PEG/phosphate ATPS were also investigated. After two-step extraction with ATPSs the purity of mAb could reach 97.6% ±0.5 with the yield of 86.8±1.0%, which was comparable to the separation with Protein A chromatography. The results indicated that the extraction with ATPSs might be a promising alternative for the purification of mAbs from the cells supernatant.

* This work was supported by National Natural Science Foundation of China and Zhejiang Provincial Natural Science Foundation of China.

(802g) Development of Aqueous Two-Phase Processes for the Recovery of High Value Biological Products

Rito-Palomares, M., Tecnologico de Monterrey
Originally presented on: 11/8/2013 14:30:00 - 14:50:00

There is an increasing need to establish selective and scalable methods of product recovery and purification that integrate effectively with upstream operations. The established processes to purify biological products are characterized by a sequence of excessive membrane and chromatography steps. It is urgent to define bioengineering strategies to establish novel processes that can alleviate the existing process constrains for the recovery of biological products.

The development of novel purification processes based on two-phase extraction will be presented in this work. In particular, the fractionation of PEGylated proteins will be critically discussed. An attractive strategy for the recovery of recombinant proteins from plants, particularly rhG-CSF, from alfalfa will be examined. Finally, challenges for the establishment of bioprocesses for the purification of stem cells will be presented. Conclusions will be drawn concerning the strategies used to develop the downstream processes that greatly simplify the current way in which bioproducts can be recovered.

(808a) CO2 Capture By Gas Hydrate Formation: Combined Effect of Thermodynamic and Kinetic Promoters

Daraboina, N., Denmark Technical University
von Solms, N., Technical University of Denmark (DTU)
Originally presented on: 11/8/2013 12:30:00 - 12:55:00

Carbon capture by gas hydrate crystallization is a promising technology currently being considered in the battle to mitigate climate change. However, the requirement of both high pressure and slow formation rates are serious limitations which need to be resolved prior to implementing this technology at demonstration scale. Various additives (e.g.: THF, TBAB, cyclopentane and others) are known to promote hydrate formation thermodynamically by reducing the hydrate formation pressure. Furthermore there are also kinetic promoters such as surfactants which can increase the hydrate crystallization rate based on the concentration. The appropriate combination of thermodynamic and kinetic promoters has the potential to move the technology towards practical application.. In this work, the hydrate-based process for carbon dioxide capture was tested in a rocking cell (RC-5; Systemtechnik, Germany) apparatus. Various combinations of surfactant (kinetic promoters: SDS and CTAB) and THF (thermodynamic promoter) in liquid water were assessed in order to find the optimum operating conditions.

(808b) CO2/Brine/Rock Interactions Under CO2 Sequestration Conditions

Soong, Y., National Energy Technology Laboratory (NETL), Office of Research and Development, Department of Energy

Howard, B. H., National Energy Technology Laboratory

Hedges, S. W., U.S. Department of Energy

McIntyre, D., National Energy Technology Laboratory

Warzinski, R. P., U.S. Department of Energy, National Energy Technology Laboratory

Irdi, G.,

Haljasmaa, I.,

Crandall, D., National Energy Technology Laboratory (NETL), Office of Research and Development, Department of Energy

Originally presented on: 11/8/2013 12:55:00 - 13:20:00

CO_2/brine/rock interactions under CO_2 sequestration conditions

Y. Soong*, B. H. Howard, S. W. Hedges, D. McIntyre, R. Warzinski, G. Irdi, I. Haljasmaa[1], and D. Crandall[1]

U.S. DOE, National Energy Technology Laboratory,

P.O. Box 10940, Pittsburgh, PA 15236

[1] URS Co.

It has indicated by Intergovernmental Panel on Climate Change Study that injection of CO_2 into confined geological formations, given their potentially massive carbon storage capacity and widespread geographic distribution, represents one of the most promising options for mitigating anthropogenically-induced global climate change. The Mount Simon sandstone in the Midwest region of the U.S is recognized as a promising candidate host reservoir for carbon sequestration. The Mount Simon is a deep saline aquifer and is a primary reservoir target for large scale carbon dioxide injection tests due to its proximity to CO_2 sources, favorable depth, thickness, permeability, porosity, and the presence of the overlying Eau Claire Formation as a seal. The Midwest Geological Sequestration Consortium (MGSC) has selected the Mount Simon as the reservoir at its major demonstration project. While the Mount Simon Sandstone has ideal reservoir characteristics in some areas, there are significant variations in porosity, permeability, and mineralogy.

Numerous studies have been conducted to investigate changes in host rock properties when exposed to CO_2. Luquot and Gouze conducted CO_2-enriched fluid flow through a carbonate core sample under in situ sequestration conditions (T = 100 °C and P = 12 MPa). Under these conditions, a decrease in permeability and porosity is seen that is linked to precipitation of Mg-rich calcite. Gouze and Luquot also conducted an X-ray microtomography characterization study of porosity, permeability and reactive surface changes during the dissolution of pure calcite in a brine-CO_2 mixture. It showed that the increase in permeability is due to the decrease of the tortuosity for homogeneous dissolution, whereas it is due to the combination of tortuosity decrease and hydraulic radius increase for heterogeneous dissolution. Jove Colon et al., studied the permeability, porosity and reactive surface area evolution during dissolution of nonfractured, clay-free Fontainebleau sandstone cores using a flow-through percolation reactor. The experiments were performed at 80°C with 0.1 M NaOH solution. The permeability was found to vary depending on where the dissolution occurred: in the pore void or the pore throat. Liu et al., conducted a simulation study of CO_2 sequestration in the Mount Simon sandstone via TOUGHREACT. They indicated that a strongly acidified zone (pH 3-5) forms in the areas affected by the injected CO_2 and consequently causes extensive secondary mineral precipitation (calcite, magnesite, ankerite, alunite and anhydrite) and dissolution of feldspars. Thus, the injection of CO_2 into a deep saline aquifer affects porosity since it can cause both mineral dissolution and precipitation in the formation. To predict if there will be any changes to the reservoir during sequestration, it is critical to understand the behavior of the authigenic minerals that line pore spaces when they are in contact with the CO_2 saturated brine.

An experimental study was conducted to assess the impacts of injected CO_2. Core samples from the Mount Simon formation were placed in CO_2 saturated brines at sequestration temperature of 85 °C and pressure of 23.8 MPa in the presence of synthetic Illinois Basin brine for a period of six months in a set of 1.3 liter high-pressure vessels (17-4PH-1150 stainless steel, 10.2 cm I.D. x 16.5 cm depth, manufactured by Thar Technologies, Inc.). One sandstone sample with a porosity of 7.9% was obtained at depth of 1,769.7 m from a well located in Vermillion County, IN. The other sample with a porosity of 1.4% was collected at depth of 2634.2 m from a well located in Knox County, IN. The major mineral composition for the first sample is 78% quartz, 15% Feldspar and with less than 2% microcline and trace minerals. The primary mineral composition of the second sample obtained from Knox County is 70% quartz, 22% Feldspar, and around 4% illite/muscovite. Micro CT, CT, XRD, SEM, petrography, porosity, and permeability analyses were conducted prior to the six month exposure. Upon completion of the experiment, the analyses were repeated.

The permeability of the Vermillion sandstone after the six month exposure to CO_2 and brine was 0.85 mD compared to 1.6 mD for the fresh sample. On the other hand, the permeability of Knox sandstone was 69 nD compared to 50 nD for the fresh sample. The results suggest that the exposure altered the pore characteristics and thus the permeability of the sample. Micro CT analysis as well as SEM backscattered images were used to probe the Vermillion and Knox Co. sandstone before and after the six month exposure to the CO_2/brine environment. It can be seen that mineral dissolution and mineral precipitation occurred after the samples were exposed to CO_2/brine for six months. Mineral precipitation occurred in the Vermillion sample pores and cracks blocking flow and resulting within the observed 50% decrease in permeability. For the case of Knox sandstone, the mineral dissolution occurred in pore though the precipitation was observed in the exterior of the sample. This might result in increasing in permeability. In addition, from the analyses of the reacted brines within the reactor, it indicated the possibly of Feldspar dissolution.

When CO_2 is injected into a saline formation, acidic conditions form in the brine in the nearby portion of the reservoir thus reducing the pH of the brine in the vicinity of the injection system. Minor minerals in the sandstone, such as K-feldspar, calcite, kaolinite, microcline etc., that are stable in the native environment but not stable at lower pH can dissolve. Eventually, the increase of cation concentration produced by mineral dissolution can supersaturate the fluids with respect to other minerals. If this occurs, precipitation of new minerals can result in a change of porosity and permeability which could affect both CO_2 storage capacity and fluid movement. The issue is further complicated by the evolution of pore geometry due to these geochemical reactions that may affect the overall hydrologic properties of the host rock and therefore the performance of the unit as an effective reservoir.

The decreasing permeability found in Vermillion sandstone was probably due to Feldspar dissolution, migration and precipitation altering the sandstone pore/crack structure. For the case of sandstone from Knox county, the permeability increased by 38%. It might be due to the Feldspar dissolution in the pore and mineral precipitation at exterior surface of the sandstone. Investigations are continuing to more clearly elucidate the mechanisms and geochemistry of this system, especially at the pore structure scale.

(808c) Novel CO2 Capture and Carbonation Process With Waste Concrete

Shuto, D.,
Yamasaki, A., Seikei University
Hongo, T., Seikei University
Iizuka, A., Tohoku University
Originally presented on: 11/8/2013 13:20:00 - 13:45:00

A novel CO_2 capture and carbonation process is developed. In this process, wastes containing calcium, such as waste cement particles or steel slug, are used as a calcium source for carbonation reaction with CO_2 emitted from point sources. The process is composed of several steps. CO_2 in the flue gas is captured through absorption process with alkaline solution such as sodium hydroxide to form sodium carbonate and sodium bicarbonate. The calcium content in the waste cement particles is extracted with acid such as nitric acid or acetic acid to form solution containing calcium. Mixing of these two solutions leads to the precipitation of calcium carbonate: CO_2 fixation via carbonation reaction. Electrodialysis is applied for the regeneration of acid and alkaline from the remained solution after filtration of the precipitant. Thus, the net input of the process is CO_2 in the flue gas and waste cement particles, and the output is calcium carbonate. The process feasibility was examined through laboratory -scale experimental studies under various operation conditions. Based on the experimental results, the process was optimized and evaluated in terms of the power consumption and cost for the CO_2 recovery, and compared with conventional processes. Net CO_2 sequestration can be achieved by the process under certain conditions.

(808d) Direct Capture and Conversion of CO_2 From Flue Gas Via Ex-Situ Mineral Carbonation Using Chemical and Biological Catalysts

Swanson, E. J., Columbia University
Brady, P. V., Sandia National Laboratories
Park, A. H. A., Columbia University
Originally presented on: 11/8/2013 13:45:00 - 14:10:00

Converting carbon dioxide from flue gas directly into a solid magnesium carbonate presents a compelling option for the mitigation of power plant emissions. The carbonate product is environmentally benign and thermodynamically stable, thus creating a safe and reliable alternative to geological storage. However, given the low partial pressure of carbon dioxide in flue gas and the low reactivity of magnesium bearing minerals, achieving the kinetics required to make such a process feasible is challenging. In addition, magnesium carbonates formed at low temperature tend to incorporate significant waters of hydration, which increases transportation requirements and limits utilization of by-products. Each of these issues was addressed in the design of a multistep process for the conversion of magnesium silicates to carbonates as a carbon capture and storage option. The mineral dissolution reaction is accelerated using a mixture of organic chelating agents to produce a concentrated magnesium solution. This acceleration appears to occur through a stabilization effect on the pores formed as the silicate dissolves. The concentrated magnesium solution produced during the dissolution of the silicate is used to drive the precipitation of carbonates directly from flue gas. Through strict control over the chemistry of a simultaneous absorption and precipitation process, the phase of magnesium carbonate formed can be controlled. In addition, the mass transfer rate of carbon dioxide from flue gas into the precipitating liquor is accelerated using biological catalysis, which is shown to provide significant savings in process equipment size and operating parameters. Finally, the process as a whole is demonstrated to have a negative overall carbon balance through life cycle assessment.

(808e) A Green Approach to Mitigate CO2 Using Mixed Microbial Culture

Raghuvanshi, S., Birla Institute of Technology & Science, Pilani
Gupta, S., Birla Institute of Technology & Science, Pilani
Mishra, S.,
Originally presented on: 11/8/2013 14:10:00 - 14:35:00

Carbon dioxide is one of the GHG's (Green House Gases) & responsible for Global warming.. As of July 2011, CO_2 in the earth's atmosphere is at a concentration of 392 ppm by volume which is 35% higher as compared to the mid-1800s. Carbon dioxide, which is estimated to contribute 50-60% of the Global warming in the last century, is majorly released during the combustion of fossil fuel. The CO_2 concentration in earth atmosphere can be reduced by biological, chemical and technological techniques available through either mitigating or sequestering emissions. In the recent years, biological route has been identified as one of the most viable options available for the mitigation of CO_2. There are six carbon fixing pathway's. Out of which, the Calvin cycle (Calvin-Benson-Bassham) is used by photoautotroph's to fix CO_2. There are microbes which are ubiquitous in nature and thrive in extreme environment classified as chemoautotrops. They also classified as Extremophiles. The requirement of industries is to develop a economic, efficient and effective system which not only fix CO_2 but also capable to withstand in the severe industrial environment. Mixed culture is proved to be a superior candidate, as the consortium of different engineered species develop a symbiotic relationship which helps each other not only to survive in harsh conditions but also curb down CO_2. The another possibilities are simultaneous waste water treatment, production of Methane, and bio-hydrogen production. The present work deals in identification of microbes present in sludge of waste water treatment plant having CO_2 fixation capabilities. It also gives the insight of the relation which exists between natural selected mixed culture and to understand the capabilities of CO_2 fixation. The batch study has been conducted to measure the capabilities of mixed culture to mitigate CO_2 for five different V/V % concentration of CO_2 4, 12, 16, 22 and 25. It was observed that CO_2 is completed consumed by microorganisms in 4 days. Mitigation rate for different concentration of CO_2, bio-mass growth rate and resulting product has been analyzed on per hour basis and per day basis. Mixed culture is found to be superior candidate than photoautotrophic organisms for CO_2 mitigation. They show a promising future to develop a low cost, compact size closed system for *in-situ* CO_2 mitigation.

(811a) Formulation and Assessment of Membrane Separation Processes for Ethanol Dehydration

Kahwaji Janho, M., Cleveland State University
Gatica, J. E., Cleveland State University
Hernandez, M. R., Universidad Nacional de Tucuman
Mele, F. D., Universidad Nacional de Tucuman
Colombo, M., Universidad Nacional de Tucuman
Originally presented on: 11/8/2013 12:30:00 - 12:55:00

Bio-ethanol production is a rapidly growing market due to the possible uses of the anhydrous product in energy generation. This production however, is weighed down by the energetic demands and the environmental effects of the dehydration step needed due to the azeotrope formed by ethanol and water.

This paper studies the formulation and use of a pervaporation module to assess the production of fuel-grade bio-ethanol in distillery installations complementing a sugar mill. A baseline condition is defined as a case-study to illustrate the effect of structural and parametric changes on membrane separation units.

The user-defined module is integrated in commercial process simulators to study the environmental and energetic impact of different process configurations. Different schemes meeting equivalent performance standards in terms of purity and recovery are examined in this paper. Particular attention is given to energetic demands as comparison metrics. Life Cycle Analysis of the optimum process configuration is further used to identify critical sources of environmental impact.

(811b) Enhancement of Membrane Pervaporation By Manipulation of Nanoparticles Within the Concentration Boundary Layer

Smart, J. L., University of Kentucky
Originally presented on: 11/8/2013 12:55:00 - 13:20:00

Dilute concentrations of volatile organic compounds (VOCs) can be effectively pervaporated from aqueous systems with use of silicone rubber hollow fibers [1,2]. A source of vacuum or sweep gas on the downstream side of a membrane can be effectively used to provide the driving force for pervaporation of these VOCs.

With use of pervaporation technology, the effect of mixing perturbations in the boundary layer adjacent to the upstream side of a silicone rubber hollow fiber was studied. With use of a sweep gas imposed on the downstream side of the hollow fiber, the downstream side resistance is removed, and overall resistance to component permeation is reduced to only two resistances. These are resistances offered by the membrane wall thickness (easily designed and controlled) and resistance of the boundary layer. Resistance between the bulk feed solution and interface of the membrane is an interplay of several competing phenomena, including effects of component diffusion, component sorption, and hydrodynamics.

Resistance of the concentration boundary layer cannot be entirely eliminated, but its effects can be reduced. Varying concentrations of ferromagnetic nanoparticles (size < 100 nm) were introduced into a dilute feed mixture of toluene/water. As the feed flowed through the inside of the hollow fiber, external magnets were used to manipulate movement of the particles. Perturbation effects upon the boundary layer were evaluated. Micromixing action provided by the tiny magnetized nanoparticles act to increase turbulence and promote component diffusion across the effectively-thinned boundary layer. Besides precise control of mixing action, another advantage of the use of magnetized particles is the ability to easily separate and collect them from the feed retentate for recycle.

The work as outlined in this study is the first known application of use of magnetic fields to enhance component flux through a membrane. Other investigators have used sonication, vibration, mechanical turbulence promoters, moving boundaries, gas sparging, and pulsating flow to enhance component flux. In many cases, these earlier studies have shown marked improvements in flux, but it is believed manipulation of magnetic particles within the boundary layer offers added advantages of better process control through manipulation of magnetic fields and the reduction of fouling layers that accumulate on the inside wall of a hollow fiber membrane.

The initial experimental apparatus consists of a single silicone rubber hollow fiber that is surrounded by a variable rotating geometric arrangement of permanent magnets. The toluene/water feed containing a low concentration of ferrofluid is pumped through the inside of the hollow fiber. A nitrogen sweep gas is fed past the outside surface of the membrane to collect the permeated species. The sweep gas is fed directly into a gas chromatograph to quantify toluene flux. A side stream of the sweep gas is also condensed into a cold trap for later quantification to evaluate membrane separation. Early efforts in this study using magnetic nanoparticles have demonstrated greater than double the flux rate over conventional hollow fiber membranes. Continuing efforts are underway to optimize flux rates by improved location and manipulation of magnetic fields surrounding the nanoparticles.

1. J. Smart, R.C. Schucker, and D.R. Lloyd, Pervaporative extraction of volatile organic compounds from aqueous systems with use of a tubular transverse flow module. Part 1. Composite membrane study, J. Mem. Sci., 143, (1998), pp. 137 - 157.
2. J. Smart, V.M. Starov, R.C. Schucker, and D.R. Lloyd, Pervaporative extraction of volatile organic compounds from aqueous systems with use of a tubular transverse flow module. Part 2. Experimental results, J. Mem. Sci., 143, (1998), pp. 159 - 179.

(811c) Novel System and Field Characterization of Self-Adaptive UF-RO Membrane Desalination of Coastal Seawater

Gu, H., University of California, Los Angeles
Gao, L., University of California, Los Angeles
Pascual Caro, X., Rovira i Virgili University
Rahardianto, A., University of California, Los Angeles
Christofides, P. D., University of California, Los Angeles
Cohen, Y., University of California, Los Angeles
Originally presented on: 11/8/2013 13:20:00 - 13:45:00

Effective feed water pretreatment in reverse osmosis (RO) membrane desalination is essential to ensure optimum performance of RO plants. Microfiltration (MF) and ultrafiltration (UF) systems have been widely used as integral processes of RO feed pretreatment. While the concept of MF/UF integration with RO is appealing and has met with a reasonable level of success, there are still technical challenges in situations where feed water quality fluctuates (with respect to time), and where it is desired to reduce the use of chemical feed additives (e.g., coagulants) and membrane cleaning chemicals (for the purpose of both cost reduction and minimizing environmental impact). In meeting the above challenges, a novel approach of self-adaptive UF operation was developed and field demonstrated. Self-adaptive control of coagulant dose and filtration duration was successfully demonstrated in extensive long-term field studies of seawater desalination at the US Naval Base at Port Hueneme California. The present approach to self-adaptive UF operation integrates a number of novel model-based control and optimization methods and RO/UF design principles that were implemented in a unique "smart" (i.e., autonomous) MF-UF-RO system. This system is capable of processing up to about 36,000 gallons per day of raw feed water at a maximum recovery of about 36%. Real-time individual UF module resistance and fouling profiles are utilized to determine the appropriate coagulant dose, backwash frequency and intensity as feed water conditions change. Assessment of the impact of various operational parameters including coagulant dose, UF feed and backwash directions and module configurations, backwash source, duration, and flux on UF fouling behavior and permeability recovery was evaluated over a field study period of about a year. Adaptive control of inline coagulation was shown to be essential for effective UF backwash in order to minimize the frequency of chemical cleaning (e.g., effective UF operation without system shutdown and chemical cleaning). Furthermore, coagulation enhanced UF was found to assist in reducing TEP (Transparent Exopolymer Particles) passage through the UF and reduce the propensity for RO membrane fouling. The potential use of the present integrated MF-UF-RO approach for distributed deployment of smart water filtration and desalination systems will be discussed with specific focus on operability and fouling mitigation.

(811d) Dialysis Coupled With Ultrafiltration Through Parallel-Flow Rectangular Membrane Modules

Yeh, H. M., Tamkang University

Originally presented on: 11/8/2013 13:45:00 - 14:10:00

Dialysis is a process for transport of molecules through the microporous membrane by the driving force of the difference in chemical potential between the retentate and dialysate. An effective approach to enhance the mass transfer in a dialyzer may be introduced by the fluid filtration so that the substances is to be carried through the membrane by convective forces. The mass transfer for the systems of dialysis with ultrafiltration in cocurrently parallel-flow rectangular membrane modules was investigated, based on mass balances with the assumption of uniform ultrafiltration flux. Considerable improvement in separation efficiency is achievable if the effect of ultrafiltration is applied, especially for the system with low mass transfer coefficient. The enhancement in separation efficiency is significantly increased with increasing ultrafiltration flux, as well as with increasing the volumetric flow rates. Furthermore, increasing the volumetric flow rate in retentate phase is more beneficial to mass transfer than increasing in dialysate phase.

(811e) UF/NF Based Processes for Wastewater Treatment and Recovery of Process Water and Tannins in the Cork Industry

Minhalma, M., Instituto Superior de Engenharia de Lisboa

Bernardo, M., Instituto Superior de Engenharia de Lisboa

Geraldes, V., Instituto Superior Tecnico

De Pinho, M. N., Instituto Superior Tecnico

Originally presented on: 11/8/2013 14:10:00 - 14:35:00

Cork has a large range of applications in many activity areas and cork industry is a very important industrial sector in Mediterranean countries. Portugal, with only 30 percent of the world's cork oak forests, accounts for more than half of the world's cork production. Prior to industrial uses, cork planks are subjected to a boiling treatment that makes the planks softer and at the same time removes undesirable compounds from the cork. Although this treatment is essential to the quality of the final products, this process consumes a large amount of water and generates a highly pollutant wastewater which is rich in tannins, sugars and salts [1]. In fact, the pollutant charge is directly related to tannins content (Fig. 1) and therefore the approach followed in this work aims not only to address the treatment of these wastewaters but also to recover the tannins, as they can constitute an asset in the tanning industry.

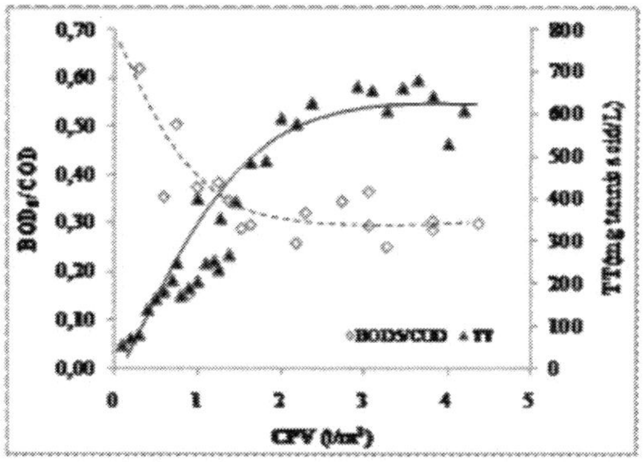

Fig. 1. Variation of Biodegradability (BOD_5/COD) and Total Tannins (TT) with the amount of cork processed in the boiling tanks (CPV, t/m^3).

Membrane processes, namely ultrafiltration (UF) and nanofiltration (NF), are playing a crucial role for the development of sustainable and cleaner processes in industries that are water-intensive users, like cork processing industries [2, 3, 4, 5].

In this work two goals are envisaged: i) water recycling (NF permeates) and ii) by-products recovery (through the production of tannin-rich concentrates) that can be used by the tanning industry. The achievement of these goals leads to the minimization of water consumption and wastewater discharge in the cork processing industry.

Ultrafiltration and nanofiltration experiments were carried out in spiral wound pilot plants with 2 m^2 of surface area. A 5 kDa polyethersulfone ultrafiltration membrane and a DS5 nanofiltration membrane were used. Total recirculation mode experiments were carried out in order to define the permeation optimal operation conditions (transmembrane pressure and feed flow rate). Three different concentrate mode experiments were carried out envisaging the tannin concentration by i) UF, by ii) NF and by an iii) UF/NF sequence. The UF/NF performance was assessed in terms of permeation flux and tannin content variation with the concentration factor. The NF permeate presents low organic content enabling its reuse as process water in the boiling tanks. The tanning power of the three tannin concentrates (UF, NF and UF/NF) was assessed (used as is and with mimosa tannin extracts) through the quantification of the temperature of contraction (Tc), Resistance to the extension of flower, Resistance to light and Tensile strength and elongation at break. These results indicated a high tanning capacity of cork extracts concentrated by ultrafiltration and nanofiltration, when used together with mimosa extracts. The good results achieved will allow significant reductions in the mimosa extract dosage. Even when tannin concentrates are used alone in the tanning process, the resulting skin is already showing a reasonable degree of stabilization. In conclusion, the tannins present in the boiling water of cork planks when subjected to concentration by UF and NF membranes can produce concentrates that have the technical feasibility for application in the Tannin industry.

[1] - ?Cork Industry Wastewater Characterization: assessment of the biodegradability, reuse and of the relationship between BOD, COD and Tannins with TOC?, Ana Santos, Marisa Bernardo, Carla Vespeira, Paula Cantinho, Miguel Minhalma, Journal of Water Reuse and Desalination, 2 (1) (2012) 33-39.

[2] ? ?Flocculation/flotation/ultrafiltration integrated process for the treatment of cork processing wastewaters?, Miguel Minhalma e Maria Norberta de Pinho, Environmental Science and Technology, 35 (2001) 4916-4921.

[3] - ?Nanofiltration of cork wastewaters and their possible use in leather industry as tanning agents?, Vítor Geraldes, Miguel Minhalma, Maria N de Pinho, S A Anil, Hasan Özgünay, Behzat O Bitlisli, Özcan Saric, Polish Journal of Environmental Studies, 18 (3) (2009) 353-357.

[4] - ?Cork Processing Wastewater Treatment/Valorisation by Nanofiltration?, J. Oliveira, M. Nunes, P. Santos, P. Cantinho, M. Minhalma, Desalination and Water Treatment, 11 (2009) 224-228.

[5] - "Cork industry wastewater partition by ultra/nanofiltration: a biodegradation and valorisation study?, M. Bernardo, A. Santos, P. Cantinho, M. Minhalma, Water Research, 45 (2011) 904-912.

(811f) Emergency Plan of Nanofiltration Integrated System for Saline Water From Tide of Qiantang River

Chen, H. L., Zhejiang University

Originally presented on: 11/8/2013 14:35:00 - 15:00:00

Qiantang River is the main drinking water source for Hangzhou area in China. Last decade, Hangzhou municipal supply drinking water system were affected seriously in view of the saline water going upstream by the tide in Qiantang River, and the wastewater excessive discharged into the river.

In order to ensure drinking water safety in tidal period, a pilot plant with 500 m^3/d of drinking water had been designed and manufactured used nanofiltration integrated technique with ultrafiltration as a pretreatment unit and reverse osmosis as a postreatment unit. The system had been set up in third drinking water company in Xiaoshan and it has been run nearly one year, The demonstration results show that the system is stability, and the quality of drinking water meet the Chinese standards for drinking water. That means this new technique and process can be scale up for the high quality of drinking water especially for where the source of water was slightly polluted.

According to the Qiantang River's water salinity level, the four emergency response grades for this kind of integrated system were set up. The suitable scale equipment and the response emergency security system for three corresponding drinking water plants respectively were designed due to water salinity by tide of Qiantang River.

(814a) Polyimide Membranes With High Fractional Free Volume for Gas Separations

Wiegand, J.,
Guo, R., University of Notre Dame
Originally presented on: 11/8/2013 12:30:00 - 12:55:00

Polyimide Membranes with High Fractional Free volume for Gas Separations

Jennifer Wiegand (jwiegan2@nd.edu), Ruilan Guo (rguo@nd.edu)

Department of Chemical and Biomolecular Engineering, University of Notre Dame

Notre Dame, IN 46556-5637

As being realized, the biggest challenge for these polymeric gas separation membranes is the trade-off relationship between gas permeability and selectivity defining the so-called "upper bound". Over the years, substantial research efforts have been directed to expanding the polymeric membrane performance envelope via designing novel polymer materials. Particularly, it has been demonstrated that polymer membranes featuring high fractional free volume, such as thermally rearranged (TR) polymers in our previous studies[1-4] and polymers of intrinsic microporosity (PIMs)[5-7], provide unprecedented combination of high gas permeability and high selectivity. Here we report a new series of high performance polyimide membranes with high fractional free volume as well as tunable microcavities that have great potential for fast and selective transport of small molecules. These novel polyimides were designed and derived via incorporating a unique set of triptycene components in their backbone structures, which are bulky, three-dimensional, shape persistent units and possess rich structural hierarchy and versatile chemistry possibilities allowing unique chain arrangement and organization mechanism to induce hierarchically controlled architectures in the resulting membranes. Molecular design, synthesis and characterization of these new polyimides will be presented with a focus on how to incorporate and tune the architecture of the microcavities via introduction of side bulky groups and/or adjusting the interlinking geometry. Preliminary pure gas permeation data (with H_2, N_2, O_2, CO_2, and CH_4) will be presented for these new polyimide membranes. The correlation between microscopic structures (primary backbone structure, chain packing efficiency, fractional free volume, etc.) of the triptycene-containing polyimides with macroscopic membrane gas transport properties will be discussed.

References:

1. R. Guo, D. F. Sanders, Z. P. Smith, B. D. Freeman, D. R. Paul and J. E. McGrath, *Journal of Materials Chemistry A*, 2013, **1**, 6063-6072.

2. R. Guo, D. F. Sanders, Z. P. Smith, B. D. Freeman, D. R. Paul and J. E. McGrath, *Journal of Materials Chemistry A*, 2013, **1**, 262-272.

3. D. F. Sanders, Z. P. Smith, C. P. Ribeiro, R. Guo, J. E. McGrath, D. R. Paul and B. D. Freeman, *J Membrane Sci*, 2012, **409-410**, 232-241.

4. Z. P. Smith, D. F. Sanders, C. P. Ribeiro, R. Guo, B. D. Freeman, D. R. Paul, J. E. McGrath and S. Swinnea, *J Membrane Sci*, 2012, **415-416**, 558-567.

5. P. M. Budd, E. S. Elabas, B. S. Ghanem, S. Makhseed, N. B. McKeown, K. J. Msayib, C. E. Tattershall and D. Wang, *Advanced Materials*, 2004, **16**, 456-459.

6. P. M. Budd, B. S. Ghanem, S. Makhseed, N. B. McKeown, K. J. Msayib and C. E. Tattershall, *Chem Commun*, 2004, 230-231.

7. P. M. Budd, K. J. Msayib, C. E. Tattershall, B. S. Ghanem, K. J. Reynolds, N. B. McKeown and D. Fritsch, *Journal of Membrane Science*, 2005, **251**, 263-269.

(814b) Membrane Gas Separation Processes: Improved Performances Through Unsteady Cyclic Operation

Wang, L., Laboratoire Réaction et Génie des Procédés LRGP- CNRS, Université de Lorraine

Corriou, J. P., Laboratoire Réaction et Génie des Procédés LRGP- CNRS, Université de Lorraine

Castel, C., Laboratoire Réaction et Génie des Procédés LRGP- CNRS, Université de Lorraine

Belaissaoui, B., Laboratoire Réaction et Génie des Procédés LRGP- CNRS, Université de Lorraine

Favre, E., Laboratoire Réaction et Génie des Procédés LRGP- CNRS, Université de Lorraine

Originally presented on: 11/8/2013 12:55:00 - 13:20:00

Membrane processes based on dense polymeric materials offer attractive performances for gas separation applications at industrial scale such as air separation, hydrogen purification, natural gas treatment or volatile organic compounds recovery [1].

Steady state operation is systematically performed for industrial installations and the separation performances are primarily governed by the membrane selectivity (a), which corresponds to the ratio of the compounds permeability through the polymer. In several cases however, the separation specifications cannot be reached due to too low membrane selectivity. In that event, the unique separation performances which can be obtained in some cases when a transient regime is applied can be of interest, especially when the differences between the diffusion coefficients of the permeating species largely exceed the differences in permeability. Paul explored this issue in a pioneering study, based on a cyclic process with synchronous valve operation [2]. We recently extended the analysis thanks to numerical simulation methods and optimization techniques which offer the possibility to explore asynchronous operation modes [3]. The concept proposed by Paul shows however a major limitation. Because the upstream composition has to remain unchanged, an infinitely low stage cut has to be applied (i.e. a very low fraction of the feed permeates through the membrane); as a consequence, the recovery ratio of the module is infinitely small.

In this study, a different approach, based on unsteady cyclic operation over long time scale (i.e. much larger than the time-lag of the system), a completely unexplored research area, will be detailed and described. The evolution of the upstream and downstream compartments pressure of a membrane module under cyclic unsteady operation have been first simulated through a tailor made computer program. The predictions obtained by simulation have been experimentally validated on a dedicated installation equipped with a hollow fiber gas separation module for air separation application (PPO fibers, Piccolo module from Parker). In a second step, the set of operating conditions which maximize the separation performances of the unsteady process have been identified through optimization. It will be shown for the first time that unsteady operation offers the opportunity to improve separation selectivity and overall recovery compared to steady operation [4].

The results of this exploratory study suggest to more systematically investigate the possibilities and limitations of gas separation modules under transient operation. A series of potential target applications (gas pairs) of industrial interest will be detailed. The possibilities of improvement of the concept and future challenges will finally be addressed.

References

[1] Baker, R, *Membrane Technology and Applications*, 2nd ed., John Wiley & Sons, New York, 2004

[2] Paul, D.R. "Membrane separation of gases using steady cyclic operation." *Ind. Eng. Chem. Process Des.*

Dev, 1971: 375-379.

[3] Corriou, J.P., C. Fonteix, et E. Favre. «Optimization of a pulsed operation of gas separation by membrane.» *AIChE Journal* 54, n° 5 (2008): 1224-1234.

[4] Favre, E., Castel, C., Corriou, J.P., Wang, L. " Procédé de séparation membranaire en régime discontinu" French patent # FR 11 60587.

(814c) Multiwalled Carbon Nanotube Mixed-Matrix Membranes Containing Amines for Acid Gas Removal From Natural Gas

Ansaloni, L., Alma Mater Studiorum - Università di Bologna

Zhao, Y., The Ohio State University

Ramasubramanian, K., The Ohio State University

Giacinti Baschetti, M., Alma Mater Studiorum - Università di Bologna

Ho, W. S. W., The Ohio State University

Originally presented on: 11/8/2013 13:20:00 - 13:45:00

Multiwalled Carbon Nanotube Mixed-Matrix Membranes Containing Amines for Acid Gas Removal from Natural Gas

Luca Ansaloni[1], Yanan Zhao[2], Kartik Ramasubramanian[2], Marco Giacinti Baschetti[1], W.S. Winston Ho[2,3]

[1]Dipartimento di Ingegneria Civile, Chimica, Ambientale e dei Materiali, Alma Mater Studiorum - Università di Bologna, Bologna, Italy

[2]William G. Lowrie Department of Chemical and Biomolecular Engineering, The Ohio State University, Columbus, Ohio 43210, United States

[3]Department of Materials Science and Engineering, The Ohio State University, Columbus, Ohio 43210, United States

Nowadays, natural gas is an important source for energy production. Depending on the original source, raw natural gas contains various impurities in composition, which must be removed before delivery to the pipelines. Removal of acid gases, namely, CO_2 and H_2S, from natural gas is therefore an important industrial gas separation process.

For CO_2 removal from natural gas, polymeric membranes compete against conventional separation processes, such as amine absorption, due to superior advantages of lower costs, lower energy consumption, and lower waste production.[1] Several attempts have been made to develop CO_2/CH_4 separation membranes that match the pipeline specifications. Glassy polymers, in particular polyimides, have been reported to show the best performance for this specific separation.[2,3] The higher condensability and smaller kinetic size make CO_2 molecules more soluble and diffuse faster in the membrane matrix than methane, resulting in good selectivity and permeability. Recently, polyimide-based polymer with intrinsic microporosity (PIM)[4] and cyclodextrine grafted co-polyimides[5] demonstrated very high CO_2 permeability at high pressure (10 atm) but low CO_2/CH_4 selectivity, less than 25. In fact, at high pressures, these membranes suffer considerably from the plasticization problem caused by high concentration of the acid gas, resulting in decreasing membrane selectivity significantly. Crosslinking can help to minimize the CO_2-induced swelling; however, an increase in separation factor of a solution-diffusion polymeric membrane always accompanies a reduction in the penetrant flow due to the well-known trade-off relationship.[6]

Facilitated transport membranes hold great potential for natural gas CO_2 removal. This type of membrane is able to achieve high CO_2 permeability accompanied with high selectivities vs. CH_4, H_2, or N_2 due to the specific, prompt reaction between amine carriers and CO_2 molecules.[7,8] This work exploited the superlative mechanical property of multi-walled carbon nanotubes (MWNTs) to reinforce the mechanical strength of amine facilitated transport membranes, from which we successfully synthesized nanostructured polymer-MWNT membranes by incorporating MWNTs (about 18 nm diameter by 180 nm) into the crosslinked polyvinylalcohol-polysiloxane network containing amine carriers.

The synthesized membranes inherited the facilitated transport mechanism and thus achieved high CO_2 permeability and selectivity vs. CH_4 at high pressures of at least 15 atm and high temperatures of more than 100°C. At 15 atm and 107°C using a feed gas of 50% CH_4 and 50% CO_2, which simulated the natural gas composition, the membrane showed a high CO_2/CH_4 selectivity of about 100 accompanied with a CO_2 permeability of about 350 Barrers. This lower permeability was due to the higher degree of carrier saturation at the higher CO_2 concentration. It should be noted that this permeability is still higher than that of commercial membrane, e.g., cellulose acetate (about 5 Barrers).

In this work, we also synthesized acid-treated MWNTs and amino-functionalized MWNTs as advanced nanofillers for amine facilitated transport membranes. The acid-treated MWNTs were functionalized with hydrophilic groups including carboxylic acid and hydroxyl groups by strong acid oxidation. The amino-functionalized MWNTs were chemically modified with 3-aminopropyltriethoxysilane. The presence of these hydrophilic groups was confirmed by FTIR spectroscopy. The membranes incorporated with the modified MWNTs demonstrated significant improvement in CO_2/CH_4 separation performance, especially in term of separation factor. A CO_2/CH_4 selectivity of 180 along with CO_2 permeability of around 1000 Barrers have been obtained at 15 atm and 107°C for a feed gas composition of 59.8% CH_4, 20.2% H_2 and 20% CO_2. These improvements are attributed to the modification of MWNTs. With the hydrophilic surface groups, MWNTs were compatible with the network of crosslinked polyvinylalcohol and polyamines. Thereby, the MWNTs were able to insert between polymer chains and disrupted polymer chain packing more effectively, resulting in increased free volumes between polymer chains. In addition, the membrane performance at higher feed pressure up to 60 atm as well as the effects of membrane composition, temperature and membrane thickness, has been investigated.

Keywords: multiwalled carbon nanotube; mixed matrix membrane; facilitated transport; carbon dioxide separation; natural gas.

References:

1. Baker, R. W. & Lokhandwala, K. Natural Gas Processing with Membranes: An Overview. *Industrial & Engineering Chemistry Research* **47**, 2109?2121 (2008).

2. Wang, L. *et al.* Gas transport properties of 6FDA-TMPDA/MOCA copolyimides. *European Polymer Journal* **44**, 225?232 (2008).

3. Wind, J. D., Paul, D. R. & Koros, W. J. Natural gas permeation in polyimide membranes. *Journal of Membrane Science* **228**, 227?236 (2004).

4. Ghanem, B. S., McKeown, N. B., Budd, P. M., Selbie, J. D. & Fritsch, D. High-Performance Membranes from Polyimides with Intrinsic Microporosity. *Advanced Materials* **20**, 2766?2771 (2008).

5. Askari, M., Xiao, Y., Li, P. & Chung, T.-S. Natural gas purification and olefin/paraffin separation using

cross-linkable 6FDA-Durene/DABA co-polyimides grafted with α, β, and γ-cyclodextrin. *Journal of Membrane Science* **390-391**, 141?151 (2012).

6. Robeson, L. M. The upper bound revisited. *Journal of Membrane Science* **320**, 390?400 (2008).

7. Bai, H. & Ho, W. S. W. Carbon dioxide-selective membranes for high-pressure synthesis gas purification. *Industrial & Engineering Chemistry Research* **50**, 12152?12161 (2011).

8. Xing, R. & Ho, W. S. W. Crosslinked polyvinylalcohol?polysiloxane/fumed silica mixed matrix membranes containing amines for CO2/H2 separation. *Journal of Membrane Science* **367**, 91?102 (2011).

(814d) CO2 Removal From High Pressure Natural Gas With Hybrid Fixed-Site-Carrier Membranes: Membrane Material Development

He, X., Norwegian University of Science and Technology
Kim, T. J., SINTEF
Uddin, M. W.,
Hagg, M. B., NTNU
Originally presented on: 11/8/2013 13:45:00 - 14:10:00

CO_2 removal from natural gas (NG) is mandatory to meet the natural gas network grid specifications as CO_2 reduces the heating value of natural gas, is corrosive, and easily forms hydrates that can clog the equipment or damage the pumps. Processing of natural gas presents a large industrial gas separation market owing to the increased worldwide natural gas consumption. Traditional chemical (amine) absorption is well known and still considered as the state-of-the-art technology. However, membrane systems show a great potential for natural gas sweetening, even though it has only 5 % of the market today. Recently, strong interest has been focused on the development of different materials, especially for the composite membranes with a very thin selective layer, to overcome the challenges of the commercial membranes in natural gas sweetening. The novel fixed-site-carrier (FSC) membranes showed a good performance for CO_2 / CH_4 separation in our previous work [1-3]. However, high pressure still challenging the membrane operation due to the membrane plasticization and compaction caused by the strong adsorption of CO_2 and heavy hydrocarbon (BTEX) in the polymer matrix. The feasible solutions are the crosslinking of the membrane materials and development of pressurization-resistant membranes. Thus, the carbon nanotubes (CNTs) reinforced polyvinyl amine (PVAm) / polyvinyl alcohol (PVA) blend FSC membranes were prepared using an optimized membrane condition in current work.

The casting solution was prepared by adding a certain amount of CNTs into the PVAm (MW 340K) /PVA (72K) aqueous solution, and completely mixed using an ultrasonic mixer. The casting solution was evenly casted on the top of the polysulfone (PSf) supports following the procedure reported by Kim et al. [4]. The membrane structure and morphology was characterized by scanning electron microscopy (SEM), and a typical selective layer thickness was found to be ~1 μm. The membrane separation performance was characterized by gas permeation testing at 30 °C with a 10 % CO_2 - 90% CH_4 and a controlled feed flow rate. A lab-scale plate-and-frame module (membrane area 19.6 cm^2) and a small pilot-scale module involves 3 sheets (each sheet is 110 cm^2) were tested in high pressure permeation rig.

The permeation testing results indicated that the membrane preparation parameters such as heat treatment temperature, duration time and the pH of casting solution will significantly influence the membrane performance, and an optimal preparation condition is found to be: 95 Â°C - 0.75h ? pH10. Moreover, the operating parameters such as feed pressure, feed CO_2 concentration, and feed flow rate were also found to influence the membrane separation performance, which should be optimized in a specific application. The experimental results indicated **that CO_2 permeance decreases with the increase of feed pressure, which may due to the less contribution from facilitated transport mechanism (carrier saturation and low water vapor content at high pressure) and membrane compaction. It was also found** that both CO_2 permeance and CO_2/CH_4 selectivity are lower when the feed gas contains a high concentration of CO_2 50 % compared to a low concentration CO_2 at the same feed pressure. In addition, a high feed flow rate presents a much high separation performance due to a better flow pattern, but a relatively lower stage-cut. Therefore, a pilot-scale membrane module with larger membrane area should be tested in future work to document the membrane performance at a relatively high stage-cut which is usually involved in an industrial process.

The dependence of the membrane separation performance on the membrane area was tested with the small pilot-scale module, and the results indicated that both CO_2 permeance and CO_2/CH_4 selectivity decrease with the increase of membrane area, large membrane area gives a high CH_4 purity in the retentate but high CH_4 losses as well. Thus, a two-stage or multi-stage membrane system may be needed to accomplish a specific separation requirement (i.e., high CH_4 purity and low CH_4 losses). The developed FSC membrane showed a nice potential application for CO_2 removal from high pressure natural gas, but the process configuration and operating condition should be well designed and controlled in a real process. The current work can be well used to guide the process simulation to evaluate the process feasibility of CO_2 removal from high pressure natural gas process.

Acknowledgements

Â The authors acknowledge the NaGaMa project (partners: Norwegian Research Council, Statoil and Petrobras) for the funding of this work. The high pressure pilot-scale module designed by PHILOS is also acknowledged. The authors also thank SHOWA DENKO K. K. company in Japan to provide the carbon nanotubes for this work.

References

[1] Deng L, Kim T-J, Sandru M, HÃ¤gg M-B. PVA/PVAm Blend FSC Membrane for Natural Gas Sweetening. Proceedings of the 1st Annual Gas Processing Symposium. Doha, Qatar; p. 247-55.

[2] Washim Uddin M, HÃ¤gg M-B. Natural gas sweetening?the effect on CO2?CH4 separation after exposing a facilitated transport membrane to hydrogen sulfide and higher hydrocarbons. J Membr Sci. 2012;423?424(0):143-9.

[3] He X, HÃ¤gg MB. Hybrid Fixed-site-carrier Membranes for CO_2/CH_4 Separation. Euromembrane 2012. London, UK.

Â [4] Kim T-J, VrÃ¥lstad H, Sandru M, HÃ¤gg M-B. Separation performance of PVAm composite membrane for CO2 capture at various pH levels. J Membr Sci. 2013;428(0):218-24.

Â ADDIN EN.REFLIST

(814e) Performance and Stability of RTIL-Membranes for Dehumidification of Methane

Sullivan-Gonzalez, F.,
Amos, R., University of Mississippi
Bae, S. K.,
Scovazzo, P., University of Mississippi
Originally presented on: 11/8/2013 14:10:00 - 14:35:00

The biogas production from organic waste—animal, sewage, landfill material - could be a valuable renewable fuel. Before this biogas—predominately bio-methane—can be converted to fuel, it must be dehumidified. Typically membrane-based gas dehumidification uses hydrophilic polymer membrane materials, such as polydimethylsiloxane (PDMS) and cellulose acetate (CA). The main problems with polymer dehumidification membranes are methane loss and the susceptibility of polymer materials to plasticization by H_2O. Room temperature ionic liquid membranes (RTIL-membranes), according to recent literature, are not plasticized by water and have large water/CH_4 selectivities. In our paper, we will build on the initial literature RTIL-dehumidification data by looking at a series of ammonium and imidazolium-based ionic liquids. One membrane based on the trifluoromethanesulfonate [TfO] anion had a water permeance of 2000 GPUs with a H_2O/CH_4 selectivity of 12 000. With such high water permeances, boundary layer and support structures might significantly contribute to the overall water transport resistance, we will, therefore, present our work to measure and minimize this effect in our data. Of greater interest to the membrane community, however, will be our finding that there is an impact on some of the RTIL-membranes from the water vapor in the feed. This impact appears to be an increase in membrane mechanical stability and H_2O/CH_4 selectivity with increasing feed gas relative humidity. In an attempt to explain these phenomena we will explore the impact of the Feed rH the on surface tension of the stabilized RTIL.

(814f) Polymerized Pyrrolidinium-Based Ionic Liquid Membranes for CO2 Capture

Wang, J.,
Baker, S. N.,
Baker, G. A., Oak Ridge National Laboratory
Originally presented on: 11/8/2013 14:35:00 - 15:00:00

Ionic liquids (ILs) are salts with melting points below 100°C. They have been utilized for CO_2 capture due to some of their unique properties like high viscosity, low vapor pressure, and thermal and chemical stability. Utilizing IL membranes for CO_2 capture highlights the both properties of ILs and advantages of membrane technologies. Currently, supported IL membranes (SILMs) are playing a major role in the area of CO_2 capture. Such SLIMs have been fabricated with supported gelled and polymerized ILs. In SILMs, the use of porous supporting membrane materials, such as PVDF membranes or other organic frameworks, decreases the contact area of CO_2 and the effective IL components, which leads to lower capture efficiency. In this work, we are reporting IL membranes forming using only polymerized ILs without support materials. These membranes are then characterized and studied for their ability to capture CO_2 and to better understand the separation mechanism especially in terms of the cation and anion components.

(817a) Numerical Simulation Investigation On Particle Loading Process of Nano-Fiber Filter

Lu, P., Institute of Process Engineering, Chinese Academy of Sciences
Hua, C., Chinese Academy of Sciences
Bai, F., Institute of Process Engineering, Chinese Academy of Sciences
Tong, R., University of Science and Technology Beijing
Zhang, X., Institute of Process Engineering, Chinese Academy of Sciences
Originally presented on: 11/8/2013 12:30:00 - 12:55:00

For worsening air pollution□fibrous filter is respected to have higher performance to capture suspended particle in industrial production and life area, so nano-fiber is considered to design next generation filter because of low pressure drop and high particle capture efficiency. In this project, theoretical analysis, numerical simulation and experimental method were used to deeply investigate effect of slip-boundary of nano-fiber on flow filed and particle loading process in single fiber model and nano-fibrous filter model respectively. The Fluent software with User-Defined Function was used for simulation. The Euler – Lagrange method was used to solve flow filed and particles trajectory, particle interface was captured by voxel method and slip-boundary condition was realized by permeable fiber assumption. The particles deposition on single nano-fiber with different mechanism (diffusion, inertia and interception) was compared with normal fiber. In addition, a mathematic model was generated for nano-fibrous filter to describe effect of dust holding capacity (DHC) on subsequent pressure drop and particle capture efficiency. The predicted DHC data of nano-fibrous filter agreed well with experimental value of hand sheet. This research will provide available theoretical analysis and numerical results for design of next generation nano-fiber filter.

(817b) Development of Carbon Nanofibers By Electrospinning for Aerosol Filtration

Chattopadhyay, S.,
Rutledge, G. C., Massachusetts Institute of Technology
Originally presented on: 11/8/2013 12:55:00 - 13:20:00

Among various air filtration technologies, nanofibrous media are of interest, since theoretical calculations predict that these materials have high efficiencies to remove sub-micrometer aerosol particles with relatively lower energy consumption. Conventional fibrous filters made of nanofibers exhibit a local minimum of fractional collection efficiency, which is called the most penetrating particle size (MPPS). This is usually observed for particles 0.1-0.5 μm in diameter. Increasing filtration efficiency in this critical size range could be targeted by altering fiber composition, size and morphology. A method called "electrospinning" to prepare fibers with sub-micrometer diameters has received considerable attention in aerosol filtration research due to the reduction in filter mass required to obtain nanoparticle filtration efficiencies comparable to currently available commercial filters. Recently, electrospun carbon fibers prepared by carbonization of precursors like polyacrylonitrile, polycarbosilane, polyvinylidene fluoride and cellulose acetate fibers have been reported for various applications. During carbonization, the fiber structure remained unaffected but the mean diameter decreased by a factor of about 3-4. Additionally, properties like high electrical conductivity, wettability and specific surface area observed in such studies may offer other advantages for air filtration performance.

To this end, we report the development of continuous cellulose acetate and carbon fibers with diameters in the range of 0.1-1 μm and large surface areas (300-3000 m^2/g) as high efficiency particle filtration media. Important filter properties such as aerosol removal efficiency at low pressure drop are studied vis-à-vis electrospinning and carbonization parameters. For the current study, engineered nanoparticles comprising monodisperse oil droplets of di-2-ethyl hexyl sebacate (0.09 to 0.35 μm in physical diameter), are generated using a condensation monodisperse aerosol generator (CMAG). This aerosolizer generates a high number concentration (10^7 counts/cc) of spherical and charge-neutral particles by condensing oil vapor onto sodium chloride nuclei, based on the *Sinclair-LaMer* principle. The size distribution, number concentration, relative humidity and temperature of the aerosol were customized in an aerosol diluter before exposure to the test filter, which is housed in a standard 13-mm aerosol filter holder. The face velocity is regulated (1-100 cm/s) by a sump operating downstream of the filter assembly, and the pressure drop across the filter is monitored using a pressure transducer. The real time number concentrations of aerosol are measured using a scanning mobility particle sizer (SMPS), which measures the electrical mobility of aerosol in the size range 0.02-0.8 μm. The ratio of particle concentration, with and without the filter material and the geometric mean diameter of the particles that pass through the filter material are defined as 'particle penetration' and MPPS, respectively. Initial results indicate that the diameter (0.1-1 μm) and cross-section shape (ribbon and cylindrical) of electrospun cellulose acetate fibers can be controlled through selection of polymer solution and electrospinning parameters. The measured filtration efficiency for fibers having ribbon and cylindrical cross sections were found to be comparable, but the filters of cylindrical fibers were easier to handle and compact. The percentage retention by these cylindrical fiber filters was altered when solutions having different concentrations of cellulose acetate were used. Solutions with lower polymer concentration (15 and 17 wt %), which are anticipated to generate smaller diameter fibers, had 1.3 times higher percentage retention than counterparts formed from solutions of 19 wt % polymer. Likewise, the MPPS of filters decreased (from 90 nm to 77 nm) with decreasing solution concentration. In general, this seems to corroborate the understanding that smaller fiber diameter increases the filtration efficiency and decreases the MPPS. However, the anticipated decrease in fiber diameter was not evident through scanning electron micrographs, which showed that the fibers vary from 0.1-0.5 μm in diameter. Studies are underway to investigate other properties like surface area and solidity that could affect filtration properties. In addition, the filtration capabilities of carbonized polymeric electrospun will be reported for their possible application in high efficiency aerosol filters.

(817c) Modeling Investigation On Distillation Process Optimization of Benzene-Clohexene-Cyclohexane System By Using Aspen Plus

Tong, R., Institute of Process Engineering, Chinese Academy of Sciences
Mi, W., University of Science and Technology Beijing
Lu, P., Institute of Process Engineering, Chinese Academy of Sciences
Bai, F., Institute of Process Engineering, Chinese Academy of Sciences
Hua, C., Chinese Academy of Sciences
Originally presented on: 11/8/2013 13:20:00 - 13:45:00

In this work, Aspen plus software was used to model and optimize distillation process of Benzene(BZ)-Clohexene(HE)-Cyclohexane(HA) system with Dimethylacetamide □DMAC□as extractant. The thermodynamic model used NTRL. The effect of important operation parameter (theoretical plate number, feed position, solvent ratio, reflux ratio, and extractant feed position and temperature) on products separation efficiency (recovery percentage of BZ, mass percentage of BZ, HE and HA) was modeled and investigated in BZ separation tower and recovery tower respectively. The modeling results indicated when distillation process was operated in following condition, the separation efficiency was best and energy consumption was the lowest.

1) In Benzene separation tower, the theoretical plates was 36, feed position of raw material on the 12nd theoretical plates, solvent ratio 3.18, and reflux ratio 2.4; extractant feed position was on the 2 ndtheoretical plates and feed temperature was 64℃.

2) In recovery tower, the best operation condition was 26 theoretical plates, feed position 10nd and reflux ratio 1.1.

(817d) A Mass Transfer Model of Absorption of Carbon Dioxide in a Bubble Column Reactor

Cheng, L., University of Cincinnati
Li, T., University of cincinnati
Keener, T. C., University of Cincinnati
Lee, J. Y., University of Cincinnati
Originally presented on: 11/8/2013 13:45:00 - 14:10:00

A bubble column reactor was tested and evaluated for carbon dioxide removal from flue gases by using magnesium hydroxide slurry. An absorption model was developed for system analysis and CO_2 removal data analysis. The overall mass transfer coefficients for CO_2-H_2O and CO_2-$Mg(OH)_2$-H_2O systems have been calculated. The chemistry of the system, temperature effects, and changes of equilibrium constants also have been discussed.

(822a) Functional Nanostructured Polymer Materials Based On Liquid Crystal Building Blocks

Noble, R., University of Colorado

Originally presented on: 11/8/2013 12:30:00 - 12:55:00

Our research program in LC-based, nanostructured polymers is divided into three main areas. The first area is the development of polymerizable *lyotropic* LCs for the construction of functional, nanostructured polymeric materials. The second area involves the design of monomers based on functional *thermotropic* LCs to control order, symmetry, and symmetry-based bulk properties in the final polymer assembly. The third area is the development of new strategies for the synthesis of LCs that exhibit functional properties and new supramolecular architectures to serve as building blocks for new materials.

(822b) Self-Assembled Superhydrophilic Polymer Membranes for Size-Exclusion Separation

Fang, Y., University of Florida
Jiang, P., University of Florida
Originally presented on: 11/8/2013 12:55:00 - 13:20:00

Here we report a simple and roll-to-roll compatible bottom-up technology for continuously fabricating macroporous polymer membranes with highly uniform and 3-D interconnected nanopores for size-exclusion separation of viruses and bacteria from water. A vertically beveled doctor blade is utilized to shear-align colloidal silica suspensions to form large-area colloidal crystals with high crystalline qualities. The thickness of the shear-aligned crystals can be easily tuned by adjusting the concentration of the colloidal suspensions and the coating speed. The self-assembled colloidal crystals can then be used as structural templates in making self-standing macroporous polymer membranes. The large voids templated from colloidal particles are interconnected through much smaller nanopores that originate from the touching sites of particles. The high porosity of the resulting membranes enables the achievement of superhydrophilic surface with an apparent water contact angle of less than 5°, promising for high-flux filtration. In addition, by using appropriate polymers, the templated macroporous films can effectively reduce bio-fouling.

(822c) Developing Tunable Ultrafiltration Membranes With Chemically-Tailored Pore Walls From Triblock Terpolymer Templates

Weidman, J., West Virginia University
Mulvena, R.,
Boudouris, B. W., Purdue University
Phillip, W. A., University of Notre Dame
Originally presented on: 11/8/2013 13:20:00 - 13:45:00

Recent reports suggest that self-assembled block copolymers are an excellent platform for producing next-generation ultrafiltration (UF) membranes. Because the self-assembly of the block copolymer results in a high density of singly-sized pores, whose diameter can be tuned via the copolymer synthesis, these membranes have the potential to provide superior performance when compared to current membranes. However, the copolymer membranes produced thus far contain pore walls that are lined by a limited number of chemical functionalities, whereas tailored pore chemistries are preferred. This collaborative work addresses this shortcoming in order to produce UF membranes that can be chemically tailored, in a facile and scalable manner, to the specific needs of a multitude of applications.

Here we will discuss efforts to make large membrane areas from a poly(isoprene-*b*-styrene-*b*-*N,N*-dimethylacrylamide) (PI-PS-PDMA) triblock terpolymer using the self-assembly and non-solvent induced phase separation (SNIPS) method. Using a simple process after membrane fabrication, the PDMA groups that line the pore walls are converted to polyacrylic acid (PAA), which can be reacted further to functionalize the membrane pores. A suite of x-ray scattering and advanced microscopy characterization techniques (SAXS, SEM, AFM) are used to explore systematically the nanostructure of the terpolymer membrane at each step of the process (*i.e.*, during the fabrication of the parent membrane, the deprotection step, and the subsequent pore wall functionalization). These characterization techniques show that the SNIPS process produces an asymmetric membrane with a thin active layer, which contains 1.85×10^{13} self-assembled pores per m^2 of membrane with diameters that can range from 5-10 nm, on top of a microporous support. The membrane maintains its nanoporous structure following the deprotection and functionalization steps. Transport tests also are conducted at each stage to quantify the hydraulic permeability and solute rejection capabilities of the membrane. This combination of material characterization and transport testing lays the groundwork for the development of structure-property-performance relationships that can guide the use of these versatile block terpolymer membranes in filtration, sensing, and drug delivery applications.

(822d) Dynamic 2D/3D Membranes From Self-Assembly of Block Copolymer Micelles: Self-Healing Membranes and Reversible Coatings

Quemener, D., University Montpellier II
Deratani, A., University Montpellier II
Originally presented on: 11/8/2013 13:45:00 - 14:10:00

In the last 5 years, we have developed new routes using self-assembly of block copolymers to prepare porous nanomaterials[1,2], coatings[3] or to control the interactions between biocomponents[4]. In a first part, we will demonstrate the possibility of making self-healing membranes from a micelle assembly of block copolymers. Those micelles of 50 nm in diameter are strongly interacting through block copolymer bridges. In case of a major failure, the membrane is able to detect and to adjust its nanostructure in order to autonomously repair the damage. In a second part, we will introduced the possibility of using macromolecular zipper to reversibly assemble micelles into 2D and 3D coatings. The zipper effect was observed from poly(n-octadecyl methacrylate) blocks thanks to the interdigitating of the molecular brushes. Micelle monolayer was zipped to a poly(methyl methacrylate) substrate through a one pot strategy including a surface functionalization with anchoring points enabling a stable zipping of the micelles. Micelle multilayers were also obtained by increasing the micelle concentration through the solvent evaporation. The concept of zipping-assembly of micelles combines a self-assembly stage following "classical" rules and a reversible lock-in of the global morphology through a zipper mechanism.

[1] P. Tyagi, A. Deratani, D. Bouyer, D. Cot, V. Gence, M. Barboiu, T. N. T. Phan, D. Bertin, D. Gigmes, D. Quemener, *Angew. Chem. Int. Ed.* **2012**, *51*, 7166.

[2] D. Quemener, G. Bonniol, T. N. T. Phan, D. Gigmes, D. Bertin, A. Deratani, *Macromolecules* **2010**, *43*, 5060.

[3] P. Tyagi, I.E Raschip, A. Deratani, D. Quemener, *Adv. Mater.* **2013**, in press.

[4] Y-J. Shih, Y. Chang, A. Deratani, D. Quemener, *Biomacromolecules* **2012**, *13*, 2849.

(822e) Mixed Matrix Nanocomposite Membranes for Water Purification

Deng, B., University of Missouri
Yin, J., University of Missouri-Columbia
Originally presented on: 11/8/2013 14:10:00 - 14:35:00

Sustainable development of human society depends on adequate supplies of water. In areas where high quality water supplies are limited, water desalination and recycling are often needed. Among various techniques for water purification, membrane technology has been increasingly explored because it can be adapted to treat a wide range of waters for different purposes. The key issue in the membrane treatment is the control of operational costs and energy consumption, which is closely tied to water permeability, salt/contaminant rejection, and increasing membrane fouling with treatment time. In this presentation, we will discuss our recent progress on the development of mixed matrix membranes with a goal to 1) increase water flux and salt rejection of the membranes, and 2) to minimize membrane fouling due to natural organic matters and microorganisms. A particular focus is on how the incorporation of nanomaterials (e.g., mesoporous carbon and silica, carbon nanotubes, Ag TiO_2) will alter membrane structure, porosity, surface hydrophobicity, roughness, and surface functionality, and how these measureable characteristics correlate with the membrane performance.

(822f) How Do Charge Density and Roughness of Surface Affect Formation of Supported Lipid Bilayers: A Coarse-Grained Molecular Dynamics Simulation

Liu, Z., Tsinghua University
Lu, D., Tsinghua University
Kong, X., Tsinghua University
Originally presented on: 11/8/2013 14:35:00 - 15:00:00

Dispersion of uni-lamellar vesicles to attractive surface is widely used to create supported lipid bilayers (SLBs), which are frequently used to embed membrane proteins for biosensors and novel artificial membranes for water or ion transport. Although vesicle spread and fusion is the most convenient and frequently used way to prepare SLBs with different composition and supporting surface in experiments, the mechanism of their formation and its behavior on surfaces with different properties are not well understood. To solve above-mentioned problem, we performed coarse-grained molecular dynamics simulation to study vesicle spreading mechanism on charged surface with special emphasis on charge type, charge density, surface roughness and also the existence of membrane protein. The liposome is composed of 677 DPPC molecules (1,2-dipalmitoyl-sn-glycero-3-phosphocholine) and 200 negatively charged DPPG molecules (1,2-dipalmitoyl-sn-glycero-3-phosphoglycerol). The model membrane protein is AqpZ, a kind of aquaporin protein from bacteria, and the charged surface is modeled by a plane of fixed particles with charges and some degree of roughness. It is shown that the liposome is only adsorbed on the surface and does not spread to from SLBs when the solid surface carries negative or zero charge. Once the surface is positively charged, the electrical interaction between the surface and the liposome enabled the liposome to overcome the bending energy barrier in the spreading process, which leads to the spread of the liposome. It is also shown that the lipids in the inner layer display several peaks in the density profile in the normal direction of the solid surface, indicating more multilayer structure forms once lipid adsorbed on the charge surface. Once the surface charge density increases, the SLBs become thinner, and more DPPG lipids appear in the inner layer. When AqpZ is embedded into liposome, the spreading process becomes more complex. It is shown that at low surface charge density, the liposome forms micelle like structure with AqpZ in it or non-flat lipid bilayer with AqpZ protuberance, rather than uniformed flat lipid bilayer with embedded protein. Only when surface possesses enough charge density, i.e., the interaction between the lipids and the surface is large enough, does lipid bilayer form flatly and AqpZ embedded in it with correct orientation. This is important for keeping fluid mosaic structure of SLBs, which is critical for functional membrane protein. Above simulation results would help us to design suitable surface for spreading lipid bilayer with membrane protein.

AUTHOR INDEX

AUTHOR INDEX

AUTHOR INDEX

AUTHOR INDEX

AUTHOR INDEX

AUTHOR INDEX

AUTHOR INDEX

AUTHOR INDEX

AUTHOR INDEX

AUTHOR INDEX

AUTHOR INDEX

AUTHOR INDEX

AUTHOR INDEX

AUTHOR INDEX

AUTHOR INDEX

AUTHOR INDEX

AUTHOR INDEX

AUTHOR INDEX

AUTHOR INDEX